A HISTORY

OF

MATHEMATICS

Jeff Suzuki
College of General Studies
Boston University

Prentice Hall
Upper Saddle River, New Jersey 07458

Library of Congress Cataloging-in-Publication Data

Suzuki, Jeff.
  A history of mathematics / Jeff Suzuki.
    p. cm.
  Includes bibliographical references.
  ISBN 0-13-019074-8
    1. Mathematics–History. I. Title.

QA 21 .S975 2002
  510′.9–dc21

2001045936

Acquisitions Editor: George Lobell
Editor-in-Chief: Sally Yagan
Vice President/Director of Production and Manufacturing: David W. Riccardi
Executive Managing Editor: Kathleen Schiaparelli
Senior Managing Editor: Linda Mihatov Behrens
Production Editor: Bob Walters
Manufacturing Buyer: Alan Fischer
Manufacturing Manager: Trudy Pisciotti
Marketing Manager: Angela Battle
Marketing Assistant: Rachel Beckman
Editorial Assistant: Melanie Van Benthuysen
Art Director: Jayne Conte
Cover Designer: Bruce Kenselaar
Cover Photo: Thomas Cole, "The Architect's Dream," 1840, oil on canvas, 53 ×
84-1/16 in., Toledo Museum of Art; Toledo, Ohio; Purchased with funds from Florence
Scott Libbey Bequest in Memory of her Father, Maurice A. Scott, no. 1949.162.
Art Studio: Marita Froimson

Printed in the United States of America
10 9 8 7 6 5 4 3 2 1

ISBN: 0-13-019074-8

Pearson Education LTD., *London*
Pearson Education Australia PTY, Limited, *Sydney*
Pearson Education Singapore, Pte. Ltd.
Pearson Education North Asia Ltd., *Hong Kong*
Pearson Education Canada, Ltd., *Toronto*
Pearson Educaciûn de Mexico S.A. de C.V.
Pearson Education - Japan, *Tokyo*
Pearson Education Malaysia, Pte. Ltd.

Contents

Introduction

The Mathematics of History

The author of a text on the history of mathematics is faced with a difficult question: how to handle the mathematics? There are several good choices. One is to give concise descriptions of the mathematics, which allows many topics to be covered. Another is to present the mathematics in modern terms, which makes clear the connection between the past and the present. There are many excellent texts that use either or both of these strategies.

This book offers a third choice, based on a simple philosophy: the best way to *understand* history is to *experience* it. To understand why mathematics developed the way it did, why certain discoveries were made and others missed, and why a mathematician chose a particular line of investigation, we should use the tools they used, see mathematics as they saw it, and above all think about mathematics as they did.

Thus to provide the best understanding of the history of mathematics, this book is a *mathematics* text, first and foremost. The diligent reader will be classmate to Archimedes, al Khwārizmī, and Gauss. He or she will be looking over Newton's shoulders as he discovers the binomial theorem, and will read Euler's latest discoveries in number theory as they arrive from St. Petersburg. Above all, the reader will experience the mathematical creative process firsthand to answer the key question of the history of mathematics: how is mathematics created?

In this text I have emphasized:

1. **Numeration, computation, and notation**. Notation both limits and guides. Limits, because it is difficult to think "outside the notation"; guides, because a good system of notation can suggest relationships worthy of further study. As much as possible, I avoid the temptation to "translate" a mathematical result into modern (mathematical) language or notation, for modern notation brings modern ways of thinking. In a similar vein, the means of computing and expressing numbers guides what discoveries one may make, and ultimately influences the direction taken by mathematicians.

2. **Mathematical results in their original form with their original arguments**. The most dramatic change in mathematics over the past

four millennia has been the standard of proof. What was acceptable to Pythagoras is no longer be acceptable today. The binomial theorem was, when Newton proposed it, nothing more than a conjecture, and while most of Euler's results on series summation are accepted today, his methods are not. But to provide a modern proof of a result would be historically inaccurate, while to omit how the these results were supported would be to neglect a vital part of the history of mathematics. To protect modern sensibilities, I will distinguish between *proofs*, where the term is used in the modern sense, and *demonstrations*, which contain some elements no longer acceptable in a modern context, and if a non-obvious result is presented without a proof, it will be labeled a *conjecture*.

3. **Mathematics as an evolving science**. The most important thing we can learn from a history of mathematics is that mathematics is created by human beings and not by semimythical demigods. The ideas of Newton, Euler, Gauss, and others originated from the mathematics they knew and the problems they saw around them; they made great contributions, but they also made mistakes, which were faithfully replicated (or caustically reviewed) by fellow mathematicians. Finally, no mathematical idea is born, fully mature and in the modern form: we will follow, where space and time permit, the development of mathematical ideas, through their birth pains, their early, formative years, and onwards to their early maturity and final, modern form.

To fully enter into all of the above for all of mathematics would take a much larger book, or even a multi-volume series: a project for another day. Thus, to keep the book to manageable length, I have restricted its scope to what I deem "elementary" mathematics: the fundamental mathematics every mathematics major and every mathematics teacher should know. This includes numeration, arithmetic, geometry, algebra, calculus, real analysis, and the elementary aspects of abstract algebra, probability, statistics, number theory, complex analysis, differential equations, and some other topics that can be introduced easily as a direct application of these "elementary" topics. Advanced topics that would be incomprehensible without a long explanation have been omitted, as have been some very interesting topics which, through cultural and historical circumstances, had no discernible impact on the development of modern mathematics.

Using this book

I believe a history of mathematics class can be a great "leveler", in that no student is inherently better prepared for it than any other. By selecting sections carefully, this book can be used for students with any level of background, from the most basic to the most advanced. However, it is geared towards students who have had calculus. In general, a year of calculus and proficiency in elementary algebra should be sufficient for all but the most advanced sections of the present work. A second year of calculus, where the student becomes more familiar and comfortable with differential equations and linear algebra, would be helpful for the more advanced sections (but of course, these can be omitted). Some of the sections require some familiarity with abstract algebra, as would be obtained by an introductory undergraduate course in the subject. A few of the problems require critiques of proofs by modern standards, which would require some knowledge of what those standards are (this would be dealt with in an introductory analysis course).

There is more than enough material in this book for a one-year course covering the full history of mathematics. For shorter courses, some choices are necessary. This book was written with two particular themes in mind, either of which are suitable for students who have had at least one year of calculus:

1. **Creating mathematics**: a study of the mathematical creative process. This is embodied throughout the work, but the following sections form a relatively self-contained sequence: 3.2, 6.4.1, 9.1.5, 9.2.3, 11.6, 13.1.2, 13.4.1, 13.6.1, 17.2.2, 20.2.4, 20.3.1, 22.4.

2. **Origins**: why mathematics is done the way it is done. Again, this is embodied throughout the work, but some of the more important ideas can be found in the following sections: 3.2, 4.3.1, 5.1, 6.1.2, 9.2.3, 9.4.5, 13.6.1, 15.1, 15.2, 19.5.2, 20.1, 20.2.5, 22.1, 23.3.2

In addition, there are the more traditional themes:

3. **Computation and Numeration**: For students with little to no background beyond elementary mathematics, or for those who intend to teach elementary mathematics, the following sections are particularly relevant: 1.1, 1.2, 2.1, 3.1, 7.1, 8.1, 8.3.1, 8.3.2, 9.1, 10.1.2 through 10.1.4, 10.2.1, 10.2.6, and 11.4.1.

4. **Problem Solving**: In the interests of avoiding anachronisms, these sections are not labeled "algebra" until the Islamic era. For students with no calculus background, but a sufficiently good background in elementary algebra, the following sequence is suggested: Sections 1.3, 2.2, 5.4, 7.4, 8.2, 8.3, 9.2, 9.3, 10.2.2, 10.3.1, 11.2, 11.3, 11.5.1, and for the more advanced students, Chapter 16.

5. **Calculus to Newton and Leibniz**: The history of integral calculus can be traced through Sections 1.4.2, 4.2, 4.3.3, 4.3.4, 5.6, 5.9, 6.1, 6.3, 10.4, 11.4.3, 12.4, 12.3.2, 13.1.2, 13.2, and 13.4.2. Meanwhile, the history of differential calculus can be traced through Sections 5.5.2, 6.6.2, 12.1.1, 13.3, 13.5, 13.6.

6. **Number theory**: I would suggest Sections 3.2, 5.8, 7.4, 8.3.5, 8.3.6, 10.2.5, 12.2, and all of Chapter 17.

Acknowledgments and Dedications

I'd like to dedicate this work to my parents, who got me started, and to Jacqui, who kept me going. Special thanks to Elizabeth McCrank and Millard Baublitz, at the College of General Studies, Boston University, and Jack and Kris Page, for looking over preliminary editions. In addition, the following reviewers provided invaluable advice during the course of writing:

David F. Anderson
University of Tennessee

Karen Z. Benbury
Bowie State University

LaDawn Haws
California State University, Chico

Victor E. Hill, IV
Williams College

Daniel J. Madden
University of Arizona

Paul C. Pasles
Villanova University

For anyone else with comments, suggestions, or criticisms, please feel free to contact me via email at `jeffs@bu.edu`.

Jeff Suzuki

Chapter 1

Egyptian Mathematics

The oldest evidence of mathematical activity is a carved wolf's bone, dating from around 30,000 B.C. and found in the Czech Republic. A series of 55 notches is carved into it, indicating that someone was keeping track of a number of objects (possibly the number of lunar months). The method of using one mark to indicate one unit is known as a **tally system**, and one form or another of tally numeration occurs in all cultures.

However, mathematics as we know it is a product of civilization: the culture of people who live in permanent cities. The oldest civilizations developed around two great river systems: the Egyptian civilization around the Nile, and the Mesopotamian around the Tigris-Euphrates. Both rivers underwent periodic floods, but Tigris-Euphrates floods tended to be sudden, destructive events that disrupted civilization, whereas Nile floods were regular, welcome events that refertilized the fields of the Egyptian farmers; it was not for nothing that the Greek historian Herodotus (fifth century B.C.) called Egypt "the Gift of the Nile." The Nile Delta, so named because it has the appearance of a Greek capital delta, Δ (and from which we get the generic term delta for a river's outlet into a sea or lake), forms Lower Egypt; Upper Egypt is formed by the region between the delta and the First Cataract, or set of waterfalls, at Aswan. According to tradition, Upper and Lower Egypt were united around 3100 B.C. by Menes, who became first pharaoh of Egypt.

1.1   Numeration

The ancient Egyptians wrote primarily on four types of materials: papyrus, made from the reeds that grow along the Nile River; leather, widely available from the numerous cattle raised by the Egyptians; cloth, manufactured from cotton or linen; and stone, which could be quarried throughout Egypt. Leather, cloth, and papyrus have many advantages. They are cheap; they are easy to write

1

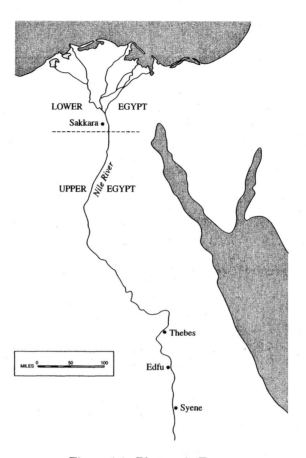

Figure 1.1: Pharaonic Egypt

on, and mistakes are easy to correct; they are flexible, so a long, unwieldy piece of material can be wrapped around a cylinder into a "roll," which means that documents written on leather, cloth, or papyrus can be transported and stored readily. However, they have one tremendous disadvantage: they decay rapidly, especially when exposed to water, insects, or sunlight.

Stone is immune to these hazards and can last thousands of years virtually unchanged. As a result, the oldest surviving documents we have are written on stone. However, the limitations of stone mean that any stone documents must be simple and concise. Consider the most common stone documents: tombstones. They generally list birth and death dates, and maybe one or two relevant facts about the deceased.

1.1.1 Hieroglyphic Numerals

By 2700 B.C., the Egyptians had developed a system of writing that adorned the stone walls of their temples and other buildings. Because the earliest known

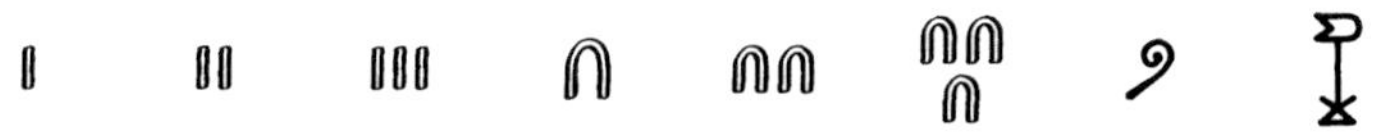
1
2
3
10
20
30
100
1000

Table 1.1: Egyptian hieroglyphic numerals

examples were found along the walls of temples, it was believed (erroneously) that the symbols had religious significance; hence, many centuries later, they were called hieroglyphics, which means "sacred writings" in Greek.

The hieroglyphic numbering system was a **tally system**. There were symbols for the unit, ten, one hundred, and all the powers of ten up to one million. Table 1.1 shows some of the hieroglyphic symbols. An early example of the use of hieroglyphics is a memorial carved during the reign of the Pharaoh Sahure, who went to war against the Libyans around 2500 B.C. Sahure returned victorious with a large amount of booty: 232,413 goats, according to the hieroglyphics along the bottom edge.

To indicate a number, these symbols would be repeated as many times as necessary.

Example 1.1. *Write two hundred thirty-one in hieroglyphic. This is two hundreds, three tens, and one unit:* 𓍢𓍢𓎆𓎆𓎆𓏺 .

This type of system, in which the value of a number can be determined by adding the values of the individual symbols used to write the number, is called **additive notation**. Since three hundred forty-two is also two units, four tens, and three hundreds, the number could also be written 𓏺𓏺𓎆𓎆𓎆𓎆𓍢𓍢𓍢, or even as 𓍢𓎆𓏺𓍢𓎆𓏺𓍢𓎆𓎆. In general, the Egyptians wrote their numbers, as we do, with the largest values first. However, since they wrote from right to left, this means that they placed the largest values in the rightmost position.

A civilization will inevitably have to deal with fractions, if only to deal with taxes! The Egyptians were no exception. Unlike ourselves, the Egyptians dealt primarily with unit fractions, those with a numerator of 1, although a symbol existed for $\frac{2}{3}$, and some evidence exists that late in Egyptian history, a symbol existed for $\frac{3}{4}$ as well. To write a fraction like $\frac{1}{13}$, first the symbols for thirteen would be written, and then a  , known as *ro* (an open mouth), would be placed above it:  . Some commonly used fractions would also have special symbols, though the only one used with any consistency was a symbol for two-thirds. During the Old Kingdom (roughly 3100 B.C. to 2200 B.C.), two-thirds was written as  but a thousand years later, this form would change into  . It would be another thousand years before this form was written 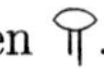.

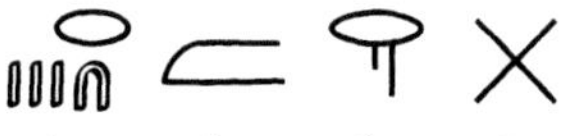
1/13
1/2
2/3
1/4

1.1.2 Hieratic Numerals

Hieroglyphics, the oldest form of Egyptian writing, obviously presents difficulties to anyone not an artist. Thus, by 2700 B.C., another form was developed, known as hieratic.

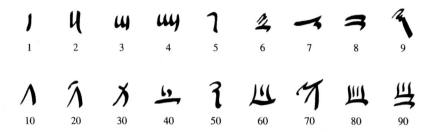

Table 1.2: Hieratic Numbers

Hieratic was, in general, a "cursive" form of the hieroglyphic, suitable for daily use. However, numbers in hieratic were written using an entirely different scheme, as Table 1.2 shows. We can see that the hieratic symbols for one, two, and three are clearly cursive forms of the hieroglyphic, but there the similarity ends. Although hieratic numeration is still a form of additive notation, it differs from hieroglyphics in an important respect: each symbol must be recognized as a whole.

To preserve the essential features of hieratic notation, we note that the hieratic symbols correspond to the numbers we write as 1, 2, 3, and so on up to 9, then 10, 20, 30, and so on up to 90, then 100, 200, and so on up to 900, and so on. Thus, we will write the number equivalents of the hieratic symbols using our own numbering system, separating the numbers by commas. We will call this method of writing pseudohieratic.

Example 1.2. *Write* 2035 *in pseudohieratic. This is two thousand, thirty, and five, or* $2000, 30, 5$.

Example 1.3. *The number* $1000, 600, 40, 3$ *is the number* 1643 *in standard notation.*

Ш λ̇
Hieratic $\frac{1}{23}$.

To indicate their fractions, the Egyptians shrank the *ro* into a dot, placed above the leading number.

The hieroglyphic *ro* could be stretched out to clearly indicate the entirety of the denominator; the hieratic · was not as flexible. For example, to write the fraction $\frac{1}{15}$, the Egyptian scribe would write down the symbols for 10, then 5, then place a · over the first term: we might read the result as $1\dot{0}, 5$. This could lead to confusion, for $1\dot{0}, 5$ could be read as $\frac{1}{10}$ and 5. It might seem that this ambiguity when dealing with fractions is an unacceptable defect of hieratic notation. However, the problems the Egyptians worked were almost always tied to some real world application. Thus, the intended value of a number could usually be determined from context.

See Problem 7.

Example 1.4. *The number of loaves required to feed one man for one week is determined to be* $1\dot{0}0, 5$*. This number can either be read as* $5\frac{1}{100}$*, or* $\frac{1}{105}$*. Given the context of the problem,* $\frac{1}{105}$ *is not a sensible interpretation, so the amount indicated must be* $5\frac{1}{100}$*.*

Shortly after the invention of hieratic and hieroglyphic numerals, the Egyptians began building pyramids, the very existence of which attests to a fairly high degree of mathematical sophistication. The step pyramid at Sakkar, near the Egyptian city of Memphis, is the oldest known; it was completed under the direction of the architect Imhotep (fl. 2650 B.C.). After Imhotep, even larger structures were built, the greatest of all being the Great Pyramid at Gizeh, finished around 2500 B.C. These structures have lasted for nearly five thousand years, to the credit of their Egyptian builders.

1.1 Exercises

1. What numbers are represented below?

   (a) ∩||

   (b) 99∩|||

   (c) ||∩9

   (d) ∩||9∩∩||9

   (e) ∩∩| ͡ᴖ

   (f) || ᗡ̃ ᗡ̃

2. A portion of a papyrus is shown in Figure 1.2, page 7. Translate the numbers along the sides of the rectangle, triangle, and trapezoidal figure.

3. Write in hieroglyphic and pseudohieratic notation:

   (a) one hundred thirty-seven

   (b) two thousand, five hundred forty-one

   (c) one thousand eleven

   (d) four hundred thirty-six

4. Write in hieroglyphic and pseudohieratic notation:

   (a) one eighth

   (b) one twenty-seventh

   (c) one fiftieth

   (d) one hundredth

5. Write in hieroglyphic and pseudohieratic notation:

   (a) Seven and one fifth

   (b) Forty-two and one ninth

   (c) Nine and a quarter

   (d) Two hundred eight, one twelfth, and one fifteenth

6. What numbers are represented by the following?

   (a) $\dot{4}0$

   (b) $\dot{2}0, 5$

   (c) $\dot{1}0, 8$

   (d) $8, \dot{1}0$

   (e) $10, 2, \dot{1}0, \dot{5}$

   (f) $10, 2, \dot{1}0, 5$

7. The following mixed numbers are written without a separator between the whole number and fractional parts. Thus, they can be read in a number of different ways. First, determine the possible interpretations of the number; then determine which interpretation is probably meant, given the context of the statement.

   (a) The pharaoh's army traveled $\dot{4}0, 5$ miles yesterday.

   (b) The feast required $\dot{1}0, 8$ loaves of bread.

   (c) After all the grain had been distributed, there was only $\dot{2}0, 5$ of a bushel remaining.

   (d) The time it takes to travel the length of Egypt is $\dot{1}0, 2$ years.

8. What are the advantages and disadvantages to the Egyptian system of numeration?

1.2   Arithmetic Operations

Around 280 B.C., a Greek-Egyptian scholar named Manetho compiled a chronicle of the history of Egypt. Manetho divided the reigns of the Egyptian pharaohs into 30 dynasties. These dynasties were divided into three groups; two intervals between the periods were later called interregnums (from the Latin words meaning "between the reigns"): the Old Kingdom (roughly 3100 B.C. to 2200 B.C.); a First Interregnum (2200 B.C. to 2100 B.C.); the Middle Kingdom (2100 B.C. to 1788 B.C.); a Second Interregnum (1788 B.C. to 1580 B.C.); and the New Kingdom (1580 B.C. to 1090 B.C.).

Around 1680 B.C., the Middle Kingdom was in the midst of a civil war when a new threat appeared. A mixed group of tribes, consisting primarily of Semites from Palestine and Hurrians from Asia Minor (modern-day Turkey), invaded Egypt. The Egyptians named them the Hyksos, meaning "Rulers of Foreign Lands." It has been suggested that the Biblical story of Joseph, who went to Egypt and achieved a position of great power in the government, was set against the background of these invasions; it seems reasonable that the ancient Hebrews comprised at least a part of the invading tribes.

The quarreling fragments of the Middle Kingdom were no match for the invaders, who founded their own dynasty. Meanwhile, the Egyptians assimilated elements of the Hyksos culture they found useful, including the horse, the chariot, and the compound bow. Thus, the Hyksos invasions represented an important turning point in Egyptian history, for the Hyksos brought the outside world to the attention of the Egyptians.

In the middle of the Hyksos period, a scribe named A'h-mosè wrote a treatise on Egyptian mathematics. Though not the oldest work of Egyptian mathematics, it is the most comprehensive that still exists as a complete work. In 1858, a portion of this treatise came into the hands of A. Henry Rhind, who purchased it near the resort town of Luxor, Egypt. Around 1900, the remaining portions of the papyrus turned up in the collections of the New York Historical Society, and

Figure 1.2: A portion of the Rhind papyrus, redrawn for clarity.

the parts of the papyrus were re-united; it is currently in the British museum. When new, the papyrus was about 18 feet long and 13 inches high. It is known variously as the Rhind papyrus, after its first European owner; the A'h-mosè papyrus, after its author; or the British Museum papyrus, after its current location. We will refer to it as the **Rhind papyrus**.

In this introduction, the scribe lists his name and says he was writing in the fourth month of the inundation season (around the beginning of Autumn), in the thirty-third year of the reign of the pharaoh A-user-Re, a Hyksos pharaoh who reigned sometime between 1788 and 1580 B.C. Thus, most scholars place A'h-mosè around 1650 B.C.

A'h-mosè claimed his work is based on an older manuscript, written during the reign of Ne-ma'et-Re, whom Egyptologists date between 1849 and 1801 B.C. This earlier text has never been found, leading some to believe that A'h-mosè actually composed the papyrus himself. Why would an author attribute his own work to someone else? In the days before movable type made printing cheap and easy, all written material had to be copied by hand: the very word **manuscript** comes from the Latin words meaning "written by hand." The laborious task of hand copying meant that only the most important works would be copied and, thus, preserved. To make it more likely a work would be copied, authors might claim their work was actually that of a great thinker of the past. Of course, this ran the risk of losing all association with a discovery, so many authors added their own names into the work, either as the "discoverer" or the "copyist"; in this way, they could obtain some measure of immortality.

1.2.1 Multiplication

It is clear from the Rhind papyrus and other sources that addition and subtraction were routinely and accurately performed by the Egyptians, though A'h-mosè gives no explanation of the techniques of addition and subtraction. The first mathematical operations actually described in the papyrus are those of multiplication and division. A'h-mosè did not explicitly explain how to multiply or divide; however, he "shows his work," allowing us to recreate the exact procedure used by the scribes. Multiplication is simply a repeated addition, and this is precisely how the Egyptians found products. The system used is now referred to as **doubling and halving**. To use this procedure, a table of the successive doubles of one factor of a product are generated; then the multiples that add up to the other factor are selected and the corresponding multiples are added to find the product.

Example 1.5. *Multiply* $10, 3$ *by* $20, 7$. *Begin by doubling one of the factors, say* $20, 7$.

$$
\begin{array}{ll}
1 & 20, 7 \\
2 & 50, 4 \\
4 & 100, 8 \\
8 & 200, 10, 6
\end{array}
$$

We note that 8, 4, and 1 together make $10, 3$, *the other factor. Mark these rows with a* \.

$$
\begin{array}{lll}
\backslash & 1 & 20, 7 \\
           & 2 & 50, 4 \\
\backslash & 4 & 100, 8 \\
\backslash & 8 & 200, 10, 6
\end{array}
$$

Now sum the multiples adjacent to 1, 4, and 8: namely, $20, 7$, $100, 8$, *and* $200, 10, 6$ *respectively. The total is* $300, 50, 1$, *which is the product.*

1.2.2 Division

The Egyptians recognized that division was the opposite of multiplication, and indeed, their method of division makes explicit use of this fact. Thus, the Egyptian scribe did not have to learn a separate procedure for division, as we do. Since the division of a by b asks how many times b must be multiplied to obtain a, the Egyptian procedure for finding the quotient was to form multiples of b and determine which set of multiples added to a.

Example 1.6. *Divide* $100, 30, 5$ *by* $10, 5$. *Since this is asking how many times* $10, 5$ *must be multiplied to obtain* $100, 30, 5$, *we begin by creating the multiple table for* $10, 5$:

$$
\begin{array}{ll}
1 & 10, 5 \\
2 & 30 \\
4 & 60 \\
8 & 100, 20
\end{array}
$$

We note that one more doubling is unnecessary, since the resulting multiple would exceed $100, 30, 5$. *Notice that* $100, 20$ *and* $10, 5$ *add to* $100, 30, 5$, *so we will indicate those rows.*

$$
\begin{array}{lll}
\backslash & 1 & 10, 5 \\
           & 2 & 30 \\
           & 4 & 60 \\
\backslash & 8 & 100, 20
\end{array}
$$

The quotient is thus 1 *and* 8, *or* 9.

To handle division with remainders, the table of multiples would have to be extended to include submultiples as well.

The Egyptians frequently began by finding $\frac{1}{2}$ or $\frac{2}{3}$ of the divisor, then halving as necessary. It is fairly simple to find $\frac{1}{2}$ of any number. However, the Egyptians more commonly found $\frac{2}{3}$ of a number, and it is reasonable to assume that the scribes learned a "2/3" times table much as we learn our ordinary times tables. One scribe was so focused on the method that to find $\frac{1}{3}$ of 3, he first found $\frac{2}{3}$ of 3, then halved the result!

To avoid the difficulties of having to distinguish between $10, 4$, $\dot{1}0, 4$, and $\dot{1}0, \dot{4}$, we will write unit fractions in nearly the form we are used to, though keep in mind that whereas we might write $\frac{1}{15}$, the Egyptians would have written $\dot{1}0, 5$.

Example 1.7. *Divide* $40, 2$ *by* 9. *We form the table of doubles.*

$$
\begin{array}{rl}
1 & 9 \\
2 & 10, 8 \\
\backslash \quad 4 & 30, 6
\end{array}
$$

We take the partial quotient 4, *which gives* $30, 6$; *an additional* 6 *is needed. By taking two-thirds of* 9, *we have:*

$$
\begin{array}{rl}
1 & 9 \\
2 & 10, 8 \\
\backslash \quad 4 & 30, 6 \\
\backslash \quad \frac{2}{3} & 6
\end{array}
$$

Thus, the quotient is $4, \frac{2}{3}$.

The first portion of the Rhind papyrus, referred to as the recto, consists of the division of 2 by the odd numbers between 3 and 101. A particularly important example is the division of 2 by 7, which A'h-mosè performed as follows. Note that A'h-mosè does not use the "obvious" submultiple, $\frac{1}{7}$; instead, he began by taking half, even though it results in fractions.

$$
\begin{array}{rl} \qquad\qquad
1 & 7 \\
\frac{1}{2} & 3, \frac{1}{2} \\
\backslash \quad \frac{1}{4} & 1, \frac{1}{2}, \frac{1}{4} \\
4 & 20, 8
\end{array}
\qquad
\begin{array}{rl}
1 & 7 \\
2 & 10, 4 \\
\frac{1}{4} \quad 4 & 20, 8
\end{array}
$$

What should we make of A'h-mosè's procedure, which produced the quotient $\frac{1}{4}, \frac{1}{28}$?

We can discern some important features of the multiple table: the indicated row shows that one fourth of 7 is $1, \frac{1}{2}, \frac{1}{4}$. To make the total equal to 2, an additional $\frac{1}{4}$ is necessary. To find $\frac{1}{4}$, A'h-mosè used an important fact: if $7 \cdot 4 = 28$, then $7 \cdot \frac{1}{28} = \frac{1}{4}$. If we "straighten out" the table, we would see:

$$
\begin{array}{cl}
1 & 7 \\
\frac{1}{2} & 3, \frac{1}{2} \\
\backslash \quad \frac{1}{4} & 1, \frac{1}{2}, \frac{1}{4} \\
2 & 10, 4 \\
4 & 20, 8 \\
\backslash \quad \frac{1}{28} & \frac{1}{4}
\end{array}
$$

Thus, the result of the division of 2 by 7 is $\frac{1}{4} \frac{1}{28}$ (i.e., $\frac{1}{4}$ and $\frac{1}{28}$).

It is from the recto that we get our best examples of how the ancient Egyptians performed computations. The purpose of the recto, however, was not to provide instruction in arithmetic. Instead, it provides, for later use, a table for the fractional decomposition of $\frac{2}{n}$ in terms of unit fractions. The need for such a table is clear. Since the Egyptian method of multiplication and division required doubling a factor or the divisor, it would occasionally be necessary to double unit fractions.

Obviously, if n is an even number, then $\frac{2}{n}$ is just a unit fraction, and thus A'h-mosè omitted quotients such as $\frac{2}{24}$ (though other tables did not). Finding $\frac{2}{n}$ is simply a matter of dividing n into 2, but there are many possible ways to accomplish this. A'h-mosè most often began by finding two-thirds of 2, but did not use any one procedure consistently. All scribes, however, seem to have used a set of "canonical" decompositions, which appear in the Rhind recto, summarized in Table 1.3.

Example 1.8. *Divide* $20, 7$ *by* $10, 1$. *The first part of the division is:*

$$
\begin{array}{cl}
1 & 10, 1 \\
\backslash \quad 2 & 20, 2
\end{array}
$$

We are 5 *units short. Taking the* $\frac{1}{11}$ *part and doubling (using Table 1.3) we obtain:*

$$
\begin{array}{cl}
1 & 10, 1 \\
\backslash \quad 2 & 20, 2 \\
\backslash \quad \frac{1}{11} & 1 \\
\frac{1}{6}, \frac{1}{66} & 2 \\
\backslash \quad \frac{1}{3}, \frac{1}{33} & 4
\end{array}
$$

The quotient is thus $2, \frac{1}{11}, \frac{1}{3}, \frac{1}{33}$.

To keep the string of fractions in a quotient from becoming too large, A'h-mosè sometimes used a reduction by the aliquot parts of a common divisor. First,

n	2 divided by n	n	2 divided by n	n	2 divided by n
3	$\frac{2}{3}$	30,7	$\frac{1}{24}, \frac{1}{111}, \frac{1}{296}$	70,1	$\frac{1}{40}, \frac{1}{568}, \frac{1}{710}$
5	$\frac{1}{3}, \frac{1}{15}$	30,9	$\frac{1}{26}, \frac{1}{78}$	70,3	$\frac{1}{60}, \frac{1}{219}, \frac{1}{292}, \frac{1}{365}$
7	$\frac{1}{4}, \frac{1}{28}$	40,1	$\frac{1}{24}, \frac{1}{246}, \frac{1}{328}$	70,5	$\frac{1}{50}, \frac{1}{150}$
9	$\frac{1}{6}, \frac{1}{18}$	40,3	$\frac{1}{42}, \frac{1}{86}, \frac{1}{129}, \frac{1}{301}$	70,7	$\frac{1}{44}, \frac{1}{308}$
10,1	$\frac{1}{6}, \frac{1}{66}$	40,5	$\frac{1}{30}, \frac{1}{90}$	70,9	$\frac{1}{60}, \frac{1}{237}, \frac{1}{316}, \frac{1}{790}$
10,3	$\frac{1}{8}, \frac{1}{52}, \frac{1}{104}$	40,7	$\frac{1}{30}, \frac{1}{141}, \frac{1}{470}$	80,1	$\frac{1}{54}, \frac{1}{162}$
10,5	$\frac{1}{10}, \frac{1}{30}$	40,9	$\frac{1}{28}, \frac{1}{196}$	80,3	$\frac{1}{60}, \frac{1}{332}, \frac{1}{415}, \frac{1}{498}$
10,7	$\frac{1}{12}, \frac{1}{51}, \frac{1}{68}$	50,1	$\frac{1}{34}, \frac{1}{102}$	80,5	$\frac{1}{51}, \frac{1}{255}$
10,9	$\frac{1}{12}, \frac{1}{76}, \frac{1}{114}$	50,3	$\frac{1}{30}, \frac{1}{318}, \frac{1}{795}$	80,7	$\frac{1}{58}, \frac{1}{174}$
20,1	$\frac{1}{14}, \frac{1}{42}$	50,5	$\frac{1}{30}, \frac{1}{330}$	80,9	$\frac{1}{60}, \frac{1}{356}, \frac{1}{534}, \frac{1}{890}$
20,3	$\frac{1}{12}, \frac{1}{276}$	50,7	$\frac{1}{38}, \frac{1}{114}$	90,1	$\frac{1}{70}, \frac{1}{130}$
20,5	$\frac{1}{15}, \frac{1}{75}$	50,9	$\frac{1}{36}, \frac{1}{236}, \frac{1}{531}$	90,3	$\frac{1}{62}, \frac{1}{186}$
20,7	$\frac{1}{18}, \frac{1}{54}$	60,1	$\frac{1}{40}, \frac{1}{244}, \frac{1}{488}, \frac{1}{610}$	90,5	$\frac{1}{60}, \frac{1}{380}, \frac{1}{570}$
20,9	$\frac{1}{24}, \frac{1}{58}, \frac{1}{174}, \frac{1}{232}$	60,3	$\frac{1}{42}, \frac{1}{126}$	90,7	$\frac{1}{56}, \frac{1}{679}, \frac{1}{776}$
30,1	$\frac{1}{20}, \frac{1}{124}, \frac{1}{155}$	60,5	$\frac{1}{39}, \frac{1}{195}$	90,9	$\frac{1}{66}, \frac{1}{198}$
30,3	$\frac{1}{22}, \frac{1}{66}$	60,7	$\frac{1}{40}, \frac{1}{335}, \frac{1}{536}$	100,1	$\frac{1}{101}, \frac{1}{202}, \frac{1}{303}, \frac{1}{606}$
30,5	$\frac{1}{30}, \frac{1}{42}$	60,9	$\frac{1}{46}, \frac{1}{138}$		

Table 1.3: The Canonical Decompositions of $\frac{2}{n}$.

the scribe would write the fractions as parts of a common divisor, then sum the parts. The sum would then be rewritten as the sum of aliquot and two-thirds parts of the divisor, and the fractions reduced to unit fractions.

> The **aliquot parts** of a number are its proper divisors. For example, the aliquot parts of 12 are 1, 2, 3, 4, and 6.

1.2 Exercises

1. Multiply.

    (a) Forty-three times eighteen

    (b) Twelve times ninety-seven

    (c) Thirteen times seventy-three

    (d) Eleven times twenty-one

2. Divide.

    (a) Forty-eight divided by twelve

    (b) Ninety-six divided by eight

    (c) One hundred fifty-three divided by nine

    (d) One hundred thirty-two divided by six

    (e) Two hundred sixteen divided by eighteen

3. Divide. Simplify your answer where possible.

   (a) $20, 5$ by $10, 3$ (b) $40, 2$ by 8 (c) $90, 8$ by $30, 1$

4. Construct a 2/3 times table for use with pseudo-hieratic.

5. Divide. Begin by taking the 2/3 part.

   (a) Twenty-six divided by twelve
   (b) Forty-nine divided by twenty-one
   (c) Eleven divided by six
   (d) One hundred sixty-one divided by forty-two

6. Perform the divisions in Problem 5, but begin by taking the 1/2 part.

7. Divide, beginning with the specified part.

   (a) $2 \div 11$, initial part $\frac{1}{6}$ (c) $2 \div 27$, initial part $\frac{2}{3}$
   (b) $2 \div 17$, initial part $\frac{1}{12}$ (d) $2 \div 31$, initial part $\frac{1}{20}$

8. Find unit fractional decompositions for $\frac{2}{103}$, $\frac{2}{105}$, and $\frac{2}{107}$.

9. There are a number of ways of decomposing $\frac{2}{n}$ into a simple sum of unit fractions. Two of these methods are:

$$\frac{2}{n} = \frac{1}{\frac{n+1}{2}} + \frac{1}{\frac{n(n+1)}{2}} \qquad\qquad \frac{2}{pq} = \frac{1}{p\frac{p+q}{2}} + \frac{1}{q\frac{p+q}{2}}$$

   Which of the decompositions in Table 1.3 can be found using these formulas?

10. The fraction $\frac{2}{3}$ is the only non-unit fraction consistently used by the Egyptians. The following suggests one reason why.

   (a) Find the unit fractional decompositions of $\frac{2}{11}$. Begin by taking halves of $10, 1$.
   (b) Find a different decomposition by first taking $\frac{2}{3}$ of $10, 1$, then taking halves.
   (c) What is the advantage to taking an initial $\frac{2}{3}$ over taking an initial $\frac{1}{2}$?
   (d) For $\frac{2}{13}$, $\frac{2}{15}$, and $\frac{2}{17}$, find decompositions by first taking an initial $\frac{2}{3}$ of the numbers and then taking halves.
   (e) Compare the decompositions. In which cases does taking $\frac{2}{3}$ of the number first and then halving produce shorter decompositions?
   (f) None of the decompositions in the Recto require more than four unit fractions. For which, if any, numbers between 3 and 101 will the decomposition by halves require more than four unit fractions?

11. Another advantage of using $\frac{2}{3}$ follows.

    (a) Prove that taking $\frac{2}{3}$ of a number n divisible by 3 always yields a two-term decomposition of $\frac{2}{n}$ into unit fractions with even denominators.

    (b) Prove that by taking successive halves of n, you will always produce a decomposition of $\frac{2}{n}$ into unit fractions with even denominators.

    (c) Under what conditions will this decomposition consist of two terms?

12. Respond to the following objection to the Egyptian system of notation: "Because of the restriction to unit fractions, the Egyptians expressed simple fractions such as $\frac{7}{8}$ as a clumsy sum of unit fractions, $\frac{1}{2} + \frac{1}{4} + \frac{1}{8}$."

13. Explain why doubling and halving works for multiplication.

14. What are the advantages to the Egyptian method of multiplication and division? What are the disadvantages?

1.3   Problem Solving

The mathematical papyri of the ancient Egyptians show that they were able to solve linear equations easily, and even some nonlinear equations were within their capabilities. These problems were stated and solved entirely without notation: in other words, every problem solved by an Egyptian scribe was a "word problem," and every solution was obtained, in effect, by talking through to the answer.

1.3.1   Linear Equations

The first problem solved in the Rhind papyrus is

Problem 1.1. *A number and its seventh make* 10, 9. *What is the number?*

In some cases, A'h-mosè solved these using the *aha* (or "heap") method, now called the **method of false position** or the **method of false solution**.

Solution. Suppose 7 is the solution. Then 7, and its seventh, will make 8, as shown:

$$
\begin{array}{rcc}
 & 1 & 7 \\
 & \frac{1}{7} & 1 \\
\text{Total} & 1, \frac{1}{7} & 8
\end{array}
$$

Since 7 results in 8, how many times larger must a number be to result in 10, 9? First, divide 10, 9 by 8.

$$
\begin{array}{cl}
1 & 8 \\
\backslash \quad 2 & 10,6 \\
\frac{1}{2} & 4 \\
\backslash \quad \frac{1}{4} & 2 \\
\backslash \quad \frac{1}{8} & 1
\end{array}
$$

Thus, multiply 7 by 2, $\frac{1}{4}$, $\frac{1}{8}$.

$$
\begin{array}{cl}
1 & 7 \\
\backslash \quad 2 & 10,4 \\
\frac{1}{2} & 3,\frac{1}{2} \\
\backslash \quad \frac{1}{4} & 1,\frac{1}{2},\frac{1}{4} \\
\backslash \quad \frac{1}{8} & \frac{1}{2},\frac{1}{4},\frac{1}{8} \\
\text{Total} & 10,5,\frac{1}{2},\frac{1}{2},\frac{1}{4},\frac{1}{4},\frac{1}{8}
\end{array}
$$

We can simplify the total to $10,6,\frac{1}{2},\frac{1}{8}$. This is then verified by multiplication.

□

A'h-mosè did not give a procedure for solving such problems; the student was presumably left to generalize from the examples. It is worth noting that every example A'h-mosè gave has a check on the solution at the end.

Example 1.9. *A third of a quantity and half of the quantity make* $10,2$. *Suppose the solution is* 6.

$$
\begin{array}{cl}
1 & 6 \\
\backslash \quad \frac{1}{2} & 3 \\
\backslash \quad \frac{1}{3} & 2 \\
\text{Total} & 5
\end{array}
$$

Since 6 *results in* 5, *how many times larger must* 6 *be to result in* $10,2$? *Divide* $10,2$ *by* 5, *and multiply the guess,* 6, *by this quotient.*

$$
\begin{array}{cl}
1 & 5 \\
\backslash \quad 2 & 10 \\
\frac{1}{5} & 1 \\
\backslash \quad \frac{1}{3},\frac{1}{15} & 2 \\
\text{Total} & 2,\frac{1}{3},\frac{1}{15}
\end{array}
$$

Multiply 6 *by* 2, $\frac{1}{3}$, $\frac{1}{15}$.

$$
\begin{array}{cl}
1 & 6 \\
\backslash \quad 2 & 10,2 \\
\backslash \quad \frac{1}{3} & 2 \\
\backslash \quad \frac{1}{15} & \frac{1}{3},\frac{1}{15} \\
\text{Total} & 10,4,\frac{1}{3},\frac{1}{15}
\end{array}
$$

In most cases, however, the *aha* method was not used; instead, the Egyptians solved problems similarly to the way we would solve an equation such as $3x = 5$: by dividing the right hand side by the coefficient of the unknown.

1.3.2 Nonlinear Equations

All of the problems in the Rhind papyrus result in simple, linear equations, though in other papyri, there is evidence that the Egyptians solved nonlinear equations as well. However, the papyri or leather rolls on which this claim is based are incomplete, which means that the texts must be interpolated, something that the unique nature of mathematics makes possible. In the Golenishchev See Problem 1. (or Moscow) papyrus, the scribe posed the problem of finding the length and breadth of a rectangle whose area is 12 and whose breadth is $\frac{1}{2}$, $\frac{1}{4}$ the length. The scribe's solution is quite clever: he divides 1 by $\frac{1}{2}$, $\frac{1}{4}$ to get $1\frac{1}{3}$, then he multiplies the area by $1\frac{1}{3}$, which makes 16. The square root of 16 gives the length, 4; $\frac{1}{2}$, $\frac{1}{4}$ of this is the breadth, 3. See Problem 4.

The Berlin papyrus (ca. 1320 B.C.) and the Kahun papyrus (ca. 1700 B.C.) both provide examples of what might be interpreted as solutions to nonlinear problems. The Kahun papyrus's problem statement is missing entirely, and the problem has been interpolated from the solution; hence, we shall not deal with it here. The Berlin papyrus, on the other hand, contains two nonlinear problems. The first is:

Problem 1.2. *A square whose area is* 100 *square cubits is equal to two smaller squares; the side of one is* $\frac{1}{2}$, $\frac{1}{4}$ *[i.e., $\frac{1}{2}$ and $\frac{1}{4}$] the side of the other. What are the sides of the unknown squares?*

The solution is found by the method of false position: assume a square of side 1 as the larger; the smaller is $\frac{1}{2}$, $\frac{1}{4}$ in size. Thus, the area of the smaller square is $\frac{1}{2}$, $\frac{1}{16}$, and in total both squares are $1\frac{1}{2}$, $\frac{1}{16}$. Take the square root, which is $1\frac{1}{4}$. The square root of 100 is 10; divide 10 by $1\frac{1}{4}$, which gives 8. This is the size of the larger square; the smaller square is thus 6.

How did the scribe extract the square root of $1\frac{1}{2}$, $\frac{1}{16}$? It is not unreasonable to suppose that the scribe had access to a table of squares and square roots, though such a table would have to be very extensive if it included the square roots of numbers such as $1\frac{1}{2}$, $\frac{1}{16}$. One possibility is that the scribes frequently considered a sum of fractions as "parts" of a common denominator; thus, $1\frac{1}{2}$, $\frac{1}{16}$ as parts of 16, is 25; the square root of which would be 4 parts of 5, or $1\frac{1}{4}$.

1.3 Exercises

1. Suggest plausible interpolations for the asterisked phrases:

    (a) "The product of 2 and *** is 11."

    (b) "The *** of 5 and 3 is 1 2/3."

    (c) "Itself and its third make ten. *** 4 ***. Solution 7 1/2."

2. Verify that $10, 6, \frac{1}{2}, \frac{1}{8}$ solves A'h-mosè's first problem.

3. Solve using the "aha" method:

   (a) Itself, its third are 5.   (c) Itself, its quarter are 7.

   (b) Itself, its fifth are 20.   (d) Itself, its third, its fifth are 12.

4. Explain the reasoning behind the scribe's method in solving the problems from the Golenishchev and Berlin papyri.

5. Solve the following area problems, using the method from the Golenishchev papyrus.

   (a) The area of a rectangle is 20,4; its length is $\frac{2}{3}$ the breadth.

   (b) The area of a rectangle is 21; its length is $\frac{1}{3}, \frac{1}{4}$ the breadth.

   (c) The area of a rectangle is 60; its length is $2, \frac{1}{3}, \frac{1}{15}$ the breadth.

6. Solve using the method in the text: the area of a square of 100 square cubits is equal to two smaller squares, one of which is $3\frac{1}{7}, \frac{1}{4}, \frac{1}{28}$ the other side. Find the sides of the squares.

7. To what form must a linear equation be reduced to in order to make the method of false position work?

1.4 Geometry

The annual flooding of the Nile, some claim, required the constant redrawing of property lines, which in turn fostered the development of geometry. Although this sounds plausible, it is important to note that the geometry needed, and thus the geometry developed by the Egyptians, was a purely practical geometry, concerning itself with quantities associated with geometric objects and not their abstract properties.

1.4.1 Slopes

The geometry of the Egyptians is, not surprisingly, filled with problems relating to pyramids. In the Rhind papyrus, A'h-mosè discussed how to find the *seked* of a pyramid.

Problem 1.3. *A pyramid is* $200, 50$ *cubits high with a base of* $300, 60$. *Find its* seked.

Solution. Take half of the base $300, 60$, which is $100, 80$. Divide $100, 80$ by its height, $200, 50$, to get $\frac{1}{2}, \frac{1}{5}, \frac{1}{50}$ of a cubit; this is its *seked*. $\qquad\square$

The exact meaning of the word *seked* is debatable. However, the scribe's method of computing the *seked* indicates it is equivalent to the **slope of a line**; more precisely, the *seked* is what we would call the reciprocal of the slope. For master builders, the *seked* was probably more useful than our slope.

See Problem 4.

Example 1.10. *If a pyramid is* 100, 50 *cubits high, and the side of its base is* 400 *cubits, what is its* seked*? Half of the base is* 200. *Divide* 200 *by* 100, 50 *to get* 1, $\frac{1}{3}$ *of a cubit, which is its* seked.

1.4.2   Areas and Volumes

The ancient Egyptians calculated areas and volumes of various figures, but did not have area or volume formulas as we do. Instead, A'h-mosé and other scribes gave worked out examples, and expected the reader to generalize from them. For a circle, A'h-mosé used:

Problem 1.4. *Find the area of a circle with a diameter of* 9 khet.

Solution. [**Area of a Circle**] Take away $\frac{1}{9}$ of the diameter, leaving 8. Multiply 8 by itself, giving the area: 60, 4 *setat*.   □

It is not known if the scribes knew their formula yielded the approximate area of a circle. From this description, we can discern that one *setat* is equal to a square one *khet* on a side. By comparing the Egyptian formula for the area of a circle and our own, we can determine the so-called "Egyptian value" for π: approximately 3.16.

See Problem 5.

Problem 51 of the Rhind papyrus, shown in Figure 1.2 on page 7, shows how A'h-mosè calculated the areas of triangles:

Problem 1.5. *A triangle has side* 10 khet *and base* 4 khet. *What is its area?*

Solution. [**Area of a Triangle**] Take half of 4 to make a rectangle [of 2 *khet* by 10 *khet*]. The area of the rectangle is the area of the triangle: 20 *setat*.   □

The procedure described by A'h-mosè gives the correct area only for right triangles, but it seems that the scribes did not know this.

Problem 52 of the Rhind papyrus deals with a cut off triangle (probably a trapezoidal figure):

See Problem 7.

Problem 1.6. *Find the area of a cut off triangle with side of* 20 khet, 6 khet *base, and* 4 khet *the cut off line.*

Solution. [**Area of a Trapezoid**] Add the base and cut off line, which is 10. Take half of 10, which is 5. Multiply 20 by 5, to get 100 *setat*, the area.   □

See Problem 8.

On the walls of a temple in Edfu (about halfway between the ancient capital of Thebes and the first cataract of the Nile, where the kingdom of Egypt had its southernmost boundary), another geometrical formula can be found. The temple wall lists gifts of land to the temple; most of the plots have four sides, and the calculation of the area suggests that the formula $(\frac{a+c}{2})(\frac{b+d}{2})$ was used, where a and c, and b and d are the lengths of opposite sides of the quadrilateral. In some cases, the plot is triangular, and one side is ignored (in our terms, d would be equal to 0).

The Egyptian procedure for finding volumes is essentially the same as our own: first, find the area of the base, then multiply the area by the height. Curiously, volume computations precede computation of areas; in fact, the aforementioned formula for the area of a circle actually appeared when A'h-mosè showed how to calculate the volume of a cylindrical granary.

What is perhaps the greatest achievement of the Egyptian geometers is a method of correctly finding the volume of the frustrum of a square pyramid (the frustrum is the lower portion of a pyramid, when a plane parallel to the base separates the pyramid into two portions). The formula is found in Problem 14 of the Golenishchev (or Moscow) papyrus, which appears to be contemporaneous with the Rhind. The bracketed words are interpolations.

Problem 1.7. *Find the volume of a truncated pyramid with a height of 6 cubits, and [square] top and bottom bases [whose sides are] 2 and 4 cubits.*

Solution. Square the base of 4 to get 10, 6; multiply the base of 2 by the base of 4 to get 8; square the base of 2 to get 4. Add 10, 6, 8, and 4 to get 20, 8. Multiply the height 6 by 1/3, to get 2. Multiply 20, 8 by 2 to get 50, 6, which is the volume.   □

This is a remarkable achievement, since if the pyramid's base is square, the result is exact. Presumably, the scribes knew some way of finding the volume of a whole pyramid, though no existing papyrus shows such a computation.

1.4 Exercises

1. Calculate the *seked* of the following pyramids:

   (a) Base of 90; height of 100

   (b) Base of 100; height of 50

   (c) Base of 40; height of 20

2. Find the areas of the following triangles, using the Egyptian formula.

   (a) Side 10, base 6   (c) Side 4, base 20

   (b) Side 20, base 8   (d) Side 6, base 24

3. Calculate the volumes of the frustrums of the following pyramids.

   (a) Lower side 8, upper side 4; height 12.

   (b) Lower side 100, upper side 2; height 50.

   (c) Lower side 40, upper side 20; height 20.

4. Why might knowing the *seked* be more useful than knowing the slope?

5. In the calculation of the seked, what is assumed about the pyramid whose *seked* is being calculated?

6. Consider the Egyptian computation for finding the area of a circle.

   (a) Convert the Egyptian method into an "area formula" for the area of a circle with radius r.

   (b) Equate the Egyptian formula to our own formula for the area of a circle, and solve for π to determine the "Egyptian value of π."

   (c) Is it proper to say that "the Egyptians approximated π as 3.16?" Explain.

7. The Egyptian method of finding the area of a triangle is only valid in certain circumstances.

   (a) One interpretation is that it gives the area of an isosceles triangle. Assume that this is the case; how accurate is the Egyptian method?

   (b) How must "side" be interpreted to make the formula correct?

   (c) The triangle in the portion of the Rhind papyrus shown on page 7 corresponds to the triangle whose area is calculated in Problem 1.5. Does the figure support this interpretation of "side"?

8. Problem 1.6, from the Rhind papyrus, deals with finding the area of a "cut off" triangle.

   (a) Under what conditions will the Egyptian method give the correct area?

   (b) The trapezoid in the portion of the Rhind papyrus shown on page 7 corresponds to the trapezoid whose area is calculated in Problem 1.6. Does the figure support the interpretation of "side" necessary to make the area formula exact?

9. Comment on the accuracy of the method for calculating the areas of quadrilaterals from the temple in Edfu.

10. Show that the procedure for calculating the volume of the frustrum of a pyramid is correct if the pyramid has a square base.

11. Mystics claim that the dimensions of the Great Pyramid were deliberately designed to precisely embody certain mathematical constants, such as π. All reputable mathematicians, Egyptologists, and historians claim that this is nonsense. Explain.

Chapter 2

Babylonian Mathematics

Mesopotamia is the name given to the region lying between the Tigris and Euphrates rivers, which corresponds roughly to modern day Iraq. Today, the Tigris and Euphrates merge about a hundred miles inland of the Persian Gulf, though in ancient times, the two rivers flowed separately into the ocean. The name Mesopotamia itself is purely descriptive, coming from the Greek words meaning "between the rivers."

Many cultures inhabited this land in ancient times, the most ancient of which were the Sumerians. By 3000 B.C., they had developed as a group of city states in Lower Mesopotamia, the region closer to the Persian Gulf. The best known of these city states was probably Ur, mentioned in the Bible as "Ur of the Chaldees." Each city state was a separate political entity; as a result of this, they were eventually conquered by their more powerful neighbors: first the Akkadians, then the Babylonians. At the same time A'hmosè was writing the Rhind papyrus, Hammurabi, a Babylonian King, had inscribed on a pillar the world's first written code of law.

2.1 Numeration and Computation

It is convenient, though not entirely accurate, to refer to the various cultures of Mesopotamia as Babylonian, a practice we will follow here. The Egyptian scribes usually wrote on papyrus, made from reeds, or on rock. The Babylonians had neither of these in abundance; instead, they had mud. Fortunately, mud could be written on with a wooden stylus, and if the resulting "page" was baked, it became a brick-hard clay tablet, much more durable than Egyptian papyri. In contrast to a handful of Egyptian mathematical papyri, over half a million clay tablets exist from Mesopotamia, many of them untranslated.

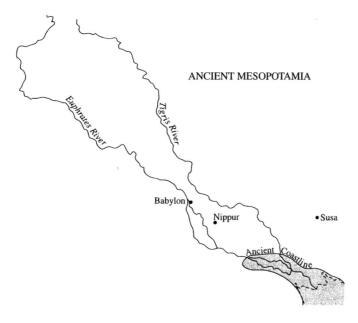

Figure 2.1: Mesopotamia (showing ancient coastline)

2.1.1 Positional Notation

As in the Egyptian hieroglyphic system, the Babylonians indicated large numbers by repeating their number symbols. Babylonian writing is known as cuneiform, which means "wedge-shaped" in Greek. Unlike Egyptian hieratic, the Babylonians only had two symbols: a units symbol, and a tens symbol (see margin). The Babylonians never developed a shorthand corresponding to Egyptian hieratic; to write a number such as forty-seven, they would repeat the tens symbol four times, and the units symbol seven times. To make up for this clumsier system of writing numbers, the Babylonians were forced to invent something the Egyptians never did: **positional notation**. The Babylonian system was sexagesimal, in which the positions differed by powers of sixty. Since our own system of notation is positional, the Babylonian system appears to be more sophisticated than the Egyptian. Since it will be useful (though not essential) to convert from base 10 to base 60, we review here the (modern) rules for base conversion:

1. Divide the base-10 number by 60, and record the remainder.

2. Divide the quotient from Step 1 by 60 again, and record the remainder.

3. Repeat the process until the quotient cannot be divided by 60 (in this case, the quotient will be 0 with a remainder of the original number).

4. The number, in base 60, will be the remainders in the reverse order.

Cuneiform 1

Cuneiform 5

Cuneiform 10

Cuneiform 47

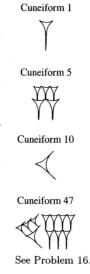

See Problem 16.

Example 2.1. *Convert* 148 *into base* 60. *The divisions yield:*

$$148 \div 60 = 2 \text{ remainder } 28$$
$$2 \div 60 = 0 \text{ remainder } 2$$

Thus, 148 *written in base* 60 *is* 2, 28.

To convert back, remember that each place is 60 times the value of the previous place.

Example 2.2. *Convert* 3, 20, 5 *into base* 10. *The number is* $3 \times 60^2 + 20 \times 60^1 + 5 \times 60^0$, *or* 12, 005.

Positional notation has many advantages, particularly when dealing with division and the inevitable fractions; as we shall see, Babylonian division, more so than the Egyptian, made use of the fact that division is simply the inverse of multiplication. Even easier is the operation of addition with fractional parts: to add $\frac{1}{10}$ and $\frac{1}{15}$ requires finding a common denominator, although if both are expressed as sexagesimal fractions, they are already expressed in terms of a common denominator: $\frac{6}{60}$ and $\frac{4}{60}$. Moreover, the operations of arithmetic have a similarity they do not have in non-positional systems. For an Egyptian scribe, the division of $\frac{1}{2}$ by 5 and the division of 5 by 5 were very different operations, but if we were to write the same problem using our modern positional notation, we would note the similarity: the first problem is $0.5 \div 5$, whereas the second is $5 \div 5$. Except for the location of the decimal point, the two problems are identical.

However, positional notation leads to several problems. For example, what should one make of the number shown in the margin? This might indicate three units, or one sixty and two units, or even one thirty-six hundred, one sixty, and one; still more interpretations are possible. To make matters even more complicated, the Babylonians did not indicate the position of the units place with a symbol (as we do using a decimal point). Thus, the number 1 might mean 1, 60, 3600, or even $\frac{1}{60}$ or $\frac{1}{3600}$. Table 2.1 shows some numbers, their sexagesimal equivalents, and how the Babylonians would have written them. (Note the use of the ; in standard sexagesimal notation to represent the location of the sexagesimal point)

As nearly as can be determined from translated cuneiform texts, the only way to read the number correctly is by context. Given the nature of the problem, only one of the interpretations is likely to make sense, so that only one of the numbers need be considered. In the oldest Babylonian texts, context alone can determine the value of a number, though by the Seleucid era (250 B.C.), a special symbol had been developed to indicate a blank space in between symbols; thus, numbers like thirty-six hundred two could be written with a blank space indicating the absence of sixties. However, this did not solve the problem completely, since the symbol was not used to indicate blank spaces at the end of a number.

Blank symbol

Cuneiform 1, 0, 2

Number (Base 10)	Number (Base 60)	Number (Cuneiform)
60	1, 0	𒁹
64	1, 4	𒁹 𒐘
3600	1, 0, 0	𒁹
3604	1, 0, 4	𒁹 𒐘
1/60	0; 1	𒁹

Table 2.1: Babylonian Equivalents of Some Numbers.

As we did with Egyptian hieratic, we should develop a pseudo-cuneiform to represent Babylonian numbers; thus, we might write

$$1 \quad 10 \quad 1 \quad 1$$
$$1 \quad 10 \quad 1 \quad 1$$

to represent two sixties and twenty-four units. We will instead write this number as a traditional sexagesimal, separating the places by commas: $2, 24$. It is important to remember, however, that when we write "2" the Babylonian would have written "1 1," repeating the symbol for units twice, and so on; moreover, a number we write as $1, 0, 4$ would have been written by the Babylonians as

$$1 \quad \begin{matrix} 1 & 1 \\ 1 & 1 \end{matrix}$$

and the correct interpretation would have to be made based on the context of the number.

2.1.2 Computation

The two major eras from which Babylonian mathematical texts exist are the Old Babylonian era, between 1800-1600 B.C. (and thus contemporaneous with the Egyptian A'hmosè), and the Seleucid era (ca. 200 B.C.). The texts fall into one of two categories: **table texts**, which consist of several parallel columns of numbers; and **problem texts**, which provide various algebraic and geometric problems to be solved. First, we shall examine the table texts; a reproduction of one made around 1350 B.C. and found in the temple library at Nippur is shown in Figure 2.2, with the front on the left and the back on the right.

What is the purpose of this tablet? In the upper left of the front (not shown), we can read the number 18. In the left column, we can read the numbers 2 through 11. In the right column, the numbers 36, 54, and 72 can be read, which are clearly the products of 18 by 2, 3, and 4. It would appear, then, that this is a table showing the products of 18 with the numbers 2 through 11. The back side of the tablet gives the products of 18 with the numbers 12 through 20, 30, 40,

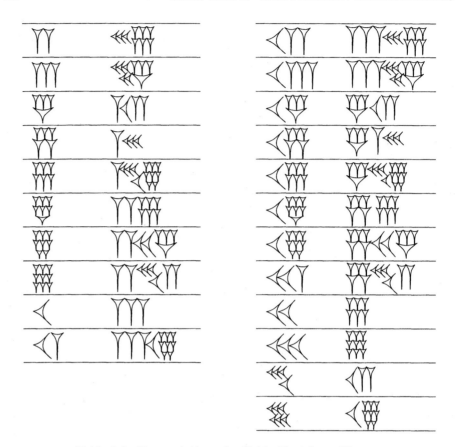

Table 2.2: Transcription of a Table Text from Nippur.

and 60. There are thousands of similar tablets, generally giving the products of a number with the numbers 1 through 20, 30, 40, 50 (not present on the Nippur tablet).

For certain multiplications, the product could be read off the table. Care must be taken in properly placing the orders of the various sexagesimals.

Example 2.3. *Multiply* 18 *by* 10. *The tablet indicates the result is* 3, *which should be read as* 3 *sixties, or* 180.

There is no hard evidence of how other multiplications were accomplished though the virtue of additive notation is that any multiplications can be performed symbol by symbol, and the results added together.

Example 2.4. *Multiply* 18 *by* 13. 18 *times* 10 *is* 3 *sixties;* 18 *times* 3 *is* 54, *so the result is* 3, 54.

A complete multiplication table for Babylonian arithmetic ought to go up to 60×60. Many existing tables go farther, however, and include such products as a number and $1, 20$, a number and $1, 15$, and so on. These additional products may seem unnecessary, until it is realized that in addition to these multiplication tables, other tablets exist, consisting of lists of *reciprocals*; one such table is reproduced on the left. Recall that the Babylonians had no method of indicating the positional value; thus, the Babylonian tablets indicate the reciprocal of 2 as 30.

2	$0; 30$
3	$0; 20$
4	$0; 15$
5	$0; 12$
6	$0; 10$
8	$0; 7, 30$
9	$0; 6, 40$
10	$0; 6$

A table of reciprocals.

It is with the table of reciprocals that the true value of positional notation makes itself apparent. Recall the Egyptians had to resort to a fairly complicated process of finding multiples and submultiples of the divisor to make the dividend. Although there exists some evidence that the Babylonians performed long division as we do, it seems the more common method was to divide by multiplying by the reciprocal of the divisor. Part of the simplicity of this method is the fact that the sexagesimal point can be shifted to obtain the correct order of magnitude.

See Problem 15.

Example 2.5. *Divide* $1, 47$ *by* 5. *This is the same as multiplying* $1, 47$ *by* $0; 12$; *the product is* $21; 24$.

Example 2.6. *Divide* $8, 16, 5$ *by* $6, 40$. *Since* $6, 40$ *is the reciprocal of* 9, *then* 9 *is the reciprocal of* $6, 40$, *so we can multiply* $8, 16, 5$ *by* 9, *where we leave the sexagesimal point to be determined later. This multiplication yields* $1, 14, 24, 45$. *Determining the correct place to put the sexagesimal point is difficult, requiring that we keep track of magnitudes: the division of a three digit number by a two digit number ought to yield a two digit number. Thus, the correct quotient is* $1, 14; 24, 45$.

The Babylonians, as attested to by the surviving texts, had greater computational abilities than the Egyptians. For example, the Babylonians were able to calculate roots to a high degree of accuracy, and even began to investigate exponentials.

See Problem 7.

2.1 Exercises

1. Convert the following numbers into base 60.

   (a) 78
   (b) 126
   (c) 38921
   (d) 1432
   (e) $216, 061$
   (f) $42, 532$

2. Convert the following sexagesimal numbers to decimal.

   (a) $1, 12$
   (b) $3, 15, 11$
   (c) $1, 1, 1$
   (d) $4, 0, 3$
   (e) $1, 4$
   (f) $1, 0, 4$

3. Translate the numbers shown in Table 2.2.

4. Write the following numbers as a Babylonian would have written them, using "1" to indicate a unit and "10" to indicate ten units. If a position is empty, leave the position blank.

   (a) $15, 35$ (c) $3, 0, 21$ (e) $0; 1$

   (b) $1, 24, 0, 47$ (d) $1, 0, 0, 0$ (f) $4, 0, 4; 0, 4$

5. The following numbers are written in Babylonian style, using our numeral symbols for ten and one. For each, write down at least five different decimal numbers they might represent.

   (a) $\begin{matrix} 10 \\ 10 \end{matrix}$ 1 1 1   (b) 1 1 10 10

6. Like the previous problem, the following numbers are written in the cuneiform style, using our modern numeral symbols for ten and one. Interpret the numbers in a manner that makes sense given the context of the problem.

   (a) A baker must make enough bread for $10, 1, 1, 1$ workmen for five days. Each workman consumes $1, 1$ loaves a day. The baker must make $1, 1, 10$ loaves.

   (b) A man walks for 10, 1 hours a day at a steady rate of 1, 1 miles an hour. After 1, 1 days, he has traveled a total of 10, 10, 1, 1, 10, 10, 1, 1 miles.

7. An Old Babylonian tablet exists that contains the following two tables:

		2	1
15	2	4	2
30	4	8	3
45	8	16	4
1	16	32	5
		1,4	6

   Suggest an interpretation of the two tables.

8. The following are computations taken from Babylonian algebra and geometry problems discussed in the next sections. For all the numbers, add placeholding zeroes or indicate the proper locations of the sexagesimal points using a ;.

   (a) Subtract 7,30 from 10,33,45 to get 3,3,45

   (b) Multiply 5 and $1\frac{1}{2}$ to get 7,30

   (c) Multiply 4,30 by 10 to get 45

   (d) Halve 7 to get 3,30

    (e) Divide 9 by 12 to get 45

    (f) Multiply 3,30 and 3,30 to get 12,15

    (g) The square root of 1,12, 15 is 8,30

9. Construct Babylonian multiplication tables for products of:

    (a) 8   (b) 12   (c) 10   (d) 20

    (e) What do the tables for 10 and 20 suggest about creating a table for 30? For 40?

10. Multiply the following numbers using the table (or one that you constructed).

    (a) 3,15 times 12   (c) 4,27,31,5 times 5

    (b) 2,11,8 times 7   (d) 8,1,25,4 times 9

11. The following table was discovered on a tablet:

$$
\begin{array}{cc}
1 & 1 \\
1,2,1 & 1,1 \\
1,2,3,2,1 & 1,1,1 \\
1,2,3,4,3,2,1 & 1,1,1,1
\end{array}
$$

    (a) Translate both sides into decimal notation.

    (b) What is the relationship between the two sides of the table?

12. Find the reciprocals in sexagesimal notation for 12, 15, 18, 20, 24, 30, 32, 36, 40, 45, 48, 50, 54.

13. Divide the following numbers using the Babylonian technique.

    (a) 18,5 divided by 6   (c) 8,3,5 divided by 1,15

    (b) 42,7,15 divided by 4   (d) 1,1,1 divided by 1,40

14. For numbers that contain prime factors other than 2, 3, or 5, the reciprocals are nonterminating sexagesimals; in these cases, the texts said that a number "does not divide." Find the first five sexagesimal places in the reciprocals of:

    (a) 7   (b) 11   (c) 14   (d) 31

15. The evidence of a "long division"-type algorithm for Babylonian division comes from a study of the errors made on a student's clay tablet. Consider the problem of dividing a number by 9. What sort of errors could occur if a "long division" algorithm was used? In what sort of errors would the standard method of Babylonian division result?

16. (Teaching Activity) Create a "clay" tablet (a mixture of flour with just enough water added to make the flour pliable will work), then write a variety of symbols (the letters of the alphabet; numbers; Babylonian cuneiform marks) on it while it is still soft. Then bake the tablet. What happens to the shapes of the symbols? Could the Babylonians have created an elaborate system analogous to the Egyptian hieratic notation? Explain.

17. Compare and contrast Babylonian numeration and computation to Egyptian numeration and computation. Some factors to consider are the ease of writing, particularly with the available material, and the possibilities for misinterpretation.

2.2   Problem Solving

The Babylonians, like the Egyptians, lacked a system of notation, so their means of handling problems was entirely verbal. Egyptian scribes rarely ventured beyond linear equations, but the Babylonians routinely solved more complicated and complex problems.

2.2.1   Simple Problems

Canals were necessities of life in Mesopotamia, so it is not surprising that many of the applied mathematics problems dealt with the digging of canals. For example:

Problem 2.1. *A canal* 5 *GAR long,* $1\frac{1}{2}$ *GAR wide, and* $\frac{1}{2}$ *GAR deep is to be dug. Each worker is assigned to dig* 10 *GIN, and is paid* 6 *SE. Find the area, volume, number of workers, and total cost.*

Here GAR, SAR, GIN, and SE are all units of quantity. Such a problem involved seven parameters: the length, width, and depth of a hole to be excavated; the assignment (the volume each worker was expected to excavate); the number of workers; the wages per worker; and finally the total cost. The problem itself is straightforward but is made more difficult by the lack of a rational set of units, as can be seen in the scribe's solution:

Solution. Multiply length and width to get 7; 30 SAR, the area. Multiply 7; 30 by depth to get 45 SAR, the volume. Multiply the reciprocal of the assignment, 6, by 45 to get 4, 30, which is the number of workers. Multiply 4, 30 by the wages to get 9 GIN, the total expenses.   □

Conversion between units is done "on the fly" and may have been so basic that the scribe did not feel it necessary to write down the appropriate conversions. Note again that the Babylonians did not indicate the place value; hence the product of 5 GAR times $1\frac{1}{2}$ GAR would have been written as $7, 30$ SAR. See Problem 1.

2.2.2 Quadratic Equations

Some of the canal problems resulted in quadratic equations, such as:

Problem 2.2. *The length and width of a canal are together* $6; 30$ *GAR; the area of the canal is* $7; 30$ *SAR. What are the length and width?*

The scribe's solution follows.

Solution*.* Take half of the sum of the length and width, which is $3; 15$. Square $3; 15$ to get $10; 33, 45$. Subtract the product of length and width, $7; 30$, from $10; 33, 45$ to get $3; 3, 45$. Take its square root, which is $1; 45$. Add it to the sum of the length and width, to get 5 GAR, the length, and subtract it from the sum, to get $1; 30$ GAR, the width. □

A second form of the quadratic is when the difference between the length and width is given. For a problem such as:

Problem 2.3. *A canal's area is* $7; 30$ *SAR, and its length exceeded its width by* $3; 30$ *GAR.*

Solution*.* Take half of the amount by which the length exceeded the width, which is $1; 45$. Square $1; 45$ to get $3; 3, 45$. Add $7; 30$ to get $10; 33, 45$. Take its square root, to get $3; 15$. Add $1; 45$ to the square root to get 5 GAR, the length. Subtract $1; 45$ from the square root to get $1; 30$ GAR, the width. □

2.2.3 Advanced Problems

The Babylonians also solved certain types of third and fourth degree equations, probably by referencing tables: Problem 5 suggests a method by which they may have solved these type of equations. The most advanced problems dealt with by the Babylonians concerned interest, and resulted in equations we would today solve logarithmically. Around 1700 B.C. a scribe posed the problem:

Problem 2.4. *One KUR is borrowed at interest. How many years before the interest and principal are equal?*

The interest rate was not stated in the problem, but most such problems used a rate of 20% per year (an unusually low rate of interest for the premodern world). The scribe's solution, with modern annotations in brackets, follows.

Solution*.* Compute the amount of interest and principal for 4 years, which is [0.0736] more than 2 KUR. How much less than 4 years to make 2 KUR? [3 years gives a total of interest and principal of 1.728, so the extra year makes an extra

0.3456. Since we need 0.0736 less, we must take $\frac{0.0736}{0.3456}$ of 1 year less than 4, or approximately 0.21296 years. A Babylonian year consisted of 12 months, so the answer would be] four years less 2; 33, 20 months.   □

This is the earliest appearance of the **method of double false position**.

2.2 Exercises

1. Determine the relationships between some of the Babylonian units using their word problems.

   (a) Since 5 GAR times $1\frac{1}{2}$ GAR is 7; 30 SAR, what is the relationship between GAR and SAR?

   (b) Since 7; 30 SAR multiplied by $\frac{1}{2}$ GAR is 45 volume SAR, what is the relationship between volume SAR and GAR?

   (c) Use the fact that 45 volume SAR divided by 10 GIN per worker gives 4,30 workers to find the relationship between volume SAR and GIN.

   (d) Use the fact that 4,30 workers multiplied by 6 SE per worker is 9 GIN to find the relationship between GIN and SE.

2. Use the Babylonian procedure to solve the following problems.

   (a) Length and width together are 10; area is 22; 45.

   (b) Length and width together are 4; area is 3; 26, 15.

   (c) Length and width together are 5; area is 6; 11, 25.

3. Use the Babylonian procedure to solve the following problems:

   (a) The difference of length and width is 4; the area is 32.

   (b) The difference of length and width is 5; the area is 1,6.

   (c) The difference of length and width is 5; the area is 18; 45.

4. Solve Problem 2.4. How accurate is the answer given by the scribe?

5. The Babylonians were able to solve certain third degree equations. In order to do this, they probably referenced tables that gave the values of $n^2 + n^3$ (or $n^2(n + 1)$). Construct such a table, then use it to solve the following equations.

   (a) $x^2(x + 1) = 1, 30$

   (b) $x^2(12x + 1) = 1; 45$ (Hint: multiply both sides by 12^2)

2.3   Geometry

The Babylonians also developed an impressive practical geometry. Their area formulas, like those of the Egyptians, tended to give approximate results. For triangles, the Babylonians multiplied half the width by one of the sides. For an isosceles trapezoid, a tablet suggests that the Babylonians used the procedure of adding the lengths of the two parallel sides, halving them, then multiplying by the length of one of the nonparallel sides; there is no indication of what would be done if the trapezoid was not isosceles. A set of tablets discovered in 1936 by French archaeologists at Susa indicates the Babylonians also knew how to approximate the areas of regular pentagons, hexagons, and heptagons.

See Problem 3.

From the same set of tablets is the determination of the radius of a circle that circumscribes an isosceles triangle with sides of 50, 50, and 60: 31, 15. The result is exact.

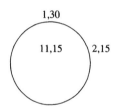

The figures in the margin are based on two cuneiform tablets. The best interpretation of the tablets is that they show the procedure for computing the area of a circle. For the first, we have a circle whose circumference is 3. Squaring 3 to obtain 9, then multiplying 9 by 5 (i.e., dividing by 12) we obtain the area, 45 (which should be read as 0; 45). The second tablet confirms that this is the sequence of operation. Thus, the Babylonian approximation for the area of a circle was worse than the Egyptian.

On the other hand, the Pythagorean theorem, of which the Egyptians appeared to be ignorant, was well known to the Babylonians. A tablet in the possession of Yale University shows a square, whose diagonal is marked with the cuneiform number equivalent to 1; 24, 51, 10. If this is interpreted as being the length of the diagonal (and there is no reason to suspect otherwise), then it shows the Babylonians were not only aware of the Pythagorean theorem for isosceles right triangles, but were also able to compute square roots to a high degree of accuracy.

How were they able to do so? One possibility is by using successive approximations. 1; 30 is too large an approximation, since the square of 1; 30 is 2; 15. To find a smaller approximation, divide 2 by 1; 30, which is the same as multiplying by 0; 40; this gives the second approximation 1; 20, which will be too small. The average of the two is 1; 25. Repeating the procedure, using 1; 25 as the initial approximation, gives a second approximation of 1; 24, 51, 10.

Even more evidence of Babylonian geometrical sophistication may be found in from Tablet 322 of the Plimpton collection at Columbia University; Table 2.3 shows a transcription of this tablet. The original dates back to between 1900 and 1600 B.C. Plimpton 322 was originally part of a larger tablet; only four columns have been incompletely preserved, and at least one column is missing entirely.

What do the numbers represent? The fourth column is, obviously, an index column. The two numbers that can be read in the first row are 1, 59 and 2, 49. After some trial and error, we find the square of 2, 49, minus the square of 1, 59, is a perfect square, namely that of 2, 0. On line 11, we find again 1, 15 squared, minus 45 squared, is 1, 0 squared, so it suggests the square of the third column,

[ ] 15	1,59	2,49	1
[ ] 58, 14, 50, 6, 15	56,7	3,12,1	2
[1, 55, 7,] 41, 15, 33, 45	1,16,41	1,50,49	3
[1,] 5[3,] 10, 29, 32, 52, 16	3,31,49	5,9,1	4
[1,] 48, 54, 1, 40	1,5	1,37	5
[1,] 47, 6, 41, 40	5, 19	8, 1	6
[ ] 43, 11, 56, 28, 26, 40	38,11	59,1	7
[1,] 41, 33, 59, 3, 45	13, 19	20, 49	8
[1,] 38, 33, 36, 36	9, 1	12, 49	9
1, 35, 10, 2, 28, 27, 24, 26, 40	1, 22, 41	2, 16, 1	10
1,33,45	45	1, 15	11
1, 29, 21, 54, 2, 15	27, 59	48, 49	12
[1,] 27, 0, 3, 45	7, 12, 1	4, 49	13
1, 25, 48, 51, 35, 6, 40	29, 31	53, 49	14
[1,] 23, 13, 46, 40	56	53	15

Table 2.3: Numbers from Plimpton 322.

minus the square of the second column, is always a perfect square. There are some problems with this conjecture, since this is not always true: in the ninth row, it is not true that $12, 49$ squared minus $9, 1$ squared is a perfect square. However, it is true that $12, 49$ squared minus $8, 1$ squared is a perfect square, so apparently, $9, 1$ is the equivalent of a typographical error in the construction of the tablet. If we except the occasional errors, we find that the second and third column are numbers b and c where $c^2 - b^2 = a^2$ for integer values of a, b and c: what we now call a **Pythagorean triple**.

What about the first column? The tenth, eleventh, and twelfth rows are complete. Taking the eleventh row, we note that $1, 15$ squared, minus 45 squared, is $1, 0$ squared, and that $1, 15$ squared, divided by $1, 0$ squared, is exactly $1; 33, 45$. Thus, the third column represents $\frac{c^2}{a^2}$. Once we have deciphered the tablet, its mathematical nature makes it possible to restore the missing numbers.

Notice that the numbers in the first column decrease at a nearly constant rate of close to $0; 2$. This means that the ratio $\frac{c}{a}$ also decreases at a nearly constant rate of $0; 1$. The work needed to construct such a table of Pythagorean triples using trial and error methods is staggering; it is more than likely that the scribes had some means of generating Pythagorean triples, corresponding to some version of the methods we will encounter later.

2.3 Exercises

1. Translate $1; 24, 51, 10$ into decimal. How accurate an approximation is it to $\sqrt{2}$?

2. Show that the radius of a circle circumscribed about an isosceles triangle with sides 50, 50, and 60 is $31, 25$. Where is the sexagesimal point located?

3. If s_n is the length of a side of a regular n- gon, and A_n is the area of the n-gon, then the Susa tablets can be interpreted to read:

$$A_5 = 1;40s_5^2 \qquad A_6 = 2;37,30s_6^2 \qquad A_7 = 3;41s_7^2$$

   (a) Comment on the accuracy of these formulas.

   (b) On the same set of tablets, the text can be read as giving the relationship between the perimeter of a regular hexagon, c_6, and a circumscribed circle, c, as $c_6 = 0;57,36c$. Comment on the validity of this formula. What does this formula suggest the Babylonians used for the ratio of the circumference of a circle to its perimeter?

4. Use the second of the two Babylonian "circle area" tablets to show that the Babylonian procedure for finding the area of a circle is equivalent to squaring the circumference and then dividing by 12.

5. Use the procedure in the text to find an approximation to $\sqrt{3}$.

6. Use lines 10, 11, and 12 of Table 2.3 to justify the conclusion that the second and third columns represent the values of b and c for a Pythagorean triple, and the first column is equal to the ratio $\frac{c^2}{a^2}$.

7. Examine Table 2.3. Reconstruct the missing numbers in lines 1, 2, and 7.

8. There are errors in Table 2.3 on lines 2, 9, 13, and 15.

   (a) Find these errors, and determine what the correct values should be.

   (b) Classify these errors as being "scribal" (arising from miscopying of numbers), "computational" (arising from computational errors), or "inexplicable."

9. Suggest possible uses for a table of Pythagorean triples $a^2 + b^2 = c^2$, where the ratio between $\frac{c}{a}$ decreases at a constant rate.

Chapter 3

Greek Arithmetic

The land we now call Greece was once part of the Minoan Empire, centered on the island of Crete. In 1500 B.C., the volcanic island of Thera exploded, causing a tsunami so devastating that the Minoans never recovered, possibly giving rise to the Atlantis legend. The Greek subjects of the Minoans revolted and invaded Crete, forming the Mycenaean Empire. By 1200 B.C., the Mycenaeans would be powerful enough to attack and destroy a city on the coast of Asia Minor (modern Turkey): Troy. But shortly after their great victory, the Mycenaean homeland was invaded by the barbarian Dorians from the north, and Greece entered a dark age about which very little is known.

The dark age lasted until about 700 B.C. By then, the population of Greece had increased tremendously. Greece itself, a mountainous land of limited fertility, could not support a large population, so many people left to form Greek colonies around the Mediterranean Sea. Because of geographic separation, the Greeks never founded a single, large, national state, as did Egypt or Babylonia; rather, they remained divided into hundreds of independent city-states, the two most important of which were Athens and Sparta.

3.1 Numeration and Computation

Around 600 B.C., the leader Solon codified the laws of the city of Athens. Before the modern era, imprisonment as a punishment was almost unheard of: punishment consisted of death, mutilation, or fines, and prison was merely a place to hold the accused until trial and the guilty until punishment. The laws of Solon were typical, giving specific fines for specific offenses. Since the fines were numerical amounts, they give us a glimpse of early Greek numeration. Solon's law codes used a form of numbers called Attic (since Athens was on the Attic Peninsula in Greece) or, sometimes, Herodianic (after Herodian, a grammarian of the second century A.D.). The Attic symbols are little more than the first letters of

Ι	ΙΙ	ΙΙΙ	Γ	Δ	Ͷ	Η	Ͷ	Χ	Ͷ
1	2	3	5	10	50	100	500	1000	5000

Table 3.1: Attic Numbers.

A	B	Γ	Δ	E	F	Z	H	Θ
1	2	3	4	5	6	7	8	9
I	K	Λ	M	N	Ξ	O	Π	ϙ
10	20	30	40	50	60	70	80	90
P	Σ	T	Y	Φ	X	Ψ	Ω	λ
100	200	300	400	500	600	700	800	900

Table 3.2: Upper Case Greek Alphabetic Numbers.

the corresponding number words (see Table 3.1).

3.1.1 Alphabetic Numeration

By 450 B.C., a new type of numbering system was in use called **alphabetic**, since each letter of the Greek alphabet was used to represent a number. On the tomb of Mausolus (one of the "Seven Wonders of the World" and the source of our word mausoleum), $\Psi N\Delta$ appears, representing seven hundred fifty-four. The Greek symbols are shown in Table 3.2. Since the standard Greek alphabet only has twenty-four letters, three additional letters were necessary to make the twenty-seven required symbols: *vau* (the symbol for 6), also called *digamma* or, even later, *stigma*; *koppa* (the symbol for 90); and *sampi* (the symbol for 900). Even later, after the invention of lower case letters, the Greeks used them to represent numbers (see Table 3.3).

The symbols are similar to the Egyptian hieratic in that each symbol represents a distinct number. To mimic the use of Greek alphabetic numerals, we might use the capital letters of the Roman alphabet and call our results pseudo-Greek numbers; the correspondence between our own letters and the numbers from 1 through 900 are shown in Table 3.4, where "?" is used for the twenty-

α	β	γ	δ	ϵ	ς	ζ	η	θ
1	2	3	4	5	6	7	8	9
ι	κ	λ	μ	ν	ξ	o	π	$\varqoppa$
10	20	30	40	50	60	70	80	90
ρ	σ	τ	υ	ϕ	χ	ψ	ω	λ
100	200	300	400	500	600	700	800	900

Table 3.3: Lower Case Greek Alphabetic Numbers.

1	2	3	4	5	6	7	8	9
A	B	C	D	E	F	G	H	I

10	20	30	40	50	60	70	80	90
J	K	L	M	N	O	P	Q	R

100	200	300	400	500	600	700	800	900
S	T	U	V	W	X	Y	Z	?

Table 3.4: Pseudo-Greek Numeration.

seventh symbol.

For one thousand, two thousand, three thousand, and so on the Greeks would use the symbol for the corresponding unit with a stroke ′ before it: thus, four thousand would be ′Δ or ′δ. In pseudo-Greek, we will write four thousand as ′D.

A disadvantage of alphabetic numerals is that the same letters used to write the language also represent numbers. To distinguish between numbers and words, the Greeks placed a bar over a letter sequence that represented a number.

Example 3.1. *Write the number five thousand two hundred thirty-four in Greek and pseudo-Greek. This would be* $\overline{′\epsilon\sigma\lambda\delta}$ *in Greek alphabetic, or* $\overline{′\text{ETLD}}$ *in pseudo-Greek.*

For even larger numbers, the Greeks would write M (the first letter in *myrioi*, their word for 10,000 and from which we get the word myriad) and above it the number of myriads.

Example 3.2. *Write 70,000. This is 7 ten thousands:* $\overset{\zeta}{\text{M}}$ *in Greek numerals or* $\overset{G}{\text{M}}$ *in pseudo-Greek.*

Example 3.3. *Write 1,347,295 in pseudo-Greek. This is 134 myriads and 7,295. 134 myriads is* $\overset{\text{SLD}}{\text{M}}$ *; 7295 is* ′GTRE*. Together,* $\overset{\text{SLD}}{\text{M}}$ ′GTRE*.*

The Greeks used common fractions, sexagesimal fractions, and unit fractions. For sexagesimal fractions, the Greeks followed the Babylonian model and simply wrote down the sexagesimals, leaving a space between them: thus $2 + \frac{13}{60} + \frac{25}{60^2}$ would be $\overline{\beta}\ \overline{\iota\gamma}\ \overline{\kappa\epsilon}$. Unit fractions would have a ′ after the denominator: thus, one-third became γ'. For common fractions, a variety of forms were used, as shown in Table 3.5.

> **Common fractions:** Fractions where the numerator and denominator can be any whole number (e.g., $\frac{15}{37}$).

Since a sequence of letters represents the same number regardless of how it is written, another disadvantage of alphabetic numerals is that they allow superstition to masquerade as mathematics. For example, by reading the letters of a name as numbers, the numerical value of the name could be calculated. The most famous instance of this occurs in the biblical Book of Revelations: the "number of the beast," 666, is believed to refer to the number value of the name of Nero, one of the early Roman emperors who persecuted Christians.

Form	Author	Date
$\overline{\gamma}$ ten $o\alpha'$	Aristarchus	300 B.C.
$\overline{\gamma\iota}\ o\alpha'$	Archimedes	250 B.C.
$\overline{\gamma}\,\overset{o\alpha}{\iota}$	Diophantus	200 A.D.
$\overline{\gamma\iota}\ o\alpha'o\alpha'$	Heron	200 A.D.

Table 3.5: Some Greek forms of the mixed number $3\frac{10}{71}$.

3.1.2 Computation

The Greeks referred to the skill of computation as **logistic**, reserving the word **arithmetic** for the study of numbers. Because the only differences between the Greek and Egyptian systems of numeration are the specific symbols used, we might suspect that the Greeks and Egyptians used similar methods of computing. Our suspicions would be incorrect, and later writers distinguish between Greek and Egyptian forms of logistic. Unlike Egyptian and Babylonian mathematics, for which we have many original sources, all known samples of Greek calculation date from the first few centuries A.D. Thus, it is entirely possible that the earliest Greek mathematicians computed (when they did so) using methods entirely different from the ones we will discuss.

Several fragments of multiplication tables exist, making it appear that the Greeks, like the Babylonians, multiplied using tables. However, a form of long multiplication, similar to our own, is recorded by Eutocius (fl. sixth century A.D.). Eutocius showed how to multiply $\prime\alpha\tau\nu\alpha$ by $\prime\alpha\tau\nu\alpha$ (1351 by 1351). The computation, in the Greek numeral system, is:

A Businessman's Letter. On Papyrus. Digitally reproduced with the permission of the Papyrology Collection, Graduate Library, University of Michigan.

$$
\begin{array}{cccc}
\prime\alpha & \tau & \nu & \alpha \\
\prime\alpha & \tau & \nu & \alpha \\
\hline
\overset{\rho}{M} & \overset{\lambda}{M} & \overset{\epsilon}{M} & \prime\alpha \\
\overset{\lambda}{M} & \overset{\theta}{M} & \overset{\alpha}{M} & \prime\epsilon\ \ \tau \\
 & \overset{\epsilon}{M} & \overset{\alpha}{M}\prime\epsilon & \prime\beta\phi\ \ \nu \\
 & & & \prime\alpha\ \ \tau\ \ \nu\ \ \alpha
\end{array}
$$

We may interpret the multiplication as follows: $\prime\alpha$ (one thousand) times $\prime\alpha$ (one thousand) is $\overset{\rho}{M}$ (one hundred myriad, or one million). Thus, the first line contains the products of $\prime\alpha$ with each of the numbers $\prime\alpha\tau\nu\alpha$. The second, third, and fourth lines consist of the successive products of the other terms.

The most difficult computations in the ancient world occurred in mathematical astronomy. Because of this, existing examples of more difficult operations, such as division and extraction of roots, used sexagesimals to represent the fractional parts of a number. To make the computation more comprehensible, we will indicate the sexagesimal point and write place-holding zeroes, but the Greeks, like the Babylonians, usually omitted them. Theon of Alexandria (fl. 375 A.D.) performed the division of $1515; 20, 15$ by $25; 12, 10$ as follows:

Solution. $25; 12, 10$ goes into $1515; 20, 15$ a little more than 60 times. From the dividend, subtract 60 times 25, 60 times $0; 12$ and 60 times $0; 0, 10$. 60 times 25 is 1500; subtracting this leaves $15; 20, 15$; we convert 15 into $0; 900$ and thus $15; 20, 15 = 0; 920, 15$. 60 times $0; 12$ is $0; 720$, and subtracting this we obtain $0; 200, 15$. Finally, 60 times $0; 0, 10$ is $0; 10$, so the remainder after the first quotient is $0; 190, 15$.

Again, divide this by 25, which gives a quotient of $0; 7$. Multiplying $0; 7$ by 25 gives $0; 175$, which we subtract from $0; 190, 15$, leaving $0; 15, 15$, or $0; 0, 915$. Multiplying $0; 7$ by $0; 12$ gives $0; 0, 84$, subtracting gives us a remainder of $0; 0, 831$. Finally, multiplying $0; 7$ times $0; 0, 10$ gives $0; 0, 1, 10$, and subtracting this leaves $0; 0, 829, 50$.   □

Theon continued for one more step to determine the second sexagesimal place, arriving at the quotient $60; 7, 33$.

3.1 Exercises

1. Create a multiplication table for pseudo-Greek numbers. Use it to perform the following multiplications:

   (a) KA times NF

   (b) YMA times KH

   (c) ?RC times WLB

   (d) ‚CYPA times ‚AXLE

2. Divide ‚CSKB by KE.

3. What are the advantages and disadvantages of the various forms of fractions used by the Greeks? Explain with regards to ease of writing, ease of interpretation, and possibility of misinterpretation.

4. Compare the Greek methods of computation with the Babylonian and the Egyptian. What are the strengths and weaknesses of each?

3.2   The Pythagoreans

The colonizing era of Greek history can lead to some confusion for students of classical geography. The shores of the Ionian Sea, to the west of modern Greece, was the original homeland of many Greeks who made their way east, towards Asia Minor. These colonies in Asia Minor became known as *Ionia*. Meanwhile, the Greek settlers who went west and settled in Italy were the first Greeks encountered by the Romans, who called southern Italy *Magna Graecia*, Latin for Greater (as in larger) Greece.

One settlement in Magna Graecia was Croton (now Crotona, Italy). Around 532 B.C., Pythagoras (580-500 B.C.) made his way there, after spending some time traveling about the Mediterranean. At Croton, Pythagoras established a secretive, mystical school that lasted about a century until it was suppressed by the authorities in the middle of the fifth century B.C. None of Pythagoras's

Figure 3.1: Magna Graecia

own work has survived; we have only what his followers claim he said. Some of the rules of the school, such as always wearing white and not eating beans, have parallels with ancient Egyptian practices, lending credence to a claim that Pythagoras spent time in Egypt.

Pythagoras and his followers were struck by the many relationships among numbers; they sought to analyze the physical world in terms of these number relations. Pythagoras is said to have established the dictum "All is number," though a later Pythagorean, Philolaus (fl. 430 B.C.), modified this to a more cautious "All can be represented by number."

3.2.1 Number Classification

If, as the Pythagoreans believed, "all is number," then by understanding number, it should be possible to understand the universe. The Pythagoreans viewed number as a collection of units; in principle, the units could be anything—line segments, squares, or three dimensional solids. In practice, the Pythagoreans used a point as the unit, and considered a number to be a collection of these points.

The first step in understanding numbers would be to classify them. The result was a vast collection of number categories, of which only a few remain in common use, such as even and odd; prime; square; and cube. There are several sources in which we may find the Pythagorean scheme of classification, though all date from centuries after Pythagoras. One such source is the *Introduction to Arithmetic* of Nicomachus of Gerasa (fl. A.D. 100), a neo-Pythagorean.

The simplest division was that into even and odd. Nicomachus actually provided three definitions of even number, but according to Nicomachus, the most ancient definition—presumably the Pythagorean one—was that an **even number** could be divided in two ways: into two equal parts, and into two

1

3

6

10

See Problem 4a.

unequal parts. The exception was the **dyad**, 2, which was the root form of the even numbers: the dyad could only be divided into equal parts.

Other number classifications came by arranging the points, whose aggregate represented a number, in a geometrical pattern; this produced a **figurate** or **polygonal number**. The simplest figure is the triangle, and hence the simplest of the figurate numbers is the **triangular number**. The first four triangular numbers are shown in the margin.

The Pythagoreans, who were interested in numbers and the relationships between numbers, were aware of the following. Since we do not know how, or even if, they proved it, we shall call it a **conjecture**; in this book, unless a proof accompanied a result when originally stated, we will refer to the result as a conjecture, regardless of whether a modern proof exists.

Conjecture 3.1. *The sum of the numbers in order produces the triangular numbers in order.*

To prove this, the Pythagoreans may have pointed to a diagram of a triangular number and showed how the next triangular number could be obtained by adding another line of points (see margin). An argument of this sort is not considered sufficient today, though it does provide a demonstration of the general principle.

Ten, the fourth triangular number, came to be known as the **tetractys** and held a special place in Pythagorean mysticism; among other things, they swore oaths "by the tetractys." Since it was the sum of the first four numbers, the Pythagoreans felt it embodied perfection and called it the perfect number; this is not our modern definition.

The number ten also influenced their natural philosophy. Philolaus created the first known system of the universe in which the Earth was itself in motion. In Philolaus's system, the Earth and five planets known to the ancients (Mercury, Venus, Mars, Jupiter, and Saturn), plus the Moon and the Sun, all orbited around a "central fire." Since the number of bodies in this system was only nine, Philolaus added a tenth, the "counter Earth," which always orbited on the other side of the central fire and was thus unobservable.

After the triangular numbers came the **square numbers**, which are still a part of our number classification system. Several properties of square numbers can be discerned by examining their figures. One relates the square numbers to the triangular numbers.

Conjecture 3.2. *Each square number is the sum of two consecutive triangular numbers.*

A second fact relates the square numbers to previous square numbers. Each square is the previous square, plus an L-shaped region the Pythagoreans called a **gnomon**. The gnomons of a square are always odd numbers (see Figure 3.2). Thus, we obtain the following:

Conjecture 3.3. *The square numbers differ by successive odd numbers.*

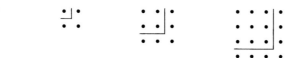

Figure 3.2: Gnomons of a Square Number.

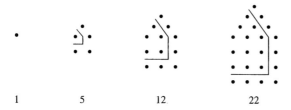

1　　　5　　　12　　　22

Figure 3.3: Pentagonal Numbers.

The early Pythagoreans probably knew several other propositions involving square numbers. Two of these are given by Theon of Smyrna (fl. A.D. 130).

Conjecture 3.4. *A square number is either divisible by three, or will be divisible after the subtraction of a unit.*

Conjecture 3.5. *A square is either divisible by four, or will be divisible after the subtraction of a unit.*

The Pythagoreans also formed regular pentagons, hexagons, and so on by beginning with the unit and placing a gnomon of the appropriate length about it to form the successive polygonal numbers (see Figures 3.3, 3.4 and 3.5). Each sequence of figurate numbers begins with 1, and the successive numbers are formed by placing a gnomon around the previous number.

Notice that the hexagonal, heptagonal, and further figurate numbers do not form "filled in" figures; the reason may have had to do with the form of the gnomons of polygonal numbers.

See Problem 2.

Perhaps because it is the last of the figurate numbers whose gnomons "fill in" the space, the Pythagoreans held the pentagon in high esteem, holding it to be a secret symbol. A story is told that a Pythagorean, dying in a distant country, was unable to repay the hospitality of his host: he told his host to

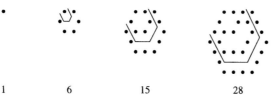

1　　　6　　　15　　　28

Figure 3.4: Hexagonal Numbers.

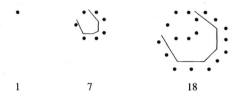

Figure 3.5: Heptagonal Numbers.

draw the pentagon on the wall, and someday a Pythagorean would come by, see the pentagon, and make inquiries. The host did, and eventually a Pythagorean came by, whence the host told the story of the dying Pythagorean, and was richly rewarded by the living one.

3.2.2 The Pythagorean Theorem

The best known of all the Pythagorean results about number properties is the **Pythagorean theorem**. Before we discuss it, we should address several myths that have grown around it. The first is that the Pythagorean theorem was Egyptian in origin, based on the notion that Egyptian "rope stretchers" routinely formed right angles by using a triangle with sides of lengths 3, 4, and 5. The available evidence argues against the Egyptians having *any* knowledge of the Pythagorean theorem; it is even questionable that they knew that the 3-4-5 triangle is right angled.

Another myth is reported by Proclus, an author of the sixth century A.D. from whom much of our knowledge of early geometry comes. According to Proclus, "Those who like to record antiquities" claim that Pythagoras sacrificed an ox when he discovered the theorem. Proclus himself sounds skeptical, and with good reason. Even if Pythagoras discovered the Pythagorean theorem, it is beyond belief he would have sacrificed an ox (or a hundred oxen in some accounts); part of the reason for the vegetarian philosophy of the Pythagoreans was the belief in the transmigration of souls.

It is certainly possible that Pythagoras proved the Pythagorean theorem; one possible Pythagorean proof will be given later. More reliably, we may attribute to the early Pythagoreans the discovery of a method for creating integer solutions to $a^2 + b^2 = c^2$. The set of integers a, b, and c is known as a **Pythagorean triple**, and Pythagoras himself is credited with discovering a special type of Pythagorean triple. According to Proclus, one began with an odd number, assumed to be the lesser of the two sides about the right angle; the second side is half of one less than the square of the number; and the third side is one more than the second side. Thus, Pythagoras's triples consisted of numbers where the smallest number was odd, and the two larger numbers differed by exactly 1.

Example 3.4. *Find the Pythagorean triple with smallest number* 7. *The second side is half of one less than the square of* 7, *or* 24. *The third side is one more than* 24, *or* 25. *Hence the triple is* 7, 24, 25.

Nothing is known about how Pythagoras arrived at this formula. However, the nature of the formula suggests that he arrived at it by considering the gnomons of square numbers. In particular, since any square plus a gnomon yields another square, all that is necessary is that the gnomon itself be a square number. See Problem 7.

A second method for finding Pythagorean triples is attributed to Plato, although it may have been an earlier discovery. Notice that the Pythagorean formula always begins with an odd number as the shorter leg of the right triangle. According to Proclus, Plato's method began with an even number as one of the sides; one less than the square of half this number gives the second side; and one more than the square of half the number gives the second third side.

Example 3.5. *Find the Platonic Pythagorean triple with smallest side* 8*. One less than the square of half of* 8 *is* 15*; one more than the square of half is* 17*. Thus, the triple is* 8, 15, 17*.*

It is possible that the method was developed by considering two gnomons placed around a square. As before, it was necessary to make the double gnomon a square number. See Problem 8.

3.2.3 Commensurable and Incommensurable Numbers

We have seen how numbers were divided into odd and even, and into many figurate forms. Several other classifications existed, based primarily on how the number could be divided. To understand the Pythagorean classifications, which will ultimately lead to the incommensurables, we need to introduce a few concepts of Greek number theory.

One number was **part** of a larger number if some multiple of the first was equal to the second; in this case, the first was said to **measure** the second. If a smaller number did not measure a larger number, the smaller number was called **parts** of the larger. Thus, 4 was part of 12, but 5 was parts of 12. In the former case, 4 was said to **measure** 12. If a number had no measures but the unit, it was said to be **prime**. Two quantities were **commensurable** if there is some quantity that will measure them both; otherwise, the quantities were **incommensurable**. All numbers in this sense are commensurable. The largest quantity to measure them both is their **greatest common measure**, which can be found using the "Euclidean" algorithm, called so because it is given in Euclid's *Elements*, Book VII, Proposition 2 (though it predates Euclid by some centuries). Nicomachus gave the algorithm as follows: See Problem 14.

Rule 3.1 (Euclidean Algorithm). *Subtract the smaller from the larger as many times as possible. Then change the order of subtraction and subtract the remainder from the smaller as many times as possible. Continue until one ends with unity, or the same number as is being subtracted; this last number will be the greatest common measure.*

Nicomachus illustrated the method using 21 and 49.

Example 3.6. *Given* 21 *and* 49. *Subtract* 21 *from* 49, *leaving* 28. *Subtract* 21 *again, leaving* 7. *You cannot subtract* 21 *from* 7, *so change the order of subtraction and subtract* 7 *from* 21, *leaving* 14; *subtracting again, you leave* 7, *which is the same number as that which is being subtracted, and hence* 7 *is the greatest common measure.*

Example 3.7. *Given* 23 *and* 50. *Subtract* 23 *from* 50, *leaving* 27. *Subtract* 23 *from* 27, *leaving* 4. 23 *cannot be subtracted from* 4, *so subtract* 4 *from* 23 *as many times as possible; the remainders are* 19, 15, 11, 7, 3. *Since, again,* 4 *cannot be subtracted from* 3, *reverse the order and subtract* 3 *from* 4, *leaving* 1. *Hence,* 1 *is the greatest common measure of* 23 *and* 50.

See Problem 10.

The Euclidean algorithm can also be applied to line segments or any other geometric figures, though in the latter case, the application is more difficult.

3.2.4   Ratio and Proportion

The Pythagoreans, in addition to giving names and classifications to the numbers, also classified the *relations* between numbers. Today, we might think of a ratio between numbers, such as the ratio of 5 to 3, as just another number that we might express as 5/3. To the Pythagoreans, the ratio was a separate type of quantity.

These ratios were probably a by-product of the Pythagorean study of music. Pythagoras himself is credited with having noted that if the lengths of two strings had small, whole number ratio to one another, the resulting music when both were plucked would sound pleasing. Some of the important ratios were: the ratio of 2 to 1, which in music is said to produce an interval of an octave; the ratio of 3 to 2, producing an interval of a fifth; the ratio of 4 to 3, producing an interval of a fourth; and the ratio of 5 to 4, producing an interval of a third.

By the time of Nicomachus, the Pythagorean classification of ratios had evolved into a vast scheme that could classify an enormous number of ratios, each with its own name. For example, the ratio of 8 to 6 was known as the **sesquitertian**, since the larger number, 8, contained the smaller number, 6, and one third of 6. The larger number by itself kept the name of the entire ratio, and the smaller number had the prefix "sub-" added: hence, in the ratio, 8 is the sesquitertian of 6, whereas 6 would be the subsesquitertian of 8. The scheme survived into the Middle Ages, when it was excised from the study of mathematics; all that remains is the ratio of a multiple (and even the submultiples have dropped out of common usage).

The next logical step is the classification of the relationship among three numbers. These relationships went under the general name of **means**, a term which was originally applied to the entire relationship, though later geometers (and modern mathematicians) now refer to the relationship as a **proportion**, and reserve the term mean for the middle term of the proportion. There were originally three types: the arithmetic, the geometric, and the subcontrary.

According to one tradition, Archytas (fl. 375 B.C.) renamed the subcontrary mean the harmonic, a name which it retains (the subcontrary was later assigned to a different mean). Archytas defined the three proportions as:

1. The **arithmetic proportion**, where the first exceeds the second by the same amount the second exceeds the third.

2. The **geometric proportion**, where the first is to the second as the second is to the third.

3. The **harmonic proportion**, where the part of the first by which the first exceeds the second, the second exceeds the third by the same part of the third.

Example 3.8. *Given the first two terms 24 and 16, complete the harmonic proportion. 24 exceeds 16 by 8, which is one-third of 24; hence 16 must exceed the third term by one-third of the third term: the third term must be 12.*

Seven other proportions and their corresponding means have followed the ratio classification into obscurity.

3.2 Exercises

1. Classify the following numbers by as many categories as apply (e.g., odd, square, triangular, etc.): 15, 25, 16, 8, 28. Be sure to explain how the number fits the definition of the category.

2. Describe the gnomons for each of the following. For example, the gnomons of the square numbers are successive odd numbers.

   (a) The gnomon of the pentagonal numbers
   (b) The gnomon of the hexagonal numbers
   (c) The gnomon of the heptagonal numbers
   (d) Make a general statement about the gnomons of the polygonal numbers.

3. Plutarch, the Greek historian, noted that eight times a triangular number plus one unit is a square number; this fact would be used by Diophantus. Demonstrate this conjecture.

4. For each of the following, draw a picture that suggests the conjecture is true.

   (a) Conjecture 3.3.
   (b) Conjecture 3.4.
   (c) Conjecture 3.5.
   (d) The sum of the first n odd numbers is the nth square number.

5. Find the Pythagorean triplet with the shortest side of:

    (a) 5   (b) 8   (c) 12   (d) 13

6. Using the Pythagorean definitions of odd and even, prove that if the side of a square is even, then the square itself is even.

7. Provide a demonstration of the Pythagorean method of finding right triangles whose sides are integers. Hint: the square of an odd number is an odd number, and successive squares differ by successive odd numbers.

8. The historian of mathematics Thomas Heath suggested that the Platonic method was developed by considering a two-rowed gnomon about a square.

    (a) Provide a demonstration of the Platonic method.

    (b) Can you generalize the method of adding n-rowed gnomons about a square to produce other Pythagorean triples?

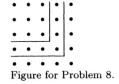

Figure for Problem 8.

9. Use the Euclidean algorithm to find the greatest common measure of the given numbers.

    (a) 36, 123   (b) 21, 184   (c) 18, 480

10. Apply the Euclidean algorithm to find the greatest common measure of an isoceles right triangle whose sides are of length 5 and a square whose sides are of length 2.

    (a) Draw the two figures, then subtract the square from the triangle.

    (b) Form the remaining pieces into a figure of equal area.

    (c) Subtract this remainder from the square.

    (d) Repeat until you have obtained the greatest common measure between the square and the triangle.

    (e) Express the square and the triangle in terms of their greatest common measure.

11. Use the Euclidean algorithm to find the greatest common measure of the indicated quantities. Express both quantities in terms of multiples of their greatest common measure.

    (a) Two squares, the side of the larger being one and a half times the side of the smaller.

    (b) Two similar right triangles, the hypotenuse of the larger being twice the size of the hypotenuse of the smaller.

(c) A rectangle with sides of length 1 and 2, with another rectangle whose sides are of length 5 and 7.

(d) A rectangle with sides of length 2 and 3, with the square of side 5.

12. Demonstrate that the isosceles right triangle measures the square on its hypotenuse.

13. For each of the following, complete the arithmetic, harmonic, and geometric proportion, given the first two numbers.

(a) 6, 3   (b) 6, 4

14. Why are all numbers, in the Pythagorean sense, commensurable?

15. What does the remark that "2 cannot be broken into unequal parts" say about the Pythagorean conception of what a number was?

16. What are the similarities and differences between the Pythagorean definition of commensurable and incommensurable and our own ideas about rational and irrational numbers? Explain.

3.3 The Irrational

Recall that Pythagoras had been of the opinion that "all is number," which later Pythagoreans softened to "all can be represented by numbers." This viewpoint would be seriously challenged if there was anything that could not be represented by number. Thus, one of the major turning points in the history of mathematics was the discovery of incommensurable quantities, which we now call **irrational numbers**, though it is important to remember that the irrational number is merely a quantity that is incommensurable with the unit number. See page 39.

Despite the great impact of this discovery, practically nothing is known about the details: the identity of the discoverer is completely unknown, and the discovery itself could have taken place any time between Pythagoras (550 B.C.) and Plato (350 B.C.). Even the exact nature of the first incommensurable to be discovered is open for debate, though there are two strong candidates: the diagonal of a square and the diagonal of a regular pentagon, which are incommensurable with the side of their respective polygon.

3.3.1 Diagonal of a Square

According to Aristotle, the proof that the diagonal of a square is incommensurable with its diagonal was shown using facts about even and odd numbers; by assuming the side was commensurable, one was able to show that odd numbers were equal to even numbers. Given this hint, it is easy to construct a proof; the following is one possible reconstruction of the Pythagorean proof.

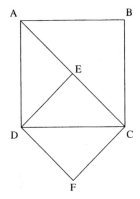

See Problem 10.

Proof. Suppose one has the square $ABCD$, with diagonal AC, and suppose that AC, DC are commensurable; further, suppose that their greatest common measure is a unit (i.e., they are reduced so that AC, DC are in lowest terms). Since the square on AC is twice the square on DC, then the square on AC is even, and so AC must be even and consequently DC odd, since it is assumed they had no common measure besides 1. Divide the even diagonal AC at E, into AE, EC, which are equal numbers. This can be formed into the square $ECFD$, with diagonal DC. Since the square on DC is twice the square on EC, then the square on DC must be even, and hence DC must be even. However, DC was supposed odd, so DC is both even and odd, which is impossible.   □

When was this fact discovered? An apocryphal story says that Hippasus of Metapontum (fl. 430 B.C.), an early Pythagorean and a contemporary of Philolaus, was expelled for revealing certain secrets of the order. According to one tradition, these secrets included revealing the existence of incommensurables. This date ties in well with a book of Democritus (fl. 430 B.C.) entitled *Two Books on Irrational Lines and Solids.* Although the book has been lost, and its very contents are a mystery, the title suggests that Democritus and his contemporaries were aware of the existence of incommensurable quantities.

A more intriguing question is *how* this fact was discovered. The proof suggested by Aristotle's comments is a very simple proof, and one that can easily be extended to prove that $\sqrt{3}$, $\sqrt{5}$, and so on are likewise irrational. Yet it would seem that some time elapsed before the next major discovery of irrationals occurred. Plato, in his *Theaetetus*, says that Theodorus of Cyrene (470-400? B.C.) demonstrated that the squares with areas of three to seventeen square feet had (except for the squares of four, nine, and sixteen square feet) sides incommensurable with the side of the square of one foot. If the proof used was that suggested by Aristotle, then why did Theodorus feel compelled to prove the incommensurability of so many quantities? Likewise, if the proofs were that simple, why did Plato add, "at which point, for some reason, he stopped," as if Plato felt there would have been some value in continuing to prove the incommensurability of further square roots?

3.3.2   Side and Diameter Numbers

An alternative proof of the incommensurability of the side and diagonal of a square uses the Euclidean algorithm and ties in with the **side and diameter numbers** of the Pythagoreans. These numbers provide successively better approximations to the ratio of the side of a square to its diagonal. Theon of Smyrna described how to obtain the successive side and diameter numbers. Beginning with a side of 1 and a diameter of 1, a new side is found by adding the original side and diameter together, making 2, and a new diameter is found by adding the original diameter and twice the original side, making 3. These become the new side and new diameter; the process can be repeated.

Example 3.9. *Given the side and diameter of 2 and 3, find the next set. The*

*next side will be the sum of 2 and 3, or 5; the next diameter will be the sum of
the original diameter, 3, and twice the original side, or 7. Thus, the new side is
5 and the new diameter is 7.*

Theon noted that the square on the diameter will alternately be greater by 1
unit than twice the square on the side, and less by 1 unit than twice the square
on the side. If the complete side and diameter relationship was known by the
time of Plato, it may represent the earliest method of approximation with a
statement of the amount of error.   See Problem 1.

How did the Pythagoreans discover the side and diameter relationship? Recall that the Euclidean algorithm began by subtracting the lesser of two quantities successively from the larger, until one reached a remainder; this remainder
was then subtracted from the smaller of the two original numbers, and so on. At   See page 43.
the end of the process, one would have the greatest common measure of the two
numbers. If the process never terminated, however, there would be no greatest
common measure.

A consideration of the side and diagonal of a square leads to precisely this
sort of unending process. From the square shown in the margin, mark off BE,
equal to AD, from the diagonal DB, leaving the remainder DE. This remainder
should be subtracted from AD. To do so, draw EF perpendicular to DB, and
note that DE, EF, and FA are all equal. Thus, to subtract DE from AD,
subtract AF from AD, leaving DF. Notice that DF, DE are the diagonal and
side of a new, smaller square, and we are about to subtract the side DE from
the diagonal DF. But we started our procedure by subtracting a side from a
diagonal, so it is clear that the successive iterations of the Euclidean algorithm
will simply result in the production of ever smaller sides and diagonals, but
never a termination to the process. Thus, the side AD and diagonal DB have
no common measure.

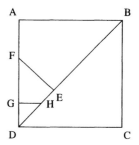

See Problem 4.

How can this be converted into the formula for the side and diameter numbers? Suppose we work from the smaller side and diagonal, DE and DF; we
want to find a relationship between the smaller side and diagonal and the larger
side and diagonal. To do this, we first notice that the new side, DA, is equal to
the old diagonal plus AF, which is equal to the old side; hence the new side, DA,
is the old side plus the old diagonal. Likewise, the new diagonal, DB, is equal
to the old side plus the new side, or twice the old side plus the old diagonal.
Thus, we obtain Theon's formula that the new side is the old side plus the old
diagonal, and the new diagonal is twice the old side, plus the old diagonal.

If this or some similar method is how Theodorus proved the incommensurability of the sides of the squares of three through seventeen square feet with the
side of the square of one foot, it is not surprising that Plato thought the achievement worth noting. The proof for each irrational is different, and the procedure
is not easy to generalize, since a new construction would be necessary for every
case, prompting Plato's wonder at why Theodorus did not continue (and perhaps
explaining why Theodorus stopped, the problem becoming tediously difficult!).

For example, to show that the side of the square of five square feet is incommensurable with the side of the square of one foot, one might draw the diagram

shown in the margin. Here, a rectangle with sides of two feet and one foot is used. Since (by the Pythagorean theorem) the square on the diagonal will be equal to the squares on the sides, then the square on the diagonal will be equal to the square of two feet, plus the square of one foot, or five square feet; hence, this will be the side of the square of five square feet.

In the diagram, BC is twice DC; CE is equal to BC. After the subtraction of the longer side BC from the diagonal, one is left with a remainder AE. Draw EF perpendicular to AC; EF and FB are equal, and a little geometry shows that AEF is half of a rectangle where again the longer side is twice the shorter side. Once again, the Euclidean algorithm will continue indefinitely, as the results of the application of the algorithm are the side and diagonal of smaller, but similar, rectangles. There are obvious constructions for proving the incommensurability of $\sqrt{10}$, $\sqrt{17}$, and similar numbers, and the diagonals of these rectangles, too, can be proven incommensurable with their sides.

Although the Euclidean algorithm can easily be applied to anything, the important feature, insofar as discovery of incommensurable quantities is concerned, is the fact that it never terminates. As we have seen earlier, while this is "obvious" in a geometric sense, it can be difficult to demonstrate, since it requires not only a proof that the side and diameter numbers never terminate, which is easy, but also that the application corresponds to the use of the Euclidean algorithm.

It is much more obvious that the properly applied Euclidean algorithm never terminates if one considers the other strong candidate for "first incommensurable quantity": the diagonal of the regular pentagon. If the diagonal of the regular pentagon was the first incommensurable quantity to be discovered, it supports a date for the discovery of incommensurables by the time of Hippasus of Metapontum. Recall he was expelled from the Pythagorean order for revealing certain secrets. One of these secrets may have been the method of constructing a regular dodecahedron in a sphere; Hippasus further compounded his guilt by claiming the construction as his own, which we might contrast with the Egyptian scribe A'hmosè claiming his work as that of someone else. What is suggestive is that the construction of a regular dodecahedron, a polyhedron consisting of twelve regular pentagons, requires the ability to construct a regular pentagon, an examination of which leads to the discovery of incommensurables.

To show that the side and diagonal of the regular pentagon are incommensurable, note that the diagonals form another regular pentagon, $A'B'C'D'E'$. Note also that $AE = ED'$, and $D'B = D'B'$. Applying the Euclidean algorithm, one can take the diagonal EB and the side AE. Subtracting AE from EB, one obtains $D'B$ as the remainder. Subtracting $D'B$ from AE, which is equal to ED', one obtains $C'D'$ as the remainder. One can continue this process by drawing the diagonals $A'C'$, $C'E'$, $E'B'$, $B'D'$, $D'A'$ inside the smaller pentagon, producing a smaller regular pentagon, $A''B''C''D''E''$. Since this process will obviously continue forever, the Euclidean algorithm will never terminate, and there will never be found a greatest common measure of the side and diagonal.

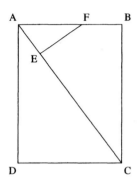

See Problem 5.

See page 48.

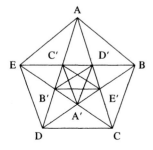

3.3.3 Effects of the Incommensurable

The existence of incommensurables may have caused a revolution in Greek thought. The diagonal of a square clearly existed, even if the ratio between the diagonal and the side could not be expressed as a ratio between whole numbers. This suggested that, in some sense, geometry was more inclusive than arithmetic. Thus, it would make more sense to use geometry as the basis of mathematics. This is precisely what would happen, and for the next two thousand years, rigorous mathematics would be based on geometric principles; indeed, key elements in the foundations of mathematics would remain geometric in nature until the nineteenth century.

3.3 Exercises

1. Refer to Theon's formula for the side and diameter numbers.

   (a) Find the first five pairs of side and diameter numbers, and then obtain five estimates for the ratio between the diagonal and side of a square (in other words, estimate $\sqrt{2}$).

   (b) Prove that the side of the diagonal is twice the square on the side plus or minus 1.

   (c) Use this fact to estimate how rapidly the ratio between the side and diameter numbers approaches $\sqrt{2}$.

   (d) Prove that the quotient of the diameter and the side approaches $\sqrt{2}$ as the number of iterations increases.

   (e) Estimate $\sqrt{2}$ to six decimal places.

2. Refer to the diagram on page 50 demonstrating the incommensurability of $\sqrt{5}$.

   (a) Find the side and diameter relationship for $\sqrt{5}$; include a statement about how the successive side and diameter numbers relate to the value of the root.

   (b) Beginning with a side and a diameter of 1, estimate $\sqrt{5}$ to six decimal places.

3. Use the side and diameter numbers to approximate the ratio of the diagonal of a pentagon to its side to six decimal places.

4. Refer to the side and diameter numbers.

   (a) Prove that DE, EF, FA are all equal, regardless of whether or not AB, AD are equal in length; in other words, the side and diameter process can work on any rectangle.

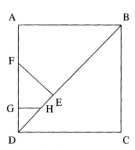

Figure for Problem 4.

(b) Show that DE, DF are the side and diagonal of a rectangle similar to $ABCD$.

(c) Draw the next two iterations of the Euclidean algorithm.

5. Draw a rectangle with sides of 3 and 4. Construct the side and diameter numbers. The process never terminates, which might suggest that the diagonal of 5 is incommensurable with the side of 4! This is obviously false. What is the flaw in the argument?

6. Find a side and diameter relationship for $\sqrt{10}$ and $\sqrt{17}$. Include a statement of how the approximations tend to the root, then estimate $\sqrt{10}$ and $\sqrt{17}$ to six decimal places.

7. Consider the diagram shown, where the diagonal is twice as long as the side, and thus the ratio between the two sides is the ratio of 1 to $\sqrt{3}$.

(a) Find the side and *side* relationship. Hint: use the fact that the diagonal is twice the shorter side. We will refer to the long side number and the short side number.

(b) In geometric terms, the square on the long side is three times the square on the smaller side. Prove that the difference between the square on the long side number and three times the square on the short side number is equal to the difference between the *first* long side number and three times the square on the *first* short side number.

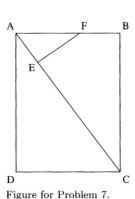

Figure for Problem 7.

8. Use either a side and diameter relation, or a side and side relation, to approximate the following quantities. Complete the relation by indicating how the successive numbers relate to the actual value (e.g., in the case of the side and diameter numbers that approximate $\sqrt{2}$, twice the square on the side is alternately one less or one more than the square on the diameter).

(a) $\sqrt{41}$ (b) $\sqrt{13}$ (c) $\sqrt{8}$

(d) For which radicals may side and diameter numbers be found? For which radicals may side and side numbers be found?

9. Refer to the proof of the incommensurability of the side and diagonal of a regular pentagon on page 50.

(a) Show that all the diagonals AC, CE, EB, BD, DA of a regular pentagon are equal.

(b) Show that the pentagon $A'B'C'D'E'$ is regular as well.

(c) Show that AE is equal to ED'.

(d) Show that $D'B$ is equal to EC'.

(e) Show that the diagonal $D'B'$ is equal to the segment EC'.

(f) Write the side and diameter relationship between the side and diagonal of a regular pentagon.

10. Plato wondered why Theodorus "stopped at seventeen." One suggestion is given by the following construction.

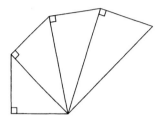

(a) Draw an isosceles right triangle. What is the length of the diagonal?

(b) On the diagonal, draw a perpendicular (see picture), and form a new right triangle. What is the length of this diagonal?

(c) Continue the process. Where is the right triangle with a hypotenuse of length $\sqrt{17}$ located?

Figure for Problem 10.

Chapter 4

Pre-Euclidean Geometry

The Greek historian Herodotus (484-430 B.C.) wrote a history of the Persian invasion of Greece titled *Histories*. Herodotus tried to establish the background for the invasion, and along the way, discussed the histories of many of the important Greek city-states and their key figures, which include Thales. Herodotus has been called the "father of history," because he is the first to apply recognizably modern historical methods to the problem of the past, as well as the "father of falsehoods," because so much of what he concludes is fantastical and hard to believe. However, the more fantastical statements were usually Herodotus quoting what someone else said, so a proper title for Herodotus might be the "father of journalism."

4.1 Thales and Pythagoras

We can speak more confidently about the mathematics of the Babylonians and the Egyptians than we can about the early Greeks, up to and including Euclid, since we have the actual mathematical treatises of the former, whereas for the latter, we must rely on copies, usually made centuries after the original. Thus, our two main sources of early Greek geometry were texts written by Proclus (sixth cent. A.D.) and by Theon of Alexandria (fl. A.D. 350). Because each copyist tended to add his (or, very occasionally, her) own commentaries, called **glosses**, to the copy, and subsequent copyists did not always distinguish between the glosses and the text, it is quite difficult to distinguish what was present in the original work and what was added later.

Ironically, the task is made easier by the fact that the copyists were rarely expert mathematicians. As a result, they often made mathematical errors, which later scribes faithfully replicated. Thus, a certain error might be common to manuscripts A, B, and C, so an historian would deduce that all three came from a common source. The problem then resembles one large logic puzzle, and by

Figure 4.1: Ionia and the Attic Peninsula.

piecing the puzzle together, one may attempt to reconstruct a lost original.

Our information about the lives of the early geometers is even sketchier than our knowledge of their work. Much of the available information is clouded with mythology and as time passes, the legend of a great mathematician grows even larger, while information about contemporary mathematicians tends to be lost into the historical background. For many early mathematicians, we know little more than their approximate dates, and in many cases, the approximate dates span a range of decades, if not centuries.

4.1.1 Thales

In some cases, a mathematician is prominent enough to be recorded in the secular histories. Thales of Miletos (ca. 624-548 B.C.) was such a mathematician, and we have a nearly firsthand account of his life from Herodotus. Miletos was a town in Ionia, on the western coast of Asia Minor (modern Turkey); like many of the Greek colonies, it was an independent city with its own army and navy. The quarreling Ionian city-states were easy prey for the expanding Lydian Empire, which conquered them, one by one, until only Miletos remained. Around the time of Thales, the Milesians realized they would be next in line for conquest, so they formally allied themselves with the Lydians and preserved some measure of independence.

None of Thales's original work has survived, though Herodotus recorded several anecdotes. One story Herodotus tells concerns a battle between the Lydians and the Medes, when the "day turned to night": in other words, a total solar eclipse occurred. Herodotus said Thales predicted the eclipse the year that it occurred. Given the astronomical knowledge of the time, it seems unlikely Thales could predict the actual *date* of the eclipse; more probably, Thales determined that an eclipse would occur in that particular year. (The eclipse—and thus, the battle—actually occurred on May 28, 585 B.C., and this battle between the Lydians and the Medes is the most ancient event whose exact date is known.)

Another story told by Herodotus about Thales concerns the invasion of Persia by the Lydian king, Croesus. Croesus's brother-in-law, Astyages, had been attacked and deposed by Cyrus of Persia around 550 B.C. To punish Cyrus, and expand his own territory at the same time, Croesus decided to go to war against the Persians. Like any head of state about to make a momentous decision, Croesus first consulted his advisors; one of these advisors was the Oracle of Delphi, supposedly possessed of the gift of prophecy. Croesus sent a messenger to the Oracle, to ask for advice. She replied that if Croesus invaded Persia, he would bring down a mighty empire. This sounded like good news to Croesus, and he sent his army into Persia.

To enter Persia, Croesus's army would have to cross the river Halys, and Herodotus recorded a belief among the Greeks that Thales helped Croesus invade Persia by diverting the river so Croesus could build a bridge across it. If this story is true, then Thales, like so many later mathematicians, was a military engineer. As for Croesus, the Oracle's prophecy was fulfilled. Unfortunately, the "mighty empire" brought down was Croesus's own, which Cyrus conquered in 547 B.C.

Thales's achievements grew into almost mythical proportions after his death. He is one of the Seven Wise Men of ancient Greece, and the only one who appears in all versions of the list. He is said to have been the first person to approach questions about the universe from a perspective that did not require supernatural intervention; to have founded a school of philosophy whose basic maxim was "know thyself"; and to have come up with the first theory of matter, believing that all things are created from a single substance, which he called "water."

Other stories about Thales are reported. Plato says that Thales once fell into a well while stargazing, thus making him the first "absent-minded professor." Aristotle reports that Thales, to prove that philosophy had a practical value, bought up all the olive presses in Miletus in a year when he determined there would be a particularly bountiful crop of olives; when the crop came in, the only presses to be had were those that Thales rented out at his price, and thus he made a small fortune.[1]

Our knowledge of Thales's mathematical achievements comes along a very indirect route. Eudemus of Rhodes (fl. 320 B.C.) wrote a history of geometry, which has since been lost. However, before it was lost, Proclus wrote a commentary on the first book of Euclid's *Elements*, and made extensive use of Eudemus's work. This makes Proclus, writing seven hundred to one thousand years after the events he described, one of our main sources on the early history of geometry. According to Proclus, Thales was the first to introduce geometry to the Greeks; Proclus implied that Thales actually learned geometry while traveling through Egypt.

---

[1] Some versions add that Thales gave back the money, because he was interested only in making a point.

Proclus said that Thales "noticed" and "asserted" four mathematical propositions; as Thales apparently gave no proof, we will refer to his discoveries as conjectures (remember, we are calling a conjecture any result originally presented without proof).

Conjecture 4.1. *A circle is bisected by its diameter.*

Conjecture 4.2. *The base angles of an isosceles triangle are equal.*

Conjecture 4.3. *The vertical angles formed by two intersecting lines are equal.*

Conjecture 4.4. *If, in two triangles, a side and the two adjacent angles of one triangle are equal to a side and the two adjacent angles of the other triangle, the two triangles are equal.*

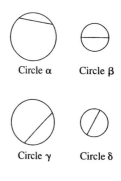

Circle α Circle β

Circle γ Circle δ

How were these conjectures generated? The most reasonable answer is that they were found inductively. Thus, early geometry resembles modern science, in that observations were made on geometric objects and some general conclusions were drawn. For example, observations of circles cut by diameter and non-diameter lines would make it "obvious" that a circle is bisected by its diameter.

There are two other conjectures associated with Thales. The first, related by Pamphile (fl. A.D. 60), a Roman philosopher, is that Thales was the first to inscribe a right angled triangle in a circle, and thus determine that any angle inscribed in a semicircle was a right angle. Pamphile goes on to tell us that Thales sacrificed an ox in honor of this great discovery, a story that sounds suspiciously familiar.

See page 42.

It is not entirely certain how Thales defined a right angle. Functionally, Thales's definition and our own would coincide. However, it is virtually certain that Thales did *not* consider a right angle to be an angle of a particular measure, since the notion of the actual measurement of an angle never arose in the theoretical geometry of the ancient Greeks. We will, for now, sidestep the issue of the definition of a right angle, though we will return to it when discussing Euclid.

Another conjecture attributed to Thales comes to us from two sources. The older account, by Hieronymus of Rhodes (290-230 B.C.), is that Thales determined the height of the Great Pyramid of Egypt by noticing that at a certain time of day, the length of a man's shadow was the same as his height, and thus the length of the pyramid's shadow was the same as the height of the pyramid. Hieronymus's account has been lost, though before it vanished, portions were quoted by Diogenes Laertius (fl. A.D. 350). The account claimed that Thales inferred that if all shadows he could observe were equal in length to the height of the object producing them, the same must hold true for the shadow of the pyramids. This is a reasonable conclusion, drawn inductively.

We can actually see "legend creep" in action, when Plutarch (A.D. 47-120) tells the same story, several centuries after Hieronymus. According to Plutarch, Thales inferred the equality of an invariant *ratio* between the shadow and the object's height. Given that the theory of proportions was, so far as we know, nonexistent in the time of Thales, this would have been an incredible achievement.

4.1.2   Pythagoras

To fully understand ancient cultures, one must take into account the prevalence of slavery, a key element of almost every culture before the present day, though generally men (and women) were slaves by virtue of military conquest, rather than by race. The Greek philosopher Aristotle defended slavery on two grounds: first, it was necessary, since machines would not work by themselves; and second, some men were, by their very nature, fit for nothing better.

Slaves worked with their hands (and backs): hence, the work done by slaves came to be known as **manual labor**, from the Latin word *manus*, "hand." Free men, on the other hand, were expected to pursue the **liberal arts**, from the Latin word *libera*, "free man." Free women, incidentally, were expected to bear children; *raising* the children was a task for slaves.

The mathematics of the Egyptians and the Babylonians, and possibly that of Thales, concerned itself with "practical" matters, whether it was computation of the height of a pyramid, or the determination of the cost of digging a canal. Because of its association with manual labor, this type of mathematics was associated with the work of slaves. *Theoretical* mathematics, on the other hand, was a pursuit worthy of free men, and Proclus gave Pythagoras (580-500 B.C.) credit for transforming mathematics into a liberal art, by studying its theorems in an "immaterial" and "intellectual" manner. This probably means that Pythagoras introduced the deductive method to mathematics, while restricting its domain to the theoretical properties of abstract objects.

Theorems of the Pythagoreans

There are five important results associated with Pythagoras or his school, the most famous of which is, of course, the Pythagorean theorem. It is not known for certain if Pythagoras even proved the theorem that bears his name; Proclus himself sounds dubious. However, it is not unreasonable to suppose that Pythagoras had a proof that the Pythagorean theorem held for *certain* right triangles.

From Plato's *Meno*, written about two hundred years after Pythagoras, we have the proof that the square of the diagonal of a square is twice the square on the sides. In the figure, $ABCD$ is a square with diagonal AC. The squares $ABHE$, $FHBK$, $KBCG$ are all duplicates of $ABCD$, and together they form the large square $EFGD$. Plato proved the equality by noting that large square $EFGD$ is four times the smaller squares individually, and hence four times $ABCD$; meanwhile, the diagonal bisects each square, so the square $AHKC$ is half the large square $EFGD$, and thus twice the square on $ABCD$.

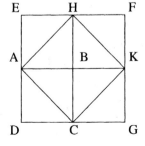

If Pythagoras knew this, and there is no reason to doubt that he did, it was a simple enough matter to extend the proof to more generalized figures. The intermediate steps are a mystery, though it is reasonable to suppose that the diagonals of other rectangles were examined first, from which a general procedure might have been found.

See Problems 4 and 5.

For the other discoveries of the Pythagoreans, Proclus not only claimed that the results were proven, but in some cases, he gave the actual proofs of the Pythagoreans; hence, we will call the Pythagorean results **propositions** and, in general, we will label as a proposition any result originally presented with a proof or, in some cases, propositions for which the proof is self-evident. The honor of "first proposition" in geometry goes to the following Pythagorean discovery:

Proposition 4.1. *The three angles in a triangle are together equal to two right angles.*

The Pythagorean proof, given by Eudemus and quoted by Proclus, is the following:

Proof. In triangle ABC, draw DE through A parallel to BC. Because DE, BC are parallel, the alternate interior angles DAB, ABC are equal; likewise, angles EAC, ACB are also equal. Hence, the angles DAB, BAC, CAE are equal to the three angles of a triangle. They are also equal to angles DAB, BAE, which together equal two right angles, so the three angles of a triangle equal two right angles. □

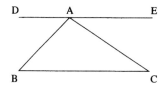

This is the proof most commonly given in modern geometry texts. The proof assumes certain things are known about parallel lines: for example, the statement that angles DAB and ABC are equal is based on the proposition that when two parallel lines are cut by a transversal, the alternate interior angles are equal. In addition, the assumption was made that the three angles on one side of the line DE are together equal to two right angles. Thus, we may infer that the Pythagoreans knew or assumed these two propositions.

Another proposition credited to the Pythagoreans is the following:

Proposition 4.2. *The angles in a polygon are together equal to the angles in a number of triangles two less than the number of sides.*

Thus, the angles in a four sided figure are together equal to the angles in two triangles; the angles in a pentagon are together equal to the angles in three triangles, and so on.

See Problem 6.

If all sides and interior angles of a polygon are equal, the polygon is said to be **regular**. The third proposition credited to the Pythagoreans, with Proclus's proof, dealt with regular polygons:

Proposition 4.3. *The space about a point can be filled with regular polygons in one of three ways: six triangles, four squares, or three hexagons.*

Proof. Each angle of an equilateral triangle is two thirds of a right angle. Thus, six such angles will make four right angles, and will fill the space about a point. In like manner, the square and regular hexagon fill the space about the point. All other regular polygons will exceed or fall short of four right angles. □

See Problems 7 and 8.

The proof, as given, is incomplete.

The filling of a space about a point also leads to a result regarding **regular polyhedra**, solids whose sides are regular polygons: there can be no regular polyhedra except those formed by triangles, squares, or regular pentagons. This may have led the Pythagoreans to the discovery of three of the five regular solids: the tetrahedron, formed by regular triangles; the cube, formed by regular quadrilaterals; and the dodecahedron, formed by regular pentagons.

See Problems 9 and 10.

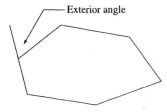

If one side of a polygon is extended, the angle between the extension and the adjacent side is called an **exterior angle**. According to Aristotle, the Pythagoreans also proved the following.

Proposition 4.4. *The exterior angles of a polygon are equal to four right angles.*

Application of Areas

Pythagoras himself is credited with the introduction of the notion of "application of areas" or, more properly, the "application of figures." Given a line, a rectilineal figure, and a parallelogram, the goal is to construct a parallelogram whose area is equal to the given figure. The three possibilities were:

1. The whole of the line was the side of the parallelogram; this was **parabole**.

2. The parallelogram "fell short" of the entire line by a parallelogram similar to the given parallelogram; this was **elleipsis**.

3. The parallelogram "exceeded" the line by a parallelogram similar to the given parallelogram; this was **hyperbole**.

In Figure 4.2, the given line is CD, the rectilineal figure is A, and the parallelogram is B. Thus, the three situations correspond to

1. Parallelogram $CDNL$, whose area is equal to that of A and with angles equal to those in B.

2. Parallelogram $CEQP$, whose area is equal to that of A; this "falls short" by parallelogram $EDRQ$, constructed on the remainder of the line and similar in shape to B.

3. Parallelogram $CFKG$, whose area is equal to that of A; this "exceeds" the original line CD by $DFKJ$, which is similar to B.

The application of areas usually involved the solution of algebraic equations, couched in geometric terms. We shall examine these problems in detail when we discuss Euclid. However, one particular problem of the application of areas is an important element in Pythagorean geometry.

The problem, which appears in Euclid, Book II, Proposition 11, is the following: given a line AB, to divide it at a point C so that the rectangle formed on the segments AB, BC, is equal to the square formed on AC; this is equivalent to

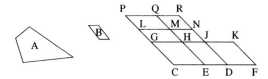

Figure 4.2: Parabole, elleipsis, hyperbole.

solving a quadratic equation. Suppose that AC is the larger of the two segments. Then the ratio of AB to AC is what came to be known as the **golden ratio**, though this name did not come into use until the Renaissance. The Greeks simply referred to the division as **the section**, since it was such an important one that it needed no other name.

Once this is done, then a regular pentagon may be formed, as outlined in Euclid, Book IV, Propositions 10 and 11. The Euclidean proposition requires See Problem 12. one to inscribe a pentagon in a given circle; the following is a simplified version, which seeks only to create a regular pentagon.

1. An isosceles triangle is constructed in which the base angles are twice the vertex angle. To do this, take the line AB and cut it at C, so that the rectangle formed by the segments AB, BC is equal to the square formed on AC.

2. Draw the circle with center at A and radius AB.

3. Locate D on the circle, so that BD is equal to AC.

4. The triangle ABD is a triangle whose base angles ABD, ADB are twice the vertex angle DAB.

5. Construct circle ABD. BD is the side of a regular pentagon inscribed in the circle ABD; the other four sides may be constructed by measuring out the points E, F, so that BE, DF are equal to BD.

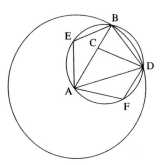

The diagonals of the pentagon can be joined to form a pentagram or a five-pointed star figure, and one finds that the ratio of the diagonals to the sides is equal to the ratio of AB to AC, and so the section reappears in the pentagon. Once a regular pentagon is inscribed in a circle, it is easy enough to find a construction for the regular 15-sided polygon; this appears in Euclid's *Elements*, Book IV, Proposition 16.

According to Lucian (A.D. 120-180), a Roman satirist and philosopher, the Pythagoreans inscribed the letters of the Greek word for health, *hygieia*, at the vertices of the pentagon. If this story is true, it may represent the first use of letters to designate points in a geometric diagram, and may have led to the general practice of labeling points with letters.

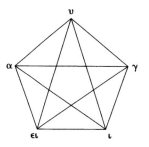

4.1.3 The Ionian Revolt

Shortly after the death of Pythagoras, the Mediterranean world would become a battleground, and progress in mathematics would be halted for half a century. After the conquest of Lydia, one of Cyrus's generals, Harpagus, was sent to conquer Ionia: this he did, though again the Milesians remained independent. In 538 B.C., the Persians conquered the Neo-Babylonian Empire established by Nebuchadrezzar, the Nebuchadnezzar of the Bible. Since Nebuchadnezzar conquered the Kingdom of Israel and destroyed the Temple of Solomon, the Jews viewed the Persian conquest as divine retribution, and centuries later the events were retold in a form that became the Biblical Book of Daniel. The Persian Empire continued to expand under Cyrus's son Cambyses, who conquered Egypt in 525 B.C.; by his death in 521 B.C., the Persian Empire included most of modern Iran, Iraq, Turkey, Israel, Lebanon, Jordan, Syria, and Egypt: it was the largest empire the West had yet seen.

The greatest of the Persian kings was Cambyses's son Darius, who became king in 521 B.C. The Persians were unusually benevolent conquerors: local traditions were respected, and Darius even allowed the Jews to return to Palestine to rebuild the Temple of Solomon. Darius also commissioned a vast system of roads, complete with military patrols to deter bandits, and established an efficient postal system, of which Herodotus wrote: "Neither rain, nor sleet, nor dark of night stays [prevents] these couriers from the swift completion of their appointed rounds," a phrase that adorns the headquarters of the United States Post Office in Washington, D.C. and is sometimes taken (incorrectly) to be the motto of the Post Office.

Despite the advantages of being part of this great empire, the Ionian city states wanted independence. Led by Miletos, they revolted around 500 B.C., and pled for help from the Greek mainland, pleas that were mostly ignored. The aid that was sent proved insufficient, and Miletos finally lost its independence when the Persians conquered it in 494 B.C. To punish the Greek city-states that had supported the revolt, Athens and Sparta, Darius organized a punitive expedition in 490 B.C., and the period in Greek history known as the Persian Wars began. It would be the first of many wars that would keep the land of Pythagoras and Thales in turmoil for three and a half centuries.

4.1 Exercises

1. We do not know how Thales defined *right angle, equal, bisect,* or any of his geometric terms. Define these terms. Be sure to define any other terms that you introduce in your definition, and avoid circularity (for example, do not define bisect as "to divide in half").

2. Demonstrate Thales's conjecture that the base angles of an isosceles triangle are equal, using the following method.

   (a) First, draw an isosceles triangle.

   (b) Fold the triangle, so the base angles overlap.

   (c) Are the base angles equal? Explain why. Note that you will have to explain what you mean by equal.

   (d) Repeat these steps for several other isosceles triangles.

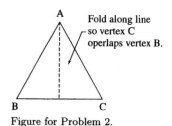

Figure for Problem 2.

3. Design an experiment to support the notion that if the length of a man's shadow is the same as the height of the man, then the length of a pyramid's shadow at the same time is the same as the height of the pyramid. Explain some of the difficulties in applying this on a practical basis.

4. Use the method of Plato to show that the diagonal of a rectangle whose width is twice the length is equal to five times the square on the length.

5. There are several speculations on how the Pythagorean theorem was proven. The simplest is based on a generalization of the method described by Plato. In the accompanying picture, AFE, BGF, CHG, DEH are all identical right triangles situated about the square $EFGH$ on their common hypotenuse. Show that this square is equal to the squares on AF, AE. Indicate what other assumptions and/or propositions are necessary to use this proof. Hint: divide the large square $ABCD$ by drawing EK and FL perpendicular to AD and AB, then determine the areas of the resulting squares and rectangles.

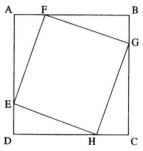

Figure for Problem 5.

6. Prove Proposition 4.2. Indicate what additional propositions or assumptions are necessary.

7. Complete the proof of Proposition 4.3.

8. Proclus implies, but does not prove, that no other regular polygon can fill the space about a point but the equilateral triangle, the square, and the equilateral, equiangular hexagon. Prove this fact.

9. Prove that regular polyhedra can only be formed out of regular triangles, regular squares, or regular pentagons.

10. Prove that there can be no more than five regular polyhedra. Note that this does not prove that the five regular polyhedra exist, but only that there are *at most* five.

11. Determine the value of the golden ratio.

12. Examine the method of creating a regular pentagon.

   (a) What are the interior angles of the isosceles triangle ABD?

   (b) Show that the ratio of AC to AB is equal to the golden ratio.

   (c) Show that BD is the side of a regular pentagon inscribed in circle ABD. Hint: find the interior angle that would intersect the arc BD.

13. Show that the ratio of the side of a regular pentagon to its diagonal is equal to the golden ratio.

14. Explain how to inscribe a regular 15-sided polygon (a quindecagon) in a circle, once you can inscribe a regular pentagon in a circle.

4.2   The Athenian Empire

In 490 B.C., the Athenians defeated a Persian army just outside Athens. A runner, Pheidippides, ran the 26 miles from the battle site to the marketplace in Athens, where he announced victory (*nike* in Greek) over the Persians at a place called Marathon. The Persians withdrew, not so much because of Greek resistance, but because Egypt had revolted against Persian rule as well, and keeping rich Egypt was worth more than punishing the Greeks. The Persians returned in 480 B.C., under Xerxes, Darius's successor. At the pass of Thermopylae (480 B.C.), 1400 Greeks, led by the Spartans, faced the entire Persian army: 200,000 soldiers, including 10,000 elite troops, known as "Immortals." Defeat was inevitable, but the Spartans stopped the Persian advance for three days (dying to the last man), buying enough time for other parts of Greece to improvise hasty defenses. It was not enough time to save Athens, and Xerxes swept across the Attic peninsula, and burned the city.

To finish off the Greeks, Xerxes would have to destroy their fleet, which had taken refuge in the Saronic Gulf, between Athens and the island of Salamis. The naval battle of Salamis (the first great naval battle in recorded history) ended in a disaster for the Persians, who lost nearly half their fleet; Greek losses were insignificant, and Xerxes withdrew—for one year. The next year, Xerxes returned with another army, suffered another defeat (at Plataea), and withdrew again. Even then, the resources of the Persian Empire dwarfed what the Greeks could field.

To fight Persia, Athens organized an alliance now known as the Delian League, because the headquarters and the treasury of the alliance were located on the island of Delos. Member states of the alliance could either contribute ships and men, or the equivalent in money. Most members chose to supply money, letting Athens build and man the ships of the fleet. Eventually the Persians gave up trying to conquer Greece.

In theory the Delian League, with no enemy to fight, ought to have been disbanded, but Athens argued, reasonably, that the League was still necessary to defend Greece from Persia. Member states grudgingly agreed to maintain what was rapidly becoming the Athenian Navy, but after a while, the threat of Persian intervention in Greece seemed a remote prospect. Tired of paying for a navy that did it no good, the inhabitants of the island of Naxos attempted to withdraw from the League. The Athenians refused to accept the withdrawal, laid siege to Naxos, and in 467 B.C., captured the city, destroyed its fortifications, and sold the inhabitants into slavery. Other member states were cowed into submission, their annual dues being converted into tribute, and the Athenian Empire was established. Heavy tribute from the "allies" went to Athens, making it one of the most beautiful cities in the world.

In 461 B.C., an ambitious aristocrat sought to expand the empire through a series of wars of conquest, sometimes called the First Peloponnesian War. The aristocrat's name was Pericles, and his era became known as the **Golden Age of Athens**, during which democracy, slavery, and imperialism existed side by side. Pericles's ambitions went far beyond the resources of Athens, and the land campaigns bogged down into a series of inconclusive battles that lasted for a generation. By 431 B.C., Athens was exhausted by a generation of warfare, which had left it no more and no less powerful than it was when Pericles came to power.

4.2.1 Three Classical Problems

During the Golden Age of Athens, between 460 and 420 B.C., three problems were posed, sometimes called the **Three Classical Problems**. They were:

1. The **trisection of an angle**: Given any angle, find an angle exactly one third as large.

2. The **squaring of the circle**: Given any circle, find a square equal in area.

3. The **duplication of the cube**: Given any cube, construct another cube with exactly twice the volume.

These problems presumably arose out of contemporary geometric knowledge. From the first, we can infer that bisecting an angle could be accomplished, and the method of dividing a given line into any number of parts was probably also well known; hence, the question of whether an angle could be trisected probably arose out of an extension of these two processes. From the second, it would appear that a procedure for squaring rectilineal figures was known, and so the question of squaring a *non*rectilineal figure arose. Finally, creating a square twice as large as an existing square was simple, so it was natural to ask if the same could be done with a cube.

The importance of these problems was not their theoretical value—no theorems required, for example, the trisection of an angle in their proof—but in how the investigation of the problems inspired new mathematics, and a re-examination of existing mathematics. From these problems came the conic sections and the foundations of Euclidean geometry.

4.2.2 Anaxagoras

It was against this background that Anaxagoras (500-428 B.C.) came to Athens. Originally from Clazomenae, in Ionia, Anaxagoras became a friend and a tutor of Pericles. Some of his scientific notions were revolutionary; among other things, he proposed that the sun was nothing more than a red hot stone, 4000 miles away, and for this he was punished for heresy.

According to Plutarch, Anaxagoras "wrote on" the squaring of the circle while in prison. Plutarch gives no details, so we do not know if this means that

Anaxagoras worked out a method of squaring the circle, or whether he worked out a method of approximating a circle with a square, or whether he simply wrote about the problem.

4.2.3   Hippias of Elis

See page 58.

Hippias of Elis (b. ca. 460 B.C.) would also make his way to Athens. Hippias was a member of a new branch of philosophy: the Sophists. Unlike the Pythagoreans, who were secretive and taught only those willing to become Pythagoreans, the Sophists were willing to teach anybody—for a price. As a result, the Sophists were often reviled by other philosophers: teaching for money seemed too much like manual labor.

It might seem that the demand for philosophy lessons would be rather limited. However, part of philosophy was the skill of debate, which in turn was essential for success in the Athenian legal system. In it, the two sides to a dispute would present their cases, and the jury's duty was to decide which party had the more persuasive argument. Learning how to debate from a Sophist or any other philosopher could prove invaluable in court.

Of Hippias's work, nothing has survived. However, Proclus noted that a curve invented by Hippias was used to trisect any angle (it is not clear whether Hippias used it for this purpose). Hippias's curve was revolutionary: it was the first curve not constructed out of straight lines or arcs of circles. Because this curve can be used to trisect an angle, it is sometimes known as the trisectrix, but it is better known for another use: since it can be used to square (or quadrate) the circle, Hippias's curve is known as the **quadratrix**.

We do not know precisely how Hippias defined the quadratrix. The earliest description we have comes from Pappus (third century A.D.), a geometer who lived in the twilight of Greek science. In his *Mathematical Collection*, Pappus described the quadratrix: construct the square $ABCD$, and with D as center and AD as a radius, draw the circular arc AYC. Let the radius DA rotate clockwise to position DC; in the same amount of time, let AB drop, parallel to itself, arriving at DC at the same time as the radius. While this motion is occurring, the radius and the line will intersect at points G, P, and so on, describing the quadratrix AGH.

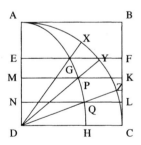

See Problem 1.

To use the curve to trisect an angle, let the given angle be GDC, which intersects the quadratrix AGH at the point G. Through G, draw EF parallel to DC, and trisect FC at points K and L. Draw NL parallel to DC; the intersection of this line and the quadratrix will be at Q. The angle QDH is one third the angle GDC. The quadratrix can actually be used to divide an angle into any number of parts. In a later section, we will look at how the quadratrix was used to square the circle.

4.2.4   Democritus of Abdera

Democritus (ca. 460-370 B.C.) came from Abdera, part of the Athenian Empire. He is credited with the following conjectures:

Conjecture 4.5. *A pyramid is one third the volume of the prism on the same base with the same height.*

If the base is a circle, then the pyramid becomes a cone and the prism becomes a cylinder; thus the second conjecture of Democritus was:

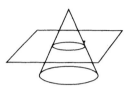

Conjecture 4.6. *A cone is one third the volume of the cylinder on the same base with the same height.*

Plutarch recorded an intriguing statement about the second conjecture: Democritus considered what would happen if a plane cut a cone into two parts, and considered the circles forming, respectively, the base of the cone and the top of the frustrum. If the two circles are unequal, then the cone is "stepped" and is no longer a cone. On the other hand, if the two circles are equal, then it would seem that every cross section of the cone was equal, and the cone was a cylinder. Democritus had no solution to this dilemma. However, the problem itself suggests that the notion of thinking of a volume as consisting of a number of cross-sectional figures had already been established by his time.

4.2.5 Oenipides of Chios

Sometime around the end of the First Peloponnesian War, Oenipides from Chios came to Athens. Chios was a member of the Delian League that managed to retain some measure of independence. It is not known what Oenipides intended to do when he came to Athens; perhaps he was attracted by the lure of the "big city," and perhaps he came on other business. In any case, Proclus credits him with solving two problems: first, he determined how to draw a perpendicular from a line to a given point; second, he determined how to construct an angle equal to a given angle.

Taken at face value, neither of these achievements is especially remarkable, and it is inconceivable that Oenipides was the *first* to construct a perpendicular, since so common an instrument as an architect's "T" could be used to do so; in fact, it was about this time that the Parthenon at Athens was being built. So why did Proclus consider Oenipides's achievement worth mention? The most reasonable guess is that it was Oenipides who showed how to perform these constructions using *only* the compass and straightedge.

After Oenipides, geometers looked for solutions that not only solved the problem at hand, but did so with the least amount of "machinery" in the form of mechanical devices. Thus, while it is possible to trisect the angle using the quadratrix of Hippias, this method of solution required whatever mechanical device was necessary to draw the quadratrix in the first place. As a method of solution to the real world problem of trisecting the angle, this was satisfactory; as a method of solution to the intellectual problem of trisecting the angle, however, this could be improved upon *if* the requirement of an additional, physical device could be eliminated.

This would eventually become the doctrine that the *only* instruments allowed in geometry were the compass and straightedge; the use of these, too, were restricted: the compass was to be used only to measure equal segments along an existing line, or to draw a circle of a given radius, whereas the straightedge could only be used to draw a line between two points.

4.2.6   Hippocrates of Chios

Hippocrates, one of Oenipides's fellow countrymen, came to Athens soon after he did. Hippocrates from *Chios* should not be confused with Hippocrates from *Cos*, the physician after whom the "Hippocratic Oath" is named. Hippocrates from Chios was originally a businessman. According to one story, a ship of his was captured by pirates and the cargo later turned up at Athens, where Hippocrates went to claim it around 450 B.C. The legal proceedings took so long, however, that he soon ran out of money, and turned to the teaching of mathematics to support himself, staying in Athens until his death around 430 B.C. Thus, Hippocrates is the first person we know to be a professional teacher of mathematics.

Proclus tells us that Hippocrates was the first to write a textbook in deductive geometry. This text, now lost, was called *The Elements of Geometry*; to distinguish it from other texts called the *Elements*, it is usually called the *Elements* of Hippocrates.

Hippocrates also made some original contributions to mathematics. According to Philoponus (sixth century A.D.), Hippocrates "consorted with philosophers" and at one point reached such a great proficiency in geometry that he tried to square the circle. He did not succeed, although he made a remarkable discovery: he was the first to show that a region bounded by curves can have the same area as a region bounded by straight lines.

See Problem 9b.

Hippocrates's discovery is the earliest theoretical achievement of Greek mathematics that has been preserved in its original form. Eudemus, whose history was so often cited by Proclus, had access to some of the works of Hippocrates; before it was lost, portions of Eudemus's work were copied "word for word" by Simplicius, a sixth century A.D. commentator on Aristotle. Although our source is thirdhand, from Hippocrates through Eudemus to Simplicius, we may reasonably reconstruct parts of Hippocrates's work.

In general, if a line cuts across a curve, the region between the line and the curve is called the **segment of the curve** and the line is the **base of the segment**: for example, the region between a chord and the circle is a **segment of a circle**. Two segments of a circle are **similar** if the central angles of the circular arcs are equal. Hippocrates dealt with regions bounded by arcs of intersecting circles, called **lunes** because they are moon-shaped.

Hippocrates is said to have **squared the lune**, a term that needs some interpretation. Literally, to square the lune would be to construct a square that was equal in area to a given lune; in practice, it was sufficient to construct any rectilineal figure that was equal to a lune, since squaring a rectilineal figure was trivial.

To effect the quadrature of these lunes, Hippocrates used several results regarding circles. These results are:

Conjecture 4.7. *Circles are to each other as the squares on their diameters.*

Conjecture 4.8. *Similar segments of circles are to each other as the squares on their bases.*

These results can be proven using the theory of proportions. However, so far as is known, the theory of proportions was not developed until the time of Eudoxus, about half a century after Hippocrates, so it seems improbable that Hippocrates had a valid proof of either statement; hence we will call them conjectures.

See Problem 3.

After these preliminary statements, Eudemus gives us Hippocrates's first proposition on the quadrature of lunes, with, presumably, the proof of Hippocrates.

Proposition 4.5. *If a semicircle is circumscribed about a right-angled isosceles triangle, and about the base a segment is described similar to the segments cut off by the sides of the triangle, then the lune formed is equal to the triangle.*

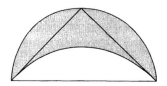

The lune referred to is shaded in the picture in the margin.

The proof, as given by Eudemus, is purely rhetorical:

Proof. Circumscribe a semicircle about a right-angled isosceles triangle, and about the base describe a segment similar to the segments of the circle cut off by the sides of the triangle. The segment about the base is equal to the sum of the segments about the sides, so the part of the triangle above the segment about the base, together with the two lunes, is equal to the triangle. Therefore the lune is equal to the triangle. □

If Simplicius is faithfully copying Eudemus, and if Eudemus is in turn faithful to Hippocrates's original work, we notice a few key facts about the geometric proofs of Hippocrates and by extension, the geometric proofs of the earliest Greek geometers: the proof is entirely verbal, and does not contain the symbolism we associate with later geometry. Naturally, this makes the proof more difficult to follow. Moreover, Hippocrates made a subtle assumption: that the similar segments did not overlap. On the other hand, it is believed that Hippocrates may have been the first geometer to indicate important points on a diagram using letters of the Greek alphabet, in which case, the lack of letters is due to Eudemus or Simplicius.

After discussing the lune in a semicircle, Eudemus gave Hippocrates's quadrature of a lune whose outer circumference is greater than a semicircle. In this case, he began with a trapezium having three equal sides, and a fourth side whose square is three times the squares on the other sides: in other words, the square of the fourth side is equal to the squares of the other three sides put together. As before, upon the long side is constructed a segment similar to the segments cut off by the short sides; the lune is indicated by shading. Eudemus shows that

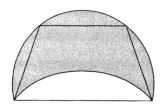

the lune's outer circumference is greater than a semicircle, but does not actually give the area of the lune.

See Problems 5 and 6.

There are at least two other lunes whose squaring Eudemus ascribes to Hippocrates. One is the lune on less than a semicircle; the other is a more complicated lune subtended by two adjacent sides of a regular hexagon. In addition, Alexander of Aphrodisias (3rd century A.D.) claimed Hippocrates squared two other lunes. However, the accuracy of Alexander's account is debatable.

See Problems 9a and 9b.

Hippocrates's other major achievement concerns the duplication of the cube. In a story told by Eratosthenes (276-196 B.C.), the problem originated when Minos, the legendary king of Crete, saw the tomb being built for his son Glaucus, and decreed that it should be doubled in size. Minos went on to order the sides of the tomb doubled in length, which, of course, would make the tomb eight times larger, not twice as large. Hippocrates made an important contribution to this problem, though he apparently did not solve it, by showing it reduced to the problem of finding two mean proportionals.

Recall that A, B, and C form a geometric proportion if A is to B as B is to C; we say that B is a **mean proportional**. The notion of mean proportionals can be extended to include more terms: if A is to B as B is to C, and B is to C as C is to D, then there are two mean proportionals, namely B and C, that have been inserted in between A and D. If D is twice A, then the first ratio, A to B, is the ratio necessary to double the cube, since the cube on a side of B will be exactly twice a cube on a side of A.

See Problem 8.

The use of mean proportionals, although it solved one problem, raised several others. In particular, the notions of ratio and proportion had been developed for numbers; to use them on geometrical objects required a more general theory. This was the task for Eudoxus; we will examine his work in the next section.

The problem of duplicating the cube is also known as the **Delian problem**, since, according to a different story, the problem of doubling the size of a cube was faced by the Delians. When confronted with a plague, they sent a delegation to the Oracle of Delphi, who said it was caused by the displeasure of the god Apollo. To appease the god, a certain cubical altar of Apollo had to be doubled in size. Unable to solve the problem themselves, they sent a delegation to Athens to seek help.

Plague was a fact of life in the ancient world. In 431 B.C., near the end of Hippocrates's life, the Second (or Great) Peloponnesian War broke out between the Athenian Empire and Sparta and her allies. Around 429 B.C., plague struck Athens, killing a quarter of the citizens, including the Athenian leader Pericles. Despite the death of Pericles and the weakening of the Imperial Party, the Great Peloponnesian War would drag on for another 25 years.

4.2 Exercises

1. Consider the quadratrix.

    (a) Using compass and straightedge alone, draw some of the points on the quadratrix.

(b) Prove that the quadratrix trisects the angle GDC.

(c) Show how you can use the quadratrix to divide an angle into any number of parts.

2. Part of the proof that angle trisection or cube duplication is impossible relies on the question of what sort of points in the Cartesian plane can be described using only compass and straightedge.

   (a) Show how you would construct a Cartesian plane, with the integer points clearly labeled, using only compass and straightedge.

   (b) Show how to locate any point (x, y), where x, y are both integers.

   (c) What lines $ax + by = c$ is it possible to construct? Note that in order to construct a line, you must locate two points on the line.

   (d) Show how to locate any point (x, y) where x, y are both *rational* numbers, by constructing two lines whose intersection point is (x, y).

   (e) What points (x, y), where one or both of x, y is *irrational*, can be constructed?

   (f) Determine what lengths a can be constructed by noting that if you can construct the points (x_1, y_1), (x_2, y_2) you can also construct the line between them, and consequently can construct a line equal to the distance between the two points.

   (g) What circles is it possible to construct? Remember that to uniquely identify a circle, you need to specify its center and its radius.

3. We can use the fact that Hippocrates proved that similar segments are to each other as the squares on their bases to give us additional information about the level of geometry in his time.

   (a) Prove that similar segments are to each other as the squares on their bases.

   (b) What additional propositions or assumptions are needed for the proof?

4. Provide commentary on Hippocrates's proof of the lune of an isosceles triangle.

   (a) First, identify the lunes, triangles, and so on, by labeling the important points of the diagram provided.

   (b) Next, explain the proof. For example, the proof claims that the segment about the base is equal to the sum of "those about the sides." Why is this true? Fill in the details that are left out.

5. Show that the trapezium, in the case of the lune whose outer circumference is greater than a semicircle, actually does mark out an arc of the circle that is greater than a semicircle. (Hint: consider the diagonal of the trapezium. Use the Pythagorean Theorem to show that the angle that subtends the long side of the trapezium must be acute, and therefore the long side subtends less than half the circle.)

6. How does the lune whose outer circumference is greater than a semicircle compare to the trapezium?

7. What problems do you run into when extending the method of Hippocrates for a general lune?

8. Recall that B is a mean proportional between A and C if the ratio A to B is the same as the ratio B to C.

    (a) Show how the determination of two mean proportionals between A and D will allow you to double the cube. (Hint: let D be twice A.)

    (b) If B, C are two mean proportionals between A, D, show that $A \cdot D = B \cdot C$. Consequently, if A, B, C, D, are considered lengths, then the rectangle on A and D is equal to the rectangle on B and C.

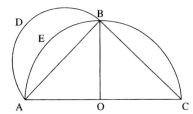

Figure for Problem 9a.   Figure for Problem 9b.

9. Alexander of Aphrodisias claimed that Hippocrates found the quadrature of two lunes, and that he squared the circle.

    (a) Let AB is the side of a square inscribed in a circle, and let ABC be the semicircle (see figure). Show that the lune $ADBE$ is equal to the triangle ABO.

    (b) Let AB and $CDEF$ be semicircles, with CF twice AB (see figure). CD, DE, EF are three sides of a regular hexagon, and CGD, DHE, EJF are semicircles with diameters CD, DE, EF, respectively. Show that the lunes CGD, DHE, EJF, and the semicircle AB together equal the trapezoid $CDEF$.

    (c) Alexander claimed that by subtracting from the trapezoid $CDEF$ a rectilineal figure equal to CGD, DHE, EJF, one has then squared the circle. Why won't this work? (It is believed the error is Alexander's, and not Hippocrates, which tends to cast doubt on the accuracy of Alexander's whole account.)

4.3 The Age of Plato

By 404 B.C., Sparta and her allies had "won" the Peloponnesian Wars. However, the Spartans were so exhausted by the war that effective leadership among the Greek city-states was taken over by Thebes. Athens, having lost her land empire, was about to establish a new one—this one far more lasting, since it was based on an intellectual and philosophical system, rather than on a military one. Indeed, modern Western culture is still part of the Athenian intellectual empire.

4.3.1 Plato

A key figure in the foundation of this empire is Plato (427-347 B.C.). In any history of mathematics, his name inevitably arises, along with a pertinent question: why? He was neither a geometer nor a teacher of geometers; he wrote no great works on mathematics, and proved no great theorems. Plato's primary contribution to mathematics was purely philosophical: he outlined what he thought mathematics should *be*.

Born to an aristocratic family in Athens, Plato originally had political ambitions. The fate of his teacher and friend, Socrates, suggested to him that politics was no place for a man with a conscience: Socrates had been ordered to commit suicide, having been found guilty on charges of "corrupting youth." After Socrates's execution in 399 B.C., Plato left Athens, and spent several years traveling about the Mediterranean, through Egypt and southern Italy, where he met Pythagoreans and other philosophers. He returned to Athens and around 387 B.C., he founded a school. Because the school held its meetings in a park devoted to the local hero Academus, the school itself came to be known as the Academy, a name that came to be applied to all similar institutions. This Academy would stay open for over 900 years, until 529 A.D., when the Christian Emperor Justinian closed it and all the other "pagan" schools in Athens.

A story is told that the entrance plaque to the Academy read, "Let No One Unversed In Geometry Come Under My Roof." The story is told by Johanes Tzetzes, a Byzantine author of the twelfth century, writing six hundred years after the closing of the Academy. Whether or not such a plaque actually existed, it cannot be denied that Plato considered geometry an essential prerequisite for true knowledge. One of Plato's goals was to establish all knowledge on a deductive basis, for which a thorough training in geometry would well prepare one's mind. Plato may have also been responsible for the notion that geometry and, by extension, mathematics in general, ought to be as impractical as possible: by studying the "real world," one studied imperfect objects, and thus obtained imperfect knowledge. According to Plutarch (ca. A.D. 46-120), Plato chided other geometers for attempting to solve the problem of duplicating the cube using mechanical devices.

It is difficult to say what Plato meant by "mechanical devices." Plato may be referring to the mechanical devices necessary to draw the curves used to duplicate the cube or solve the other problems. In his view, geometry ought to be performed without any instruments and consist solely of the contemplation of ideal figures.

See page 43

It is not known if Plato actually discovered the method of finding Pythagorean triples usually attributed to him; it seems likely that it was an earlier discovery. Eutocius (sixth century A.D.) gave what he claimed to be Plato's method of duplicating the cube, although the attribution is dubious: no other ancient authors seem to be aware of the Platonic method of duplicating the cube, and the method cited by Eutocius involved the use of the mechanical devices that Plato found so abhorrent. Plato might have invented the term **mathematics** in its modern sense. The *mathema* were the three liberal arts worthy of study: arithmetic, geometry, and astronomy.

There are, in fact, only two mathematical achievements that can be unambiguously attributed to Plato. First, he defined two terms used by later geometers: analysis and synthesis. In **analysis**, as defined by Plato, one begins with a proposition to be proven, and deduces from the proposition other statements, until one arrives at a statement known to be valid. **Synthesis** is the procedure of reversing the steps, so the proposition may be proven.

See page 69.

For example, consider the second lune of Hippocrates, which began with a trapezium having three equal sides, and the fourth side having a square three times the squares on the other side. To determine how to construct such a trapezium, begin by assuming the construction already done, with trapezoid $ABCD$ in circle ABC with center G, where the square on AB is three times the square on AD, DC, or CB, all assumed to be equal to some length MN. We note that once D is found, the trapezium can be constructed.

Example 4.1. *The trapezoid $ABCD$ in the circle $ABCD$ has sides AD, DC, CB equal, and the square on side AB is three times the square on any of the other sides. Construct the perpendicular bisectors of DC, CB. They meet at G, which will be in the center of the circle, since the perpendicular bisector of a chord is the diameter, and the intersection of two diameters is the center. Extend the diameter until it bisects the chord AB at H, and draw KD perpendicular to AB. Thus, D is located.*

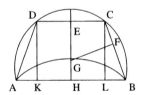

Reversing the steps: bisect AB at H. Construct KH equal to half of MN; construct KD perpendicular to AB; likewise, construct HL equal to half of MN, and LC perpendicular to AB. Construct a circle with center at A and radius of MN; let the intersection with the perpendicular KD be at D. Construct DC parallel to AB and equal to MN. CB is equal to MN (proof omitted; See Problem 4). The center is found by bisecting CB, DC, and drawing the perpendiculars EH, FG, which intersect at G, the center.

Plato's other achievement was the Academy itself. Because of Plato's emphasis on the deductive method, his Academy attracted a great number of geometers, and many of the most important mathematical discoveries of the fourth century B.C. came from the Academy in Athens.

One of the earliest geometers at the Academy was Archytas (428-347 B.C.), born in Tarentum a year after the plague began in Athens. Plato visited Archytas around 388 B.C., and it is probable that Archytas was the one who influenced Plato to turn toward mathematical ways of thinking. Archytas eventually moved to Athens and lived the rest of his life there as a member of the Academy. He was, apparently, the first to solve the problem of duplicating the cube, using a complex, three-dimensional solution involving the intersection of several space curves; it will not be reproduced here.

4.3.2 Theaetetus

Another member of the Academy was Theaetetus (415-369 B.C.), who died in 369 B.C. from wounds received during a battle near Corinth. Thaeatetus classified various types of irrationals in a form that ultimately appeared in Euclid's *Elements* as Book X, and probably showed how some types could be reduced to other types. The work is an early example of what modern mathematicians would call a **classification theorem**, in which a large group of seemingly different objects is broken down into a number of categories, all of which have similar properties. If two lines were commensurable, or if the squares on the lines were commensurable, the two lines were said to be **rational straight lines**. Given two rational straight lines:

Definition. *A **binomial** line is formed by the sum of two rational lines, commensurable in square only; the **apotome** is formed by the difference of the two rational lines. The side of the square equal to the rectangle whose sides are the two rational lines is a **medial** line.*

Thus, the line formed by adding the diagonal of a square to one of its sides is a binomial, whereas the line formed by removing from the diagonal the side of the square is an apotome, and if a square is constructed equal to the rectangle whose sides are the side of the square and its diagonal, the new square's side is a medial line. We will see the importance of these classifications when we deal with the Greek solution to the quadratic equation. See page 123.

Theaetetus himself was the first to write a treatise on the five regular solids. Three of these solids, the tetrahedron, the cube, and the dodecahedron, were definitely known to the Pythagoreans. There is no reason to doubt that the Pythagoreans were aware of the existence of two others, namely the octahedron and the icosahedron, although Theaetetus was apparently the first to give theoretical constructions of them; in fact, he is credited with being the first to be able to construct regular octahedrons and icosahedrons in a sphere. The five solids together are known as the **Platonic solids** because of Plato's discussion of them in his dialog *Timaeus*.

4.3.3 Antiphon of Athens and Bryson of Heraclea

Antiphon (fl. 450 B.C.), like Hippias of Elis, was a Sophist. Like Hippocrates and Anaxagoras, he worked to square the circle, and introduced a powerful new concept to geometry. Antiphon's attempt to square the circle began with a circle

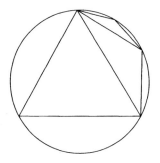

The first three stages of
Antiphon's quadrature

See Problem 2.

in which an equilateral triangle was inscribed. The equilateral triangle, of course, could easily be squared. On each side of the triangle, he constructed an isosceles triangle, with the vertex on the circle; this process was repeated indefinitely. The inscribed figure, being bound by straight lines, could be squared using known procedures.

Antiphon apparently believed that the constructed polygon would eventually equal the circle. However, for this to occur, it would be necessary for the side of one of the isosceles triangles to coincide with the arc of the circle between two adjacent vertices. This assumption was criticized by Aristotle and others, on the grounds that a curved segment would never coincide with a straight segment.

Bryson (fl. 450 B.C.), a student of Socrates, built upon Antiphon's notion of inscribed polygons, and may have been the first to consider both inscribed and circumscribed polygons; instead of beginning with an equilateral triangle, Bryson began with a square (Simplicius also claims that Antiphon began with a square). According to a very critical account by Alexander of Aphrodisias, who called Bryson's work "captious sophistry," Bryson inscribed a square in a circle, circumscribed a square about a circle, then constructed a third square "in between" the two, saying that the square was equal to the circle.

Even though the work of Bryson and Antiphon did not give exact results, they both provided a means of *estimating* the area of the circle that was more powerful than any that existed before. What it lacked was a theory that could quantify the relationship between the circles and the squares in an organized, axiomatic way; the new theory could then be applied to a number of other curvilineal figures. The geometer who would take the work of Bryson and Antiphon to new heights was Eudoxus of Cnidos.

4.3.4   Eudoxus

Unusual for an early geometer, the life of Eudoxus (408-355 B.C.) is fairly well documented. He was born in Cnidos, in Asia Minor, and studied geometry with Archytas in Tarentum, and medicine with Philistion of Locri. At 23, he traveled to Athens and attended the Academy, in 385 B.C. He was so poor, according to one story, that he could not afford to live in Athens, but instead took residence in Piraeus, the port section (near the site of the previous century's battle of Salamis), and walked to Athens every day to attend the Academy, a distance of about ten kilometers each way (uphill, at least in the morning). He spent only two months there, and continued his travels to Egypt, where a sacred bull licked his cloak, which meant (according to the priests) that he would become famous but die young. He paused in Cyzicus, founding a school there; after a while, he returned to Athens briefly and was greeted warmly by Plato. Finally, he returned to Cnidos and died there, at the not-so-young age of fifty-three. The fact that we are discussing Eudoxus more than two thousand years after his death shows the second part of the prediction, at least, to be true.

Like so many ancient geometers, none of his work has survived. According to an anonymous commentator on Euclid, "some say" the theory of ratios and proportions in Books V and VI of Euclid is the work of Eudoxus, "the teacher of Plato." The anonymous commentator is believed to be Proclus, perhaps basing his comment on Eudemus.

A **magnitude** is some geometrical object, such as a line, or a point, or a two- or three-dimensional figure. To **multiply a magnitude**, one would replicate the magnitude some number of times; the result would be the multiple of the magnitude. Two magnitudes can have a **ratio** to each other if either can be multiplied to "exceed" the other. For example, a circle and a square can be said to have a ratio, since the circle can be multiplied to exceed the square, and vice versa. On the other hand, a line and a square cannot have a ratio, since the line cannot be replicated to cover the square. This avoids the question of whether a circle and a square can be equal: they can have a ratio.

To make use of these ratios, you must be able to compare them. Euclid recorded the Eudoxian definition of equality of ratios:

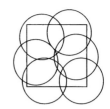

Six circles exceed the square: a circle can have a ratio to a square

The lines cannot exceed the square: a line may not have a ratio to a square

Definition. *Magnitudes have the same ratio, the first to the second and the third to the fourth, if for any equimultiples whatever of the first and third, and any equimultiples whatever of the second and fourth, the former equimultiples alike equal, alike exceed, or alike fall short of the latter equimultiples.*

To interpret the definition, consider the following: suppose you have four quantities, and the ratios A to B, and C to D. Take any multiple m of A, and the same multiple of C. Take any other multiple n of B, and the same multiple of D. Then compare the multiples of mA and nB, and the multiples of mC and nD. If the ratios are equal, then the multiples of A have the same relationship to the multiples of B that the multiples of C have to the multiples of D.

If the two ratios are not equal, then one is greater.

$A : B = C : D$ if, for any whole numbers n, m, $mA \gtreqless nB$ whenever $mC \gtreqless nD$.

Definition. *When, of the equimultiples, the multiple of the first exceeds the multiple of the second, but the multiple of the third does not exceed the multiple of the fourth, the first ratio is greater than the second.*

Thus, if there was some m and n for which mA exceeds nB, but mC does not exceed nD, then the ratio $A : B$ is greater than the ratio $C : D$. These definitions were most often used in the negative, to show when two ratios are *not* equal, and furthermore, to determine the inequality between them.

If there is an m, n, for which $mA > nB$, but $mC \not> nD$, then $A : B > C : D$.

Example 4.2. *Show that the ratio between the numbers 5 to 2, and the ratio between the circle and the square on its diameter, are not equal, and determine the inequality between them. Take the "1" multiple of 5 and the circle, which is, respectively, 5 and one circle; take the "2" multiple of 2 and the square, which is, respectively, 4 and two squares. 5 is greater than 4, but one circle is not greater than the two squares on its diameter. Thus, the ratio 5 to 2, and the ratio between the circle and the square on its diameter, are not equal. Since the multiple of the first, 5, exceeded the multiple of the second, 2, and the multiple of the third, the circle, did not exceed the multiple of the second, the square on the diameter, then the ratio of 5 to 2 is greater than the ratio between the circle and the square.*

The use of ratios and proportions avoids the problem of trying to compare directly the areas of rectilineal and curvilineal figures. Instead of comparing a square and a circle directly, you could compare their ratios or, in the form that the theory of proportions was ultimately used on this problem, you could compare the ratio between two circles to the ratio between two squares.

Using the theory of proportions, Eudoxus would have been able to place the notions of Antiphon and Bryson on an axiomatic basis, and prove results about circles and, indeed, other types of nonrectilineal figures. The method used by Eudoxus was named the **method of exhaustion** in 1647, by Gregory of Saint-Vincent. The Greeks themselves never gave it a special name. Using the method of exhaustion, Eudoxus proved:

1. Two circles are to each other as the squares on their diameters.

2. Two spheres are to each other as the cubes on their diameters.

3. A pyramid is one-third the prism of the same height on the same base.

4. A cone is one-third the cylinder of the same height on the same base.

Note that Eudoxus was able to prove three conjectures of previous geometers: the first was that of Hippocrates; the last two were the conjectures of Democritus. We will delay discussing the method of exhaustion until we have more thoroughly discussed Euclidean geometry.

4.3.5   Menaechmus

Menaechmus (fl. 350 B.C.) was also a member of Plato's Academy. He may have been a student of Eudoxus; later historians say he was one of the teachers of Alexander the Great. An old story says that Alexander was having difficulty understanding a proposition in geometry, and asked if there was an easier way; Menaechmus supposedly replied that while kings and princes often trod easier paths than the common people, there was no royal road to geometry.

Menaechmus was the first geometer we know to have studied conic sections systematically. In Menaechmus's time, the **conic sections** were formed by cutting a right circular cone with a plane perpendicular to a side. The sections were named whether the vertex angle was acute, right, or obtuse. Later, the **section of an acute-angled cone** would be renamed the ellipse; the **section of a right-angled cone** would become the parabola; and the **section of an obtuse-angled cone** would become the hyperbola (see Figure 4.3). We shall use these anachronistic names here in place of the more historic ones.

Menaechmus apparently discovered a key property of the ellipse, parabola, and hyperbola. These properties are called **symptoms**. Suppose one has a fixed line, DE, and a line AB, and some curve AC. Draw the perpendicular BC at any point along the curve (see Figure 4.4). If the rectangle formed by the lines DE, AB, and the square formed by BC, are everywhere equal, then the curve AC is a parabola. Alternatively, given an area E, line AB, curve HOJ, draw

See Problem 8.

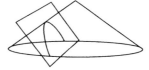

Figure 4.3: Sections of a Cone

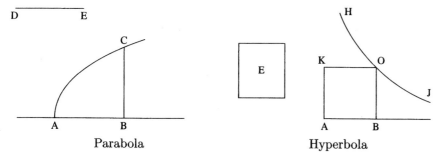

Parabola   Hyperbola

Figure 4.4: Symptoms of the Conic Sections.

the perpendicular BO. If the rectangle formed by lines AB, BO is equal to the area E everywhere along the curve, then HOJ is a hyperbola. We do not know how Menaechmus discovered these properties, or if he proved them, though a proof is relatively easy to find.

See Problem 8, page 111.

Menaechmus used these symptoms to duplicate the cube. Recall that Hippocrates had shown the problem reduced to finding two mean proportionals. Menaechmus's solution, as quoted by Eutocius, provides excellent illustrations of the use of both analysis and synthesis. Menaechmus let A, D be the lines between which two mean proportionals are to be found. The analysis begins by assuming the mean proportionals, B and C, have already been found.

See page 70.

Problem 4.1. *Given A, D, to place two mean proportionals between them.*

Solution. Given line EF, place EG equal to C and GH perpendicular and equal to B (see Figure 4.5). Because B is the mean proportional between A and C, then the rectangle on A, C is equal to the square on B; hence the rectangle on EG and the fixed line A is equal to the square on GH, and thus H is on

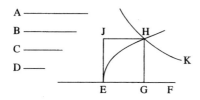

Figure 4.5: Menaechmus's Solution to Duplicating the Cube

some parabola EH. Next, because B, C are two mean proportionals, then the rectangle on A, D, which forms a fixed area, is the same as the rectangle on B, C. But B, C are equal to GH, EG respectively, so H is also on some hyperbola HK. Thus, H is at the intersection of a parabola and a hyperbola. We note that A, D are known quantities (with A twice D), and hence the rectangle with sides A, D is also a known quantity.

The synthesis is as follows: construct hyperbola HK with the property that the rectangle on GH, EG is equal to the rectangle on A, D. Construct parabola EH with the property that the square on GH is equal to the rectangle on EG and A. Let their point of intersection be H. Then the two mean proportionals are EG, GH. $\square$

See Problem 10.

4.3.6 Dinostratus

Dinostratus was the brother of Menaechmus, and he is credited with having squared the circle, using the quadratrix of Hippias. Given the quadratrix AFH, Dinostratus's quadrature relies on knowing the position of the point H, which itself was a point of controversy to Greek geometers. Assuming this point was known, Dinostratus claimed that the arc of the circle AKC was to AD as AD was to DH. The proof is a good example of a proof by contradiction.

See Problem 13.

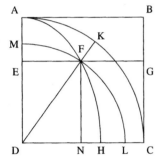

Proposition 4.6. *In the quadratrix AFH, arc AKC is to AD as AD is to DH.*

Proof. Suppose otherwise. Then AKC is to AD as AD is to DL, where DL is a line either greater or less than DH. Suppose it is greater. Draw the circle MFL with center D and radius DL, and FN perpendicular to DC. AKC is to MFL as AD is to DL. Since (by assumption) AD is to DL as AKC is to AD, then AKC is to MFL as AKC is to AD, and hence MFL, AD are equal.

By the property of the quadratrix (see page 66), AKC is to KC as AD is to FN. AKC is to KC as MFL is to FL, so MFL is to FL as AD is to FN. Finally, since MFL, AD are equal, then AD is to FL as AD is to FN, so FL, FN are equal. However, this is absurd.

See Problem 11.

A similar proof shows that DL cannot be less than DH. Hence, AKC is to AD as AD is to DH. By finding a line such that PQ is to AD as AD is to DH, then PQ is equal to AKC, and hence PQ is one fourth the circumference of the circle AKC. $\square$

The determination of PQ, however, does not actually square the circle, since all that has been found is a line equal to one-fourth the circumference. The fact that Dinostratus was given credit for having squared the circle implies that the relationship between the circumference and the area of the circle was known by the time of Dinostratus.

See Problem 12.

4.3.7 Aristotle

The greatest thinker of the ancient Greeks was probably Aristotle (384-322 or 321 B.C.), from Stagira, a nondescript town in the backwoods of Greece. Aristotle came to Athens, studied under Plato, and may have stayed in Athens for a while, but apparently life in the "big city" was not for him, and he returned to his native Stagira.

Aristotle was a one-man encyclopedist, writing on anything that interested him, including biology, physics, government, ethics, and comedy. His scientific writings had a peculiar effect on the development of science: where he was most wrong, as in his assertion that heavier bodies fell faster than lighter ones, he was most religiously adhered to, but when he was most correct, as when he identified dolphins as mammals because the unborn dolphin was nurtured in a placenta, he was most often ignored. This has the unfortunate effect of making Aristotle almost a villain who retarded the progress of science, though the fault lay not with him, but with those who would rely on blind authority instead of their own observations.

Although Aristotle made no original contributions to mathematics, he often used mathematical examples to illustrate proof and reasoning. As a result, though we do not have the works of many early geometers, we have Aristotle's references to their works, and we can, through Aristotle, attempt to reconstruct some of the details of early geometry. Moreover, he codified and clarified the rules of logic, and defined many terms we use today. An **axiom** was a principle that could not be demonstrated, such as the notion that if equals were subtracted from equals, the remainders were equal. A **hypothesis** was something believed to be true, but a **postulate** is something whose truth was assumed as part of the study of a science.

Aristotle returned to his native Stagira after his stay in Athens. Shortly thereafter, he found employment in the royal household of Philip of Macedon, who had built his kingdom by unifying a number of warring tribes. Philip hoped to pass on the kingdom to his son, but he feared that unless his son was well trained in both the martial and the philosophical arts, the son would destroy the kingdom. Hence Philip hired Aristotle to teach the elements of philosophy to young Prince Alexander.

After Philip's death, Alexander became king of Macedon, and embarked on a career of conquest hitherto unmatched in the classical world. History remembers him as Alexander the Great. The Persians had attempted to conquer Greece, and failed. Alexander conquered first Greece, then the Persian Empire, and made his way to India, all by the time he was 30. En route, he helped maintain local control by encouraging his men to marry local women; moreover, he founded cities—all named Alexandria—where Greek and local cultures could mix. He was on the borders of India in 326 B.C. when his generals, who had faithfully followed him from Macedonia and across the width of the Persian Empire, balked at yet another campaign.

While returning through Babylonia, Alexander contracted a disease (probably malaria), and died in 323 B.C. The young king had not yet designated a successor, and as he was dying, his generals clustered around him, hoping to be designated Alexander's successor (and fearing that their greatest rival would receive the honor). Who, they wanted to know, would be Alexander's heir?

To whom would the empire go? Leaning close, they heard his reply: "To the strongest." Thus, yet another round of wars wracked the land of Greece, and yet another generation would pass before progress was made in mathematics.

4.3 Exercises

1. Use analysis and synthesis to solve the problem $5x + 7 = 22$.

    (a) First, assume that you have a solution, namely $x = a$. Apply the appropriate operations until you arrive at $5x + 7 = 5a + 7$. Record the operations you had to perform.

    (b) Remember that $5x + 7 = 22$, so you can also state that $22 = 5a + 7$. These first two steps are the analysis.

    (c) Reverse the steps you went through in the analysis to find the solution. This is the synthesis.

2. Consider Bryson's method of quadrature. According to Alexander of Aphrodisias, Bryson found a square "between" the inscribed and circumscribed squares.

    (a) One interpretation of "between" is either the arithmetic or geometric mean of the areas of the two squares. Determine these means; how closely do they compare to the actual area of the circle?

    (b) A second interpretation of "between" is the square whose side is either the arithmetic or geometric mean of the sides of the two squares. Determine the side of such a square, its area, and how closely its area relates to the area of the circle.

3. Determine if the two magnitudes in each pair can have a ratio.

    (a) The numbers 8 and 3

    (b) A cube and a square

    (c) The surface of a cube and a circle

    (d) The quadratrix and the circle

    (e) A circle and the number 3.14159

4. The Eudoxian definition of equal ratios implies that you can compare the equimultiples of the first and second, and the equimultiples of the third and fourth. Why is this guaranteed?

5. Show that the Eudoxian definition of equal ratios is equivalent to noting that, given any rational number q, the relationship of $A : B$ to q is the same as the relationship of $C : D$ to q.

6. Determine the inequality of the following ratio pairs (i.e., determine which is greater and which is less) using the Eudoxian method.

    (a) The ratio between a circle and square on its diameter, compared to the ratio between a square and the square on its diameter.

    (b) The ratio between the diagonal of a square and its side, compared to the ratio between the numbers 3 and 2.

    (c) The ratio between a circle and the square on its diameter, compared to the ratio between the side of a square and its diagonal.

7. In Example 4.1, prove that the third side CB of the trapezoid is equal to MN in the constructed lune.

8. Show that Menaechmus's symptom of a parabola coincides with the modern algebraic equation for a parabola. What is the equation of the corresponding parabola?

9. To what algebraic curve does Menaechmus's symptom of a hyperbola correspond?

10. Show that the intersection of the parabola and hyperbola finds the two mean proportionals.

11. Show that in the quadratrix, AKC is to AD as AD is to DL where DL is less than DH also leads to a contradiction.

12. Carefully examine the use of the quadratrix to square the circle.

    (a) What propositions regarding proportions were used?

    (b) What other geometric propositions were used?

    (c) What is the relationship between the quarter circumference and the area of the circle?

    (d) Complete the problem of squaring the circle, using the quadratrix.

13. The quadratrix raised one of the earliest questions regarding limiting values. To square the circle, it was necessary to find the location of the point H. Explain why this is a problem.

Chapter 5

The *Elements*

Alexander's generals quarreled among themselves during the Wars of the Successors, but in the end, an uneasy stalemate developed between two of them: most of the Persian Empire, except for Egypt, went to Seleucus, while Egypt ended in the hands of Ptolemy. The Seleucid Empire, although larger, was less stable, for it contained a vast number of subject peoples, always ready to rebel. Egypt, which had not existed as a separate nation for centuries, was much easier to control, and during the reign of the Ptolemies, Egypt underwent a great cultural renaissance.

Ptolemy made his capital at Alexandria (founded by Alexander the Great in 332 B.C.). Besides being a military man, Ptolemy was also a man of learning, and the following suspiciously familiar story is told: at one point, while Ptolemy was trying to learn geometry, he had great difficulty with a proposition. In exasperation, he asked his teacher if there might be an easier way to learn geometry. The teacher replied, "Sire, there is no royal road to geometry." The teacher was Euclid.

5.1 Deductive Geometry

In his capital at Alexandria, located at the mouth of the Nile River, Ptolemy established a great institution of learning, whose main entrance housed a statue of the Muses, the patron goddesses of the arts and sciences. Because of this statue, the institution came to be known as the Museum of Alexandria. It combined the research and teaching aspects of a university with the activities we associate with museums today. It was also a state-supported institution, first by the Ptolemaic kings of Egypt and then, after the Roman conquest of Egypt, by a grant from the Roman emperor himself.

The museum contained a botanical garden, a zoo, and a library, which came to be the best known part of the museum; indeed, most people speak of the Library of Alexandria and forget that it was part of a larger institution. One of

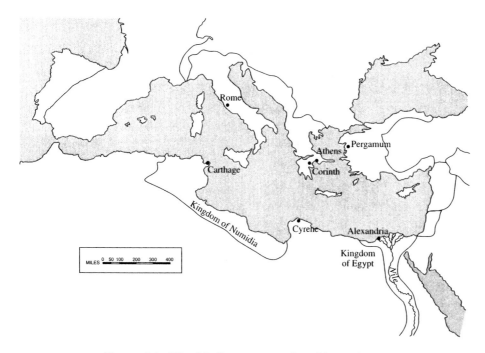

Figure 5.1: The Mediterranean after Alexander.

the goals of the library seems to have been the collection of everything that had been written in Greek, from the most ancient manuscripts to the most recent. At one point, the library may have held half a million papyri. It is said that in order to augment its collection of papyri, ships entering the harbor had to either sell their manuscripts to the library, or allow the library to take them so copies could be made; the *copies* would then be returned to the owners. The library grew so large that a library annex was established at the nearby Temple of Sarapis in 235 B.C.

5.1.1 Euclid

About the only certain information about Euclid's life is that around 300 B.C., he was living in Alexandria. Euclid may have founded a school there, akin to Plato's Academy. A story told about Euclid is that one day, a student wanted to know the value of geometry. Euclid is said to have given him a coin, then dismissed him, chiding him for requiring that knowledge have tangible benefits. Most accounts say he was a great teacher, though his actual contributions to mathematics were probably minimal: there is not a single new mathematical discovery that can definitely be attributed to Euclid. However, Euclid would change the shape of geometry forever.

In the library, Euclid had access to three hundred years of Greek geometrical development. The manuscripts dealing with geometry available to him might be divided into three broad categories. The first were works that focused on one particular topic. These treatises were meant for scholars who already had a strong background in geometry and were usually difficult for the typical student to understand, so a second type of writing also existed: the **commentary**, which took the work of another and explained and elaborated it. Euclid's *Data* was a concentrated study of certain aspects of geometry necessary for the advanced student, whereas this book is an example of a commentary.

The third type of work available to Euclid was the introductory texts, such as the *Elements* of Hippocrates. At one point, Euclid decided to write his own *Elements*. Whether it was because he was dissatisfied with the work of his predecessors, or he just wanted his own personalized version of the geometry of the Greeks, the result was the best selling textbook of all time, still in print after more than two thousand years.

The impact of the *Elements* cannot be overstated. The study of Euclid's geometry, and especially the first six books, made up an essential part of the medieval quadrivium, the course of study that corresponded to a modern bachelor's degree. Euclidean geometry was the benchmark against which mathematical rigor was measured as late as the eighteenth century, and it was not until the nineteenth century that the first serious mathematical challenges to Euclid arose. Because of the importance of the *Elements* to the next two thousand years of mathematics, we will examine the *Elements* and its contents in some detail.

See Chapter 21.4.

It is difficult to determine what Euclid actually wrote. Consider Figure 5.3, which appears to be classroom notes giving the first ten definitions in Euclid's *Elements*. This is one of the two earliest known copies of the contents of the *Elements* (the papyrus itself does not specifically refer to Euclid's work), yet it was written in the third century A.D., more than five hundred years after Euclid lived and wrote! Careful cross-checking of manuscripts leads historians to believe almost all existing copies the *Elements* come from one copied by Theon of Alexandria around A.D. 350. Thus, our primary knowledge of what Euclid actually wrote comes from a manuscript written over six hundred years after his death.

The *Elements* is divided into thirteen books, each about the length of a modern chapter. Although it is ordinarily considered a geometry textbook, one should remember that to the Greeks, "geometry" was the all-encompassing mathematical science. A rough outline of the *Elements* is:

1. Book One: the geometry of rectilinear figures

2. Book Two: algebraic propositions

3. Book Three: the geometry of circles

4. Book Four: polygons inscribed and circumscribed about circles

5. Book Five: theory of proportions

6. Book Six: similar figures

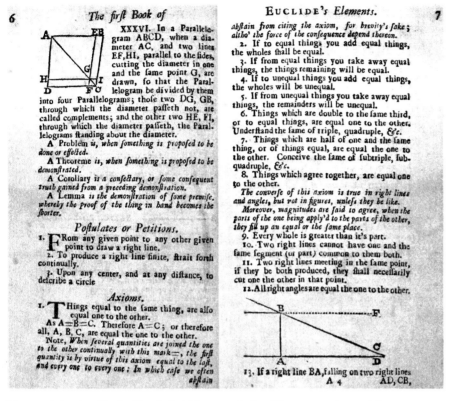

Figure 5.2: An Early English Edition of the *Elements*, that of Isaac Barrow. Note that Barrow's ordering and placement of the axioms are different from that of Euclid. Boston Public Library/Rare books Department. Courtesy of the Trustees.

7. Books Seven through Ten: number theory

8. Books Eleven and Twelve: solid geometry

9. Book Thirteen: algebraic propositions

Thus, only slightly more than half of the *Elements* deals with geometry.

5.1.2 Definitions from the *Elements*

The *Elements* begins with the definitions of many geometrical objects: points, lines, polygons, circles, and so on. The first seven of these definitions are:

Definition. *A **point** is that which has no parts.*

Definition. *A **line** is a length without width.*

Definition. *The **ends of a line** are points.*

Figure 5.3: The first ten definitions from the *Elements*, on papyrus from a third century A.D. Greek source. Digitally reproduced with the permission of the Papyrology Collection, Graduate Library, University of Michigan.

Definition. *A **straight line** is one that lies evenly with its points.*

Definition. *A **surface** is that which has length and width only.*

Definition. *The **ends of a surface** are lines.*

Definition. *A **plane surface** is one that lies evenly with its straight lines.*

Most of these definitions are unclear and not very useful. Indeed, these first seven definitions are of such a different quality from the remaining definitions that some scholars believe they were added later, possibly by Heron of Alexandria. In this case, Euclid may have left terms such as "point", "straight line" and "plane" undefined, thus anticipating nineteenth century geometers by more than two thousand years.

See page 188.

The next few definitions deal with plane angles.

Definition. *A **plane angle** is the inclination between two lines that intersect and do not lie in a straight line.*

Definition. *If two straight lines intersect and make the adjacent angles equal, the equal angles are **right angles**, and the lines are said to be **perpendicular** to each other.*

Note that Euclid's definition of a right angle has nothing to do with the actual measure of the angle; instead, an angle is a right angle by virtue of some property that it has with respect to the line on which it is located. As we do, Euclid defined obtuse and acute angles as angles that are, respectively, greater or less than a right angle.

More difficult to understand are Euclid's definitions relating to circles.

Definition. *A **circle** is a plane figure contained by a line so that all the straight lines falling on the line from a point, called the **center**, are equal.*

Euclid's definition needs to be explained a little. The fact that Euclid felt compelled to define a *straight* line is an acknowledgment that there are other types of lines. Thus, to Euclid, a circle is a plane figure contained by *some* line, not necessarily straight, so that all the lines from the center to the line are equal to one another.

See Problem 5.

The circle illustrates one of the peculiar blind spots of the Greek geometers; despite the fact that modern science and mathematics are filled with technical terminology which derive from Greek root words (such as "technical" and "mathematics"), the Greeks were, themselves, not very good about naming things. Thus, while Greek geometers would speak of Euclid's "line drawn from the center of a circle to the circle," or the more concise "line drawn from the center," they did not use the term **radius** (which is Latin in origin).

Definition. *A **diameter** of a circle is a straight line drawn through the center and ending on the circumference of the circle; the diameter bisects the circle.*

This definition is not a proper one, for it actually includes a proposition, that of Thales: a diameter bisects the circle. While this statement may seem self-evident, if Euclid is to construct a proper, axiomatic system, he must *explicitly* indicate all of his assumptions. This is one of many places where Euclid assumed more than he claimed he assumed.

See page 57.

Euclid then defined a number of plane figures: semicircles; equilateral and isosceles triangles; right-, obtuse-, and acute angled triangles. These definitions are the same as our own, and will not be listed here. His classification of quadrilaterals is, however, different from our own.

Definition. *A **square** is a quadrilateral that is right angled and equilateral. An **oblong** is right angled but not equilateral. A **rhombus** is equilateral but not right angled. A **rhomboid** has opposite sides equal, but is neither right angled nor equilateral. All other quadrilaterals are called **trapezia**.*

See Problem 1.

Notice Euclid has no specific definition for a *trapezoid*.

Euclid's last definition, before he lists his postulates and common notions, concerns parallel lines:

Definition. *Parallel straight lines are straight lines in the same plane that do not meet, however far they may be extended.*

5.1.3   Postulates and Common Notions

After his list of definitions, Euclid gave a list of five postulates, which relate to geometrical objects, and five common notions, which relate to logical deduction. From the definitions, postulates, and common notions, every proposition in the *Elements* was to be derived.

Because all of geometry is to be derived from the postulates and common notions, Euclid had to choose carefully. First, enough postulates must be chosen to allow a rich and full geometry to be developed. However, the list of assumptions should not be overly long. Even more importantly, the set of postulates had to be consistent: one postulate could not contradict another, nor could *any* two deductions drawn from the postulates contradict each other. Euclid's real brilliance was his choice of postulates. Euclid's five postulates are:

1. A straight line may be drawn between any two points.

2. A straight line may be extended indefinitely in a straight line.

3. About any point, a circle of any radius may be constructed.

4. All right angles are equal to one another.

5. If a straight line falling on two straight lines makes the interior angles on one side less than two right angles, the two lines, if extended, meet on the side where the interior angles are less than two right angles.

We will use the word radius instead of "line drawn from center."

The first three postulates set the stage for Euclidean geometry, and may be a response to an objection of Aristotle. Aristotle felt that, all too often, mathematicians defined objects, without worrying about their existence. For example, See page 66. consider the quadratrix of Hippias. The quadratrix could be used to trisect an angle, *provided* one had the curve to begin with. But, Aristotle would have objected, merely defining the curve did not guarantee its existence. If geometry is to be a complete deductive system, then any curve that geometry considers must have its existence proven or explicitly assumed.

What figures did Euclid assume the existence of? The first postulate assumes that a straight line can be drawn between any two points; the second assumes that the straight line can be extended as far as one likes. Thus, the first two postulates assume the existence of straight lines of arbitrary lengths and positions. The third postulate assumes that about any point, and with any radius, a circle can be drawn; thus, circles with arbitrary centers and radii exist. Euclidean geometry is limited by these three postulates: it concerns itself with lines and circles, and the figures formed from them.

See Problem 2.

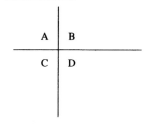

Euclid's fourth postulate, concerning right angles, might seem unnecessary, especially if we think of right angles as "angles with a 90° measure." But Euclid defined two lines to be at right angles if the adjacent angles are equal. Thus, if two adjacent angles A and B are equal, they are right angles; likewise, if two other adjacent angles E and F are equal, they, too, are right angles. The fourth postulate made explicit the assumption that A and E were likewise equal. We might make the following analogy: if one pie is cut into four equal pieces, and a second pie is cut into four equal pieces, there is nothing to guarantee that the "equal pieces" of the one pie are the same as the "equal pieces" of the other pie.

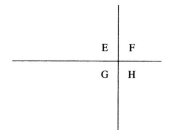

Of all Euclid's postulates, it was the fifth, usually called "the parallel postulate" that came under the most intense scrutiny, and generated the most controversy. All the other postulates are simple to state, and easy to comprehend. The fifth, however, is a long, complex statement, and is the only one that is not immediately apparent. Proclus said it ought to be removed completely from the list of postulates, and almost immediately after Euclid, other geometers attempted to find a means of avoiding it or proving it from the other four postulates.

For now, let us try and interpret the fifth postulate, as Euclid wrote it. Suppose line L falls on lines M and N, as shown in the diagram. Four interior angles are formed, which are indicated as A, B, C and D. Suppose angles A and B are together less than two right angles. By the second postulate, we can extend M and N indefinitely, and the fifth postulate claims that the lines will meet on the side containing the angles A and B. Alternatively, if it is the angles C and D that are together less than two right angles, the lines M and N, if extended indefinitely, will meet on that side.

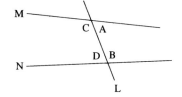

After the postulates, Euclid lists his five "common notions," which dealt not with geometric objects, but with reasoning in general. These common notions are:

1. Things equal to the same thing are also equal to each other.

2. If equals are added to equals, the results are equal.

3. If equals are subtracted from equals, the results are equal.

4. Things that coincide are equal.

5. The whole is greater than the part.

The first three are generally considered part of the rules of algebra. However, Euclid's notion of "things" is much more general: he will apply these common

See Problem 4.

notions to rectilinear figures, circles, lines, angles, and all manner of geometric objects.

The fourth common notion is an important concept. Although Thales, Pythagoras, and Hippocrates dealt with the notion of "equal figures," Euclid is the first whose surviving work tells us, in precise terms, what is meant by "equal figures": they are figures that *coincide*. In the next section, we will examine the Euclidean geometric method by looking at the propositions in Euclid's geometry.

5.1 Exercises

1. Consider Euclid's classification of quadrilaterals.

    (a) Reclassify Euclid's quadrilaterals, in terms of our modern classification system.

    (b) We define a square as a rectangle with four equal sides. Is Euclid's definition of a square compatible with our own, or could there be figures that one set of definitions calls "square" and the other does not? Explain.

    (c) What are some of the advantages and disadvantages of the Euclidean classification scheme?

2. Which of the figures discussed in the previous chapter can be included in Euclidean geometry? Which cannot be? Explain.

3. The second and the fifth postulates are both connected with the notion of parallel lines.

    (a) What is the connection between the second postulate and parallel lines?

    (b) What would be the consequences if the second postulate was that a line could *not* always be extended indefinitely?

4. How is the geometric formulation of the common notions more general than the algebraic? How is it more restricted?

5. Compare and contrast the way Euclid uses "line" with the way modern mathematicians use "curve."

6. Consider the geometry on the surface of a sphere, where the "straight line between two points" corresponds to the great circle arc through the two points. If the two points are the north and south poles, then a line of longitude is a straight line between the two points, and a line of latitude corresponds to a circle. How must the postulates be changed to reflect the geometry on the surface of a sphere?

5.2 Rectilineal Figures

We will begin our examination of Euclid's propositions with Book I. In this book, Euclid dealt with the geometry of figures formed by straight lines; specifically, triangles and quadrilaterals. Thematically, Book I is divided into two sections, the first dealing with triangles, and the second dealing with parallel lines.

Modern mathematicians distinguish among lemmas, propositions, theorems, and corollaries. The Greek geometers tended to make no distinction, except for corollaries, which they called "porisms." As Euclid's propositions are so important to the later development of geometry, we will designate them using just the book and proposition number; hence I-47 will refer to the 47th proposition of Book I of Euclid.

Of course, not all propositions are equally important. Proclus noted that many of Euclid's propositions were actually methods of solving a particular problem. Thus, Proclus divided Euclid's propositions into "problems" and "theorems." Euclid's first three propositions are problems, according to Proclus's classification. Euclid's first proposition is:

Proposition 5.1 (*Elements*, I-1). *To construct an equilateral triangle about a given line.*

In this case, the problem is that of constructing an equilateral triangle about a given line; the solution is *how* to construct the required triangle.

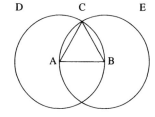

Proof. Given the line AB; construct circle BCD about A, with a radius of AB. Likewise, construct a circle ACE about B, with a radius of AB. The two circles meet at the point C; draw the lines AC and BC and construct the triangle ABC. It is an equilateral triangle, since AC is equal to AB, and BC is equal to BA, and thus all three sides are equal. ☐

The proof of I-1 assumes that two circles that share a radius will meet in at least one point; hence Euclid should have included this assumption as an additional postulate. From the beginning, we see that Euclid's proofs often rely on assumptions beyond the five postulates and five common notions.

The next few propositions are problems: to place a line at a given point equal to a given line (I-2); and to cut off, from a given line, a line equal to a given line (I-3). The first theorem, according to Proclus's classification, is

Proposition 5.2 (*Elements*, I-4). *If two triangles have two sides and the included angle equal, then the remaining sides and angles are equal, and the triangles are equal.*

This full well deserves the title of proposition, as it is an important component of many later proofs.

Proof. Given triangles ABC, DEF, where AB is equal to DE and AC equal to DF, and angle BAC is equal to angle EDF. Apply the triangle ABC to the triangle DEF by placing the point A at D, and the line AB on the line DE.

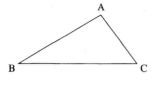

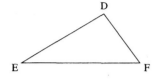

Since AB and DE are equal, then B also coincides with E. Likewise, since the angle BAC is equal to the angle EDF, then the line AC lies on the line DF. The equality of AC and DF then means that C and F coincide. Since BC is drawn between the points B and C, and EF is drawn between the points E and F, and the points B and D and the points C and F coincide, then BC must be equal to EF. Thus, the triangle ABC will coincide with the triangle DEF, and be equal to it. Hence the angles ABC and DEF are equal, as are the angles ACB and DFE. □

Once again, in the course of the proof Euclid introduced an additional assumption. Since the proof involves "moving" triangle ABC onto triangle DEF, it is necessary that, if AB is equal to DE in the original position, then AB will still be equal to DE once it has been moved onto DE: in other words, lengths (and angles) remain unchanged under translation and rotation (now referred to as **rigid transformations**). Since this is a very formidable assumption, I-4 is today treated as a postulate in its own right.

Another important point to note is that Euclid's statements of the five postulates are slightly different from how he actually used them. The proof of I-4 relied on the fact that, since B and E coincide, and C and F coincide, then the line BC and the line EF coincided as well. Euclid's first postulate was that a line could be drawn between any two points, but in I-4, Euclid assumed that the line that could be drawn was *unique*.

This analysis of Euclid's proof of I-4 might seem like nitpicking. After all, what is the difference between saying that *a* line could be drawn between two points, and that a *unique* line could be drawn between the two points? In practice, very little, since every time Euclid used the first postulate he interpreted it to mean that a unique line could be drawn between two points. The consistency of interpretation means that the precise form of the postulate is unimportant, in the same way that if *everyone* interprets "red light" as go and "green light" as stop, then traffic will flow normally.

However, if one chooses to venture beyond the geometry of Euclid, the precise statement of the postulates becomes an issue. By stating the postulate in the form it is actually used, that a *unique* line could be drawn between two points, one is more easily able to see the logical alternatives: that *more than one* line could be drawn between two points, or that *no* lines could be drawn between two points. Rather than being overly critical, a careful analysis of I-4 shows new territory into which geometers may venture.

Next is one of Thales's conjectures.

Proposition 5.3 (*Elements*, I-5). *In an isosceles triangle, the angles opposite the equal sides are equal, and if the equal sides are extended, the angles beneath the equal angles are equal.*

Frequently, Euclid will prove both a proposition and its converse. He began this practice with I-6.

Proposition 5.4 (*Elements*, I-6). *If, in a triangle, two angles are equal, the sides opposite them are equal.*

The proof is the first, of several hundred, in which Euclid used proof by contradiction: he began by assuming the opposite of what he wants to prove, then showed that it led to a contradiction.

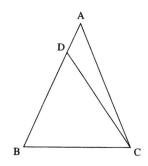

Proof. Suppose in triangle *ABC*, angle *ABC* equals angle *ACB*. If *AC* is not equal to *AB*, then one is greater; let it be *AB*. From *AB*, cut *DB* equal to *AC*. Draw *DC*. Then since *DB* is equal to *AC*, and angle *ABC* equal to angle *ACB*, and *BC* is common, then the triangles *DBC*, *ACB* have two sides and the included angle equal, and so they are equal. But *ACB* contains the triangle *DBC*, so it cannot be equal to it. Thus, *AC* cannot be unequal to *AB*, so it must be equal to it. □

Euclid included several important relationships about the sides of triangles. One was:

Proposition 5.5 (*Elements*, I-18). *In any triangle, the greater side is opposite the greater angle.*

This is proven by contradiction. The **triangle inequality** appeared as I-20:

Proposition 5.6 (Triangle Inequality, *Elements*, I-20). *In any triangle, two sides taken together are greater than the third side.*

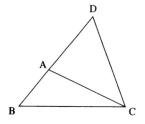

Proof. Let *ABC* be any triangle. Extend *BA* to *D*, so that *AD* and *AC* are equal. Join *DC*. Then, since *AD*, *AC* are equal, *ADC* is isosceles and angles *ADC*, *ACD* are equal. Thus angle *BCD* is greater than angle *ACD*, and thus greater than angle *BDC*. Since *BCD* is a triangle having angle *BCD* greater than angle *ADC*, then *BD* is greater than *BC*. But *BD* is *BA*, *AD*, and so *BD* is *BA*, *AC*. Thus, the sides *BA*, *AC* are greater than the side *BC*. Similarly, *BA*, *BC* is greater than *AC*, and *AC*, *BC* is greater than *AB*. □

Today, the triangle inequality is generally stated as an axiom.

5.2 Exercises

1. Carefully examine the proof of I-4. Indicate the postulate required for each claim.

2. Prove I-5. Let *ABC* be an isosceles triangle, with *AB*, *AC* equal.

   (a) Extend the sides and make *AG*, *AF* equal.
   (b) Show that triangles *FAC*, *GAB* are equal.
   (c) Consequently, show that triangles *BFC*, *CGB* are equal.
   (d) Finally, show that the angles under the equal angles are equal.
   (e) A proof given by Pappus (discussed later) relied on showing *ABC* and *CBA* were equal. Construct such a proof.

3. Refer to the proof of I-6. What assumptions besides the five common notions and five axioms is Euclid making?

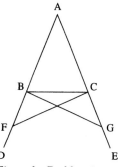

Figure for Problem 2e.

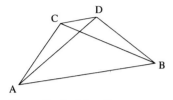

Figure for Problem 5.

4. Prove that if two oblongs have three adjacent sides equal, they are equal. (In other words, if three adjacent sides of one oblong are separately equal to three adjacent sides of a second oblong, the two oblongs are equal.) Do not use any of the properties of parallel lines.

5. Prove I-7: Given lines AC, BC, drawn from AB, there cannot be two lines AD, BD, where AC, AD are equal, and BC, BD are equal, but C and D do not coincide.

6. Prove I-8: If two triangles have three sides equal to three sides, then the triangles are equal, and the angles contained by equal sides are equal.

7. Prove I-15: if two lines intersect, the vertical angles formed are equal.

8. Solve the following problems. Use synthesis and analysis where possible. Remember that part of the solution is the proof that your construction is valid.

   (a) I-9: To bisect a given angle

   (b) I-10: To bisect a given straight line.

   (c) I-11: To draw a straight line at right angles to a given line, from a point on the given line.

   (d) I-12: To draw a straight line at right angles to a given line, from a point *not* on the given line.

   (e) I-22: To construct a triangle given three lines.

9. The solution to the following problem was known by Euclid, though the first surviving proof is that of Heron of Alexandria (see page 188.)

   (a) Prove, under the assumption that light travels in the path of least distance, that the angle of reflection off a plane mirror is equal to the angle of incidence. Hint: consider the diagram in the margin, with $CE = EF$, and CF perpendicular to DE. Consider the two paths, ABC, AGC, where G is located at the intersection between AF and DE.

   (b) Write and solve the above proposition as a calculus optimization problem.

   (c) Prove that if DE is an arc that curves everywhere away from DE, then the proof of 9a still holds.

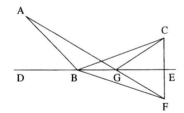

Figure for Problem 9a.

10. Restate the problem propositions as theorems (i.e., restate them so they claim a certain theoretical result).

11. Restate the theorems so that they are problems.

12. Examine how Euclid uses the "problem" propositions in later proofs. Why do you think Euclid made no distinction between problems and theorems?

5.3   Parallel Lines

The second section of Book I, beginning with Proposition 27, concerns parallel lines and figures bounded by parallel lines (i.e., parallelograms), and culminates in the Euclidean proof of the Pythagorean theorem.

Euclid was probably aware that the fifth postulate would generate a great deal of controversy; thus, he avoided using it whenever possible. One way to avoid it was to use the proposition:

Proposition 5.7 (*Elements*, I-16). *Any exterior angle of a triangle exceeds the two remote interior angles.*

With this, Euclid proved I-27 without using the parallel postulate.

Proposition 5.8 (*Elements*, I-27). *If two lines are cut by a third line and the alternate interior angles are equal, then the two lines will be parallel to each other.*

Proof. Let EB falling on DF, AC make alternate interior angles FEB, EBA, equal to each other. The lines are parallel. Suppose not. Then extending indefinitely, they will meet, either in the direction of F, C, or in the direction of D, A. Suppose they meet in the direction of F, C, at the point H. Then for triangle HEB, the exterior angle EBA is greater than the interior angles FEB, EHB. But this is impossible, since FEB, EBA are equal. Thus the lines will not meet in the direction of F, C. Similarly, they cannot meet in the direction of D, A. Hence, they cannot meet, and are parallel. □

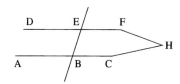

The first proposition that the fifth postulate is actually used for is:

Proposition 5.9 (*Elements*, I-29). *If a straight line crosses parallel lines, the alternate interior angles equal, and the interior angles on the same side of the crossing line equal two right angles.*

Proof. Suppose DF, AC are parallel, and GH falls upon them. If the alternate interior angles DEB, EBC are not equal, then one of them is greater. Let it be DEB. Add the angle FEB to each. Then DEB, FEB are equal to two right angles, and greater than FEB, EBC, so FEB, EBC are less than two right angles, and the lines, if extended indefinitely, meet towards F, C. But this is impossible, since the lines are parallel; hence, DEB cannot be unequal to EBC. The rest of the proof is Problem 3. □

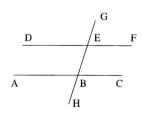

Euclid spoke of **parallelogrammic areas**: quadrilaterals with opposite sides parallel, and what we now call **parallelograms**. An important proposition, used in many later proofs, is:

Proposition 5.10 (*Elements*, I-34). *The opposite sides and angles of parallelograms are equal, and the diameter bisects the figure.*

I-35 begins a sequence of propositions that highlight the difference between Euclidean geometry and what we call geometry today. Recall the Egyptians and Babylonians could compute areas and volumes, sometimes approximately, and other times exactly. The branch of practical mathematics that dealt with such computations was called **geodesy** by the Greeks, although by Roman times, it was renamed **mensuration** (from the Latin word for measurement). Geometry was a liberal art; geodesy was not.

See page 58.

Since the *Elements* is a work on theoretical, not practical, geometry, we can expect Euclid to focus on the abstract relations between figures, and not on the practical problem of *computing* the area of a given figure. Thus in Euclid, we have general theorems about the equality of two figures. First, Euclid proved:

Proposition 5.11 (*Elements*, I-35). *Parallelograms on the same base and between the same parallels are equal.*

Proof. Consider parallelograms $ABCD$, $FEBC$, on the same base BC between the same parallels BC, AF. Since $ABCD$ is a parallelogram, the opposite sides AD, BC are equal; likewise, since $FEBC$ is a parallelogram, EF, BC are equal. Hence AD, EF are equal. Adding DE to both, AE equals DF. Since $ABCD$ is a parallelogram, then AB, DC are equal. Finally, since AB, DC are parallel, and FA is a line drawn across them, then the exterior angle FDC and the interior angle on the same side DAB are equal. Hence the triangles EAB, FDC have two sides equal to two sides, and the enclosed angle equal, and thus the triangles are equal. Subtracting DGE from both, the trapezia $ADGB$, $FEGC$ are equal. Adding triangle BGC to both, the parallelogrammic areas $ABCD$, $FEBC$ are equal.   □

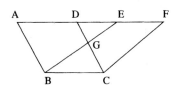

The proof relies on common notion four: things which coincide are equal. Like his use of the postulates, Euclid used the common notions in a slightly different way than he phrased them. In this case, the two parallelograms do *not* coincide, but rather, they can be *made to* coincide, by the appropriate decomposition and reconstruction.

Proposition 5.12 (*Elements*, I-36). *Parallelograms on equal bases between the same parallels are equal.*

Together with I-35, this provides an area formula for a parallelogram, since one could then compare the area of one parallelogram, anywhere in space, with the area of a rectangle. As we shall see, this is not quite a complete area formula, and considerably more mathematics is required before the area formula as we know it can be derived. The corresponding proposition for triangles is:

Proposition 5.13 (*Elements*, I-38). *Triangles on equal bases between the same parallels are equal.*

An important concept first appears in I-43. If $ABCD$ is a parallelogram with diagonal AC, then the parallelograms $DEFG$ and $JFHB$ are the **complements about the diameter**; I-43 then states:

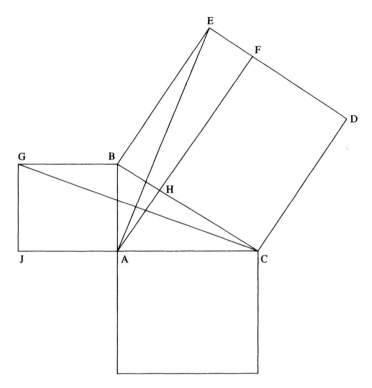

Figure 5.4: Proof of the Pythagorean Theorem.

Proposition 5.14 (*Elements*, I-43). *In a parallelogram, the complements of the parallelogram about the diameter are equal.*

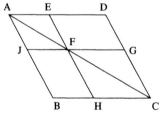

The final few propositions of Book I lead up to the proof of the Pythagorean theorem. Over two hundred proofs of the Pythagorean theorem exist. Euclid proved the Pythagorean theorem by showing how the squares on the legs of a right triangle can be added to form the square on the hypotenuse. The proof is believed to be Euclid's own, and hence may be one of Euclid's few original contributions to mathematics.

Proposition 5.15 (Pythagorean Theorem, *Elements*, I-47). *In a right-angled triangle, the square on the side opposite the right angle is equal to the squares on the sides containing the right angle.*

Proof. Given triangle ABC, with angle BAC a right angle. Construct the squares on BC, AB, AC. Draw AF parallel to BE; join AE, GC. BG is equal to BA; CB is equal to EB. Angles ABG, EBC are right angles and are equal; adding angle CBA to both, we find that angles CBG, EBA are equal. Thus, triangles CBG, EBA have two sides equal two sides, and the included angle equal, and they are equal. Triangle CBG is between parallels GB, AC, and on the same base GB as the square $GBAJ$; hence, triangle CBG is half the square. Likewise, triangle EBA is between the parallels BE, FA, and on the same base as the rectangle $BEFH$, so it is half the rectangle. But triangle EBA equals triangle CBG, so half the rectangle equals half the square, so the rectangle $BEFH$ equals the square $GBAJ$. Likewise, the rectangle $CHFD$ equals the square on AC. Thus, the square on the side opposite the right angle equals the squares on the sides containing the right angle.   □

The converse of the Pythagorean theorem concludes the first book of the *Elements*.

Proposition 5.16 (*Elements*, I-48). *If in a triangle the square on one of the sides is equal to the squares on the other two sides, the angle contained by the remaining two sides is a right angle.*

5.3 Exercises

1. Euclid proved I-16 without using the fifth postulate. Construct such a proof. Hint: in triangle ABC, bisect AC at E; extend BE to F so $BC = CF$. Show that angles EAB, ECF are equal.

2. Prove I-28: If a line falls on two lines so that the exterior and the interior angles on the same side are equal, or the interior angles on the same side are equal to two right angles, the lines are parallel. Do not use the fifth postulate.

3. Complete the proof of I-29.

4. Euclid's proof of I-35 contains an important, unstated assumption.

   (a) Identify the assumption.

   (b) Find a new proof of I-35 that does not require this assumption.

5. Prove I-36. (Hint: draw BE, CF, and consider parallelograms $ABCD$, $EBCF$, then $EBCF$, $FEGH$.)

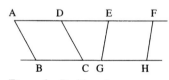

Figure for Problem 5.

6. I-41 states, "If a parallelogram is on the same base as a triangle, and is between the same parallels, the parallelogram is twice the triangle."

   (a) Prove this proposition.

   (b) How does this proposition translate into an area formula for a triangle?

7. Prove that of all parallelograms on a given base between the same parallels, the rectangle has the least perimeter.

8. Consider the problem of squaring the circle. What difficulties arise when trying to prove a given square is equal to a given circle?

5.4 Geometric Algebra

Book II is the shortest of the thirteen books of the *Elements*, containing only fourteen propositions. However, in many ways, it is the most important, for it is concerned with what we call **geometric algebra**, and serves as the key to much of the geometry as developed by Euclid and his successors.

Two new definitions appear in Book II. One involves the notion of **containment**:

Definition. *Any rectangle is said to be **contained** by its two sides.*

We will refer to the rectangle contained by the sides A and B as **rect. A, B** or, if the two sides are the same, we can speak of **sq. A**.

The other definition is that of a gnomon:

Definition. *The **gnomon** of a parallelogram is either of the parallelograms about the diameter, with its two complements.*

We have run into gnomons before, with Pythagorean number theory; here now is the geometric equivalent.

In the diagram, recall that the parallelograms $DEFG$, $HBJF$ were the complements about the diameter AC. In addition, the parallelograms $AEFJ$, $FGCH$ are referred to as the **parallelograms on the diameter**. If we take the two complements and either of the parallelograms on the diameter, we have a gnomon. The gnomons are named after a point on each of the sides of the parallelograms; thus, we will speak of the gnomon KLM, formed by the parallelograms $DEFG$, $HBJF$, and $AEFJ$.

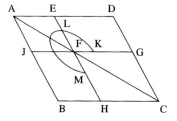

5.4.1 Identities

Most of the propositions in Book II of Euclid are now considered in the realm of algebra. To make Euclid's very general statements of the propositions easier to understand, we will state them as they are proven. The very first proposition in Book II is:

Proposition 5.17 (*Elements*, II-1). *Given two straight lines AB, AC, with AB cut at D, E. Then the* rect. AB, AC *is equal to* rect. AD, AC, rect. DE, AC, *and* rect. EB, AC.

Proof. Construct the rectangle $ACHB$ contained by AB, AC. Construct the perpendiculars FD, GE. Then rect. AB, AC is equal to rect. AD, AC, rect. DE, FD, and rect. EB, GE. Since CH is parallel to AB, then AC, FD, GE are all equal, hence rect. AB, AC is equal to the rectangles contained by AC and the segments AD, DE, and EB. □

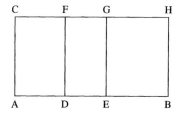

This is the geometric equivalent of the distributive property, $a(b+c) = ab + ac$, which was actually used by Euclid in the proof of the Pythagorean theorem, but not proven until Book II.

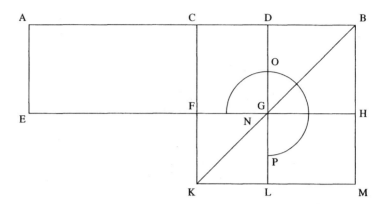

Figure 5.5: Proof of II-5.

The next few propositions deal with lines cut at random; in other words, any division of a line into two segments. Although these all translate into algebraic identities, it is important, for the proper understanding of Greek geometry, to view them primarily as *geometric* statements, and to prove them in a geometric manner, not an algebraic one.

Proposition 5.18 (*Elements*, II-4). *If AB is cut at C, then the square on AB is equal to the square on AC, together with the square on CB, plus twice the rectangle on AC, CB.*

This is equivalent to the algebraic proposition that $(a + b)^2 = a^2 + b^2 + 2ab$.

Many propositions in geometric algebra involve the notion of a line being cut into equal and unequal segments; II-5 is the first of many such propositions.

Proposition 5.19 (*Elements*, II-5). *If the straight line AB is cut into equal segments at C and unequal segments at D, then the rectangle on AD, DB and the square on CD is equal to the square on CB (see Figure 5.5).*

See Problem 5.

Proof. Draw the square $CKMB$, and the diagonal BK. Note that any parallelogram formed on the same diagonal BK, such as $DBHG$, is also a square.

Draw DL parallel to BM. The parallelograms $CDGF$, $GHML$ are complementary, and thus they are equal. Add the parallelogram $DBHG$ to both, and thus $CBHF$, $DBML$ are equal. Since AC, CB are equal, then parallelogram $ACFE$, $CBHF$ are also equal. Thus, $ACFE$ is equal to $DBML$. Add $CDGF$ to both. Thus, the gnomon NOP is equal to the parallelogram $ADGE$, which is the same as rectangle AD, DG; since $DBHG$ is a square, then DG, DB are equal, so gnomon NOP is equal to rectangle AD, DB. Add the square $FGLK$ to both. Gnomon NOP and $FGLK$ forms the square $CBMK$, which is thus equal to the rectangle on AD, DB, plus the square $FGKL$. But the square $FGKL$ is the square on CD; hence, rectangle AD, DB, plus square CD, is equal to square CB.   □

5.4.2 Square Roots

An important question that arises is whether or not the Greeks used these propositions in the same way we use their algebraic equivalents. The answer seems to be yes. For example, Theon of Alexandria explicitly referred to II-4 when he gave the procedure for extracting a square root. To find the square root of 144, he noted that 10 was the largest of the tens whose square was less than 144; hence he partitioned a square (supposed of size 144) into a smaller square (of size 100) and a gnomon. The gnomon must be of size 44. As the two complementary rectangles are equal, then the side of the small square (10) times the width CB, times two, must be less than 44: thus, 20 times CB must be less than 44, so CB is at most 2, which we obtain by dividing 20 into 44. In this case, the gnomon is exactly 44, and the square root is exact. If not, additional approximations can be made; Theon used sexagesimals when the calculation extended into fractions.

Example 5.1. *Approximate the square root of 740. Note the square on 20 is less than 740, whereas the square on 30 is more, so begin by removing a square of side 20, leaving a gnomon of size 340. Twice 20, times CB, must be no more than 340; thus, we might begin with $CB = 8$. However, in this case the gnomon would be 384, which is too large; thus, we let $CB = 7$. The gnomon is thus 329, and the remaining area is 11. To continue, we want a gnomon of size 11 to be added to a square of side 27. Thus, we repeat the procedure: the side of the square AB is 27, so twice this is 54. Dividing this into the remaining area, 11 (converted into the sexagesimal $0; 660$ for convenience) we get $0; 12$, and a corresponding gnomon of $10; 50, 24$, leaving a gnomon of size $0; 9, 36$ to be found.*

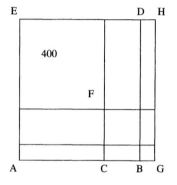

The procedure can obviously be continued as far as desired.

Example 5.2. *Find the square root of 600 to the first sexagesimal place. The largest tens whose square is less than 600 is 20; hence suppose AB is the square on 20 and so the square AB is 400. This leaves a gnomon of size 200. The rectangle on twice 20 must be less than 200; thus, suppose DE is 4 (since 200 divided by 40 is 5, but this would make the gnomon larger than 200). If DE is 4, the gnomon is of size 176.*

This leaves a second gnomon of size 24 to be found. The side, CE, is now 24, so the rectangle on 48 must be less than 24. We suppose EF to be $0; 29$ (as if we try EF to be $0; 30$, the gnomon is too large), so the gnomon will be of size $23; 26, 1$, leaving a remainder of $0; 33, 59$. If we wished to continue, we would need the rectangle on twice CF, which is $24; 29$, to be less than $0; 33, 59$.

5.4.3 The Section

An important proposition is *Elements*, II-11. Recall that the Pythagoreans may have discovered how to form a regular pentagon; the construction required dividing a line so that the rectangle on the whole and one of the segments is equal to the square on the remaining segment. This produces what writers of the See page 61.

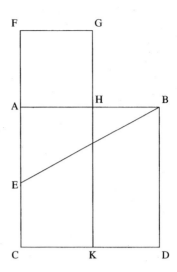

Figure 5.6: Proof of Proposition II-11.

Renaissance called the **golden section**, though the Greeks simply called it the **section**. Euclid's proof is the following. Notice that it is stated as a problem.

Proposition 5.20 (*Elements*, II-11). *Given a line, to cut it so the square on one segment is equal to the rectangle on the whole line and the other segment.*

Given a, find x so that $x^2 = a(a - x)$.

Proof. Given line AB; we want to find H so that the square on AH is equal to the rectangle on AB, BH. Construct square $ABDC$ (see Figure 5.6). Bisect AC at E, and draw BE. Draw EF, equal to BE, and make square $FGHA$. The square on AH is equal to the rectangle on AB, HB. Complete the proof as problem 9.   □

See page 61.

This will be used, in Book IV, for the construction of the regular pentagon.

5.4.4   Sides of Triangles

See page 98.

The next two propositions, II-12 and II-13, deal with the lengths of the sides of acute- and obtuse-angled triangles. We might say they are trigonometric propositions corresponding to the **law of cosines**. However, the measurement aspects of trigonometry were part of the *manual* art of geodesy. The *liberal* art of geometry concerned itself with the relationship between the sides of the triangles.

Proposition 5.21 (*Elements*, II-12). *In the obtuse-angled triangle ABC, with angle BAC obtuse, let perpendicular DB meet AC at D. Then the square on BC is equal to the square on AB, together with the square on AC, plus twice the rectangle on CA, AD.*

Proof. Since CD is cut at random at A, square CD is equal to square CA and square AD, together with twice rectangle CA, AD. Add square BD to both. Then the square CD, together with square BD, is equal to the squares CA, AD, and BD, plus twice rectangle CA, AD. But BDA, BDC are right triangles, so the squares CD, BD are together equal to square BC; likewise, the squares AD, BD are together equal to square AB. Hence, square BC is equal to squares CA, AB, plus twice rectangle CA, AD. $\square$

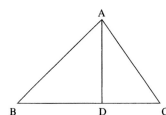

II-13 is the corresponding result for acute-angled triangles:

Proposition 5.22 (*Elements*, II-13). *Let ABC be an acute-angled triangle with perpendicular AD. Then the square on AC is equal to the square on AB, together with the square on BC, less the rectangle on BC, BD.*

5.4 Exercises

Note: Although most of the following problems can be proved algebraically, it is important, for the full understanding of Greek geometric methods, that problems be worked using the foregoing geometric methods.

1. Use Theon's procedure to evaluate the square roots indicated. Carry your results to the first sexagesimal place.

   (a) The square root of 38   (c) The square root of 325

   (b) The square root of 72   (d) The square root of 783

2. Express the following algebraic statements in a geometric manner. Then prove the statements geometrically.

   (a) $(a + b)(a - b) = a^2 - b^2$ (b) $a(b - a) = ab - a^2$

   (c) $(a + b)(c + d) = ac + ad + bc + bd$

3. What are some key differences between the algebraic equivalent of II-4 and the geometric theorem II-4?

4. To what algebraic proposition is II-5 equivalent? Explain the differences between the algebraic form and the geometric form.

5. In the proof of II-5, prove that any parallelogram formed on the diagonal KB is also a square. In general, prove that in a parallelogram $ABCD$, any parallelogram formed on the diagonal AC will have the same angles as the parallelogram $ABCD$. Do not use the properties of similar figures.

   In Problems 6 to 8, identify what algebraic propositions each of the following are equivalent to. Then prove them *geometrically*.

6. II-6: If a straight line AB is bisected at C, and AB extended through B to D, then the rectangle on AD, DB, together with the square on AC, is equal to the square on CD.

7. II-7: If a straight line AB is cut at random at C, the square on AB and the square on CB is equal to twice the rectangle on AB, CB, plus the square on AC.

8. II-8: If a straight line AB is cut at random at C, four times the rectangle on AB, CB, plus the square on AC, is equal to the square whose side is AB plus CB.

9. Complete the proof of II-11. Note: the proof should not rely on the theory of proportional figures, which Euclid has not yet established.

10. Complete the proof of II-13 as follows: given the acute-angled triangle ABC, draw perpendicular AD. Prove that the square on AC is less than the squares on BC, AB by twice the rectangle contained by BC, BD.

11. Explain how II-12 and II-13 are equivalent to the law of cosines. Why was it necessary for Euclid to state II-12 and II-13 as two separate propositions?

12. Discuss the advantages and disadvantages of geometric algebra. Things to consider: how much more "obvious" are algebraic propositions such as $(a + b)^2 = a^2 + 2ab + b^2$? What difficulties are encountered dealing with subtracted quantities, like $(a - b)^2$, and how are they dealt with?

5.5   Circles

Book III of the *Elements* deals with circles. More than half of the proofs are done by contradiction: Euclid first assumed the opposite of what he wanted to prove, and then showed it led to a contradiction. This is in sharp contrast to Book II, which has no proofs by contradiction, and Book I, which has only a few.

5.5.1   Chords

The first proposition is to find the center of a circle. Euclid then proved:

Proposition 5.23 (*Elements*, III-2). *If two points are taken on the circumference of a circle, the chord between them lies within the circle.*

Proof. Given circle ABC. Suppose the straight line between A, B, falls outside, as AEB. Find the center of circle ABC, let it be D. Draw AD, BD, and DE, which (by assumption) cuts the circle at F. Since DA is equal to DB, then the angles DAE, DBE are equal. Since DEB is an exterior angle of the triangle DAE, it is greater than the angle DAE. But angle DAE is equal to angle DBE, so angle DEB is greater than angle DBE. The greater angle subtends

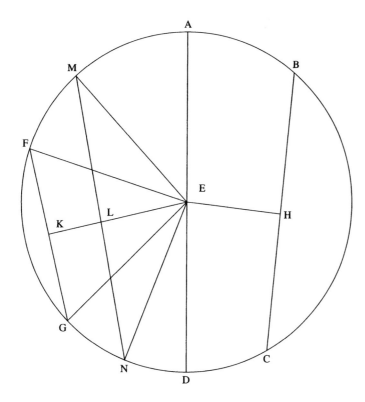

Figure 5.7: Proposition III-15.

the greater side, so DB is greater than DE. But DF is equal to DB, so DF must be greater than DE, which is absurd. Thus the line AEB cannot fall outside the circle. Similarly, the line AEB cannot fall on the circumference itself; hence, it must fall within the circle.   □

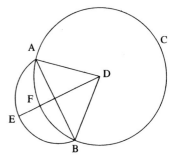

An important property about the diameter and the chords is:

Proposition 5.24 (*Elements*, III-3). *If the diameter is perpendicular to a chord, it bisects it, and conversely.*

Later in the book, Euclid proved several propositions about chords in a circle. The first important one is III-15:

Proposition 5.25 (*Elements*, III-15). *Of all chords in a circle, the diameter is the longest. Given two chords, the chord closer to the center is longer than the chord farther away.*

For Euclid, the distance between a chord and the center is the length of the perpendicular between the center and the chord. Thus in circle $ABCD$ with center E and two chords BC, FG, the perpendiculars EH, EK are the distance between the center of the circle and the chord (see Figure 5.7).

III-35 and III-36 are an interesting pair of propositions that deal with the intersection of two lines and the segments cut off on them by a circle. Given a circle and two intersecting lines, Euclid considered two cases: first, the intersection point may be inside the circle (III-35); or the intersection point may be outside the circle. In the latter case, both lines may be tangent (in which case, it is easy to prove the lines are equal in length), or one might be tangent (III-36); Euclid did not deal with the case in which neither was tangent.

Since geometric proof requires that certain objects be constructed, it is necessary that certain degenerate cases be dealt with *before* the general case. Thus, III-35, which deals with chords in a circle, must first deal with the case that the two chords are the diameters.

Proposition 5.26 (*Elements*, III-35). *If two chords intersect in a circle, then the rectangle formed by the segments of one chord is equal to the rectangle formed by the segments of the other chord.*

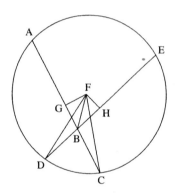

Proof. In the circle $AECD$, let the two straight lines AC, ED cut one another at B. We claim rectangle AB, BC is equal to rectangle EB, BD. If AC, ED cut each other at the center of the circle, then it is obvious that rectangle AB, BC, and rectangle EB, BD, are equal.

Otherwise, suppose the center is at F. Draw FG, FH at right angles to AC, DE; also draw FB, FD, FC. FG, FH bisect AC, ED. Thus, AC is cut into equal segments at G and unequal segments at B, so rectangle AB, BC, and square BG is equal to square CG. Add square GF to both, so rectangle AB, BC, and squares BG, GF are equal to squares CG, GF. The squares CG, GF are together equal to square FC, and squares BG, GF are together equal to square FB, so rectangle AB, BC, together with square FB, is equal to square FC. Likewise, rectangle EB, BD and square FB are equal to square FD.

But FD, FC are equal, so rectangle AB, BC, with square FB is equal to rectangle EB, ED, with square FB. Hence, rectangle AB, BC is equal to rectangle EB, ED.   □

If two lines intersect outside a circle, and one of them is tangent, then:

Proposition 5.27 (*Elements*, III-36). *If a tangent and a chord meet outside the circle, the square on the tangent is equal to the rectangle formed by the whole of the line, and the segment of line outside the circle.*

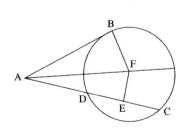

In other words, if AB is tangent to the circle and AC is another line, then sq. AB is equal to rect. AC, AD.

5.5.2   Tangents and Angles

The Greek geometers considered a wide variety of angles, of which the rectilineal angle was the simplest. Remember that the Greeks used the word line in much the same way modern mathematicians use the term curve. Thus, Euclid's definition of a rectilineal angle specifically involves straight lines. The only other type

See problem 5 on page 92.

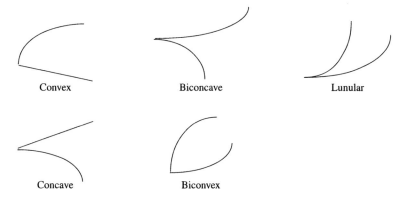

Figure 5.8: Angles

of line in the *Elements* is the circle; together, the straight line and circle form what Proclus called **simple lines**. The intersection of these simple lines on the surface of a plane can, in the classification of Proclus, form one of six types of angles.

1. The **rectilineal angle**, formed by the intersection of two straight lines.

2. The **concave** and **convex** angles, formed by the intersection of a circle with a straight line.

3. The **biconcave**, **biconvex**, and **lunular** angles, formed by the intersection of two circles.

These are illustrated in Figure 5.8. Of these, the *Elements* dealt almost entirely with rectilineal angles. However, the concave and convex angles receive passing notice in III-16, where Euclid dealt with tangents.

Proposition 5.28 (*Elements*, III-16). *The straight line drawn perpendicular to the diameter of a circle from a point on the circumference will be tangent to the circle, and no other straight line may be placed within the space between the tangent and the circumference; further, the angle in the semicircle is greater, and the remaining angle less, than any acute rectilineal angle.*

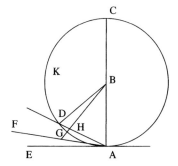

Proof. Given the circle AKC with diameter AC and center at B. Suppose the line AD, drawn at right angles to the diameter, lies inside the circle. Then BD is equal to BA, and thus angles BDA, BAD are equal; since BAD is a right angle by construction, they must both be. Thus, the triangle BDA has two right angles, which is impossible. Similarly, AD cannot fall on the circumference of the circle, and therefore, AD must fall outside the circle.

Next, suppose we can interpose a line AF between AE, which is perpendicular to AC, and the circumference; hence angle BAG is less than a right angle. Draw BG perpendicular to AF; it cuts the circle at H. By construction, AGB is a right angle, and BAG is less than a right angle. Hence, BA must be greater than BG. But BA, BH are equal, and BG is greater than BH, so BA is less than BG, and thus BA is both greater and less than BG, which is impossible. Therefore, there cannot be interposed a line AF.

Finally, if there is any acute rectilineal angle greater than that contained by the line AB and the circumference, and thus a rectilineal angle less than that between the line AE and the circumference, then a line can be interposed between AE and the circumference, which was proven impossible. Hence the angle between the line AE and the circumference is less, and the angle between AB and the circumference is greater than any acute rectilineal angle.   □

The angle between the line AE and the circumference was referred to, by the Greeks, as the **horn-like angle**, and was specifically referred to by Proclus as an example of a concave angle. The angle formed by the diameter AC and the circumference at the point A is called the **angle of a semicircle**, and it was the consideration of these two angles that gave rise to a variety of controversies between the thirteenth and seventeenth centuries regarding the nature of what we would now call limits.

See problem 4.

5.5 Exercises

1. Prove III-2 directly. Hint: consider any point E on the line between A and B. Show that DE must be less than DA.

2. Prove III-10: two circles cannot intersect in more than two points.

3. Prove III-15 (see Figure 5.7).

   (a) First, prove that if EL, EH are equal, so are BC, MN, and conversely.

   (b) Euclid continues the proof of III-15 by claiming without proof that angle MEN is greater than angle FEG. Show that this is true, and complete the proof.

4. Refer to the proof of Euclid III-16. Consider a sequence of circles, all with centers on AC and with A as the endpoint of a diameter. All these circles will have AE as a tangent.

   (a) What is the relationship between the horn-like angle in each case, and the diameter of the circle? Prove this.

   (b) State and prove the corresponding proposition regarding the angle of a semicircle.

   (c) What paradox does the preceding proposition lead to? Hint: consider that the horn-like angle can be increased indefinitely.

5. Prove III-20: Given circle BDC with center O and arc BC, and a point A on the circle. Then the central angle BOC of the arc is twice the angle BAC.

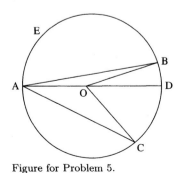

Figure for Problem 5.

6. Prove III-21: Given arc BC on a circle and two points A, E located anywhere else on the circle. Then the angles BAC, BEC are equal.

7. Prove III-36. Hint: consider also the line AF from the point through the center of the circle.

8. Use III-35 to prove the symptom of a parabola, as defined in the time of Menaechmus. See page 79.

9. Prove that given a point and a circle, the two tangents that can be drawn from the point to the circle are equal.

10. State and prove a result corresponding to III-35, if the two lines meet outside the circle.

11. Prove the following proposition: of all rectangles of a given area, the square has the least perimeter, and so conversely. Hint: consider the chords in a circle.

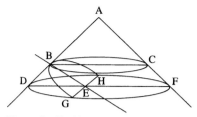

Figure for Problem 8.

5.6 Ratio and Proportion

Most historians of mathematics believe that most of Euclid, Book V, especially the definitions of ratio, and equality and inequality of two ratios, was originally developed by Eudoxus, although the arrangement of the propositions is believed to be Euclid's work. Two important technical terms are antecedent and consequent. In the ratio of A to B, A is the **antecedent**, and B is the **consequent**.

See page 77.

5.6.1 Ratios

One of the difficulties in dealing with Euclid, Book V, is that the language of ratio and multiple is not familiar to us. On the other hand, rephrasing it in modern arithmetic language would trivialize the theory of ratio and proportion. We will take the middle ground, and state the propositions in the form they are proven: that is to say, with regards to a particular set of magnitudes. The proofs will be given in the original form, but will have modern annotations in brackets. The first proposition in Book V is thus:

Proposition 5.29 (*Elements*, V-1). *Given any magnitudes A, B, where A is some multiple of C, and B the same multiple of D, then the combined magnitude A,B is the same multiple of the combined magnitude C,D.*
[*If $A = nC$, and $B = nD$, then $A + B = n(C + D)$.*]

We will use A, B to indicate the quantity of the combined magnitude A, B.

Proof. Suppose A is some multiple of C, and B is the same multiple of D. Then there are as many magnitudes in A equal to C as there are magnitudes in B equal to D. Divide A into magnitudes G, H, ... each equal to C; divide B likewise into magnitudes K, L, each equal to D. Since C, G, are equal, and D, K, are equal, then G, K is equal to C, D. Likewise, H, L is equal to C, D. Therefore, as many magnitudes as there are in A equal to C, there are magnitudes in A, B equal to C, D.

[Let $A = \overbrace{G + \ldots + H}^{n \text{ times}}$, $B = \overbrace{K + \ldots + L}^{n \text{ times}}$. Then it follows that $A + B = \overbrace{(G + K) + \ldots + (H + L)}^{n \text{ times}}$.] $\square$

An important proposition is V-12.

Proposition 5.30 (*Elements*, V-12). *Given magnitudes A, B, C, D, E, F, and suppose A is to B as C is to D, and C is to D as E is to F. Then A is to B as A, C, E is to B, D, F.*
[If $\frac{A}{B} = \frac{C}{D} = \frac{E}{F}$, then $\frac{A}{B} = \frac{A+C+E}{B+D+F}$]

This transformation of ratios is called **summing the antecedents and consequents**. Euclid's method illustrates the nature of the equality of ratios.

Proof. Take any multiple of A, making G; take the same multiples of C, E, making H, F. Take any multiple of B, making L; take the same multiples of D, F, making M, N. Since A is to B as C is to D as E is to F, then if the multiples of A exceed, equal, or fall short of the multiples of B, then so do the multiples of C, E alike exceed, alike equal, or alike fall short of the multiples of D, F, according to the definition of equal ratios. Thus if G exceeds, equals, or falls short of L, then G, H, K each alike exceed, alike equal, or alike fall short of L, M, N, so together, G, H, K alike exceeds, equals, or falls short of L, M, N.

[Let $nA = G$, $nC = H$, $nE = F$; let $mB = L$, $mD = M$, $mF = N$. Then if $nA > mB$, so is $nC > mD$, $nE > mF$, and so on. Thus $nA + nC + nE > mB + mD + mF$, so $G + H + K > L + M + N$. Likewise for $nA < mB$ or $nA = mB$.]

But G and G, H, K are the same multiples of A and A, C, E, and L and L, M, N are the same multiples of B, and B, D, F. Thus, if G, H, K exceeds, equals, or falls shorts of L, M, N, then the multiples of A, C, E exceed, equal, or fall short of B, D, F, according to whether the multiples of A exceed, equal, or fall short of the multiples of B; hence A, C, E and B, D, F have the same ratio as A and B.

[$G = nA$, and $G + H + K = n(A + C + E)$, likewise, $L = mB$ and $L + M + N = m(B + D + F)$, so if $nA > mB$, so is $n(A + C + E) > m(B + D + F)$, and thus A is to B as A, C, E is to B, D, F.] $\square$

Several propositions involve transforming one equality of ratios into a different equality of ratios. The first is:

Proposition 5.31 (*Alternating, Elements*, V-16). *If A is to B as C is to D, then A is to C as B is to D.*
[If $\frac{A}{B} = \frac{C}{D}$, then $\frac{A}{C} = \frac{B}{D}$.]

Proposition 5.32 (*Separating, Elements,* V-17). *If* AE, EB *is to* EB *as* CF, FD *is to* FD, *then* AE *is to* EB *as* CF *is to* FD.
[If $\frac{A+B}{B} = \frac{C+D}{D}$, *then* $\frac{A}{B} = \frac{C}{D}$.]*

Proposition 5.33 (*Combining, Elements,* V-18). *If* AE *is to* EB *as* CF *is to* FD, *then* AE, EB *is to* EB *as* CF, FD *is to* FD.
[If $\frac{A}{B} = \frac{C}{D}$, *then* $\frac{A+B}{B} = \frac{C+D}{D}$.]*

Proposition 5.34 (*Converting, Elements,* V-19). *If* AE, EB *is to* CF, FD *as* AE *is to* CF, *then* EB *is to* FD *as* AE *to* CF.
[If $\frac{A+B}{C+D} = \frac{A}{C}$, *then* $\frac{B}{D} = \frac{A}{C}$.]*

In the most significant indicator of the shift in mathematical thinking, the ratio proposition that is easiest and most obvious to us is the ratio proposition that would have been most difficult and least obvious to Euclid:

Proposition 5.35 (*Compounding, Elements,* V-20). *If* A *is to* B *as* D *is to* E, *and* B *is to* C *as* E *is to* F, *then* A *is to* C *as* D *is to* F.
[If $\frac{A}{B} = \frac{D}{E}$, *and* $\frac{B}{C} = \frac{E}{F}$, *then* $\frac{A}{C} = \frac{D}{F}$.]*

5.6.2 Similar Figures

The ratios of Book V are put to immediate use in Book VI, which deals with similar figures. Again, the use of the definition of equal ratios is an important part of the Euclidean theory of proportions.

Proposition 5.36 (*Elements,* VI-1). *Triangles and parallelograms with the same height are to each other as the bases.*

Proof. Given triangles ABC, ABD, on bases BC, BD and under the same height. We claim as BC is to BD, so will triangle ABC be to triangle ABD.

Let BD be extended in both directions to H, L, and let BH be divided into segments CG, GH, and so on, equal to CB; likewise, let BL be divided into segments DK, KL equal to BD. The triangles ABC, ACG, AGH are equal, as are the triangles ABD, ADK, AKL. If BH exceeds, equals, or falls short of BL, then triangle ABH exceeds, equals, or falls short of triangle ABL. Thus the ratio of BC to CD is the same as the ratio of ABC to ACD. Likewise for parallelograms. □

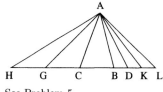

See Problem 5.

Recall Proposition I-35 allowed a given parallelogram to be compared to a given rectangle. This is not quite the area formula for a parallelogram, since it does not indicate how the area of one parallelogram compares to the area of a standard unit square. In order to make the final connection, VI-1 must be invoked. Thus, a parallelogram is compared (via I-35) to a rectangle on the same base, between the same parallels. Then VI-1 is used twice, first to compare this rectangle to another with a unit base, and then to compare this last rectangle to a unit square.

See page 98.

Proposition VI-1 is used to prove VI-2.

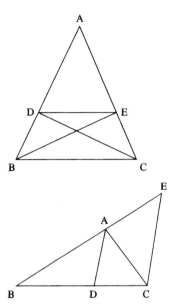

Proposition 5.37 (*Elements*, VI-2). *A line drawn parallel to the side of a triangle cuts the sides proportionally, and conversely, if a line cuts two sides of a triangle proportionally, it is parallel to the third side.*

Proof. Draw DE parallel to BC. We claim that BD is to AD as CE is to AE. Triangle BDE is equal to triangle CDE, since they are between the same parallels and on the same base. Thus BDE is to ADE as CDE is to ADE. But BDE is to ADE as BD is to AD, and CDE is to ADE as CE is to AE; hence BD is to AD as CE is to AE. The proof of the converse is Problem 6.   □

An important proposition used by Archimedes is (VI-3):

Proposition 5.38 (*Elements*, VI-3). *In triangle ABC, if angle BAC is bisected by AD, then BD is to DC as AB is to AC.*

After this follows a number of proportions relating to similar triangles. The first one, establishing the proportionality of the sides in similar triangles, was:

Proposition 5.39 (*Elements*, VI-4). *In triangles whose corresponding angles are equal, the sides opposite the equal angles are proportional.*

5.6.3   Geometric Arithmetic

Proposition VI-2 has far-reaching implications, for there are a number of constructions in Euclidean geometry that correspond to arithmetic operations. The most obvious are the addition and subtraction of two lines, although as we have seen, the problem of finding the greatest common divisor is also amenable to geometric techniques. An important tool in **geometric arithmetic** is finding lines proportional to other lines. The problem of finding proportional lines occurs in Book VI, beginning with:

Proposition 5.40 (*Elements*, VI-9). *To cut off a given part from a given line.*

In other words, cut off half, a third, a fourth, or any other specified part of a given line. Euclid's method is identical to that used in modern elementary geometry, and corresponds to dividing a given *real* number by a whole number.

Proposition 5.41 (*Elements*, VI-11). *Given lines AB, AC, find CE so AB is to AC as AC is to CE.*

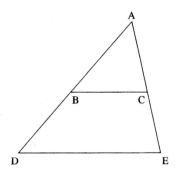

Proof. Extend AB to D, and AC to E, so that BD is equal to AC. Join BC, and draw DE parallel to BC. Since BC, DE are parallel, then AB is to BD as AC is to CE. Hence AB is to AC as AC is to CE, and CE is the desired third proportional.   □

This proposition corresponds to finding the third of a sequence of terms in geometric proportion, given the first two terms. Two other cases are dealt with by:

Proposition 5.42 (*Elements*, VI-12). *Given lines A, B, C, find D so A is to B as C is to D.*

which is solved using similar triangles, and

Proposition 5.43 (*Elements*, VI-13). *Given lines AB, BC, find BD so that AB is to BD as BD is to BC.*

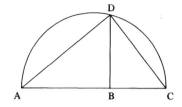

which is solved by constructing the semicircle *ADC* with perpendicular *BD*; *BD* is the geometric mean.

All of these geometric procedures correspond to arithmetic operations. If the first term, *A*, is taken to be a unit, then VI-11 is equivalent to squaring a given real number *B*; VI-12 is equivalent to multiplying two real numbers; and VI-13 is equivalent to taking the square root of a given real number.

5.6.4 Proportionality of Figures

A pair of propositions that will be important when dealing with the areas of circles are VI-15 and VI-19.

Proposition 5.44 (*Elements*, VI-15). *In two triangles with one equal angle, if the triangles are equal, the sides about the equal angle are reciprocally proportional, and so conversely.*

Proof. Given equal triangles *ABC*, *ADE*, where angle *BAC*, *DAE* are equal. Then we claim that *CA* is to *AD* as *AE* is to *AB*. Place the triangles so *CA* is in line with *AD*, and thus *EA* is in line with *AB*. Since triangles *ABC*, *ADE* are equal, and triangle *BAD* is another area, then the ratio of triangle *ABC* to triangle *BAD* is equal to the ratio of triangle *ADE* to triangle *BAD*. But triangles under the same height are to each other as their bases; hence triangle *ABC* is to triangle *BAD* as *CA* is to *AD*; and triangle *ADE* is to triangle *BAD* as *AE* is to *AB*, so *CA* is to *AD* as *AE* is to *AB*. □

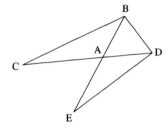

Proposition 5.45 (*Elements*, VI-19). *Similar triangles are to each other as the duplicate ratios on their corresponding sides.*

Remember that if *A* is to *B* as *B* is to *C*, then *A* is said to have a duplicate ratio to *C* that it has to *B*.

Proof. Given triangles *ABC*, *DEF* similar, so angle *B* is equal to angle *E*. Construct a third proportional *BG* so that *BC* is to *EF* as *EF* is to *BG*. Join *AG*. Since *ABC*, *DEF* are similar, *AB* is to *BC* as *DE* is to *EF*. Alternating, we have *AB* is to *DE* as *BC* is to *EF*. By construction, *BC* is to *EF* as *EF* is to *BG*, so *AB* is to *DE* as *EF* is to *BG*. Since triangles *ABG*, *DEF* share an angle and the sides about the angle are reciprocally proportional, then triangle *ABG* is equal to triangle *DEF*. Since *BC* is to *EF* as *EF* is to *BG*, then *BC* has the duplicate ratio to *BG*. But the ratio of *BC* to *BG* is equal to the ratio between triangle *ABC* to *ABG*, since they are under the same height; hence

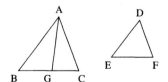

triangle ABC to triangle ABG, and thus to triangle DEF, has the duplicate
ratio of BC to EF.   □

5.6 Exercises

As with previous sections, most of the following problems can be solved using
algebra, although to better appreciate Greek geometry, and to understand its
methods more thoroughly, these problems should be done using the methods of
this chapter.

1. Prove V-14: if A is to B as C is to D, then if A exceeds, equals, or falls
   short of C, then B likewise exceeds, equals, or falls short of D.

2. Prove V-15: if A is the same multiple of B that C is of D, then A is to B
   as C is to D.

3. What assumption is necessary for the alternation of ratios, in V-16?

4. Why would compounding a ratio V-20 be conceptually difficult for a Euclidean geometer?

5. Consider VI-1.

   (a) Complete the proof regarding the proportionality of parallelograms.

   (b) Euclid assumes but does not prove that if two triangles or parallelo-
       grams under the same height are on unequal bases, the greater base
       will produce the greater triangle or parallelogram. Prove this; do not
       use the area formula for parallelograms or triangles.

6. Complete the proof of VI-2.

7. Prove VI-3. Draw CE parallel to AD, then apply VI-2.

8. Prove VI-4. Let ABC, DCE be the equiangular triangles, with angle A
   equal to angle D, angle B equal to angle C, and angle C equal to angle E.
   Extend AB, DE until they meet at F, and apply VI-2.

9. Explain how VI-11,VI-12, and VI-13 are respectively equivalent to the
   arithmetic operations of squaring, multiplying, and taking the square root
   of a number.

10. Prove the converse of VI-15: if two triangles have an equal angle and the
    sides about the equal angle are reciprocally proportional, then the triangles
    are equal in area.

11. How are duplicate and triplicate ratios expressed today? In other words,
    if A, C have a duplicate ratio, then how would we express the relationship
    between A, C?

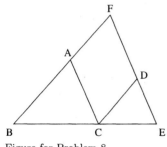

Figure for Problem 8.

5.7   The Quadratic Equation

As we have seen, the solution to the problem of cutting a line into "extreme and mean" ratios is equivalent to solving a particular quadratic equation; this was done in Euclid, Book IV as a prelude to the construction of the regular pentagon. In Book VI, Euclid tackled the general problem of application of areas, which is equivalent to solving quadratic equations.

See page 60.

First, a few important techniques must be established. In VI-25, Euclid showed how to construct a rectilineal figure equal (in area) to another figure and similar to a third. VI-26 then shows that if one parallelogram is similar to another, and shares an angle, then the same diameter will include both. Then:

Proposition 5.46 (*Elements*, VI-27). *Of all parallelograms on a given straight line and deficient by a parallelogram similar to the parallelogram on half the line, the greatest is the parallelogram on half the line and similar to the defect.*

What Euclid meant by this rather complex statement is the following: on a given a straight line, AB, which is bisected at C, form *any* parallelogram $CBED$. Cut the line AB at any other point, such as K. The parallelogram $KAGF$ is on the given line, and is deficient by the parallelogram $KBHF$, which is similar to $CBED$; it is smaller than the parallelogram $ACDL$, which is on half the given line and similar to the defect $CBED$.

Proof. Bisect AB at C, and create the parallelogram $ACDL$, which is on half the line and deficient by the figure on half the line, namely $CBED$. We claim that the parallelogram $ACDL$ is greater than the parallelogram $KAGF$, which is on the line and deficient by the figure $KBHF$, similar to $CBED$. Since $KBHF$, $CBED$ are similar, they are on the same diameter DB. Recall that the complements about the diameter in a parallelogram are equal; thus, $KCNF$, $MEHF$ are equal. By adding $KBHF$ to both, we find that $CNHB$, $KMEB$ are equal. But $CNHB$ is equal to $AGNC$, so $AGNC$ is equal to $KMEB$. Add $KCNF$ to both. Thus, $KAGF$ is equal to gnomon PRQ, which is less than $CBED$. But $CBED$, $ACDL$ are equal; hence $KAGF$ is less than $ACDL$.   □

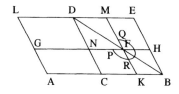

This problem is used, by Euclid, to establish a necessary condition for the existence of a solution to a quadratic equation. It also solves an optimization problem.

See Problem 2.

5.7.1   Application of Areas

With these propositions, Euclid can now solve one type of quadratic equation.

Proposition 5.47 (*Elements*, VI-28). *Given a line, a parallelogram, and a rectilineal figure. To construct on the line a parallelogram equal to the rectilineal figure and deficient by a parallelogram similar to the given parallelogram. It is necessary that the given figure must be less than half the figure on the straight line and similar to the defect. (See Figure 5.9.)*

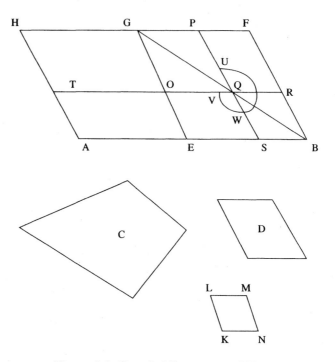

Figure 5.9: Proof of Proposition VI-28.

This and VI-29 form the Greek geometric solution to the quadratic equation. Notice that Euclid imposed a condition (called a **diorism**) on the problem, corresponding to a statement that certain problems of this type are unsolvable.

Proof. Given the line AB, the figure C, and the parallelogram D, we seek to find a parallelogram $ATQS$ which is equal to C, and deficient by the parallelogram $SQRB$, which is similar to D. Bisect AB at E, and construct $EGFB$ similar to D. If $AEGH$ is equal to C, then the problem is solved.

Otherwise, suppose $AEGH$ is greater than C, and thus $EGFB$ is greater than C. Create $KLMN$, similar to D and equal to the excess of $EGFB$ over C. Since $KLMN$, $EGFB$ are both similar to D, they are similar to each other; let KL correspond to GE, LM to GF. Since $EGFB$ is equal to C and $KLMN$ together, then GB is greater than LN, and GE is greater than KL. Make GO, GP equal to KL, LM respectively. Then $GPQO$ is equal to $LMNK$; since it is similar to $EGFB$, they are both on the same diameter GB. Note that the gnomon VWU is equal to C.

The complements $QPFR$, $EOQS$ are also equal. Adding $SQRB$ to both, we have that $SPFB$ is equal to $EORB$, which is equal to $ATOE$. Add $EOQS$ to both; thus the parallelogram $ATQS$ is equal to the gnomon VWU. But this was equal to C, so the parallelogram $ATQS$ is equal to C, which is what was desired.   □

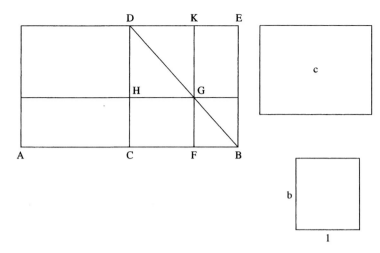

Figure 5.10: Solving Quadratic Equations Using Proposition VI-28.

How does this translate into the solution of a quadratic equation? To simplify matters, suppose all the figures involved are rectangles (see Figure 5.10). Given a line AB of length a, we seek to find a rectangle with an area c, where base of the rectangle is less than the whole line and the rectangle itself falls short of the rectangle on the whole line by another rectangle, whose sides are in ratio b to 1. Thus, we can see that this proposition gives the solution to a quadratic equation $bx(a - x) = c$, with a, b, c all positive. The Euclidean procedures corresponds to:

1. Divide AB in half at C.

2. Construct the rectangle $CDEB$, where $DC : CB = b : 1$.

3. Construct $HDKG$, equal to the excess of $CDEB$ over c.

4. The rectangle contained by AF, FG is the desired rectangle.

Example 5.3. *Given a line 8 feet in length, a figure with an area of 14 square feet, and a rectangle whose sides have a ratio of 2 to 1, find the rectangle on the line. Divide the line of 8 feet in half; construct on the half a rectangle of 8 feet by 4 feet, so the sides of the rectangle are in the ratio of 2 to 1 (see Figure 5.11). This rectangle has an area of 32 square feet, which exceeds the 14 square feet by 18 square feet; if it did not exceed the given 14 square feet, the problem would be unsolvable. A rectangle of 18 square feet with a 2 to 1 side ratio must be 6 feet by 3 feet; applying this to the rectangle on 8 feet by 4 feet in the manner indicated leaves 1 foot left on the end; hence a rectangle of 7 feet (8 feet minus 1 foot) by 2 feet as our solution.*

Example 5.4. *Given a line of 12 inches, an area of 70 square inches, and a rectangle whose sides have a ratio of 3 to 2, find the rectangle on the line. Divide*

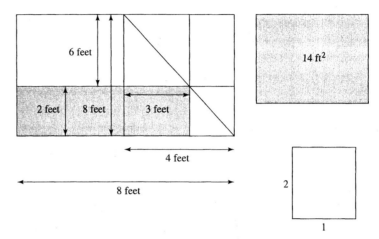

The shaded areas are equal; thus, on the line of 8 feet has been constructed a parallelogram equal to 14 square feet, deficient by a parallelogram whose sides are in ratio of 2 to1.

Figure 5.11: Sample Quadratic

the line in half, making 6 *inches; a rectangle with sides in ratio* 3 *to* 2 *would be* 9 *inches by* 6 *inches, which has an area of* 54 *square inches. Since this does not exceed the given area, the problem is unsolvable.*

Example 5.5. *Given a line of* 12 *inches, a figure with an area of* 30 *square inches, and a rectangle with sides in ratio* 3 *to* 2. *As in the previous problem, the rectangle on half has an area of* 54 *square inches, which exceeds* 30 *square inches by* 24 *square inches. The rectangle with sides in ratio* 3 *to* 2 *is* 6 *inches by* 4 *inches. Thus, the desired rectangle must be* 3 *inches by* 10 *inches.*

VI-28 solves only one type of quadratic equation with positive coefficients. Another type is solved by:

Proposition 5.48 (*Elements*, VI-29). *Given a line, a parallelogram, and a rectilineal figure. To construct on the line a parallelogram equal to the rectilineal figure and exceeding the line by a figure similar to the given parallelogram.*

Proof. Given the line AB, a rectilineal figure C, and a parallelogram D, it is desired to construct on AB a parallelogram $RAPO$ equal to C, and exceeding AB by a figure $BPOQ$ similar to D (see Figure 5.12). Bisect AB at E, and describe parallelogram $FEBL$ similar to D. Draw $GKJH$ similar to D, and equal to $FEBL$ together with C. Let KJ correspond to FL, KG to FE. Since $GKJH$ is greater then $FEBL$, therefore KJ is greater than FL, and KG greater than FE. Make $NFMO$ equal to $GKJH$. The gnomon VWX is equal to C

The complements about the diameter OF, namely $BLMP$, $NEBQ$, are equal. Since $NEBQ$ is equal to $RAEN$, then $BLMP$, $RAEN$ are equal. Add

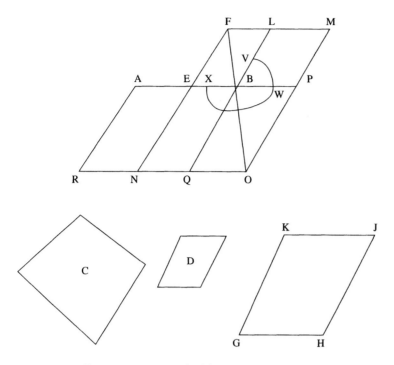

Figure 5.12: Proof of Proposition VI-29.

$NEPO$ to both. Thus, $RAPO$ is equal to the gnomon VWX, which is equal to C, as was desired. □

Again, suppose all the figures involved are rectangles (see Figure 5.13). Given the line AB of length a, area c, and rectangle D whose sides are in ratio $b : 1$, then the procedure is

1. Divide AB in half at E.

2. Construct the rectangle $FEBL$, where $FE : EB = b : 1$.

3. Add $FEBL$ to c, and create $GKJH$, where $KG : GH = b : 1$.

4. Apply this rectangle to $FNOM$, where $FN : NO = b : 1$. The desired rectangle is $RAPO$.

Which quadratic equation does this solve? Given line AB of length a, the problem seeks a rectangle of area c, where the base of the rectangle is greater than the line and the rectangle exceeds that on base AB by a rectangle whose sides are in ratio $b : 1$. Thus, the proposition solves quadratics of the form $bx(a + x) = c$. Again, a, b, and c must all be positive numbers.

Example 5.6. *To a line of length 4 feet, apply a rectangle of area of 15 square feet, so that the excess is similar to a rectangle whose sides are in ratio 3 to 1.*

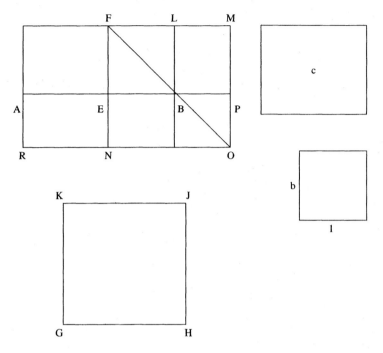

Figure 5.13: Solving Quadratic Equations Using Proposition VI-29.

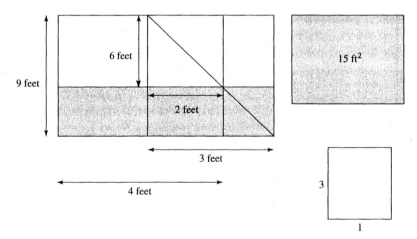

The shaded areas are equal; thus, on the line of 4 feet has been
constructed a parallelogram equal to 15 square feet, exeeding the
line by a figure whose sides are in ratio of 3 to1.

Figure 5.14: Sample Quadratic

Bisect the line, and construct a rectangle of 2 feet by 6 feet; the area is 12 square feet (see Figure 5.14). Add it to the given 15 square feet, which has a total of 27 square feet. The rectangle similar to this would have sides of 9 feet and 3 feet; apply this to the line. The result is a rectangle of 4 feet plus 1 foot, by 3 feet, which has an area of 15 square feet, and exceeds the given line by a rectangle whose sides are in ratio 3 to 1.

Example 5.7. *To a line of length 6 inches, apply a rectangle of 32 square inches, exceeding the line by a rectangle whose sides are in ratio of 2 to 1. Cut the line in half and construct a rectangle 3 inches by 6 inches; add this area, 18 square inches, to the given area, 32 square inches, to produce an area of 50 square inches; this can be produced by a rectangle of sides 10 inches and 5 inches. Applying this to the rectangle, we obtain a rectangle of 6 inches plus 2 inches, by 4 inches, which has an area of 32 square inches, and exceeds the line by a rectangle with sides in ratio 2 to 1.*

5.7.2 Rational and Irrational Lines

Did the Greek geometers actually solve quadratic equations in this fashion? Evidence in favor is found in Book X, which includes over a hundred propositions for reducing certain irrational lines to other irrational lines; it is, in fact, the longest book of the *Elements,* and perhaps the least accessible of all thirteen books. The reductions really serve no purpose unless one assumes that the Greeks did, in fact, solve quadratic equations in the manner outlined by Book VI of the *Elements.*

See page 75.

Consider a quadratic equation of the form solved by VI-29, which we would now write as $x^2 + 2\beta x = \alpha^2$, where x, β, and α represent the lengths of lines. In the simplest case, α and β are commensurable. Note that this does not mean they are rational numbers: to the Greeks, the quantities we would express as $\sqrt{8}$ and $\sqrt{18}$ are commensurable. The positive solution to this equation is $x = \sqrt{\beta^2 + \alpha^2} - \beta$; we may interpret x to be the line formed by taking the difference of two lines: the first is the side of a square equal to the combined squares on α and β, and the other is the line β.

This line might or might not be commensurable with the lines α, β. If it was incommensurable, then it formed an apotome, in the classifications of Theaetetus, since the two lines, $\sqrt{\beta^2 + \alpha^2}$ and β, were commensurable in square. Euclid named this type of irrational the **first apotome**. Likewise, the solution of a quadratic of the form solved by VI-28 gave rise to a **first binomial**. Together, these two types form what are now called **quadratic irrationals**, since they arise from the solutions to quadratic equations with rational coefficients.

See page 75.

See Problem 9.

Next, suppose α and β were commensurable in square only. In some cases, such as $x^2 + \sqrt{5}x = 11$, the solution is once again an apotome. But in other cases, the result was a new type of irrational, neither binomial nor apotome: for example, $x^2 + x = \sqrt{5}$. These gave rise to other types of binomials and apotomes. Sometimes, these binomials and apotomes might only appear to be different: thus, $\sqrt{7 + 4\sqrt{3}}$ is actually a first binomial, $2 + \sqrt{3}$. The classification and reduction schemes of Book X extend to 115 propositions, making it the longest book in the *Elements*.

5.7 Exercises

1. Prove VI-24: In any parallelogram, the parallelograms about the diameter are similar to the whole and to one another. Remember that two parallelograms are similar if they are both equiangular and the sides about the equal angles have an equal ratio to each other.

2. Use VI-27 to show that given any triangle ABC, the parallelogram $BDEF$, with BD, BF on the sides AB, BC of the triangle, will have maximum area if BD, BF are both half of AB, AC.

3. Use VI-27 to solve the following optimization problem: A parking lot currently charges \$7 for parking, and has 4000 customers. It estimates that for every \$1 increase in the price, it loses 800 customers, while it will gain 800 customers for every \$1 decrease. How much should it charge to maximize its revenue?

4. Determine whether the following quadratics are solvable, using VI-27.

    (a) A rectangle on a side of 9 feet, equal to an area of 25 square feet, and falling short by a rectangle with sides in ratio 4 to 1.

    (b) A rectangle on a side of 6 feet, equal to an area of 20 square feet, and falling short by a rectangle with sides in ratio 2 to 1.

    (c) A rectangle on a side of 12 feet, equal to an area of 40 square feet, and falling short by a rectangle with sides in ratio 3 to 1.

5. Why does Euclid not need a diorism for VI-29?

6. VI-28 and VI-29 only solve two types of quadratic equations.

    (a) Given that a, b, and c must always be positive, how many different types of quadratic equations are there?

    (b) Of these types, which two did Euclid solve?

    (c) Why did Euclid not solve the other type(s)?

7. Show that cutting a given line into an extreme and mean ratio is equivalent to solving a special case of VI-29.

8. Solve the following quadratic equations using either VI-28 or VI-29, whichever is appropriate.

   (a) $x^2 + 4x = 60$ (c) $4x^2 - 5x = 21$

   (b) $x^2 - 6x = 27$ (d) $2x^2 = 3x + 77$

9. Consider VI-28.

   (a) Show it solves quadratic equations of the form $2\beta x - x^2 = \alpha^2$.

   (b) Show that, if α, β, are commensurable, then solutions are either commensurable with the α, β, or are binomial straight lines.

5.8 Number Theory

Geometry has been defined as the science of magnitudes, which is quite evident in Books V and VI, in which Euclid developed the theory of proportion and ratio. Book VII returns to the science of whole number, which the Greeks called arithmetic. Euclid defined **number** as a collection of units; a **unit** was that which numbered the quantity one. *Measure, part,* and *parts* were defined as they were by the Pythagoreans. A **prime number** is measured by the unit alone; all other numbers are **composite**. Two numbers are **prime to one another** (our "relatively prime") if there is no number that measures them both.

See page 43.

Like the Pythagoreans, Euclid divided numbers into several categories, including even, odd, square, and cube. In addition, he further classified numbers as:

Definition. *A number is **even-even** if it can be measured by an even number an even number of times; **even-odd** if it can be measured by an even number an odd number of times; and **odd-odd** if it is measured by an odd number an odd number of times.*

It is interesting to note that, in contrast to Euclid's classifications of quadrilaterals, where, for example, a square was not an oblong, Euclid's classification of numbers allowed a single number to be a member of several categories. Thus, 36 was both even-even (as 6 measured it 6 times) and even-odd (as 12 measured it 3 times).

Our modern definition of a **perfect number** comes from Euclid:

Definition. *A number is **perfect** if it is equal to the sum of its parts.*

In Latin, *perfectus* simply means complete. Two additional classifications were added later, by Theon of Smyrna: overperfect numbers (what we now call **abundant** numbers), whose parts are greater than the number itself, and **deficient** numbers, whose parts are less than the number. For example, 12 is overperfect, for its parts are 1, 2, 3, 4, and 6, and their sum is 16. Meanwhile

10 is deficient, for its parts are 1, 2, and 5, whose sum is 8. Finally, 6 is perfect, since its parts are 1, 2, and 3, and the sum of the parts is equal 6.

Number theory in the *Elements* begins with VII-1, the Euclidean algorithm, and continues onward through books VIII and IX. Many of the propositions are duplicates of those in Books V and VI. For example

See page 43.

Proposition 5.49 (*Elements*, VII-12). *If A is to B as C is to D and as E is to F, then A is to B as A, C, E is to B, D, F.*

But for the change from "magnitude" to "number," this proposition is identical to *Elements* V-12.

An important proposition, utilizing what Fermat would later popularize as the **method of infinite descent**, is VI-31:

Proposition 5.50 (*Elements* VII-31). *Any composite number is measured by some prime.*

Proof. Suppose A is composite. Therefore some number measures it, say B. Either B is prime or it is composite; if it is prime, we are done. If B is composite, then some number, C, measures it, and since C measures B and B measures A, then C measures A. If C is prime, we are done; otherwise, we may find another number, D, which measures C. In this way, we produce a sequence of decreasing numbers, which must end with a prime number that measures A. Otherwise, we have an infinite sequence of decreasing numbers, which is impossible. □

This is part of the **fundamental theorem of arithmetic**, though Euclid never states the complete theorem (which is that every whole number has a unique decomposition into prime factors).

Book IX considers the properties of the whole numbers. One of Euclid's most important results is IX-20:

Proposition 5.51 (Prime Number Theorem, *Elements*, IX-20). *The number of primes exceeds any given number.*

Today, we state that the number of primes is *infinite*; however the Greeks, who were uncomfortable with the idea of an actual infinite amount, stated the result in the above form. Euclid's proof is the same as our modern one: if A, B, C are primes, then there are more primes than these, since the number $ABC + 1$ is either prime, or is measured by some prime other than A, B, or C.

An important proposition concerns the sum of a finite geometric series:

Proposition 5.52 (*Elements*, IX-35). *Given any number of numbers in continued proportion, the second minus the first is to the first as the last minus the first is to the sum of all except the last.*
[Given a geometric sequence b, b^2, b^3, ..., b^n, then $(b^2 - b) : b = (b^n - b) : (b + b^2 + b^3 + \ldots + b^{n-1})$.]

Euclid's proof supposed there were only four terms in continued proportion, though it is easy to generalize.

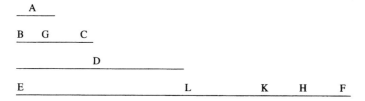

Figure 5.15: Proof of Proposition IX-35.

Proof. Let A, BC, D, EF be numbers in continued proportion, with A the first and EF the last (see Figure 5.15). From BC, EF, take BG, FH, equal to the first, A, leaving GC, EH. Let $FK = BC$, $FL = D$. Since $FH = A$ and $FK = BC$, then $HK = CG$. Since A, BC, D, EF are in continued proportion, we have

$$EF : D = D : BC = BC : A$$

and thus

$$EF : FL = FL : FK = FK : FH$$

Separating the ratios, we have

$$EL : FL = LK : FK = KH : FH$$

Thus, EL, LK, KH are to FL, FK, FH, as KH is to FH. Thus EH is to D, BC, A as CG is to A, which was to be proven. □

The last proposition of Euclidean number theory concerns the generation of perfect numbers.

Proposition 5.53 (*Elements*, IX-36). *If any number of numbers, beginning with the unit, are in double proportion, and their sum is prime, then the product of the sum and the last number is perfect.*

For example, the sequence $1, 2, 4$ has a sum of 7, which is prime; hence $7 \times 4 = 28$ is perfect; until the middle ages, only four perfect numbers were known. Perfect numbers of this type are called **Euclidean perfect numbers**; it is known that all *even* perfect numbers must be of this form, but it is unknown, even today, whether any odd perfect numbers exist.

5.8 Exercises

1. Is one a number? Explain in terms of Euclidean number theory.

2. Use *Elements*, IX-36 for the following.

   (a) Find the first four perfect numbers.

   (b) Based on these four, make a conjecture regarding the distribution of perfect numbers.

   (c) Find the fifth perfect number.

   (d) Is the conjecture in Problem 2b validated?

5.9   The Method of Exhaustion

Recall that Euclid used the term equal figures in two senses. The first sense was
that two figures are equal if they are identical, and thus Euclid speaks of equal
triangles, equal parallelograms, and so on. The other meaning of equality of
figures is that two figures are equal if their component parts are equal in the first
sense; thus, Euclid does not hesitate to say that two parallelograms are equal if
a proper dissection of one can yield the other.

These two definitions can be applied to any rectilineal figures, but only the
first makes sense when applied to circles or regions bound by arcs of circles.
Thus, while Euclid can and does speak of equal circles, he cannot speak of the
equality of a circle and a rectangle, since they are neither identical figures, nor
can one figure be dissected and reformed into the other. In fact, he cannot even
compare two circles of different sizes in any useful manner, other than to say
that one is larger than the other. Thus, to discuss the areas of circles, Euclid
resorts to the equality of ratios, as developed in Book VI.

Book XII of the *Elements* deals primarily with the volumes of cylinders,
cones, and spheres, using the method of exhaustion. The method relies on:

Proposition 5.54 (*Elements*, X-1). *If, from two unequal magnitudes, there is
subtracted from the greater a quantity more than its half, and from the remainder
a quantity more than its half, and if the process is continued, there will eventually
be a remainder less than the smaller of the two original magnitudes.*

Proof. Suppose AB is greater than C. C can be multiplied to exceed AB. Let
C be multiplied to DE, greater than AB, and divide C into DF, FG, and so on,
all equal to C. Let AB be divided at H so that BH is greater than half of AB;
let AH be divided at K so KH is greater than half of AH, and so on until AB is
divided into parts equal in number to the parts of DE. Now, DE is greater than
AB, and if we subtract GE, which is less than half of DE, from DE, we obtain
DG; likewise, if we subtract HB, which is greater than half of AB, we obtain
AH; DG is thus greater than AH. Repeating, we have DF greater than AK.
But DF is equal to C. Hence, if we subtract from the larger, AB, a magnitude
greater than its half, and repeat, we eventually come to a magnitude, AK, which
is less than the smaller magnitude C.   □

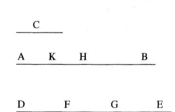

An obvious but unstated requirement for the proof is that the two magnitudes
have a ratio to one another. Although Euclid provided a proof, the proposition
is today generally considered an axiom, and called the **axiom of Archimedes**.

5.9.1   Circles

Euclid began Book XII by showing similar polygons inscribed in a circle are to
each other as the squares on the diameters; this requires only the geometry of
similar figures from Book VI, and not the method of exhaustion. Then Euclid
proved the following.

See Problem 2.

Proposition 5.55 (*Elements*, XII-2). *Circles are to each other as the squares on their diameters.*

This is, perhaps, the most elegant theorem in all of the *Elements*.

The key to the theorem is the use of the theory of proportions and a double application of a proof by contradiction to show that the ratio can neither be greater, nor less, than the claimed amount.

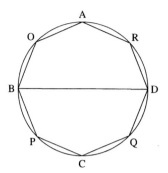

Proof. Given circles $ABCD$, $EFGH$, with diameters BD, FH respectively. We claim that the square on BD is to the square on FH as the circle $ABCD$ is to the circle $EFGH$.

Suppose not. Then the square on BD is to the square on FH as the circle $ABCD$ is to some area S, either greater or less than circle $EFGH$.

Suppose S is less than the circle $EFGH$. Consider now the two magnitudes, namely the amount that the area S is less than the circle, and the circle itself. Inscribe in $EFGH$ the square $EFGH$, which is greater than half the circle. Bisect the arcs EF, FG, GH, HE at K, L, M, N, and join to form the polygon $EKFLGMHN$. The triangles EKF, FLG, GMH, HNE are greater than half the remaining portion between the square and the circle. Continuing this process, we will eventually obtain a polygon such that the difference between the polygon and the circle is less than the difference between S and the circle. Call this polygon $EKFLGMHN$; this polygon exceeds S.

Inscribe in $ABCD$ polygon $AOBPCQDR$ similar to $EKFLGMHN$. By assumption, circle $ABCD$ is to S as the square on BD is to the square on FH. But similar inscribed polygons are to each other on the squares of the diameters, so the square on BD is to the square on FH as $AOBPCQDR$ is to $EKFLGMHN$. Hence the circle is to S as $AOBPCQDR$ is to $EKFLGMHN$. Alternating the ratio, we have circle $ABCD$ is to polygon $AOBPCQDR$ as S is to $EKFLGMHN$. Since the circle $ABCD$ exceeds the polygon $AOBPCQDR$, the area S must likewise exceed that of polygon $EKFLGMHN$. But it is also less, which is impossible. Thus, the square on BD cannot be to the square on FH as the circle $ABCD$ is to some area S less than the circle $EFGH$; in general, the square on the diameter cannot be to the square on the diameter of another circle in a ratio of the one circle to some area less than the other circle.

Similarly, it can be shown that the circle $EFGH$ to any area less than the circle $ABCD$ is not as the square on FH is to the square on BD.

Suppose next that the square on BD is to the square on FH as the circle $ABCD$ is to an area T greater than the circle $EFGH$. Therefore, the square on FH is to the square on BD as the area T is to the circle $ABCD$. But as the area T is to the circle $ABCD$, so the circle $EFGH$ is to some area smaller than the circle $ABCD$. Therefore as the square on FH is to the square on BD, so is the circle $EFGH$ to some area less than the circle $ABCD$.

But, by the first half of the proposition, the circle $EFGH$ to any area less than the circle $ABCD$ cannot be as the square on FH to the square on BD. Hence, it cannot be that the square on BD is to the square on FH as the circle $ABCD$ is to an area T greater than the circle $EFGH$. Since it cannot be to an area T greater than $EFGH$, nor an area S less than $EFGH$, it must be equal to the circle $EFGH$, and hence circles are to each other as the squares on their diameters.   □

5.9.2 Solid Figures

The next few propositions involve pyramids. As this involves the consideration of three dimensional figures, a few definitions from Book XI, where Euclid introduces solid geometry, are necessary. In the following, it is important to realize that when Euclid speaks of "planes," he means a finite plane figure.

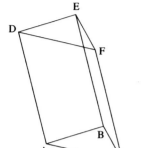

Definition. *Similar solid figures are those contained by similar planes, that are equal in number.*

Definition. *Equal solid figures are those contained by similar planes, that are equal in number and magnitude.*

Definition. *A **pyramid** is a solid figure, contained by planes, from one point to a plane.*

Definition. *A **prism** is a solid figure that is contained by two equal parallel planes.*

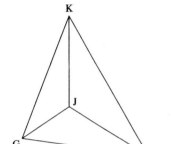

We shall refer to prisms by naming the equal parallel figures. Thus, the prism between the two triangular bases ABC, DEF will be spoken of as the prism ABC, DEF. Pyramids will be named by referring to their base and their vertex; thus the pyramid on base GHJ with vertex at K will be the prism GHJ, K.

With this in mind, Euclid's next proposition, after proving that circles are to each other as the squares on their diameter, is:

Proposition 5.56 (*Elements*, XII-3). *Any pyramid on a triangular base can be divided into two pyramids and two prisms, with the pyramids equal and similar to one another and the whole pyramid, and the two prisms greater than half the whole pyramid.*

What is the importance of this proposition? Recall that the use of the method of exhaustion requires that a quantity "greater than half" be subtracted from the larger of two given quantities. The foregoing shows that by subtracting the two prisms from the pyramid, one has subtracted from the pyramid a quantity greater than its half. This is necessary for:

Proposition 5.57 (*Elements*, XII-5). *Pyramids on triangular bases and with the same height are to each other as their bases.*

Proof. Given pyramids ABC, G and DEF, H of the same height; suppose they are not to each other as their bases, but instead ABC is to DEF as ABC, G is to K, either greater or less than DEF, H (see Figure 5.16). Suppose K is

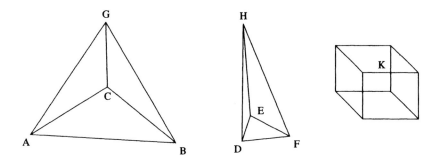

Figure 5.16: Proof of Proposition XII-5.

less than DEF, H. Divide DEF, H into two pyramids equal and similar to it, and two prisms greater than its half. Take these pyramids, and divide them as well, continuing until the pyramids are together less than the difference between DEF, H and K.

Let ABC, G be divided a similar number of times into similar figures, so as ABC is to DEF, the prisms in ABC, G are to the prisms in DEF, H. Since ABC is to DEF as ABC, G is to K, then ABC, G is to K as the prisms in ABC, G are to the prisms in DEF, H or, alternately, ABC, G is to the prisms in ABC, G as K is to the prisms in DEF, H. But ABC, G is greater than the prisms in ABC, G, so K must likewise be greater than the prisms in DEF, H. But K was assumed less than the prisms in DEF, H. Therefore, ABC to DEF cannot be as ABC, G to any volume less than DEF, H.

Likewise, K cannot be greater than ABC, G. □

See Problem 5.

The immediate corollary, which Euclid felt important enough to warrant its own proposition, is the following.

Proposition 5.58 (*Elements*, XII-6). *Pyramids of the same height with polygonal bases are to each other as their bases.*

An important fact about the proofs of XII-5 and XII-6 is that they required the use of the method of exhaustion, which mars the mathematical beauty of the proof. We might make the following analogy: I-35 showed that the equality of two parallelograms could be proven using only decomposition of figures and, by extension, any two rectilineal figures could be shown equal, greater, or less using decomposition. It is, of course, impossible to do this if one of the figures is curvilineal (except in the case of identical figures): thus, it is reasonable for Euclid to have used the method of exhaustion to compare two circles.

But XII-5 concerns the comparison of two pyramids, which are rectilineal figures. Conceptually, there is no reason why decomposition cannot be used in this case; however, Euclid was unable to find a proof using only decomposition. One might ask whether it is possible to prove the equality of two pyramids, using only decomposition? Over two thousand years would pass before this question was answered.

The volume formula for a pyramid comes from the following.

Proposition 5.59 (*Elements*, XII-7). *Any prism with a triangular base can be divided into three equal pyramids.*

Unlike the previous pair of propositions, Euclid draws the corollary as an immediate porism; namely, that pyramids on polygonal bases are one third the prisms on the same base; from this, our volume formula of a pyramid can be derived. Euclid also proved that the cone was one-third the cylinder (XII-10); the volumes of cones and cylinders on the same base were to each other as their heights (XII-14); and spheres were to each other in triplicate ratio of their diameters (XII-18).

5.9 Exercises

1. Explain, by means of an example, why Euclid requires in X-1 that the amount being subtracted is than half the whole.

2. Prove XII-1: similar inscribed polygons are to each other as the squares on the diameters.

3. Consider XII-2.

   (a) Show that the triangles EKF, FLE, GMH, HNE are greater than half the remaining portion between the polygon and the circle.

   (b) Explain why XII-2 suggests the existence of the numerical constant π.

   (c) How does XII-2 translate into the area formula for a circle?

   (d) How does Euclid's use of the Method of Exhaustion in Book XII foreshadow the development of calculus?

4. Prove XII-3, using the following steps. Given pyramid ABC, D; bisect AB, BC, CA, AD, DB, DC at E, F, G, H, L, K.

   (a) Show that triangles AEH, AHG, AGE, GHE are equal and similar to triangles HKD, HDL, HLK, LDK respectively, and thus pyramid AGE, H is equal to the pyramid with HLK, D.

   (b) Show that triangles HKD, HDL, HLK, LDK are similar to triangles ABD, ADC, ACB, CDB respectively, and thus the pyramid with HLK, D is similar to the pyramid ACB, D.

   (c) Show the prism BKF, EHG is equal to the prism GFC, HKL.

   (d) Finally, show that the prisms BKG, EHG and GFC, HKL are together greater than half the pyramid ABC, D.

5. The proof of XII-5 requires XII-4, "Given two pyramids of equal height on triangular bases, if each is divided into similar and equal pyramids and

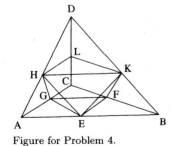

Figure for Problem 4.

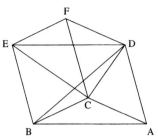

prisms greater than half the whole, then as the base of one pyramid is to the base of the other, so will all the prisms in one pyramid be to all the prisms in the other pyramid." Prove this statement.

6. Complete the proof of XII-5 by the method of exhaustion.

7. Prove XII-7 by first joining DB, EC, CD, then showing that the pyramids that result are equal.

Figure for Problem 7.

8. Prove the corollary to XII-7: any prism is three times the pyramid with the same (polygonal) base and the same height.

9. What hidden assumption is Euclid making in his definition of equal solid figures? Explain by providing a counterexample of two figures, "equal" under Euclid's definition in Book XII, but not equal in the sense of Book I (i.e., they do not coincide).

5.10 Book XIII

The thirteenth and final book of Euclid's *Elements* concerns itself with the construction of the five regular solids. As we have already noted, Theaetetus was probably the first to write a systematic treatment of the five regular solids, although his treatment has been lost.

See page 75.

Two of the propositions in Book XIII are important for their later use for the table of chords developed by Ptolemy. These are:

Proposition 5.60 (*Elements*, XIII-9). *If the side of a regular hexagon and a regular decagon inscribed in the same circle are combined to form a line, the line will be cut into extreme and mean ratios, with the larger segment the side of the hexagon.*

Proof. In circle ACB with center E, let BC, CD be equal to the sides of a regular decagon and a regular hexagon inscribed in the circle, and let them be joined in a straight line. From the center E, join EB, EC, ED, and carry BE to A. Then triangle EBD is similar to triangle CBE (See Problem 1). Hence DB is to BE as EB is to BC. Since BE, CD are equal, and EB, CD are also equal, then DB is to CD as CD is to BC, and thus the straight line DB has been cut into extreme and mean ratios at C. Moreover, since DB is greater than DC, then DC must be greater than CB, and hence the side of the hexagon is the longer segment. □

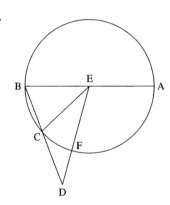

Immediately following is:

Proposition 5.61 (*Elements*, XIII-10). *If a regular pentagon is inscribed in a circle, the square on the side of the pentagon is equal to the square on the side of the regular hexagon inscribed within the same circle, together with the square on the side of the regular decagon, also inscribed in the same circle.*

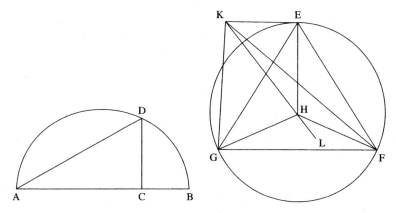

Figure 5.17: Proof of Proposition XIII-13.

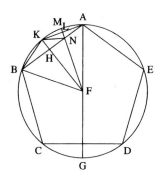

Proof. Given circle $ABCDE$ with regular pentagon $ABCDE$ inscribed. From A through center F draw diameter AG. Join FB, draw FH perpendicular to AB; extend FH to K. Join AK, KB. Draw FL perpendicular to AK; extend FL to M. Then triangle AFB is similar to triangle FNB, and thus AB is to BF as FB is to BN, and so rectangle AB, BN is equal to square BF. Also, triangle KBA is similar to triangle NKA, and thus BA is to AK as KA is to AN, and thus rectangle BA, NA is equal to square KA. Hence square AB is the sum of the squares on BF and on KA, and thus the square on the side of the regular pentagon is equal to the square on the side of the regular hexagon, together with the side on the square of the regular decagon.   □

After establishing a few key theorems, Euclid then showed how all five regular solids may be constructed. Euclid's procedure for constructing a regular solid whose faces are equilateral triangles is:

Proposition 5.62 (*Elements*, XIII-13). *To inscribe a regular tetrahedron in a sphere, and prove the square on the diameter of the sphere is one and a half times the square on the side of the tetrahedron.*

Proof. On semicircle ABD, cut C so that AC is twice CB (see Figure 5.17). Draw CD perpendicular to AB, and join DA. Draw circle EFG with center H and radius equal to DC; inscribe equilateral triangle EFG inside circle EFG. From H set HK equal to AC and perpendicular to the plane of circle EFG. Join KE, KF, KG.

The triangles EFG, KEF, KFG, KEG are all equal.

Extend KH to L, so that HL is equal to CB. Since AC is to CD as CD is to CB, and AC is equal to KH, and CD is equal to HE, then KH is to HE as HE is to HL. Since angles KHE, EHL are right, then the semicircle on KL will also pass through E. If KL remains fixed and the semicircle revolved around it, it will also pass through F, G as well, since KHF, FHL are also right angles, as are GHK, GHL, and the pyramid will thus be contained within the sphere with diameter KL.

Finally, the square on the diameter of the sphere is one and a half times the square on the side of the pyramid. □ Complete the proof: problem 4.

5.10 Exercises

1. Fill in the missing steps of the Euclidean proof of XIII-9 by showing triangle EBD is similar to triangle CBE.

2. Complete the missing steps in XIII-10, by showing:

   (a) AK is the side of a regular decagon, and AK is twice KM.

   (b) Triangle AFB is similar to triangle FNB. Triangle KBA is similar to triangle NKA.

   (c) The square on AB is equal to the square on BF, together with the square on KA.

3. Prove XIII-12: The square of the side of an equilateral triangle inscribed in a circle is three times the square on the radius.

4. Complete the proof of XIII-13.

   (a) Prove KE, KF, KG are equal to each other.

   (b) Prove DA is equal to KE, and consequently DA, KE, KF, KG are all equal.

   (c) Prove the square on AD is triple the square on DC. (This Euclid proves as a lemma following the proposition), and thus AD is equal to EF, and thus all six edges of the polyhedron are equal.

   (d) Prove that the triangles EFG, KEF, KFG, KEG are all equal, and thus all four faces of the polyhedron are equal, and equilateral triangles; hence the pyramid is a regular tetrahedron.

5. Prove XIII-14: To inscribe a regular octahedron in a sphere, and prove the square on the diameter of the sphere is twice the square on the side of the octahedron.

Chapter 6

Archimedes and Apollonius

Shortly after Euclid wrote the *Elements*, the Roman Republic began to expand southward. In 282 B.C., the Romans reached Magna Graecia, in southern Italy, and the city-state of Tarentum called on Pyrrhus of Epirus to save them from the Roman onslaught. In 279 B.C., Pyrrhus met the Romans at Ausculum, driving them away, but at such a cost to Pyrrhus's own forces that he was reputed to have said, "Another such victory and I am lost": it was from this battle we get the phrase "pyrrhic victory." In 275 B.C., Pyrrhus was forced to leave Italy. Of Sicily, he said, "What a battleground I am leaving for the Romans and Carthaginians."

Carthage was a city-state on the coast of North Africa (in modern day Tunisia). The First Punic War (so called because Carthage was a Phoenician city, hence "Punic") between Rome and Carthage occurred between 264 and 241 B.C., and Rome gained control of the island of Sicily, which became its first Province. Rome ruled the western half of Sicily directly and gave the eastern half to one of Rome's allies, Hieron II, king of Syracuse. Carthage, deprived of one Mediterranean empire by the Romans, built a new one in Spain. Her success would provoke a jealous Rome into waging the Second Punic War.

6.1 Circles

If one were to make a list of the greatest mathematicians of all time, near or perhaps at the top of the list would have to be the figure of Archimedes (287-212 B.C.). While some of the achievements attributed to him are of a legendary nature, there is much in his surviving work to credit him with some of the most important discoveries in mathematics and the applications of mathematics. Archimedes, a relative of King Hieron II, was born in Syracuse. He studied in Alexandria, making many acquaintances to whom he addressed much of his work, and eventually returned to Syracuse. Archimedes achieved fame as a scientist and an engineer. As a scientist, he is known as the discoverer of the principles of hydrostatics, and in a particularly famous experiment, he discovered how to measure the volume of an irregular object by measuring the volume of water the submerged object displaces.

6.1.1   The Second Punic War

Archimedes's life was caught up in the events of the Second Punic War. Rome had exacted from Carthage a promise not to cross the Ebro River, and not to attack the city of Saguntum (a Greek colony in Spain). The Sagantines were aware of Carthage's promise, so they repeatedly attacked Carthaginian territory, believing themselves to be untouchable. This proved wrong, and in 219 B.C., Hannibal led a punitive force to attack and destroy Saguntum. This began the Second Punic War.

Hannibal realized that the strength of Rome was her allies, many of whom had been recently conquered by Rome, and who might reasonably be expected to bear some resentment toward their conquerors. Thus in 218 B.C. Hannibal entered Italy (the exact point of crossing is still a matter of debate) with 50 elephants and 26,000 men, intent on destroying the Roman confederation by convincing Rome's allies to abandon her. He did so by being merciless to Roman soldiers and Roman property, while at the same time sparing the men and property of the allies; he hoped to make clear his quarrel was with Rome alone.

The strategy began to work, and Rome's allies began to waver. The ranks of Hannibal's army swelled with new recruits. But Hannibal needed a great military victory if he wanted to break the Roman confederation, for the allies would not completely desert Rome until it became clear that Rome was losing the war. Thus, in 218 B.C., Hannibal met a Roman army of 85,000 men at Cannae. Hannibal had only 50,000 men with him, but used them masterfully; by day's end, 50,000 Roman soldiers were dead, against 2000 easily replaceable mercenaries on Hannibal's side. To the present day, no army in history has suffered more casualties in a single day than the Romans at Cannae.

Cannae was the victory Hannibal needed, and Rome's allies began to desert her. In Syracuse, Hiero II felt honor-bound to uphold the alliance, but he died in 215 B.C. and his successor and grandson, Hieronymus, abandoned the alliance and joined with the Carthaginians. However, Hannibal was facing problems at home: the rulers of Carthage feared what a popular general might do when he got home, so they withheld money and support. Moreover, the core allies of the Roman Republic proved more attached to Rome than Hannibal imagined, and Rome rebounded. Soon the Romans gained the upper hand and conquered Carthage itself in 201 B.C.

In the meantime, the Romans turned their attention to their rebellious allies. In 213 B.C., Roman legions under Marcellus besieged Syracuse. The siege of Syracuse took an unusually long time for a Roman siege, and in part it was due to Archimedes. To help defend his native city, Archimedes was reputed to have developed fantastic war machines, such as catapults that could hurl enormous stones to sink Roman ships, or a giant mirror that could set them on fire. The Roman soldiers grew to be so terrified of Archimedes's machines that every time a rope descended from the walls of the city, it is said that they fled in terror, thinking it to be a precursor to some more fearsome attack.

Eventually, the Romans stormed the city. Recognizing the value of the mind of Archimedes, Marcellus gave strict orders that he was not to be harmed. But in the confusion of the battle, Archimedes was killed. The exact details of his death are unknown, and Plutarch, in his life of Marcellus, gives us three versions of Archimedes's death. Plutarch says that Archimedes was, during the siege of Syracuse, working on some problem, and did not even notice the city had been taken. In one version, Archimedes refused to go along with the Roman soldiers until he finished the problem. The soldiers, having no patience, killed him.

This seems rather unlikely, given Marcellus's order. Plutarch's third version of Archimedes's death seems the most probable: Archimedes was en route to meet Marcellus, and carried with him various astronomical and mathematical equipment. Some soldiers mistook him for a rich citizen who might have something worth stealing, so they killed him. Marcellus "regarded him that killed him as a murderer," which probably means the soldier was executed. Marcellus also did his best to treat Archimedes's surviving relatives with great honor and kindness.

6.1.2 *Measurement of a Circle*

Almost all of what we consider to be the geometry of circles, spheres, and cones is due to Archimedes. His *On the Measurement of the Circle* gives the formula for the area of a circle, as well as the approximation $\frac{22}{7}$ for the ratio of the circumference of a circle to its diameter. The work as we have it appears to be a much later copy of the Greek original, and might only be a fragment of a larger work. The first proposition gives the area formula for a circle.

Proposition 6.1 (Area of a circle). *The area of a circle is equal to that of a right triangle with sides equal to the circumference and radius of the circle.*

To prove this, Archimedes used inscribed and circumscribed circles, as Bryson and Antiphon did. First, Archimedes assumed the areas were not equal, and that the triangle was the lesser area.

See page 75.

Proof. Suppose $ABCD$ is the given circle, and K a right-angled triangle with one side equal to the circumference and the other side equal to the radius of the circle, and suppose that the area of circle $ABCD$ is not equal to the area of K (see Figure 6.1). Then it is either greater or less.

Suppose circle $ABCD$ is greater than K. Inscribe in $ABCD$ the square $ABCD$; bisect its sides, etc., until a regular polygon $AEBFCGDH$ is found that is greater than K. Let AE be any side of the polygon, and ON the perpendicular between that side and the center of the circle at O. ON is less than the radius, and hence ON is less than the side of K that is equal to the radius. The perimeter of the polygon is less than the perimeter of the circle, and hence the perimeter

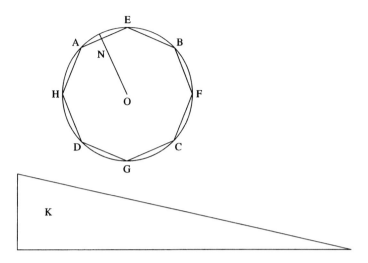

Figure 6.1: The Area of a Circle

of the polygon is less than the side of K equal to the perimeter of the circle. Hence the area of the polygon is less than K. But it was also greater, which is impossible. ☐

Archimedes then used circumscribed polygons to show that K cannot be greater than the circle.

See Problem 1.

The unreliability of early Greek sources is evidenced by the next proposition, which appears in the commentary by Eutocius (500 AD):

Proposition 6.2 (Flawed). *A circle is to the square on its diameter as* 11 *to* 14.

The text of the second proposition is unclear; in any case, it is flawed and is likely the interpolation of a later copyist.

Archimedes was the first person to find a value of π in its modern sense as the ratio of the circumference of a circle to its diameter:

Proposition 6.3. *The ratio of the circumference of a circle to its diameter is less than* $3\frac{1}{7}$ *but greater than* $3\frac{10}{71}$.

To find this value, he estimated the perimeter of a 96-sided regular polygon inscribed and circumscribed about a circle.

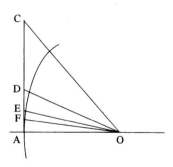

Proof. Let AO be the radius of a circle, and AOC one third of a right angle.

Draw AC tangent to the circle at A. Then:

AO to AC is greater than 265 to 153.

CO to AC is equal to 2 to 1, or 306 to 153.

Draw OD bisecting angle AOC.

CO to OA is equal to CD to DA [by *Elements*, VI-3].

Thus CO and OA is to OA as CD and DA is to DA, or CA to DA. Alternating, we have CO and OA is to CA as OA is to DA.

Thus, OA is to DA in a ratio greater than $306 + 153 = 571$ to 153. So OA^2 is to DA^2 in a ratio greater than $571^2 : 153^2$. So OA^2 and DA^2 is to DA^2 in ratio greater than $571^2 + 153^2$ to 153^2; hence OD^2 is to DA^2 in a ratio greater than $349{,}450$ to $23{,}409$.

Hence OD is to DA in a ratio greater than $591\frac{1}{8}$ to 153. $\qquad\square$

After this, Archimedes let OE bisect the angle AOD at E; then let OF bisect AOE at F, and OG bisect AOF at G, and obtained the ratios:

Proof, continued. OE to EA greater than $1172\frac{1}{8}$ to 153.
$\quad$ OF to FA greater than $2339\frac{1}{4}$ to 153.
$\quad$ OG to GA greater than $4673\frac{1}{2}$ to 153. $\qquad\square$

Since the original angle, AOC, was one third of a right angle, and was bisected four times, the angle AOG is $\frac{1}{48}$th of a right angle, and hence GA is half the side of a regular 96-sided polygon circumscribed about the circle. Archimedes concluded the perimeter of the circumscribed 96-sided regular polygon is to the diameter AB in a ratio less than $4673\frac{1}{2}$ to 14688, which is in turn less than the ratio of $3\frac{1}{7}$ to 1, and thus the circumference of the circle is to the diameter in a ratio less than $3\frac{1}{7}$.

The most remarkable feature of this proposition is the series of rational approximations to irrational numbers, which Archimedes introduced without comment; there has been much speculation on how Archimedes arrived at his results. One likely candidate is through consideration of side and diameter numbers, suggesting that the side and diameter numbers were not limited to the approximation to $\sqrt{2}$, but were in fact used routinely in Greek computational mathematics.

See Problem 4.

Next, to find the lower bound, Archimedes considered the polygon of 96 sides inscribed in the circle.

Proof, continued. Let AB be the diameter, and draw AC so that CAB is one third of a right angle (see Figure 6.2). Then AC is to CB in a ratio less than 1351 to 780.

Draw AD so that it bisects angle CAB, and meets CB at d. The angles at D, C are right angles, and angle BAD is equal to angle dAC and angle dBD. Hence triangle ADB is similar to triangle BDd, and to triangle ACd. Hence AD is to DB as BD is to Dd, and as AC is to Cd.

But by *Elements* VI-3, we have AC is to Cd as AB is to Bd.

Hence AD is to DB as AC is to Cd, and as AB is to Bd. Summing the antecedents and consequents, we have AD is to DB as AC and AB is to Cd and Bd, or AD is to DB as AB and AC is to BC. Since $AC : BC < 1351 : 780$, $AB : BC = 1560 : 780$, so AB and $AC : BC < 2911 : 780$, so $AD : DB$ is less than $2911 : 780$.

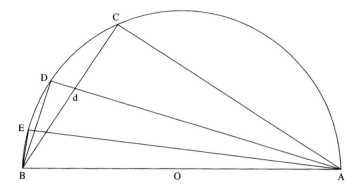

Figure 6.2: Approximating the circumference of a circle.

Thus, AD^2 is to DB^2 in a ratio less than 2911^2 to 780^2, AD^2 and DB^2 is to DB^2 in a ratio less than $2911^2 + 780^2$ to 780^2, so AB^2 is to DB^2 in a ratio less than 9082321 to 608400. Thus, $AB : DB < 3013\frac{3}{4} : 780$.

$\square$

Again, bisecting the angle DAB by AE, then EAB by AF, then FAB by AG, and repeating the procedure, Archimedes obtained the ratios:

AB is to BE in ratio less than $1838\frac{9}{11}$ to 240.

AB is to BF in ratio less than $1009\frac{1}{6}$ to 66.

AB is to BG in ratio less than $2017\frac{1}{4}$ to 66.

Since BAG is the fourth bisection of the angle CAB, which was one third of a right angle, then BG will be the side of a regular inscribed polygon of 96 sides. Hence the perimeter of the inscribed 96-sided regular polygon is to the diameter AB in a ratio greater than 6336 to $2017\frac{1}{4}$, which is greater than $3\frac{10}{71}$ to 1, or less than $3\frac{1}{7}$ but greater than $3\frac{10}{71}$.

6.1 Exercises

1. Refer to the Archimedean proof of Proposition 6.1.

    (a) How does Archimedes use of "equal" compare with Euclid's use of "equal"?

    (b) In order to use the method of exhaustion, it is necessary that the segments subtracted be greater than half the difference between the two quantities. In the case of the inscribed circles, it is therefore necessary that the triangles AEB, BFC, CGD, DHA are greater than half the difference between the square $ABCD$ and the area K. Complete the first part of the proof by showing this is true.

    (c) To complete the proof, Archimedes used circumscribed polygons, and showed that doubling the number of sides also satisfies the requirements of the method of exhaustion; namely that the subtracted area is greater than half the distance. Begin with the circumscribed square, touching the circle at points A, B, C, D, then bisect each of the arcs AB, BC, CD, DA at E, F, G, H, and show that the triangles formed are greater than half the difference between K and the circle, and thus the method of exhaustion can be applied.

(d) Complete the proof.

2. In the proof of Proposition 6.3, show $265 : 153 < AO : AC$, and $AC : CB < 1351 : 780$.

3. Prove the following: the perimeters of circles are to each other as their diameters.

4. Archimedes arrived at two approximations to $\sqrt{3}$, without explanation. Refer to the side-to-side relationship for a rectangle whose diagonal is twice as large as the shorter side in Problem 7, page 52.

   (a) Let the two sides have a starting value of 1; use this relationship to arrive at the approximation $\sqrt{3} \approx \frac{265}{153}$.

   (b) Let the longer side have a starting value of 2; use this relationship to arrive at the approximation $\sqrt{3} \approx \frac{1351}{780}$.

   (c) To assign the direction of the inequalities to the two approximations, you could square both ratios. Find a simpler method.

6.2   Spheres, Cones, and Cylinders

Euclid, in the *Elements*, defined spheres, cones, and cylinders by rotation of a plane figure:

Definition. *If a semicircle is rotated about its diameter until returns to its starting position, the resulting figure is a* **sphere**, *and the diameter is the* **axis** *of the sphere.*

Definition. *If a right-angled triangle is rotated about one of the sides about the right angle until the it returns to its starting position, the resulting figure is a* **cone**, *and the fixed side is the* **axis** *of the cone. The cone is* **right-angled**, **obtuse-angled**, *or* **acute-angled** *depending on whether the fixed side is equal, lesser, or greater than the other side about the right angle.*

Definition. *If one side of a rectangle is fixed while the rectangle rotates about the line until it returns to its starting point, the resulting figure is a* **cylinder**, *and the fixed straight line is the* **axis** *of the cylinder.*

As with circles, most of what we consider the geometry of spheres, cylinders, and cones can be traced to Archimedes. *On the Sphere and Cylinder* was dedicated to his friend Dositheus. According to Archimedes's introduction, he had earlier sent to Dositheus his researches concerning the section of a right-angled cone, showing that a segment of it was equal to four thirds the triangle with the same base and equal height, a result we will discuss later. Since then, he had *discovered* other properties. Archimedes noted these properties were "inherent in the figures," an interesting early contribution to the philosophy of mathematics.

6.2.1 Circumscribed and Inscribed Figures

Recall that Euclid's use of the method of exhaustion relied only on inscribed figures, whereas the method of exhaustion originally developed by Bryson and Antiphon used inscribed and circumscribed figures. Why did Euclid restrict himself so? Part of the problem was that while there was no question that the perimeter of the circle was greater than that of the inscribed polygon, the same could not be said about a circumscribed polygon and the perimeter of the circle. Thus in *On the Measurement of a Circle*, Archimedes assumed that a circumscribed polygon necessarily had a greater perimeter than the circle.

See page 129.

To address this issue, Archimedes began *On the Sphere and Cylinder* with the idea of **concavity**:

Definition. *A line is **concave in the same direction** if all straight lines joining any two points on the line fall on the same side of the line, or some fall on the line and the rest on one side.*

Similarly, a surface could be concave in the same direction if all lines joining two points of the surface were on the same side of the surface.

Archimedes then gave a list of explicit axioms. The first two are:

Axiom. *Of all lines between two points, the straight line is the shortest.*

Axiom. *Given two lines between the same points and both concave in the same direction. If one is wholly contained by the other and the straight line between the same points, or partly included by and partly in common with the other, it is the shorter.*

These allowed Archimedes to prove two statements that had previously been assumed:

Proposition 6.4. *If a polygon is inscribed in a circle, the perimeter of the inscribed polygon is less than the circumference of the circle.*

Proposition 6.5. *If a polygon is circumscribed about a circle, the perimeter of the circumscribed polygon is greater than the circumference of the circle.*

Another important axiom is now called the **axiom of Archimedes**.

Axiom (Axiom of Archimedes). *Of unequal lines, surfaces, or solids, the greater exceeds the lesser by a quantity that, when added to itself, can be made to exceed any other quantity comparable to it.*

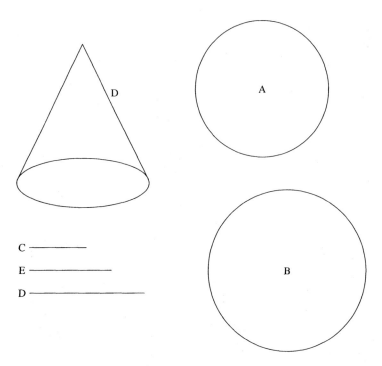

Figure 6.3: The area of a cone.

In other words, no matter how small the difference between two comparable quantities, so long as they are unequal, the difference may be made as large as desired.

6.2.2 Surface and Volume of a Sphere

The first key proposition for proving that the surface of a sphere is four times the greatest circle contained by it is:

Proposition 6.6 (Area of a Cone). *The surface of an isosceles cone, excluding the base, is equal to a circle whose radius is the geometric mean between the side of the cone and the radius of the circle.*

> Given a right circular cone of height h and base a circle of radius r, the surface area is $\pi r \sqrt{r^2 + h^2}$.

Proof. Let A be the base of the cone. Draw C equal to the radius of A, and let D be the side of the cone. Let E be the geometric mean proportional between C and D. Let B be a circle of radius E. We claim circle B is equal to the surface S of the cone, excluding the base (see Figure 6.3). Suppose they are not, and let B be less than S. Circumscribe a polygon about B and inscribe a similar polygon such that the area of the outer polygon is to the inner polygon in a ratio less than that of S to B. Describe around A a similar polygon, and set up a pyramid with the same apex as the cone. Then:

Polygon around A is to the polygon around B as the square on C is to the square on E.

Hence polygon around A is to the polygon around B as C is to D.

Polygon around A is to the polygon around B as the polygon around A is to the surface of the pyramid without its base.

Thus, the surface of the pyramid without its base is equal to the polygon around B.

But the polygon around B is to the polygon inside B in a ratio less than that of S to B. Thus, the pyramid is to the polygon inside B in ratio less than S to B. But the pyramid is greater than S, and the polygon in B is less than B, which is impossible. Thus, the circle B cannot be less than the surface S of the cone. Likewise, the circle B cannot be greater than the surface S. □

This is an important step in determining the area of a frustrum of a cone (less its bases):

Proposition 6.7 (Area of a Frustrum). *If an isosceles cone is cut by a plane parallel to its base, the surface of the frustrum is equal to a circle whose radius is a geometric mean between the line on the portion of the cone between the parallel planes, and the line equal to the sum of the radii of the circles forming the two bases on the frustrum.*

The next important proposition is:

Proposition 6.8. *Let a regular polygon with an even number of sides be inscribed in a circle with diameter AA'; and let BB' and other lines parallel to it, such as CC', DD', ..., be drawn joining the vertices of the polygon (see Figure 6.4). Then $(BB' + CC' + DD' + ...) : AA' = A'B : BA$.*

Proof. Let BB', CC', DD', ... meet AA' at points F, G, H ...; let CB', DC', ... be drawn, meeting AA' at points K,L, ... CB', DC', ... are all parallel to each other and to AB. Hence:

$$A'B : BA = BF : FA$$
$$= B'F : FK$$
$$= CG : GK$$
$$= C'G : GL$$
$$\vdots$$

Summing the antecedents and consequents we have $A'B : BA = (BB' + CC' + ...) : AA'$. □

Next, Archimedes proved a rather complex proposition regarding the surface area of a solid of revolution.

Proposition 6.9. *Let a regular polygon $AB...A'B'$ with a number of sides equal to a multiple of 4 be inscribed in the great circle of a sphere, and let BB' subtending two sides be joined. Draw all other lines parallel to BB' and joining*

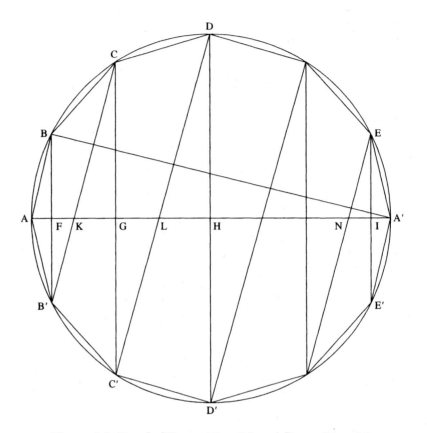

Figure 6.4: Proof of Proposition 6.8 and Proposition 6.9.

pairs of vertices (see Figure 6.4). If the figure is revolved about the diameter AA', the surface of the figure is equal to a circle, the square of whose radius is equal to rect. $BA, BB' + CC' + \ldots + HH'$.

The next proposition is key, for it is the proposition that allows Archimedes to invoke the method of exhaustion.

Proposition 6.10. *The surface of the previous figure is less than four times the great circle in the sphere.*

The **great circle** in a sphere is one that shares the same center as the sphere.

Proof. Let R be a circle such that the square on the radius is equal to rect. BA, $BB' + CC' + \ldots + HH'$. Now $BB' + CC' + \ldots$ is to AA' as $A'B$ is to AB, thus rect. $AB, BB' + CC' + \ldots + HH'$ is equal to rect. $AA', A'B$, and thus the square on the radius of R is equal to rect. $AA', A'B$. This is less than the square on AA'. Therefore the surface of the inscribed figure is less than four times the great circle in the sphere. $\qquad\square$

Finally, we arrive at the proposition relating the surface of a sphere to the greatest circle contained by it.

Proposition 6.11 (Area of a Sphere). *The surface of a sphere is equal to four times the greatest circle contained in it.*

Proof. Let C be a circle equal to four times the greatest circle in the sphere. If C is not equal to the surface of the sphere, it is either greater or less.

Suppose C is less than the surface of the sphere. Find two lines, β, γ, where β is the greater, so that $\beta : \gamma$ is less than the sphere to C. Find the mean proportional δ, between β and γ. Inscribe similar polygons with a number of sides equal to a multiple of four inside and outside the circle so that the ratio between the exterior and interior sides is less than $\beta : \delta$. Let the polygons and the circle revolve around the diameter, describing a solid of revolution as before. Then

$$\begin{pmatrix} \text{surface area} \\ \text{outer solid} \end{pmatrix} : \begin{pmatrix} \text{surface area} \\ \text{inner solid} \end{pmatrix} = \begin{pmatrix} \text{sq. on} \\ \text{outer side} \end{pmatrix} : \begin{pmatrix} \text{sq. on} \\ \text{inner side} \end{pmatrix}$$

$$< \quad \text{sq. } \beta : \text{sq. } \delta$$

$$= \quad \beta : \gamma$$

$$< \quad \text{sphere} : C$$

But this is impossible, since by the previous, the surface area of the inner solid is less than C, while the surface area of the outer solid is greater than the sphere. Thus, the C cannot be less than the surface of the sphere.

Likewise, C cannot be greater than the surface of the sphere. $\square$

Archimedes also determined the volume of a sphere.

Proposition 6.12 (Volume of a Sphere). *Any sphere is four times the cone whose base is equal to the greatest circle in the sphere and whose height is equal to the radius of the sphere.*

Of all Archimedes's results, it would seem that he was proudest of this one and its corollary, sometimes called the Theorem of Archimedes:

Corollary (Theorem of Archimedes). *Every cylinder whose base is the greatest circle in a sphere and whose height is the diameter of the sphere is $\frac{3}{2}$ the sphere, and its surface together with its bases is $\frac{3}{2}$ the surface of the sphere.*

He requested that a diagram, showing the cylinder and sphere, be part of his gravestone. Apparently his request was granted (possibly by Marcellus). When the Roman orator Cicero (106-43 B.C.) was *quaestor* (auditor) in Sicily, he found the tomb of Archimedes, much neglected, and ordered it to be restored; this was done, but since the time of Cicero, the tomb has vanished.

6.2 Exercises

1. Consider Proposition 6.5.

   (a) Prove it, using Archimedes's second axiom.

   (b) Archimedes's result is true for all polygons, but a proof that a circumscribed regular polygon has a greater perimeter than the circle is possible, using only Euclidean geometry and the method of exhaustion. Prove this. Hint: given circle O, and a circumscribed regular polygon $ABCD$, suppose the perimeter of $ABCD$ is not greater than the circle. Inscribe a similar polygon in O. Show that it is possible to create an inscribed polygon with perimeter greater than a similar circumscribed polygon, which is impossible.

2. Consider Proposition 6.6.

   (a) Show how it is equivalent to the modern formula for the surface area of a cone.

   (b) Explain why the square on C is to the square on E as C is to D.

   (c) Prove that C is to D as the polygon around A is to the pyramid without the base.

3. In the proof of Proposition 6.10, Archimedes stated, without proof, that since the square on the radius of R is less than the square on the diameter AA', the surface of the inscribed figure is less than four times the great circle. Prove this statement.

4. Consider Proposition 6.11.

   (a) Complete the proof, showing that C cannot be greater than the surface of the sphere.

   (b) Archimedes used (and proved) several propositions we did not state. Determine what propositions they are, and provide a proof.

5. How does Proposition 6.12 translate into the modern volume formula for a sphere?

6. Prove the corollary. Do not use the modern volume formulas.

6.3   Quadratures

The general problem of squaring a figure came to be known as the **quadrature problem**, from the Latin word *quadratus*, square; occasionally, one speaks of the **cubature problem**, corresponding to determining the volume of a given figure, and in general, a figure is said to be **rectifiable** if quadrature or cubature (whichever is appropriate) is possible; the Greek geometers almost never considered the problem of measuring the length of a curve. In Euclidean geometry, quadrature of rectilineal figures is straightforward, and using other curves like the quadratrix it was possible to square the circle. However, this exhausted the list of rectifiable figures known prior to Archimedes.

6.3.1 *Quadrature of a Parabola*

Thus, Archimedes was the first geometer in history to exactly determine areas and volumes of figures not formed by circles and straight lines; nowhere is his method more clearly illustrated than in *Quadrature of the Parabola*. *Quadrature* is a model of the use of the Greek method of exhaustion, showing both its power and its limits.

Recall that conic sections were known as far back as the time of Menaechmus. Euclid and others wrote comprehensive treatises on the subject. Unfortunately, as Euclid's *Elements* swept away earlier texts in plane geometry, Apollonius's *Conics* would sweep away earlier texts on the conic sections, and *Quadrature* is one of the few places where we may find a pre-Apollonian treatment of conics. See page 78.

Archimedes's terminology is different from ours: indeed, he still calls the parabola by its ancient name, the section of a right-angled cone. Whereas we are used to parabolas having a single vertex and a single diameter, in the work of Archimedes and Apollonius we find these terms are relative to a given situation. In particular, the **diameter** of a parabola was a line parallel to the axis, and the **vertex** was the point of intersection of a diameter with the parabola itself. The **segment of a parabola** was the region between the parabola and a chord intersecting the parabola in two places.

The first three propositions of *Quadrature* were, according to Archimedes, already proven in the *Elements of Conics* (probably the treatise by Euclid). These propositions are:

Proposition 6.13. *Given any point P on a parabola, and straight line QQ' parallel to the tangent to the parabola at P, then the line PV parallel to the axis of the parabola will bisect QQ'. Conversely, if a chord QQ' is bisected at V by line PV parallel to the diameter, then QQ' is parallel to the tangent to the parabola at P.*

Proposition 6.14. *Given any point P and any chord QQ' parallel to the tangent to the parabola at P. Draw PV parallel to the axis, intersecting QQ' at V, and let the tangent at Q be QT, intersecting PV at T. Then PV is equal to PT.*

Proposition 6.15. *Given any point P and any line PV parallel to the axis, and through any other points Q, R, draw lines parallel to the tangent to the parabola at P and intersecting PV at V, S. Then $PV : PS = QV^2 : RS^2$.*

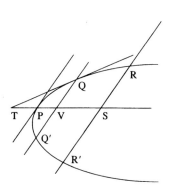

Lines QV and RS were the **ordinates**; note that ordinates are not necessarily perpendicular to the diameter, but instead cross it at a fixed angle.

Propositions 4 through 17 consist of showing how a parabolic segment could be "weighed" and, using the principles of mechanics, the quadrature may be

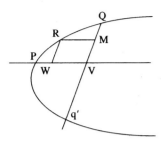

effected. Since this first proof involved the use of a mechanical principle, namely that of the lever, Archimedes also included a proof based on sound, geometric principles.

The key propositions in the geometrical quadrature are:

Proposition 6.16. *Let chord Qq of a parabola be bisected at V by diameter PV; and let RM be a diameter bisecting QV at M, and let RW be the ordinate from R to PV. Then PV is $\frac{4}{3}RM$.*

Proof. By the property of a parabola, $PV : PW = QV^2 : RW^2$; since $QV = 2MV = 2RW$, then

$$PV : PW = QV^2 : RW^2 = 4RW^2 : RW^2$$

Hence $PV = 4PW$, and since $WV = RM$, then $PV = \frac{4}{3}RM$. □

Proposition 6.17. *If Qq is the base of a parabolic segment with vertex P, then the triangle PQq is greater than half the segment.*

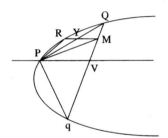

The **vertex of a segment** was the vertex corresponding to the diameter that bisected the chord.

Proposition 6.18. *Let chord Qq be the base of a parabolic segment with vertex P, and let R be the vertex of the parabolic segment with base PQ. Then triangle PQq is eight times triangle PRQ.*

Essential to the Archimedean quadrature method were a number of series summation formulas. To set up the geometric sum, Archimedes proved:

Proposition 6.19. *If there is a sequence of areas A, B, C, ..., each of which is four times the next, and if A, the largest, is equal to triangle PQq inscribed in a parabolic segment with base Qq and vertex P, with the same base and equal height, then $A + B + C + \ldots$ is less than the area of the segment PQq.*

To complete the quadrature, Archimedes then proved the geometric sum formula:

Proposition 6.20 (Sum of a Geometric Series). *If A, B, C, D, $\ldots Z$ is a sequence of areas, with A the largest and each four times the next, then*

$$A + B + C + \ldots + \frac{1}{3}Z = \frac{4}{3}A$$

Finally, Archimedes proved:

Proposition 6.21 (Area of a Parabolic Segment). *The area of a segment of a parabola is equal to $\frac{4}{3}$ the triangle with the same base and vertex.*

Proof. Given segment PQq, with vertex P and base Qq. Suppose K is 4/3 the triangle PQq. If the segment is not equal to K, then it is greater or less. Suppose it is greater. Then if we inscribe in the segments cut off by PQ, Pq triangles with the same base and height as the segments, i.e., triangles with the same vertices R, r as the segments, and in like manner inscribe in segments RQP, rqP triangles with the same base and height as those segments, and so on, we shall finally have segments remaining whose sum is less than the difference by which the PQq exceeds K. Therefore the polygon inscribed must have an area which exceeds K, which is impossible, for $A + B + C + \ldots + Z < \frac{4}{3}A$, where A is the area of triangle PQq. $\qquad\square$

6.3.2 *On Spirals*

One of the difficulties encountered by the Greek geometers was creating new curves: all known curves had their origin in the straight line and the circle. Moreover, with the exception of the quadratrix, the curves were "static": for example, the conic sections existed by virtue of an intersection. Archimedes produced a new curve that, like the quadratrix, existed by virtue of motion. *On Spirals* introduced and completed the study of spirals, which Archimedes defined by:

Definition. *Let a straight line revolve at a uniform rate about a fixed point, called the **origin**. Designate the initial position of the line the **initial line**. Let a point move from the fixed point along the moving straight line at a uniform rate; the moving point describes a **spiral**. The length along the moving line traversed by the point during the first revolution is the **first distance**; the area enclosed by the spiral and the first distance is the **first area**, and the circle with radius equal to the first distance is the **first circle**; likewise for the second distances, etc.*

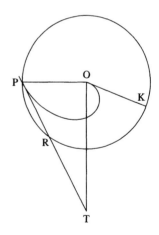

Archimedes gave two propositions that solved the tangent and quadrature problems for the spiral.

Proposition 6.22 (Tangent to a Spiral). *Let P be any point on the first turn of the spiral with origin O and initial line OK; draw OT perpendicular to OP, and let the tangent at P meet the perpendicular at T; let circle KRP be drawn with radius OP. Then OT is equal to the arc KRP.*

Proposition 6.23 (Area of a Spiral). *The first area of the spiral is one-third of the first circle.*

6.3.3 *The Method*

The proofs of the Archimedean results on quadratures are difficult enough, though with some perseverance, they can be understood. But how did Archimedes arrive at the results in the first place? This is a key problem of theoretical mathematics: how can one *find* a result to prove? Archimedes recognized this difficulty, and wrote a letter, *The Method of Treating Mechanical Problems*, to explain how he arrived at some of his conclusions. The letter was addressed to Eratosthenes (276-196 B.C.), a friend Archimedes met in Alexandria. Eratosthenes was nicknamed by his contemporaries "beta," after the second letter of the Greek alphabet: he was second best in everything. The "everything" included astronomy, geometry, history, and physics, so "beta" is as much of a compliment as a criticism. Eratosthenes is best known for having measured the size of the Earth.

See Problem 7.

The method communicated to Eratosthenes by Archimedes was a method of discovery, not of demonstration. Archimedes recognized this, and warned Eratosthenes that the propositions still needed to be proven geometrically. The method was a mechanical one, and according to the introduction, the very first theorem he discovered using this method was that found in *Quadrature of a Parabola*: recall the first part of this work showed how one may "weigh" a parabolic segment and show that it is 4/3 the triangle contained by the base and vertex of the segment. The key property used is that of the lever: if weights A, B are placed at distances a, b from a fulcrum, then the lever will be in equilibrium if $A : B = b : a$.

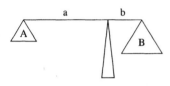

Using the method, Archimedes discovered

Conjecture 6.1. *The volume of a sphere is four times the cone whose base is a great circle of the sphere, and whose height is equal to the radius of the sphere; the cylinder whose base is a great circle of the sphere and whose height is equal to the diameter of the sphere is $1\frac{1}{2}$ times the sphere.*

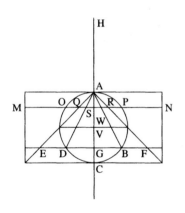

Demonstration. Let $ABCD$ be a great circle in a sphere with center W. Let MS, SN, AC be equal, and let MN be perpendicular to diameter AC. Let MN be the diameter of a circle perpendicular to AC; let this circle be in a cylinder with height AC. Let YZ perpendicular to AC at W be a diameter of a great circle in the sphere $ABCD$, and let the great circle YZ be the base of a cone with height AW. Extend AY through E and AZ through F, to make a larger cone whose base is a circle with diameter MN and whose height is AC. Let CH be a bar of balance, with A as midpoint. $\qquad\square$

Note that the initial comparison will be between the sphere and a cylinder whose base is a circle with twice the diameter of the great circle in a sphere, which we will call the large cylinder, and a cone whose base is the same as the cylinder and whose height is equal to the diameter of the sphere, which we will call the large cone.

Demonstration. [Continued] Now, $MS = AC$, $QS = AS$, and thus $MS \cdot QS = AC \cdot AS$. Since $AS : AO = AO : AC$, then $AC \cdot AS = AO^2$; also (by the Pythagorean Theorem), $AO^2 = OS^2 + AS^2$. Hence $MS \cdot QS = OS^2 + AS^2 =$

$OS^2 + QS^2$. Since $HA = AC$, then we have

$$
\begin{aligned}
HA : AS &= AC : AS \\
&= MS : QS \\
&= MS^2 : QS \cdot MS \\
&= MS^2 : OS^2 + QS^2 \\
&= \left(\begin{array}{c} \text{Circle,} \\ \text{diameter } MN \end{array} \right) : \left(\begin{array}{c} \text{Circle,} \\ \text{diameter } OP \end{array} \right) + \left(\begin{array}{c} \text{Circle,} \\ \text{diameter } QR \end{array} \right)
\end{aligned}
$$

Thus, the circle in the large cylinder placed at S will be in equilibrium with the circle in the sphere and the circle in the large cone, placed at H.

Since this is true regardless of the position of MN, then the large cylinder, in its place, will be in equilibrium with the sphere and the large cone. Now the center of gravity of the large cylinder is at W. Thus the large cylinder is twice the sphere and large cone together. But the large cylinder is three times the large cone; hence the large cylinder is six times the sphere. But the large cylinder is four times the cylinder whose base is equal to a great circle of the sphere and whose height is the diameter; hence such a cylinder would be 3/2 the sphere.

Alternatively, since the large cone is eight times the cone with base a circle YZ and height AW, then the sphere is four times this latter cone. $\qquad \square$

Unfortunately for the progress of mathematics, little was made of the method of Archimedes, and by the Renaissance, every written copy of *The Method* had vanished, leading Renaissance geometers to speculate the Greeks deliberately hid all but the final results and the proof, to make their discoveries more impressive.

If every written copy of *The Method* has vanished, how do we know of its contents? Before the invention of paper, which could be made cheaply from rags, writing material was expensive: the most widely available substance was vellum, made from sheep skins. The greater durability of vellum offset its higher cost, for a piece of vellum could be written on and later, the writing could be washed off and the vellum reused. The recycled piece is called a **palimpset**. A tenth century copy of *The Method* suffered this fate, and was recycled to be used for a book of prayers and sermons for the Eastern Orthodox Church; this copy remained in Constantinople (soon to become Istanbul, Turkey). The erasure was incomplete, however, and clear traces of a work of mathematical nature could still be seen. In 1906, the Danish scholar J. Heiberg, hearing of a palimpset that contained a mathematical work, sought it out, and recovered the original text of Archimedes.

6.3 Exercises

1. Prove Proposition 6.17. Why is this proposition necessary?

2. To prove Proposition 6.20, Archimedes began by creating a set of auxiliary quantities, b, c, ... z, where $b = \frac{1}{3}B$, $c = \frac{1}{3}C$, and so on, then noting

$B + C + D + \ldots + Z + b + c + \ldots + z = \frac{1}{3}(A + B + C + \ldots + Y)$. Complete the Archimedean proof.

3. Prove the second half of Proposition 6.21, that the area of the segment cannot be less than K.

4. In the statement of the law of the lever, why would it be improper to say A is to b as B is to a?

5. Use the method to show that if a triangle BLC is on base BC of a rectangle $ABCD$ and has the same height as the rectangle, then the triangle is half the rectangle.

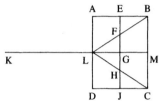

Figure for Problem 5.

   (a) Let the figure be drawn as shown, and let KM be a bar of balance, with middle point L. Show that the line JE of the rectangle, placed at G, will balance the line HF of the triangle, placed at K, and thus the triangle BLC, placed at K, will balance the rectangle $ABCD$, in its location.

   (b) Show that this implies the rectangle is twice the triangle.

6. (Teaching Activity) Cut out a parabolic segment and a triangle with the same base that shares the same vertex as the segment. Construct a beam of balance, and find the fulcrum point so the parabola and triangle balance. Use this to conjecture the relationship between the area of the segment and the area of the triangle. Repeat the process with other figures.

7. Eratosthenes determined the circumference of the Earth by using the following facts: on the day of the summer solstice, the noonday sun was directly overhead in Syene, but made an angle of about 7° to the vertical in Alexandria; the north-south distance between Syene and Alexandria was 5000 stadia.

   (a) Determine the circumference of the Earth.

   (b) One stadia is believed to be about 0.1 miles, and the circumference of the Earth is approximately 24,900 miles. How accurate was Eratosthenes measurement?

   (c) Suppose the Earth was flat. What alternative explanation could be made for the inclination of the sun to the vertical? What information would this give you? Hint: see Anaxagoras, page 65.

8. Discuss the following claim: calculus was invented by the ancient Greek geometers. Explain what you mean by "calculus"; be sure to justify your definition.

6.4 Large Numbers

The "very large" tends to blend into the infinite in popular conception: "innumerable as the stars in the sky" say the poets, who might not realize that only about 2000 stars are visible on a typical cloudless night. All too frequently the word "infinite" is used to mean "very large." Archimedes, no doubt, heard this misuse of language once too often, and wrote a letter, now referred to as "The Sand-Reckoner."

6.4.1 Octad Notation

In ancient times, there was no way of expressing very large numbers, in part because there was no need to, so "uncountable" might literally mean that. However, there is a difference between the very large and the infinite, and Archimedes, to point out the difference, calculated the number of grains of sand it would take to fill the known universe. To express the number, Archimedes invented a system called **octad notation**, which will have an interesting historic echo two thousand years later for counting the *infinite*. The largest named Greek number was the **myriad**, ten thousand. One could say "two myriad," "three myriads," etc., up to "myriad myriad"; Archimedes called this sequence the **numbers of the first order**. Let a myriad myriad be the **unit of the second order**, which for convenience we can call M.

The numbers "one M," "two M," "three M," and so on, up to "$M\ M$" formed the **numbers of the second order**, and the last was the **unit of the third order**, which again for convenience we might call N. The **numbers of the third order** are created in a similar manner, beginning with "one N," "two N," "three N," and so on up to "$M\ N$" (*not* $N\ N$), which was **unit of the fourth order**. Continuing up to the Mth order, Archimedes designated by P last number of the Mth order, and called the numbers from 1 to P the **numbers of the first period**. We can again speak of P, $2P$, $3P$, etc., up to $M\ P$, which forms the **first order of the second period**; we can continue through the second, third, fourth, etc., orders of the second period, up to the Pth order of the second period, or P^2, and continue this way for as long as we like.

By estimating the size of a sand grain to be far smaller than it actually is (Archimedes's grains of sand were more like specks of dust), and by estimating the size of the universe to be even larger than his contemporaries suggested (though far, far smaller than is believed today), Archimedes was able to show that even if the universe was filled with sand, the number would not exceed ten million units of the eighth order of numbers.

6.4.2 The Cattle Problem

Archimedes is also associated with the first recorded "challenge" problem in history, posed to his younger contemporary Apollonius of Perga. The problem was to find the number of oxen of Helios (the Greek sun god); it is not certain, however, that Archimedes actually posed the problem to Apollonius or to anyone else, and the problem probably dates from centuries afterwards.

The problem is to find the numbers of each of four types (white, black, yellow, and dappled) of bulls and cows: thus, the problem gives rise to a system of equations with eight unknowns. The restrictions are:

1. The number of white bulls is one-half plus one-third the number of black bulls, plus the number of yellow bulls.

2. The number of black bulls is one-fourth plus one-fifth the number of dappled bulls, plus the number of yellow bulls.

3. The number of dappled bulls is one-sixth plus one-seventh the number of white bulls, plus the number of yellow bulls.

4. The number of white cows is one-third plus one-fourth the total number of black cattle.

5. The number of black cows is one-fourth plus one-fifth the total number of dappled cattle.

6. The number of dappled cows is one-fifth plus one-sixth the total number of yellow cattle.

7. The number of yellow cows is one-sixth plus one-seventh the number of white cattle.

Since there are seven equations and eight unknowns, there are an infinite number of solutions. Obviously, only whole number solutions are acceptable; the smallest of these is:

1. White bulls 1,217,263,415,886; white cows 846,192,410,280.

2. Black bulls 876,035,935,422; black cows 574,579,625,058.

3. Yellow bulls 487,233,469,701; yellow cows 638,688,708,099.

4. Dappled bulls 846,005,479,380; dappled cows 412,838,131,860.

To make the problem even more difficult, Archimedes (or whoever actually posed the problem) added a few more conditions:

1. The number of white bulls, plus the number of black bulls, is a square number.

2. The number of yellow bulls, plus the number of dappled bulls, is a triangular number.

The first condition was vaguely expressed in the original problem, and might either mean that the number was square or, possibly, composite. The smallest solution to the complete problem was not found until 1880; the smallest answer (for the size of the herd) is a number that begins with 7766 and has 206,541 more digits. It seems highly unlikely that Archimedes had the solution when he posed the problem!

6.4 Exercises

1. Examine the development of numbers in Archimedes's octad notation.

    (a) What are the values of M and P in modern notation?

    (b) How many sand grains would it take to fill the universe, according to Archimedes's calculation?

    (c) Modern cosmology estimates there are about 10^{68} atoms in the universe. Compare Archimedes's number to this one.

    (d) Why is the name octad appropriate? (*Oct* is the Greek prefix meaning "eight")

2. The cattle problem stated that the white and black bulls together stood in a formation "as wide as it was deep." Explain why this condition is vague, and could be interpreted as meaning either the total was a square number, or it was a composite number.

6.5   Vergings and Loci

The number of general curves known to the Greeks was very limited. Two important classes of curves were vergings and loci. Both originated from the requirement that a curve satisfy a certain set of properties. **Vergings** or **inclinations** dealt with the following problem: given two lines (one or both of which might be straight) and a fixed point, to place between them a straight line of a fixed length so that it points to a fixed point. The general form of a verging was a **locus** (plural, **loci**), which was a curve that satisfied a given property.

6.5.1   Nicomedes

One of the earliest vergings known was invented and studied by Nicomedes (fl. 250 B.C.) who introduced a new curve, called by him the cochloid. His treatise on the cochloid has been completely lost, and we know of the curve of Nicomedes only through the description of Pappus (third century A.D.), who gave it the name **conchoid**. According to Pappus, there were four types of conchoids, but he described only the first:

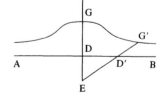

Definition. *Given a **canon** AB, a **pole** E, and EG perpendicular to AB at D; let **interval** DG be a fixed line. Let line GE move about E describing the **conchoid** GG′ where D′G′ is everywhere equal to DG.*

Nicomedes apparently described a mechanical device for drawing the conchoid, and discovered many of its properties and uses, including the fact that GG' is asymptotic to the canon in both directions. Moreover, he described how it could be used to solve two of the classical problems: trisection of an angle, and duplication of the cube.

Author and Title	Availability
Euclid, *Data*	Complete
Apollonius, *Cutting off of a Ratio*	Complete
Apollonius, *Cutting off of an Area*	**LOST**
Apollonius, *Determinate Sections*	**LOST**
Apollonius, *Tangencies*	**LOST**
Euclid, *Porisms*	**LOST**
Apollonius, *Neuses*	**LOST**
Apollonius, *Conics*	Seven out of eight books
Aristaeus, *Solid Loci*	**LOST**
Euclid, *Loci on Surfaces*	**LOST**
Eratosthenes, *On Means*	**LOST**

Table 6.1: The domain of analysis.

6.5.2 Apollonius

Pappus, writing five hundred or more years after the events he described, is one of our best sources of geometry in the era between 300 and 200 B.C. In his *Mathematical Collection*, he described the **domain of analysis**: those works beyond the *Elements* that were necessary to learn how to solve geometrical problems. In modern terms, we might call them the core readings for the "graduate study" of mathematics. Pappus arranged the works from most elementary to most complex; Table 6.1 summarizes Pappus's choices, as well as the works that survive to the present day. Of the 12 works in 32 books, only 10 books survive: Euclid's *Data* (one book); and Apollonius's *Cutting off of a Ratio* (two books) and *Conics* (seven out of eight books).

Notice the majority of the works were by Apollonius of Perga, the third and last giant of Greek geometry. If more of Apollonius's works had survived the passage of time, his reputation might surpass that of Archimedes, for what we have suggests a geometer of high caliber. In a refrain that has become, perhaps, tiresome, little is known about him. He was perhaps twenty-five to forty years younger than Archimedes, which would place his birth around 260 B.C.; the dates of 262-190 B.C. have been suggested for his life. Apollonius was born in Perga in Pamphilia (southern Asia Minor), and spent some time in both Alexandria and Pergamum. During Apollonius's lifetime, the Seleucid Empire was collapsing, and Pergamum, under its leader Attalus, broke free of the Seleucids and established itself as an independent kingdom in 230 B.C. Like Eratosthenes, Apollonius received a Greek letter nickname, "epsilon," for its resemblance (ϵ) to the Moon: Apollonius was also active in working out the theory of the Moon (i.e., the determination of its future position).

See page 84.

Apollonius's *Cutting off of a Ratio* survives only in an Arabic translation. The problem it deals with is the following: given two lines, AB, CD, and a point on each line, such as E, F, and a point G not on the lines, and a given ratio $m : n$, to draw through G a line GH, intersecting the given lines at K, L, so that $EK : FL = m : n$. The two books deal with each of the cases that could possibly arise, and proceed by the classic method of analysis and synthesis.

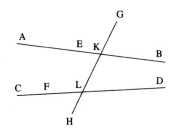

Tangencies apparently dealt with the following problem: given three things, be they points, lines, or circles, to draw a circle through the given points (if any) and tangent to the given lines or circles. There are ten cases; the two simplest (three points, or two points and a line) were dealt with by Euclid, but the remaining eight cases were solved by Apollonius. The problem is a difficult one to carry out even using the tools of modern analytic geometry, and the fact that Apollonius solved this problem using compass and straightedge alone says something of his mastery of geometry.

Pappus described another lost work of Apollonius, *Plane Loci*, which consisted of two books with 147 theorems and 8 lemmas. Pappus's description of *Plane Loci* was sufficiently detailed to allow seventeenth century mathematicians, notably Fermat, to fruitfully attempt reconstructions of the work. Thus, though it is lost, *Plane Loci* was the most influential of all Apollonius's works, for its reconstruction led to our system of coordinate geometry.

Before discussing what is known about *Plane Loci* in detail, a few terms need to be introduced. As mentioned before, a locus is a figure whose points satisfy a certain relationship: for example, a circle is the locus of points equidistant from a given point. A point, line, or circle is **given in position** if it has a specified location in space; if one end of a line is given in position, this means that the line itself passes through a specified location in space; if a line, circle, or other figure is **given in magnitude**, it has a specified size but is otherwise free to move about in space (and the figure's shape is also free to change within the constraints of the definition of the figure itself).

Pappus's description of *Plane Loci* began with a single, rather complex, proposition summarizing the first book. Then Pappus gave three propositions, enunciated by "Charmandrus," about whom this single reference in Pappus is all that is known; presumably, Pappus wanted to give the reader an idea of what locus problems were solved before Apollonius. These first three propositions are: See Problem 5.

Proposition 6.24. *If one end of a straight line, given in magnitude, is fixed, the other will be on a circle given in position.*

Proposition 6.25. *If, from two points, straight lines are drawn and meet at a given angle, the vertex of the angle will be on a circle given in position.*

Proposition 6.26. *If a triangle is given in magnitude and its base is a line given in position and magnitude, the vertex of the triangle will be on a line given in position.*

One of the locus problems dealt with by Apollonius was:

Proposition 6.27. *If a straight line is given in magnitude and parallel to a line given in position, and one end of line given in magnitude is on another line given in position, then the other end of the line given in magnitude will also be on a straight line given in position.*

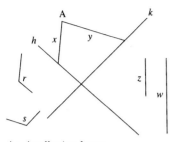

An Apollonian locus.

To make the sixth locus problem of the first book easier to understand, it will be stated in terms of a specific figure:

Proposition 6.28. *Let lines h, k be given in position, and let angles r and s be given in magnitude. Let a : b be a given fixed ratio. If, from A, two lines x and y are drawn to the given lines h and k, making with them angles r and s so x : y = a : b, then A is on a line given in position (see margin). Alternatively, if x + z = w, where z : y = a : b, and w is another line, given in magnitude, the same is true.*

It is possible to see the origin of analytic geometry in this locus statement, for suppose h and k are perpendicular, and angles r and s right angles. Then in the first case, A will be on the line $y = \frac{b}{a}x$. The second case corresponds to the graph of a line not through the origin.

The second book contains more complex propositions about loci.

We also know that Apollonius wrote a work, *Quick Delivery*, which discussed rapid means of computation and gave a method for computing the ratio of the circumference of a circle to its diameter that resulted in a value for π more accurate than that of Archimedes (though we do not know what the value was). Apollonius also created a system for expressing large numbers based on the myriad and called **tetrad notation**, since the numbers up to a myriad could be expressed in four or fewer symbols. It was, perhaps, in response to this "tetrad" system of Apollonius that Archimedes created his own "octad" notation.

6.5 Exercises

1. Prove the conchoid GG' is asymptotic to the canon AB. Do not use the Cartesian equation of the curve.

2. Design and build a device for drawing the conchoid.

3. Suggest possible variations that would make the three other types of conchoids of Nicomedes.

4. Find a circle solving the tangency problem, given:

    (a) Three points.   (b) Three lines.

    (c) Two points and a line. Hint: Use *Elements* III-36, given in Problem 7, Section 7.

5. What can be inferred about Charmandrus given the nature of the loci attributed to him by Pappus? Explain.

6. Prove that the specified figure solves the locus problem in the three propositions attributed to Charmandrus.

7. Consider Proposition 6.28.

   (a) To which lines do the two forms of the locus proposition correspond?

   (b) Prove that if P, Q, R are points that satisfy the locus, then they lie on the same line.

   (c) From this, prove that the equation $Ax + By = C$, in Cartesian coordinates where A, B, C are positive constants, is the equation for a line.

8. A line can be given in position, in magnitude, or both. Explain why a circle given in position is also a circle given in magnitude, using a Euclidean construction; include a proof. (In other words, given a circle in position, show that any other circle having the same position must necessarily be the same circle)

9. Discuss the fact that no work of Archimedes appeared in the domain of analysis. What does this suggest about the work of Archimedes?

6.6 Apollonius's *Conics*

If we consider the works in Pappus's domain of analysis to be the core reading of the graduate study of mathematics in the Greek era, then Apollonius's *Conics* would correspond to the mathematical work required in the second or even third year of graduate school. Yet, like most graduate texts, Apollonius built on the work of his predecessors, and though he may have added his own discoveries, the bulk of the results were known by his time.

6.6.1 Conic Sections Before Apollonius

Unfortunately for the historian, Apollonius's work was so successful that, like Euclid's *Elements*, it replaced all prior works on the subject, and thus we have no surviving treatises on conic sections before Apollonius. There are two lost treatises that are of particular interest.

The first is the *Conics* of Euclid, in four books. According to Pappus, Euclid's *Conics*, like his *Elements*, was meant to systematize and organize the knowledge on conics. Apollonius then "filled out" Euclid's work and added four more books of his own. Thus, it sounds as if the first four books of Apollonius's *Conics* are essentially similar to the four books of Euclid's *Conics*. However, though Apollonius based the first four books on *Conics* on Euclid, he actually broke new ground in the first four books by examining the conic sections in a manner that was original with him; then he continued the work with more advanced material, most of it original. Thus, though the *results* in Euclid's work have probably been preserved, the *form* has not, and the loss is a particularly acute one to the historian.

An even more significant loss was Aristaeus's *On Solid Loci*, in five books, which apparently included some original results. What is especially interesting about this work is that the title suggests the treatment of conic sections as the solutions to locus problems. Unfortunately, the title is the only part of the work that has survived.

Of the eight books of Apollonius's *Conics*, the eighth book no longer exists, and only the first four books exist in Greek; books five through seven exist only in Arabic translations made during the Middle Ages. Most of the books have an introduction (Book III is the exception). The introduction to the first two books are notes to Eudemus of Pergamum, and from them we may discern some additional biographical information about Apollonius. He had a son (or student); he spent some time with Eudemus in Pergamum, and Eudemus and Apollonius spent time in Alexandria, working with the geometer Naucrates.

6.6.2   Generalization of Conic Sections

Apollonius claimed he had studied the conic sections in greater generality than any of his predecessors. Recall that as late as the time of Archimedes, the conic sections were formed by the intersection of a right circular cone with a plane perpendicular to the side of the cone. Might the intersection of an oblique cone with a plane not perpendicular to the side produce other curves? The answer, as Apollonius proved, was no: if a cone with a circular base is cut by a plane, the only possible sections are a point, two intersecting lines, a circle, an ellipse, a parabola, or a hyperbola.

Apollonius defined a cone by:

Definition. *Given a point, called the* **vertex**, *and a circle not in the same plane as the point, called the* **generating circle**. *Let a straight line be drawn from the vertex to the circumference of the circle, and extended indefinitely in both directions, and while the point remains fixed, let the line trace along the circumference until it returns to its starting point. The two surfaces generated are the* **conic surface**. *and the straight line drawn from the vertex to the center of the generating circle is the* **axis**.

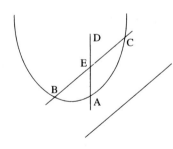

Definition. *Given a curve and a straight line. The line AD is a* **diameter** *of the curve if any line BC parallel to the given line is bisected by AD. The intersection of the diameter and the curve is the* **vertex** *of the curve, and the lines parallel to the given straight line are said to be drawn* **ordinatewise to the diameter** *with BE (or CE) the* **ordinate**. *If the diameter cuts the lines drawn ordinatewise at right angles, it is the* **axis** *of the curve.*

Note that the diameter and vertex of a curve are relative to the given line; they do not, in general, correspond to our modern conception of the diameter and vertex of a conic section. Moreover, in general, the ordinate of a point on a conic section only rarely will correspond to what we think of as one of the coordinates.

A key concept in Apollonius is that of the **axial triangle**, which is the section formed by cutting a cone with a plane that passes through the axis of the cone. A key property of the axial triangle is that the intersection of the axial triangle with the generating circle is a diameter of the circle. In all subsequent sections of the cone, Apollonius assumed that the intersection of the cutting plane and the plane of the generating circle is a line perpendicular to the base of the axial triangle. If this occurs, Apollonius proved, then the line formed by the intersection of the cutting plane and the plane of the axial triangle is a diameter of the section.

See Problem 16.

See Problem 3.

6.6.3 Properties of Conic Sections

To replace the terms "section of an acute-angled cone," "section of a right-angled cone," and "section of an obtuse-angled cone," Apollonius used the terms ellipse, parabola, and hyperbola, drawn from Euclidean geometry. Recall that these terms appeared in connection with the application of areas; Apollonius showed that the sections of a cone would, depending on how the cutting plane intersected the cone, result in a relationship between the ordinates that could be classified as falling short, equaling, or exceeding another area. These properties were referred to as the **symptoms** of the curve.

See page 60.

Apollonius's language is very difficult to follow, since he must refer to several cutting planes, some of which must have certain required properties. As we did with the *Elements*, Book VI, we will state the propositions as they relate to specific figures.

Proposition 6.29 (Symptom of a Parabola). *Suppose the cone with vertex A and generating circle BDC is cut by plane to produce the section DFE; let the intersection of the plane with the generating circle be DE, and suppose that axial triangle ABC has BC perpendicular to DE; suppose further than the diameter FG is parallel to one side of the axial triangle AC. From any point on the section, such as K, draw KL parallel to DE (see figure in margin). Then square KL will equal* rect. *FH, FL, where FH : FA =* sq. *BC :* rect. *BA, AC.*

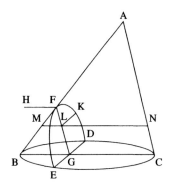

Symptom of a parabola.

Since *BC, BA,* and *AC* are all lines of constant length for a given cone, the length of *FH* is fixed for any particular intersection of a plane and the cone; moreover the intersection of the plane and the cone determines *FA,* which is another constant. If we think of *FL, KL* as corresponding to the *y* and *x* coordinates of a parabola, and *FH = k,* then this Apollonian proposition is equivalent to saying that $x^2 = ky$.

Apollonius demonstrated, once again, the peculiar blind spot of Greek geometers when he did not give *FH* a specific name but instead referred to it as "the straight line to which the straight lines drawn ordinatewise to the diameter *FG* are applied in square." This convoluted phrase became the **parameter** of the conic section. Apollonius also noted *FH* was also called the "upright side," which in Latin is **latus rectum**.

For the hyperbola and the ellipse, Apollonius gave

Proposition 6.30 (Symptom of a Hyperbola). *Given a cone with vertex A and axial triangle ABC, where the cutting plane intersects the generating circle on DE perpendicular to BC; let the axis of the section be FG and let this axis*

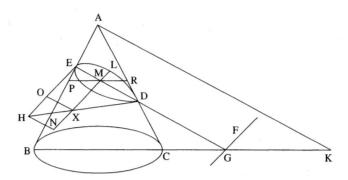

Figure 6.5: Symptom of an ellipse.

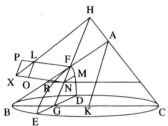

Symptom of a hyperbola.

*meet the side of the axial triangle at H beyond the vertex A (see figure in margin). Draw AK parallel to FG and intersecting the base BC of the axial triangle at K. Let FL be chosen so that FH : FL = sq. KA : rect. BK, KC, and draw FL perpendicular to FG; let NX be perpendicular to FG and intersect HL as shown; let LO, PX be parallel to FG and intersect FL, NX as shown. Let MN be drawn parallel to DE. Then the square on MN is equal to the parallelogram FX, applied to FL with FN as its base and exceeding by the parallelogram LPXO similar to the rectangle contained by HF and FL. The section DFE will be called a **hyperbola**; FL is the **parameter** of the hyperbola, or its upright side; FH will be called the **transverse side**.*

Proposition 6.31 (Symptom of an Ellipse). *Given a cone with vertex A and axial triangle ABC, where the cutting plane intersects both sides of the axial triangle, and meets the extension of BC at G so that the intersection FG between the cutting plane and the plane of the generating circle is perpendicular to BG; let the diameter of the section be ED (see Figure 6.5). Draw EH perpendicular to EG, and AK parallel to ED such that sq. AK : rect. BK, KC = DE : EH. Draw ML parallel to FG. Then the square on ML is equal to the parallelogram EX, which is the area applied to EH, having EM as its width and deficient by a figure similar to the rectangle contained by DE, EH. The section DLE will be called an **ellipse**. EH is the **parameter** of the ellipse, or its upright side; ED will be called the **transverse side**.*

Apollonius was the first to recognize both branches of the hyperbola; in *Conics* I-14 he showed that both branches of the hyperbola would have the same diameter and same parameter; he called the two sections together the **opposite** and reserved the name hyperbola for when he dealt with a single branch.

For the ellipse and hyperbola, Apollonius defined the center:

Definition. *The **center** of an ellipse or hyperbola is the midpoint of its diameter or transverse side.*

Notice that a center is specific to a diameter, and that the uniqueness of the center is not guaranteed by the definition.

See Problem 15.

A useful relationship for the ellipse and hyperbola can be found in:

Proposition 6.32. *In hyperbola DA with parameter AC, diameter BE with vertex A and transverse side AB, if from F, D, lines are dropped ordinatewise to the diameter, as FG, DE, then sq. FG : rect. AG, GB = AC : AB; moreover, sq. FG : sq. DE = rect. AG, GB : rect. AE, EB. Likewise, the same result is true if DA is an ellipse or a circle with diameter AB.*

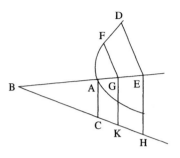

Proof. Join BC, and draw EH, GK parallel to AC. By the property of the hyperbola, sq. FG = rect. KG, GA, and by similar triangles, $KG : GB = CA : AB$. Hence rect. KG, GA : rect. $GB, GA = CA : AB$, and thus sq. FG : rect. $GB, GA = AC : AB$, which was to be proven. □

The fact that a diameter of a section bisects all lines parallel to some given line is quite useful. From it, one can immediately prove that a line drawn from the vertex and parallel to a line drawn ordinatewise is tangent to the conic section. It also allows a very simple method of constructing the tangent to a parabola at a point. The first part of the construction is I-33:

See Problem 9.

Proposition 6.33 (Tangent to a Parabola). *Given a parabola ACD with diameter AE and vertex A. If a line is dropped ordinatewise from C to meet the diameter at E, and AE = AF, then FC does not cross the parabola at C.*

Proof. Let the construction be made as indicated. Suppose FC falls inside, as FH. Extend GH ordinatewise to D on the parabola. Obviously, sq. GD : sq. CE > sq. GH : sq. CE; because ACD is a parabola, then sq. GD : sq. CE = $GA : AE$; by similar triangles, $GH : CE = FG : FE$, and thus sq. GH : sq. CE = sq. FG : sq. FE. Since sq. GD : sq. CE > sq. GH : sq. CE, and sq. GH : sq. CE = sq. FG : sq. FE, then sq. GD : sq. CE > sq. FG : sq. FE. But sq. GD : sq. $CE = GA : AE$, so $GA : AE$ > sq. FG : sq. FE.

See Problem 10.

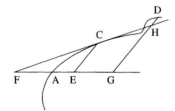

Also, $GA : AE = 4$ rect. GA, FA : 4 rect. AE, FA. Hence 4 rect. GA, FA : 4 rect. AE, FA > sq. FG : sq. FE; alternating, we have 4 rect. GA, FA : sq. FG > 4 rect. AE, FA : sq. FE.

By construction, $FA = AE$, and thus 4 rect. AE, FA is 4 sq. FA, which is equal to sq. FE. Hence 4 rect. GA, FA > sq. FG. But this is impossible, since A is not the midpoint on FG. Hence, FC cannot fall within the parabola, and so FC does not cross the parabola. □

The second part involved proving that no straight line through C could be interposed between FC and the parabola.

The difficulty is finding a diameter suited to a particular problem. In the second half of Book I, Apollonius proved a number of key propositions, the details of which are far too complicated to give here, but in the summary following *Conics* I-51, Apollonius noted that in a parabola, every line parallel to a diameter is also a diameter, while in hyperbolas, ellipses, and the opposite sections, every line drawn through the center is a diameter.

6.6 Exercises

1. What are the differences between Euclid's definition of a cone and Apollonius's?

2. Show that the definition of diameter of a curve is consistent with the definition of the diameter of a circle.

3. Prove part of *Conics* I-7: If a plane cuts a cone so that the intersection of the cutting plane and the generating circle is a line perpendicular to the base of the axial triangle, then the intersection of the cutting plane and the axial triangle is a diameter of the section. In the figure, let the intersection of the cutting plane and the generating circle be the line DE, which is perpendicular to the base BC of the axial triangle ABC; prove that the intersection FD of the cutting plane and the axial triangle is a diameter of the conic section EFG. Note: in general, FG will *not* be perpendicular to the lines drawn ordinatewise, such as LK.

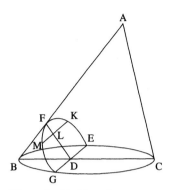

Figure for Problem 3.

4. Prove Proposition 6.29. Begin by showing that if MN is parallel to BC, then MKN are on a circle in a plane parallel to the plane of the generating circle.

5. How do Propositions 6.30 and 6.31 translate into the Cartesian equations for a hyperbola and an ellipse?

6. Show that the ancient definitions of the section of a right, obtuse, and acute angled cone correspond to the Apollonian parabola, hyperbola, and ellipse, respectively.

7. Why did other geometers before Apollonius not recognize the two branches of the hyperbola?

8. Justify the names parabola, hyperbola, and ellipse.

9. Prove *Conics* I-17: given a conic section and a diameter, if a straight line is drawn from the vertex that is parallel to an ordinate, that line will be tangent to the conic section at the vertex.

10. Prove *Conics* I-20: if two lines are dropped ordinatewise to the diameter of a parabola, then the squares on the ordinates will be to each other as the straight lines cut off by them on the diameter beginning from the vertex. In the diagram in the margin, AF is a diameter with vertex A; show that the square on CE is to the square on DF as AE is to AF. Remember that the diameter does not correspond to the axis of the conic.

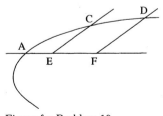

Figure for Problem 10.

11. The Greek concept of the tangent to a curve also required that if a line FC touched a curve, no straight line could be interposed between the line and the curve at C. Prove the second half of Proposition 6.33: if a line touches a parabola, no straight line can be interposed between the line and the parabola.

12. Prove that, given a segment AC and a point B on AC but not the midpoint, the square on the whole must exceed four times the rectangle on AB, BC.

13. Prove *Conics* I-46, which is equivalent to saying that all diameters of a parabola are parallel. Remember that the lines drawn ordinatewise to any diameter are parallel to the tangent to the curve at the vertex of the diameter.

14. Given a conic section, explain how to solve each of the following problems.

    (a) How to find a diameter (*Conics* II-44).

    (b) How to find the center of an ellipse or hyperbola (*Conics* II-45).

    (c) How to find the axis of a parabola (*Conics* II-46) or an ellipse or hyperbola (II-47).

    (d) How to find the tangent to the conic at a given point (II-49).

15. Apollonius's definition of the center of an ellipse or a hyperbola raises the possibility that such a figure might have more than one "center."

    (a) *Conics* I-30 states that in an ellipse or the opposite sections, if a straight line is drawn from the center of a diameter, it will be bisected by the diameter. Prove this.

    (b) *Conics* I-47 states that if a line is tangent to a hyperbola, ellipse, or circle and it meets the diameter, then the line drawn from the point of contact to the center of that diameter will bisect the straight lines drawn in the section parallel to the tangent; in other words, the line drawn from the point of tangency to the center is also a diameter. Prove this.

    (c) *Conics* I-48 is the corresponding result for the opposite sections. Prove this.

    (d) Explain why these results prove that the center of an ellipse, hyperbola, or circle is unique.

16. Prove that given *any* intersection of a cone with a plane that does not contain the vertex, an axial triangle may be found where the intersection of the cutting plane with the plane of the generating circle is perpendicular to the base of the axial triangle.

6.7 Loci and Extrema

Today, conic sections are usually introduced as solutions to locus problems, relating to the distances between a fixed point, called the **focus**, and a fixed line, called the **directrix**. For example, the parabola is defined as the locus of points P so that the distance between the point and the focus, PF, is equal to the distance between the point and the directrix, PD. Another way to express this is to say that the ratio PF to PD is equal to unity. If PF to PD is a fixed number greater than 1, the locus forms a hyperbola, while if PF to PD is less than 1, the locus forms an ellipse. In all cases, the ratio of PF to PD is called the **eccentricity** of the conic section.

6.7.1 The Focus-Directrix Property

As we shall see, Apollonius was clearly aware of the relationship between conics and the solution to locus problems. Thus it is especially striking that *Conics* nowhere mentions the directrix or the focus of a parabola. The focus of a hyperbola or an ellipse appears, but as a consequence of the properties of the conic sections, and not as their defining property.

If, as intimated by Pappus, Apollonius based his work on Euclid's *Conics*, it suggests that Euclid did not derive the conic sections from the focus-directrix property. We may speculate that Euclid, desiring to create a work that rested on a minimum of assumptions, avoided the locus definition of a conic section because that would entail the additional assumption that such a locus *existed*.

See Problem 12.

However, earlier geometers were clearly aware of the focus-directrix property, as evidenced by what is known about Euclid's lost *Loci on Surfaces*. In Pappus's sequence, this work immediately followed Aristaeus's *Solid Loci*; hence, it might be assumed that Euclid's *Loci* was a more advanced work. The section by Pappus seems to be added as an afterthought, for in his introduction, he expressed his intention to discuss the works only so far as Apollonius's *Conics*; certainly, the clarity and comprehensibility of the short section on *Loci on Surfaces* that has survived make it appear to be more of a rough draft than a completed study of the work.

From Euclid's *Loci on Surfaces*, Pappus proved the following locus property:

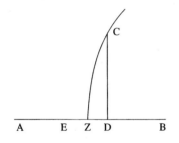

Proposition 6.34. *Given two points, A, B, fixed in position, and a line CD perpendicular to AB and intersecting it at D. If the ratio of* sq. *AD to* sq. *CD, together with* sq. *DB, is a constant, then C is on a conic section; if the ratio is greater than 1 to 1, the conic section is a hyperbola; if the ratio is less, the conic section is an ellipse; and if the ratio is equal, the conic section is a parabola.*

Pappus proved the result using analysis and synthesis. For the parabola, the proof is:

Proof. Suppose the ratio is that of equality; i.e., sq. AD is equal to sq. CD together with sq. DB, and let CZ be the locus. Let DE be made equal to BD. Then rect. BA, AE equals sq. CD. Let AB be bisected at Z; since AE, EZ are together equal to ZD, DB, let ZD be added to both; hence AE, DE are together equal to twice ZD, and DB; but DE, DB are equal. Hence AE is twice ZD. Hence rect. BA, AE is twice rect. BA, ZD, or the rect. on twice BA, ZD. Hence, the sq. CD is equal to rect. on twice BA, ZD, and thus C is on a parabola with parameter twice BA.   □

This statement translates into the focus-directrix property. It is thus worth noting that the focus-directrix property was *not* the means by which the conic sections were studied, and it was not until the time of Kepler, nearly two thousand years after Apollonius, that conic sections were re-examined in terms of their foci and directrices.

See Problem 11.

6.7.2 Conics and Loci

Although Apollonius did not define the conic sections in terms of a locus, he did know that certain locus problems could be solved using conic sections; indeed, he takes credit for being the first to solve three- and four-line locus problems. These problems may be concisely stated as follows:

Problem 6.1 (Three-Line Locus Problem). *Given three lines AD, DC, AC and three angles r, s, t, find the locus of points H so that if lines are drawn from H cutting AD, DC, AC so that the angles HJA, HLD, and HKA are equal, respectively, to r, s, t, then* sq. *HK is equal to* rect. *HJ, HL.*

The four line locus problem is similar, but instead of the square on HK equal to rect. HJ, HL, the locus rests on the equality of two rectangles.

The solutions to the locus problem are embodied in the last propositions of Book III. As they are very complicated, we will discuss only one of them, *Conics* III-54, which deals with the parabola.

Proposition 6.35. *Given two intersecting tangents of a parabola, AD, DC and any other point on the parabola, such as H. If AF, CG are parallel to DC, AD respectively, and meet CH, AH in F and G, respectively, and if E is the midpoint of AC, then* rect. *AF, CG* : sq. *AC as* sq. *EB* : sq. *BD, compounded with the ratio* rect. *AD, DC* : rect. *AE, EC.*

$$[I.e., \ \frac{AF \cdot CG}{AC^2} = \frac{EB^2}{BD^2} \cdot \frac{AD \cdot DC}{AE \cdot EC}]$$

Thus, the parabola solves the three-line locus problem.

See Problem 1.

6.7.3 Extrema

Perhaps the most interesting, and original, contributions of Apollonius were contained in Book V of the *Conics*, which discussed the shortest and longest lines that could be drawn to a conic from a given point. Though Apollonius noted the subject is one of those that "seem worthy of study for their own sake," we will see another possible reason for his examination of minimum and maximum lines.

For all the sections, the location of the minimum line depends on the parameter or, more specifically, half the parameter. Since the parabola has the simplest relationship between the ordinate and the parameter, the results about the minimum lines drawn from a point to the parabola are, correspondingly, the most comprehensible.

The general problem is: given any point and a conic section not through the point, to find the shortest (or longest) line that can be drawn from the point to the conic section. To solve this problem, Apollonius began by examining simpler problems, such as the case where the point is on the axis of a conic section.

Proposition 6.36. *Given parabola CBA with axis CE, and CZ half the parameter. Of all the lines that can be drawn from Z to the parabola, e.g., HZ, BZ, AZ, CZ, we claim CZ is the shortest of all these lines. Moreover, given two lines, such as HZ, BZ, the line closer to the vertex will be the shorter.*

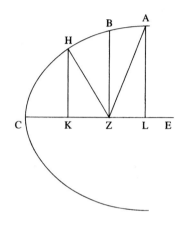

Proof. Draw HK, AL, etc. perpendicular to CE. Since CZ is half the parameter, then twice CZ is the parameter; by the symptom of a parabola, then, rect. twice CZ, CK is equal to sq. HK. Hence rect. twice CZ, CK plus sq. KZ equals sq. HK together with sq. KZ. Hence sq. HZ is rect. twice CZ, CK, and sq. KZ.

Since CZ is CK and KZ, then rect. twice CZ, CK is twice sq. CK, plus twice rect. KZ, CK. Hence rect. twice CZ, CK and sq. KZ is twice sq. CK, plus twice rect. KZ, CK, plus sq. KZ, which is sq. CK plus the square on CK, KZ, or sq. CZ. Hence sq. HZ is equal to sq. CK plus sq. CZ, and therefore HZ greater than sq. CZ. Thus, HZ is greater than CZ. □

After proving results about minimum lines when a point is on the axis, Apollonius generalized the results. The final result for a parabola is:

Proposition 6.37. *Suppose AB is a parabola, with axis BH and C any point on the interior of the parabola. Construct CD perpendicular to BH, and let DE be measured, equal to half the parameter, towards the vertex. Construct ET perpendicular to BH, and describe the hyperbola through C with asymptotes TE, EH; let the intersection of the hyperbola and the parabola be A. CA is the shortest line that can be drawn from C to the parabola.*

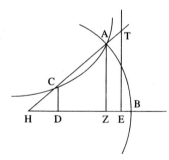

See Problem 5.

Apollonius also proved corresponding results for the minimum lines in ellipses and hyperbolas.

Why did Apollonius study minimum lines? Propositions V-27 and V-28 may give us some insight as to the reason. In Proposition V-27, Apollonius proved that, given a parabola, a minimum line, and the tangent at the point where the minimum line crosses the parabola, the tangent and the minimum lines are perpendicular. In other words, the minimum line is a **normal to the curve** at the point where it crosses the curve. One of the problems is that defining a tangent line is problematic for figures like parabolas and hyperbolas: one cannot define a tangent to a parabola as a line that intersects the figure exactly once, since many nontangent lines have this property. Apollonius's use of normals allowed him to make use of another property of the tangent: it divides the plane into two regions, one of which contains the entire curve, and the other of which contains no part of the curve.

6.7.4 The Rise of Rome

After Apollonius, over two centuries would pass before any significant progress was made in mathematics. This gap can be attributed to the vast upheavals soon to occur in the Mediterranean world as the old kingdoms were swept away and a new empire was built: the Roman Empire.

Even after the Second Punic War, Rome felt Carthage was a threat. The anti-Carthaginian party was led by Cato the Elder, a prominent public figure. After every public speech, regardless of the topic, Cato would add, "And I am also of the opinion that Carthage must be destroyed."

Under the terms of the treaty, Carthage was not permitted to wage war without Rome's permission. Rome promised the King of Numidia (now in Algeria), Masinissa, support if he attacked Carthage. When Masinissa attacked Carthage, the Carthaginians appealed to Rome, which did nothing. When the Carthaginians dared to defend themselves without Rome's permission, Rome declared Carthage in violation of the treaty, and went to war in 149 B.C. Shortly after the war began, both Masinissa and Cato died (they were in their eighties); we can only hope that Cato's last days were full of anguish, for the war he had worked so hard to start began with a series of Carthaginian victories. However, defeat was inevitable, and in 146 B.C., Rome destroyed Carthage. Shortly thereafter, Rome added Greece to her expanding empire.

One wonders how the Greeks felt: for the first time since the Persian Wars, 350 years before, Greece was at peace. Greek culture flowed into Rome: it was a mark of distinction to have a Greek philosopher teach your children. The teacher was, as often as not, a purchased slave, but soon, Greek became the language of high society.

Meanwhile, the Kingdom of Pergamum had been willed to Rome by its king, Attalus III, and by 100 B.C., the Roman Republic included most of Spain, southern France, Italy, Greece, Asia Minor, and Palestine. By then, the Alexandrine successor states were gone, with one exception: the Ptolemaic kingdom of Egypt.

6.7 Exercises

1. To show that the parabola solves the three-line locus problem, first suppose the given lines are AD, DC, AC, and the given angles are DAC, ACD, and CDA (i.e., the angles of intersection of the lines).

   (a) Suppose H is on a parabola. Explain why rect. AF, CG must have a constant size.

   (b) Draw DE through the midpoint of AC; from H, draw HX, HY, and HZ, parallel to DE, AD, and DC, respectively. Show that the ratio between rect. CZ, AY and rect. ZH, YH is constant.

   (c) Show that the ratio between rect. ZH, YH and sq. HX is constant.

   (d) Hence, the two ratios compounded is constant; so the ratio between rect. CZ, AY to sq. HX is constant, and thus the sq. HX is to rect. HL, HJ in some constant ratio.

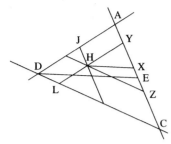

Figure for Problem 1.

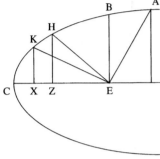

Figure for Problem 4.

(e) How would the parabola have to be constructed so the square on HX will equal rect. HL, HJ?

(f) Generalize the solution so that HX, HL, HJ can make arbitrary (but fixed) angles with the given lines.

2. Prove the second half of Proposition 6.36.

3. Prove that if the point on the diameter of a parabola is closer to the vertex than half the parameter, the minimum line will be the line to the vertex.

4. Prove *Conics*, V-8: Given the parabola ABC, with diameter EC and vertex C. Let EZ be equal to half the parameter, and draw ZH perpendicular to EC. Of all the lines that can be drawn from E to the parabola, the shortest will be the line drawn to the intersection of the perpendicular ZH with the parabola.

5. Prove *Conics* V-12: In any conic AB, if AC is the minimum line from a point on the axis to the curve, and D is any point on the line, then DA is shorter than any line from D to the conic section.

6. Prove *Conics* V-24: From any point on any conic, there can be only one minimum line drawn from the point to a point on the axis of the conic.

7. Prove *Conics* V-27: Given a minimum line from a point to a parabola, the tangent to the parabola at the point on the parabola will be perpendicular to the minimum line. Note that the point does not have to be on the axis of the parabola.

8. Given a parabola and any point on the parabola, explain how to use the propositions from Book V to find the tangent to the parabola at the given point.

9. Suppose CDE is concave towards A (see page 143). Prove that if AD is the longest line that can be drawn from A to the curve, then DF, perpendicular to AD, cannot meet the curve, and thus is tangent.

10. Complete the proposition from *Plane Loci*, to show that if the ratio exceeds unity, the conic section must be a hyperbola, and if the ratio is less than unity, the conic section must be an ellipse. Determine the parameter in each case, and prove your result.

11. Explain how Proposition 6.34 translates into the modern focus-directrix properties for a conic section. How do the ratios in the proposition relate to the eccentricity?

12. Compare and contrast the logical difficulties of defining the conic sections as the solution to a locus problem involving a focus and a directrix, as compared to defining conic sections as the intersection of a plane and a cone. What assumptions must be made in either case?

13. Prove, using *only* the axioms and propositions from the *Elements*, that the locus of points equidistant from two points, given in position, is a line. Note that you will have to prove that the line exists, and also that any point on the line satisfies the locus.

Chapter 7

Roman Era

By 100 B.C., the Roman Republic had acquired a vast empire. This was a disaster for the middle and lower classes, for their products (including their own labor) could be obtained more cheaply overseas. Thus, from 100 B.C. to 48 B.C., Roman fought Roman in a terrible civil war between the *optimates* ("best people"), consisting of the aristocrats, and the *populares* ("rabble rousers"), consisting of the middle and lower classes bankrupted by the Roman conquest of the Mediterranean. Gaius Pompey led the military forces of the *optimates*. Pompey's record was impressive. In 67 B.C., pirates made it nearly impossible for Rome to get food shipments from overseas; in six months, Pompey cleared the seas and piracy was no longer a problem. Between 66 and 64 B.C., Pompey swept through Asia Minor, and even conquered the troublesome kingdom of Judaea, adding more territory to Rome's vast empire.

The optimates had proven military leadership; moreover, their soldiers were rich enough to afford the best arms and armor. It was against these formidable opponents the populares had to fight, and all the odds seemed against them. But leading the forces of the *populares* was a "class traitor": a wealthy aristocrat, and Pompey's father-in-law. His name was Julius Caesar.

7.1   Numeration and Computation

Roman numerals were one of the last form of additive notation to be invented and used in the West. Originally, the numerals were abstract symbols: | for the unit, ∨ for five; × for ten. However, the similarity to letters forced the two forms together, and the final forms became:

I	V	X	L	C	D
One	Five	Ten	Fifty	One Hundred	Five Hundred

Like all additive systems, Roman numerals expressed numbers by writing down the appropriate symbol the required number of times: two hundred thirty six would be two C, three X, one V, and one I: CCXXXVI. Order does not matter, and one finds IIIX and VIX to represent thirteen and sixteen because in Latin, the words for these numbers are literally "three-ten" and "six-ten" (as they are in English, "thirteen" and "sixteen" being archaic forms of "three-ten" and "six-ten"). The **subtractive principle**, where the symbol for a smaller number precedes the symbol for a larger number and is read as the larger *minus* the smaller, was almost never used by the Romans and only became common practice during the fifteenth century.

For higher numbers, the Romans used a number of different techniques. The Latin word for "thousand" is *MILLE* (literally, since the Romans did not use lower case letters), and by the first century A.D., M began to be used to represent a thousand, but this was by no means standard. Other methods involved modifying the unit: thus, two thousand was (I)(I). Alternatively, a bar might be written over the number of thousands: $\overline{\text{II}}$. By the Middle Ages, M began to see common use to represent one thousand. Finally, the symbol ∞ was also used, possibly because it was a "cursive" M (alternatively, it was a variation of (I)). Even larger numbers might be written with a bar atop: forty-seven thousand, eight hundred sixty-three was $\overline{\text{XXXXVII}}\text{DCCCLXIII}$.

7.1.1 The Abacus

Before the invention of paper, material to write upon was either very expensive (vellum or papyrus) or very bulky (clay or wax tablets). Thus, when extensive numerical computations had to be done, they were usually done using some sort of mechanical device, the most common of which was the **abacus**. The word abacus comes from a Semitic word meaning "dust," suggesting one "cleared" the table by wiping off the dust. Various forms of abaci were known to Roman, Greek, Egyptian, Chinese, and Indian civilizations as early as the seventh century B.C., and there is no reason to suspect a common origin.

Originally, the abacus might have just been a series of grooves drawn in the dirt. Each groove represented a particular place value; by setting markers into the appropriate grooves, a number was indicated. The most convenient markers were pebbles, or *calculus* in Latin, from which we get words such as "calculate." For example, to represent the number three thousand, two hundred, and six, three pebbles would be placed in the thousand (M) column; two pebbles in the hundred (C) column; and six in the units (I) column.

M	C	X	I
○	○		○
○	○		○
○			○
			○
			○
			○

Three thousand, two hundred six.

Addition was trivial: to add three hundred twenty-five to three thousand, two hundred six, add three pebbles to the C column, two to the X column, and five pebbles to the I column, as shown in Figure 7.1: on the abacus, addition proceeds as rapidly as the numbers can be set down. Since there are more than ten pebbles in the I column, ten of these may be exchanged for a single unit in the X column, giving the final answer: three M, five C, three X, and one I, which can almost immediately be recorded in Roman numerals: MMMCCCCCXXXI, or MMMDXXXI.

Since Roman numerals have a symbol for five, fifty, and five hundred, an

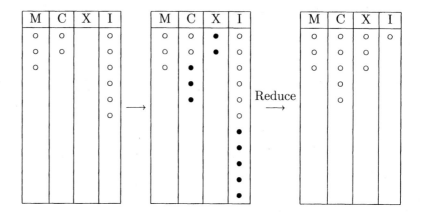

Figure 7.1: Adding MMMCCIIIIII and CCXXIIIII.

additional set of columns (usually placed above the main columns) were added. Later on, the abacus took the form of a hand-held mechanical device. The only difference here is the "pebbles" have been replaced with actual beads, which can slide along a wire; moreover, there is a division between the top set (each bead representing the "fives": V, L, and D) and the bottom set (which represent the "ones": I, X, C, M). When a bead is adjacent to the center bar (or touching a set of beads touching the center bar), it counts; otherwise, it is ignored.

7.1.2   The Julian Calendar

At Pharsalus, Greece, in 48 B.C., Caesar's army, consisting mainly of poorly armed and armored footsoldiers, met Pompey's army, consisting mainly of well armed and armored cavalry. Caesar demonstrated the true nature of military genius: understanding his opponents. He instructed his footsoldiers to point their spears not at the cavalry horses, which was the standard tactic, but at the riders. The riders, all aristocrats, had a choice: they could win the battle, but be disfigured; or they could lose. They chose to lose.

Pompey fled to Egypt, hoping to recruit a new army. The regent, Pothinus, was in the midst of a civil war. His charge, Ptolemy XII, was fighting his sister for control. Not wanting to have trouble with Rome, Pothinus had Pompey killed as he disembarked from his ship. Pothinus showed the body to Caesar and suggested that as he, Pothinus, had helped Caesar by eliminating a rival, Caesar should help Pothinus by eliminating *his* rival, Ptolemy's sister. The sister was Cleopatra (the seventh of that name). She met and charmed Caesar so much that what was intended to be a brief stay in Egypt in the pursuit of Pompey turned into a much longer one. One of the important consequences of Caesar's stay in Egypt was the reform of the Roman calendar.

We may describe the **calendar problem** as follows: given a small number of different whole numbers, to find a set consisting of these whole numbers whose mean approximates a given real number to any desired degree of accuracy. The solar year is 365.242199 days long, and it is the solar year that is relevant to the seasons, which in turn is relevant to agriculture, and thus of vital importance for everyone. Using 365 and 366 day years, the Egyptians had solved the calendar problem by making three out of every four years 365 days, with the fourth year 366 days: this gave a mean of 365.25 days. A complicating factor that the Romans tried to incorporate was the synodic period of the Moon, the time it takes the Moon to run through a full set of phases. This is approximately 29.5306 days, so the Roman calendar year consisted of 12 months, alternating between 29 and 30 days.

The traditional Roman calendar began in March (hence, December was actually the tenth month, as its name suggests) and ended in February. Since the Roman calendar year was a poor approximation for the solar year, the beginning of the year (and the start of planting) had to be announced; in fact, our word "calendar" comes from the Greek word meaning "proclamation." Since incorrectly determining the date of the start of the year would have disastrous consequences, in most cultures the job of determining the actual date of the start of the year was left to priests, who could devote their full time and effort to the task. In Rome, these priests were elected officials, and they frequently delayed or accelerated the start of the calendar year, depending on whether their party was in power or not. As a result, by the time of Caesar, the Roman calendar had little relationship to the seasons. When Caesar finally returned to Rome, he brought Sosigenes, an Egyptian of Greek descent, to help reform the Roman civil calendar.

Based on Sosigenes's recommendations, Caesar instituted a number of reforms. First, to ensure the year began in spring, Caesar ordained 45 B.C. would last 445 days: 45 B.C. came to be known as "The Year of Confusion." Next, Caesar made the month lengths fixed: beginning with March, the months alternated 31 and 30 days with the exception of February. February was an unpopular month: all bills came due in February. Thus, Caesar shortened it to 29 days. Finally, to make the average calendar year approximate the solar year, every fourth year was a leap year, in which February would have 30 days. Note that in the system originally established by Caesar, the months of September, October, November, and December had 31, 30, 31, and 30 days.

Shortly after Caesar's proclamation of the calendar in 45 B.C., he was assassinated (the two events were probably not related!). Marc Antony, Caesar's friend, suggested the fifth month be named July, in honor of Caesar's family, the Julians. Caesar's death was followed by another civil war. The major contenders were Caesar's nephew Octavian, and Marc Antony. Marc Antony went to Egypt, to seek help from the rich (independent) kingdom of Egypt, while Octavian stayed in Rome and turned the Roman public against Marc Antony. It was not too difficult: Marc Antony had abandoned his wife and was clearly having an affair with Cleopatra.

Marc Antony and Cleopatra's forces suffered a disastrous naval defeat at Actium in 31 B.C. Marc Antony committed suicide. Cleopatra held out hope that she could entrance Octavian as she did Marc Antony and Caesar, but Octavian (something of a stern moralist) refused, offering her only a position in his triumph at Rome—as a captive. She killed herself instead. Octavian swept up the remnants of Ptolemaic Egypt and emerged undisputed master of the Mediterranean.

To allay the fears of the Senate, he styled himself "First among equals" and gave himself the title of *princeps*, which means "First Citizen"; this became the word prince, and the Roman world became known as the *Principate*. Octavian took on the name Augustus ("exalted one") and the title of Caesar (which, until Julius Caesar, had simply been a name); in time, "Caesar" came to be a title for many ruling monarchs: kaiser, tsar, czar, and (possibly) shah all stem from the word Caesar. (So does the word Jersey, from a group of islands off the north coast of France spotted by Julius Caesar when he ventured into Britain.)

Augustus made two changes to the calendar, which put it almost into its modern form. First, Augustus appropriated the sixth month for himself, naming it August. Then, since August would only have 30 days under the original, Julian scheme, a day was taken from February (reducing it to its current 28 days, 29 in leap years). But since the month of September had 31 days, this would leave three 31-day months in a row, so Augustus switched the month lengths to their current amounts.

7.1 Exercises

1. Show how the following computations can be performed on the abacus.

   (a) Four hundred twenty-six plus thirty eight

   (b) Two thousand eighty-five plus four hundred seven

2. Compare and contrast the numbering systems of the Egyptians, Babylonians, Greeks, and Romans. What are the strengths and weaknesses of each? Pay particular attention to the methods of computation used, and how well suited the notation is for the particular method of logistic.

3. It is said that the Roman system of numeration and computation is clumsy and ill-suited for advanced mathematics. Comment.

4. How well does the Julian calendar solve the calendar problem?

5. Find a solution to the calendar problem as described in the text.

7.2   Geometry and Trigonometry

Greek contributions to trigonometry were scattered: some propositions attributed, on dubious authority, to Thales; the law of cosines in the Book II of Euclid's *Elements* and the results on similar triangles in Book VI. However, the development of trigonometry as a practical branch of mathematics was left to the Roman era.

A key figure in both the history of astronomy and mathematics was Hipparchus of Nicaea (ca. 180-ca. 125 B.C.), about whom we know very little.

Hipparchus may have been responsible for the systematic division of the circle into 360 parts. From Theon of Alexandria, we know Hipparchus wrote twelve books on the chords in a circle, which probably included a table of chord lengths, the precursor to the modern table of sines. Unfortunately, this work (and almost all of Hipparchus's other works) has been lost; shortly after Hipparchus's death, the Roman Republic underwent the long and painful transition to Principate and then Empire.

After Augustus's death in 14 A.D., the rulers of Rome were weak, incompetent, insane, or all three. This was the era of Caligula and Nero, and of the most gory excesses of Rome. The chaos of the first century ended in 96 A.D., when Nerva, a respected Senator, was elected Emperor. The reign of Nerva began a period of peace and prosperity that would last for nearly a century. Before his death (of old age) a few years later, Nerva recognized Trajan, a Spanish general, to be competent and conscientious, and appointed him his successor. This started a trend, and the "Five Good Emperors" (Nerva, Trajan, Hadrian, Antoninus Pius, and Marcus Aurelius) reigned in peaceful succession until 180 A.D., each appointing as his successor a man he felt was competent and conscientious.

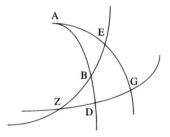

7.2.1 Menelaos of Alexandria

Menelaos of Alexandria (ca. 100 A.D.) lived during the peace of the reign of Trajan. According to Theon of Alexandria, Menelaos wrote six books on chords in a circle. Unfortunately, these have vanished. However, his *Sphaerics* survives in an Arabic translation. The first book deals with spherical triangles, an essential part of mathematical astronomy, and the first extension of the triangle concept that has come down to us:

Definition. *The triangle on the surface of a sphere is the figure formed from three arcs of great circles, each less than a semicircle.*

Definition. *The angles of the triangle are the angles between the planes of the great circles.*

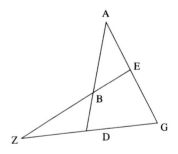

An important discovery of his is known as Menelaos's theorem:

Theorem 7.1 (Menelaos's Theorem). *If two great circle arcs GE, BD intersect at a point A, and from the two points G, B are drawn great circle arcs GD, BE which intersect at Z, then chord GE to chord EA is equal to the compounded ratio of chord GZ to chord ZD, and chord BD to chord BA.*

The proof of this theorem relies on a corresponding result for a plane triangle *ADG*, namely that in a plane triangle, the ratio of side *GE* to *EA* is the ratio *GZ* to *ZD*, compounded with the ratio *DB* to *BA*. It is in this form that Menelaos's theorem is usually stated.

7.2.2 Ptolemy

Claudius Ptolemy (fl. 125-151) lived during the reign of Trajan's successor, Hadrian. Trajan himself made one of the last additions to the Roman Empire when he sent a force across the Danube into Dacia, which became so Romanized that even today, though the land is surrounded by Slavic culture and languages, a romance (Latin-based) language is spoken. Even Dacia's modern name reminds us of its history: Romania.

The Roman Empire of Hadrian's time included most of western Europe, Asia Minor, Palestine, Egypt, and North Africa. However, Hadrian realized that Rome's borders were insecure. In the northernmost part of the Empire, England (first visited by Caesar, but not added to the Empire until a century later), Hadrian had built a wall, now known as Hadrian's Wall, to secure England from marauding Picts.

About all we know for certain about Ptolemy's life is that he made astronomical observations from Alexandria between the years 125 and 151. He would, however, write a work that, like Euclid's *Elements*, incorporated the work of his predecessors in so successful a fashion that tracing the history of trigonometry before Ptolemy is quite difficult.

Ptolemy called his work the *Mathematical Collection*. It might be called a text on mathematical astronomy, for it begins with the mathematical prerequisites for understanding and using the astronomical system established by the Greeks. Later commentators called it the *Great Collection* ("great" in the sense of large, to distinguish it from "lesser," or smaller, collections), which in Greek is *Megale Syntaxis*. The Arabs would combine their definite article *al* with the superlative form *Megistos*, and thus the book returned to Europe titled *Almagest*, by which the work is generally known today. Aside from its practical bent, the mathematics of Ptolemy follow the strict Euclidean form of proposition and proof, showing, at least, that the axiomatic tradition had not yet vanished from western mathematics.

The *Almagest* highlights one of the greatest distinctions between mathematics and the sciences. The *Almagest* is known primarily as a work on astronomy, not mathematics; however, the astronomy of the *Almagest* depicts a universe centered around the Earth, and is considered erroneous today. The mathematics, on the other hand, is just as valid today as in Ptolemy's time.

Chords in a Circle

One of the key issues in mathematical astronomy is determining the lengths of chords in a circle (assumed to be a great circle in a sphere). Hipparchus and Menelaos made tables of chords before Ptolemy, but these have not survived to the present. Hence, Ptolemy's table is the oldest known. To create the table, Ptolemy first divided the circumference of the circle into 360 parts, and expressed his intention to set, "side by side" at intervals of a half a part, the chords of the corresponding arcs: in other words, he would create a table of chords whose central angles measured from $0°$ to $360°$, at intervals of $\frac{1}{2}°$. Ptolemy then divided the diameter of the circle into 120 parts, which would be the units for the length of the chords themselves. To distinguish these parts from the parts of the circumference, Ptolemy designated them with a superscripted P: 120^P. In his computations, he expressed an intention to use sexagesimal notation, "because of the difficulty of fractions."

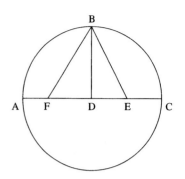

To begin finding the chord lengths, Ptolemy began by introducing what is perhaps the simplest method of constructing a regular pentagon inside a circle, far simpler than that used by Euclid in Book IV of the *Elements*:

Proposition 7.1. *Given semicircle ABC on diameter AC and center D, with perpendicular DB. Let DC be bisected at E, and BE joined. Make EF equal to EB, and join FB. Then FD is the side of a regular decagon inscribed in circle ABC, and BF is the side of a regular pentagon inscribed in ABC.*

Proof. Since *DC* is bisected at *E*, and *DF* added to it in a straight line, then rect. *CF, FD*, together with sq. *ED*, is equal to sq. *EF*, which is equal to sq. *BE*. But sq. *ED*, together with sq. *DB*, is also equal to sq. *BE*. Hence rect. *CF, FD* is equal to sq. *DB*, or sq. *DC*. Thus *CF* is cut into extreme and mean ratio at *D*. But since the side of the regular inscribed hexagon and the regular inscribed decagon, when they are on the same line, also cut the line into extreme and mean ratios, and *DC* is the side of the regular hexagon, then *DF* is the side of the regular decagon.

See *Elements*, XIII-9.

Likewise, *BF* is the side of a regular pentagon inscribed in a circle. □

Since *FD* is the side of a regular decagon, it will also equal the length of the chord whose central angle is 36° (we will, from now on, use Ptolemy's phraseology, and simply refer to this as the "chord of 36°," or crd. 36°). To find *FD*, we note that *DB* is a radius, so $DB = 60^P$; *DC* is bisected at *E*, so $DE = 30^P$; hence the Pythagorean theorem can be used to find *BE*, and thus *EF*, and thus *FD* can be found: $37^P 4' 55''$ (where ′ and ″ represent 1/60 and 1/3600 of a part of the diameter; note that they do *not* correspond to our modern notion of minutes and seconds of arc). A second application of the Pythagorean theorem allows *BF*, the side of a regular pentagon, to be found: *BF* and hence crd. 72° is $70^P 32' 3''$.

See Problem 2.

Ptolemy then noted that knowing the chord of an arc allowed the determination of the chord of the supplementary arcs, since the squares on the chords added to the square on the diameter.

Example 7.1. *Suppose AC is $70^P 32' 3''$ (i.e., AC is crd. 72°). Find DC (which corresponds to crd. 108°). By the Pythagorean theorem we have*

$$\text{sq. } AC + \text{sq. } DC = \text{sq. } AD$$

where AD is the diameter, 120^P; hence

$$DC = 97^P 4' 56''$$

As for the remaining chords, Ptolemy introduced a proposition which, because his is the oldest surviving record, is known as Ptolemy's theorem, though it almost certainly predates him.

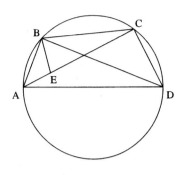

Proposition 7.2 (Ptolemy's Theorem). *If quadrilateral ABCD is inscribed in a circle,* rect. *AC, BD is equal to* rect. *BC, AD together with* rect. *AB, CD.*

Proof. Construct angle *ABE* equal to angle *DBC*; thus angles *ABD, EBC* are equal; since angles *BDA, BCE* subtend the same arc, they are also equal. Therefore triangle *ABD* is similar to triangle *EBC*, and thus *BC* : *CE* as *BD* : *AD*, so rect. *BC, AD* is equal to rect. *BD, EC*. Likewise, triangle *ABE* is similar to triangle *DBC*, so *AB* : *AE* as *DB* : *DC*, so rect. *AB, DC* is equal to rect. *BD, AE*. Thus, rect. *BC, AD* together with rect. *AB, DC* is equal to rect. *BD, EC* together with rect. *BD, AE*. But rect. *BD, EC* with rect. *BD, AE* is rect. *BD, AC*, which was to be proven. □

Consequently, Ptolemy noted, if *AD* was the diameter, and the chords *AB, AC* were of known lengths, then the length of the chord *BC* could also be determined. It is sometimes claimed this discovery of Ptolemy's is equivalent to a formula for $\sin(a - b)$. This follows from supposing the center of the circle is at *O*; then chord *BC* is a chord subtended by a central angle equal to the difference between *AOC* and *AOB*.

Example 7.2. *Suppose AB is* 60^{P} *(i.e., AB is* crd. $60°$*) and AC is* $70^{\mathrm{P}}32'3''$ *(i.e., AC is* crd. $72°$*). Find BC (which corresponds to* crd. $12°$*). By Ptolemy's theorem, and noting AD is the diameter of the circle, we have*

$$BC \cdot AD + AB \cdot DC = BD \cdot AC$$

From Example 7.1, we found $DC = 97^{\mathrm{P}}4'56''$*. In a similar fashion, we find* $BD = 103^{\mathrm{P}}55'23''$ *(which we note is equal to* crd. $120°$*). Thus*

$$BC \cdot (120^{\mathrm{P}}) + (60^{\mathrm{P}})(97^{\mathrm{P}}4'56'') = (103^{\mathrm{P}}55'23'')(70^{\mathrm{P}}32'3'')$$

Solving for BC, we obtain

$$BC = 12^{\mathrm{P}}32'36''$$

which is also crd. $12°$.

Since not all angles can be determined using Ptolemy's theorem alone, it is necessary to find some other method of generating chords. Thus, Ptolemy showed how, given a chord, one can find the chord of half the angle. First, it is necessary to show:

Proposition 7.3 (Half-Chord Theorem). *Given semicircle ABC, on diameter AC, and CB a chord of known length. Bisect arc BC at D. Join AB, BD, DC, and AD; draw DF perpendicular to AC. Then CF is half of the difference between AC and AB, and thus the chord DC on half the angle may be found.*

Proof. Draw AE equal to AB, and join ED. Then triangle EDC is isosceles, with equal sides DC, DE. Since CE is the difference between AC and AB, then CF is half the difference.

Since triangle ACD is similar to triangle DCF, then $AC : CD$ as $DC : CF$, and thus rect. AC, CF is equal to sq. DC. But AC is the diameter, so CF can be found using the previous proposition, and DC, the chord on half the angle, can be found. $\square$

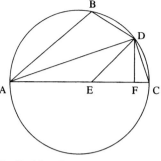

See Problem 5.

Example 7.3. *Suppose BC is 60^{P} (i.e., BC is* crd. *$60°$). Then AB is $103^{\mathrm{P}}55'23''$, and thus CF is half the difference between AB and diameter AC; hence CF is $8^{\mathrm{P}}2'19''$. Hence*

$$AC \cdot CF = \text{sq. } DC$$
$$(120^{\mathrm{P}}) \cdot (8^{\mathrm{P}}2'19'') = \text{sq. } DC$$
$$31^{\mathrm{P}}3'31'' = DC$$

Note that DC is crd. *$30°$.*

From the half-chord proposition, the chord of $1\frac{1}{2}^{°}$ could be found.

Ptolemy also proved the formula for the chords of a sum.

Proposition 7.4 (Sum of Chords). *Given circle $ABCD$, on diameter AD and center at F. From A, cut arcs AB, BC of known lengths; join the chords. The chord AC may be found.*

The proof of the proposition shows how to find the chord AC.

See Problem 6.

From this point, the construction of a useful table of sines is an exercise in endurance, for all the necessary theoretical foundations are present. However, for practical purposes, Ptolemy's methods are unable to manufacture a table of chords separated by half degree increments, as Ptolemy desires. It is necessary to find the chord of $1°$, something that is impossible given the purely geometric propositions above.

See Problem 9.

Instead, a numerical procedure must be used. The basis of Ptolemy's numerical approximation is:

Proposition 7.5. *In a circle, if arc AB is less than arc BC, then $BC : AB$ is less than* arc $BC :$ arc AB.

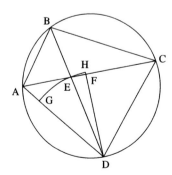

Proof. Let AB be less than BC. Let angle ABC be bisected by BD; join AC, AD, CD. Then CD is equal to AD, and CE is greater than AE (by Euclid, VI-3). Drop DF perpendicular to AC. Then AD is greater than DE, and DE is greater than DF. Draw the circle with center D and radius DE, through G, H. Then the sector DEH is greater than the triangle DEF; likewise, the triangle DEA is greater than the sector DEG. Hence

triangle $DEF :$ triangle $DEA <$ sector $DEH :$ sector DEG

But triangle DEF : triangle DEA as EF : AE, while sector DEH : sector DEG as angle FDE : angle EDA. Hence

$$EF : AE < \text{angle } FDE : \text{angle } EDA$$

And thus (combining the ratios)

$$AF : AE < \text{angle } FDA : \text{angle } EDA$$

Doubling the antecedents

$$AC : AE < \text{angle } CDA : \text{angle } EDA$$

Separating the ratios:

$$CE : AE < \text{angle } CDB : \text{angle } EDA$$

But $CE : AE$ is equal to $BC : AB$ (by *Elements*, VI-3); also, angles BDA, EDA are the same. Finally, angle CDB : angle BDA as arc BC : arc BA. Hence

$$BC : AB < \text{arc } BC : \text{arc } BA$$

which was to be proven.   □

Ptolemy used this fact to find the chord of $1°$. First, let the arc AB subtend a central angle of $\frac{3}{4}°$, and the arc AC subtend an angle of $1°$. Then, by the preceding,

$$AC : AB \quad < \quad \text{arc AC} : \text{arc } AB$$

and thus crd. AC is less than $\frac{4}{3}$crd. AB. But crd. AB can be found; it has a measure of $0^P 47' 8''$, so crd. AC is less than $1^P 2' 50''$.

Likewise, if arc AD subtends an angle of $1\frac{1}{2}°$, crd. AD is equal to $1^P 34' 15''$, so crd. AC must be greater than $1^P 2' 50''$. Thus, crd. AC must be greater and less than the same number, so the best approximation is $1^P 2' 50''$. This means that crd. $\frac{1}{2}°$ is $0^P 31' 25''$.

With the values for the chord of $\frac{1}{2}°$ and $1°$ in hand, Ptolemy can create his table, primarily by applying the arc sum and arc difference results, as well as the values already determined. The value of the chords were found to the second sexagesimal place, giving an accuracy of 1 part in 216,000, or about five decimal places.

See Problem 7.

A feature added by Ptolemy extended both the usefulness and the reliability of the table. In addition to the actual chord lengths, he included an extra column that gave one thirtieth of the difference between successive chords, carried out to three sexagesimal places. Table 7.1 shows some of the values. The purpose of this extra column is two-fold. Since the successive arcs differ by $\frac{1}{2}°$, and the column entries correspond to 1/30th the difference between successive arcs, the column entries approximate the difference between chords 1/60th of $1°$ apart;

Arc	Chord			Sixtieths			
°	P	′	″	P	′	″	‴
20	20	50	16	0	1	1	51
$20\frac{1}{2}$	21	21	11	0	1	1	48
21	21	52	6	0	1	1	45
$21\frac{1}{2}$	22	22	58	0	1	1	42

Table 7.1: A Section of Ptolemy's Table of Chords

thus, the table, with interpolation, provides the chords of arcs differing by 1′ of arc.

Ptolemy noted a secondary purpose. He was well aware that his book would be hand copied by scribes who might or might not have a mathematical background. As a result, errors would inevitably creep into the table, especially since it is a long table of numbers rather than words. Thus, if any chord on the table was suspect, the additional column provided a means of checking, since it represented the difference between successive chords on the table. This was the earliest appearance of a **checksum**, a means by which the accuracy of a copy can be verified.

Euclid's Fifth Postulate

Ptolemy also considered Euclid's Fifth Postulate, in a work (now lost) entitled, *That Lines Produced from Angles Less Than Two Right Angles Meet Each Other.* Recall that Euclid's first four postulates were simple, comprehensible statements, such as, "A line can be drawn between any two points." The fifth postulate, on the other hand, represented a major departure from the simplicity of the other four, and Proclus even says of it, "This ought to be struck from the postulates altogether." Proclus and many other geometers felt that the fifth postulate should be provable by the use of the others, and for two thousand years after Euclid, geometers attempted to do so. Ptolemy's was among the first important attempts.

Ptolemy's proof is, however, flawed (a fact that Proclus was aware of), so we will call it a "demonstration" and note that, in fact, it is invalid. According See Problem 15. to Proclus, Ptolemy began with a number of propositions equivalent to the first part of Euclid, Book I (remember that the first half of Book I of Euclid did not rely on the parallel postulate, and indeed, in some cases where the parallel postulate would have greatly simplified the proof, Euclid deliberately avoided it). Then Ptolemy proved

Proposition 7.6. *Let lines AB, CD cross EH at F, G, so that angles BFG, DGF are together equal to two right angles. Then the two lines do not meet.*

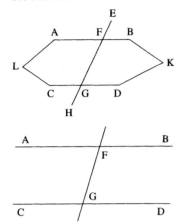

See Problem 14.

Proof. Suppose the lines meet at K. Since angles AFG, FGC are also equal to two right angles, they must also meet, suppose at L. Hence the lines AB and CD enclose an area, which is impossible. Thus, the lines cannot meet. $\square$

After this, Ptolemy claimed

Proposition 7.7. *If two lines are parallel, the interior angles on the same side of the transversal are equal to two right angles.*

Ptolemy's proof, not considered valid, is:

Flawed proof. Let AB, CD be parallel and cut by transversal FG. Suppose angles BFG, DGF are not equal to two right angles. Then they are either greater or less. If they are greater, then the angles AFG, CGF must be less. But this is impossible, since AF, CG are no more parallel than BF, GD, and hence if AFG, CGF are less than two right angles, so must BFG, DGF, which is impossible. Hence, angles BFG, DGF must be equal to two right angles. $\square$

From this, Ptolemy concluded the fifth postulate.

Proof. Suppose a line falling on two given lines makes the interior angles on one side less than two right angles. The two given lines must meet on the side where the interior angles are less than two right angles. If they do not, then they certainly do not meet on the side where the interior angles are greater than two right angles, and thus the lines are parallel. But if they are parallel, then the interior angles must be equal to two right angles, which is impossible. Hence, the lines must meet on the side where the interior angles are less than two right angles. $\square$

7.2.3   The Decline of Rome

In 161, Marcus Aurelius became emperor. Aurelius was intensely interested in learning, and his *Meditations* is considered one of the greatest works on philosophy ever written. In 176, in an attempt to revive classical learning in Athens, he established four endowed chairs of philosophy, as well as at least one chair in rhetoric.

Unfortunately, during the reign of Marcus Aurelius, the empire soon found itself under siege from both within and without, from enemies both human and inhuman. In 166, a plague, possibly the bubonic plague, struck the empire, causing enormous casualties. On the empire's frontiers, the Marcomanni, a Germanic tribe, crossed the Danube and threatened Roman territory. Although both of these threats were dealt with, Marcus Aurelius's reign foreshadowed the chaos to come.

7.2 Exercises

1. What is the relationship to the length of the chord and what we now call the sine of the central angle?

2. Show that the chord of 36° on a circle whose diameter is 120 parts has a length of approximately $37^P 4' 55''$, and the chord of 72° has a length of approximately $70^P 32' 3''$.

3. Find the chords of 90° and 120°. Round your answer to the second sexagesimal place.

4. Use Ptolemy's theorem to find the chords of: 15°, 9°, 12°, 78°, 3°.

5. Complete the proof of Ptolemy's method of finding half chords.

    (a) Show triangle EDC is isosceles, with equal sides DC, DE.

    (b) Show that triangle ACD is similar to triangle DCF.

6. Complete the proof of Ptolemy's statement that, given chords AB, BC of known length, then the chord AC may be determined.

    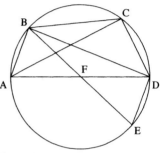

    Figure for Problem 6.

    (a) Begin by drawing the diameter BFE, and the chords BD, DC, CE, DE.

    (b) Note that $BCDE$ is a quadrilateral inscribed in a circle; use this to find CD.

    (c) Show how CD may be used to determine AC.

7. Determine the chord of $1\frac{1}{2}°$ and $\frac{3}{4}°$, and note what chords had to be computed to find this chord.

8. Determine, using one or more of Ptolemy's theorems and results, the chords with central angles:

    (a) 120°   (b) 81°   (c) 18°   (d) 6°

9. Using Ptolemy's theorem, the half-chord theorem; and the propositions regarding the sines of sums and differences, determine the central angles θ that may have their chords determined as precisely as one wishes. For example, the chord of 15° can be found using the difference of sines, as the sine of 45° − 15°. Use these values to explain why you cannot find the chord of exactly 1°.

10. Approximate the circumference of a circle with a diameter of 120 parts by using the value for the chord of 1°, and again for the value of the chord of $\frac{3}{4}°$. How accurate are these values?

11. Find the chords of $1\frac{1}{2}°$, 2°, $2\frac{1}{2}°$, 3°. Also find, as Ptolemy did, one thirtieth the difference between the successive chords (the chord of 0° can be taken to be 0), carried out to the third sexagesimal place. Verify your results by comparing them with the sines of the corresponding angles.

12. Use Table 7.1 to find the chords of the following. Verify the result using the sines of the corresponding angles. How accurate are the results obtained?

    (a) $20°1'$ (b) $20°36'$ (c) $21°12'$

13. Consider the major practical applications of trigonometry. For which of these applications is it more practical to use a table of chords than to use a modern sine table? Explain your reasoning.

14. Ptolemy asserted in his proof of Proposition 7.6 that two lines cannot enclose an area. Prove this statement.

15. Carefully examine Ptolemy's proof of the fifth postulate. What is Ptolemy assuming that is unwarranted?

7.3   Heron of Alexandria

Heron's dates are very uncertain. In 1896, a copy of Heron's *On Measurement* was discovered, which referred to Archimedes, Apollonius, and a work "on chords in a circle." The earliest known work on chords in a circle was that of Hipparchus, so Heron probably lived after 150 B.C. On the other hand, Pappus refers to Heron's work, so Heron lived before 250 A.D. Heron mentioned an eclipse he observed, which some have identified as being that of March 13, 62 A.D. But Heron also described a "well known" method for measuring the distance between two places on the Earth with the same latitude, whereas Ptolemy claimed he was the first to find such a method. Most historians place Heron around 200 A.D.

By then, a dark change had come over the empire. Marcus Aurelius, the last of the "Five Good Emperors," chose his son Commodus to be his successor. Besides being incompetent, Commodus was insane, believing himself to be the god Hercules. He entered gladiatorial games, killing wild animals in the arena, and even engaged in combats with opponents (the combats were, presumably, rigged). Tired of his excesses, a group of conspirators (including his mistress, Marcia) arranged to have Commodus strangled in 192 A.D. by a professional wrestler.

Commodus's successor, Pertinax, was a good man who sought to replenish the treasury by cutting expenditures: in particular, he refused to bribe the Praetorian Guard, the personal bodyguard of the emperor and the most powerful military force in the city of Rome. They killed him, and sold the empire to the highest bidder, a Senator named Marcus Didius Julianus. But Lucius Septimius Severus, a general commanding the Danubian legions, marched on Rome with the intent of declaring himself emperor; the Praetorians, knowing they stood no chance against real soldiers, deposed Julianus as Severus entered Rome. Julianus was executed after holding the throne for only two months, protesting to the end that he had done nothing wrong. If the Praetorians thought their support of Severus would save them, they were wrong: Severus disbanded the guard and established a precedent of military rule. Peace broke out: the peace of the sword. It would last less than one generation.

It was during this generation of enforced peace that Heron probably lived. Heron was a masterful engineer, and his *Pneumatics* was on the uses of compressed air and steam to run devices. An early form of the steam engine is attributed to him: a ball, set on an axle, which could be filled with water. When heated, the water would jet out through two nozzles, and set the ball spinning. This invention should have merited Heron's inclusion in a list of master engineers compiled by the Roman architect Vitruvius, writing around 14 B.C.; the omission of Heron adds weight to the conjecture that Heron lived some time after Ptolemy.

Besides his work in engineering, Heron also produced several works of mathematics, such as a very popular commentary on Euclid. Indeed, it is possible that it was through Heron's work that the first seven definitions of the *Elements* appeared in the form we have them. However, his works on mathematics were See page 88. distinctly inferior to the great works of the previous generations of Greek geometers. The *Geometry* begins with a compilation of rules for finding areas and perimeters. By Heron's time, "geometry" had taken on new meaning as "surveying," and thus, because of the practical nature of this work, the formulas were given without explanation, and no distinction was made between approximate and exact results. Like the ancient Egyptians, Heron calculated a specific area or perimeter, though he also gave a summary of the relevant rule:

Rule 7.1. *To find the area of a circle: square the diameter, multiply by* 14; *divide by* 11. *To find the circumference: multiply the diameter by* 22, *then divide by* 7.

Heron cited Archimedes as his source for both of these formulas.

In the eighth problem of the first book of *Measurement*, Heron introduces what is now called **Heron's formula** for the area of a triangle. The description is entirely verbal:

Problem 7.1. *Find the area of a triangle with sides* 7, 8, 9.

Solution. [Heron's Formula] Add the sides, making 24. Take half, which is 12. From this subtract 7, leaving 5; subtract 8, leaving 4; subtract 9, leaving 3. Multiply 12 by 5, which is 60; multiply this by 4, which is 240; multiply this by 3, which is 720. The square root is the area. □

This is one of the few cases where Heron actually supplied a proof.

Proof. In triangle *ABC*, inscribe circle *DEZ* with center at *H*; join *AH*, *BH*, *CH*, and radii *DH*, *ZH*, *EH* (see figure in margin). *AB* times *DH* is twice triangle *ABH*; likewise, *AC* times *ZH* is twice triangle *ACH*, and *BC* times *EH* is twice triangle *BCH*; thus the perimeter of *ABC* times *EH* is twice triangle *ABC*. Extend *CB* to *T*, so that *BT* = *AD*. Then *CT* is half the perimeter of the triangle (since if two tangents are drawn to a circle from a point, the tangents are equal), and so *CT* times *EH* is the area of the triangle *ABC*. Now,

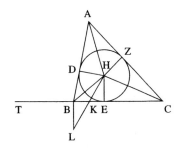

$\frac{a}{b} = \frac{c}{d}$ implies $\frac{a+b}{b} = \frac{c+d}{d}$.

CT times EH is the root of the square on CT times the root of the square on EH, so the product of the squares on CT, EH is the square of the area of the triangle.

Draw HL, BL perpendicular to CH, BC respectively; join CL. Thus, a circle can be circumscribed about quadrilateral $CHBL$, and angles CHB, CLB are together equal to two right angles. But angles CHB, AHD are together equal to angles AHC, DHB, and all four are together equal to four right angles, so angles CHB, AHD are equal to two right angles, and thus angle CLB, AHD are equal. Also, the right angles CBL, ADH are equal to each other, so triangles CBL and ADH are similar.

Thus, BC is to BL as AD is to DH. Since AD is equal to BT and DH is a radius and equal to EH, we have BC is to BL as BT is to EH. Alternating, this gives us BC is to BT as BL is to EH. But triangle KBL is similar to triangle KEH, so BL is to EH as BK is to KE, and thus BC is to BT as BK is to KE. Hence CT is to BT as BE is to KE.

Hence, the square of CT is to the product of BT, CT as the product of BE, EC is to the product of KE, EC. But the product of KE, EC is equal to the square of EH, so the square of CT is to the product of BT, CT as the product of BE, EC is to the square of EH. Thus, the square of CT, multiplied by the square of EH, is equal to the product of CT, TB multiplied by the product of CE, EB. But CT, TB, CE, EB are all known: CT is half the perimeter; TB is half the perimeter less side BC; CE is half the perimeter less side AB; and EB is half the perimeter less side AC. Hence, the area can be found.   □

7.3 Exercises

1. Comment on Heron's area formulas. How accurate are they?

2. Heron uses the proposition that if each diagonal of a quadrilateral is perpendicular to a side, a circle can be circumscribed about the quadrilateral. Prove this.

3. Comment on Heron's proof. What features would make it unacceptable to a Euclidean geometer?

7.4   Diophantus

Of all the figures in the history of mathematics, none is more enigmatic than Diophantus of Alexandria. Even the exact century of his life is unknown. Most of the evidence is negative: Diophantus quotes Hypsicles (150 B.C.), and is in turn quoted by Theon of Alexandria (350 A.D.); hence he lived sometime in the five hundred years between the two.

Until recently, the best positive evidence for Diophantus's date has been a letter of Michael Psellus (eleventh century century A.D.), a Byzantine historian. According to this letter, Anatolius, the Bishop of Laodicea, wrote a book on the Egyptian method of calculation. This book was dedicated to Diophantus. Anatolius wrote his work around 278-9 A.D.; hence, this suggests Diophantus lived during the third century A.D., and the date that has usually been given to him is thus 250 A.D.

This tentative dating has recently been called into question. Based on a study of a work called *Definitions of Terms of Geometry*, which was originally attributed to Heron of Alexandria, historian of mathematics Wilbur R. Knorr suggests this work was actually written by Diophantus. The style of the text very closely mimics the style of Diophantus; Knorr suggests that the dedication of the work of Anatolius referred to by Psellus was for a different Diophantus. Thus, it is possible that Diophantus was a contemporary of Heron of Alexandria, or possibly even Heron's predecessor.

Diophantus's contribution to mathematics was mostly in the field of what we would now call algebra; his *Arithmetic* is, in fact, the only known Greek algebra to have survived to the present day. Appropriately enough, an algebra problem, in the form of an epigram (a short, tombstone inscription), commemorates Diophantus's life, though the problem, given below, is far simpler than those Diophantus normally dealt with.

Problem 7.2. *For one-sixth of his life, Diophantus was a boy. A twelfth part later, he grew a beard, and married after a seventh. Five years later had a son. The son lived only half the father's life, and Diophantus died four years later.*

The *Arithmetic* is addressed to Dionysius, a friend of Diophantus who had expressed interest in learning the "science of numbers." Beginning with simple foundations and working through to more difficult concepts, Diophantus hoped to instruct Dionysius well, and expressed faith that Dionysius would learn rapidly, because he was both enthusiastic about the subject and had a good teacher (Diophantus himself).

Diophantus concluded his introduction by noting that the *Arithmetic* was divided into 13 books. Only six books are known to have survived in Greek; their survival is attributed to the commentary on these books by Hypatia. The remaining seven books were considered irretrievably lost, though several Renaissance mathematicians claim to have seen a complete manuscript in Rome.

In 1968, an Arabic manuscript in a library in Meshed, Iran, was found claiming to be a translation by Qusta ibn Luqa (d. 912) of Books IV through VII of Diophantus's *Arithmetic*. There is some question as to whether the books are authentic Diophantus, as the problems contained tend to be much simpler than those in the extant copies of the *Arithmetic*. It is possible that they are added material, possibly written by Hypatia as part of her commentary on the *Arithmetic*.

The placement of the lost books, assuming they are part of Diophantus's original manuscript, is problematic, since the six Greek books have already been designated books one through six. Based on internal evidence, it seems the four "lost" books should go between books three and four of the Greek text. We will use the traditional designations of Books I through VI for the Greek books, and

Symbol	Diophantus's Term	English	Modern Notation
ς	arithmos	number	x
Δ^Y	dynamos	power	x^2
K^Y	cubos	cube	x^3
$\Delta^Y\Delta$	dynamodynamos	square square	x^4
ΔK^Y	dynamocubos	square cube	x^5
$K^Y K$	cubocubos	cube cube	x^6

Table 7.2: Diophantus's Notation for Positive Powers.

The sequence of extant books is thus: I, II, III, A, B, C, D, IV, V, VI.

A through D for the four Arabic books. Book IV is Book IV of the Greek edition, while Book A is what Diophantus would have considered the fourth book.

Diophantus returned to the notion of number as a collection of units, and though he speaks of square and cube numbers, these are divorced from their geometric meanings. Hence, Diophantus will freely speak of adding squares together to make cubes, and consider squares of squares and other concepts which a Euclidean geometer would find absurd.

7.4.1 Syncopated Notation

Diophantus and those who followed him for the next thousand years would consider "species" of the unknown quantity, corresponding to our "degrees." In Diophantus, we see the beginning of a form of algebraic notation. It is a form known as **syncopated**: essentially, a set of abbreviations. To indicate the unknown quantity itself, the number which is "an indeterminate multitude of units" he used a Greek terminal 's', written ς, possibly because it was the last letter in the Greek word for number, *arithmos*, but more likely because, alone among the Greek letters, it was not used to represent a number. The various species of the unknown are summarized in Table 7.2.

In Diophantus's system, these symbols preceded the coefficient; thus, where we would write $5x^2$, Diophantus would have written $\Delta^Y \epsilon$. To indicate reciprocals, Diophantus used a χ, placed as a superscript; thus $K^{Y\chi}$ would indicate $\frac{1}{x^3}$.

To indicate the units, Diophantus used the symbol $\overset{\circ}{\mathrm{M}}$. Because Greek numerals were additive, Diophantus felt no need, as we do, to separate the terms of a polynomial expression using an explicit addition symbol; he simply wrote them in sequence. To indicate subtraction, Diophantus used a truncated Ψ turned upside down, $\wedge$, and wrote all terms to be subtracted after it. Thus, for $8x^3 - 5x^2 - 3x + 10$, Diophantus would have written $K^Y \eta \overset{\circ}{\mathrm{M}} \iota \quad \wedge \quad \Delta^Y \epsilon \varsigma \gamma$

To remain true to Diophantine notation, we will use the following convention: we shall indicate the species of numbers with x for unknown, S for the square of the unknown, C for the cube, SS for the square square (i.e., the fourth power), SC for the square cube (or fifth power), and CC for the cube cube (or sixth power); pure numbers will be indicated by U (for "unit"). We will indicate the reciprocals using the Diophantine $^\chi$, and thus $\frac{1}{x^3}$ would be $1C^\chi$. Coefficients will be written as we are used to, in front, and M (for "minus") will be used to set off the subtracted terms. Thus, $3x^3 - 5x^2 + 2x - 10$ becomes $3C\ 2x\ M\ 5S\ 10U$.

Example 7.4. *Express $2x - 5$ and its square, in Diophantine notation. $2x - 5$ is two unknowns minus 5 units, or $2x\ M\ 5U$; the square is 4 squares and 25 units minus 20 unknowns, or $4S\ 25U\ M\ 20x$.*

Diophantus's work gives no general solutions, and is thus similar to the Egyptian or Babylonian style of writing mathematics; indeed, Diophantus has been called the "flower of Babylonian algebra." Diophantus outlined his general method: the equations need to be reduced so that one term of one species is equal to one term of another species, adding subtracted terms to both sides as needed.

7.4.2 Equation Solving

The *Arithmetic* begins with

Problem 7.3. *To divide a given number into two, having a given difference. Given number* 100, *given difference* 40.

Solution. Let $1x$ be one of the numbers, $1x\ 40U$ the other. Hence $2x\ 40U$ is 100, and thus x is 30. □

This first problem illustrates one of Diophantus's key techniques. Having only one symbol for a variable, he is careful to express the other unknown quantities so they satisfy one of the conditions of the problem. In this case, by making the other unknown $1x\ 40U$, the condition that the two unknowns had a difference of 40 was satisfied.

Problem 7.4. *To find two numbers in a given ratio and such that their difference is also given. Given ratio* 5 : 1; *given difference* 20.

Solution. Let the numbers be $5x$, $1x$. Therefore $4x$ is 20, and x is 5. The numbers are 25, 5. □

7.4.3 Diophantine Equations

The problems in Diophantus rapidly become more difficult. Many have an infinite number of rational solutions, and give rise to what are now called **Diophantine equations**. Though today, the solutions to Diophantine equations are limited to whole numbers, Diophantus simply solved the equations for a single positive, rational value. Without a doubt the most famous Diophantine problem is the eighth problem of Book II:

Problem 7.5. *To divide a given square number into two squares. Given square number* 16.

Solution. Let $1S$ be one of the squares; the other is $16U\ M\ 1S$, which must be equal to a square. Let it be equal to the square of any number of x, less the square root of 16, for example, $2x\ M\ 4U$. Thus, $16U\ M\ 1S$ is equal to $4S\ 16U\ M\ 16x$. Thus, $5S$ is equal to $16x$, and thus x is $\frac{16}{5}$. The required squares are thus $\frac{256}{25}$ and $\frac{144}{25}$. ☐

This would lead, eventually, to "Fermat's Last Theorem."

A more complex illustration of how Diophantus took great care in choosing his unknown quantities is:

Problem 7.6. *To find two numbers so that the square of either minus the other is a square.*

Solution. Suppose the numbers are $1x\ 1U$, $2x\ 1U$, satisfying one property.

[Note that

$$(x+1)^2 - (2x+1) = x^2$$

hence the square of the first, minus the second, is a square, satisfying one property.]

To satisfy the other, it is required that $4S\ 4x\ 1U$ minus $1x\ 1U$, or $4S\ 3x$ be a square. Suppose this square is $9S$.

[Assume

$$(2x+1)^2 - (x+1) = 9x^2$$

which will ensure that the square of the second, minus the first, is also a square.]

Thus, $5S$ is $3x$, and thus x is $\frac{3}{5}$, and the required numbers are $\frac{8}{5}$, $\frac{11}{5}$. ☐

An elegant, if complex, example of the use of the method of false position is given by the following problem.

Problem 7.7. *To find three numbers so that the sum of any two times the third is a given number. Suppose the sum of the first and second, times the third, is equal to 35; the sum of the second and third, times the first, is equal to 27; and the sum of the third and first, times the second, is equal to 32.*

Solution. Let the third number be $1x$. Then the first and the second together must be $35x^\chi$.

[The product of the sum of the first and second with $1x$ must be 35, so the sum of the first and second must be $\frac{35}{x}$.]

Suppose the first is $10x^\chi$ and the second is $25x^\chi$. The sum of second and third, times the first, is $250S^\chi\ 10U$, which is equal to $27U$. The sum of the third and first, times the second, is $250S^\chi\ 25U$, which is equal to $32U$. But these equations are inconsistent.

$$\frac{250}{x^2} + 10 = 27$$
$$\frac{250}{x^2} + 25 = 32$$

So we need to divide 35 into two parts so that the difference is 5. This can be done, and the two parts are 20, 15. Thus, taking $15x^\chi$ for the first and $20x^\chi$ for the second, we have $300S^\chi\ 15U$ equal to $27U$, and thus x is 5 and the solutions are 3, 4, 5. ☐

A lemma preceding problem the fifteenth problem of Book VI deals with the so-called **Pell equation** of the form $Ax^2 - B = y^2$.

Lemma 7.1. *Given two numbers and a square. If one of the numbers times the square minus the other number is a square, another square larger than the first square can be found that satisfies the same property.*

Example 7.5. *Given the numbers 3, 11, and the square of 5; these satisfy the first requirement, since 3 times 25 minus 11 is 64, a square. Let the square of $1x$ $5U$ be the new square. Then $3S$ $30x$ $64U$ is a square, say the square on $8U$ M $2x$, or $4S$ $64U$ M $32x$. Thus, $3S$ $30x$ $64U$ is $4S$ $64U$ M $32x$, or $1S$ is $62x$, so x is 62. Thus the side of the new square is 67, and the square itself is 4489.*

7.4.4 Adequating

In two problems Diophantus used a procedure his Latin translators would call *adequatio* (adequating). The simpler problem is the ninth problem of Book V:

Problem 7.8. *To divide unity into two parts such that, if the same given number be added to either part, the result will be square. Given number 6.*

Attached to the problem is a condition. The first part reads "The given number cannot be odd." The text of the second part has been lost to the passage of time, so it is not clear what Diophantus required, though we may reconstruct it using modern methods of number theory.

See Problem 7.

In any case, Diophantus's procedure is the following:

Solution. If we divide 13 into two squares the difference of which is less than 1, we solve the problem.

Take half of 13, or $6\frac{1}{2}$, and add to it a small square that makes it square. Or, multiplying it by 4, we have to make $1S^x$ $26U$ a square, or $26S$ $1U$ a square. Suppose it is the square of $5x$ $1U$, whence x is 10. That is, in order to make 26 a square, we must add $\frac{1}{100}$, and so to make $6\frac{1}{2}$ a square, we must add $\frac{1}{400}$, and $6\frac{1}{2} + \frac{1}{400} = \left(\frac{51}{20}\right)^2$.

Therefore, we must divide 13 into two squares whose sides are as nearly as possible equal to $\frac{51}{20}$. Thus, we seek two numbers such that 3 minus the first is very nearly $\frac{51}{20}$, so that the first number is very nearly $\frac{9}{20}$, and 2 plus the second is very nearly $\frac{51}{20}$, so that the second is very nearly $\frac{11}{20}$. We write accordingly the squares of $11x$ $2U$, $3U$ M $9x$ for the required squares.

[Since $3^2 + 2^2 = 13$, find x so that $(3 - 9x)^2 + (2 + 11x)^2 = 13$.]

The sum is then $202S$ $13U$ M $10x$ equal to 13, and thus $x = \frac{5}{101}$, and the sides are $\frac{257}{101}$ and $\frac{258}{101}$. Subtracting 6 from the squares of each, we have, as parts of unity, $\frac{4843}{10201}$, $\frac{5358}{10201}$. □

7.4.5 Negative Numbers

In at least one case (Book V, Problem 2), Diophantus's method originally led to a negative value, which he rejected.

Problem 7.9. *Find three numbers in geometrical progression so that each of them, added to a given number, gives a square. Given number 20.*

Solution. Take a square that, added to 20, gives a square: say 16. Let this be one of the extremes; let the other be $1S$, so the mean is $4x$ (remember that if three terms are in geometric progression, the product of the extremes is the square of the mean). Thus it is required that both $1S\ 20U$ and $4x\ 20U$ be squares. Their difference is $1S\ M\ 4x$, or $1x$ times $1x\ M4U$. The usual method gives $4x\ 20U$ equal to 4, which is absurd, because $4x\ 20U$ ought to be greater than 20.   □

The "usual method" Diophantus is referring to stems from the fact that the difference of two squares is the product of the sum and difference of the roots: assuming the sum and difference to be, respectively, $1x$ and $1x\ M4U$, then the smaller square is the square of half the difference.

See Problem 10.

7.4 Exercises

1. Solve the epigram to determine how long Diophantus lived.

2. Write the following polynomials in Diophantine notation.

    (a) $x^3 + 4x^2 - 6x - 5$ (b) $3x^2 - 4x - 5$

3. For each of the following polynomials, write the expression inside the parentheses in Diophantine notation, then expand it and write the expansion in Diophantine notation.

    (a) $(2x - 3)^2$ (c) $(4 - 2x)^2$
    (b) $(5x + 2)^2$ (d) $(x^2 + 3x - 5)^2$

4. Solve the following problems using Diophantus's techniques.

    (a) Book I, Problem 3: Separate a given number so that one is a given ratio to the other plus a given difference. Given number 80, given ratio $3 : 1$, and given difference 4.

    (b) Book I, Problem 7: From the same number, subtract two given numbers so the remainders have a given ratio; find the number. Given numbers to be subtracted, 100, 20, and given ratio $3 : 1$.

    (c) Book I, Problem 9: From two given numbers, subtract the same number so the remainders have a given ratio; find the subtracted number. Given numbers 20, 100; given ratio $6 : 1$.

(d) Book I, Problem 11: To the same number, add and subtract given numbers so the results have a given ratio; find the number. Given numbers 20 and 100, and given ratio 3 : 1.

5. Solve the following problems using Diophantus's techniques.

   (a) Book II, Problem 22: To find two numbers such that the square of either added to the sum of both gives a square.

   (b) Book II, Problem 23: To find two numbers such that the square of either minus the sum of both gives a square.

   (c) Book A, Problem 3: Find two square numbers whose sum is a cube.

6. Most of Diophantus's problems have an infinite number of solutions, differing primarily by the choice of the form of the unknown. Find other solutions for the following problems.

   (a) In Problem 7.5, Diophantus assumed that $16U$ M $1S$ is equal to the square of $2x$ M $4U$. Find other squares $16U$ M $1S$ could be equal to, then find the corresponding solutions.

   (b) In Problem 7.6, Diophantus assumed that $4S$ $3x$ was equal to the square $9S$. Find other squares $4S$ $3x$ could be equal to, then find the corresponding solutions.

7. Examine Lemma 7.1. Diophantus's solution states that the side of the new square 67, and the square itself 4489. In other words, $3(67)^2 - 11$ is a square.

   (a) Determine the square.

   (b) This is the square of $8U$ M $2x$, but Diophantus does not evaluate this quantity. What difficulties would arise if Diophantus did evaluate the quantity?

   (c) If Diophantus let the square be $2x$ M $8U$, what problems would arise in the solution?

   (d) Is Diophantus justified in calling his result a "lemma"? In other words, is this a rigorous proof? Explain.

8. Use the method in Lemma 7.1 to solve the following equations, given one solution.

   (a) Given the numbers 2, 1, and the square of 5. (In other words, find other numbers x, y, so that $2x^2 - 1 = y^2$, where $x = 5$, $y = 7$ is already a known solution. Note that $x = 5$, $y = 7$ is a pair of side and diameter numbers. See Problem 4.)

   (b) Given the number 3, 12, and the square of 4.

   (c) Given the numbers 5, 76 and the square of 7.

9. Provide a commentary on Problem 7.8.

   (a) The complete condition for the lemma, as determined by Fermat, is "The given number must not be odd, and the double of it increased by one, when divided by the greatest square which divides it, must not be divisible by a prime number of the form $4n - 1$." Prove this condition.

   (b) Given the number 6, why does Diophantus begin by dividing 13 into two squares each greater than 6? Explain the validity of Diophantus's transformation of the equation $6\frac{1}{2} + \frac{1}{x^2} = $ a square into $x^2 + 26 = $ a square.

   (c) Find another solution.

10. In the solution to Problem 7.9, Diophantus assumed that if the difference of two squares is a product $1x$ times $1x$ M $4U$, the lesser square is the square of half the difference of the two factors.

    (a) Does this assumption work? In other words, if the difference of two squares can be written as a product, can the method be used to find two squares whose difference is the given product?

    (b) Is this assumption valid? In other words, given the difference of two squares that can be written as the product of two factors, will this method always give you one of the two squares?

11. Diophantus completed solving Problem 7.9 by letting the square $1x$ $9U$ be one of the extremes; by adding this to 16 one should obtain a square, which Diophantus assumed was the square of $1x$ M $11U$. Complete the solution.

12. Diophantus has been called the "father of algebraic notation." Discuss this claim.

7.5   The Decline of Classical Learning

A generation after Severus, the empire was again torn apart by civil war. Finally, a dour general named Diocletian became emperor in 284. Diocletian made two lasting changes to the empire. To avoid the fate of his predecessors, who reigned an average of two years before being murdered, Diocletian took on the trappings of an Oriental monarch, where the ruler was more god than man. One consequence was that if the emperor was a god, it was treason not to worship him. In particular, this meant the new sect of Christianity, whose adherents refused to worship any god but their own, was a threat to the safety of the empire. Diocletian began a massive persecution of Christians—the last the Roman world would see.

The other reform of Diocletian was to divide the empire into two halves, an eastern and a western half, to be ruled by *Augusti*. The co-emperors would each have co-rulers, called *Caesars*; the four rulers formed the *tetrarchy* (from the Greek word for "four rulers"). In theory, the Caesars would become emperors, after having gained experience governing. Diocletian retired in 305, and persuaded his fellow *Augustus* Maximian to do likewise. Diocletian's reward: he lived until 313, and was one of a small handful of Roman emperors to have died of natural causes.

To help stabilize the empire further, Diocletian mandated that all occupations were to be hereditary: sons had to follow in the professions of their fathers. Certain professions were forbidden entirely; in one famous edict, Diocletian forbade the practice of **mathematics**. Diocletian was joined in his distaste for mathematicians by the Christian thinker Augustine of Hippo. However, the mathematics and mathematicians criticized by Diocletian and Augustine were charlatans: astrologers, numerologists, and other practitioners of useless superstition.

What we now call mathematics was called **geometry**, and a practitioner was a geometer; indeed, the Diocletian edict goes on to praise the work of **geometers**, who provide useful services to the state. As late as the eighteenth century, mathematics and mathematician were terms of insult; we will retain the more general terms mathematics and mathematician, and note that no slur is intended with the use of the word!

7.5.1 Pappus of Alexandria

Though the Roman world produced some great advances in practical mathematics, primarily through the work of Heron and Ptolemy, theoretical mathematics had languished since the time of Apollonius. Centuries after Apollonius's death, Pappus of Alexandria attempted to rekindle interest in theoretical mathematics.

Very little is known about Pappus's life; even his dates are uncertain, though according to a marginal note in a copy of one of his works, Pappus lived during the reign of the Roman Emperor Diocletian. If so, he may have been the last mathematician to have full access to the Library of Alexandria, established in the time of Euclid. In 294 , Egypt revolted from Roman rule. Diocletian sent an army to deal with the rebels. The army besieged and eventually captured Alexandria but during the siege, a fire was started (no one knows by whom) that consumed half the city, and destroyed much of the main library. The smaller library at the Temple of Sarapis remained, but the Great Library never recovered.

Pappus's greatest work was undoubtedly his *Mathematical Collection*, in eight books. The *Collection* comes to us from a tenth-century Greek manuscript, which is missing Book I, the first half of Book II, and the end of Book VIII. There is some internal evidence that the *Collection* was meant to contain twelve books, but whether the extra four books have been lost or never were cannot be determined.

The intent of the *Collection* is clear: it is a handbook of Greek geometry. It was not designed as a textbook; rather, it was apparently meant to be used in conjunction with the extant Greek texts, to help students understand Greek geometry by, in many cases, filling in the details, or by offering alternative proofs. At least part of the motivating factor, judging by some of the non-mathematical comments interspersed throughout the work, was that the state of mathematical education was at a particularly low ebb, and in several places, Pappus lamented the lack of proficiency in both his colleagues and their students.

One such place is in Book III of the *Collection*. The introduction addressed Pandrosion, a woman teacher of mathematics. Nothing is known about her aside from this mention in Pappus, but the name was an extremely uncommon one in the classical world: Pandrosos was the (legendary) daughter of Cecrops, a king of Athens; the pandroseion, site of the sacred olive tree of Athens, was named in her honor (hence, it is possible that Pandrosion is a pseudonym).

See page 65.

Pappus apparently encountered some of Pandrosion's students, who claimed to be able to solve some of the classical problems from antiquity, such as the trisection of an angle, using compass and straightedge methods alone. The solutions were invalid, and Pappus, noting that Pandrosion should take heed, proceeded to give the actual solutions to the problems. At the beginning, he classified geometric problems into three types: **plane**, so called because the solutions involved curves from plane geometry (i.e., circles and straight lines); **solid**, so called because the curves originated in solid figures (i.e., the conic sections); and **linear**, a catch-all phrase for problems requiring curves other than plane or solid (e.g., the quadratrix or the conchoid).

It was desirable to solve a problem using plane methods; if these proved unsuccessful, solid methods were the next best choice; and only if neither plane nor solid solutions could be found was the geometer to use the linear solutions.

Isoperimetric Figures

Book V is, perhaps, the most interesting of the extant works of the *Collection*, for it deals with figures of equal perimeter, or **isoperimetric figures**. Pappus addressed the work to Megethius, probably a colleague. Pappus began by noting even bees could appreciate mathematics, since honeycombs were hexagonal: of all the figures that could fill the space about a point, the hexagonal prism contained the most volume for the amount of wax required for its sides. Man, Pappus continued, is possessed of a greater reasoning power than the bees, and consequently can (and should) discern something more: that of all the regular polygonal plane figures of the same perimeter, the one with the greater number of sides has the greater area, and the area of the circle is greater than that of any regular polygon of equal area.

The first part follows the work of another mathematician named Zenodorus (first cent. A.D.?), and concluded by showing that the circle has an area greater than that of any polygon of equal perimeter.

First, Pappus proved

Proposition 7.8 (Area of a Sector). *A circle is to a sector of it as the circumference of the circle is to the arc of the sector.*

Next, he proved

Proposition 7.9. *If ABC is a right angle, and AD is the arc of a circle with center C and radius AC, then sector ADC is to region ADB in a ratio greater than that of a right angle to angle BCA.*

This proposition is necessary to prove

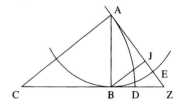

Proposition 7.10. *If ABC is a right angle, and AD is the arc of a circle with center at C and radius AC, then sector ACD is to ABD in ratio greater than that of a right angle to angle ACD.*

Proof. Draw circle ZCE with center A and radius AC, with AE perpendicular to AC and Z on the extension of AB. Since ZCE and AD are equal sectors, and angle ACD is greater than angle CAE, then sector ACD is greater than sector CAE.

Thus, sector ACD is to triangle ABC in ratio greater than that of sector CAE to triangle ABC, which in turn has a ratio greater than that of sector CAE to sector CAZ.

Angle EAC is to angle CAZ as sector CAE is to sector CAZ; hence sector ACD is to triangle ABC in ratio greater than that of angle EAC is to angle CAZ. Inverting the ratio, we have triangle ABC is to sector ACD in ratio less than angle CAZ to angle EAC; combining the ratio we have region ABD is to sector ACD in ratio less than angle EAZ is to angle EAC; inverting the ratio again we have sector ACD is to region ABD in ratio greater than the right angle EAC to the angle EAZ, which is also angle ACD, which was to be proven. □

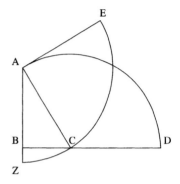

Finally, he proved that the arc enclosing a semicircle encloses the largest segment of all arcs of the same length.

Proposition 7.11. *If AC, DZ are two arcs of equal length, and AC is the arc of a semicircle, then the semicircle is greater than the segment enclosed by DZ.*

Proof. Suppose DZ is less than a semicircle, and suppose the centers of the circles are H, J. Draw HB perpendicular to AC, and EJ perpendicular to LM so arc DZ is bisected at K. $LJ : AH$ as arc $LE :$ arc AB; since $AB = DE$, then $LJ : AH =$ arc $LE :$ arc ED. But sector $LJE :$ sector $EJD =$ arc $LE :$ arc ED, so $LJ : AH =$ sector $LJE :$ sector EJD.

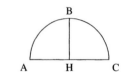

Also, sector $LJE :$ sector $AHB =$ sq. $LJ :$ sq. AH, so the ratio of sector LJE to sector AHB is the duplicate ratio of sector LJE to sector EJD; hence sector DJE is the mean proportional between sectors LJE and AHB. Thus sector LJE is to sector DJE as sector DJE is to sector AHB.

By Proposition 7.9, sector $DJE :$ region EDK in ratio greater than that of a right angle to angle DJE. But the ratio of a right angle to angle DJE is the same as that of sector $LJE :$ sector DJE.

Thus, sector $DJE :$ region EDK is in ratio greater than that of sector $DJE :$ sector AHB, and thus sector AHB is greater than region DKE; likewise for their doubles. Hence the segment ABC enclosed by the semicircle is greater than the segment DEZ enclosed by the equal arc on less than a semicircle.

If the arc DZ is greater than a semicircle, then the segment will likewise be smaller than the semicircle. □

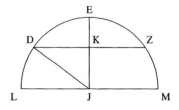

See Problem 2.

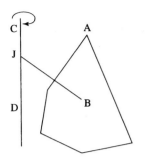

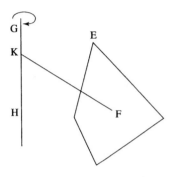

The Pappus-Guldin Theorem

Pappus did not, in general, indicate what contributions were those of other geometers, and which of those were his own. There is at least one instance, however, where Pappus claimed a result as his own. The claim appears as part of a scathing criticism of his contemporaries, in which he accuses them of occupying themselves with only the most elementary mathematics. He claimed to have proven a very general theorem about volumes. Pappus's language is difficult to follow; we will state Pappus's result as it relates to a specific case:

Proposition 7.12 (Pappus-Guldin Theorem). *If figure A with center of gravity B is rotated about the axis CD, and figure E with center of gravity F is rotated about the axis GH, and lines are drawn from the axis to the center of gravity so that the lines make equal angles with the axis (i.e., angles BJD, FKH are equal), then the ratio of the volumes of the solids so generated will be the product of the ratios between the lines drawn at similar angles from the axis to the centers of gravity, and the ratio of the figure A to the figure E.*

If the rotation is incomplete, then the volumes so generated will have a ratio to one another that is the product of the ratios between the arcs swept out by their centers of gravity, and the ratio of the figure A to the figure E.

Though Pappus claimed this result was his own, he did not give a proof for it. The result is perfectly general, though it seems that, at best, Pappus could have proven it only for figures bounded by straight lines, arcs of circles, or, possibly, curves of one of the conic sections.

The result is similar to that discovered by Paul Guldin (1577-1643), and Pappus's work may have influenced Guldin; on the other hand, it may have been an independent discovery by Guldin. For this reason, the result is usually known as the Pappus-Guldin theorem.

7.5.2   Theon and Hypatia

Diocletian's tetrarchy lasted only one generation, before a new round of civil wars began. It ended when Constantine became sole emperor in 313. Before the Battle of Milvian Bridge (312), Constantine was said to have seen a flaming cross, with the words, "With This Sign, Conquer." Constantine had his troops paint the symbol on their shields, and won the day. Though it is claimed that Constantine also converted to Christianity after the battle, he was not baptized (if he ever was) until many years later; however, he did soften the official stand on Christianity. In 313, he proclaimed the Edict of Milan, which officially ended the persecution of Christianity, and in 325, Constantine sponsored the Council of Nicaea, which formally established Christian doctrine. Christianity became a powerful force within the empire, and the ancient Greek learning, because it was pagan, came under increasing attack.

Our best sources of early geometry come from this twilight era of Greek scholarship. Theon of Alexandria (fl. 375) was one of the more important commentators. His edition of Euclid's *Elements*, written around 350, seems to be the primary source for almost all later copies. Theon also wrote commentaries on other works, and our knowledge of Greek logistic comes from his works.

Little is known for certain about Theon's daughter Hypatia (d. 415). The factual information is further clouded by the fact that no commentator was objective: she was either the devil incarnate or a martyr who died because of her pursuit of learning, and in both cases, the myth has overshadowed the reality. The facts about her life and work are these: she wrote a commentary on Diophantus's *Arithmetic*, and it is believed that the six books of Diophantus available in Greek are, in fact, copied from her now lost commentary. Theon credits her with helping him in his commentary on Ptolemy, which suggests she was quite learned in the mathematics and science of the time.

A portentous event occurred in 391, when Christian mobs destroyed the remaining portion of the library of Alexandria at the Temple of Sarapis. It was a dangerous time to be a pagan, and even more dangerous to be a female and a scholar; Hypatia's murder (about the only thing the legends agree upon) was practically inevitable. In March, 415 she was set upon by a fanatical mob of Christians who dragged her into the street, stripped her naked, then flayed the flesh from her body with shells (or tiles) before burning her, still alive.

7.5.3   The End of Classical Learning

If fifth century Alexandria was no longer a place where classical scholarship could survive, Athens, at least, retained a semblance of its pre-eminence, and the pagan Academies in Athens were still accepting students. Proclus (410-485), whose commentary on the first book of Euclid provides us with most of our information about the early history of geometry, was a teacher at one of the private academies in Athens, possibly the Academy of Plato. Fewer and fewer students were interested, however, for it was becoming clear that paganism was on the decline, and Christianity would be the religion of the future.

By Proclus's time, the capital of the empire had moved to Constantinople (modern Istanbul, Turkey), a city founded by Constantine. Italy itself was ruled by the Ostrogoths, nominally for the Roman emperor, but in fact as an independent kingdom. Still, Roman traditions were preserved. Anicius Manlius Severinus Boethius (ca. 480-524) was a consul in Rome, but got into a disagreement with Theodoric, the King of the Ostrogoths; Theodoric had him executed. While in prison, Boethius wrote *The Consolation of Philosophy*. Though Boethius himself was not a Christian, this work became one of the cornerstones of medieval Christian philosophy.

Like many Romans, Boethius had great admiration for the works of the Greeks. But most Romans could not read Greek, so Boethius began translating Greek works into Latin. In addition to the works of Aristotle and Plato, he also translated some of the works of the Greek mathematicians, and for this reason earns a minor place in the history of mathematics. However, the translations of Boethius point to the depths to which the study of mathematics had fallen:

Figure 7.2: The Tower of Knowledge: A Renaissance conception. At the top of the tower is theology or metaphysics. On the next tier down are Pythagoras (music), Euclid (geometry), and Ptolemy (astronomy), representing three of the quadrivial studies. Boston Public Library/Rare books Department. Courtesy of the Trustees.

inferior as they were, they still formed an essential part of the mathematical curriculum of the European Middle Ages.

Boethius gave partial translations of two major mathematical works into Latin. The first of these, and the only one which we have complete editions, was Nicomachus's *Arithmetic*. Boethius intended the translation to be a handbook, the first of four, one for each of the "mathematical sciences": arithmetic, music, geometry, and astronomy. In describing these four sciences, Boethius used the word **quadrivium** for the first time. By analogy, the three rhetorical subjects, grammar, rhetoric, and logic, became known as the **trivium**, from which we get our word "trivial."

The *Arithmetic* is a more or less complete translation of Nicomachus, though it does omit some of the more difficult sections. The second handbook, *Geometry*, no longer exists as a complete work, but surviving fragments show the work to consist of only the most elementary sections of Euclid. The definitions from Book I are included; some of the propositions (without proof) from Books III and IV; some practical geometry, such as area computations; and a discussion on computation using the abacus.

In 529 the Emperor Justinian closed the schools of pagan learning in Athens and elsewhere, and passed an edict that banned pagans from teaching, public office, and military service. Learning in the West declined precipitously, and western Europe entered a "Dark Age." Elsewhere in the world, and particularly in the Eurasian continent, learning and scholarship existed and thrived, and we now turn to the contributions of the Chinese, the Indians, and, most importantly, the medieval Arabs. It was in Arabia, where western geometrical concepts met eastern numerical concepts, and mingled in an atmosphere conducive to learning and scholarship, that modern mathematics was born.

7.5 Exercises

1. Prove Proposition 7.8. Do not use the formulas for the areas of the sector of a circle or of the length of an arc; use instead the definition (from Euclid) of the equality of ratios. Pappus's proof is broken down into two parts: where the arc is commensurable with the circumference, and where the arc is incommensurable with the circumference.

2. Prove the second half of Proposition 7.11: if the arc is greater than a semicircle, the segment on a semicircle is still larger.

Chapter 8

China and India

The earliest, non-legendary ruler of the land we now call China was the
Emperor Yao (ca. 2300 B.C.), whose architect, Kung Kung, built dikes along
the Yellow River to help control flooding. Eras of strong, centralized power
alternated with eras of disruption. Around 1000 B.C., the Chou barbarians
conquered China but found themselves so attracted by Chinese imperial culture
that they quickly assimilated the ideas and culture of the empire. But by the
fifth century B.C., Chou authority had so declined that the various kingdoms
under their rule rebelled, and China entered the Period of Warring States.

8.1 Chinese Numeration and Computation

By the Period of Warring States, the Chinese system of writing and numbering
had already been established. Chinese writing consists of ideographs, symbols
that encapsulate an idea. Some of these ideographs obviously originate in pic-
tures, such as the symbols for the numbers one through three (see Table 8.1).
The Chinese system of numbering, more so than any other system of numbering
we have yet encountered, utilizes the notion of decimal place value, and there
are plausible suggestions that the Hindu-Arabic numbers actually originated in
China. Indeed, the form of notation is so similar to our own that we will not
hesitate to use a Hindu-Arabic equivalent for a Chinese numerical expression.
There are symbols for the units one through nine, and to write out a number
larger than ten, the number of tens, hundreds, and so on, is written with a place
value symbol; for these, we will use "T" for tens, "H" for hundreds, "Th" for
thousands, "TTh" for ten thousands, "HTh" for hundred thousands, and "M"
for millions.

Example 8.1. *The number four hundred twenty-five would be 4 H 2 T 5.*

Example 8.2. *The number forty seven thousand, eight hundred sixty three would
be 4 TTh 7 Th 8 H 6 T 3.*

一 二 三 四 五 六 七 八 九
One　Two　Three　Four　Five　Six　Seven　Eight　Nine

十 十一 十二 十三 二十 三十
Ten　Eleven　Twelve　Thirteen　Twenty　Thirty

百 二百 六百四十五 千
Hundred　Two hundred　Six hundred forty-five　Thousand

Table 8.1: Chinese Numeration.

|	||	|||	||||	|||||	T	TT	TTT	TTTT
One	Two	Three	Four	Five	Six	Seven	Eight	Nine

Table 8.2: Rod Numbers.

Another system of writing numbers existed as well: the rod numerals (Table 8.2). As each group of rods corresponds to a single number symbol, and it is the *position* that determines the actual numerical value, we will write the rod numbers in our own form: thus, ||| will be written as the number 3. To solve the problem of an empty space, rod numerals in adjacent positions were rotated or flipped: thus, 33 was ≡ |||, and 303 was ||| |||, while | T was 66.

8.1.1　Counting Board Arithmetic

The rod numerals themselves were, like Roman numerals, especially suited for computation using a counting board, essentially a board divided into rows and columns to allow the placement of the numbers in their appropriate place positions. The rods would then be actual sticks, made of whatever cheap material was available. Computation with the rod numerals was very straightforward. For example, to multiply 83 by 27 , the last digit of the second number would be placed below the first digit of the first number:

		8	3
	2	7	

The multiplication would then proceed as follows: 2 times 8 is 16; the ones digit would be written above the 2, and the tens would be written in the next space over. Next, 7 times 8 is 56, and again, the ones digit was written above the 7, while the tens digit was written in the next space over. Since the 8 was "done with," it would then be crossed out (on the counting board, the rods representing the 8 would be removed). The 16 and 56 would be added in the appropriate columns, and the 27 moved over one space (so the last digit is below the 3); with rods, it is simply a matter of shifting the rods over.

		8̸	3
1	6		
	5	6	
	2	7	

$\xrightarrow{\text{add}}$

			3	
	2	1	6	
		2	7	

$\xrightarrow{\text{shift}}$

			3
2	1	6	
		2	7

Repeating the procedure: 2 times 3 is 6, written above the 2; 7 times 3 is 21, written above the 7. Adding 21 and 6 (i.e., 60) and erasing the 3, we have:

			8̸
		6	
		2	1
	2	1	6
		2	7

$\xrightarrow{\text{add}}$

	2	2	4	1
			2	7

Hence, the product is 2241.

Division is equally simple. To divide 18815 by 71, begin, as in multiplication, by placing the numbers atop one another. Since 71 does not go into 18, we shift it over.

1	8	8	1	5
7	1			

$\xrightarrow{\text{shift}}$

1	8	8	1	5
	7	1		

To divide 71 into 188, we can subtract 71 repeatedly, keeping track of how many times we have subtracted it by placing a rod in column of the 1 in 71.

1	8	8	1	5
	7	1		

$\xrightarrow{\text{subtract}}$

		1		
1	1	7	1	5
	7	1		

$\xrightarrow{\text{subtract}}$

		2		
	4	6	1	5
	7	1		

Thus the quotient is 2 with a remainder of 46. Shifting 71 over and again subtracting, we find 71 can be subtracted 6 times from 461.

	2		
4	6	1	5
7	1		

$\xrightarrow{\text{shift}}$

	2		
4	6	1	5
	7	1	

$\xrightarrow{\text{subtract}}$

	2	6	
	3	5	5
	7	1	

Shifting the 71 over one space, we find 357 divided by 71 is 5 with no remainder.

	2	6	
	3	5	5
	7	1	

$\xrightarrow{\text{shift}}$

	2	6	
	3	5	5
		7	1

$\xrightarrow{\text{subtract}}$

	2	6	5
		7	1

Thus, the quotient is 265.

8.1.2 Root Extractions

Procedures for the extraction of roots also existed. Some of these methods are very similar to those described by Theon of Alexandria and the medieval Indians, though there is no reason to suspect a connection between any of them. All rely on the fact that finding the nth root of K was equivalent to solving $(a+b)^n = K$, where $a^n \approx K$, a variation of a procedure sometimes called **Horner's method**. Thus, it was natural that the determination of binomial coefficients would enter into Chinese mathematics, and as early as 1050 A.D., in a work of Chia Hsien, the so-called **Pascal triangle** appeared in Chinese texts; Chia Hsien even included the fact that any entry of the triangle was the sum of the two entries immediately above it. The Pascal triangle appeared even earlier, in Indian sources of the tenth century.

By the thirteenth century, an ingenious algorithm for finding the nth roots of a number was known to Chinese mathematicians; the earliest appearance was in the *Hsiang Chieh Suan Fa* of Yang Hui (fl. thirteenth cent. A.D.), about whom little is known aside from his approximate date and his birthplace in Ch'ien-t'ang. Yang Hui himself claimed the work was taken from an earlier source. To find the fourth root of $1,336,336$, Yang Hui used the following procedure: first, set down the number in the *shih* column of a counting board, and place one rod in fourth space over (for a cube root, the rod should be in the third space over, and so on); this is the *hsia fa*, and the intervening columns are called *hsia lien*, *shiang lien*, and *li fang*; finally the "answer" column, adjacent to the *shih*, Yang Hui referred to as the *shang shang*. For convenience, we will refer to these by their initials; the initial counting board set-up is thus:

HF	HL	SL	LF	S	SS
1				$1,336,336$	

Find an initial value for the fourth root: in this case, 30, presumably obtained by trial and error, since $30^4 < 1,336,336 < 40^4$; write 30 in the SS column. Now, multiply SS by HF, and write the product (30) in the HL; multiply SS by HL, and write the product (900) in SL; multiply SS by SL and write the product, (27,000) in LF. Finally, multiply SS by LF and subtract this from S. We will call this sequence of steps the **first procedure**.

HF	HL	SL	LF	S	SS
1	30	900	$27,000$	$1,336,336$	30
				$-810,000$	
				$\overline{526,336}$	

Next comes a sequence of steps we will call the **second procedure**. First, take the SS and multiply it by the HF, and write the answer (30) in the HL and add it to the previous HL; multiply the total (60) by the SS, and write the product (1800) in the SL, and add it to the previous SL to obtain a new total, 2700. Finally, multiply the SL by the SS and write the answer (81,000) in the LF, and add it to get the new total, 108,000.

HF	HL	SL	LF	S	SS
1	30	900	27,000	1,336,336	30
	30	1800	81,000	−810,000	
	60	2700	108,000	526,336	

Again, multiply SS by HF and write the product in the HL, adding to get a new HL (90). Multiply SS by HL and write the product in SL, adding to get a new SL. Finally, multiply SS by HF and add the result in HL. Note that the multiplications performed in each repetition stop one column short of the previous repetition.

HF	HL	SL	LF	S	SS
1	30	900	27,000	1,336,336	30
	30	1800	81,000	−810,000	
	60	2700	108,000	526,336	
	30	2700			
	90	5400			
	30				
	120				

Finally, the **third procedure** begins by finding a new SS (again by trial and error); in this case, it is 4. Multiply the SS by the HF and add it to the HL; multiply the SS by the HL and add it to the SL; multiply the SS by the SL and add it to the LF; multiply the LF by the SS and subtract it from the S to get 0, the remainder.

HF	HL	SL	LF	S	SS
1	30	900	27,000	1,336,336	30
	30	1800	81,000	−810,000	4
	60	2700	108,000	526,336	
	30	2700	23,584	−526,336	
	90	5400	131,584	0	
	30	496			
	120	5896			
	4				
	124				

See Problem 6.

Neither Yang Hui nor any other Chinese author offered a justification for the procedure.

8.1 Exercises

1. Consider numbers such as sixty-five, thirteen, fifty-three thousand, and so on. What advantages are there to writing (and speaking) numbers in the Chinese system?

2. Consider the rod numbers. What problems of positional notation are solved by the rod numbers? What problems are not solved?

3. Write the following numbers in rod numerals.

    (a) Four hundred sixty-three

    (b) Two thousand, eight hundred sixty-seven

    (c) Fifty three thousand, four hundred twenty-nine

4. Use the rod system for finding the following products.

    (a) Twenty-six times thirty-nine

    (b) Four hundred thirty-seven times sixty-one

    (c) Two thousand, four hundred eighty one times forty-six

5. Use the rod system for calculating the following quotients.

    (a) Three thousand nine hundred thirty-four divided by seven

    (b) Five thousand, six hundred eighty-one divided by thirteen

    (c) One hundred seventy-nine thousand, seven hundred twelve divided by two hundred thirty-four

6. Justify Yang Hui's procedure by showing that the *Second Procedure* generates the binomial coefficients in the expansion of $(30 + x)^4$, and the *Third Procedure* generates the corresponding terms in the expansion of $(30 + 4)^4$.

7. Use the method of Yang Hui to determine $\sqrt[4]{50,625}$.

8. Explain how the method of Yang Hui would have to be modified to find nth roots.

9. Use the results from Problem 8 to find the following.

    (a) $\sqrt[3]{8921}$ (b) $\sqrt[5]{7,962,624}$ (c) $\sqrt[3]{50,653}$

10. (Teaching Activity) Note that the numbers subtracted in the S column come from

$$(30 + 4)^4 = \underbrace{30^4}_{810,000} + \underbrace{4(30)^3(4) + 6(30)^2(4)^2 + 4(30)(4)^3 + 4^4}_{526,336}$$

Vary the method to find the expansion of $(a + b)^4$.

8.2   Practical Mathematics in China

In 221 B.C. the state of Ch'i surrendered to King Cheng of Ch'in, ending the Period of Warring States. King Cheng took the title Shih Huang Ti (First Emperor Ti) and, on the advice of astrologers, renamed the land he ruled Ch'in-a: China. By joining previously existing fortifications, Shih Huang Ti completed the Great Wall of China; he also ordained a common system of writing and currency across China, and standardized weights and measures in a remarkably rational system: the basic unit of length was the *chi*, generally between 25 and 35 centimeters depending on the era. For the most part, a larger unit was ten times the size of the next smaller unit; thus, there were 10 *cun* in one *chi*, and 10 *chi* in one *zhang*. Areas and volumes used the same names: thus, a square one *chi* by one *chi* had an area of one *chi*.

This passion for order had a dark side. Shih Huang Ti declared, "History beings with Ch'in" and ordered the destruction of virtually every Chinese text written before the formation of the Ch'in state. Shih Huang Ti, who wanted to live forever, spared only medical texts. The empire was to last for ten thousand years. It lasted for just fourteen. Shih Huang Ti's attempt to live forever failed, and he died from a stroke in 210 B.C. He was buried underneath a mountain, with 6000 terra cotta "soldiers"; the discovery of his tomb in 1974 plays the role in Chinese history that the discovery of the tomb of Tutankhamen plays in Egyptian history. Shih Huang Ti's son proved incompetent, and a rebellion engulfed the empire, leading to the destruction of the Ch'in dynasty.

In 202 B.C., Liu Pang, a former Ch'in official, succeeded in overcoming the last rival to the throne and, as Emperor Kao Tsu, founded the Han dynasty. One of Kao Tsu's first acts was to seek out scholars who had memorized the books that had been destroyed, and have them recreate the works from memory. It was during the Han dynasty that Chinese culture reached its definitive form, and, though the Han dynasty itself eventually fell, the Chinese called themselves the Men of Han for millennia afterward.

8.2.1   The *Nine Chapters*

The earliest surviving works of Chinese mathematics date back to the Han dynasty. The most advanced of these early works is the *Computational Prescriptions in Nine Chapters*, probably written sometime around 150 B.C.; the author is unknown. Although it is primarily a work on applied, not theoretical, mathematics, the *Nine Chapters* has a place in Chinese mathematics similar to that of the *Elements*: it is the mathematical classic to which many later texts refer, and it is very likely a compilation of the work of many earlier texts. One of these earlier texts, the *Suanshu shu*, was unknown until 1985, when a tomb, dating to around 175 B.C., was unearthed and copies of the *Suanshu shu* were discovered.

The *Nine Chapters* covers a variety of mathematical problems with solutions clearly expressed in terms of rules. For example:

Rule 8.1. *To find the area of a circular field: multiply half the circumference* *by half the diameter. Or multiply the diameter by itself; multiply the result by 3 and divide by 4. Or multiply the circumference by itself, and divide the result by 12.*

The Pythagorean theorem and rules for dealing with similar triangles are correctly introduced and used. Another work, *Sunzi's Computational Canon* (ca. 550 A.D.), gave other formulas for volumes and areas, some accurate and others not. For example, the surface area of a spherical segment is given as the product of half the circumference of the base of the segment times half the diameter.

Excess and Shortfall

Chapter Seven of the *Nine Chapters* is called "Excess and Shortfall," where we find the first systematic appearance of the method of **double false position**. The explanation in the *Nine Chapters* is rather complicated, and it comes at the *end* of a section of problems that rely on it; we will explain it using one of the earlier problems:

Problem 8.1. *There is a wall 90 cun high. A melon grows from the base of the wall, creeping upward and becoming 7 cun taller each day. A gourd grows from the top downward, becoming 10 cun longer each day. When do they meet?*

Solution. After 5 days, the two are 5 *cun* apart, after 6 days, they are 12 *cun* past each other. Write down the excess and shortfall, with their corresponding trial numbers:

$$5 \quad 5$$
$$6 \quad 12$$

Cross multiply and add the products: hence $5 \cdot 12 + 5 \cdot 6 = 90$. Use this sum as the dividend. Add the excess and shortfall; use it as the divisor: $5 + 12 = 17$. Divide the dividend by the divisor; this will be the number of days: $\frac{90}{17} = 5\frac{5}{17}$.

[Since a change of 1 in the guess leads to a change of 17 in the distance, while a change of 5 is what is desired, then the guess must be changed by $\frac{5}{17}$.] □

Matrix Algebra

Chapter Seven of the *Nine Chapters* deals with problems like

Problem 8.2. *Some men buy an item together. If each pays 8 coins, there are 3 coins too many; if each pays 7, there are 4 coins too few. How many men are there, and how much is the item?*

which is a problem in two unknowns (the number of people and the total cost of the item). To solve this and similar problems, the author of the *Nine Chapters* gives the rule:

Rule 8.2. *Put down the assumed amounts, and under them the corresponding excess and shortfall. Cross multiply; add the products and use the sum as the dividend. Add excess and shortfall; it is the divisor.*

From the assumed amounts, find the difference, and divide the dividend by the difference to find the price; divide the divisor by the difference to find the number of people.

This is a version of **Cramer's rule**. Applying this to the preceding problem gives the solution:

Solution. The amounts assumed were 8 and 7, which gave an excess of 3 and a shortfall of 4, respectively. Writing these down, we have

$$8 \quad 3$$
$$7 \quad 4$$

Cross multiplying and adding, we have $8 \cdot 4 + 7 \cdot 3 = 53$, which is the dividend. The excess plus the shortfall is $3 + 4 = 7$, the divisor. The difference between the assumed amounts is $8 - 7 = 1$; divide this into the dividend, 53, to get the total price, 53; and divide 1 into the divisor, to get the number of people, 7. $\square$

Corresponding rules existed if both assumed amounts resulted in an excess, a shortfall, or if one was exact and the other was an excess or a shortfall.

Matrix algebra is even more apparent in Chapter Eight, subtitled "Rectangular Tabulation" (in Chinese, *fang chang*).

Problem 8.3. *Three sheaves of a good harvest, 2 sheaves of a mediocre harvest, and 1 sheaf of a bad harvest yield a profit of 39 tous. Two sheaves of a good harvest, 3 sheaves of a mediocre harvest, and 1 sheaf of a bad harvest, yield a profit of 34 tous. One sheaf of a good harvest, 2 sheaves of a mediocre harvest, and 3 sheaves of a bad harvest yield a profit of 26 tous. How much is the profit from each sheaf of good, mediocre and bad harvest?*

A technique very similar to Gaussian elimination is used. The Chinese would begin by writing down the amounts of sheaves and the profits, forming the table:

1	2	3	good
2	3	2	mediocre
3	1	1	bad
26	34	39	

Thus, whereas we would write the coefficients in rows, the Chinese wrote them in columns, and arranged them from right to left (thus, the "first column" is the rightmost). However, this is a difference that is more apparent than real: Chinese is written from top to bottom, and from right to left, so it is natural they would organize the *fang shang* table in this fashion. Thus, we will set up the table in our familiar form, which is the correct form in the context of the written language.

Solution. Writing down the amounts of sheaves and the profit, we form the rows

good	mediocre	bad	
3	2	1	39
2	3	1	34
1	2	3	26

Multiply the second and third rows by 3, the leftmost number of the topmost row.

good	mediocre	bad	
3	2	1	39
6	9	3	102
3	6	9	78

Now subtract the numbers in the first row as many times as possible from the numbers in the second and third rows. Subtracting the first row twice for the second row, we get

good	mediocre	bad	
3	2	1	39
	5	1	24
3	6	9	78

where we leave a blank on the left of the second row, since there is nothing left. Subtracting the first row once from the third row, we obtain

good	mediocre	bad	
3	2	1	39
	5	1	24
	4	8	39

Again, multiply the third row by the number at the front of the second row.

good	mediocre	bad	
3	2	1	39
	5	1	24
	20	40	195

Subtracting now the second row four times from the third row, we are left with

good	mediocre	bad	
3	2	1	39
	5	1	24
		36	99

Divide 99 by 36 to obtain the value of the bad harvest; then use this value to find the value of the mediocre harvest; finally, use it to obtain the value of the good harvest. $\square$

The author of the *Nine Chapters* was fond of beginning problems by noting that a quantity was *less* than a certain amount but, when an additional quantity was added, the total was *equal* to a given amount. Thus he gave:

Problem 8.4. *One military horse cannot pull a load of* 40 dan; *neither can* 2 *ordinary horses, nor* 3 *inferior horses. But* 1 *military and* 1 *ordinary horse can pull the load, as can* 2 *ordinary horses and* 1 *inferior horse, or* 3 *inferior and* 1 *military horse. How much can each horse pull?*

Note that this problem results in coefficients of zero which, on a counting board, would be indicated by leaving a space empty.

Also in the *Nine Chapters* are quadratic equations (solved in a manner similar to that of the Babylonians), and indeterminate equations. A later work, *Suanzi's Computational Canon* (ca. 400 A.D.) incorporated much of the work of the *Nine Chapters* and added new problems. A key problem in number theory, now called the Chinese remainder problem, appeared for the first time in *Suanzi's Computational Canon*, though no explanation of the method of solution was given. In the *Shushu jiuzhang* of Qin Jiushao (1202-1261) we find a variation of the method of Yang Hui used to solve an equation we would write as $x^4 = 15245x^2 - 6262506.25$.

See page 209.

Negative Numbers

The *fang chang* method, which requires one row to be repeatedly subtracted from another row, occasionally results in negative numbers, which the Chinese handled with no difficulty whatsoever. To indicate a negative number on the counter board, the Chinese used black rods (positive numbers were represented with red rods), which we will indicate using boldfaced numbers: hence 1 is a positive number, but **1** is a negative number. One problem was:

Problem 8.5. *Five sheaves of a good harvest, less* 11 shang, *are equal in value to* 7 *sheaves of a poor harvest. Seven sheaves of a good harvest, less* 25 shang, *are equal in value to* 5 *sheaves of a poor harvest. How much is each sheaf worth?*

A *shang* is a monetary amount. To solve this problem, the author rewrote the two conditions as:

1. Two sheaves of a good harvest less 7 sheaves of a poor harvest are worth 11 *shang*.

2. Seven sheaves of a good harvest less 5 sheaves of a poor harvest are worth 25 *shang*.

essentially transposing the negative terms. Thus, the *fang chang* method would begin by setting down

$$
\begin{array}{ccc}
5 & \mathbf{7} & 11 \\
7 & \mathbf{5} & 25
\end{array}
\quad \xrightarrow{\text{Multiply by 5}} \quad
\begin{array}{ccc}
5 & \mathbf{7} & 11 \\
35 & \mathbf{25} & 125
\end{array}
$$

The first row would have to be subtracted seven times from the second row, requiring the rule:

Rule 8.3. *For subtraction: with same type, subtract. With different types, add. For addition; with different types, subtract. With same type, add.*

Thus, since **7** and **25** are the same type, and we are subtracting, we subtract 7 from 25 seven times. The first three subtractions are:

$$
\begin{array}{ccc}
5 & \mathbf{7} & 11 \\
30 & \mathbf{18} & 114
\end{array}
\rightarrow
\begin{array}{ccc}
5 & \mathbf{7} & 11 \\
25 & \mathbf{11} & 103
\end{array}
\rightarrow
\begin{array}{ccc}
5 & \mathbf{7} & 11 \\
20 & \mathbf{4} & 92
\end{array}
$$

Subtracting another **7** is impossible, requiring the additional rule:

Rule 8.4. *If you are not adding, black [negative] becomes red [positive] and red [positive] becomes negative [black].*

What this means is the following: from the **4** we must subtract 7. First, subtract 4, leaving nothing. Then subtract 3 more, leaving a positive 3.

$$
\begin{array}{ccc}
5 & \mathbf{7} & 11 \\
15 & \mathbf{3} & 81
\end{array}
$$

Now the **7** and the 3 have different signs, so now 7 is added.

$$
\begin{array}{ccc}
5 & \mathbf{7} & 11 \\
10 & \mathbf{10} & 70
\end{array}
\rightarrow
\begin{array}{ccc}
5 & \mathbf{7} & 11 \\
5 & \mathbf{17} & 59
\end{array}
\rightarrow
\begin{array}{ccc}
5 & \mathbf{7} & 11 \\
 & \mathbf{24} & 48
\end{array}
$$

Hence the solution is that 1 sheaf of a poor harvest is 2 *shang*, while 1 sheaf of a good harvest is 5 *shang*.

8.2.2 The *Sea Island Mathematical Manual*

Around 263 A.D., the *Sea Island Mathematical Manual* of Liu Hui appeared. It is essentially a work on surveying, and shows that the Chinese had developed a high degree of sophistication. The author, Liu Hui, lived in an era when the Han Dynasty was collapsing. Among his other achievements, Liu Hui approximated the area of a circle by inscribing regular polygons in a circle with a radius of one foot. The polygonal areas could be computed; by using a 192-sided polygon, Liu Hui obtained an estimate of 3.141024 for the area.

The *Sea Island Mathematical Manual* shows us, among other things, the highly practical nature of Chinese mathematics. Although Liu Hui apparently gave proofs of his results, these proofs have not been retained. What comes down to us is the statement and solution of various problems in practical surveying. The first problem concerns measuring the distance and height of an island in the ocean (hence the name of the manual):

Problem 8.6. *Two poles are erected, 5 bu high and 1000 bu apart and in line with the top of a sea island. Viewed from the ground level 123 bu behind the front pole, the top of a sea island coincides with the top of the pole. Viewed from the ground level 127 bu behind the rear pole, the top of the sea island coincides with the top of the pole. Find the height of the island and the distance from the poles.*

Solution. Multiply the distance between the poles by their height, giving the dividend 5000. Take the difference in distances from the pole to the respective points of observation as the divisor 4. Divide the two, making 1250, and add this to the height of the pole to get the height of the island, 1255 *bu.*

To find the distance, multiply the distance between the poles by the distance between the first pole and the first point of observation to get the dividend 123,000. Divide by the difference in distances from the pole to the respective points of observation 4 to get the distance to the island, 30,750 *bu*   □

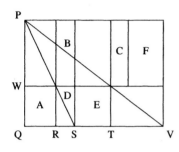

No explanation of the procedure is given. The problem can be solved through a straightforward use of similar triangles. But another method, called the **out-in method**, seems to have been used. The out-in method relies on the decomposition of rectangular figures. In this case, the procedure might have been the following: if QP represents the height of the island, and the poles are located at R, T and sighted from S, V, then complete the rectangles as shown in the figure.

Remember that the complements about the diagonal in a parallelogram are equal. Let rectangle C equal rectangle B; since PS, PV are diagonals, then A, B are complementary rectangles and are thus equal; likewise, A, D, E is equal to C, F, so D, E is equal to F. But the area of D, E is the height of the pole multiplied by the distance, so it is a known value; meanwhile, the area of F will be the difference in distances multiplied by the height PW. Thus the height PQ of the island can be found by dividing the area D, E by the difference in distances, then adding the height of one of the poles.

To find the distance, Liu Hui could have noted that the equivalence of areas meant 123 times 1250 was 5 times the distance to the island, but instead, to make the computation easier, he relied on the fact that 1250 was obtained by multiplying 1000 by 5, then dividing by 4. Hence

$$123 \cdot 1250 = 5l$$
$$123 \cdot \frac{1000 \cdot 5}{4} = 5l$$
$$123 \cdot \frac{1000}{4} = l$$

and so the distance to the island could be found more simply.

8.2 Exercises
Solve Problems 1 to 5 using an appropriate method from the *Nine Chapters.*

1. Some chickens are bought. If everyone pays 9, there is an excess of 11; if everyone pays 6, there is a shortfall of 16. How many people are there and how much are the chickens?

2. Seven sheaves of a good harvest and 2 sheaves of a mediocre harvest yield a profit of 11 *tous.* Two sheaves of a good harvest and 8 sheaves of a mediocre harvest yield a profit of 9 *tous.* How much is the profit from each sheaf of a good and mediocre harvest?

3. Five cattle and 2 sheep are worth 10 *liang*. Two cattle and 2 sheep are worth 8 *liang*. How much are one bull and one sheep worth?

4. Two sheaves of a good harvest and 1 sheaf of a mediocre harvest yield a profit of 1 *tou*. Three sheaves of a mediocre harvest and 1 sheaf of a poor harvest yield a profit of 1 *tou*. One sheaf of a good harvest and 4 sheaves of a poor harvest yield a profit of 1 *tou*. How much is one sheaf of each?

5. Solve Problem 8.4. Note that the problem assumes that one military horse and one ordinary horse can pull *exactly* 40 *dan*.

6. Explain how Cramer's rule is equivalent to Rule 8.2.

7. Explain how the distance to the Sea Island (Problem 8.6) can be determined using the out-in method.

8.3 Indian Problem Solving

The Chinese were able to achieve a very high level of mathematical proficiency; however, it is difficult to determine what, if any, influence they had on the development of mathematics as a whole. The same cannot be said about the Indians, whose influence upon the development of Islamic mathematics was significant. A key factor was that the Indians did not, as the Greeks did, treat arithmetic as a specialized case of geometry; instead, their arithmetic, and consequently their method of solving equations, used purely numerical procedures. An important consequence of this was the free use of zero, irrational, and even negative quantities. We see this liberation from the constraints of Greek geometry in the Bhakhshālī manuscript, discovered in 1881. It is purportedly a copy of a seventh century A.D. text written by the "son of Chajaka," and shows that a very sophisticated algebra had developed in India by then, which included operations on signed numbers, root approximation techniques, and the rule of single false position.

8.3.1 Aryabhata

Unfortunately, the history of Indian mathematics is very difficult to trace. Unlike China, India was divided into many independent states until recently, and it is not known for certain how much intercommunication there was between various states. Unlike Greece, India is a subcontinent, twenty times larger than all the Greek city-states put together. In many cases, we know little more than the name of a medieval Indian mathematician—which is frequently the same as that of another Indian mathematician from a different time and place, so we cannot even distinguish which of several made a key discovery. A good example is the life and work of Aryabhata (476-550), who was born in "Kusumapura," whose exact location is uncertain, though it is believed to be near Pataliputra (Patna). Enough confusion existed about Aryabhata's identity that the Persian historian

Figure 8.1: Medieval India.

al-Biruni believed there were two Aryabhatas living in sixth century India, and it was not until 1926 that it was shown the two Aryabhatas were, in fact, the same person.

In 499, Aryabhata completed his *Aryabhatiya*, a collection of mathematical problems and procedures consisting of 123 verse stanzas divided into three books. The first ten stanzas deal with astronomical concepts; the next thirty-three deal with mathematical ideas, and the remainder with the reckoning of time.

There is an intriguing statement in the beginning of the second book that suggests positional notation was already in use in India at the time of Aryabhata. First, he gave the names for the numbers 1, 10, 100, and so on up to a billion, then noted that the numbers are, "from place to place each ten times the preceding." Earlier, in the first book, Aryabhata discussed how to render numbers into alphabetic sequences; again, the implication is that a place value system is already in existence. Thus, the word *cayagiyinusuchlr*, taken syllable by syllable, represents the number 57,753,336. The syllables represent the numbers 6, 30, 300, 3000, and so on, numbering from smallest to greatest. Aryabhata also included formulas for the sums of arithmetic series, series of squares, and series of cubes.

8.3.2 Brahmagupta

Other important contributions were made by Brahmagupta (598-670), part of the school of mathematics established at Ujjain. *The Opening of the Universe* was Brahmagupta's major work: it consists of ten original chapters, plus an additional fifteen chapters that are essentially commentary on the first ten. In it, he gave an explanation of zero as the result of subtracting a number from itself, and gave the correct rules for adding, subtracting, and multiplying by zero. For division by zero, Brahmagupta claimed zero divided by zero was zero, and that any number divided by zero was a fraction with zero as a denominator (which is simply saying $a \div 0 = \frac{a}{0}$). More importantly, he described the rules of operating on *fortunes* (positive numbers) and *debts* (negative numbers), giving such rules as, "A debt subtracted from zero is a fortune" (i.e., subtracting a negative number from zero resulted in a positive number).

8.3.3 Quadratic Equations

The average medieval person was probably even less aware of advanced mathematics than the average person of today. A businessman might be expected to know a little bit more, and **commercial arithmetic** was an important part of learning business. Much of Indian mathematics involved commercial arithmetic. This factor played an important role in the development of mathematics, since in pure geometry, negative quantities do not exist and zero quantities require special handling. Commercial arithmetic, on the other hand, deals very easily with negative quantities, usually as debts; zero quantities, being treated from a purely numerical point of view, posed no special problems. Although the negative quantities were originally treated as *different* from positive quantities, in the same way that a debt is different from a credit, it was a step in the direction of generalizing the number concept.

Problems in money-lending with simple interest generate some of the quadratic equations that appear in Indian sources:

Problem 8.7. *One hundred is loaned for one month, and the interest received is loaned for six months. The total of the interest and the interest on the interest is 16. Find the interest on the principal.*

The problem results in a quadratic equation; Aryabhata's solution is:

Solution. Add the interest on the principal and the interest on this interest, and multiply this by the time and by the principal: the result is 9600. Add this to the square of half the principal, 2500, making 12100. Find its square root, 110, then subtract half the principal and divide by the time, obtaining 10, which is the interest on the principal. □ See Problem 1.

8.3.4 The Rule of Three

One of the most important methods of commercial arithmetic, and probably the most advanced mathematics used by the average person before modern times, was the **rule of three**, as well as the lesser known rules of five, seven, and so on. These were first enunciated by the medieval Indians, and provide a formulaic method of solving simple problems in proportion.

The rules were stated poetically: the "fruit" was multiplied by the " desire" and divided by the "measure," which would give the "fruit of the desire." This rather poetic way of stating the rule of three is characteristic of medieval Indian mathematics. Aryabhata gave the rule but no examples; we take one from *Trisatika* of Sridhara (ca. 870-930?). Like many other Indian mathematicians, very little biographical information is known about Sridhara. One theory places his birthplace in Bengal; the other in southern India, a thousand miles away. There are also differing theories about when he lived, and all that is known for certain is that it was some time between the seventh and eleventh centuries. Based on the mathematical knowledge presented in his treatises, it is currently believed he lived at the end of the ninth century.

Sridhara's problem, like many Babylonian problems, used an inconsistent set of units; if the conversions are done, however, the problem is:

Problem 8.8. *If one and one quarter* pala *of sandalwood cost ten and a half* pana, *how much does nine and a quarter* pala *of sandalwood cost?*

The desire is the nine and a quarter *pala* (note that the desire is *not*, as might be expected, the unknown); the fruit of this desire is the cost of sandalwood and the unknown quantity to be found. The measure is one and a quarter *pala*, while the fruit is ten and a half *pana* of sandalwood. Thus, to solve this problem, multiply the fruit by the desire, and divide by the measure: the solution is $79\frac{7}{10}$.

Corresponding rules exist for five, seven, nine, and eleven. Brahmagupta described the most general form of the rule:

Rule 8.5 (Rule of Five, Seven, etc.). *Transpose the fruit. Multiply the larger set of terms, and divide it by the product of the smaller set of terms.*

An example from Bhaskara (see later) is:

Problem 8.9. *One hundred is lent for one month and the interest is five. What will the interest of 16 be for 12 months?*

To solve these problems, an Indian mathematician would have written down the numbers in parallel columns. There is occasional use of a symbol, $\bigcirc$, to indicate the unknown, though if the formatting of the columns is done properly, the use of the symbol for the unknown is superfluous.

Solution. Transpose the fruits:

100	16	Principal
1	12	Time
5	$\bigcirc$	Interest

transpose
fruits
$\longrightarrow$

100	16	Principal
1	12	Time
$\bigcirc$	5	Interest

Multiply the larger set of terms, 16, 12, and 5, to obtain 960. Multiply the smaller set of terms to obtain 100. Divide the larger set by the smaller set, to obtain the fruit: the interest is 9 3/5. $\square$

8.3.5 Bhaskara

Bhaskara Atscharja (1114-1185) was probably the most sophisticated of the Indian mathematicians. Atscharja is an honorific, meaning "Teacher." To distinguish him from an earlier Bhaskara, he is usually referred to as Bhaskara II; his two important works were the *Lilavati* (The Beautiful) and the *Vija Ganita* (Seed Counting). A story told by Bhaskara's Persian translator, writing in 1587, was that *Lilavati* was the name of Bhaskara's daughter. Bhaskara computed, according to astrological precepts, the only time when she could be happily married, but when the appointed time approached, a seed fell into the waterclock and stopped it. The indicated time passed before anyone realized the clock had stopped, and Lilavati (according to the story) never married. To console her, Bhaskara gave her name to his mathematical treatise. We may suspect the effort at consolation failed.

The *Vija Ganita* is a more advanced work and shows us, perhaps, the pinnacle of the mathematics of medieval India. Bhaskara gave a lucid explanation of positive and negative quantities, including the rules for arithmetic operations with signed numbers. For example:

> The square [root] of affirmative is sometimes affirmative and sometimes negative, according to difference of circumstances... But if anyone asks the root of 9 negative I say the question is absurd, for there never can be a negative square as has been shown.[1]

The existence of both positive and negative square roots meant that Bhaskara would be forced to deal with solutions that seemed impossible, as in

Problem 8.10. *One fifth of a troop [of monkeys] less three, squared, went into a cave. One monkey was left outside. How many were there?*

Bhaskara found the answers to be 50 and 5, but rejected 5 as a valid solution: an early example of the rejection of **extraneous solutions**.

Although the zero had been known in India for some centuries, some confusion still existed over its arithmetical properties. Bhaskara gave, correctly, the rules for multiplication by zero and division of zero by another number, and even noted:

> If a number is the dividend and zero the divisor the division is impossible, for by whatever number we multiply the divisor, it will not arrive at the dividend, because [the product] will always be zero. [2]

Later on, however, he noted that if the quotient of a number divided by zero is then multiplied by zero, the result is the number: in other words, $\frac{a}{0} \cdot 0 = a$.

Vija Ganita also includes solutions to quadratic equations, as well as to some degenerate cubic and biquadratic equations, and multilinear and nonlinear equations in several variables. One problem involved finding the three sides of a right

---

[1]Bhaskara, *Bijaganita*. Translation by Edward Strachey. W. Glendinning, 1813 (?). Page 15.

[2]Ibid., 16-7

See Problem 4.

triangle, given the perimeter was 56 and the product of the three sides was 4200; Bhaskara solved this using a combination of geometric and algebraic rules.

Medieval Indian mathematicians showed a particular genius for solving indeterminate equations. A passage in the *Aryabhatiya* discussed how to find integral solutions to indeterminate systems of linear equations, which suggests that the Indians were solving indeterminate equations not long after Diophantus. The use of the same examples as Diophantus strongly suggests a connection though the Indians, particularly Brahmagupta, exceeded Diophantus in several respects. In particular, whereas Diophantus was concerned with finding only a single rational solution, the Indians gave methods of finding an infinite number of whole number solutions. For example, Brahmagupta solved $61x^2 + 1 = y^2$, and found the smallest whole number solution: $x = 226153980$ and $y = 1766319049$.

The problems were stated and solved completely without notation, making even the statements of the problems difficult to follow. From Bhaskara's *Vija Ganita*, we have

Problem 8.11. *Given a multiplicand and an augment, find a lesser square, such that when it is multiplied by the multiplicand and to the product the augment is added, the sum will be a larger square.*
[Given a, b, find whole numbers x, y so $ax^2 + b = y^2$.]

Bhaskara did not initially solve this problem; instead, he gave a method of finding other solutions from a given one.

Rule 8.6. *Given two roots of squares and the augment that satisfy the relationship, write them down on a horizontal line; write them down again. Cross multiply the lesser and greater roots, and add the products to find the new lesser root. Multiply the product of the two lesser roots by the multiplicand; add this to the product of the two greater roots to form the new greater root. The new augment will be the product of the two augments.*

Example 8.3. *Find a square, such that when it is multiplied by eight, and one is added, the result will be a square.*
[Find x, y integers so $8x^2 + 1 = y^2$.]
We note that 1 satisfies this relationship, since the square of 1, multiplied by 8, plus 1, is 9, the square of 3. The lesser root is 1, the greater root is 3, and the augment is 1. Write these down twice.

Lesser Root	Greater Root	Augment
1	3	1
1	3	1

Cross multiplying and adding, we have $3 \cdot 1 + 1 \cdot 3 = 6$, the new lesser root. Multiply the product of the two lesser roots by the multiplicand, 8, making 8; multiply the two greater roots, making 9, and add the two products together, making 17, which will be the new greater roots. Finally, the product of the two augments, $1 \cdot 1$, will be the new augment.

Lesser Root	Greater Root	Augment
1	3	1
1	3	1
6	17	1

Repeating the procedure again, the new lesser root is $1 \cdot 17 + 3 \cdot 6 = 35$; *the new greater root is* $1 \cdot 6 \cdot 8 + 3 \cdot 17 = 99$, *and the new augment is* $1 \cdot 1 = 1$.

Lesser Root	Greater Root	Augment
1	3	1
1	3	1
6	17	1
35	99	1

This procedure solves Pell equations of the form $ax^2 + 1 = y^2$, given a single solution. Of course, Bhaskara's method would be difficult to employ unless he also gave a method of *finding* the first solution. To do this, he found a solution to $ax^2 + b = y^2$, where b is arbitrary, then with this solution, found a transformation that gave a solution to the equation $ax^2 + 1 = y^2$.

See Problem 6.

8.3.6 The Arithmetic Triangle

Pascal's triangle appeared in tenth century Indian texts, as well as eleventh century Chinese texts. However, Indian and Chinese authors concerned themselves primarily with the uses of the triangle, and except for a few scattered results, did not deal with the properties of the numbers in the triangle itself.

The main use of the arithmetic triangle was finding binomial coefficients. However, another use presented itself in a new branch of mathematics, developed by the Indians: **combinatorics**, which seeks to determine the number of possible combinations of a finite set of objects. The earliest combinatorial statement can be traced back to a sixth century B.C. medical text of Susruta, an Indian physician, who noted there were 63 possible combinations of six tastes (bitter, sour, salty, astringent, sweet, and hot). The size of the problem is small enough that it is not unreasonable to suppose that the problem was solved by simple enumeration.

Varahamihira (505-587), probably born in Kapitthaka but worked and taught in Ujjain (where Brahmagupta would later study), discussed the problem of finding the number of perfumes that could be made by choosing four substances out of sixteen: he gave the answer, 1820. To find this number, Varahamihira wrote down an array

$$\vdots$$
$$1$$
$$1$$
$$1$$
$$1 \quad 1 \quad 1 \quad 1 \quad \ldots$$

The remaining entries in the array were found by adding the number to the left and the number below an entry. This produces

$$
\begin{array}{ccccc}
\vdots & \vdots & \vdots & \vdots & \\
1 & 4 & 10 & 20 & \ldots \\
1 & 3 & 6 & 10 & \ldots \\
1 & 2 & 3 & 4 & \ldots \\
1 & 1 & 1 & 1 & \ldots
\end{array}
$$

which is the familiar arithmetic triangle. To find the number of combinations of 16 objects taken 4 at a time, Varahamihira found the entry in the sixteenth row (counting upwards from the bottom), fourth column (counting from the right).

Of course, finding the number of combinations this way is not only tedious, but any arithmetical error in the table renders every entry above and to the right of the error incorrect. Mahavira (800-870), born in Mysore in southern India, gave the correct formula for determining the number of combinations of n objects taken m at a time; the rule is quoted by Bhaskara.

8.3 Exercises

1. Suppose r is the rate of interest, P is the principal amount, t_1 the original period of the loan and t_2 the extension period. Let I_1 and I_2 be the interest accrued during the periods t_1 and t_2; remember that I_2 is the interest on the interest I_1 but *not* any further interest on the principal.

    (a) Write down the equation that would have to be solved to find I_1.

    (b) To what quadratic equation does Aryabhat's solution correspond? Note that Aryabhata does not solve directly for the interest rate, but for the amount of interest during the time period t_1; i.e., he solves for PrI_1.

2. Compare Aryabhata's solution to the quadratic equation with the Babylonian solution.

3. Use the rule of false position to solve the following problems.

    (a) Out of a heap of lotus flowers, a third, a fifth, and a sixth were offered respectively to Shiva, Vishnu, and Surya, and a quarter was presented to Bhavani. The remaining six were given to the venerable preceptor. Determine the number of lotus flowers. (Bhaskara)

    (b) The third part of a necklace of pearls, broken in a lover's quarrel, fell to the ground; its fifth part rested on the couch; the sixth part was saved by the wench, and the tenth part taken by her lover; six pearls remained strung. How many pearls composed the necklace? (Sridhara)

4. Recreate Bhaskara's solution to the following problem from the *Bija Ganita*: the perimeter of a right triangle is 56 units, and the product of its three sides is 4200.

   (a) Set up the solution in this way: let the hypotenuse be the unknown, and thus the unknown is 4200 divided by the rectangle on the two sides. Then use the fact that the square of a sum of the two sides is equal to the sum of the squares of the sides plus twice the rectangle on the two sides; also, the sum of the two sides is 56 minus the hypotenuse.

   (b) Solve the problem.

   (c) Why would a traditional Greek mathematician (e.g., Euclid) be taken aback by the problem?

5. Solve Problem 8.10. Is Bhaskara's rejection of 5 as a solution valid? Explain.

6. Bhaskara's solution to the general Pell equation involved finding, first, a solution to $ax^2 + 1 = y^2$.

   (a) To find an initial value of $11x^2 + 1 = y^2$, Bhaskara first found that $11(1)^2 + 5 = (4)^2$. Apply Bhaskara's method as outlined in the text to find that $11(8)^2 + 25 = 27^2$.

   (b) Use the fact that $11(8)^2 + 25 = 27^2$ to find a rational solution to $11x^2 + 1 = y^2$.

   (c) Find rational solutions to $5x^2 + 1 = y^2$. Note that $5(2)^2 + 5 = 5^2$.

   (d) Find rational solutions to $61x^2 + 1 = y^2$.

8.4   Indian Geometry

The Persian historian al-Biruni (973-1048) once compared the mathematics and astronomy of the Medieval Indians to a "mixture of pearl shells and sour dates, or of pearl and dung, or of costly crystals and common pebbles." Nowhere is this criticism more valid than for Indian geometry. In the *Aryabhatiya*, valid propositions alternate with poor approximations, with no indication that the Indians knew (or cared) about the distinction between the two.

Rule 8.7. *The area of a triangle is half the base times the perpendicular. Half this area, times the height, is the volume of the pyramid.*

Rule 8.8. *Half the circumference times half the diameter is the area of the circle. This area multiplied by its square root is the exact volume of the sphere.*

Rule 8.9. *The area of any plane figure is the product of two of its sides. The chord of one-sixth the circumference of a circle equals the radius.*

Aryabhata also gave:

Rule 8.10. *Add four to one hundred, multiply by eight and add sixty-two thousand. The result is the approximate value of the circumference of a circle with diameter of twenty thousand.*

which can be interpreted to give a value of π. Brahmagupta gave:

Rule 8.11. *The diameter multiplied by three is the practical circumference, and the square of the radius multiplied by three is the practical area. The square root of ten times the square of the diameter, and the square root of ten times the square of the radius, are the exact circumferences and areas.*

Another pearl mixed with dung is Brahmagupta's formula for finding the area of quadrilateral:

Rule 8.12. *The product of half the sides and opposite sides is the rough area of a triangle of quadrilateral. Half the sum of the sides set down four times and each lessened by the sides being multiplied together—the square-root of the product is the exact area.*[3]

This is an extension of Heron's formula.

A key element in mathematical astronomy is the development of a table of chords. Unlike Ptolemy, who dealt with the whole chord, the Indians began to deal with the *jiva*, half the chord of half the angle, and the clear precursor to our modern sine values. In Book I of the *Aryabhatiya*, Aryabhata gave a sequence of differences between successive values of the half-chords; these differences are 225, 224, 222, 219, 215, 210, 205, 199, 191, 183, 174, 164, 154, 143, 131, 119, 106, 93, 79, 65, 51, 37, 22, 7. These values represent the differences between successive *jiva* for central angles from 0° to 90°, differing by 225 minutes of arc $\left(3\frac{3}{4}^\circ\right)$.

To find these values, Aryabhata used the following procedure:

Rule 8.13. *Take the amount the last determined difference is less than the first difference (always 225). Divide the sum of all the preceding differences by the first difference. This is how much the next difference is less than the first.*

Example 8.4. *To find the second difference, note that the last difference found is also the first difference, and the difference is thus 0. Divide the sum of the preceding differences (225 only in this case) by the first difference, 225, giving 1. The sum of 1 and 0, or 1, is is how much the next difference is less than the first difference, 225. Thus, the second difference is 224.*

Example 8.5. *To find the third difference: the last difference is 224, calculated above, which is 1 less than the first difference. Divide the sum of all the preceding differences, or 449, by the first difference, 225, to obtain 2 (to the nearest whole number). The sum of 1 and 2, or 3, is how much the next difference is less than the first difference; thus, the third difference is 222.*

See Problem 1.

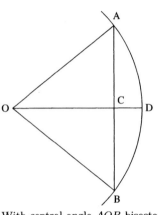

With central angle *AOB* bisected by *OD*, the chord is *AB* and the *jiva* is *AC*.

See Problem 2.

---

[3]Kaye, G. R., *Indian Mathematics*, 1915. Thacker, Spinc and Co. Page 48.

The procedure, followed precisely, does not yield exactly the numbers of Aryabhata; the fractional values were presumably redistributed somehow, though no one is certain of how the redistribution was done.

8.4 Exercises

1. Consider Rules 8.10 and 8.11.

    (a) What value of π is Aryabhata giving?

    (b) Determine the area of the circle according to Aryabhata and Brahmagupta.

    (c) Use Rules 8.10 and 8.8 to determine the volume of a sphere with diameter 20,000. How incorrect is the value?

2. Consider Aryabhata's table of differences.

    (a) The values correspond to $r \sin \theta$. Determine the value of r, given that $r \sin(3\frac{3}{4}^\circ) = 225$. Round your answer to the nearest whole number.

    (b) Determine the value of $\sin(3\frac{3}{4}^\circ)$. How accurate is Aryabhata's value?

    (c) Comment on the validity of Aryabhata's method in general.

3. Calculate the 24 differences according to the rule stated. How different are the values calculated, using the recursion method given in the *Aryabhatiya*, from the values given in the first book?

4. Translate Brahmagupta's Rule 8.12 into an area formula for a quadrilateral. For which quadrilaterals does Brahmagupta's formula give the correct area?

Chapter 9

The Islamic World

In 612, in the Arabian peninsula, a new religion was born, and the founder was Mohammed (570-632), a merchant. The religion, the youngest of the three main world religions, would become known as Islam ("submission," i.e., submission to the will of God), and its adherents are called Muslims. Mohammed united the various Arabian tribes, who soon conquered all of Arabia in a great war of religion, called a *jihad*. After Mohammed, a succession of leaders, called *caliphs* (which simply means "successor"), led the Muslims; the first "dynasty" of caliphs were the Umayyads, named after one of the founders. By 700, the territory controlled by the Umayyad caliphate included Iraq, Palestine, Egypt, and North Africa.

In 711, the Muslims moved into Spain, easily conquering it. A Muslim army crossed the Pyrenees into France where they finally suffered a defeat when they met an army led by Charles Martel at Tours in 732. Martel's horsemen used an invention new to European armies (but well known to the Arabs and the Chinese): stirrups. In *The Song of Roland*, a medieval epic poem, the battle becomes the annihilation of the Muslim army and the saving of all of Christian Europe. The reality was that a civil war had erupted within the caliphate, and the armies withdrew for the richer prize. By 750, Umayyad power existed mainly in Spain, while the remainder of the caliphate was controlled by a new family, the Abbasids.

9.1 Numeration and Computation

Our modern system of numeration can be traced back to the Hindus of medieval India. Hence, it should properly be called Indian or Hindu numerals, though since they arrived in Europe via the Arab world, they are usually called Hindu-Arabic numerals. The earliest appearance in the West was a reference by the Bishop Severus Sebokht in 662. After Justinian closed the schools of pagan

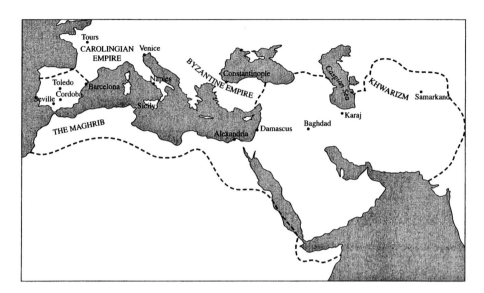

Figure 9.1: The Islamic World

learning, some of the scholars made their way to the Muslim world, which still appreciated classical scholarship, particularly Greek learning. Sebokht, living in Keneshra on the Euphrates, felt his colleagues were a little *too* appreciative of Greek learning, and wanted to point out that others made valuable contributions to scholarship as well: he mentioned the great discoveries of the Hindu astronomers, as well as their *nine* signs for numeration.

It is worth noting that Sebokht was unaware of the tenth sign, which is what made the Indian numeration system workable. Recall that the Babylonians, who also expressed numbers using positional notation, had no symbol to indicate an empty place. The '0' symbol can be traced back to an inscription dating from around 876, at Gwalior, India: the symbol itself was referred to as the *sunya*, "void." The Arabs translated this as *aṣ-ṣifr*, which became *zephirum* as it entered medieval Europe, from which we get the word zero, as well as cipher.

There were actually two forms of Indian numerals; our form stems from the so-called Gobar numerals. Through a quirk of fate, the Gobar numerals entered Europe through the work of al Khwārizmī, who in turn probably based his work on a (lost) work of Brahmagupta. At the same time, another form, later known See page 220. as East Arabic, was entering the Arab world and soon would be the form used by the Arabs themselves. Thus, many of the Arabs of the medieval period did not use what Europeans of the medieval period were calling Arabic numerals!

Like the Indians, the Arabs wrote the numbers from smallest value to largest: thus 1 thousand, 2 hundred, 3 ten and 4 units would be written beginning with the '4'. However, because Arabic is written right to left, this meant that the number itself would be written from right to left; thus, the smallest value is the

Brähmi (500 A.D.)	—	=	≡	Ɣ	⋏	φ	⁊	⅂	⁊
Bhakshäli	∩	3	3	ౙ	ⅴ	⅄	⁊	✗	ℚ
Arabic	Ι	⋗	⋓	⋛	∂	Ч	Ѵ	∧	9
Spain (976)	Ι	Ⴐ	⅀	⅙	Ч	Ⴑ	⁊	8	9
Italy or France (1400)	1	2	3	8	Ч	⅁	∧	8	9

Figure 9.2: The Development of Hindu-Arabic Numerals.

rightmost digit. Thus, European translators, reading the number as a single symbol, wrote down numbers with the largest place value first.

One of the earliest primary sources we have on computation with the Hindu-Arabic numerals is that of Kūshyār ibn Labbān (fl. ca. 1000), whose *Principles of Hindu Reckoning*, if its title correctly identifies its contents, gives us a glimpse of medieval Indian computational methods. Very little is known about ibn Labbān, though "al-jīlī" is sometimes added to his name, suggesting he was from a region in northern Iran, south of the Caspian Sea.

9.1.1 Multiplication

For multiplication, ibn Labbān used the following procedure. To multiply 325 by 243, first place the smallest digit of one below the largest digit of the latter:

```
          3  2  5
    2  4  3
```

Multiply 2 by 3, to get 6, and write the product above the 2.

```
    6     3  2  5
    2  4  3
```

Multiply 4 by 3, to get 12; write the units (2) above the 4 and add the tens (1) to the 6:

```
    7  2  3  2  5
    2  4  3
```

Now multiply 3 by 3 to get 9; write this above the 3, erase the 3 of the first factor, and move the factor 243 over one space. Notice that ibn Labbān's procedure has much potential for confusion.

```
    7  2  9  2  5
       2  4  3
```

Repeat the procedure, multiplying 2 by 2, 4, and 3, writing the numbers in the appropriate places, and move the factor 243 over to the last space.

$$
\begin{array}{ccccc}
7 & 7 & 7 & 6 & 5 \\
  &   & 2 & 4 & 3
\end{array}
$$

Finally, multiply the last digit 5 by 2, 4, and 3.

$$
\begin{array}{ccccc}
7 & 8 & 9 & 7 & 5 \\
  &   & 2 & 4 & 3
\end{array}
$$

Thus, the product is 78975.

Other methods of multiplication and division seemed to have come out of India, through Arab writers, and into Western Europe. Some methods are identical to our own form of long multiplication. The **grating method** for the product of 123 by 456 is shown in the margin, and has some advantages over our own form of long multiplication, since there is no need to "carry" intermediate sums. The split in the grating reminded Italians of shutters which, through a strange twist of etymology, were called *gelosia*, the Italian word for jealousy; as a result, this method is sometimes called the **method of gelosia**.

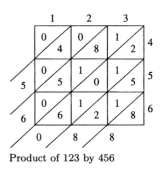

Product of 123 by 456

9.1.2 Division

For division, ibn Labbān used the following procedure. To divide 5625 by 243, he first wrote the highest rank of the dividend under the highest rank of the divisor.

$$
\begin{array}{cccc}
5 & 6 & 2 & 5 \\
2 & 4 & 3 &
\end{array}
$$

Dividing the first digit of the divisor, 2, into the first digit of the dividend, 5, he obtained 2; this he wrote above the 3 (the lowest rank of the divisor).

$$
\begin{array}{cccc}
  &   & 2 &   \\
5 & 6 & 2 & 5 \\
2 & 4 & 3 &
\end{array}
$$

Now, multiply 2 times 2 and subtract the product from 5, leaving 1.

$$
\begin{array}{cccc}
  &   & 2 &   \\
1 & 6 & 2 & 5 \\
2 & 4 & 3 &
\end{array}
$$

Multiply 4 times 2, and subtract the product from 6 (actually 16):

$$
\begin{array}{cccc}
  &   & 2 &   \\
  & 8 & 2 & 5 \\
2 & 4 & 3 &
\end{array}
$$

Finally, multiply 3 by 2 and subtract the product from 82; move the divisor over one space.

$$
\begin{array}{ccc}
 & 2 & \\
7 & 6 & 5 \\
2 & 4 & 3
\end{array}
$$

Repeat the procedure: 2 into 7 is 3, which is again written above the last ranking number.

$$
\begin{array}{ccc}
2 & 3 & \\
7 & 6 & 5 \\
2 & 4 & 3
\end{array}
$$

Multiplying 2, 4, and 3, and subtracting the result from 765, we obtain

$$
\begin{array}{ccc}
2 & 3 & \\
3 & 6 & \\
2 & 4 & 3
\end{array}
$$

Thus, the result is 23 and $\frac{36}{243}$.

The "ship" or **galley method** of division is, perhaps, the most striking visually. The method appears in both Chinese and Islamic mathematical works, so a common origin in India seems likely. It takes its name from the shape of the final configuration, which with a little imagination looks like a sailboat. In the division of 83514 by 37, begin by writing down the two numbers, with the greatest rank of each in the first column.

$$
\begin{array}{ccccc|}
8 & 3 & 5 & 1 & 4 \\
3 & 7 & & &
\end{array}
$$

Divide 3 into 8, obtaining the quotient 2. Write this down to the right of the bar as the first digit of the quotient. Then multiply 2 times 3, to obtain 6, and subtract this from 8, the number above 3. Cross out 8 and record the remainder; also cross out 3.

$$
\begin{array}{ccccc|c}
2 & & & & & \\
\not8 & 3 & 5 & 1 & 4 & 2 \\
\not3 & 7 & & & &
\end{array}
$$

Now multiply 2 times 7, to obtain 14; subtract this from 23, leaving 9 and crossing out the 2, 3, and the 7. Recopy the divisor, 37, one space over (notice the "7" is placed immediately under the 5 of the dividend).

$$
\begin{array}{ccccc|c}
\not2 & 9 & & & & \\
\not8 & \not3 & 5 & 1 & 4 & 2 \\
\not3 & \not7 & 7 & & & \\
 & 3 & & & &
\end{array}
$$

Repeat the procedure. 3 into 9 is 3, but this will be too much, so record 2 as the next digit of the divisor.

```
2   9
8   3   5   1   4 | 2   2
3   7   7
        3
```

Multiply 2 by 3 and subtract the product from 9; multiply 2 by 7 and subtract the product 14 from 35, crossing out the appropriate numbers. Again, recopy the 37 one space over.

```
        3                               2
2   9                                   3
8   3   5   1   4 | 2   2    →    2   9   1
3   7   7                         8   3   5   1   4 | 2   2
    3                             3   7   7   7
                                      3   3
```

The next digit of the quotient is 5, and the table becomes

```
        2   2
        3   6
2   9   1   6
8   3   5   1   4 | 2   2   5
3   7   7   7   7
    3   3   3
```

The last digit of the quotient is 7, giving the final table

```
        2   2
        3   6   5
2   9   1   6   5
8   3   5   1   4 | 2   2   5   7
3   7   7   7   7
    3   3   3
```

and thus the quotient is $2257\frac{5}{37}$. Some manuscripts showing this method of division often rounded it off by drawing a ship around it though, of course, this last step is optional!

Divisions often end with remainders, which were usually expressed as common fractions (as in the previous example). Alternatively, the remainder might be expressed as a sexagesimal fraction, which retains some of the key elements of place value. Like the Greeks and Babylonians, the Islamic mathematicians did not separate the place values, nor did they indicate the sexagesimal point, though we will do so using a semicolon. Ibn Kūshyār performed the long division of 49 degrees 36 minutes by 12 degrees 25 minutes as follows: first, set dividend and divisor side by side in two rows, the first row representing the degrees and the second row the minutes:

```
| 4   9 | 1   2
| 3   6 | 2   5
```

Consulting a table of sexagesimal products, ibn Kūshyār determined the quotient of $49; 36$ and $12; 25$ to be 3; record 3 in the top row and from $49; 36$ subtract 3 times $12; 25$, or $37; 15$, leaving $12; 21$. Finally, shift the divisor $12; 25$ down one row.

$$
3 \begin{array}{|cc|cc}
1 & 2 & 1 & 2 \\
2 & 1 & 2 & 5 \\
 & & & \\
\end{array}
\overset{\text{shift}}{\longrightarrow}
\quad
3 \begin{array}{|cc|}
1 & 2 \\
2 & 1 \\
 & \\
\end{array}
\begin{array}{cc}
1 & 2 \\
2 & 5 \\
\end{array}
$$

Next divide $12; 25$ into $12; 21$. Since each row represents a 60-fold increase in value, the quotient is $0; 59$. Write this in the left column and multiply $12; 25$ by $0; 59$ and subtract from $12; 21$, obtaining:

$$
\begin{array}{cc}
 & 3 \\
5 & 9 \\
 & \\
\end{array}
\begin{array}{|cc|}
0 & 0 \\
0 & 8 \\
2 & 5 \\
\end{array}
\begin{array}{cc}
 & \\
1 & 2 \\
2 & 5 \\
\end{array}
$$

Clearly the procedure can be continued indefinitely, obtaining ever more precise sexagesimal approximations to the quotient.

9.1.3   Decimal Fractions

However, this leads to the difficulty of having one system of place value for whole numbers, where the places differ by a power of 10, and another for the fractional parts, where the places differ by powers of 60. An "obvious" solution is to use the same system for both: our modern form of decimal fractions is precisely such a solution. This obvious solution took several hundred years to develop.

The earliest surviving examples of the use of decimal fractions come from the *Book of Chapters on Hindu Arithmetic*, written in Damascus around 952 by Abu'l-Ḥasan al-Uqlīdisī (920-980). Al Uqlīdisī's work comes down to us through a single copy, made in 1157. "Al-Uqlīdisī" is Arabic for "The Euclidean scribe": he was someone who made his living by making copies of the *Elements* for students of geometry. This is all the biographical information known about al-Uqlīdisī. According to the preface, the *Book of Chapters* is a compilation of the best techniques of working with the Hindu system of numeration, particularly without using the "dust board" or abacus. Instead, the computational techniques are all suited for an important technological innovation: paper.

According to one tradition, paper was invented in China around 105 A.D. by the eunuch Tsai Lun, although archaeological evidence suggests it was already in use for wrapping and packing more than two and a half centuries before. Paper is made from rags and wood pulp, and bears only a superficial similarity to papyrus, made from reeds; however, the similarity was enough to cause the new material to be named after the old. By the dawn of the ninth century, the Arabs had learned the art of paper making from captured Chinese prisoners of war, and a flourishing paper industry grew up in Samarkand, now in modern-day Uzebekistan. The existence of a cheap writing material made computation easier: a dust board had to be set up on a level surface, and in any case took up a minimum of space. Pen and paper computations, on the other hand, could be done anywhere. There were other reasons to switch to the new medium: the dust board (and dusty hands) were the hallmarks of astrologers (the "mathematicians" censured by Diocletian and Augustine). By using pen and paper, an honest calculator could distinguish himself from a charlatan.

To take half of a unit, ibn Kūshyār and others used sexagesimals: since $1 = \frac{60}{60}$, then half of 1 would be $\frac{30}{60}$. Al-Uqlīdisī described another technique, clearly recognizable as that of decimal fractions: half of 1 was 5 in the "place before." Again, since Arabic is written from right to left, this meant that the 5 would be placed in the space to the right, just as we do today. Other Arab writers, particularly al Samaw'al, further developed decimal notation. However, Europeans remained ignorant of these Arabic developments, and over six hundred years would pass before the European arithmeticians invented a decimal notation comparable to that of the Arabs.

See page 260.

9.1.4   Root Extraction

Ibn Kūshyār's arithmetic included a method for finding square roots, and methods of extracting square and cube roots were known to Indian mathematicians. One of the earliest treatises about finding roots higher than the cube was a work of Mohammad Abu'l Wafā al-Buzjani (June 10, 940-July 15, 998), *On Obtaining Cube and Fourth Roots and Roots Composed of These Two*. Nothing is known about the work except its title. Finding a fourth root might have simply been a matter of noting that $\sqrt[4]{x} = \sqrt{\sqrt{x}}$ (hence the fourth root was "composed of" two square roots in succession). Umar al Khayyami, whose work will be discussed later, claimed he was the first to find fourth, fifth, and sixth roots of a number, and that he wrote a treatise on the subject. However, his treatise, too, has been lost, so the earliest example we have of finding fifth roots in the Islamic world comes from *The Calculator's Key*, written in 1427 by Ghiyāth al-Dīn Jamshīd al-Kashī (1380-June 22, 1429).

See page 255.

Al-Kashī found $\sqrt[5]{44,240,899,506,197} \approx 536\,\frac{21}{414,237,740,281}$; the result is accurate to ten decimal places. We will illustrate al-Kashī's procedure and find $\sqrt[5]{15,443,753}$. First, the paper is divided into several regions, labeled "Row of the Result," "Row of the Number," "Row of the square-square" (one less than the fifth power), "Row of the second of the number," "Row of the cube," "Row of the third of the number," "Row of the square," "Row of the fourth of the number," "Row of the root," "Row of the fifth of the number." Since the distinction between the "Row of the square-square" and the "Row of the second of the number is not important, we will save space and simply label the sections "Result," "Number," "Second," "Third," "Fourth," and "Fifth." Next, the number itself (written in the "Row of the Number") would be broken into 5-digit groups al-Kashī called cycles (see Figure 9.3). This was because the fifth powers of one-digit numbers had five or fewer digits; the fifth powers of two-digit numbers had six to ten digits, and so on.

To begin the procedure, find the highest number whose fifth power is less

Result								
Number	1	5	4	4	3	7	5	3
Second								
Third								
Fourth								
Fifth								

Figure 9.3: Finding $\sqrt[5]{15,443,753}$: Set-up.

Result		2		← Step 2				
Number	1	5	4	4	3	7	5	3
		3	2	← Step 6				
	1	2	2	← Step 7				
Second								
		1	6	← Step 5				
Third								
			8	← Step 4				
Fourth								
			4	← Step 3				
Fifth								
			2	← Step 1				

Figure 9.4: Finding the first digit of the root.

Result		2						
Number	1	5	4	4	3	7	5	3
		3	2					
	1	2	2					
Second								
		8	0	← Step 14				
		6	4	← Step 13				
		1	6					
Third								
		3	2	← Step 12				
		2	4	← Step 11				
			8					
Fourth								
		1	2	← Step 10				
			8	← Step 9				
			4					
Fifth								
			4	← Step 8				
			2					

Figure 9.5: Once up to the Square-Square: First Part.

Result		2						
Number	1	5	4	4	3	7	5	3
		3	2					
	1	2	2					
Second								
		8	0					
		6	4					
		1	6					
Third								
		8	0	← Step 19				
		4	8	← Step 18				
		3	2					
		2	4					
			8					
Fourth								
		2	4	← Step 17				
		1	2	← Step 16				
		1	2					
			8					
			4					
Fifth								
			6	← Step 15				
			4					
			2					

Figure 9.6: Once up to the Square-Square: Second Part.

than 154: this number is 2. Write down 2 in the row of the fifth, and also in the row of the result (Step 1 and Step 2 in Figure 9.4). Then multiply the number in the row of the fifth (2) by the most recently obtained digit (also 2) and write the result (4) in the row of the fourth (Step 3). Multiply the result (4) by the most recently obtained digit of the root (still 2), and write the product (8) in the row of the third (Step 4); continuing, we fill out the powers of 2 in the first column (Step 5 and Step 6). Subtract the last, 32, from 154, leaving a remainder of 122 (Step 7); al-Kashī used a line to indicate where an operation had been performed.

Next begins a procedure called "Once up to the row of the square-square." Take the most recently obtained digit, 2, and add it to the number in the row of the fifth of the number (2), obtaining 4 (Step 8 in Figure 9.5). Multiply 4 by the most recently obtained digit (again 2), and write the result (8) in the row of the fourth of the number (Step 9). Add 8 to 4 to get 12 (Step 10); multiply this by 2 to get 24 (Step 12), and add this to the number in the row of the third of the number to get 32 (Step 13). Multiply this sum by 2 to get 64, and add it to the number in the row of the second of the number to get 80 (Step 14).

Again, take the number in the row of the fifth of the number (4), add the most recently obtained digit (2) (Step 15 in Figure 9.6) again to get 6, and multiply it by the most recently obtained digit (2) (Step 16). Add the product (12) to the number in the row of the fourth to obtain 24 (Step 17). Multiply this by 2 (Step 18), add the result (48) to the number in the row of the third to get 80 (Step 19).

Result		2						
Number	1	5	4	4	3	7	5	3
		3	2					
	1	2	2					
Second		8	0					
		6	4					
		1	6					
Third		8	0					
		4	8					
		3	2					
		2	4					
			8					
Fourth		4	0	← Step 22				
		1	6	← Step 21				
		2	4					
		1	2					
		1	2					
			8					
			4					
Fifth		1	0	← Step 23				
			8	← Step 20				
			6					
			4					
			2					

Figure 9.7: Once up to the Square-Square: Third and Fourth Part.

Once more, take the number in the row of the fifth of the number (6), add 2 to get 8 (Step 20), multiply by 2 (Step 21 in Figure 9.7) and add the result 16 to the number in the row of the fourth of the number to get 40 (Step 22). Finally, take the number in the row of the fifth of the number and add 2 to get 10 (Step 23).

Once the numbers are obtained, shift the number in the row of the second one space; the number in the row of the third two spaces; the number in the row

Result		2				7		
Number	1	5	4	4	3	7	5	3
		3	2					
	1	2	2					
Second			8	0				
Third				8	0			
Fourth					4	0		
Fifth						1	0	

Figure 9.8: Shift.

Result		2				7				
Number	1	5	4	4	3	7	5	3		
		3	2							
	1	2	2							
	1	1	1	4	8	9	0	7		← Step 30
		1	0	9	4	8	4	6		← Step 31
Second		1	5	9	2	7	0	1		← Step 29
			7	9	2	7	0	1		← Step 28
			8	0						
Third			1	1	3	2	4	3		← Step 27
				3	3	2	4	3		← Step 26
				8	0					
Fourth					4	7	4	9		← Step 25
					7	4	9			← Step 24
				4	0					
Fifth						1	0	7		

Figure 9.9: Finding Second Digit of Root.

of the fourth three spaces, and the number in the row of the fifth four spaces. The shifted numbers become the new numbers in the "Row of the second," "Row of the third," etc. (see Figure 9.8).

The first digit of the root was 2 which, because of its position, represents 20; the procedure of al Kashī has already subtracted 20^5 from the original number, leaving 12243753. The next digit, which we might call b, would make the partial root $20 + b$, and it is necessary that the fifth power not exceed the original number; in other words, we need to find the largest b so

$$(20 + b)^5 - 20^5 < 12243753 \tag{9.1}$$

Al-Kashī found the next digit of the root to be 7, so he wrote 7 in the row of the result (at the top of the second column) and as the last digit of the number in the row of the fifth of the number. This number, 107, is then multiplied by the last digit found (7) (Step 24 in Figure 9.9), and the product (749) added to the number in the row of the fourth of the number to make 4749 (Step 25). This sum is multiplied by 7 (Step 26) and the result, 33243 is added to the number in the row of the third of the number, 80000, to make 113243 (Step 27). This is multiplied by 7 (Step 28) and the product, 792701 is added to the number in the row of the second of the number, 800000, to make 1592701 (Step 29). Finally, this number is multiplied by 7 (Step 30), and the product 11148907 is subtracted from 12243753, to obtain 1094846 (Step 31) in the row of the number itself. The end result of all the steps so far is to have determined $15443753 - 27^5 = 1094846$.

What is the procedure of al-Kashī equivalent to? If the left hand side of Equation 9.1 is expanded, we obtain

$$800,000b + 80,000b^2 + 4000b^3 + 100b^4 + b^5 < 12243753$$

Result	2			7				
Number	1	5	4	4	3	7	5	3
		3	2					
	1	2	2					
	1	1	1	4	8	9	0	7
		1	0	9	4	8	4	6
Second		2	6	5	7	2	0	5
		1	0	6	4	5	0	4
		1	5	9	2	7	0	1
			7	9	2	7	0	1
		8	0					
Third			1	9	6	8	3	0
				4	4	7	5	8
			1	5	2	0	7	2
				3	8	8	2	9
			1	1	3	2	4	3
				3	3	2	4	3
			8	0				
Fourth					7	2	9	0
						8	4	7
					5	5	4	7
						7	9	8
					4	7	4	9
						7	4	9
					4	0		
Fifth						1	3	5
						1	2	8
						1	2	1
						1	1	4
						1	0	7

Figure 9.10: Once up to the Square-Square: All steps.

the first four coefficients of which are the same as the numbers in the various rows in Figure 9.8. Of course, al-Kashī could evaluate the left hand side directly, but instead he used a simpler method, equivalent to noting that the left hand side can be written as

$$b(800,000 + b(80,000 + b(4000 + b(100 + b)))) < 12243753 \qquad (9.2)$$

and thus the problem is to find the highest b for which the left hand side is less than the remaining portion of the root; b equal to 7 satisfies this requirement.

By making b the last digit of the number in the row of the fifth, al-Kashī determined $100 + b$. Multiplying this by the last digit obtained produces $b(100 + b)$. Adding this amount to the number in the row of the fourth gives $4000 + b(100 + b)$, which is then multiplied by b to produce $b(4000 + b(100 + b)))$. Thus, the left-hand side of Equation 9.2 is found.

Al Kashī certainly knew this procedure could be used to find additional decimal places in the value of the root, but he employed a different algorithm to find a rational approximation. He found the approximation by completing the procedure of "One step up to the square-square" one more time (Figure 9.10). To find the rational approximation, al-Kashī first added the numbers at the top of each of the rows, plus 1. In this case, that would produce:

$$2657205 + 196830 + 7290 + 135 + 1 = 2861461$$

This is the denominator of a fraction whose numerator is the remaining portion of the root; hence $\sqrt[5]{15443753} \approx 27\,\frac{1094846}{2861461}$.

The extraction of higher roots highlights the difficulties of untangling the history of the era and region. There are clear similarities between the method of al-Kashī and the earlier method of Yang Hui, so it is easy to suspect the Islamic mathematicians learned from the Chinese. But al Khayyami's lost work suggests Islamic mathematicians predated Yang Hui in their discovery of root extraction methods, making it plausible to suppose that it was the Chinese who learned from the Persians. However, Yang Hui himself credited an earlier source, which has not been found. The *existing* evidence is that the Chinese predated the Persians in their discovery of a method of root extraction, but the *anecdotal* evidence leaves the question open. Thus, whether the Chinese learned from the Persians, the Persians from the Chinese, or whether the marvelous method of root extraction was a simultaneous and independent invention cannot be definitively answered until more evidence surfaces. With the loss of so many sources, it is possible that the final answer may never be found. See page 209.

9.1.5 Combinatorics

Combinatorics, first studied in detail by the Indians, was further developed by Muslim mathematicians. In the Islamic world, the most advanced work in the subject was probably that of Ahmad al-Ab'dari ibn Mun'im, who lived in the Maghreb (Morocco) during the twelfth century. Ibn Mun'im considered the problem See page 225.

Problem 9.1. *Given ten different colors of silk, how many different bundles, consisting of one, two, three, etc., different colors, can be made?*

To solve this problem, ibn Mun'im began by noting that the bundles of two colors were obtained by taking each bundle of one color, and adding a second color; the bundles of three colors were obtained by taking the two color bundles and adding a third color, and so on. First, he set the ten possibilities for one-color bundles as a row of ones:

				Color						
1st	2nd	3rd	4th	5th	6th	7th	8th	9th	10th	
1	1	1	1	1	1	1	1	1	1	One color

To determine the number of two-color combinations in an orderly fashion, he noted that the second color could be combined with only the first one; the third color could be combined with either of the previous two; the fourth color could be combined with any of the previous three, and so on. By writing down these numbers, he obtained the two color combinations:

Color

1st	2nd	3rd	4th	5th	6th	7th	8th	9th	10th	
									1	Ten color
								1	9	Nine color
							1	8	36	Eight color
						1	7	28	84	Seven color
					1	6	21	56	126	Six color
				1	5	15	35	70	126	Five color
			1	4	10	20	35	56	84	Four color
		1	3	6	10	15	21	28	36	Three color
	1	2	3	4	5	6	7	8	9	Two color
1	1	1	1	1	1	1	1	1	1	One color

Table 9.1: Ibn Mun'im Combinatorial Table.

Color

1st	2nd	3rd	4th	5th	6th	7th	8th	9th	10th	
	1	2	3	4	5	6	7	8	9	Two color
1	1	1	1	1	1	1	1	1	1	One color

For the three color combinations, one could take the third color and combine it with the one previous two-color combination; the fourth color and combine it with the three previous two-color combinations, and so on. Proceeding in this way, ibn Mun'im formed Table 9.1, which is familiarly known as Pascal's triangle. Since a given bundle of silk, say red-blue-yellow, is the same regardless of the order the colors were chosen, ibn Mun'im's table gave the number of combinations.

Because of Mun'im's means of generating the triangle, he was able to conclude that every entry was the sum of the entries on previous entries on the line below it; this was one of the few *theoretical* contributions to the study of the arithmetic triangle before Pascal. Ibn Mun'im proceeded to use the table to solve a number of combinatorial problems, ultimately proceeding to his problem of interest: to determine the number of "words" that can be formed using the a number of letters.

Since two different words might use the same letters, but in a different order, ibn Mun'im's problem is one of determining the number of permutations of a set of n objects. Ibn Mun'im's analysis proceeded as follows: a one-letter word A has, obviously, only one possible permutation of the letters. A two letter word has two possible permutations, AB or BA. A third letter could go in one of three possible positions in each of the two permutations: before the two letters; between the two letters; or after the two letters. Thus, there are $3 \times 2 = 6$ possible permutations. A fourth letter may appear in any one of the four positions in each of the six permutations: at the beginning; between first and second; between second and third; between third and fourth; or at the end. Thus, there are $6 \times 4 = 24$ possible permutations. Thus, ibn Mun'im concluded:

Rule 9.1. *To find the number of permutations of the letters of a word with no* *repeated letters, multiply 1, by 2, by 3, and so on, up to the number of letters in the word.*

He also gave the correct rule if letters were repeated.

9.1 Exercises

1. Multiply, using both the method of Kūshyār ibn Labbān and the grating method.

   (a) 389 multiplied by 7   (c) 8915 multiplied by 73

   (b) 8916 multiplied by 47   (d) 10634 multiplied by 415

   (e) Compare the methods. What are each method's advantages and disadvantages?

2. Divide, using both the method of Kūshyār ibn Labbān and the galley method.

   (a) 127 divided by 7   (c) 8916 divided by 17

   (b) 2895 divided by 42   (d) 389215 divided by 195

   (e) Compare the methods. What are each method's advantages and disadvantages?

3. Use the method of al-Kashī to approximate the indicated roots.

   (a) $\sqrt[5]{3216589}$ (b) $\sqrt[5]{389238913}$ (c) $\sqrt[5]{44240899506197}$

4. Vary the method of al-Kashī to approximate the indicated roots

   (a) $\sqrt[3]{8921}$.   (b) $\sqrt[4]{1336336}$.   (c) $\sqrt[7]{49702849382}$.

5. (Teaching Activity) Consider the method of gelosia. Each diagonal corresponds to a certain order of magnitude in the product of two numbers.

   (a) Modify the method so the product of two numbers can be estimated rapidly.

   (b) Modify the method to find the product of two polynomials.

   (c) Reverse the product procedure and use the grating to determine the *quotient* of two polynomials.

6. Compare al-Kashī's procedure for finding roots with Yang Hui's. What are their similarities and differences?

7. Compare the notations and operations using Chinese numerals with those of the Hindu-Arabic system as described by Indian and Arab mathematicians. Comment on the possibility that the Hindu-Arabic system originated in China.

8. Show
$$(n^k + r)^{1/k} \approx n + \frac{r}{(n+1)^k - n^k}$$

Then show that al-Kashī's method of estimating the fractional part of the root is equivalent to this approximation.

9. Examine the methods of arithmetic used by Kūshyār ibn, Labbān, and al-Kashī, as well as the grating and ship methods. Which seem better suited for pen and paper computation? Which seem better suited for computation on the counting board? Explain your reasoning.

9.2   Al Khwārizmī

By the time of caliph al Ma'mun (809-833), Abbasid power was well established in the Muslim world. The original Abbasid capital was at Merv (now in Turkmenistan); al Ma'mun moved it to Baghdad, closer to the geographic center of the caliphate. There he established a House of Wisdom, similar in intent to the great Museum of Alexandria. He was said to have been inspired by a dream in which Aristotle appeared to him; this spurred him to try and obtain translations of as many ancient Greek manuscripts as he could obtain, including the works of Euclid and Ptolemy.

Among the scholars at the House of Wisdom was Abu Ja'far Muhammad ibn Musa al Khwārizmī (ca. 780-850). "Abu Ja'far" means "father of Ja'far"; "Ibn Musa" is "son of Musa" (the Arabic form of Moses); "al Khwārizmī" is "the man from Khwarizm" (a region near the Aral Sea, now in Uzbekistan and Kazakstan); thus, his name can be translated as "Mohammed, father of Ja'far, son of Moses, the man from Khwarizm." He is usually referred to as al Khwārizmī. As the Shari'ah (the body of Islamic law) prohibits the making of images for fear the drawing of a living object will promote idol worship, there are very few authentic portraits of medieval Islamic mathematicians, and none of al Khwārizmī.[1]

See page 220.

Al Khwārizmī was a notable scholar who wrote on a variety of subjects, including a work on the numbering system of the Hindus, probably based on the work of Brahmagupta. A Latin translation of al Khwārizmī's *On the Indian Numbers* became one of the means by which Europe learned of the new system of numeration. The translations were not always of the highest quality. One fifteenth century work, *The Craft of Numbering*, turned al Khwārizmī into Algor, a king of India! More accurately, the author went on to say that the craft

---

[1] One consequence was that art and geometry united to defeat emptiness: to fill in blank spaces, Islamic artists often drew elaborate geometric designs. Europeans later called the designs *arabesque*.

of computation was named after Algor: **algorithm**, which is a misunderstanding of the name al Khwārizmī.

Around 825, al Khwārizmī wrote one of the first books on the procedures for solving equations. His book is entitled *The Condensed Book of Completion and Restoration*. In Arabic, "completion" is *al-jabr*, from which we get **algebra**.

Al Khwārizmī began by distinguishing three types of numbers: squares, roots, and simple numbers, akin to Diophantus's classification of species. Like Diophantus, he had no difficulty comparing numbers of different types:

> A number belonging to one of these three classes may be equal to a number of another class; you may say, for instance, 'squares are equal to roots,' or 'squares are equal to numbers,' or 'roots are equal to numbers.' [2]

The algebra of al Khwārizmī represents in many ways a step backward from the work of Diophantus. Notation is nonexistent, and even the numbers are spelled out. The problems are generally simpler than those posed and solved by Diophantus and the ancient Babylonians.

This should not detract from a far more important feature. Diophantus, and before him the Babylonians, solved their problems using a variety of clever methods, each specific to a type of problem. Al Khwārizmī and his successors approached problem solving by *reducing* a problem to one of just a few fundamental types, the solutions of which are straightforward. Thus, the work of al Khwārizmī shifted the emphasis from solving equations, to transforming equations using what we would now call the **rules of algebra**.

9.2.1 Quadratic Equations

After disposing of the three simple equations (squares equal to roots; squares equal to numbers; roots equal to numbers), al Khwārizmī proceeded to more complicated "compound" equations, combining squares, roots, and numbers. There are three types:

1. Squares and roots equal to numbers;

2. Squares and numbers equal to roots;

3. Squares equal to roots and numbers.

His method of solution is familiar; in his work, "dirhems" were used to indicate quantities that are pure number:

> One square, and ten roots of the same, amount to thirty-nine dirhems ...The solution is this: you halve the number of roots, which in the present instance yields five. This you multiply by itself; the product is twenty-five. Add this to thirty-nine; the sum is sixty-four. Now take the root of which, which is eight, and subtract it from half the

---

[2]Rosen, Frederic, *The algebra of Mohammed ben Musa*. J. Murray 1831. Page 6.

number of the roots, which is five; the remainder is three. This is the root of the square which you sought for; the square itself is nine.[3]

Note that, like the Rhind papyrus, example and rule are combined, and generalization was left to the student. For problems like "two squares and ten roots are equal to forty-eight dirhems," al Khwārizmī reduced the problem to "one square and five roots are equal to twenty-four dirhems," and then solved it as above.

The second type of problem, squares and numbers equal to roots, can be solved:

Problem 9.2. *A square and twenty-one in numbers are equal to ten roots of the same square.*

Solution. Take half the number of roots, which is five. Multiply this by itself, making twenty-five. Subtract this from twenty-one (the number with the square), which is four. Find the root, two. Subtract this from half the roots, leaving three. This is the root of the square, and the square is nine. Or add the root to half the roots, which is seven, and the root of the square, whose square is forty-nine.   □

Al Khwārizmī gave two demonstrations of the validity of the solution method in the case of a square and ten roots equal to thirty-nine dirhems. In both cases, he began with a square the sides of which are unknown; key to the procedure is the process of adding the ten roots. To accomplish this, he noted that adding ten roots was the same as adding a rectangle, one of whose sides was the side of the square, and the other side of which was the number of roots to be added.

In the second case, which is directly related to the method of solution, he added the ten roots by halving ten, making five, then constructing the rectangles C, D, which have as one side the side of the square and the other side the length of five. The resulting gnomon is the square (AB) and ten roots (C and D); it is short of a complete square by the square E, which must be the square of side five. Hence E is twenty-five.

If the square E is added to the square and ten sides, the result is a square; numerically, the square and ten sides are equal to thirty-nine, so all together the square, ten sides, and the square E are sixty-four. The large square FG is thus equal to sixty-four; its side is eight, and subtracting five, the side of E, one finds the side of the original square is three.

Obviously, the geometry is the same regardless of the number of roots added or what the quantity of squares and roots is equal to. A more important question is the validity of adding the "side of five" and how, precisely, one could justify adding ten sides, which should be a larger side, to a square, which is an area. We will see the attempts of later mathematicians to justify this procedure.

The case of a square and number equal to roots is more complicated. The demonstration of the solution, "a square and twenty-one dirhems are equal to ten roots," begins with a square AD, again of unknown side length. To the unknown, add rectangle HB, representing twenty-one dirhems, with sides AB,

---

[3]Ibid., 8

NB. The complete figure HD thus represents a square and twenty-one dirhems, which is also equal to ten roots. Since a root is the side of the square AD, such as CD, then the side HC must be ten.

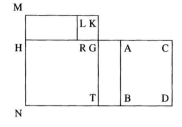

Divide HC into equal parts at G, and draw the perpendicular GT. Add GK, equal to the difference between CG, GT (in other words, equal to GA), making TK equal to KM, and thus MT is a square. Since HG is half of ten, then MT is the square of five, or twenty-five. From KM, take a piece KL equal to GK, so TG and ML are equal, and KL is equal to GK. Thus the rectangle MR is equal to the rectangle TA.

Now, since HB represents twenty-one numbers, so do the rectangles HT, MR. Since MT is twenty-five, then the square KR is four, and thus RG is equal to two, as is GK and consequently GA. Since GC is half of ten, or five, and GA is two, then AC must be three, which is the side of the original square.

9.2.2   Higher Degree Equations

One of al Khwārizmī's more interesting problems is:

Problem 9.3. *Subtract three roots from a square, then multiply the remainder by itself, and the square is restored.*

In modern terms, $(x^2 - 3x)^2 = x^2$. Since the problem deals with squaring a square, it results in a nongeometric entity. Diophantus or the Babylonians would have found a clever method of solving this specialized type of problem; al Khwārizmī instead shows how the rules of algebra can be used to reduce it to a previously solved type:

Solution. The remainder must also be a root, so the square must equal four roots. Hence the side is four.   □

In effect, al Khwārizmī reduced the problem by taking the square root of both sides, and solved $x^2 - 3x = x$. Equations involving radicals also appear in al Khwārizmī's work, and except for the complete lack of notation, the methods used by al Khwārizmī are the same as those used today.

9.2.3   The Arithmetization of Algebra

After this, al Khwārizmī proceeded to explain how to multiply quantities including the unknown. This was an early step in the **arithmetization of algebra**: the application of the techniques of arithmetic to the unknown quantities in algebra. First, he gave an example of how to multiply two numbers:

> Ten and one to be multiplied by ten and two. Ten times ten is a hundred; once ten is ten positive; twice ten is twenty positive, and once two is two positive; this altogether makes a hundred and thirty-two.[4]

---

[4]Ibid., 22

For cases where subtraction is involved, he demonstrated an understanding of the products of signed numbers:

> Ten less one, to be multiplied by ten less one, then ten times ten is a hundred; the negative one by ten is ten negative; the other negative one by ten is likewise ten negative, so that it becomes eighty; but the negative one by the negative one is one positive, and this makes the result eighty-one.[5]

By analogy, the process was extended to algebraic quantities. Here, "thing" represents the unknown quantity.

> Ten and a thing to be multiplied by itself, then ten times ten is a hundred, and ten times thing is ten things; and again ten times thing is ten things; and thing multiplied by thing is a square positive, so that the whole product is a hundred dirhems and twenty things and one positive square.[6]

9.2.4   Other Problems

Most of the problems in the first part of the book appear to be pure mathematics; it is only later that al Khwārizmī dealt with commercial arithmetic, where he used the Rule of Three (but not the Rules of Five, Seven, etc.). Many of the problems originate in the complex inheritance laws of Islam: the procedure for determining shares of an estate owed to each survivor was known as *'ilm al-farā'iḍ*. The appearance of so many inheritance problems implies they are of great importance, though the fourteenth century Islamic scholar ibn Khaldūn noted that in most cases, the division was straightforward and criticized authors for inventing problems that required very complex mathematics but would have only occurred in very rare cases.

See page 221.

Judging by the section on geometry, al Khwārizmī was completely ignorant (or chose to ignore) Euclid's *Elements*, then being translated into Arabic by his contemporaries. It would appear that al Khwārizmī instead drew from a very early Hebrew geometry, *Mishnat ha Middot*, written around 150 A.D.: a literal translation of *Mishnat ha Middot*, with only minor rearrangements, comprises the entirety of the geometry in the *The Condensed Book of Completion and Restoration*. After stating that the diameter multiplied by three and one-seventh gives an approximation to the circumference, al Khwārizmī cited two "other methods" used by geometers:

Rule 9.2. *To find the area of a circle: Multiply the diameter by itself, then by ten; take the square root.*

Astronomers, he said, used the following method:

---

[5]Ibid., 22-3
[6]Ibid., 24

Rule 9.3. *Multiply the diameter by sixty-two thousand eight hundred and thirty-two, then divide by twenty thousand.*

Both of these appear in Hindu texts. Al Khwārizmī gave a third formula for the area of a circle, identical to the correct one given by Aryabhata. The Pythagorean theorem also made an appearance, complete with a proof; however, al Khwārizmī's result is valid only for isosceles right triangles.

See page 227.

9.2 Exercises

1. Al Khwārizmī's first diagram explaining the procedure for solving "a square and ten roots is equal to thirty-nine" involved adding rectangles on all four sides of the figure.

   (a) The four rectangles are equal, and each has the side of the square as one side. What must the length of the other side be?

   (b) Explain how to complete the square for this figure.

   (c) What sort of variant method of solution can be derived from the figure?

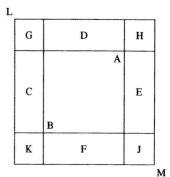

Figure for Problem 1.

2. Solve the following problems from al Khwārizmī's, using his method.

   (a) Two squares and ten roots are equal to forty-eight dirhems.

   (b) Half a square and five roots are equal to twenty-eight dirhems.

3. Explain why the case of "squares and roots equal to numbers" must always have a solution (in the geometrical sense).

4. Under what conditions will the case "squares and numbers equal roots" not have a solution?

5. Write an algorithm for solving the case "roots and numbers equal to squares." Al Khwārizmī used the example: "Three roots and four simple numbers are equal to a square."

6. Multiply.

   (a) Ten minus thing to be multiplied by ten minus thing

   (b) Ten minus thing to be multiplied by ten and thing

7. Create a geometric justification of the procedure for solving "roots and numbers equal to squares." Use the case of "three roots and four simple numbers are equal to a square." Construct square AD, the side of which is unknown. Draw HR parallel to DC, so the rectangle HD is equal to three roots, and HB is equal to four simple numbers; thus, the square (AD) is equal to three roots (HD) and four simple numbers (HB). Halve CH at

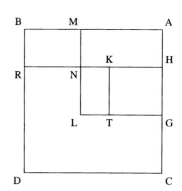

Figure for Problem 7.

G, and construct the square HT. Add to GT the quantity LT, equal to AH, so GL is equal to AG, and AL is a square.

   (a) Explain how the diagram justifies the procedure for solving the case "roots and numbers equal squares."

   (b) Explain why the case of "roots and numbers equal squares" always has a solution.

8. Solve the problems from al Khwārizmī.

   (a) A number, one-third of which and one dirhem multiplied by one-fourth of it and two dirhems restore the number, with a surplus of thirteen dirhems. (In other words, the product equals the original number plus thirteen.)

   (b) A dirhem and a half is divided among one person and some number of persons, so the share of one person be twice as many dirhems as there are other persons. (Find the number of persons.)

   (c) A square; remove one root from it, and if you add to this root a root of the remainder, the sum is two dirhems. (Find a number so that its square, less the number, plus the square root of the whole, is equal to 2.)

9. (Teaching Activity) Use al Khwārizmī's procedure for finding the product of 10 minus 1 by 10 minus 1 to justify the product of two negatives is a positive.

9.3   Other Algebras

Al Khwārizmī's work refers to itself as a "condensed" work, which suggests that larger works existed. However, only a fragment of a contemporary algebra survives: a section entitled "Logical Necessities in Mixed Equations" from the algebra of 'Abd al-Hamid ibn Wasi ibn Turk al Jili, usually referred to as ibn Turk ("son of the Turk"). Very little is known about ibn Turk; even his origin is uncertain, variously ascribed to Iran, Afghanistan, or Syria.

The style of the surviving fragment is very similar to al Khwārizmī's, suggesting that al Khwārizmī's place in the development of algebra is similar to that of Euclid's place, in other words, preparing a textbook for a science that had already reached a high level of development (and, unfortunately like Euclid's *Elements*, causing the other works to be neglected). Ibn Turk's work is superior to al Khwārizmī's from a theoretical standpoint, as he analyzed in greater detail the existence of solutions.

See Problem 1.

9.3.1 Abu Kamil

A surviving algebra of high theoretical sophistication was that of Abu Kamil (850-930). Abu Kamil means "father of Kamil," and he was sometimes referred to as the "Reckoner from Egypt," which is almost all the biographical information we have about him. Abu Kamil's work is primarily a commentary on al Khwārizmī—many of the problems are the same—but Abu Kamil takes greater care in proving algebraic identities. For example, a problem posed by al Khwārizmī was:

> I have divided ten into two parts; and have divided the first by the second, and the second by the first, and the sum of the quotient is two dirhems and one-sixth.[7]

Without proof, al Khwārizmī introduced the identity:

> If you multiply each part by itself, and add the products together, then their sum is equal to one of the parts multiplied by the other, and again by the quotient which is two and one-sixth.[8]

$$\boxed{\text{If } \frac{a}{b} + \frac{b}{a} = c, \text{ then } a^2 + b^2 = abc.}$$

Abu Kamil attempted to provide a rigorous proof. First, Abu Kamil proved:

Proposition 9.1. *If A divided by B is G, then A times A divided by B times A is also G.*

In other words, $\frac{A}{B} = \frac{A^2}{AB}$. Abu Kamil's proof relied on the Euclidean definition of number as a collection of units:

Proof. Let A times A be H, and B times A be D. Since A divided by B is G, then B is contained in A as many times as there are units in G. Since A multiplied by A is H, and B multiplied by A is D, then D is in H as B is in A, and B is in A as many times as there are units in G.

[Since $\frac{A}{B} = G$, then $A = BG$. Since $AA = H$ and $BA = D$, then $\frac{H}{D} = \frac{A}{B} = G$.] □

Then Abu Kamil proved:

Proposition 9.2. *Given two numbers A and B, so that A divided by B is D and B divided by A is Z; A multiplied by itself is G; B multiplied by itself is H, and A multiplied by B is E. If the sum of G and H is divided by E, it comes to the sum of Z and D.*

In other words, if $\frac{A}{B} = D$ and $\frac{B}{A} = Z$, then $\frac{AA+BB}{AB} = D + Z$.

Proof. A divided by B is D, which by the previous proposition is equal to G divided by E. B divided by A is Z, which again by the previous proposition is equal to H divided by E. Thus the sum of G and H, divided by E, is the sum

---

[7]Ibid., 44
[8]Ibid., 44-5

al Karaji	Modern	Text	al Karaji	Modern	Text
cube-cube	x^6	cc	part-cube-cube	x^{-6}	pcc
square-cube	x^5	sc	part-square-cube	x^{-5}	psc
square-square	x^4	ss	part-square-square	x^{-4}	pss
cube	x^3	c	part-cube	x^{-3}	pc
square	x^2	s	part-square	x^{-2}	ps
thing	x	t	part-thing	x^{-1}	pt
unit	1	u			

Table 9.2: Al Karaji's Powers

of D and Z. Or when we multiply the sum of D and Z by E, we get the sum of G and H.

[By the previous proposition, $\frac{A}{B} = \frac{A^2}{AB} = D = \frac{G}{E}$. Likewise, $\frac{B}{A} = \frac{B^2}{AB} = Z = \frac{H}{E}$. Thus $D + Z = \frac{G}{E} + \frac{H}{E} = \frac{G+H}{E}$. Or $E(D + Z) = G + H$.]   □

Besides his work in algebra, Abu Kamil also solved indeterminate equations in *Book of Rare Things in the Art of Calculation*. The work is noteworthy for having been written before Diophantus had been studied in depth by the Arabs, so it represents an independent development of methods of solving indeterminate equations.

9.3.2 Al Karaji

About Abu Bakr al Karaji (April 13, 953-ca. 1029), we know very little, not even his name: "Abu Bakr" simply means "father of Bakr," presumably a more famous son, while "al Karaji" sometimes appears as "al Karkhi," and in either case indicates a place of origin: the former means "the man from Karaj," a city in modern day Iran, while the latter indicates he was from Karkh, a suburb of Baghdad. We will use "al Karaji," as being more likely correct. Al Karaji wrote a work on algebra, *The Marvelous*, and dedicated it to the local governor, Fakhr al Mulk. The governor might not have been pleased: soon afterwards al Karaji left Baghdad for "the mountain countries," where history lost track of him.

Al Karaji took an important step in the development of algebra, and systematically dealt with the algebra of higher powers of the variable than the cube. Al Karaji named the higher powers as shown in Table 9.2. To retain the flavor of al Karaji (and later Arab authors), we will use the abbreviations in the "text" column; hence $6x^5 + 3x^2 + \frac{1}{x^2}$ would be read "six square-cubes plus three squares plus one part-square," and written "6 sc plus 3s plus 1 ps." This was an early step in the development of modern algebraic notation.

Al Karaji continued the process of arithmetization of algebra, begun by al Khwārizmī, by noting that the higher degree terms in a polynomial could be treated exactly like the digits of a multidigit number. Thus, to add two polynomial expressions, such as "three cube-cubes plus four squares plus three units" and "two square-cubes plus three square-squares plus two-squares plus five things plus two units," al Karaji lined up the terms of the same power (just like one would line up the units in performing an addition), then added the coefficients. This seemingly simple but crucial step meant polynomial expressions could be added, subtracted, multiplied, and divided, using the same algorithms that had been developed for working with numbers. Al Karaji failed to provide a complete system of arithmetic only because there was no generalized concept of a negative number; in particular, the procedure of subtracting a negative from a negative (i.e., $-a - (-b)$) had no analog among arithmetical algorithms.

9.3.3 Umar al-Khayyami

It is convenient to speak of "Arabic" mathematics, but this obscures the fact that many mathematicians of the era, though they lived in the Arab controlled part of the world, were not Arabs, either by birth or by culture. Al Karaji and al-Kashī were both Persians, as was Umar ibn Ibrahim al-Khayyami ("Omar, the son of Abraham the Tentmaker") (May 18, 1048-December 4, 1131). Khayyami is better known Omar Khayyam, author of the Rubaiyat, a series of short, four-line poems. Lines from the Rubaiyat include, "a loaf of bread, a jug of wine, and thou," as well as, "The moving finger writes, and having writ, moves on," though how much of the Rubaiyat was actually written by Khayyami, and how much existed before him, is unknown. According to al-Bayhaqī, one of his biographers, al-Khayyami once read a book seven times to memorize it, and when he returned home, copied out the whole work from memory: a comparison with the original revealed very few errors.

His major mathematical work was *On Completion and Restoration*, a work that might be seen as the culmination of Islamic algebra. According to al-Khayyami, Abu Abdallah Muhammad ibn Isa al-Māhānī (820-880), a fellow Persian, came upon an equation involving cubes, squares, and numbers, which no one could solve. The problem originated from Eutocius's commentary on Archimedes's *On the Sphere and Cylinder, Part II* and is equivalent to the equation

$$x^2(a - x) = b^2 c$$

where a, b, and c are given positive quantities. The problem was first solved using conic sections by Abu Jafar al Khazin (d. ca. 965). Further progress in solving cubics was made by other geometers. Al-Khayyami had hoped to write a book discussing all twenty-five types of cubic equations, but other duties kept him from doing so until the present work was written.

Al-Khayyami's work was probably the most rigorous of all the Arabic algebras, for he based his proofs solidly in geometry. Indeed, he noted that the work could not possibly be understood by those who did not understand Euclid's works thoroughly, both the *Elements* and the *Data*, as well as the first two books of Apollonius's *Conics*—a hefty prerequisite, considering that the *Data* and *Conics* were part of Pappus's domain of analysis! See page 158.

There are four quantities to be considered: pure numbers; roots; squares; and cubes. Al-Khayyami noted other quantities, such as the square of the square, might be used "in problems of measurement," but was not itself a "real" quantity: in other words, it did not admit of geometric interpretation.

Al-Khayyami, unlike most of his contemporaries, was concerned with the **problem of homogeneity**: how could one add, for example, the side of a square to the square itself, when the former represented a line segment and the latter a plane figure? To solve this problem, he converted all the figures involved to quantities of the same type. Thus, in order to equate a number and a rectangle, the number itself was treated as a rectangle with one side of unit length. Likewise, if a number was equal to a cube, this meant that a cube was equal to a rectangular parallelepiped with a base equal to the square of unity and whose height was equal to the given number.

Like his predecessors, he began with the problem, "A square and ten roots equal the number thirty-nine." However, his solution relied on Euclid, Book II; in fact, his demonstration, given next, is the same as that given by ibn Qurra. It clearly shows how thoroughly he expected his readers to understand Euclid.

Let the square AC be the unknown; let ten times its root represent the rectangle CE, so DE is ten. Then:

Elements, II-6.   See Problem 6, Section 5.4.

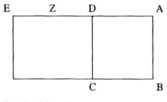

See Problem 3.

Solution. Divide DE into equal parts at Z and extend it by AD. Since the straight line DE has been divided into equal parts and a straight line DA added to it in a straight line, then the rectangle on EA, AD, which is equal to the rectangle BE, plus the square DZ, equals the square ZA. The square of DZ, or half the number of roots, is known. The area of rectangle BE, which is equal to the given number, is known; hence the square of ZA is known, and thus the line ZA is known. Subtracting ZD from ZA gives the line AD.   □

Why would al-Khayyami use such a roundabout means of justifying the solution to this simple equation? One possible reason is that the same Euclidean theorem that demonstrates the solution of a square and roots equal to a number can be applied to the different problem of a number and roots equal a square.

Two Mean Proportionals

An important element in many of al-Khayyami's solutions to cubic equations is the ability to find two mean proportionals between two given numbers. We have already seen how Menaechmus used the intersection of a parabola and a hyperbola to insert two mean proportionals; al-Khayyami gave a different method involving the intersection of two parabolas.

Given the two quantities AB, BC, construct the right angle ABC. Construct parabola BDE with axis BC, vertex at B, and parameter BC; also construct parabola BDZ with axis AB, vertex B, and parameter AB. Let the parabolas intersect at D. Draw DT, DH perpendicular to AB, BC respectively. Then AB, BH, BT, BC are in continued proportion.

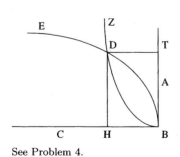

See Problem 4.

See Problem 5.

Finding the sequence of continued proportionals is essential to solving all the cubic equations in the manner of al-Khayyami. The simplest is that of the cube equal to a number. For the problem of cube and sides equal to a number, al-Khayyami used the following procedure: let AB equal the side of the square equal to the number of roots; let BC be the height of a solid whose base is the square on AB and equal to the number. In other words, for the equation $x^3 + p^2x = p^2q$, let $AB = p$, $BC = q$. Draw BC perpendicular to AB.

Extend AB to Z, and make a parabola HBD with vertex at B, axis BZ, and parameter AB. Describe the semicircle with diameter BC; the two must intersect; call the point of intersection D. Drop DZ, DE perpendicular to AZ, BC respectively.

Since DZ is an ordinate of the parabola, then the square on DZ is equal to the rectangle ED, AB, and thus $AB : DZ$ as $DZ : ED$ or, since DZ is equal to BE, then $AB : BE$ is equal to $BE : ED$. But $BE : ED$ is equal to $ED : EC$, and thus AB, BE, ED, EC are in continued proportion. Hence the square on AB is to the square on BE as $BE : EC$. Thus the solid whose base is the square AB and whose height is EC is equal to the cube on BE. Add to both the solid whose base is the square AB and whose height is BE. Thus the cube BE, plus the given number of sides, is equal to the given number, and the solution to the equation is the side BE.

See Problem 6.

Cube and Number Equal to Sides

Al-Khayyami solved the cube and number equal to sides as follows: let AB be the side of the square equal to the number of roots; let BC be the height of a solid equal to the number, with base the square on AB. Describe parabola DBE with vertex at B, axis AB and parameter AB. Construct hyperbola ECZ with vertex C, axis BC perpendicular to AB, and whose parameter and transverse side are both equal to BC. If they do not meet, then there is no solution; obviously, this is a possibility. Suppose they do meet; call the intersection point E, and draw the perpendiculars EH, ET.

See Problem 10.

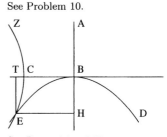

See Proposition 6.32.

Since ET is the ordinate of the hyperbola, the square on ET is to the rectangle contained by BT, TC as the parameter is to the transverse side. Hence the square on ET is equal to the product of BT and TC, and thus $BT : TE$ as $TE : TC$. Also, since DBE is a parabola, the square on BT is equal to the product of BH and BA, and thus $AB : BT$ as $BT : BH$. But BH is equal to TE, and hence $AB : BT$ as $BT : TE$, so AB, BT, TE, TC are in continued proportion. Thus the square on AB is to the square on BT as BT is to TC.

Thus, the cube on BT is equal to the solid whose base is the square on AB and whose height is TC. Add the solid whose base is the square on AB and whose height is BC to both. Thus we have cube and number equal to the sides, and the problem is solved, with the side BT being the side of the cube.

Al-Khayyami dealt with the remaining cases, also using the intersections of conic sections. In some cases, such as that of cube, squares and sides equal to numbers, a solution is guaranteed to exist, since the conics (in this case, a hyperbola and a circle) are situated so they are guaranteed to intersect. In other cases, such as that of a cube, squares, and numbers equal to sides, a solution might not exist, since the conics (two hyperbolas, in this case) might not intersect. In most cases, the solutions are hardly practical, since they require

See Problem 7 and 8.

the construction of conic sections satisfying certain properties; it is not clear how these constructions were to be accomplished.

It would be much better, then, if an algebraic method of solution could be found. Al-Khayyami admitted he was unable to find one, and further doubted that it was possible: he claimed that problems involving cubes, squares, roots, and numbers could only be solved using conic sections.

Other Equations

Al-Khayyami also discussed equations in which the unknown was raised to a negative power; his terms for the powers were almost identical to al Karaji's. Though he stopped with the "cube cube" and corresponding "part of a cube cube," al-Khayyami recognized the table of al Karaji could be extended indefinitely in both directions. But he felt no need to do so because no methods existed for solving equations involving terms beyond the cube or part of the cube.

Equations involving parts could be dealt with in one of two ways. First, terms could be replaced with their analogous terms: part of a square was analogous to a square, and so on. Thus, the problem, "A part of a square is equal to half the part of a root" could be transformed into the equation, "A square is equal to half its root." Alternatively, one could replace terms with proportional succeeding terms, and thus "a root equal to unity and two parts of a root" became "a square equal to a root and two units," replacing each term with the next higher term so the "parts" were eliminated. These transformations are equivalent to replacing $\frac{1}{x}$ with v and multiplying through by x^n, respectively.

In summary, al-Khayyami noted that any equation that "spanned" four terms in proportion could be solved, since it was one of the twenty-five equations whose solutions he found, but if the equation spanned more than four terms, a solution was impossible.

Ratios and Magnitudes

Al-Khayyami, like many other Islamic mathematicians, disliked the purely formal nature of ratios as presented in Euclid's *Elements*. In particular, the definition of equality provided no *constructive* means of determining whether or not two ratios were equal.

Al-Māhānī suggested an alternative: ratio is the mutual behavior of two magnitudes when they were compared using the Euclidean algorithm; we might think of this mutual behavior as being the partial quotients. For example, applying the Euclidean algorithm to the numbers 28, 10, one obtained:

1. Ten can be subtracted **2 times** from 28, leaving 8.

2. Eight can be subtracted **1 time** from 10, leaving 2.

3. Two can be subtracted **4 times** from 8, leaving nothing.

and thus the sequence of partial quotients is 2, 1, 4.

Two pairs of ratios were proportional, A to B as C to D, if the sequence of partial quotients was the same. Thus, since the ratio of 14 to 5 also produced the partial quotients 2, 1, 4, then $28 : 10$ is proportional to $14 : 5$.

Al-Khayyami extended the idea still farther and built an entire theory of ratios based on the Euclidean algorithm. He proved that al-Māhānī's definition of equality was equivalent to that of Euclid, given in the fifth book of the *Elements*; thus, ratios, even incommensurable ratios, could be based solidly in *number*. It was a step in the direction of joining arithmetic and geometry which, unfortunately, would have no effect on the overall development of mathematics, for al-Khayyami's work remained unknown to European mathematicians until the nineteenth century.

9.3.4 The Crusades

Not long after al Ma'mun established his House of Wisdom, the Islamic world underwent a series of internal wars that brought it to a new low; the Caliphate at Baghdad was torn apart. The study of science declined as well; in the tenth century, the scholar al Sijzi complained that in his location (which he did not specify), it was legal to kill mathematicians! Presumably, the "mathematicians" al Sijzi referred to were the same fortune tellers and charlatans that Diocletian outlawed.

The internal chaos in the Muslim world coincided with the era of the Crusades. "Crusade" comes from the Latin *crux*, cross; a crusade is a Holy War, identical in spirit to the Muslim *jihad*. At the end of the eleventh century, the Muslim world was so badly divided that western Europeans were, for the first and only time, able to capture Palestine during the First Crusade (1096-1099). Squabbling among the Franks (as the Muslims called the Europeans, most of whom were French) led to the Muslim recapture of Palestine. Additional Crusades were organized, but the Franks found it easier to fight among themselves than to unite for a common cause.

The Fourth Crusade was perhaps the greatest disaster ever to befall classical scholarship. Constantinople, which had a continuous link to the world of Archimedes and Apollonius, held enormous collections of Greek manuscripts. In the ninth century, the Emperor Leo VI re-opened the University of Constantinople. Among other things, he had the works of Archimedes prepared, based on manuscripts found in Constantinople. These made their way to Europe, via Sicily, and form the core of what we have of the work of Archimedes. Other ancient manuscripts were undoubtedly present in the city.

In 1204, the Crusaders made their way to Venice, but to transport the Crusaders to the Holy Land, the Venetians demanded $85,000$ marks. The Crusaders were unable to raise the money, so Venice offered a trade: if the Crusaders were to attack Venice's trade rival, Zara (now in Croatia), they would waive the fee. The Crusaders did so, destroying Zara (1202), then went on to sack Constantinople itself (1204). In the sack, a quarter of the city was burned, and besides the atrocities that inevitably accompany an invasion, countless treatises of antiquity were lost. The Crusaders never made it to Palestine. The remaining Crusades were likewise epics of incompetence. By 1244, it was all over: the Muslims reconquered Jerusalem in that year, and Palestine would remain in Muslim hands until 1917.

9.3.5   Al Samaw'al

Jews and Christians, as fellow "People of the Book" (the Bible), were tolerated by the Muslims and allowed to practice their own religions. Islamic tolerance was neither universal nor unlimited: pagans were mercilessly hunted down, and the legal system gave Muslims priority over non-Muslims. Moreover, special taxes were levied against those who could not serve in a Muslim army, which meant that non-Muslims (who were forbidden to serve) had a great financial motive to convert. Others were genuinely attracted to the religion. For al Samaw'al ben Yahya ben Yahuda al Maghribi (1130-1180), a Jew born in Baghdad, his conversion was a result of careful thought. After having examined the basis for the three main faiths, Samaw'al decided that Islam was the best supported, and he converted on November 8, 1163; he later wrote a book, *Decisive Refutation of the Christians and Jews*, supporting his conclusions.

Samaw'al, like his uncle, became a physician and made contributions to the field of medicine. Unlike many of his contemporaries, Samaw'al believed astrology was nonsense, and thus he wrote *The Exposure of the Errors of the Astrologers*. Our interest in Samaw'al is, of course, his work in mathematics. By Samaw'al's time, the golden era of Islamic mathematics had passed. The state of mathematics education was so poor he could find no one to teach him more than the first few books of Euclid, so he studied the works of the ancient Greek geometers on his own. He completed the arithmetization of algebra by giving procedures for finding polynomial quotients, as well as explaining how to handle the subtraction of two negative quantities.

These rules appeared in his *Shining Book of Calculations*. The title of the work might seem odd, but in English, we have the word lucid, which comes from the Latin word meaning "to shine," so the title is not that peculiar; indeed, the fact that we use "clear" to mean both transparent and understandable shows that we still equate visibility with comprehensibility. In his book, al Samaw'al gave the first examples of a polynomial division, that of 20cc plus 2sc plus 58ss plus 75c plus 125s plus 196t plus 94u plus 40pt plus 50ps plus 90pc plus 20pss, divided by 2c plus 5t plus 10pt, or

$$\frac{20x^6 + 2x^5 + 58x^4 + 75x^3 + 125x^2 + 196x + 94 + \frac{40}{x} + \frac{50}{x^2} + 90\frac{1}{x^3} + 20\frac{1}{x^4}}{2x^3 + 5x + 10\frac{1}{x}}$$

In a simpler example of Samaw'al's procedure, consider the quotient of 2ss plus 3c plus 5s plus 8t plus 5u plus 1pt, divided by 1ss plus 2u plus 1pt (i.e., $2x^4 + 3x^3 + 5x^2 + 8x + 5 + \frac{1}{x}$ divided by $x^2 + 2 + \frac{1}{x}$). First, write down the powers in their "natural order." Then, below them in the appropriate column, write the coefficients of the dividend, and below them, the coefficients of the divisor, starting in the leftmost column. If a term is missing, write '0'. Thus our division would begin by setting down:

ss	c	s	t	u	pt
2	3	5	8	5	1
1	0	2	1		

Samaw'al's procedure was meant to be performed on a counting board, not paper; hence, it may seem more complicated when reduced to print. First, the leading term of the divisor, 1, is divided into the leading term of the dividend, 2; since this is really 1s into 2ss, the quotient is 2s. This is written at the top of the 's' column; then 2s is multiplied by the divisor, and the product subtracted.

		2			
ss	c	s	t	u	pt
2	3	5	8	5	1
2	0	4	2		
		3	1	6	

Now the divisor is written down starting with the leftmost column, and the whole procedure is repeated. Notice the process of writing down the divisor again is exactly analogous to the procedure performed in the division of two whole numbers. The final result is

		2	3	1	
ss	c	s	t	u	pt
2	3	5	8	5	1
2	0	4	2		
		3	1	6	
		3	0	6	3
			1	0	2
			1	0	2	1
			0	0	0	0

so the quotient is 2s plus 3t plus 1u.

Samaw'al had no difficulty handling negative expressions, using rules identical to our own. Most interesting of all is his example of dividing 20s plus 30t by 6s plus 12u. Just as with whole numbers, some divisions do not "come out even," and one ends with an infinite repeating decimal; Samaw'al performed the division of the two polynomials using the division algorithm and obtained

$$3\frac{1}{2} + 5\text{pt} - 6\frac{2}{3}\text{ps} - 10\text{pc} + 13\frac{1}{3}\text{pss} + 20\text{psc} - 26\frac{2}{3}\text{pcc} - 40\text{pssc}$$

where he stopped, calling this the "approximate answer"; clearly, he recognized that the division would never terminate. ("pssc" is "part square-square-cube"). Al Samaw'al was also responsible for introducing a form of decimal notation, centuries before its appearance in Europe; however, there seems to be no direct connection between the two, and decimal notation would have to be invented anew.

9.3 Exercises

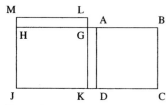

Figure for Problem 1.

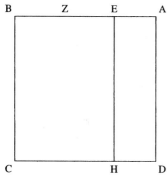

Figure for Problem 3.

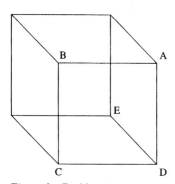

Figure for Problem 5.

1. Prove, geometrically, that the case of "a square and thirty numbers equals ten of its roots" is unsolvable, using the figure in the margin. Let AC be the side of the square which is to be determined; let AJ represent thirty numbers. Then rectangle BJ represents a square and thirty numbers; hence HB must be ten. Show that this is not possible.

2. An important proposition proven by Abu Kamil is that $\frac{A}{B}$ times $\frac{B}{A}$ is equal to 1. Prove this, using the Euclidean theory of numbers.

3. Prove, using Euclidean geometry, the method of solution for "roots and numbers equal a square." Use the problem five roots and six numbers equal to a square. Let the square AC equal five times its root and the number six. Let rectangle AH represent the number six; thus BE is five. Bisect BE at Z.

   (a) Show that the rectangle on BE, EA, with the square on EZ, is equal to the square on ZA.

   (b) How does this translate into the solution of the problem?

4. Complete al-Khayyami's problem of inserting two mean proportionals by showing that AB, BH, BT, BC are in continued proportion.

5. Solve the problem "a cube is equal to a number" in the following manner: given the solid $ABCD$ equal to the given number; hence AC is the unit square and DE is equal to the given number. Between AB, DE fit two mean proportionals, E and Z, and construct the cube with side equal to E. Show that E solves the problem.

6. In many of the solutions to the cubic using conic sections, it is necessary to resort to the following proposition: if a, b, c, d are in continued proportion, then the square on the first is to the square on the second as the second is to the fourth; in other words, a^2 is to b^2 as b is to d. Prove this geometrically.

7. To solve a cubic equation using the method of al-Khayyami, it is necessary to construct the particular conic section. The case of cubes and sides equal to a number requires the location of the intersection of a parabola and a circle.

   (a) Show that all parabolas are similar. In particular, find how to transform the parabola with a parameter unit parabola into a parabola with a parameter of AB.

(b) Because of the similarity of all parabolas, and the constructibility of the circle, this means that the case of a cube and sides equal to a number can be solved in all cases, once a single parabola is obtained. Explain how to do this. Specifically, explain how you could solve the cubic $x^3 + ax = b$, given the parabola $y = x^2$, by identifying the circle of diameter d with one endpoint of the diameter on the vertex of the parabola.

8. The case of cubes, squares, and sides equal to a number requires locating the intersection of a hyperbola with known asymptotes and a circle. Show that all hyperbolas of form $y = \frac{a}{x}$ are similar to the hyperbola with the form $y = \frac{1}{x}$; note the asymptotes of this hyperbola are perpendicular.

9. To solve the case of cubes, squares, and sides equal to numbers, al-Khayyami used the following method: let BE be the side of a square equal to the number of sides; BC the height of a solid whose base is the square on BE and equal to the number; let BD be equal to the number of squares. Extend CB to D, and use DC as the diameter of semicircle DZC. Complete the rectangle BK, and construct on C a hyperbola with asymptotes BE, EK.

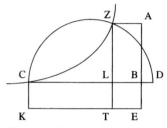
Figure for Problem 9.

(a) Let the intersection of the hyperbola and the circle be Z; draw ZT, ZA perpendicular to EK, EA. Show that the solid whose base is the square on EB and whose height is BC is equal to the cube on BL, plus the solid whose base is the square on BL and whose height is BD, plus the solid whose base is the square on EB and whose height is BL.

(b) Explain how to use the standard hyperbola to solve the general problem of the cube, squares, and sides equal to a number. Specifically, explain how to solve the cubic equation $x^3 + bx^2 + cx = d$ using the hyperbola $y = \frac{1}{x}$ by identifying the length and position of the diameter DC.

10. Examine al-Khayyami's solution to "cube and number equal to sides."

(a) To what types of cubic equations does the case "cube and number equal to sides" correspond?

(b) How are the sides AB, BC related to the constants in the cubic equation?

11. Solve the problem of cube equal sides and number as follows. Let AB be the side of the square equal to the number of sides, and BC be the height of the solid whose base is the square AB and equal to the number; draw AB, BC perpendicular. Construct the parabola with vertex B and parameter AB, and hyperbola with vertex B and parameters both equal to BC. Show that HB is the side of the desired cube.

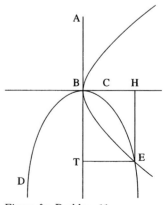
Figure for Problem 11.

12. Perform al Samaw'al's division of 20cc plus 2sc plus 58ss plus 75c plus 125s plus 196t plus 94u plus 40pt plus 50ps plus 90pc plus 20pss divided by 2c plus 5t plus 10pt.

13. Discuss the strengths and weaknesses of the system for expressing exponents of al Samaw'al and al Karaji.

9.4 Geometry and Trigonometry

The contributions of the Islamic world to mathematics were not limited to algebra and the transmission of the Hindu system of numeration. Indeed, it is clear that mathematics in the Islamic world reached heights Europe would not attain for centuries, though the influence of any particular discovery is, in many cases, difficult to determine. However, many of the geometrical discoveries of the Arabs and Persians did have an impact, particularly in the ultimate development of non-Euclidean geometry and trigonometry.

9.4.1 Thabit ibn Qurra

No one can deny that the Arabs and Persians were translators of the works of the Greeks, and among the greatest of the translators was Thabit ibn Qurra al-Harrani, "Thabit from Harran, son of Qurra," (826-February 18, 901). Harran is in modern Turkey, but Thabit made his way Baghdad, the cultural center of the Islamic world. Archimedes's *Book of Lemmas* and his *[Inscribing a Regular] Heptagon in a Circle* come to use from Arabic translation, done by ibn Qurra; Books Five through Seven of Apollonius's *Conics* also survive because of Arabic translations, which can be traced back to one done by ibn Qurra. He also greatly improved the translation of Ptolemy's *Almagest* done by Ishaq ibn Hunain (d. 930). But ibn Qurra was no mere translator; he also mastered the work of the ancient Greek geometers, suggesting extensions and improvements.

See page 150.

In *Quadrature of a Parabola*, Archimedes relied on inscribed triangles to show that the area of the parabolic segment was $\frac{4}{3}$rds the triangle with the same base and vertex. Ibn Qurra proved the same result, though he relied on both inscribed and circumscribed figures. Moreover, ibn Qurra's approach is similar to our own: unlike Archimedes, who partitioned the region into triangles whose bases were the sides of other triangles, ibn Qurra did the equivalent of partitioning the domain at points in geometric proportion, then summing the areas of the corresponding rectangles. Elsewhere, ibn Qurra attempted to find the volumes of solids of revolution generated by parabolic segments; though he was unsuccessful, he inspired al-Qūhī (fl. 970-1000) to attempt the same problem. The problem was finally solved by ibn al-Haytham (965-1039). Unfortunately, however advanced these works might have been, European mathematicians seemed to have been unaware of them until they had invented or reinvented the methods used by the medieval Arabs. For example, ibn Qurra's division of the domain

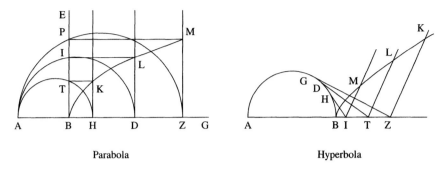

Figure 9.11: Parabola and Hyperbola Construction.

into points in geometric proportion was reinvented by Fermat—eight hundred years later.

In algebra, ibn Qurra combined the geometric algebra of Euclid to the developing algebra of the Arabs, showing how the Euclidean propositions in Book II were the geometric equivalents to the algebraic operations used to solve equations. His proof of the solution for a square and roots equal to a number anticipated that of Umar al-Khayyami.

See page 256.

9.4.2 Ibn Sinan

Further contributions to mathematics were made by other Arab geometers, who were particularly interested in the conic sections. Several treatises about them were written by Ibrahim ibn Sinan (908-946), the grandson of Ibn Qurra (in fact, a more complete name is Ibrahim ibn Sinan ibn Thabit ibn Qurra, "Ibrahim [Abraham], son of Sinan, son of Thabit [Tobit], son of Qurra," thus giving the family history for four generations). Ibn Qurra's rather lengthy work on the quadrature of the parabola was reworked by al-Māhānī into a shorter, more readable version. Ibn Sinan took this as a challenge to his family's intellectual honor. In response, ibn Sinan wrote *On the Drawing of the Three Conic Sections*, which included procedures for pointwise construction of the parabola, hyperbola, and ellipse using compass and straightedge methods.

The drawing of a conic section had more than theoretical interest, for it solved a very practical problem: that of creating accurate sundials. Suppose a stick is placed in the ground; since the position of the shadow relates to the position of the sun, it can be used to determine the **local solar time** (which clock time approximates). Moreover, since the sun traces a very nearly circular path across the sky each day, the line between the position of the sun and the tip of the stick traces out, in an Apollonian fashion, a double cone, with the stick's tip as the cone's vertex. The plane of the ground cuts both sides of this cone, and thus the path traced out by the stick's shadow is a hyperbola.

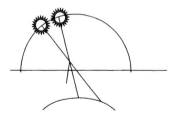

See Problem 1.

To construct a parabola, ibn Sinan began with the fixed line AG, fixed segment AB, and arbitrary points H, D, Z, ... (see Figure 9.11). Construct BE perpendicular to AG. Draw semicircles ATH, AID, APZ with diameters AH,

AD, AZ respectively; let the perpendicular BE intersect the semicircles at T, I, P. Draw TK, IL, PM parallel to BG, and KH, DL, ZM parallel to BE. Then the points of intersection B, K, L, M are on the parabola with vertex B, axis BG, and parameter AB.

Ibn Sinan gave three methods for pointwise construction of a hyperbola. The simplest is: given semicircle AGB with diameter AB, extend the diameter to Z and pick arbitrary points G, D, H on the semicircle; draw the tangents GZ, DT, HI. Draw IM at an arbitrary angle, and TL, ZK parallel to IM, with $IM = IH$, $TL = TD$, $ZK = ZG$ (see Figure 9.11). Then B, M, L lie on a hyperbola with AB as a diameter and whose parameter and transverse sides are equal to AB.

9.4.3   Jabir ibn Aflaḥ

One of the more important Islamic geometers was Jābir ibn Aflaḥ (ca. 1100-ca. 1160), about whom very little is known except that he lived in Spain, probably Seville, during the twelfth century. His influence is exerted through a single book, his *Correction of the Almagest*. The work dealt extensively with spherical trigonometry, proving a number of key theorems. Menelaus's theorem, involving as it did six quantities, was difficult to apply in practice. In its place, Jābir offered the "Rule of Four Quantities," which allowed problems in spherical trigonometry to be solved more simply. The rule was known to Islamic astronomers by 1000, but it was through the work of Jābir, translated by Gerard of Cremona, that the discoveries of the Arabs and Persians in spherical trigonometry reached the West. The work in plane trigonometry was centered around the practical problem of solving plane triangles. Though the methods used by Jābir could be found in Ptolemy, Jābir's work had a direct influence on the development of European trigonometry, owing to its incorporation in the work of Regiomontanus.

See Theorem 7.1.

9.4.4   The Fifth Postulate

Several noteworthy attempts to prove Euclid's fifth postulate originate with the Arabs. Abu Ali al Hasan ibn al Haytham (965-1039), known to the west as Alhazen, presented a proof using an assumption that the locus of points generated by a line moving so its endpoint was equidistant from a fixed line was a line parallel to the given line. This attempt was criticized by al-Khayyami, who argued that the concept of motion, a concept from the physical world, should have no place in mathematical reasoning. In any case, Alhazen's assumption is itself equivalent to the fifth postulate.

Al-Khayyami's own attempt, *Commentaries on the Difficult Postulates of Euclid*, involved the construction of a quadrilateral with two equal sides perpendicular to a base; the other two angles could either be acute, right, or obtuse. He disposed of the acute and obtuse cases by appealing to a principle of Aristotle that two converging lines necessarily intersect—again, an assumption equivalent to the fifth postulate.

Perhaps the most sophisticated attempt was that of Nasir Eddin al Tusi (February 18, 1201-June 26, 1274); "al Tusi" indicates he was born in Tus, now in modern Iran. Again, his proof depended on an assumption equivalent to Euclid's fifth postulate. Nasir Eddin's assumption was that if a line u was perpendicular to a line w at A, and another line v crossed w obliquely at B, then lines perpendicular to u and intersecting v are less than AB on the side on which v, w make acute angles, and the lines are greater on the other side.

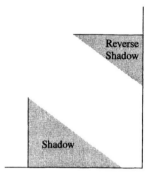

9.4.5 Trigonometric Quantities

Trigonometry as we study it today was largely an invention of the Arab and Persian mathematicians of the Islamic era. Recall that Ptolemy's trigonometry was primarily concerned with the measurement of chords in a circle. The Hindu mathematicians dealt the "half-chord"—the predecessor to the sine—but it was the Arabs who created and used the five remaining trigonometric quantities. The tangent, secant, and cosecant function was first used by Abu'l Wafā, and he created a table of sine and tangent values at $15'$ intervals, accurate to eight decimal places.

There remained some differences between our modern trigonometric functions and the values given in trigonometric tables. Until the seventeenth century, the sine (and the other trigonometric functions) were generally considered to be the *length* of a line segment. Two quantities originally referred to the lengths of shadows (*miqyas* in Arabic) of rods, which were either perpendicular to the ground (the "shadow") or perpendicular to a vertical wall (the "reverse shadow"); note that these lengths were, like the conic sections, important for the construction of sundials. However, by the time of Nasir Eddin, these lines were also defined in terms of the lengths of lines relating to a circle. In Arabic, the *jiva* of the Hindus became *jiba*, and still meant half-chord (of half the angle). This is not quite our modern sine, since it is a length, not a ratio; to distinguish between the two, we will indicate the half-chord of the half-angle as Sine.

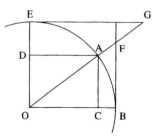

Consider circle BAE with center O and radii OB and OE perpendicular to each other; let BF and EG be tangent to the circle at B and E, respectively. Then the Arabs dealt with six trigonometric quantities relating to the arc AB:

1. Sine AB, which is the line AC perpendicular to the radius OB.

2. Sine of the complementary arc AE, which is the line AD.

3. The tangent line BF.

4. The tangent line EG.

5. The hypotenuse of the shadow, OF.

6. The hypotenuse of the reversed shadow, OG.

The existence of six (instead of one) trigonometric quantities made many problems easier to deal with.

See Problem 8.

9.4.6   The Law of Sines

Nassir Eddin also stated and prove the **law of sines**, dealing with the three cases separately. The case of the obtuse angle is:

Proposition 9.3. *In triangle ABC with obtuse angle ABC, $AB : AC =$ Sine C : Sine B.*

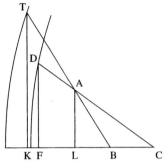

Here, Sine B means the sine of the arc with central angle B, according to the Arabic definition: it corresponds to the half-chord of half the angle. Nassir Eddin tabulated the values of sine using a circle with a radius of 60 units, so his proof is:

Proof. Extend CA to D and BA to T so that BT, CD are each 60 units in length; draw the perpendiculars TK, DF. Then $TK = $ Sine $ABL = $ Sine B and $DF = $ Sine C. Draw AL perpendicular to CK. Then $AB : AL = TB : TK$; likewise, $AL : AC = DF : DC$. Since $TB = DC = 60$, then (compounding the proportions) we have $AB : AC = DF : TK = $ Sine C : Sine B.   □

In 1206, a few years after Nasir Eddin's birth, the Mongolian chieftain Temujin united the various Mongol tribes and was declared Very Mighty King: *Genghis Khan*. Under his leadership, the Mongols began a campaign of conquest unmatched to the present day, either in speed or extent, from the east coast of China to Poland. In 1256, Hulagu, grandson of Temujin, was sent by his brother Mangu into the caliphate with an army and a mission of conquest.

Islam was by then divided into two main branches: the Sunni, who believed the caliph should be chosen in some fashion, and the Shi'i, who believed the caliph had to be a descendant of Mohammed. The caliphate at Baghdad supported the Sunni. The Nizari, who were a branch of the Isma'ili sect of Shi'ism, established a stronghold in Alamut and other places in the mountains of northern Persia, and expressed their dissatisfaction with the caliphate by periodically murdering key figures in the government and military. One rumor, of dubious authority, is that before going out on their missions of terrorism, the agents of the Nizari would smoke hashish to induce visions of the Paradise they would surely enter. As a result, they became known as *hashshashin*: assassins.

For reasons that are not entirely clear, Hulagu perceived his first task was to exterminate the band of terrorists from Alamut: one story (which seems doubtful) is that they sent a delegation of 400 assassins to murder the Great Khan in Karakorum. A more likely story is that the local governor, seeing that the caliphate was unable to stop the depredations of the assassins, called for Mongol help. Hulagu systematically destroyed the assassin strongholds, until only Alamut was left.

In 1256, the Mongols arrived at Alamut. Nasir Eddin was there: he was Shi'i himself, and had been invited by the leader of the Isma'ili to join them in their castle strongholds. What happened afterward is not clear, but the facts are these: Alamut was taken, the last of the assassins was killed, and Nasir Eddin joined the Mongol forces and was present when they smashed the caliphate and sacked

Baghdad in 1258. Shortly afterward, Hulagu gave Nasir Eddin all the support he needed to build an observatory in Maragheh. The Mongols themselves went on to Egypt, where they were finally stopped (in an odd echo of the Battle of Tours) at Ain-Jalut, by the rulers of Egypt, the Mamluks. Meanwhile, in Europe, the tale of a mighty army that defeated the Arabs in a time when the Arab armies threatened the Holy Land led to the legend of a Christian king in the far East: Prester ("Priest") John.

The Mongol Empire itself was divided into khanates, ruled by a khan who theoretically owed allegiance to the Great Khan in the capital at Karakorum then, later, after the Mongols conquered China, at the new capital: Peking. The khans who ruled Persia, known as the Il-Khans, soon became decadent and self-absorbed. In 1369, the Il-Khan Chagatay was deposed by his prime minister, Timur. A battle early in his career injured his leg so that he limped for the rest of his life, and thus became known as Timur Lang (the Lame): Tamerlane. It was his intent to restore the great Mongol Empire, and he embarked on a career of conquest. Timur soon conquered most of Persia and sent expeditions as far east as India, as far north as Moscow, and as far west as Asia Minor. He was a brutal conqueror who used terror as a routine weapon: a lesson he learned from the Mongols. The most characteristic feature of Timur's conquests was a tower of heads: in one massacre at Isfahan, 70,000 heads were piled up to serve as a warning to those who would defy his authority. He was en route to China to restore the Mongol Empire there when he died in 1405. However, the Timurid Empire remained.

9.4.7 Ulūgh Beg

In contrast to his cruelty in warfare, Timur was a great patron of the arts, and his massacres frequently spared skilled artisans who, as often as not, were commanded to rebuild and beautify the cities Timur destroyed. His son was equally friendly to artists and scholars, and Timur's grandson, Ulūgh Beg (1393-October 27, 1449), carried on the family tradition, establishing in the capital at Samarkand a *madrasah*, a center for higher education, which opened in 1420. He invited al-Kashī and about sixty other scholars there to be lecturers. Later, al-Kashī dedicated *The Calculator's Key* to Ulūgh Beg.

But Ulūgh Beg was more than merely a patron; he was an accomplished scholar himself. Al-Kashī regarded him as one of the better mathematicians in Samarkand (second only to himself, in fact). This was more than mere flattery: it was Ulūgh Beg himself who discovered errors in Ptolemy's trigonometric tables, causing him to order the creation of newer, more accurate tables of sines and tangents. Ulūgh Beg and al-Kashī collaborated closely in their construction. Other trigonometric tables had been created during the Islamic era. Around 1000, ibn Labbān constructed a table in which the successive angles were one minute of arc apart. Ibn Labbān also created a table of tangent values with angles differing by one degree; the tables were accurate to three decimal places. A more accurate but less comprehensive table was assembled by Abu Arrayhan

Muhammad ibn Ahmad al-Bīrunī (September 15, 973-December 13, 1048) a few years later: the sines were calculated at intervals of 15 seconds, but the values were accurate to four decimal places.

The fundamental principles of constructing a table of trigonometric values were established by Ptolemy: from the chord of any single angle, straightforward formulas gave a means of computing the sine of half the angle; the addition formula allowed other values to be computed easily. Since the trigonometric functions for some angles were rational numbers (e.g., $\sin 30° = \frac{1}{2}$), then constructing an accurate table of sines is "merely" a matter of endurance.

There are two difficulties that arise. No matter how complete a sine table is, it must omit some values; hence, some method of interpolating between known values must exist. Ptolemy's method of including the differences between successive arcs was one that made linear interpolation easy, but even more accurate methods existed in the Islamic world. Unfortunately, the authors never explained the conceptual basis, so we do not know how they found their methods of interpolation, or what the methods of interpolation were based upon.

See page 183.

The other difficulty is that it is impossible, using the trigonometric identities, to find an exact value for sine of 1°; hence, one cannot construct a "useful" table of sines, in which the angles differ by one degree. Ptolemy's approximation was one way out of this difficulty, but it had limited accuracy, since it bounded the solution between the values of $\sin \frac{3}{4}°$ and $\sin 1\frac{1}{2}°$: although these two bounds could be found as accurately as desired, all that was known was that the actual value of $\sin 1°$ fell between them. Al-Kashī gave a method that offered unlimited accuracy in *Treatise on the Chord and Sine*. Al-Kashī's method relied on the relationship (expressed in sexagesimals)

$$\text{Sine } (3\theta) = 3 \text{ Sine } \theta - 0;0,4\,(\text{ Sine } \theta)^3 \qquad (9.3)$$

where again Sine θ represents the half-chord length for a circle of radius 60.

This relationship holds in general. If $\theta = 1°$, then the left side is Sine 3°, whose value can be determined as precisely as one wants. Thus, finding Sine 1° becomes a matter of finding a solution to a cubic equation. Using a method of successive approximations, al-Kashī was able to find Sine 1° to nine sexagesimal places.

The high degree of accuracy in approximating Sine 1° could also be applied to finding the circumference of a circle; in *Treatise on the Circle*, al-Kashī used a procedure, similar to that of Archimedes, to find the perimeter of an inscribed $805,306,368$-sided polygon, formed by beginning with a hexagon and doubling the number of sides 27 times. If the circle has a radius of 1, then al-Kashī determined the polygon's circumference, in sexagesimals, is

$$6; 16, 59, 28, 1, 34, 51, 46, 14, 50$$

which al-Kashī also converted to the decimal fraction

$$6.2831853071795865$$

which gives the correct value of π to sixteen decimal places: a world record at
the time, far surpassing the efforts of the Greeks (three decimal places) and the
Chinese (six decimal places).

After al-Kashī's death in 1429, Ulūgh Beg produced an even better table of
trigonometric values in his star catalog of 1437. The sine and tangent values were
computed at intervals of $1°$, making it somewhat less useful than al-Kashī's own,
but the values were accurate to 8 decimal places, and Ulūgh Beg's computation of
$\sin 1°$ was accurate to an incredible 14 decimal places. Other tables, completed in
1440, were accurate to 5 decimal places, though they included angles at intervals
of one minute of arc.

Ulūgh Beg's contemporaries did not take kindly to the khan being preoccu-
pied with intellectual pursuits: they revolted, and his son, 'Abd al-Laṭīf, had
him executed. Ulūgh Beg's death was unfortunately indicative of the changes
that had come across the Islamic world since the time of the Caliph al Mam'un,
when the Islamic world was at the forefront of intellectual progress; by the time
of the death of Ulūgh Beg, Europe was ready to begin intellectual pursuits once
more.

9.4 Exercises

1. Explain why the path traced out by a stick's shadow is (approximately) a
   hyperbola. Why is it not precisely a hyperbola?

2. Consider ibn Sinan's pointwise constructions of the parabola and the hyperbola.

   (a) Prove that $BKLM$ is a parabola with diameter BG, vertex B, and
       parameter AB, by showing that if K is not on the parabola with di-
       ameter BG and parameter AB, then the parabola must pass through
       some other point on KH, say N. Prove that this results in a contra-
       diction.

   (b) Prove $BMLK$ is a hyperbola with diameter AB and parameter and
       transverse axis AB.

3. Show that al-Khayyami's assumption that two converging lines intersect is
   equivalent to Euclid's Fifth postulate.

4. Show Nasir Eddin's assumption is equivalent to Euclid's fifth postulate.

5. To which of the six modern trigonometric functions do each of the six lines
   in a circle correspond? Explain.

6. Complete Nassir Eddin's proof of the law of sines by showing the law also
   holds for the case in which all three angles are acute.

7. Consider Equation 9.3.

   (a) What is Sine θ equal to in modern terms?

    (b) Prove Equation 9.3.

    (c) Suppose the circle has a radius of 60^n. How would Equation 9.3 be modified?

8. Suggest practical uses for each of the six trigonometric quantities. Keep in mind that the trigonometric quantities before modern times always indicated *lengths*.

Chapter 10

Medieval Europe

One of the most important figures during the Middle Ages was Charlemagne, king of the Franks, and grandson of Charles Martel. Charlemagne ruled over what is now France and western Germany, as well as portions of northern Italy. During his lifetime, the so-called Carolingian Renaissance occurred, and an attempt was made to revive scholarship and education in the west. An important innovation made during the Carolingian Renaissance was a modification of typography. Latin was written with all capital letters (called the Roman alphabet) and dots between words: LIKE·THIS·EXAMPLE. During the Carolingian Renaissance, a new type of alphabet, known as *minuscule*, was developed, corresponding to our lower case letters, and spaces between words began to replace the dots.

Charlemagne was fascinated by education; as an adult, he learned to read, though lacked the dexterity to learn to write passably. In 781, he met Alcuin of York (ca. 735-804), and persuaded him to go to Aachen, Charlemagne's capital, to help establish schools of education across the realm. In 787, Charlemagne decreed that every monastery was to have a school associated with it. These would develop into the "Cathedral" schools that would form the basis of primary education in western Europe for the next thousand years. However, the time was not yet ripe for a true rebirth of education and scholarship.

10.1 The Early Medieval Period

The period following the final decline of the civilization of the Greeks and Romans is called the **medieval period**, which is the Latin translation of its more common English name: the Middle Ages. During this time, the Chinese, Indians, and Muslims were advancing mathematics, while Europeans spent several centuries contemplating religious questions. Great minds flock to great questions, and during the medieval period, the great questions asked in Europe were about

Christianity. Thus it is not surprising that the few European advancements in mathematics were made by churchmen.

10.1.1   Bede

The earliest of these churchmen was Bede (ca. 673-735), who lived in England; he was such a distinguished scholar he is known as the Venerable Bede. He wrote a number of scientific treatises, including one on the use of the astrolabe and one on the method of finger counting, a means of representing the numbers one through one thousand by various positions of one's fingers; apparently, the representation was used to record intermediate results. Bede's is the only existing treatise on finger counting from the Middle Ages. Bede also suggested that the years be counted from the birth of Christ, which popular astrological superstition placed during the year we now call 1 A.D.: January 1, 1 A.D. began on a Sunday, with a new Moon, and was a year following a leap year. In fact, if the anecdotal evidence in the Bible has any validity, then the historical Christ could not have been born later than 4 B.C.

In 325, the Council of Nicaea set the date of the vernal equinox (the first day of spring) to be March 21, and from this date all moveable feasts, such as Easter, were calculated. The date of Easter, for example, is the first Sunday after the first full Moon on or after the vernal equinox. The problem is that the Julian calendar, with a leap year every four years, gave a calendar year somewhat longer than the actual year. Thus, by Bede's time, March 21 was actually a few days after the vernal equinox, a fact Bede noted. The problem of rectifying the "official" date of Spring with the actual date of the vernal equinox would occupy astronomers and mathematicians for another ten centuries.

10.1.2   Hindu-Arabic Numerals

One of the most important events of the medieval period was the introduction of the Hindu-Arabic system of notation to Europe. The earliest appearance of Arabic numerals in the West occurred in Spain, around 976. Gerbert of Aurillac (ca. 940-1003), who later became Pope Sylvester II, may have learned of them while studying in Barcelona and wrote an introduction to them. Gerbert's introduction, one of the first European treatises, was incomplete, for '0' was not yet known in western Europe.

To make the numbering system more comprehensible, Gerbert also introduced a variation of the Roman abacus. Gerbert's abacus consisted of up to twenty-seven columns. These columns were broken down into groups of three, and labeled (from left to right) C, D, and S for *centum* (Latin for "hundred"), *decem* (Latin for "ten"), and *singularis* (Latin for "single"). What made Gerbert's abacus different from the Roman version was that instead of using unmarked counters, he had several hundred counters inscribed with the Hindu numerals from 1 to 9. As a mechanical computing device, Gerbert's abacus was a failure, for the abacus is ideally suited for an additive system of numeration, like Roman numerals; moreover, to add two numbers involved not only remembering that

"3" plus "2" was "5" (instead of the obvious addition, on the Roman abacus, that • • • plus • • makes • • • • •), but searching through the counters until you found the "5."

10.1.3 Division by Differences

Gerbert's abacus was an attempt to solve one of the major disadvantages of the Hindu-Arabic system: numerical computations are more difficult in a positional system. Additive notation made addition and subtraction trivial, and multiplication, when treated as a repeated addition, was likewise simple. Division, treated as a repeated subtraction, is also easier in additive notation, particularly when a mechanical device like an abacus is available. Only when computations are done "by hand" do the advantages of the Hindu-Arabic system of notation appear.

One interesting new form of division made its appearance during the Middle Ages, **division by differences**. The earliest appearance is in an unsigned treatise, written around 1200. The manuscript was found with others by "Gerlandi de Abaco," indicating that it might have been a copy of one of Gerbert's works; another possible author is Adelard of Bath (see later). To divide 7800 by 166, the author noted that $166 = 200 - 34$. The divisor (166), difference (34) and dividend (7800) were placed on a counting board:

M	C	X	I	
		3	4	Difference
	1	6	6	Divisor
7	8			Dividend

First, 200 is divided into 7000 to obtain a partial quotient of 30 (probably since 7 divided by 2 is 3). The product of 200 and 30, or 6000, is subtracted from 7000, leaving 1000 (we will cross out the number, though on the abacus, the number would just be eliminated). Then the quotient 30 is multiplied by the difference. Thirty times 30 is 900, written in the C column; 30 times 4 is 120, written in the C and X column. Finally, all the numbers are added to produce the new dividend.

M	C	X	I	
		3	4	Difference
	1	6	6	Divisor
7̸	8			Dividend
1				
	9			Product of 3 times 3
	1	2		Product of 3 times 4
2	8	2		Sum and New Dividend
		3		Partial Quotient

The process is repeated on the new dividend. 200 into 2000 is 10; the product of 10 and the difference is 300 and 40.

M	C	X	I	
		3	4	Difference
	1	6	6	Divisor
2	8	2		Dividend
	3			Product of 1 times 3
		4		Product of 1 times 4
1	1	6		Sum and New Dividend
		3		Partial Quotient
		1		Partial Quotient

Now 200 is divided into 1000 to get 5.

M	C	X	I	
		3	4	Difference
	1	6	6	Divisor
1	1	6		Dividend
	1	5		Product of 5 times 3
		2		Product of 5 times 4
	3	3		Sum and New Dividend
		3	5	Partial Quotient
		1		Partial Quotient

Finally, 200 is divided into 330 to get 1.

M	C	X	I	
		3	4	Difference
	1	6	6	Divisor
	3	3		Dividend
	1			Remainder
		3		Product of 1 times 3
			4	Product of 1 times 4
	1	6	4	Sum and New Dividend
		3	5	Partial Quotient
		1	1	Partial Quotient
		4	6	Total Quotient

Thus, the result of the division is 46 with a remainder of 164.

10.1.4   Negative Numbers

The Hindu and Chinese mathematicians worked with negative numbers freely, though they generally discarded negative solutions as extraneous. In the Islamic tradition, there is only a single reference to a negative number, which appeared in Abu'l-Wafa's treatise, *Book on What is Necessary from the Science of Arithmetic for Scribes and Businessmen.* Sometime before the tenth century an anonymous European wrote:

> Nonexistent seven overcomes existent three and leaves of itself non-existent four.

If "nonexistent" is read as "negative," we have a statement that $-7 + 3 = -4$. Thus, by the year 1000, negative numbers had reached Europe.

10.1.5 The Century of Translation

No matter how devoted to scholarship they may have been, Bede, Alcuin, and Gerbert all faced one enormous disadvantage: schools of higher learning in the West were almost nonexistent, so reviving scholarship meant, first of all, establishing a system of higher education. It would be much easier to bring about a rebirth—a renaissance—of learning by using existing centers of learning and letting the knowledge and scholarship spread from there.

The Islamic world had several great advantages. First, the Koran required that every male be able to read the Koran, so widespread literacy—and by extension, education—was part of the cultural background. The other great advantage of the Islamic world was its diversity: the political rulers in the Islamic world were generally Arabs, but they ruled over Persians, Turks, and other non-Arabs, as well as many non-Muslims, such as Jews and Christians. The Arab rulers realized that the best way to rule over a diverse population was to tolerate differences, and encourage their subjects to do the same. In most of the lands they conquered, the population of non-Muslims was small, but in two places—Sicily and Spain—the three cultures of Judaism, Islam, and Christianity mingled to form something that was greater than any one of them. This was extremely critical, for the important works of the ancient Greek geometers were in the hands of the Arabs and would return to Europe through the hands of the translators; in many cases, the works of the ancients are known *only* through Arabic copies.

The historian of mathematics Carl Boyer referred to the twelfth century as the "century of translation," because much of the century's major achievements revolved around the translation of works of mathematics from Arabic (a language few Europeans could understand) into Latin (a language all educated Europeans could read). It is an apt designation, though it properly extends over two centuries, between about 1100 until about 1300.

Adelard of Bath (fl. 1116-1142) was the earliest of the important translators. His life followed a pattern that is typical for many of the important scholars of all centuries. Born in Bath, England, he traveled to France and studied at Tours and taught at Laon. From there, he went to Sicily, which was probably where he first encountered Islamic culture. Intrigued, he made his way through Cilicia (in Asia Minor), Syria, Palestine, and possibly as far as Spain, before returning to Bath in 1130.

Adelard translated a number of original Arabic works into Latin, including the astronomical tables of al Khwārizmī; he also wrote a work on computation using Hindu-Arabic numerals, probably based on al Khwārizmī's work. In addition, making use of existing Arabic copies, he produced the first good translation of Euclid into Latin. For the first time since the Roman era, the fundamental work of Greek geometry could be read by European scholars.

Perhaps the greatest of the translators was Gerard of Cremona (1114-1187), born in Cremona, Italy, though he eventually settled in Toledo, Spain. He produced a translation of Euclid that was better than that of Adelard; he also translated Ptolemy's *Almagest* from Arabic copies. In addition to reintroducing Greek learning, Gerard also transmitted the original contributions of the Arabs, translating the *Correction of the Almagest* by Jābir and the *Algebra* of al Khwārizmī.

A somewhat earlier and more popular version of al Khwārizmī's *Algebra* had been done by Robert of Chester, around 1125. This translation is not of the entirety of the work; in a Christian Europe, Robert of Chester thought it best to leave out the introductory paragraphs, which effusively praised Islam and Caliph al Ma'mun. In addition, he omitted the last half of the book, which dealt with practical geometry and algebraic problems arising from the complex set of inheritance laws of Islam. Finally, many of the problems were also left out. Altogether, Europeans reading Chester's translation saw perhaps a third of what Al Khwārizmī wrote.

Robert of Chester's work shows one of the difficulties of translation. The Arabs used the Hindu word *jiva* for the half-chord of Ptolemy; in Arabic, this became *jiba*. Arabic has letters to indicate vowels, but in general, words are written without them (the Koran is the only major exception: every vowel in the Koran must be indicated). Thus, *jiba* was written in Arabic letters as "j b." In his translations, Robert of Chester read this as *jaib*, which is Arabic for "bay"; hence, in his translation, he used the Latin word for bay: *sinus*. This became our word **sine**.

Another edition of Euclid was produced by Campanus of Novarra (ca. 1225-1296), born in Novarra, Italy, and considered by contemporary Roger Bacon (ca. 1214-1294?) to be one of the four best scholars of the time. Campanus's work was more a book *on* Euclidean geometry than a translation *of* Euclid, and it rapidly became the most studied edition of Euclid in the Middle Ages. The West was not yet ready to produce original mathematics, but the foundations were being laid.

Two other translators are noteworthy. The first was William of Moerbeke (1220-1286), probably born in Moerbeke, Belgium. He provided the first translations into Latin of many of the works of Aristotle, and better translations of the works of Archimedes. The other translator was John of Holywood (1200-1256). In a trend that would soon become popular, he Latinized his last name, and "Holy Wood" became Sacrobosco, by which he is better known today. Though probably educated at Oxford, he began teaching at the University of Paris around 1220 and stayed there for the rest of his life. Sacrobosco wrote a work on computation using the new Hindu numerals, *Algorithms*, which soon became one of the most widely read treatises on the methods of arithmetic. By 1389, candidates for the bachelor's degree at the University of Vienna were required to have read the *Algorithms*, which indicated, among other things, that no disdain of the practical art of logistic was inherent in the university system at that time.

Thus, over a two-century period, the key elements of modern mathematics entered Europe. From Greek geometry came the deductive proof. From Arabic algebra came a powerful tool for handling variable quantities. From India came a bold spirit that broke with the tradition of geometrically comparable magnitudes. Besides this reintroduction of mathematics to the west, centers of learning were being established as well: universities, originally a term meaning a coalition of people gathered together for any single purpose, were established at Bologna (ca. 1050), Paris (ca. 1170), Cambridge (1209), Naples (1224), Oxford (1264), and elsewhere.

10.1 Exercises

1. Divide using the method of differences:

   (a) 900 divided by 17

   (b) 382 divided by 47

   (c) 34000 divided by 275

   (d) 321532 divided by 47.

2. Compare and contrast the types of division thus far presented. What are the advantages and disadvantages of each?

10.2 Leonardo of Pisa

Leonardo Pisano Bigollo (ca. 1180-1250), or Leonardo of Pisa, was the son of a merchant. Since Leonardo once referred to an ancestor of his, Bonaccio, he became known in the nineteenth century as Fibonacci (son of Bonaccio), though this name was never used during his lifetime. Leonardo's father (*not* Bonaccio) was a merchant, and father and son traveled about the Mediterranean, wherever their business took them. During these travels, which were often into Arab-controlled lands, Leonardo became acquainted with the mathematics of Islam, and when he returned to Europe, he began to write a number of important mathematical works.

The book for which Leonardo is most famous is his *Book of the Abacus*, first completed in 1202, with a second edition completed by 1227. It is a literally weighty tome, several hundred pages in length with many examples for every concept, no matter how basic. Despite its title, the book does not discuss the abacus, so perhaps a better title would be *Book of the Calculator*, where "calculator" is a medieval title for a person who performs computation.

Leonardo stated his purpose was to familiarize merchants with the Hindu-Arabic numerals. Most of the book discussed commercial arithmetic. Thus, a great deal of space is devoted to solving algebraic problems that merchants would have to deal with, including conversion between various forms of currency.

10.2.1 Computation

In *Book of the Abacus* Leonardo used sexagesimal, unit, and common fractions. Fractional values preceded the whole number; thus, whereas we would write $13\frac{2}{3}$, Leonardo wrote $\frac{2}{3}13$. Leonardo invented his own system of fractions, which

Figure 10.1: Europe in the Middle Ages.

he used throughout the *Book of the Abacus*. A key feature was that common fractions were written as sums of products; thus, $\frac{1\ 5\ 7}{2\ 6\ 10}$ represented $\frac{1}{2\cdot6\cdot10} + \frac{5}{6\cdot10} + \frac{7}{10}$. The value of this system was two-fold. The primary reason can be illustrated: a quantity amount using multiple units, such as 5 yards, 2 feet, 8 inches, could be represented compactly as $\frac{8\ 2}{12\ 3}$ 5 yards.

Another reason is that it simplified division. Leonardo's procedure of division involved rules, consisting of a factored divisor; he devoted several pages to finding rules. For example, to divide by 75, the rule was $\frac{1\ 0\ 0}{3\ 5\ 5}$ since $3\cdot5\cdot5$ was 75. To divide 749 by 75:

Divide 749 by 3, obtaining 249 and remainder 2; put this 2 above the 3, and divide 249 by 5, the number following 3, making 49 and remainder 4; put this remainder above the 5, then divide 49 by the second 5, leaving 9 and remainder 4; put this 4 above the 5, and put 9 after the fraction, and thus we have the result of the division, namely $\frac{2\ 4\ 4}{3\ 5\ 5}9$.[1]

In our terms, the result would be $9 + \frac{4}{5} + \frac{4}{5 \cdot 5} + \frac{2}{3 \cdot 5 \cdot 5}$.

Example 10.1. *Divide 326 by 24. Since $24 = 3 \cdot 8$, the rule might be $\frac{1\ 0}{3\ 8}$. Divide 3 into 326, and the result is 108 with remainder 2, which should be written above the 3. Divide 8 into 108, and the result is 13 with remainder 4, which is written above the 8. The result is thus $\frac{2\ 4}{3\ 8}13$.*

Multiplication of mixed numbers was handled in a manner similar to our own. To multiply $13\frac{2}{3}$ by $24\frac{5}{7}$, Leonardo first converted the numbers into the improper fractions $\frac{41}{3}$ and $\frac{173}{7}$. The product of 173 and 43 is 7093, which was then divided, as above, by 3 and 7, resulting in $\frac{1\ 5}{3\ 7}337$.

10.2.2 Algebra

The debt of European mathematicians to the Arabs is particularly evident in the algebraic portions of *Book of the Abacus*: many of the problems are taken directly from the work of Abu Kamil. Leonardo dealt with familiar problems such as

See page 253.

Problem 10.1. *The Lion, The Leopard, and the Bear: A lion ate a sheep in 4 hours, a leopard in 5 hours, and a bear in 6: if one sheep is thrown to them, in how many hours will it be devoured?*

Leonardo's solution is interesting, for it avoids dealing with fractions until the very end.

Solution. Suppose they devour sheep for 60 hours. In that time, the lion devours 15, the leopard 12, and the bear 10 sheep, for a total of 37 sheep. Thus in 60 hours, they ate 37 sheep. How long to eat one sheep? Multiply 1 by 60, and divide by 37, which results in $\frac{23}{37}1$ hours. □

This is a variation on single false position. An important innovation used by Leonardo appeared first in the problem:

Problem 10.2. *Suppose 100 rotuli are worth 260 soldi; how much is 1 soldi worth?*

Though this problem can be solved using the rule of three, Leonardo used **double false position**, first used systematically by the Chinese but probably introduced to Leonardo by the Arabs. Leonardo's first example shows some of the problems associated with merchanting in the Middle Ages: a plethora of

---

[1]Leonardo of Pisa, *Scritti di Leonardo Pisano*, B. Boncompagni, 1857. Vol. I, page 41.

different currencies required additional conversions to be made, both between the currencies of different city-states, as well as the different denominations of the same currency. To simplify the problem, all currency amounts have been converted into a common set of units.

Solution. Suppose 1 rotuli is worth 1 soldi; thus 100 rotuli are worth 100 soldi. But it was supposed to be 260 soldi; thus the first guess is false and differs from the correct value by 160. Next, we suppose a rotuli is worth 2 soldi, and as before, this gives a value of 200 soldi, which is also false, differing by 60. Now the first guess was short by 160, and the second by 60. The difference between the first and second guesses is 1 soldi, and the corresponding difference is 100 soldi; 60 more are needed. Thus multiply 1 by 60 and divide by 100; the result is added to the second guess, 2, to make $\frac{3}{5}2$ soldi.   □

Quadratic equations also appear, many taken from Arabic algebras. Other problems are solved using a variety of methods, ranging from the tedious to the extremely clever. By far the most famous problem in *Book of the Abacus* is:

Problem 10.3. *How many pairs of rabbits can be bred from one pair in a year, if the pairs produce a pair every month and begin to breed in the second month after their birth?*

This gives rise to the **Fibonacci sequence**. Leonardo's solution is:

Solution. In the first month, the first pair breeds one pair, so there are 2 pairs. In the second month, the first pair breeds another pair, so there are 3 pairs. Of these, two pairs will be fertile in the third month, so in the third month, there will be 5 pairs, of which 3 will be fertile in the fourth month, so there will be 8 pairs in the fifth month. Of these, 5 pairs will be fertile, making 13 pairs in the fifth month. Continuing in this fashion, we find there will be 377 pairs in a year.   □

Month	Pairs
	1
First	2
Second	3
Third	5
Fourth	8
Fifth	13
Sixth	21
Seventh	34
Eighth	55
Ninth	89
Tenth	144
Eleventh	233
Twelfth	377

Leonardo goes on to note that the number of pairs in each month can be found by adding the number of pairs in the two preceding months. Note that in the modern formulation of the problem, the pair is *born* in the first month, whereas in Leonardo's statement of the problem, the pair *breeds* the first month.

In another problem, Leonardo described a number of men who had certain amounts of money; the problem was to find the amounts each man had. No solution was possible unless, Leonardo stated, one of the men was assumed to have a debt. Thus, he restated the problem with one man having a debt, and solved it under this assumption. This was the typical medieval method of dealing with negative numbers: they were treated as quantities different from positive numbers, as debts rather than credits. Thus, unlike the Greeks, who would have denied the existence of a solution entirely, Leonardo and others began to consider the possibility of a generalization of the number concept.

10.2.3 Diophantine Equations

The *Book of the Abacus* also includes the solutions to many indeterminate equations. Unlike Diophantus, who gave only a single solution, Leonardo discussed how to generate more solutions, suggesting a possible link between Leonardo and the Hindu mathematicians, probably via the Arabs.

> Find a number which, upon being divided by 2, 3, 4, 5, 6, always leaves a remainder of 1; but is divisible by 7. Thus I add 1 to a number that can be divided by these given numbers...such a number is 60, which divided by 7 leaves 4, which needs to be 6. Thus, double 60, or triple it, or multiply it by some number so that when it is divided by 7, the result leaves 6; multiplying 5 by 60, obtaining 300, then adding 1, will be 301, and that is the desired number. Likewise, if 420, which can be divided by all of the above number, is added to 301 once, or as many times as you wish, it will create a number which is evenly divisible by 7 and when divided by all of the others leaves a remainder of 1.[2]

In another case, Leonardo considers a number that, upon division by 2, 3, 4, 5, or 6, leaves remainders of 1, 2, 3, 4, or 5. In this case, the trick is to subtract 1; Leonardo again begins with 60; his solution is 119. For many Diophantine problems, Leonardo gave both a rational solution and found ways to transform it into a whole number solution.

Leonardo also dealt with the problem of writing a given number as the sum of two squares, another form of the Pell equation. Given that $4^2 + 5^2 = 41$, Leonardo's procedure for finding two other squares that added to 41 was to first find two squares that added to a square; for example, $3^2 + 4^2 = 5^2$. Then, since $(4 \cdot 5)^2 + (5 \cdot 5)^2 = 41 \cdot 5^2 = 1025$, the problem then becomes to find two squares that add to 1025. To find them, Leonardo's procedure is:

> Put the roots that made 25 one below the other; before them put those that make 41; and multiply 3 by 4, which is before that 3, and which was one of the roots whose squares added to 25; and 4 by 5, which is before it, and you will have 12 and 20. Now multiply the roots across from each other, that is the 3 by 5 and the 4 by 4, which is 15 and 16; add these to make 31, and subtract 12 from 20, leaving 8; and thus you have two roots, 31 and 8, whose squares, that is 961 and 64, add to 1025. Now divide these other numbers, 31 and 8, by 5, by which we multiply the roots above, namely 4 and 5, making the quotients 6 and $\frac{1}{5}$, and $\frac{3}{5}1$ whose squares, if added, make 41.[3]

$$\begin{array}{cc} 4 & 3 \\ 5 & 4 \end{array}$$

In other words, since $31^2 + 8^2 = 41 \cdot 5^2$, then so will $\left(\frac{31}{5}\right)^2 + \left(\frac{8}{5}\right)^2 = 41$.

---

[2] Ibid., 281
[3] Ibid., 402-3

10.2.4 The Wonder of the World

The empire established by Charlemagne was renamed the Holy Roman Empire by Frederick I (reigned 1152-1190), known as Barbarossa ("red beard"). The emperor was chosen by a group of electors: at first, the most powerful leaders in the empire, but later a specified group of princes and bishops. One of the greatest of the medieval emperors was Barbarossa's grandson, Frederick II (reigned 1211-1250).

Frederick was a remarkable monarch. The First Crusade, the only one to achieve its objective of freeing Palestine from Muslim occupation, was followed by a series of bungled attempts that accomplished nothing—or less. Perhaps the most disturbing was the Children's Crusade, where two groups, of 30,000 and 20,000 children, sought to recover Palestine by love. The children made their way to Marseille, France, where they boarded ships and were probably sold into slavery. The episode may have been the origin of the story of the Pied Piper.

After the failure of the second, third, fourth, and fifth crusades to take Palestine by force, Frederick hit upon a clever new strategy: negotiation. In 1229, Frederick won a ten-year truce, unhindered passage for pilgrims, and in general, accomplished by diplomacy everything that previous crusaders had attempted to do by force. Frederick's reward: the Pope, who had been seeking to increase his power in the empire, excommunicated Frederick, declared his titles forfeit, and promised great rewards in heaven and Earth for anyone who would depose him. As a result, when Frederick returned from Palestine, he had to fight off a papal army. Frederick crushed the papal forces, but wisely swore to support the papacy in the future, thereby averting future difficulties.

Frederick was both diplomat and scholar. He wrote a book on falconry (one of his hobbies), and took a keen interest in scientific pursuits. He collected about him a remarkable entourage that included a number of prominent scholars; thus his court and entourage came to be known as the "Wonder of the World" (*stupor mundi* in Latin).

10.2.5 *Book of Squares*

Leonardo eventually joined Frederick's court. Another member was Master John of Palermo: master is a title that comes from the Latin *magister*, "teacher." It is from this use of the word Master that we obtain the academic degree of the Master of Arts, though perhaps a better modern equivalent to the medieval title is the academic Doctor. Master John posed the following problem to Leonardo:

Problem 10.4. *To find a square number that, if five is added or subtracted, gives a square number.*

Leonardo's investigation and solution of the problem form *The Book of Squares*, dedicated to Frederick II and forming the core of what was known in Europe of indeterminate analysis during the Middle Ages.

First, Leonardo noted that the squares originate in the odd numbers, since the sums of the odd numbers in sequence produce the square numbers in se-

Figure 10.2: Proof of Proposition 10.2.

quence. This was stated, without proof, as the first proposition; Leonardo's proof appeared later. Based on this proposition, Leonardo solved the problem:

Problem 10.5. *To find two squares equal to a square.*

Solution. Begin with any odd square. The sum of the odd numbers from unity to the odd number preceding the square will be a square number. The two squares add to a third square. □

For example, 9 is an odd square; the sum of the odd numbers from 1 to 7 is 16; so 9 and 16 together make a square. Leonardo then proved

Proposition 10.1. *Successive square numbers differ by the sums of their roots.*

After explaining a second method for finding two squares whose sum is another square, Leonardo proved his first conjecture, that the squares were the sums of the consecutive odd numbers.

Proposition 10.2. *The sum of the odd numbers from 1 is a square number.*

His proof is typical of the proofs in the *Book of Squares*; the proofs are, in general, very difficult to follow, and we shall present only this one. Leonardo used line segments to represent numbers; the line segment from a to b was written $.ab$.

Proof. To the unit $.ab.$ add any number of consecutive numbers $.bg.$, $.gd.$, $.de.$, $.ez.$, and so on; let $.ag.$ be the number $.t.$, $.bd.$ equal the number $.k.$, $.ge.$ equal the number $.l.$, $.dz.$ equal the number $.m.$, and so on (see Figure 10.2).

First, $.t.$, $.k.$, $.l.$, $.m.$, and so on, are odd numbers, for each is the sum of two consecutive numbers, one odd, one even. They are additionally consecutive odd numbers. $.n.$ is the sum of $.ez.$ and $.zi.$, while $.m.$ is the sum of $.ez.$ and $.de..$ Thus the difference between $.n.$ and $.m.$ is the difference between $.zi.$ and $.de..$ But $.zi.$ exceeds $.ez.$ by one, and $.ez.$ exceeds $.de.$ by one, hence $.zi.$ exceeds $.de.$ by two. Likewise, $.m.$ exceeds $.l.$ by two, and $.l.$ exceeds $.k.$ by two, and $.k.$ exceeds $.t.$ by two. But since $.t.$ exceeds unity by two, then the numbers unity, $.t.$, $,.k.$, $.l.$, $.m.$, and so on, are the consecutive odd numbers.

Now the squares of consecutive numbers differ by the sum of the roots. Hence the square of the number $.zi.$ exceeds the square of the number $.ez.$ by the sum of the roots, namely $.n..$ Similarly the square of the number $.ez.$ exceeds the square of the number $.de.$ by $.m.$, and so on, until we have the square of the number $.bg.$ exceeds the square of $.ab.$, or unity, by $.t.$, which is 3. Therefore if to the square of unity, which is 1, is add $.t.$, which is 3, the square of the number

See Problem 5.

.bg. is obtained. And if to this is added the *.k.*, the square of *.gd.* is obtained, and so on. Hence the squares will be the sum of the consecutive odd numbers, starting from 1.   □

Leonardo gave several remarkable series summation formulas (series summation formulas also appear in *Book of the Abacus*). For example:

Proposition 10.3. *If the squares of the numbers from 1 are summed, the product of the last number, the number following, and the sum of the two, is six times the sum of the squares.*

In other words

$$6 \left(1^2 + 2^2 + 3^2 + \ldots + n^2\right) = n(n+1)(n+n+1)$$

Leonardo proved this by induction, first showing that the product of a number, the number following, and the sum of the two numbers is equal to six times the square of the number, plus the product of the number, the number preceding, and the sum of the two, or

$$n(n+1)(n+n+1) = 6n^2 + n(n-1)(n+n-1)$$

To solve Master John's problem, Leonardo began with:

Proposition 10.4. *If there are a number of numbers above, and an equal number below, a given number, and if each of the larger numbers exceeds the given number by the same amount one of the smaller numbers falls short of the given number, then the sum of all the smaller and larger numbers will be the product of the number of given numbers and the given number.*

Leonardo proved this proposition by pairing the numbers above and below the given number, and showing that the sum was just twice the given number; hence the sum of all the numbers above and below the given number was the same as the number of numbers times the given number.

Next is a Diophantine problem:

Proposition 10.5. *To find a number that, added to a square number and subtracted from a square number, yields a square number.*

Leonardo's proof is very complex, but the essential element is that since the square numbers are the sums of the consecutive odd numbers, then:

$$\text{Sum of } m \text{ odd numbers} + c = \text{Sum of } n \text{ odd numbers}$$
$$\text{Sum of } n \text{ odd numbers} + c = \text{Sum of } p \text{ odd numbers}$$

Hence, c can be written as the sum of consecutive odd numbers in two different ways. To solve the problem in whole numbers, it is necessary that c be a multiple of 24; Leonardo called these numbers **congruous numbers**, and the squares associated with the congruous number are added, the **congruent squares**. For example, the congruous number 24 has congruent squares 1^2 and 5^2, since $1^2 + 24 = 5^2$, and $5^2 + 24 = 7^2$. Leonardo explained how to find the congruous numbers.

Note that in this case, $5^2 \pm 24$ is a square. Hence, the problem of finding two squares to which 24 added or subtracted make a square is trivial. What is necessary to solve Master John's problem is an appropriate transformation so that adding or subtracting 5 makes a square; Leonardo, ever adept at the transformation of solutions, proved: See Problem 7.

Proposition 10.6. *If a congruous number and its congruent squares are multiplied by a square, the result will be another congruous number and its congruent squares.*

What is necessary to solve the problem of Master John is to find a congruous number that is also a square multiple of 5; it is easy to see that 720 satisfies this requirement. Thus:

Problem 10.6 (Master John's Problem). *To find a square which, increased or decreased by 5, makes a square.*

Solution. Take a congruous number that is 5 times a square number; 720 is such a number, which is 5 times 144. The congruent numbers are 961, 1681, and 2401, whose roots are 31, 41, and 49. Thus the square of $\frac{41}{12}$, decreased by 5, is the square of $\frac{31}{12}$, while if it is increased by 5, is the square of $\frac{49}{12}$.   □

10.2.6   Hindu-Arabic Numerals

Leonardo stated that his purpose in writing *Book of the Abacus* was to introduce the Indian number system to the West. Others had done so before Leonardo, and others would continue to do so after Leonardo. One particularly charming work was by one of Leonardo's contemporaries, Alexander de Villedieu (d. 1240), who wrote a long poem, "Ode to Algorithm," which began:

> Here follows the art of algorithm, which is
> Born from the twice five figures of India
> 0 9 8 7 6 5 4 3 2 1
> The first signifies one, the second indeed two
> The third signifies three; thus leftwards proceed
> Until the end, the one which "cipher" is called...[4]

and continues onward, going through the essential arithmetic operations of addition, subtraction, duplation, mediation, multiplication, division, and extraction of roots. Note that "first" indicates the rightmost digit, suggestive of an Arabic influence on the ordering.

It might seem quaint that the rules of arithmetic were stated poetically; however, this was a very practical device. In the days before printing, all texts

---

[4]Steele, Robert, *The Earliest arithmetics in English.* Oxford University Press, 1922. Page 72

had to be hand copied, and were thus quite expensive. It was a rare (and wealthy) individual who had more than a few books. At universities, the reference books were so valuable they were frequently chained to the desk on which they rested. Putting the rules of arithmetic into poetic form made them easier to remember, just as rhyming advertising jingles are more memorable than a prose description of a product.

Acceptance of the Hindu-Arabic numerals would be a long time in coming. In 1299, the city of Florence banned the use of Hindu-Arabic numerals in account books, and other cities followed suit. In 1348, the University of Pavia required that the prices of books be written out in "clear letters," not in "ciphers": in other words, in Roman, and not Arabic, numerals. Spelling out the amount of a check in addition to writing the numerical amount is a vestige of this policy. The restrictions might seem to be a case of conservatism, though there were very real legal issues involved. The Roman numerals could not easily be confused for one another; on the other hand, a hastily scrawled 1 might appear to be a 7, while a 0 could easily turn into a 6 or 9. That we write 7 with a bar across the middle to distinguish it from 1 shows that this was a legitimate concern; inadvertent mistakes, not to mention deliberate fraud, were a very real possibility, and as late as the sixteenth century, bankers around Europe were using Roman numerals to keep account records.

10.2 Exercises

1. Find a useful rule for each of the following numbers: 36, 48, 250.

2. Find a rule, then perform the following divisions:

   (a) 178 divided by 48   (c) 389215 divided by 105

   (b) 32654 divided by 270   (d) 1893214 divided by 2475

3. Consider the rabbit problem (Problem 10.3). Solve the modern version of the problem, where the first pair of rabbits is *born* in the first month.

4. Find a number that, when divided by the numbers from 2 through 10, leave remainders of 1, 2, 3, 4, 5, 6, 7, 8, 9, but is divisible by 11.

5. Prove Proposition 2 of the *Book of Squares*: consecutive squares differ by the sum of the roots. Show this geometrically. Hint: use one of Euclid's propositions from Book II of the *Elements*.

6. Prove Proposition 10.3.

   (a) Show that

   $$n(n+1)(n+n+1) = 6n^2 + n(n-1)(n+n-1)$$

   and thus $n(n+1)(n+n+1) - n(n-1)(n+n-1) = 6n^2$.

(b) Hence show

$$6\left(1^2 + 2^2 + \ldots + n^2\right) = n(n+1)(n+n+1)$$

7. Given a congruous number, explain how to find its congruent squares. Hint: to find the congruent squares for 24, note that $11 + 13 = 3 + 5 + 7 + 9$, and the square of n is the sum of the first n odd numbers.

8. (Teaching Activity) Make up a poem describing how to perform an arithmetic operation. Be sure the poem is a complete lesson in performing the operation, and not just a description of a few specific cases.

9. (Teaching Activity) The rhyme and scansion of most poems make them easy to sing. Set a poem about arithmetic operations to music.

10.3   The High Middle Ages

The people living in medieval times did not think they were living in the "middle ages," and certainly did not consider their era to be the "Dark Ages" (another popular term for the period). It is, however, convenient to divide history into periods, and most historians agree there is a distinction between the Early Middle Ages and the High Middle Ages. What they do not agree upon is the dividing line between the two. For no reason other than its convenience, we will suppose the transition to be in Leonardo's time, for after Leonardo's time, theoretical mathematics began its revival in Europe. Leonardo himself was a key transitional figure: his *Book of the Abacus* is a work on purely practical mathematics, while his *Book of Squares* shows a more theoretical bent.

Further evidence of interest in theoretical mathematics was the increasing popularity of a game called **rithmomachy**, invented in the eleventh century in France or the Holy Roman Empire. In rithmomachy, both players had a set of numbered tiles (different numbers for the different sides). These moved on a board in accordance to simple rules, akin to chess or checkers. An opponent's piece could be captured (and then used by the other player), as in chess or as in *go* (where a piece is captured if it is completely surrounded by opposing pieces), but other methods of capture existed, based on arithmetical properties. If a piece was surrounded by opposing pieces that added or multiplied to the tile's value, it was captured; another form of capture was that a piece numbered m could capture a piece numbered mn if the two pieces were n spaces apart.

10.3.1   Jordan de Nemore

The work of Jordan de Nemore (or, in Latinized form, Jordanus Nemorarius) was perhaps the most theoretical of the period. About all that is known for certain about Jordan de Nemore is that he was a French contemporary of Leonardo of Pisa and probably taught in Paris around 1220. Whoever Jordan de Nemore was, he wrote two works on mathematics that concern us. The first, called

On Arithmetic, is a work on arithmetic far more sophisticated than Boethius's simplified version: Nemore's work was grounded firmly in the Euclidean model of definition-axiom-proposition.

The other important work of Nemore is his *Regarding the Given Numbers*, which we would consider a work on algebra. Like the Arabic algebras that preceded it, *Regarding the Given Numbers* is a long compilation of types of equations with instructions on how to reduce each type to a few fundamental forms. The first form is:

Proposition 10.7. *If a given number is separated into two parts with a given difference, the parts can be found.*

In modern terms, given $N = x + y$ and $M = x - y$, then x and y can be found.

This is a key problem, for many of the remaining problems in the first book of *Regarding the Given Numbers* reduce to finding the difference between two parts of a given number. Note that Nemore's states that the solution *can* be found, and not what the solution is; if we resort to Proclus's division of propositions into theorems and problems, then all the propositions in *Regarding the Given Numbers* are theorems. Today we would call the proposition an **existence theorem**. The actual determination of the solution is left to the proof:

See page 93.

Proof. The smaller plus the difference equals the larger. Thus the smaller with itself and the difference equal the given number. Hence subtract the difference from the given number, giving twice the smaller; half of this is the smaller number. □

After this, an example is given.

Proposition 10.8. *If a given number is separated into two parts with a given product, the parts can be found.*

In other words, given $N = x + y$ and $M = xy$, x and y can be found.

In his proof, Nemore introduced a system of notation. It was an early attempt, barely an improvement upon the syncopated notation of the Arabs, but it was a step in the right direction. Letters were used to represent the variables, and the barest beginnings of symbolic expression are discernible in Nemore's work. The geometric inspiration is clear: in the following, Nemore let abc represent the given number, which was separated into two parts, ab and c.

Proof. Let the given number abc be separated into ab and c, and let ab times c be the given product d. Let abc times itself be e.

[Given $N = x + y$ and $M = xy$; let $e = N^2$.]

Sum the quadruple of d, which we will call f, and subtract this from e leaving g, which is the square of the difference between ab and c. Therefore take the root of g and call this b. Thus a given number is separated into two parts with a given difference, and the solution can be found.

[Let $f = 4M$. Then $N^2 - 4M = (x - y)^2 = g$. Thus $\sqrt{g} = x - y$. Thus $N = x + y$ is divided into two parts with a given difference, $\sqrt{g}$, and the two parts can be found according to Proposition 10.7.] □

Proposition 10.9. *If a number is separated into two parts with a given difference, and the product is a given number, then the number can be found.*

Proof. Let b be the difference and d be the product. Let the double of d be e. Let h be the square of the difference, and add twice e and let the sum be f, which is the square of the number. Hence the number can be found. □

Books II and III of *Regarding the Given Numbers* deal with solving problems involving proportions. In Book IV, Nemore included solutions to the three types of quadratic equations.

Proposition 10.10. *If the sum of the square of a number and the product of a given number and the number is a given number, the number can be found.*

Proof. Let the square of the unknown number be a, the root b, and let b be multiplied by the given number cd. Let c, d each be half of cd and let b times cd make e, and let ae be the given number.
[Given M and N, we want to solve $x^2 + Mx = N$. Let $a = x^2$, $b = x$, $M = c + d$ with $c = d = \frac{1}{2}M$. Let $e = Mx$.]
Now, bcd multiplied by b is ae. Add the square of d to ae to make aef. Since aef is now known, bc can be found, since it is the root of aef. Subtract c to find b, which is the unknown number.
[$x + M$ multiplied by x is $x^2 + Mx$, which is equal to N. Add d^2 to both sides; then $x + c = \sqrt{N + d^2}$. Thus, to solve for x, subtract c.] □

Though his method is the same as that of the Islamic mathematicians for "a square and its roots equal a number," the notation makes the exposition more difficult to follow. Nemore also included the other two cases, recognizing the double solution in the case of a square and a number equal to roots; he did not, however, note that this particular case might not have a solution. See Problem 5.

The only time Nemore ventured beyond the second degree is in his last problem, which is a degenerate quartic:

Proposition 10.11. *If the square of a number of squares is equal to a number of roots, the root can be found.*

10.3.2 Ben Gershon

One of the most powerful techniques available to the modern mathematician is the method of **mathematical induction**. An early user of this technique was Rabbi Levi Ben Gershon (1288-?). Rabbi is an honorific (roughly equivalent to the medieval "Master"); "Ben" is like the Arabic "ibn," and indicates "son of"; thus he might be called Master Levi [Leo], son of Gershon [Gershwin] and his name is often Latinized to Leo the Jew. In 1321, he finished *The Art of* *the Calculator.* We know the exact year because Ben Gershon wrote down the Jewish calendar date on which he finished the work. The second half of the work concerned itself with practical arithmetic. The first half, however, developed a theoretical arithmetic, and several propositions were proven using mathematical induction. After proving

Proposition 10.12. *The product of one number and the product of two numbers is the same as the product of any two of the numbers, and the remaining number.*

which is a combination of the associative and the commutative laws of multiplication. Ben Gershon then proved

Proposition 10.13. *The product of one number, and the product of three numbers is the same as the product of any of the numbers, times the product of the remaining numbers. In general, the product of one number, and the product of any number of numbers is the product of one of the numbers, times the product of the remaining numbers.*

Ben Gershon's proof proceeds as follows: if we consider the product of one number, a, times the product of three numbers, bcd, then the product of three numbers b, c, and d, is the product of any of these numbers, say b, times the product of the two remaining numbers, cd, which is itself a number. Thus, the product of one number times the product of three numbers can be considered the product of one number a times the product of two numbers, b and cd. By the previous proposition, this is the product of any one of the numbers (b), times the product of the other two (a and cd), one of which is the product of two numbers: thus the theorem is proven. Proceed this way "step by step" (in Hebrew, *hadragah*, an adjective referring to the appearance of cliffs that appeared to rise from the plains like steps) to infinity.

These two propositions together contain most of the elements of a modern proof by induction: the proof of the first case, $a(bc) = b(ac)$; the proof of how another case derives from one of the proven cases, namely how the product $a(bcd)$ can be thought of as the product $a(b \cdot cd)$ and reduced to the first case; finally, a logically defensible claim that the same argument could be made for *every* case. In a like manner, Ben Gershon proved other propositions, including several on combinatorics, such as

Proposition 10.14. *If the number of permutations of a given number of things is some number, then the permutations of a set consisting of one more thing will be the number of permutations times one more than the given number of things.*

[If there are t permutations of m things, then the permutations of $m+1$ things will be $(m + 1)t.$]

Proof. Suppose there are t permutations of the m elements; let an additional element be added. Take any permutation, and place the new thing first. There are t permutations of the remaining m elements. Now take any of the other elements, and place that first. There are t permutations of the remaining elements. Repeating this for all $m + 1$ elements, we find there are $(m + 1)t$ permutations. Proceed in this way "step by step" to infinity.   □

In Ben Gershon's *Astronomy* (part of a much larger treatise, *Wars of the Lord*), Ben Gershon referred to a mathematical procedure he called "heuristic reasoning" (in Hebrew, *heqqesh tahbuli*). This is another form of the law of double false position. Since there was, as yet, no systematic way of dealing with negative numbers, Ben Gershon had to express his method as a set of rules for different situations:

Rule 10.1. *If a first quantity substituted into an equation results in an amount that exceeds the constant by a second quantity, and a third quantity results in an amount that falls short of the constant by a fourth quantity, then choose a fifth quantity between the first and third so the difference between the first and the fifth is to the difference between the first and third as the second is to the sum of the second and the fourth.*

A marginal note on the manuscript gives an example of the use of the rule.

Example 10.2. *If the first quantity* 10 *makes* 30, *which is* 6 *too much and thus the second quantity is* 6; *and the third quantity* 6 *makes* 15, *which is* 9 *too small and thus the fourth quantity is* 9, *then the fifth is* 8; 24 *[=* $8\frac{24}{60}$, *since* 1; 36 : 4 = 6 : 15*]*.

Note that the example does not state the actual problem, and that Ben Gershon's method depends *only* on the known values of the equation for the different values of the proposed solution.

Example 10.3. *An expression is equal to* 12. *If* 5 *substituted into the expression makes* 14, *which is* 2 *more, and* 1 *substituted into the expression makes* 6, *which is* 6 *less, then we have:*

First quantity $= 5$	Second quantity $= 2$
Third quantity $= 1$	Fourth quantity $= 6$
Difference, First and Third $= 4$	Sum, Second and Fourth $= 8$

Let the difference between the fifth quantity and the first quantity be Q *(thus the fifth quantity will be* $5 - Q$*). We need* $Q : 4 = 2 : 8$; *thus the difference is* 2 *and the fifth quantity is* 3.

Note that we have approximated a solution to $f(x) = 12$, when all we know is $f(5) = 14$ and $f(1) = 6$.

If the two amounts both results in an excess (or if both amounts result in a shortfall), Ben Gershon stated his rule as:

Rule 10.2. *If a first quantity substituted into an equation results in an amount that exceeds the constant by a second quantity, and if a third quantity results in an amount that exceeds the constant by a fourth quantity less than the second quantity, choose a fifth quantity so the third is between the first and fifth and the difference between the first and fifth is to the difference between the first and third as the second is to the difference between the second and fourth.*

Again, a marginal note gives an example.

Example 10.4. *If the first quantity 10 makes 36, which is 12 too much and thus the second quantity is 12; and a third quantity 8 makes 30, which is 6 too much and thus the fourth quantity is 6; then the fifth quantity will be 6 [since 4 : 2 = 12 : 6].*

Example 10.5. *If the first quantity 6 makes 24, which is 12 too little and thus the second quantity is 12; and a third quantity 8 makes 30, which is 6 too little and thus the fourth quantity is 6, then the fifth is 10 [since 4 : 2 = 12 : 6].*

Ben Gershon also included a concise statement of the rule of single false position:

Rule 10.3. *If a first quantity substituted into an equation results in an amount differing from a third quantity by a second quantity, choose a fourth quantity so the fourth is to the first as the third is to the second.*

Example 10.6. *If the first quantity 10 makes 30, the second quantity, when we want it to make 24, the third quantity, then we suppose a fourth quantity 8, since 8 is to 10 as 24 is to 30.*

Using these rules, Ben Gershon approximated Sine $\frac{1}{4}^{\circ}$ as $0; 15, 42, 28, 37, 2$, which is accurate to the third sexagesimal place.

10.3 Exercises

1. Use one of Nemore's propositions to solve:

   (a) Ten is separated into two parts whose difference is 2.

   (b) Ten is separated into two numbers whose product is 21.

2. Nemore's fourth problem is: If a given number is separated into two parts the sum of whose squares is known, then each of the numbers can be found. Derive Nemore's solution, using the following steps.

   (a) Let the sum of the squares be B, and E be twice the product of the two parts. Explain how to find E.

   (b) Subtract E from B to obtain H, which is the square of the differences. Explain why.

   (c) Explain how to use the square root of H, which Nemore calls C, to find the solution.

3. Consider Nemore's Proposition 10.9.

   (a) What algebraic equation is being solved?

   (b) Explain why the procedure works.

4. Nemore's Problem I-6 is: if the difference of two parts of a number is known, and if the sum of their squares is also known, the number can be found. Reconstruct Nemore's method.

5. Consider Nemore's Proposition 10.10. Explain how Nemore's solution of the quadratic is equivalent to the Arab solution.

6. Example 10.3 was translated into, "Find an approximate solution for $f(x) = 12$ if $f(5) = 14$ and $f(1) = 6$." Translate Example 10.4 and Example 10.5 in the same way.

7. For each of the following, translate the problem statements as in Problem 6. Then find an approximate solution.

   (a) A quantity is to be equal to 40. 10 makes 36 and 16 makes 48.

   (b) A quantity is to be equal to 15. 8 makes 27 and 9 makes 20.

   (c) A quantity is to be equal to 25. 4 makes 12 and 6 makes 20.

   (d) A quantity is to be equal to 30. 12 makes 36 and 10 makes 20.

8. To find Sine $\frac{1}{4}^{\circ}$ (equivalent to our $60 \sin \frac{1}{4}^{\circ}$), Ben Gershon used the following procedure.

   (a) Find the value of Sine $1\frac{1}{2}^{\circ}$ (see Exercise 7 of Section 7.2).

   (b) Find the value of Sine $15 + 1\frac{1}{2}^{\circ}$.

   (c) Use the half-angle formulas to find the value of Sine $\frac{1}{4} + \frac{1}{128}^{\circ}$.

   (d) Beginning with Sine 15°, use the half-angle formulas to find Sine $\frac{1}{4} - \frac{1}{64}^{\circ}$.

   (e) Now find an approximate solution to Sine $\frac{1}{4}^{\circ}$. Hint: let the *Sines* be the first and third quantities, and let the desired quantity be $\frac{1}{4}^{\circ}$.

9. (Teaching Activity) Create a board game (like chess or checkers) that, like rithmomachy, has captures based on arithmetical properties. Note: one of the difficulties of playing rithmomachy is that there are several different methods of capture. For ease of play, you might want to focus on two or three, with at least one of the capture methods be as in chess or checkers.

10.4 Nicholas Oresme

Like many other prominent intellectual figures of the times, Nicholas Oresme (fl. 1340-1382) was a Church scholar, obtaining his Masters of Theology around 1355. He was friends with the future Charles V of France, and between 1370 and 1377, he worked on translating several important classical works into the French language. In 1377, Charles appointed him the Bishop of Lisieux, where he spent the rest of his life.

Oresme was trained in and supported the classical Aristotelian system of the world, though he did allow for some modifications. The supporters of Aristotle, called **scholastics**, held that the world was motionless; Oresme suggested that, while the center of mass of the Earth was at rest, changes in the distribution of matter, caused by geological disturbances and variations in climate, meant the Earth itself underwent slight motions. In other aspects of Aristotelian physics, Oresme demonstrated his free-thinking spirit: he opposed the popular view that the velocity of an object was proportional to the distance the object fell and believed instead the velocity was proportional to the *time* of fall.

Oresme lived during one of the most devastating periods in French history. In 1337, Edward III of England invaded France to recover lost territories, which began a series of wars that would rage, on and off, for over a century: the Hundred Years' War. All the battles were fought on French soil, causing untold devastation. Then in 1347, another disaster struck. A ship entered Genoa, Italy, its crew dying from a new disease: the bubonic plague or, because its symptoms included black pustules forming on the victim's arms and legs, the Black Death. No disease in all of history has been more devastating: more than 90% of those who caught the disease died within a *week*; entire towns were wiped out. It has been estimated that one-third of the population of Europe died from the plague. To add to France's disasters, the peasantry, who paid the bulk of the *gabelle* (a tax on salt, instituted in 1341) and other taxes levied to raise money to fight the English, revolted in 1358. Since "Jacques" was a common French peasant name, this and subsequent peasant revolts came to be known as *jacqueries*.

10.4.1   The Configuration Doctrine

In mathematics, Oresme wrote two important works. In the 1340s, he wrote *Questions on the Geometry of Euclid*, which he expanded into the three-part *Treatise on the Configuration of Qualities and Motions*. To these works we can trace the origin of the modern notion of a graph.

The first part of the *Treatise* dealt with the **figuration of qualities**. A **quality** was something like velocity, "hotness," or "whiteness," that could be measured. The notion of quantitative comparison of qualities was in the air. Scholars at Oxford and at Paris discussed what would happen when two bodies of water of different "hotness" were added, or what would happen if, somehow, the "hotness" of one body of water could be transferred to another body of the same "hotness." Oresme's great contribution was applying geometry to quantitative comparison.

First, Oresme boldly argued that *all* qualities could be measured quantitatively, and thus the changing intensity of that quality could be represented geometrically. This was done by drawing a line, representing the intensity of the quality, *perpendicular* to what Oresme called the "extension" of the quality, usually interpreted to be time. The intensity of the quality was referred to as the **longitude** of the quality, while the extension of the quality was designated the **latitude** of the quality.

The longitudinal lines were assumed to be in proportion, so that if at some point the quality was twice as intense as at another point, the line representing the quality would likewise be twice as long.

Example 10.7. *The extension of a quality is AB; suppose that at C the intensity of the quality is twice what it was at A, and at B the intensity is three times that at A. Thus, the perpendicular drawn at C should be twice the perpendicular drawn at A; likewise, the perpendicular drawn at B should be three times the perpendicular drawn at A.*

Finally, the **quantity of quality** was the figure whose base was the extension of the quality, and whose altitude at any point was the intensity of the quality. This clearly suggests a Cartesian graph; however, it is important to realize that Oresme was concerned primarily with the quantity of the quality, as represented by the plane figure. In a sense, Oresme skipped Cartesian graphs and went straight to the definite integral. For example, we may suppose the quality to be velocity, and the extension the time over which the velocity is held. Then velocity at any particular instant would be represented by a line perpendicular to the extension, and the whole figure would represent the quantity of the quality velocity. If the velocity is constant, then all the longitudinal lines are the same height, and the quantity of quality would be represented by a rectangle.

Oresme dealt only with quantities that were, in his terms, **uniform** (constant), or **uniformly difform** (increasing so that the intensity was proportional to the extension). He noted the existence of a third category, that of the **difformly difform**, in which the intensity was changing at a rate that was not proportional to the extension. These latter he divided into sixty-six cases, most of which were unrectifiable by the tools of the time. Only the cases where the configuration was a combination of rectilineal figures and regions bounded by circular arcs could the quantity of quality be determined.

Oresme extended the configuration notion to more than two dimensions. A **surface quality** could be imagined by a solid situated on a base; if the quality was uniform, then the summit of the solid figure would be plane parallel to the base, while if it was uniformly difform, the summit would be a plane that was not parallel; finally, if the quality was difformly difform, the summit would be some curved surface. Oresme boldly envisioned a **corporeal quality**, whose base was the *solid* from which the longitudinal lines were drawn; this would generate a four-dimensional figure.

A possible objection to Oresme's scheme was the use of lines to represent other objects. After all, lines were geometric objects, and might be called upon to represent other geometric objects, or qualities that were not even geometrical, such as velocity or time. Oresme had disposed of this objection in his *Questions on the Geometry of Euclid*. In the Scholastic tradition, he first cited Aristotle, who had imagined time to be a line; then he added that through the use of lines, the problem of understanding changing qualities is much easier: hence their use is justified pedagogically.

The second part of *Treatise* is a long, philosophical discussion of little mathematical value, though Oresme periodically returns to the physical and mathematical concepts that make him a key figure in the history of mathematics. Oresme distinguished between the velocity of motion, which was measured by the distance an object traveled, and the velocity of descent, which was measured by an object's downward motion. He noted that two objects might have equal velocities, yet different motions. This distinction is equivalent to noting that velocity is a vector quantity.

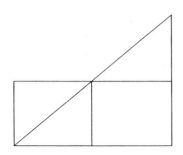

It is the third part that is the most interesting, for it deals with the determination of the quantity of qualities—particularly the quantity of velocity. The main function of the configuration doctrine was that it allowed for the comparison of changing qualities. For example, suppose a body is moving so that its velocity begins at zero and is proportional to the time traveled. The quantity of the quality of velocity will be represented by a triangle. This quantity of quality, however, is equal to a rectangle whose height is equal to the velocity at the middle point of the triangle. Thus, the quantity of quality in the two instances is the same. In the latter case, the quantity of quality is that of a body moving with a *uniform* velocity, equal to half the final velocity of the body in the first case.

This is the so-called **Merton rule**, formulated by Oresme's contemporaries at Merton College at Oxford, which states that the distance traveled by an object whose velocity is proportional to the time is equal to the distance traveled by an object of constant velocity equal to the "middle velocity" of the first object. In the case where the configuration of velocities is a simple, rectilineal figure, the determination of the quantity of quality can be done simply, by partitioning the figure into triangular, rectangular, and trapezoidal regions.

Oresme went beyond the Merton rule, however, and the *Treatise* gave a number of instances where either the final velocity or the total extension (in time) was infinite. "Common sense" would suggest that in either case, the quantity of quality (i.e., the distance) would also be infinite, but Oresme showed that this was not so. In particular, he showed:

1. It is possible for an object to increase to an infinite velocity, yet travel a finite distance.

2. It is possible for an object to travel for an infinite amount of time, yet to travel a finite distance.

For the first, Oresme begins by claiming

Proposition 10.15. *A finite surface can be made as high as desired.*

Proof. Beginning with an area of one square foot and base AB, take another, similar and equal to it on base CD. Divide the latter into an infinity of parts in ratio 2 to 1, by dividing the base CD in the same way. Designate these parts as E, F, G, ... Place E on AB with its end over B. Place F on E, with its end towards B, and again place G on F, and so on.

Imagine AB divided into parts in continued proportion of 2 to 1, towards B. Thus, on the first proportional part of AB there will be a surface one foot high.

On the second proportional part there will be a surface two feet high. On the third proportional part, there will be a surface three feet high, and so on. Yet the whole surface is only two square feet, or four times the figure on the first proportional part on *AB*. □

After this, Oresme gave an example using velocity.

Example 10.8. *If an object were to move with a certain velocity during some period of time, say an hour, and twice the velocity during the half the time, and three times the velocity during a quarter the time, and so on, the total quantity of velocity would be four times the quantity total velocity of the first part, so in the whole hour, the object would travel four times the distance what it traversed in the first half. For example, if in the first half hour it traveled one foot, it would travel four feet during the whole hour.*

It is interesting to note that Oresme discussed the specific example of velocity after solving the general problem of determining the quantity of quality; it was an implicit recognition that the quantity of quality was dependent solely on the configuration of the quality.

Parisian scholars adopted the configuration doctrine almost immediately, and Giovanni di Casali, in a work that predated the *Treatise* but drew heavily from the *Questions*, spread the configuration doctrine to Italy.

10.4.2 France and England

If we look at the prominent figures in mathematics during the thirteenth and fourteenth centuries, we note that many of them were French or English: Robert of Chester and Sacrobosco from England, and Nemore and Oresme from France. Thriving universities were established at Paris, Cambridge, and Oxford. But the next major spurt in the development of mathematics came from Italy and Germany, not England and France. The Hundred Years' War played a major role: most of the war was fought on French territory, and between the war's devastation, the Black Death, and the *jacqueries*, it is no wonder that France did little besides concentrate on internal affairs for the next century.

Meanwhile, the English faced the fact that they lost the Hundred Years' War despite some spectacular victories. The most spectacular was at Agincourt, in 1415, where 25,000 French knights met a force of 5700 exhausted, sick, and starving Englishmen, led by King Henry V. The French knights charged across a narrow front, got mired in a field of mud, and were picked off by the thousands by Henry's men, using the longbow. The "flower of French chivalry"—over 8000 knights—died at Agincourt. Henry lost 400 men and gained the throne of France.

By 1429, all had changed, due to an illiterate peasant girl who managed to rally the French forces: Jeanne Darc, known in English as Joan of Arc from a misinterpretation of "Darc" (a last name of no consequence) as "d'Arc" ("of Arc," a name implying noble descent). Joan's boost of French morale should not be underestimated, but equally important was French utilization of a new weapon: artillery. But the English attributed French success to witchcraft, and through their Burgundian allies, had Joan imprisoned, tried, and eventually burned at the stake. It did no good: England soon lost all its French territory, and afterward underwent a series of civil wars, known as the War of the Roses.

10.4 Exercises

1. Sketch the configuration of velocities for the following objects.

   (a) A body that has a constant velocity for a time, and then doubles its velocity for twice the time.

   (b) A body that has a constant velocity for a time interval, and then over an equal interval, increases at a uniform rate so that its final velocity is twice the original velocity.

   (c) A body that starts with some velocity, and over some time interval, triples its velocity at a uniform rate.

   (d) A body that starts with some velocity, and over some time interval, slows to a stop at a uniform rate.

2. For the bodies in Problem 1, find the uniform velocity another body would have to travel at for the same time interval, so that the quantity of quality of the uniformly moving body is equal to the quantity of quality of the difformly moving body.

3. Suppose that an object travels for a certain time at a certain velocity. Then it travels for the same amount of time at half the velocity, and so on, each time traveling at half the velocity for the same amount of time.

   (a) Sketch a figure showing the configuration of velocity.

   (b) Show that the object travels a finite distance. What is the finite distance?

4. In Chapter III, Part ix, Oresme gives another example. Suppose the extension AB is divided into parts in proportion 4 to 1 (thus, if the whole was 64, then the first part would be 48, the second part 12, the third part 3, and so on), and a body were to move on the first part for a constant speed, on the second part for twice the speed; on the third part for four times the speed, and so on.

   (a) Sketch the configuration of the velocity.

   (b) What is the quantity of quality?

Chapter 11

The Renaissance

Europe slowly began to recover from the Black Death. To slow the spread of the disease, Italian port cities began to confine an arriving ship's passengers and crew for a period of time, so that if they showed signs of the plague, they could be prevented from spreading it to the general population. The period of confinement came to be standardized at forty days: forty is *quaranta* in Italian, from which we get our word quarantine.

Many of the survivors of the plague were enriched by the deaths of relatives. Some used their new-found wealth to patronize artists and scholars, in the fashion of the great medieval kings, such as Frederick II. Others, seeing temperance and abstinence no proof against the disease, took to living life as if there were no tomorrow—as well there might not be. These two trends characterized the 1400s: the rebirth, or Renaissance, of classical scholarship; and the drunken orgies and week-long parties. Boccaccio's *Decameron* illustrates the two aspects. Both would have fateful consequences.

The Renaissance had the more immediate effect. In architecture, the works of Greek and Roman builders were revered and copied, while the beautiful cathedrals built during the Middle Ages were derided as "gothic," in the erroneous belief they were built by the "uncultured" Goths. Except for architecture, the Renaissance focused on the humanities: literature, sculpture, and painting. What advances there were in science and mathematics were incidental by-products of the revival of classical scholarship, two elements of which were belief in the potential of man, and the study of Greek.

11.1 Trigonometry

One of the impediments to progress in mathematics was that the great mathematicians of antiquity wrote in Greek, a language few Western scholars could read. In any case, scholars had, at best, Latin translations of Arabic transla-

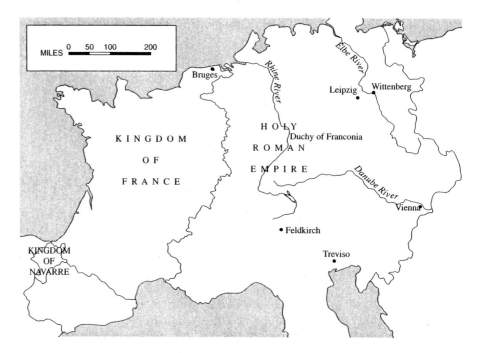

Figure 11.1: Central Europe During the Renaissance.

tions of Greek texts, which in turn were copies of originals that had long since vanished. Since the copies and translations were rarely done by mathematicians, the more difficult parts were omitted, and errors crept into the rest.

The Renaissance, which revived interest in all things of Greek and Roman antiquity, spurred the study of Greek. One of the most influential in promoting the cause of translating classical Greek texts from original sources was George Trebizond, better known as Cardinal Bessarion (1403-1472). Bessarion was himself Greek, and made a number of contacts in Europe during attempts to forge an alliance between the remnants of the Byzantine Empire and various European states to thwart the Turkish advance. These attempts at forming alliances with the West failed, and in 1453, Constantinople fell to the Turks under Mehmet the Conqueror, and the last vestiges of the Byzantine Empire were conquered shortly thereafter.

11.1.1 Regiomontanus

On May 5, 1460, Cardinal Bessarion arrived in Vienna. At the University of Vienna, he met Georg Peurbach and his student and colleague, Johann Müller (June 6, 1436-July 8, 1476). Müller is best known as Regiomontanus, after the Latin form of the name of his birthplace: Konigsburg, in Franconia. Bessarion convinced Peurbach and Regiomontanus to work on a translation of Ptolemy's *Almagest*. Unfortunately, Peurbach died the next year, before the translation could be finished; before dying, he had had Regiomontanus pledge to complete the work. Regiomontanus finished the task, and later began to translate other works from Greek originals into Latin. He hoped to provide translations of Archimedes, Heron, Apollonius, Ptolemy's *Geography*, and a number of other works; during his search for Greek original manuscripts he claimed to have seen a complete copy of Diophantus's *Arithmetic*, containing fourteen books, one more than the number claimed by Diophantus. Assuming that Regiomontanus was not completely mistaken, the extra book may have been from another work of Diophantus.

In 1475, Regiomontanus was called to Rome by Pope Sixtus IV to reform the calendar. Since the time of Bede, the inaccuracies in the Julian calendar had been known, but nothing had been done about them. By the fifteenth century, the inaccuracies in the Julian calendar meant that the vernal equinox was occurring about ten days before March 21. But before Regiomontanus could do anything, he died during one of Rome's frequent plagues. See page 274.

The death of Regiomontanus meant the reform of the calendar had to wait another century. In 1582, Pope Gregory XIII assembled a team of astronomers and mathematicians to solve the problem of the calendar. The result, known as the **Gregorian calendar**, is our modern one. There were two significant changes: first, century years not divisible by 400 were nonleap years (in the Julian system, they would be leap years). Second, to make the date of the vernal equinox March 21 (as established by the Council of Nicaea), ten days were dropped: in a Papal decree ordaining the new calendar, the day following October 4, 1582 would be October 15, 1582.

The dropping of ten days was met with some resistance, not all of which was irrational: the loss of ten days of October meant debts would come due sooner, while rents were still owed for a full month. More seriously, by then half of Europe had split from the Catholic Church, and Protestant countries refused to accept a calendar instituted by a pope. England continued to use the Julian calendar until 1752, which meant that a date in England was a different date in Spain. A curious consequence was that William Shakespeare and Miguel Cervantes (the author of *Don Quixote*) died on the same *date*, April 23, 1616, but ten days apart.

Regiomontanus's most important work was probably his *On Triangles* (1464). So much of the text was taken without attribution from Jābir's *Correction of the Almagest* that a century later, the mathematician Cardano would accuse Regiomontanus of plagiarism. Certainly Regiomontanus was well aware of Jābir's work, having copied it for his own reference twice before 1460, and a significant portion of Book IV is nearly identical to Jābir's Book I. See page 266.

The fact that so much of *On Triangles* was plagiarized should not detract from an important fact: Regiomontanus's work did for trigonometry what the *Elements* did for geometry. Though not original, it summarized the knowledge of trigonometry in an easily readable form and soon became the standard text in the subject. Book I dealt with plane trigonometry. The law of sines appears, but most of the book deals with the problem of solving a triangle (i.e., determining all three sides and all three angles of a triangle):

Proposition 11.1. *Given a triangle with three unequal sides, and the perpendicular to one of the sides, the lengths that the perpendicular divides the base can be determined.*

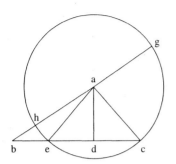

Proof. Given triangle *abc*, with perpendicular *ad*; suppose *ac* is the shortest side. Draw circle *hec*, with radius *ac* and center *a*. Then *be* is the difference between the segments *bd*, *dc*. Also, the rectangle on *gb*, *bh* is equal to the rectangle on *cb*, *be*, and *gb* = *ab* + *ag*, or *ab* + *ac*, and *bh* = *ab* − *ac*. Hence, the difference *be* between the segments of the base divided by the perpendicular may be found, and thus the segments themselves may be found.   □

After this, Regiomontanus gave a procedure:

Rule 11.1. *Multiply the sum of the two sides by the difference of the two sides, and divide by the base. This gives the difference between the segments. Subtract this from the base, or the base from this amount; half the remainder is the shorter side.*

Example 11.1. *ab is 25; ac is 17, bc is 28; let the perpendicular to bc be drawn. The sum of the two sides is 42; the difference is 8; the product is 336. Divided by the third side is 12; subtract this from 28 leaves 16, so the smaller side is 8, and the larger side is 20.*

Some of Regiomontanus's problems required solving quadratic equations; he did not give explicit solutions, and clearly expected his readers to be capable of solving them.

On Triangles quickly became the standard text on trigonometry, and as such played a valuable role in the history of science. In 1577, a "new star" appeared in the constellation of Cassiopeia. Based on his observations, the Danish astronomer Tycho Brahe (1546-1601) wrote *Regarding the New Star*; new in Latin is **nova**, and this name has been given to an entire class of stars that suddenly appear.[1] Using the triangulation methods from *On Triangles*, Brahe showed the new star was farther away than the Moon, in direct contradiction to Aristotle's claim that the regions beyond the Moon were eternal and unchanging. This was one of the first significant cracks in the Ptolemaic system of the universe, and a key step toward the modern heliocentric system.

11.1.2   The Printing Press

Before the modern era, mathematics and science grew slowly. There are many reasons for this, but certainly one of the more important is that ideas develop best when they are developed by many people: the free interchange of knowledge

---

[1]Nova do not actually represent new stars, but stars that were originally too dim to be seen that underwent a sudden increase in brightness. Modern astronomers distinguish between *recurrent* nova, which, as the name suggests, can brighten and dim repeatedly, and *supernova*, where a star literally explodes, destroying itself in the process.

is essential. Thus, we have seen many centers of scientific and mathematical development: Athens, Alexandria, Baghdad, and Sicily.

One of the most important events in world history occurred in 1454. In that year, Johann Gutenburg of Mainz published a *printed* Bible. Printing presses existed before Gutenburg. The Chinese had been using them for a thousand years, and European presses existed before the thirteenth century. These printing presses, however, used a method known as **block printing**: little more than a very large ink stamp, on which a design had to be laboriously carved into some hard material, then stamped onto the paper. Gutenburg's innovation was **movable type**, whereby a printed page could be assembled, letter by letter.[2] The press was infinitely variable, and could be used again and again. It is incorrect to say that Gutenberg invented the printing press (his exact contributions are a matter of some debate)—but he was certainly the first to make use of a *practical* press.

Books became cheap, and because they carry ideas further and more reliably than the spoken word, the interchange of ideas was facilitated. Equally important was that printing became easy. With block printing, only the most important— and shortest—ideas were worth the effort required to carve the blocks. With movable type, more ideas with more detail could be distributed to more people: the perfect environment for an explosive growth in science and mathematics.

The first printed mathematics text is a commercial arithmetic published in 1478 in the city of Treviso (not far from Venice); hence it is usually referred to as the *Treviso Arithmetic*. It is a work on commercial arithmetic, sharing many similarities in spirit with Leonardo's *Book of the Abacus*. Indeed, nearly three hundred years after Leonardo, the anonymous author found it necessary to explain the mysteries of Hindu-Arabic numeration to Europeans.

The printed word could spread other ideas as well. The Renaissance popes participated fully in the two trends that characterized the Renaissance: some were notable patrons of the arts, some were notorious libertines, and many were both. The most debauched of the Renaissance popes was Alexander VI, a member of the infamous Borgia family. In 1494, to settle claims between the Portuguese and the Spanish, he mediated the Treaty of Tordesillas, which divided the entire world between Spain and Portugal (the rest of the world was not consulted!). One result is that most of the Americas fell into Spanish hands (and is consequently Spanish speaking today). However, one parcel of land, discovered by Pedro Cabral in 1500, fell into the Portuguese side of the dividing line and today speaks Portuguese: Brazil.

The corruption and arrogance of the papacy were especially galling in regions that had not yet recovered from the Black Death. In particular, the middle and lower classes of central and eastern European felt their donations to the Church went to support the excesses of the Roman popes. Early calls for reform came from John Wycliff in England, and Jan Hus in Bohemia. Perhaps the most telling tenet of the Wycliffites, also known as Lollards, and the Hussites was the

Example of Multiplication and Division, from Regiomontanus's *On Triangles*. Boston Public Library/Rare books Department. Courtesy of the Trustees.

---

[2]The Chinese also used movable type, but since the Chinese language was and is still written using thousands of ideographs, movable type had only a slight advantage over block printing.

demand to translate the Bible into the vernacular, the language of the people, rather than leave it in Latin, a language understood only by the clergy (and sometimes not even by them). Wycliff died in 1384, but his body was dug up and burned after being condemned. Hus was promised safe conduct to the Council of Constance (1415) to air his views; there, he was arrested, tried, then burned at the stake.

Both Wycliff and Hus attempted reform *before* movable type printing, and they failed. A new round of reform began in 1517, when Martin Luther, a professor at the University of Wittenburg, posted a list of 95 theses for discussion. The Church attempted to defuse the issue, but failed: the printing press made the theses available for everyone to read, and the Reformation had begun. Luther's followers were originally known as Lutherans, and only later was the name Protestant applied to the reformed churches.

11.1.3   Copernicus and Rheticus

In 1543, Nicholas Copernicus (February 19, 1475-May 24, 1543), a canon (church lawyer) from Poland, published *On the Revolutions of the Heavenly Orbs*, which contained a good deal of trigonometric theory. Copernicus seemed to have developed his trigonometric theory before he was acquainted with Regiomontanus's work, though the final published form of trigonometry in *On the Revolutions* was influenced by *On Triangles* to no small extent.

On the Revolutions is better known for starting the scientific revolution by proposing the modern, heliocentric theory. Copernicus realized that many computations of the positions of planets would be significantly easier if one assumed that the sun, not the Earth, was the center of the solar system; gradually he became convinced that this might really be true. Copernicus gave his student, Georg Joachim von Lauchen (February 15, 1514-December 4, 1576), permission to publish a preliminary version of Copernicus's work in 1540. Von Lauchen, like Regiomontanus, took a Latin named based on his birthplace, in Feldkirch, Austria. During the Roman Empire, Feldkirch was part of the province of Raetica, so von Lauchen took on the name Rheticus.

Contrary to popular belief, the Catholic Church was not initially opposed to the heliocentric theory. It was instead the Protestants, and Lutherans in particular, who were the first to prohibit the study and teaching of the new astronomy. In 1542 Rheticus was teaching at the University of Wittenberg, where the Reformation began. His heliocentric views made him unpopular, and he was forced to leave and accepted a position at the University of Leipzig.

Rheticus combined the work of Regiomontanus and Copernicus to produce the most comprehensive trigonometric work to date, the *Palatine Work on Triangles*, named in honor of its sponsor, Frederick IV, Count Palatine of the Rhine, and one of the seven electors of the empire. Rheticus based his trigonometric tables on the sides of a triangle, rather than on the chords of a circle, and was the first author to speak of the sine of an *angle* in our modern sense. For the sine and cosine values, he used a right triangle with a hypotenuse with a value of $10,000,000$ and angles differing by $10''$, giving him the equivalent of seven decimal place accuracy; for the remaining functions, he used a right triangle with a base of $10,000,000$. Before he died, he began to construct a table of the values of tangent and secant with a base of 10^{15}, which would have given him an astounding fifteen decimal place accuracy, but died before he finished. The work was completed by a student of his from Wittenberg, Valentin Otho (ca. 1550-1605) in 1596. By then, the study of triangles had its modern name, **trigonometry**, first used by Bartholomaeus Pitiscus (1561-1613), in a work of the same name in 1595.

11.1 Exercises

1. For right triangle ABC with sides 3, 4, 5, how does the perpendicular to the hypotenuse divide the hypotenuse?

2. Show that the Gregorian calendar is a better solution to the calendar problem than the Julian calendar.

11.2   The Rise of Algebra

Renaissance medicine was filled with astrological superstitions. Vestiges of this remain: influenza, from the belief that the planets "influenced" (caused) disease on Earth; lunacy from the belief the Moon causes madness; there was even a disease called "Planet", which is listed as being a cause of death in many Renaissance mortality tables. Because it was believed necessary to cast a patient's horoscope before a proper cure could be prescribed, most physicians were also competent *mathematicians*, in the sense of Diocletian and Augustine.

See page 199.

11.2.1   Nicholas Chuquet

Some physicians were also competent mathematicians in the modern sense: we note in passing that both Eudoxus and al Samaw'al were physicians and made important contributions to mathematics. A third mathematically inclined physician was Nicholas Chuquet (fl. 1484), who received his baccalaureate in medicine from the University at Lyons and lived in Paris; little else is known about his life. His *Science of Numbers in Three Parts*, known as the *Triparty*, appeared in 1484. Only a single handwritten copy of the *Triparty* is known to exist, but large parts of it were included (without attribution) in the more widely read *printed* arithmetic of Master Etienne des la Roche, published in 1520, so the work was not without influence.

The first part of the *Triparty* dealt with numbers. Chuquet introduced the Arabic numerals and operations upon them, and was responsible for originating some of our names for higher numbers, though not the actual values. **Million** was already in use, stemming from the "large thousand" (*millione*) used by Italian merchants. For a million million, Chuquet invented the word bi-million or billion; a million million million was a tri-million or trillion, and so on. This corresponds to the use of these terms in England and in Germany but not, oddly enough, in France, or in the United States.

Chuquet gave several means of solving algebraic problems; the rule of three appears, as do single and double false position. Chuquet's own original contribution is his **rule of mean numbers**. Chuquet explained the method's conceptual basis: $\frac{1}{2}$ is the "first and beginning" of the fractions; from it arise two sequences:

$$\frac{1}{2} \quad \frac{2}{3} \quad \frac{3}{4} \quad \frac{4}{5} \quad \cdots$$

$$\frac{1}{2} \quad \frac{1}{3} \quad \frac{1}{4} \quad \frac{1}{5} \quad \cdots$$

the first consisting of steadily increasing fractions, while the second consists of steadily decreasing fractions. For all other fractions:

Rule 11.2. *To find a number between two fractions, add numerator to numerator and denominator to denominator.*

With this, Chuquet solved the problem

Problem 11.1. *To find a number which, multiplied by itself and added to itself, makes* $30\frac{13}{81}$.

Solution. The solution is between 5 (which, multiplied by itself and added to itself, makes 30) and 6 (which makes 42). A number between 5 and 6 is $5\frac{1}{2}$. This second guess is too low, so we use the increasing progression. The next fraction is $\frac{2}{3}$, and we find $5\frac{2}{3}$ is still too small. The next fraction in the progression is $\frac{3}{4}$, but $5\frac{3}{4}$ is too small; $5\frac{4}{5}$ is too large, so the solution is between the two. Adding numerator to numerator and denominator to denominator, our next guess is $5\frac{3+4}{4+5} = 5\frac{7}{9}$, which turns out to be the solution. $\square$

If the next guess was not the solution, the process of interpolation could continue indefinitely.

11.2.2   Notation

Algebraic and arithmetic notation as we know it was invented during the Renaissance. For $5x^2$, Chuquet wrote 5^2, and wrote expressions involving negative exponents, such as $7x^{-2}$, as $7^{2\bar{m}}$. He even wrote 12^0 to indicate the number 12, which indicates an understanding of the rule $x^0 = 1$, and gave explicit directions for multiplying exponential terms. Roots were indicated with a stylized $\mathfrak{R}$, with the index written as a superscript. Finally, grouping was accomplished with an underline: thus, $\sqrt{14 - \sqrt{180}}$ was written by Chuquet as $\mathfrak{R}^2\underline{14\bar{m}\mathfrak{R}^2180}$.

Complete Business Calculation (1489) by Johann Widmann of Leipzig (b. ca. 1460) introduced the now-familiar $+$ and $-$ symbols; *The Coss* (1524) of Christoph Rudolff (1499-1545) used $\sqrt{}$ to indicate square roots. In England, Robert Recorde (1510-1558) wrote a number of works on introductory arithmetic and algebra; his *Ground of Artes* (1543) was probably the most popular. The *Whetstone of Witte* (1557) was more sophisticated, and marked the introduction

Symbol	Year	Author
$+, -$	1489	Widman
$\sqrt{}$	1525	Rudolff
$=\!=\!=$	1557	Recorde
$<, >$	1631	Thomas Harriot
$\times$	1631	William Oughtred
$\div$	1659	Johann Heinrich Rahn

Table 11.1: The dates and inventors of modern arithmetical symbols.

of the modern symbol for equality: two parallel lines, chosen because "no two things can be more equal." Recorde's original symbol, $=\!=\!=$, was soon shortened to our modern $=$ symbol. Other symbols and their inventors are shown in Table 11.1.

As algebra became less and less rhetorical, it was important to develop grouping symbols to distinguish between expressions such as "The square root of 12, plus the square root of 140" and "The square root of 12 plus the square root of 140." To distinguish between these two, authors like Rudolff used a $\cdot$ to indicate the grouping: hence $\sqrt{\cdot}\,12 + \sqrt{140}$ meant $\sqrt{12 + \sqrt{140}}$. We note in passing that classical Latin inscriptions, so close to the heart of the Renaissance, used the $\cdot$ to separate words, which may have inspired Rudolff and others to use the $\cdot$ to separate symbol groupings. The modern parentheses and bracket symbols began to be used during the fifteenth and sixteenth centuries. Remarkably, the rules for the **order of operations** were not agreed upon until the twentieth century: $24 \div 4 \times 6$ might be 36 or 1, depending on the author, and $3 \times 4 - 2$ might mean 6 or 10. See page 273.

11.2.3   Algorist versus Abacist

The competition between the **algorists**, who used Hindu-Arabic numerals and computation algorithms, and the **abacists**, who used Roman numerals and the abacus, was still going strong. Conservatism and resistance to change can account for part of the slow acceptance of Hindu-Arabic numerals, but a better reason is that Hindu-Arabic numerals are ill-suited for computation using the abacus, the only available method of mechanical computation; a competent abacist could win a competition against even the best algorist. Indeed, as late as the 1950s, Japanese abacists routinely beat competitors using adding machines. Indeed, the fifteenth century saw a resurgence in the use of Roman numerals; a **German form**, using lowercase instead of uppercase letters, began to be common. We see this form in the prefaces to books. See page 288.

In Figure 11.2, the man on the left is using the Hindu-Arabic numerals, while the man on the right is using a form of an abacus known as a **counting board**. The counting board was an essential tool of merchants: as a customer made his or her purchases, the merchant would record the costs by placing markers in the

Figure 11.2: A Race Between an Abacist and an Algorist, Overseen by the Spirit of Arithmetic. Boston Public Library/Rare books Department. Courtesy of the Trustees.

appropriate spots. When the customer was done, the total could be found easily, and the customer's goods would be brought "up to the counter."

11.2 Exercises

1. Prove Chuquet's Rule 11.2 always produces a fraction intermediate between

two given fractions.

2. Use Chuquet's "rule of mean numbers" to solve the following problems.

   (a) The square of a number and the number equal to $13\frac{37}{121}$.

   (b) The square of a number minus twice the number equal to $11\frac{37}{49}$.

3. Use Chuquet's "rule of mean numbers" to approximate a solution to, "The square of a number and the number equal to 1."

11.3 The Cossists

Italian translations of al Khwārizmī existed as early as 1464. The Italian word for unknown, "thing," is *cosa*: hence algebraists became known as **cossists**, and algebra became the **cossick** art. Thus, Rudolff named his work on algebra *The Coss* and the title of Recorde's *Whetstone of Witte* is a pun: the Latin word for whetstone is "cos." One of the earliest significant Italian algebras was the See page 308. *Summary of Arithmetic, Geometry, Proportion, and Proportionality* (1487) of Luca Pacioli (1445-1514), which provided an overview of various mathematical disciplines in a textbook format. The work also included detailed instructions for a revolutionary method of keeping business accounts: double entry bookkeeping.

Pacioli's notation was primarily syncopated: for the arithmetical operations of addition and subtraction, Pacioli used "p" and "m"; the unknown was "co," an abbreviation for *cosa*; "ce" was *census*, indicating the square of the unknown, and "ae" for *aequalis*, equal; thus $x^2 - 5x = 4$ would be "ce m 5 co ae 4." Pacioli, like al-Khayyami, believed the cubic to be unsolvable algebraically. An interesting feature of Pacioli's *Summary* was the presentation of the solutions to the quadratic equation in verse form.

Rafael Bombelli (January 1526-1572) also wrote an algebra, though it was not published until the year of his death, and even then, the last three books were not published until 1929, so they probably had little influence. While in Rome, Bombelli came across a copy of Diophantus, and with Antonio Maria Pazzi at the University of Rome, began a translation (which he did not live to complete). Bombelli, like Reigomontanus, claimed to have seen a complete copy of Diophantus's *Arithmetic* when in Rome, but this copy has never been found.

11.3.1 Tartaglia

The 1400s were the golden age in Italy; in fact, the Renaissance there is known as the *quattrocento*, the "four hundreds." The Medici, a family of wealthy bankers, made the city of Florence the epitome of Renaissance culture. Patronage reached its height under Lorenzo "The Magnificent." But not all were pleased with the growing obsession with material goods, and the monk Savonarola began to preach against it, warning that the wrath of God was coming to strike down the proud. In 1494, he convinced the Florentines to expel the Medici, and destroy

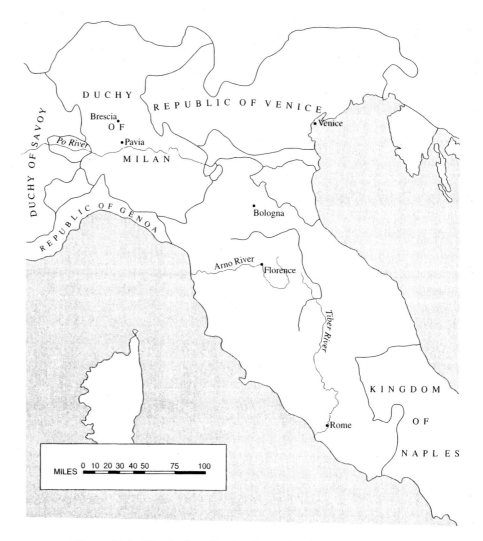

Figure 11.3: The Italian Peninsula in the Sixteenth Century.

their works of art in a great "bonfire of the vanities." Fortunately, not everyone participated.

The wrath of God—or rather, the greed of man—arrived in 1494, when Charles VIII of France brought an invading army to Italy. Charles had a flimsy claim to the throne of Naples, in southern Italy; en route, he captured Florence and Rome. An alliance, known as the Holy League, was hastily assembled to expel the French from Italy. The league's main members were Holy Roman Empire, Venice, and the Papal States; the battles between the Holy League and the French would sweep across the Italian peninsula for half a century, resulting in untold devastation and suffering.

In 1512, French armies under Gaston la Foix stormed Brescia and murdered many of its citizens. One survivor, who hid in a church but was badly wounded by the French, was a young boy named Niccolo Fontana. The wounds left him with a speech impediment, and he became known as "The Stammerer": Tartaglia (1499-December 13, 1557).

According to Tartaglia himself, he was taught to write letters as far as the letter "K," when his family fell upon hard times, and could no longer afford his education; Tartaglia was self-taught thereafter. In 1534, he moved to Venice where he lived the rest of his life except for a brief return home to Brescia, between 1548 and 1549. Tartaglia was quarrelsome, and occasionally presented the work of others, such as the translations of William of Moerbeke, as his own.

In the Renaissance, the most certain way to establish yourself as a man of learning was to challenge (and defeat) another man of learning in a public forum. These debates were a cross between mud-slinging political arguments and erudite discussions, and were as eagerly attended by the local citizens and magnates as sporting events are today. Thus, in 1535, Tartaglia was challenged by Antonio Maria Fiore, from Venice: each would pose thirty questions, to be answered by the other. The prize: thirty banquets, to be paid for by the loser.

Tartaglia asked a variety of questions, but Fiore posed variations of a single type: the so-called "cosa and cube equal to a number." In modern terms, these are equations of the form $x^3 + ax = b$, where a, b are positive numbers. Tartaglia wrestled with the problems, and on the night between February 12 and 13, inspiration struck, and he hit upon a method that allowed him to solve all thirty of Fiore's questions, while Fiore had been unable to solve most of Tartaglia's. Tartaglia declined the prize, saying that victory was reward enough.

Niccolo Tartaglia. Boston Public Library/Rare books Department. Courtesy of the Trustees.

11.3.2 Cardano and Ferrari

The method of solving cubic equations is now called **Cardano's method**, after Giralamo Cardano (September 24, 1501-September 21, 1576), born in Pavia. Whereas Tartaglia's early life had been one of hardship, Cardano's was one of privilege. His father, a friend of Leonardo da Vinci, encouraged young Cardano's study of the classics and mathematics; in 1520, Cardano entered the university at Pavia.

Pavia became the site of one of the most important battles in history. A series of shrewd dynastic marriages arranged by his grandfather, Maximilian I, left the Holy Roman Emperor Charles V in control of more territory than any monarch in modern European history. Through one set of grandparents, Ferdinand of Aragon and Isabella of Castile, Charles inherited Spain and her colonies in the New World. Through Maximilian I, Charles controlled most of Germany, Austria, and the Netherlands. He also inherited the Holy League's mandate to expel the French from Italy. Spanish and French forces clashed at Pavia on February 25, 1525, resulting in the capture of the French King, Francis I, who was taken away to Spain as a prisoner.

By the terms of the Treaty of Madrid (January 14, 1526) Francis surrendered all French claims to Italy, though as soon as he made his way back to France, he renounced the treaty, saying that it had been obtained under duress and therefore was not binding. To us, this seems reasonable, but to Francis's contemporaries, it was reprehensible behavior. By then, it became clear that the Italians had merely masters, and the League of Cognac was formed to expel Charles V. But the League was no match for the Spanish and German mercenaries under Charles's command, and in 1527, Rome was captured and sacked; the Pope found himself a virtual prisoner, under Charles's "protection." One important consequence was that when Henry VIII of England petitioned the Pope for an annulment of his marriage to Catharine of Aragon, the Pope refused: Catharine was Charles V's aunt. As a direct result, England broke away from the Catholic Church.

Cardano missed the battle of Pavia, for he was at the University of Padua and in 1526, received his doctorate in medicine. Around 1539, Cardano met Zuanne de Tonini da Coi who, like Tartaglia, was from Brescia. Da Coi informed Cardano of the victory of his fellow townsman Tartaglia. Cardano was quite interested, for he had believed Luca Pacioli's statement of the impossibility of solving the cubic equation; however, da Coi knew no specifics. Thus, Cardano visited Tartaglia. At first, Tartaglia refused to talk: by preserving the secrecy of his method of solution, he could emerge victorious from any debate, and establish his reputation as the foremost mathematician in all of Europe. But Cardano flattered Tartaglia, and eventually Tartaglia gave in. After making Cardano swear an oath never to reveal the secret, Tartaglia explained his method.

There the matter may have rested, but for Ludovico Ferrari (February 2, 1522-October 1565). Ferrari entered Cardano's household as a valet (a gentleman-in-training) but Cardano recognized the brilliance of the young man and proceeded to teach him Latin, Greek, and mathematics. Soon, Ferrari became Cardano's secretary — one entrusted with his master's "secrets." In 1540, Ferrari gained a position as lecturer in mathematics at the University of Milan. Shortly thereafter, da Coi posed the following problem to Cardano: to divide 10 into three parts in continued proportion so the first, times the second, is 6. Cardano replied that he could solve the problem, though this was (at the time) an empty boast: the problem results in a fourth degree equation, which no known technique could solve. In the terminology of the times, the equation that resulted was

Problem 11.2 (Da Coi's Problem). *Sixty things equal to one square-square plus 6 squares plus 36.*

Cardano gave the problem to Ferrari, who eventually solved it. Cardano wanted to include the solution in his upcoming work on algebra, *The Great Art*, but found himself in a quandry: Ferrari's solution required the ability to solve a cubic. If Cardano published Ferrari's solution, he would also have to publish Tartaglia's work, yet he had sworn that he would not reveal the solution to the cubic equation. What could be done?

Cardano recalled the debate between Fiore and Tartaglia. Fiore must have known how to solve the cubic, for if he did not, he would not risk posing to

Giralamo Cardano. Boston Public Library/Rare books Department. Courtesy of the Trustees.

See Problem 7.

Cardano	Modern	Text
qd.qd.	x^4	sq.sq.
cub.	x^3	cube
quad.	x^2	sq.
pos.	x	thing
quant.	y	quant.
p:	$+$	plus
m:	$-$	minus

Table 11.2: Cardano's Abbreviations.

Tartaglia questions that he himself could not solve. Fiore was the student of Scipione (del) Ferro (February 6, 1465-ca. November 7, 1526). Ferro was dead, but his nephew, Annibale dalla Nave, had his papers and his position at the University of Bologna. Around 1542, Cardano and Ferrari went to Bologna, where dalla Nave showed them his uncle's unpublished manuscripts.

There, Cardano claimed, he found evidence that Ferro knew how to solve cubic equations before Tartaglia. Though none of Ferro's writings have survived to the present day, it is believed that he was able to solve at least two types of cubic equations: both $x^3 + px = q$ and $x^3 = px + q$, where p, q are positive constants. Because of Ferro's prior discovery, Cardano felt that the oath to Tartaglia was no longer binding, and thus he published the solution to the cubic in *The Great Art* (1545). Cardano gave Tartaglia credit for having independently discovered the solution to some types of cubic equations.

Tartaglia was furious: by the standards of the sixteenth century, no mitigating circumstances allowed for the violation of an oath. Tartaglia challenged Cardano to a public debate. Cardano declined, but Ferrari took up the challenge; on August 10, 1548, the two met in the Church of Santa Maria del Giardino dei Minori Osservanti in Milan, with town citizens and various dignitaries, including the Mayor of Milan, attending. We do not know exactly what happened, but afterwards, Tartaglia returned to Brescia, and Ferrari was offered a number of positions, including the job of tutoring Charles V's son Philip in mathematics. Ferrari chose to stay in Italy, and took a position with Ercole Gonzaga, the Cardinal of Milan. Hence, we may infer that Tartaglia lost the debate.

Solution to the Cubic

Cardano's notation is primarily syncopated, with quantities being expressed verbally with abbreviations for some of the more common ones. We will emulate this practice (see Table 11.2). To solve the cubic equation, Cardano began with the three "thing and cube" equations; that is to say, equations involving only the cube of the unknown and its first power. These three are:

1. The cube and things equal the number. [$x^3 + px = q$]

2. The cube equal to the things and number. [$x^3 = px + q$]

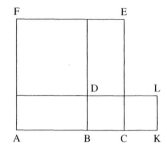

See Problem 1.

See Problem 2.

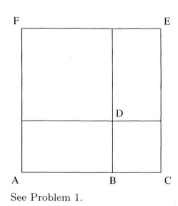

See Problem 1.

3. The cube and number equal to things. $[x^3 + q = px]$

To solve the case of the cube and things equal to a number, such as "a cube GH and six sides of GH equals 20", Cardano used the following procedure:

Solution. Take two cubes with bases AE, CL, and suppose the cubes differ by 20, and the product of side AC with side CK is 2, one-third the number of things. Let $CB = CK$. Then $AB = GH$, the side of the cube, and the solution is AB. □

Hence, Cardano's solution is to let $u^3 - v^3 = 20$, the constant term, and $uv = 2$, one-third the coefficient of the first power term; the solution is then $x = u - v$. First, Cardano proved that AB would be the solution to the cubic equation. To find the actual solution:

Rule 11.3 (Cardano's Rule for $x^3 + px = q$). *Cube one third the number of things. To this add the square of half the number, and take the square root of the whole. Work with this twice: first add half the number, and second subtract half the number. This will give you a binomial and an apotome, respectively. Find the cube root of the binomial, and subtract from it the cube root of the apotome, and the result is the value of the thing.*

Example 11.2. *A cube and* 6 *things make* 20. *One third the number of things is* 2; *cube this, making* 8. *Add this to the square of half the number, i.e., the square of* 10, *making* 108. *Take the square root, giving* $\sqrt{108}$. *Add and subtract half the number, giving the binomial* $\sqrt{108} + 10$ *and apotome* $\sqrt{108} - 10$. *The difference between the cube roots of the binomial and apotome is the thing. Thus, thing is* $\sqrt[3]{\sqrt{108} + 10} - \sqrt[3]{\sqrt{108} - 10}$.

This may have been the *only* type of cubic equation whose solution was known to Tartaglia; if so, then Cardano and Ferrari greatly advanced mathematics by finding the solutions for all types of cubics, with the exception of $x^3 + bx^2 + cx + d = 0$, where b, c, d are all positive.

Though Cardano discussed the use of negative numbers, for him all equations were expressed in terms of positive coefficients. Thus, the case of "cube equal to things and number" was distinct from the "cube and things equal to number," and required a different solution.

Solution. Let the base CF of a cube be divided into squares DC and DF, so the sum of the cubes on bases DC and DF equals the constant, and the product of the sides AB and BC equals one third the number of things. Then AC is the value of the thing. □

Again, Cardano described the solution in a purely rhetorical form.

Rule 11.4 (Cardano's Rule for $x^3 = px + q$). *Take the cube of one third the number of things, subtract it from square of half the number, then take the square root. Add it to half the number, producing a binomial; subtract it from half the number, producing an apotome. The sum of the cube roots of the binomial and the apotome will be the thing.*

Example 11.3. *The cube is equal to six things and the number 40. One third the number of things is 2; cubed is 8. Subtract 8 from half the number squared, or 400, making 392; take its square root: $\sqrt{392}$. Add it to and subtract it from half the number, 20, to make the binomial $20 + \sqrt{392}$ and the apotome $20 - \sqrt{392}$. The thing is thus $\sqrt[3]{20 + \sqrt{392}} + \sqrt[3]{20 - \sqrt{392}}$.*

The last equation is that of the cube and number equal to things. Here Cardano made the discovery that the solutions to cube equal to things and number were related to the solutions to cube and number equal things, namely that the two positive solutions to the cube and number equals things, when added together, formed the solution to the cube equal to the same number of things and the same number.

Of course, this would be useful if Cardano already had the solutions for a cube and number equal to things, for he would then be able to find the solutions to the cube equal to things and number immediately. Unfortunately, he did not, and the problem remained of how to separate the known solution to the cube equal to things and number, into the two solutions of the cube and number equal to things.

> Suppose r_1 and r_2 are two real solutions to $x^3 + q = px$. Then $r_1 + r_2$ is a solution to $x^3 = px + q$.

Rule 11.5 (Cardano's Rule for $\mathbf{x^3 + q = px}$). *First, find the solution to the cube equal to the same number of things and the number. Take three times the square of half the solution, and subtract it from the number of things; then take its square root.*

[Given the equation $x^3 + q = px$, first solve $y^3 = py + q$. Then find $\sqrt{p - 3(\frac{y}{2})^2}$. The solutions to the original equation are $x = \frac{y}{2} \pm \sqrt{p - 3(\frac{y}{2})^2}$.]

Cardano's example is interesting, for the solution he finds to the original equation is not the one that would be found applying any of the rules for solving the cubic equation.

Example 11.4. *If the cube and the number 3 are equal to 8 things, find the solution to the cube equal to 8 things and the number 3. The solution to this is 3. Three times the square of half of this is $\frac{27}{4}$; subtracting this from the number of things, 8, leaves $\frac{5}{4}$. Add or subtract the square root of this to half the solution of the cube equal to things and number, which is $\frac{3}{2}$. Thus, the solutions are $\frac{3}{2} + \sqrt{\frac{5}{4}}$ and $\frac{3}{2} - \sqrt{\frac{5}{4}}$.*

See Problem 4.

Cardano's solutions to cubics involving quadratic terms relied on transforming them into cosa and cube equations. To solve the equation, "A cube is equal to six squares, plus one hundred numbers" (in our terms, $x^3 = 6x^2 + 100$), Cardano, supposing AC is the solution to the cubic, lets AB be a new unknown, with $BC = 2$ (hence $x = y + 2$). Thus, the equation is transformed into AB^3 equal to 12 sides AB and 116.

Biquadratic Equation

At least part of Cardano's rationale for publishing Tartaglia's solution and break-ing his oath was to include Ferrari's solution to the biquadratic equation. Fer-rari's solution, reprinted by Cardano, proceeded in a very geometric fashion, despite the square-square being a very non-geometric entity. Remember the Tonini problem resulted in the equation 60 things equal to 1 square-square plus 6 squares plus 36 numbers.

The conceptual basis of Ferrari's method is to add a quantity that will make both sides of the equation perfect squares. In general, given $x^4 + bx^2 + c = dx + f$, complete the square on the left hand side, so that it is of the form $(x^2 + p)^2$. Then add $gx^2 + h$ to both sides, so that the left hand side is a square of the form $(x^2 + p + q)^2$, and the right hand side will be a square of the form $(rx + s)^2$. Taking the square root of both sides gives you $x^2 + p + q = rx + s$, which can be solved using standard techniques. The problem reduces to determining g and h so that both sides remain perfect squares.

Ferrari's solution to the Tonini Problem is thus:

Solution. A sq.sq. and 6 sq. and 36 are equal to 60 things. Adding 6 sq. to each, we have 1 sq.sq. plus 12 sq. plus 36 equal to 6 sq. plus 60 things, which does not have a root.

[From

$$x^4 + 6x^2 + 36 = 60x$$

complete the square on the left hand side, making

$$\left(x^2 + 6\right)^2 = 6x^2 + 60x$$

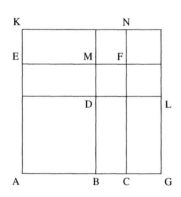

Since the right hand side is not a perfect square, we must add some number to make both sides perfect squares.]

Let there be quant. squares added, so both sides are perfect squares. As you can see from the figure, adding 2 quant. squares plus 12 quant. plus the square of quant. will result in a larger square. Thus, the total number of squares will be 6 plus 2 quant.; the total number of things will be 60, and the number will be 12 quant. plus the square of quant.

[Now add y inside the parentheses on the left, to obtain

$$\left(x^2 + 6 + y\right)^2 = (2y + 6)x^2 + 60x + (12y + y^2)]$$

Now a trinomial will be a perfect square if the product of the first and third terms is equal to the square of half the second term. Thus, we must have the square of half the second, 900, be equal to the product of the first and third, which will be 2 cubes plus 30 squares plus 72 quant.

[A trinomial $ax^2 + bx + c$ is a perfect square if $acx^2 = \left(\frac{bx}{2}\right)^2$. Ferrari's example actually used the coefficients of the terms, and not the terms themselves; thus we need $(2y + 6)(12y + y^2) = \left(\frac{60}{2}\right)^2$.]   □

Since this is a cubic equation, it can be solved using the previously outlined methods; the solution will give the value of y. Then, once y is found, the value of the original unknown may be determined.

Imaginary Numbers

Cardano was forced to deal with the square roots of negative numbers, which he called **impossible roots**, for they often appeared in the solutions to the cubic. Cardano's method of handling them is an early example of **formalism**, the viewpoint that all algebraic operations may be performed on any symbolic quantity.

Cardano first encountered impossible roots in the problem: Divide 10 into two parts whose product is 40. Using the procedures in *The Great Art*, the two numbers are $5 + \sqrt{-15}$ and $5 - \sqrt{-15}$. Cardano then proved that the solution was correct: "putting aside" the "mental tortures" involved, he multiplied $5 + \sqrt{-15}$ by $5 - \sqrt{-15}$ to obtain 40, as desired; their sum is obviously 10.

> For example, a formalist "proof" is that since $\frac{a}{b} \cdot b = a$, then $\frac{a}{0} \cdot 0 = a$. Formalism is generally frowned upon, though it has its value, particularly in combinatorics.

Numerical Analysis

The *practical* value of exact solutions to the cubic and biquadratic equation was nonexistent: methods of finding approximate solutions to equations of any degree existed, and Cardano included two of them in *The Great Art*. The first method is now called the **method of secants**, and is based on the rule of double false position. To solve the problem $x^4 + 3x^3 = 100$, Cardano noted that if the unknown was 2, the left hand side would be 40, while if the unknown as 3, the left hand side would be 162. Hence an increase of 1 in the guess resulted in an increase of 122 in the final result. Since an increase of $\frac{60}{122}$ of 122 was desired, one would add $\frac{60}{122}$ of 1 to the initial guess 2, and obtain the new solution, $2\frac{30}{61}$.

Repeating this procedure using $x \approx 2\frac{30}{61}$, then $x^4 + 3x^3$ is approximately 85. Since $x = 3$ gave $x^4 + 3x^3 = 162$ and $x = 2\frac{30}{61}$ gave $x^4 + 3x^3 = 85$, then an increase of $\frac{31}{61}$ in x yielded an increase of 77; to make 162 into 100, the desired value, an decrease of 62 is needed, and thus the approximate solution 3 should be decreased by $\frac{62}{77}$ of $\frac{31}{61}$, or $\frac{1922}{4697}$. This is the amount that the approximate solution 3 should be decreased by, so the second approximate solution is $2\frac{2775}{4697}$. Cardano noted if the approximate solutions was not accurate enough, the procedure could be repeated a third time, which would surely result in an "insensible difference" between the approximate and the actual solution.

Cardano also provided an alternative that involved varying the coefficients of the equation. For $x^2 + 20 = 10x$, Cardano let $x = 7$, which solved the equation $x^2 + 20 = 9\frac{6}{7}x$; $x = 8$ was, similarly, the solution to the equation $x^2 + 20 = 10\frac{1}{2}x$. In this case, an increase of 1 in the assumed value led to a change of $\frac{9}{14}$ in the coefficient of x, whereas an increase of $\frac{1}{7}$, which is $\frac{2}{9}$ of $\frac{9}{14}$, is actual needed; thus, the approximate solution 1 should be increased by $\frac{2}{9}$ of 1, the change in the assumed value of x.

Example 11.5. *Solve $x^2 + 5 = 5x$. We note $x = 3$ is the solution to $x^2 + 5 = \frac{14}{3}x$, and $x = 4$ is the solution to the equation $x^2 + 5 = \frac{21}{4}x$. Thus, an increase of 1 in the value of x led to an increase of $\frac{7}{12}$ in the coefficient of x; an increase of $\frac{1}{3}$ was necessary, which is $\frac{4}{7}$ of $\frac{7}{12}$. Thus, the second approximation should be 3, the original guess, increased by $\frac{4}{7}$ of 1, the difference between the guesses: $3 + \frac{4}{7}$.*

A key innovation was the treatment of the coefficients of an equation as just another quantity that could be varied.

Curiously, misfortune seemed to follow those who worked on the cubic and biquadratic equations. In 1560, Cardano's elder son was executed for having poisoned his wife, and in 1570, Cardano himself was imprisoned by the Inquisition for suggesting that even Christ was subject to astrological influences. Ferrari died in 1565, poisoned, according to one account, by his sister or her lover. Tartaglia had hoped to use his victory at the 1548 contest to obtain a well-paying position, but instead, he returned to Venice, where he spent the rest of his life in relative obscurity.

11.3 Exercises

1. Prove Cardano's geometric solutions.

    (a) Prove that if $x^3 + px = q$, $u^3 - v^3 = q$ and $uv = \frac{p}{3}$, then $x = u - v$ is a solution.

    (b) Prove that if $x^3 = px + q$, $u^3 + v^3 = q$, and $uv = \frac{p}{3}$, then $x = u + v$ is a solution.

2. What does the inadmissibility of a cubic of the form $x^3 + bx^2 + cx + d = 0$, where b, c, d are all positive, tell you about Cardano's algebra?

3. Cardano's method for solving $x^3 + px = q$ involved finding two quantities, u and v, such that $u^3 - v^3 = q$ and $uv = \frac{p}{3}$.

    (a) Solve this system, and explain how its solution corresponds to Cardano's Rule.

    (b) Cardano's solution is reminiscent of the following problem: the difference of two numbers is 20, and their product is 8. Solve this problem using a Diophantine procedure. (Hint: let one of the numbers be thing minus 10.) How does your solution compare to Cardano's Rule?

    (c) Show that if r_1, r_2 are two solutions to the cubic equation $x^3 + q = px$, where p and q are positive constants, then $r_1 + r_2$ is a solution to $x^3 = px + q$.

4. Cardano began solving $x^3 + 3 = 8x$ by "solving" the equation $x^3 = 8x + 3$. Solve this equation, using Cardano's method for the "cube equal to things and number." Explain what difficulties would be involved.

5. Show how the equation $x^3 = 6x^2 + 100$ is transformed into the equation $y^3 = 12y + 116$, using the substitution $x = y + 2$.

6. Find the appropriate substitutions to eliminate the square term in the following equations. Then solve the equation.

   (a) A cube and 6 squares equal 100.

   (b) A cube and 64 equals 18 squares.

   (c) A cube, 6 squares, 20 things equals 100.

   (d) $r^3 - 63r = 162$.

   (e) Explain how you would, in general, eliminate the square term from a cubic equation. Can you generalize this procedure? (In other words, how can you eliminate the cubic term from a biquadratic, the fourth degree term from a quintic, etc.)

7. In the problem posed by Tonini, let x be the middle number of the proportional.

   (a) Show the Tonini problem is equivalent to $60x = x^4 + 6x^2 + 36$.

   (b) Solve the problem.

8. Apply Cardano's method to solving the following equations.

   (a) $x^3 + 6x = 100$. $x = 4$ and $x = 5$ are approximate solutions.

   (b) $x^3 = 6x + 20$. Use $x = 3$, $x = 4$ as approximate solutions, and vary the coefficient of x^3.

   (c) $x^4 + 6x^2 + 36 = 60x$. $x = 1$ and $x = 2$ are approximate solutions.

9. Explain what is "formal" about Cardano's solution $5 + \sqrt{-15}$, $5 - \sqrt{-15}$, to the problem of finding two numbers that add to 10 and multiply to 40.

10. (Teaching Activity) Write a poem explaining how to solve the three main types of cubic equations.

11.4   Simon Stevin

In 1556, worn out by years of campaigning to unite Christian Europe in a crusade against the Muslim Turks, Charles V retired. Spain and the Netherlands (which included Belgium and Luxembourg) went to his son, Philip II of Spain. It was a grave mistake. The Dutch tolerated Charles V, who grew up in Flanders and even announced his abdication in Brussels. However, the Protestant Dutch detested the Catholic Philip. The treatment of the Dutch by the Spanish governors did little to foster goodwill, and in 1568, the Dutch revolted.

In 1578, Alessandro Farnese, the Duke of Parma. became governor of the Netherlands. Farnese was a brilliant military commander, *and* a shrewd statesman. Of the thirteen provinces in revolt, the southern six were inhabited mostly by Catholic, French-speakers known as Walloons, who might be reconciled to a Catholic King. By promising to restore the old political freedoms (which were

Simon Stevin.   ©Bettmann/
CORBIS.

See page 236.

not very significant in the first place), Farnese quelled the revolt in the south. The six provinces form the core of modern Belgium.

The seven northern provinces were more intractable. In 1579, they concluded the Union of Utretcht, and announced their independence from Spain, under the leadership of William of Orange. William was assassinated a few years later, and leadership fell upon the shoulders of his son, Maurice of Nassau.

Maurice of Nassau was also a serious student of the sciences. His tutor was Simon Stevin (1548-1620), about whom very little is known. Stevin was born around 1548, purchased a house in Den Haag in 1612, had children in 1612, 1613, and 1615 (and a fourth child somewhat later), and was married in 1616. He died around 1620. Stevin wrote several texts that Maurice carried with him while on campaign. Perhaps worried that the manuscripts would be lost or destroyed, Stevin had them published; one unverified story is that Maurice paid for the publication of the works.

11.4.1   Decimal Fractions

Stevin, who almost always signed his works "Simon Stevin of Bruges" (a town in the Walloon part of the Netherlands, now in Belgium), was primarily responsible for popularizing decimal fractions through his work *The Tenths*. The writing of whole numbers in Hindu-Arabic numerals was by then well established, but the Arabic use of decimal fractions was not yet known in Europe. To indicate the place, Stevin used circled numbers. Thus, 3.141 would be 3 ⓪ 1 ① 4 ② 1 ③. Arithmetical operations were handled just as we handle them today, with the added feature of indicating the explicit place value using the ◯s.

Stevin was fond of the ◯. Not only did he use it to indicate higher roots, for example writing the cube root as $\sqrt{③}$, but he also used the same symbols in algebra to represent the powers of the unknown. Thus, in his *Arithmetic*, he wrote "the square [of a number] 1 ① added to -12 makes 1 ② -12."

11.4.2   Number

Another area in which Stevin was ahead of his time was in his concept of number. In his *Arithmetic*, Stevin defined number to be that which measures the quantity of any thing. The Greeks were divided on whether 1 was a number; Stevin boldly declared that it was, because it represented the quantity of one thing. Moreover, he declared that the distinction between "irrational" and "rational" was entirely artificial and meaningless. Stevin's views were advanced for his time, and over a century later, mathematicians would still debate over whether 1 should be considered a number.

11.4.3   Infinitesimal Analysis

Stevin's system of decimal notation lasted only a few years before it was replaced with our modern one, and his system of writing algebraic expressions seems not to have been used by anyone outside of the Netherlands. It was in the field

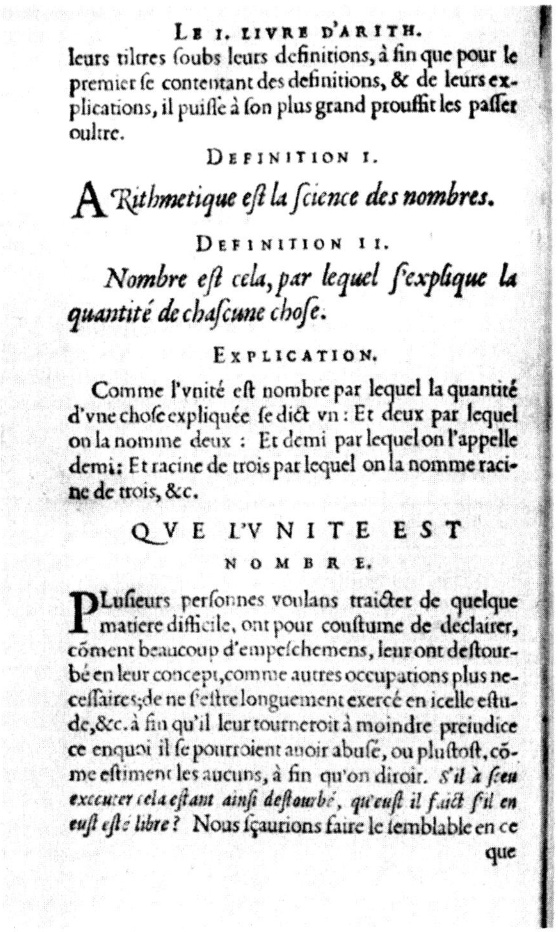

LE I. LIVRE D'ARITH.

leurs tiltres soubs leurs definitions, à fin que pour le premier se contentant des definitions, & de leurs explications, il puisse à son plus grand prouffit les passer oultre.

DEFINITION I.

ARithmetique est la science des nombres.

DEFINITION II.

Nombre est cela, par lequel s'explique la quantité de chascune chose.

EXPLICATION.

Comme l'vnité est nombre par lequel la quantité d'vne chose expliquée se dict vn: Et deux par lequel on la nomme deux: Et demi par lequel on l'appelle demi: Et racine de trois par lequel on la nomme racine de trois, &c.

QVE L'VNITE EST NOMBRE.

PLusieurs personnes voulans traicter de quelque matiere difficile, ont pour coustume de declairer, cõment beaucoup d'empeschemens, leur ont destourbé en leur concept, comme autres occupations plus necessaires; de ne s'estre longuement exercé en icelle estude, &c. à fin qu'il leur tourneroit à moindre preiudice ce enquoi il se pourroient auoir abusé, ou plustost, cõme estiment les aucuns, à fin qu'on diroit. *S'il à sceu executer cela estant ainsi destourbé, qu'eust il faict s'il en eust esté libre?* Nous sçaurions faire le semblable en ce que

Figure 11.4: The opening of Stevin's *Arithmetic*, where Stevin boldly proclaims, "THAT ONE IS A NUMBER" [QUE L'UNITE EST NOMBRE]. Boston Public Library/Rare books Department. Courtesy of the Trustees.

of determining centers of gravity that Stevin was to make his mark. His work relied on infinitesimal methods, and, though a step down in rigor from the work of Archimedes, was a step upwards in generality.

In 1586, Stevin published *The Elements of the Art of Weighing.* Stevin took the step, unusual in that time, of publishing it in Dutch, rather than Latin, the language normally used for publication of a "serious" work. It would be almost two hundred years before Latin was replaced by the vernacular on a regular basis.

Using many of the same definitions given by Archimedes in *On the Equilibrium of Planes*, Stevin gave a formal definition of the center of gravity of an object:

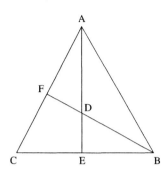

Definition. *The **center of gravity** of a solid is the point such that if the solid is suspended from that point, it will be at rest.*

None of Stevin's results are truly groundbreaking, and most were known to Archimedes and proven in *On the Equilibrium of Planes I* and *II* using proper Euclidean methods. What distinguished Stevin's work, though, was the free and easy transition to the infinite, in a manner that would become a critical part of integral calculus. A key proposition occurred in Book II.

Proposition 11.2. *The geometric center of a plane figure is also its center of gravity.*

Stevin demonstrated the case for a triangle and for a parallelogram.

Demonstration. Given equilateral triangle *ABC*, with geometrical center *D*. Draw *AE* to the midpoint *E* of *BC*; and *BF* to midpoint *F* of *AC*. If the triangle is suspended from its center line *AE*, it will balance, for *AEB*, *AEC* are equally large, similar, and of the same form (i.e., they are situated in the same way). Likewise, if the triangle is suspended with center line *BF*, it will also balance. Hence, the center of gravity must be on the intersection of *AE* and *BF* at *D*, which is also the geometrical center. Likewise for a parallelogram. □

The extension to a general triangle is

Proposition 11.3. *The center of gravity of a triangle is on the line drawn from an angle to the middle point of the opposite side.*

The proof used a concept that would be crucial in the developing calculus: the infinitesimal rectangle. Though the proof used the non-mathematical concept of a weight, this is not critical to its structure, and thus it may be called an actual proof of the preceding proposition.

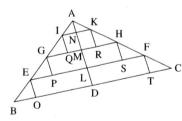

Proof. Given triangle *ABC*, we seek to prove the center of gravity is on line *AD*, drawn to the midpoint *D* of *BC*. Draw lines *IK*, *GH*, *EF*, parallel to *BC*; draw *IQ*, *KR*, *GP*, *HS*, *EO*, *FT*, parallel to *AD*. Then parallelograms *IKRQ*, *HSPG*, *FTOE* have their centers of gravity on *AD* [and will have equal weights on either side of the line]. Hence the center of gravity of figure *IKRHSFTOEPGQ* will also be on line *AD* [and the weight of *NKRHSFTD* will equal the weight of *NIQGPEOD*].

Now we may inscribe an infinite number of quadrilaterals in *ABC*, and the center of gravity of the inscribed figure will, for the same reason, be on *AD*. For if we draw lines parallel to *BC* through the middle points of *AN*, *NM*, *ML*, *LD*, the difference between the new figure constructed and the triangle will be half the difference between the original figure and the triangle. Thus, we may inscribe within the triangle a figure such that the difference between the figure and the triangle is less than any given figure. Thus, the apparent weight of the part *ADC* will differ less than the apparent weight of the part *ADB* by any plane figure that may be given.

This is because:

1. Between any two different weights, there may be found a weight less than their difference.

2. Between the weights ADB, ADC, there may not be found a weight less than their difference.

Hence, the weights ADB and ADC are equal. □

Since the center of gravity is on the line drawn from the vertex to the midpoint of the opposite side, called the **median**, this means that the center of gravity of a triangle will be at the intersection of the two medians. Stevin went on to show how the centers of gravity of other figures could be determined.

For a prism, Stevin proved

Proposition 11.4. *The center of gravity of a prism is at the midpoint of its axis.*

The proof is a straightforward extension of the results used to find the center of gravity of a triangle.

11.4 Exercises

1. Prove Proposition 11.2 for a parallelogram.

2. Prove *The Art of Weighing*, Book II, Proposition X: The center of gravity of any parabolic segment is on its diameter. Note: Stevin is using the term "diameter" in the sense of Archimedes, and not in the modern sense.

3. Prove *The Art of Weighing*, Book II, Proposition XVI: The center of gravity of any pyramid is on its axis.

11.5 François Viète

The life of François Viète (1540-December 16, 1603) spanned one of the most turbulent periods in French history. The struggle between France and the rest of Europe finally ended in 1559, with the Peace of Cateau-Cambresis. By then, France had serious internal difficulties. French Protestants, known as **Huguenots** (a name of uncertain origin) found themselves at war with French Catholics. Prominent among the Catholic Party were Henry, the Duke of Guise, and Catherine de Medici, the mother of the French king Charles IX. In 1572, a wedding was arranged between Charles's sister, Margaret, and Henry of Navarre, a Protestant stronghold in southern France. Though Henry of Navarre was a Protestant, the marriage was acceptable because he was a descendant of Louis IX (Saint Louis), and in line for the throne.

Protestants from all over France would be in Paris to celebrate the wedding and this seeming victory of the Protestant forces. On August 23, 1572, the night before the wedding, radical anti-Huguenots struck, murdering over two thousand Protestants in what would become known as the St. Bartholomew's

François Viète.   ©Bettmann/
CORBIS.

Day massacre. Henry of Navarre was spared only by an immediate conversion to Catholicism. Though her exact role in the planning of the massacre is uncertain, Catherine de Medici shed no tears over the deaths of thousands. But violence breeds violence, and for the next generation, France would be torn apart by religious warfare.

Charles, never in good health, died in 1574, and his brother became King Henry III. Henry summoned Viète to Paris. Viète, a lawyer, had made a name for himself by dispensing good advice to local Parliaments around France, and Henry may have felt the need of someone who could find a way to tread the dangerous path between the Catholic Party and the Huguenots. Soon, Viète became one of the king's most trusted advisers, being sent on important and confidential missions.

Unfortunately, all Viète's good advice could not save Henry. Because the French Parliament (known as the Estates General) would not pay for a war of extermination against the Huguenots, Henry was unable to eliminate French Protestantism. The Catholic Party took this as a sign that Henry was *unwilling* to annihilate the Huguenots, so the Pope and Philip II of Spain organized a new Holy League, which aimed to place Henry of Guise on the throne. In 1584, to appease the Catholic Party, Henry revoked all the concessions he had made to the Hugenots, leading to the War of the Three Henrys: Henry III, the king; Henry of Guise; and Henry of Navarre, who had recently renounced his conversion to Catholicism.

Guise entered Paris, forcing Henry III to flee; in 1588, Henry III had Guise murdered. Naturally, this placed Henry's own life in great danger, and, oddly enough, the one place he could find support and safety was with Henry of Navarre. Unfortunately, nothing could protect the king from fanatics, and in 1589, a monk, Jacques Clement, murdered King Henry. Ironically, this gave the throne to the Protestant Henry of Navarre, though the Catholic Party refused to acknowledge Henry's accession.

Henry of Navarre recognized Viète's value, and offered Viète the same position he had had with King Henry III; Viète accepted. Some of Viète's earliest work was in **cryptanalysis**, the making and breaking of "secret messages," which has since become one of the more practical pursuits of mathematics. While still in the employ of Henry III, Viète decoded messages to Alessandro Farnese, then commander of the forces of the Holy League.

In 1589, some messages from Philip of Spain fell into Henry of Navarre's hands; he turned them over to Viète to see what he could do. One of the messages, to the Spanish Ambassador Moreo, contained important details, including a desperate plea for 6000 men, from the Spanish Netherlands (modern Belgium), to help the Holy League's cause in France, lest it suffer a disastrous defeat. This information would have been of vital strategic importance — except by the time Viète gave the decoded message to Henry, he had already dealt the League a serious defeat.

Viète continued to read the dispatches of the Spanish, and made no secret of it, being confident that however they changed their codes, he could always

Modern	Variable	Homogeneous Quantity
x	A side	B side
x^2	A square	B plane
x^3	A cube	B solid
x^4	A square-square	B plano-plane
x^5	A square-cube	B plano-cube
x^6	A cube-cube	B cubo-cube

Table 11.3: Viète's Notation.

decipher the messages. Philip, having discovered Viète could read messages he thought inviolable, complained to the Pope, claiming that the only way Viète could do so was to use black magic. The Pope, whose own cryptanalysts had been reading Spanish communications for years, ignored the complaint.

The war dragged on, and Henry of Navarre was shrewd enough to realize that the Catholic Party would cause him no end of trouble. Thus, Henry converted back to Catholicism, reputedly saying, "Paris is worth a mass." But to retain the support of the Protestants, Henry, now King Henry IV, announced the Edict of Nantes on April 15, 1598, which granted the Huguenots the right to practice their religion without persecution. Unlike the previous promises, this one was kept—for 87 years.

11.5.1 *Introduction to the Analytic Art*

Viète started a revolution in mathematical notation. The Greeks had developed geometry as the all encompassing mathematical science. In part, this was because it could provide more generality: a reference to a triangle ABC was perfectly general, and could refer to any triangle. Viète's contribution was to generalize algebra in the same way, by letting letters represent any quantity.

Viète allowed letters to substitute for the unknowns: the vowels A, E, I, O, U, and Y. His notation still contained elements of geometry: the unknowns were generally viewed as line segments; the product of two lines was a plane; the product of a plane and a line was a solid; the product of two planes was a plano-plane, and so on (Table 11.3). In some sense, Viète's notation was a throwback, for he wrote out the words, not even relying on the abbreviations and symbols used by others. Even arithmetic operations like "plus" and "times" were spelled out, though later Viète used Widmann's + symbol. Viète often dropped the "plane" as being understood, just as we do not ordinarily write the exponent of 1 in x^1.

None of this was particularly new; the advantage of A, E over Cardano's *cosa* and *res* is minimal. Viète's A cube has a slight advantage over Cardano's *cubus*, since you can distinguish more easily the cubes of two different variables, such as A cube from E cube, but it was hardly revolutionary.

What was revolutionary was Viète's next step: he allowed even the constants to be represented by letters, using the consonants B, G, D, and so on (note the

"Greek" ordering of the letters). Viète was the first mathematician to whom the idea of a "general equation," such as $ax + by = c$, would make sense.

Of course, revolutions are almost never the work of a single man, and the revolution in notation was no exception. Viète kept the idea of homogeneity: one could not add a plane and a solid, nor a plane to a side. Thus, to Viète, an equation like A square plus C was nonsense, since you were trying to add a square to a side. To get around this, Viète would call C a plane number; hence the proper notation would be A square plus C plane.

To determine what type of constant was needed, Viète included a number of multiplication rules, such as

1. Side times side is plane.

2. Side times plane is solid.

3. Side times solid is plano-plane.

Example 11.6. *Write down, "The product of two squares equals a given number." Designating the unknowns as A and E, and the given number as B, we note we are multiplying a square by a square, to make a square-square (as a variable) or a plano-plane (as a number). Hence A square times E square equals B plano-plane.*

The constants are given whatever dimensionality is required to allow the addition or equality to exist.

Example 11.7. *A cube and some sides equal a number. Let the unknown be A; let the number of sides be B and G. We must add a cube to a solid, and the result must be a solid. To make some sides a solid, we multiply it by a plane. Thus the equation is A cube $+ A B$ plane equals G solid.*

The replacement of even the constants with letters was Viète's key advance. Since the Rhind papyrus, students of mathematics learned by looking at examples and generalizing from them; with Viète, it became possible to provide a generalized example, and to write down general formulas for the solutions of an equation.

A good example of the value of Viète's contribution occurred in his posthumous work, *Two Treatises on the Understanding and Amendment of Equations.* Viète gave three formulas for solving quadratic equations. The first formula is:

Rule 11.6 (Quadratic Formula). *If A square $+ B$ times $2A$ is equal to Z plane, let $A + B$ be E. Therefore E square is equal to Z plane $+ B$ square. Therefore $\sqrt{Z \text{ plane} + B \text{ square}} - B$ is equal to A.*

11.5.2   Zetetics

Viète designated the process by which equations were created *zetetics*. In *Two Treatises on Zetetics*, he stressed the value of understanding the *origin* of equations. Viète still considered geometry to be the supreme mathematical science,

De reductione quadratorum adfectorum ad pura.

Formulæ tres.

I.

Sı A quad. -+ B 2 in A, æquetur Z plano. A -+ B esto E. Igitur E quad., æquabitur Z plano -+ B quad.

Confectarium.

Itaque, ✓ z̄p̄lām̄ -+ B̄ q̄uād. — B fit A, de qua primum quærebatur.

Sit B 1. Z planum 20. A 1 N. 1 Q -+ 2 N, æquatur 20. & fit 1 N. ✓ 21 — 1.

II. Si

Figure 11.5: The First Appearance of the Quadratic Formula. Boston Public Library/Rare books Department. Courtesy of the Trustees.

however, reminding his readers that the roots of equation were nothing more than line segments in continued proportion. For example, the equation

$$A \text{ square } + BA \text{ equal to } Z \text{ square}$$

came from the problem of finding the extremes, given the mean of three proportionals and the difference of the extremes, with A the smaller extreme. Hence if Z was the mean, B the difference between the extremes, then the extremes were A, $A + B$, and hence the given equation followed. Equally important from the theoretical standpoint was that the geometric problem, and thus the algebraic, had a unique solution.

Viète also investigated the relationship between what he called **correlative equations**: two equations were *correlative* if they had the same structure and coefficients, though by Viète's meaning of the term, the coefficients were considered independent of whether the corresponding terms were added or subtracted. Hence Viète recognized three types of correlative equations. Viète used "parabola," "grade," "power," and "homogeneous" to indicate the variables or constants raised to definite but arbitrary (whole number) powers. The three types of correlative equations were:

1. **Ambiguous**, where the coefficients on the same variables with the same powers are the same:

   $$B \text{ parabola } A \text{ grade } - A \text{ power } = Z \text{ homogeneous}$$
   $$B \text{ parabola } E \text{ grade } - E \text{ power } = Z \text{ homogeneous}$$

   are ambiguous equations.

2. **Contradictory**, where the variable terms are, in one equation, added, and in the other, subtracted.

   $$B \text{ parabola } A \text{ grade } + A \text{ power } = Z \text{ homogeneous}$$
   $$B \text{ parabola } E \text{ grade } - E \text{ power } = Z \text{ homogeneous}$$

are contradictory.

3. **Inverse**, where the variable terms are subtracted in opposite orders.

$$B \text{ parabola } A \text{ grade } - A \text{ power } = Z \text{ homogeneous}$$
$$E \text{ power } - B \text{ parabola } E \text{ grade } = Z \text{ homogeneous}$$

are inverse.

Since the constant terms are equal, the two equations, whether ambiguous, contradictory, or inverse, can be equated, and the relationship between the two solutions A and E can be determined. In the case of the homogeneous equations

$$BA - A \text{ square } = Z \text{ plane}$$
$$BE - E \text{ square } = Z \text{ plane}$$

Viète showed $B = A + E$, and $Z = EA$, which is the result that the sum of the roots of a quadratic are equal to the coefficient of the linear term, and the product of the roots is equal to the constant term.

11.5.3   Trigonometry

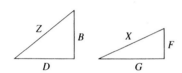

Viète also made pertinent contributions to trigonometry in *Preliminary Notes on Symbolic Logistic*. Given A, B, then the two sides $2AB$ and $A^2 - B^2$ form a right triangle with hypotenuse $A^2 + B^2$. To distinguish between the sides of the triangle, Viète refers to one side as the base and the other as the perpendicular. Given two right triangles, one with base D, perpendicular B, and hypotenuse Z, and the other with base G, perpendicular F, and hypotenuse X, two right triangles may be constructed. The first has base $BF - DG$, perpendicular $BG + DF$ and hypotenuse ZX; this triangle Viète called the **synaeresic triangle**. The other triangle has base $BF + DG$, perpendicular $BG - DF$, and hypotenuse ZX, which Viète called the **diaeresic triangle**.

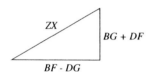

The names of these two triangles come from grammar, where *synaeresis* is the process of joining two vowels to make a single sound, such as the "a-i" in *rain*. Diaeresis is the process of pronouncing two adjacent vowels as distinct syllables, as the "o-o" in *cooperate*. Viète probably chose these names because the synaeresic triangle corresponds to a right triangle with the base angle equal to the sum of the base angles of the two given triangles, while the diaeresic triangle corresponds to the right triangle with base angle equal to the difference of the two base angles.

See Problem 3.

If the two triangles are the same, say with base D, perpendicular B, and hypotenuse A, then diaeresis is, of course, impossible, but through synaeresis we may produce what Viète calls the triangle of the double angle, with base $D^2 - B^2$, perpendicular $2BD$, and hypotenuse A^2. Repeating this procedure, one can find the triangle of the triple, quadruple, and other angles, and in general, Viète claimed:

Conjecture 11.1. *If any power of a binomial root is taken, and the resulting terms are separated into two groups, and if the terms in each group are alternately positive, then negative, the base will be the first group of terms, the perpendicular the second, and the hypotenuse will be like the power.*

If the base of the right triangle is D, the perpendicular B, and the hypotenuse A, then the expansion of $(D + B)^n$ will determine the new base, perpendicular, and A^n will be the new hypotenuse.

Example 11.8. *The triple angle. Expanding $(D + B)^3$, we have*

$$
\begin{array}{rcccc}
(D + B)^3 & = & D^3 \quad 3D^2B & 3DB^2 & B^3 \\
Base & & D^3 & -3DB^2 & \\
Perpendicular & & 3D^2B & & -B^3
\end{array}
$$

Thus, the base is $D^3 - 3DB^2$, the perpendicular is $3D^2B - B^3$, and the hypotenuse is A^3.

In modern terms, what Viète has done is to express $\cos n\alpha$ in terms of $\cos \alpha$ and $\sin \alpha$, namely that

$$
\cos n\alpha = \cos^n \alpha - \binom{n}{2} \cos^{n-2} \alpha \sin^2 \alpha + \binom{n}{4} \cos^{n-4} \alpha \sin^4 \alpha + \ldots \tag{11.1}
$$

and

$$
\sin n\alpha = n \cos^{n-1} \alpha \sin \alpha - \binom{n}{3} \cos^{n-3} \alpha \sin^3 \alpha + \binom{n}{5} \cos^{n-5} \alpha \sin^5 \alpha + \ldots
$$
$$
\tag{11.2}
$$

See Problem 4.

11.5.4 The Problem of Adrianus Romanus

In 1593, Adrianus Romanus, a Belgian mathematician, posed the following problem to the mathematicians of Europe:

Problem 11.3 (Adrianus Romanus's Problem). *If two terms have a ratio, the first to the second, of 1 to 45 ①−3795 ③+95634 ⑤ −1138500 ⑦+7811375 ⑨ −34512075 ⑪+105306075⑬ −232676280 ⑮ +384942375 ⑰ −488494125 ⑲ +483841800 ㉑−378658800 ㉓ +236030652 ㉕ −117679100 ㉗ +46955700 ㉙ −14945040 ㉛ +3764565 ㉝ −740459 ㉟ +111150 ㊲ −12300 ㊴ +945 ㊶ −45 ㊸ +1㊺, to find the two terms.*

Romanus used the notation of Stevin, where an expression like 45 ① meant $45x^1$.

Henry IV called on Viète, who solved the problem overnight. Viète explained his procedure, which relied implicitly on the properties of trigonometric functions, in a 1595 tract, *Response to a Problem Posed by Adrianus Romanus.* We will not solve Romanus's problem in its entirety, but simply outline the essential steps.

Key to Viète's procedure was a pair of theorems:

Conjecture 11.2. *Given a series of right triangles with equal hypotenuses, and whose base acute angles are B, 2B, 3B, 4B, and so on; construct a sequence of lines in continued proportion, the first of which is half the common hypotenuse, and the second of which is the base of the first triangle. Between the successive proportionals and the successive bases of the triangles there exist the following equalities:*

1. *The third continued proportional, minus twice the first, will be equal to the base of the second triangle.*

2. *The fourth proportional, minus three times the second, will be equal to the base of the third triangle.*

3. *The fifth, minus four times the third, plus twice the first, will be equal to the base of the fourth triangle.*

and so on. Moreover, if the first term of the sequence [i.e., half the hypotenuse] is 1, then the equalities may be expressed as:

$1Q - 2$	*base of second*
$1C - 3N$	*base of third*
$1QQ - 4Q + 2$	*base of fourth*
$1QC - 5C + 5N$	*base of fifth*
$1CC - 6QQ + 9Q - 1$	*base of sixth*
$1QQC - 7QC + 14C - 7N$	*base of seventh*
$1QCC - 8CC + 10QQ - 16Q + 2$	*base of eighth*
$1CCC - 9QQC + 17QC - 30C + 9N$	*base of ninth*

Notice that Viète is here using N as an abbreviation for the first power of the unknown, Q as an abbreviation for the square, C for the cube, QQ for the square-square, and so on. This theorem is the equivalent of expressing $\cos n\theta$ in terms of members of the sequence $1, 2\cos\theta, 4\cos^2\theta, 8\cos^3\theta, \ldots$.

To understand Viète's work, consider a sequence of right triangles, each with hypotenuse 2, and base angles $\theta, 2\theta, 3\theta, \ldots$. Let the base of the first triangle be N (which is $2\cos\theta$, although Viète would never have expressed it in this form). Then the base of the second triangle would be $1Q - 2$.[3] The base of the third triangle is $1C - 3N$, and so on. To make use of Viète's table, note that if, for

---

[3]Since the base of the second triangle is

$$\begin{aligned} 2\cos 2\theta &= 2\left(\cos^2\theta - \sin^2\theta\right) \\ &= 2\left(2\cos^2\theta - 1\right) \\ &= 4\cos^2\theta - 2 \\ &= 1Q - 2 \end{aligned}$$

Remember that in Viète's notation, Q is $N \cdot N$.

example, $1Q - 2$ is equal to some number, that number will be the base of the triangle with hypotenuse 2 and base angle 2θ, and the unknown is the base of the triangle with hypotenuse 2 and base angle θ. It is then a matter of reconstructing the original triangle and finding the base of the triangle with base angle θ and hypotenuse 2.

The next theorem is the corresponding result if the first two terms of the sequence of proportionals are half the hypotenuse and the altitude of the first triangle.

Conjecture 11.3. *Given a series of right triangles with equal hypotenuses, and whose base acute angles are B, 2B, 3B, 4B, and so on; construct a sequence of lines in continued proportion, the first of which is half the common hypotenuse, and the second of which is the altitude of the first triangle. Between the successive proportionals and the successive bases of the triangles there exist the following equalities*

1. *Twice the first, minus the third, is equal to the base of the second triangle.*

2. *Three times the second, minus the fourth, is the altitude of the third triangle.*

3. *Twice the first, minus four times the third, plus the fifth, is the base of the fourth triangle.*

and so on. Moreover, if the first term of the sequence [i.e., half the hypotenuse] is 1, then the equalities may be expressed as:

$2 - 1Q$	*base of second*
$3N - 1C$	*altitude of third*
$2 - 4Q + 1QQ$	*base of fourth*
$5N - 5C + 1QC$	*altitude of fifth*
$2 - 9Q + 6QQ - 1CC$	*base of sixth*
$7N + 14C + 7QC - 1QQC$	*altitude of seventh*
$2 - 16Q + 10QQ - 8CC + 1QCC$	*base of eighth*
$9N - 30C + 27C - 9QQC + 1CCC$	*altitude of ninth*

These two theorems allowed certain expressions to be reduced to simpler trigonometric expressions.

Example 11.9. *Solve the equation $3N - 1C = \sqrt{2}$. The left hand side is the altitude of the third triangle in the sequence of right triangles, where half the hypotenuse is 1 and the altitude of the first triangle is $1N$; this third triangle will have a base angle of three times the base angle of the first triangle. Hence, the first altitude (and thus $1N$) is $\sqrt{2 - \sqrt{3}}$.*
[Solve

$$3x - x^3 = \sqrt{2}$$

Let $x = 2\sin\theta$, so $3x - x^3 = 2\sin 3\theta$, and thus

$$2\sin 3\theta = \sqrt{2}$$

Hence $2\sin\theta = \frac{\sqrt{6}}{2} - \frac{\sqrt{2}}{2}$.

To solve the problem of Adrianus Romanus, Viète noted that it could be reduced to the sine of 45θ; using trigonometric tables, he found not just one, but twenty-three positive real solutions.

11.5.5   Area of a Circle

One of Viète's most celebrated accomplishments is his determination of a value for π. Since Archimedes laid the essential *theoretical* foundations, finding a value for π was merely a matter of endurance. Viète made an important contribution in his *Responses to Various Mathematical Questions*, for the formulation of Archimedes gave no easy method to obtain more accurate approximations.

If BD is the side of a regular polygon inscribed in a circle, and BDC a semicircle, then Viète referred to DC as the **apotomic side**. Viète proved:

Proposition 11.5. *If two regular polygons are inscribed in a circle, where one has double the number of sides of the other, then the area of the first to the second is as the apotomic side of the first to the diameter of the circle.*

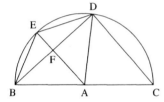

Proof. Given circle with center A, diameter BC, and BD the side of regular polygons inscribed in a circle. Bisect arc BD at E; join AE and let it intersect chord BD at F. Join BE, ED, DC. BE is the side of a regular polygon with twice as many sides inscribed in a circle; note AE is perpendicular to BD. We claim the polygon with side BD is to the polygon with side BE as DC is to BC.

Now, the regular polygon with side BD consists of triangles equal to BAD, while the regular polygon with side BE consists of trapezia equal to $BEDA$, which is equal to triangle BAD and triangle BED. Triangles BAD and BED are on the same base, BD; hence triangle BAD is to triangle BED as the altitude AF is to FE. Combining the ratio, we have triangle BAD is to trapezia $BEDA$ is AF is to AE, or as AF is to AB (since AE, AB are both radii of the circle). But triangles AFB, CDB are similar; hence AF is to AB as DC is to BC. Hence triangle BAD is to trapezia $BEDA$ as the apotome is to the diameter; hence the regular polygon with side BD is to the regular polygon with side BE as the apotome of the first is to the diameter of the circle, which was to be proven.   □

Thus:

Proposition 11.6. *If an infinite sequence of regular polygons is inscribed in a circle, each having twice as many sides as the other, then the area of the first to the area of the last is as the apotomic of the first is to the diameter, compounded with the apotomic side of the second to the diameter, compounded with the apotomic side of the third to the diameter, and so on.*

As a corollary, Viète noted:

Corollary. *The ratio of an inscribed square to a circle is the product of the ratio of the side of the square to the diameter, times the ratio of the apotomic side of the octagon to the diameter, times the ratio of the apotomic side of the hexadecagon to the diameter, and so on, each apotomic side being that of a figure with twice the number of sides.*

Without proof, Viète went on to note that

If a circle has a diameter of 2, the side of the square inscribed in the circle is $\sqrt{2}$, and its square is 2. The apotomic side of the octagon is $\sqrt{2+\sqrt{2}}$. The apotomic side of the hexadecagon is $\sqrt{2+\sqrt{2+\sqrt{2}}}$.

The apotomic side of the 32-sided polygon is $\sqrt{2+\sqrt{2+\sqrt{2+\sqrt{2}}}}$.

The apotomic side of the 64-sided polygon is $\sqrt{2+\sqrt{2+\sqrt{2+\sqrt{2+\sqrt{2}}}}}$, and the progression continues.

Let the diameter of the circle be 1. Let the [area of the] circle be $1N$. As $\frac{1}{2}$ is to $1N$, so is the product of $\sqrt{\frac{1}{2}}$ by $\sqrt{\frac{1}{2}+\frac{1}{2}\sqrt{\frac{1}{2}}}$

by $\sqrt{\frac{1}{2}+\frac{1}{2}\sqrt{\frac{1}{2}+\frac{1}{2}\sqrt{\frac{1}{2}}}}$ by $\sqrt{\frac{1}{2}+\frac{1}{2}\sqrt{\frac{1}{2}+\frac{1}{2}\sqrt{\frac{1}{2}+\frac{1}{2}\sqrt{\frac{1}{2}}}}}$ by

$\sqrt{\frac{1}{2}+\frac{1}{2}\sqrt{\frac{1}{2}+\frac{1}{2}\sqrt{\frac{1}{2}+\frac{1}{2}\sqrt{\frac{1}{2}+\frac{1}{2}\sqrt{\frac{1}{2}}}}}}$ [and so on].[4]

Since the area of a circle with diameter 1 is $\frac{\pi}{4}$, we can interpret Viète's last statement to mean that

$$\frac{2}{\pi} = \sqrt{\frac{1}{2}}\sqrt{\frac{1}{2}+\frac{1}{2}\sqrt{\frac{1}{2}}}\sqrt{\frac{1}{2}+\frac{1}{2}\sqrt{\frac{1}{2}+\frac{1}{2}\sqrt{\frac{1}{2}}}}\sqrt{\frac{1}{2}+\frac{1}{2}\sqrt{\frac{1}{2}+\frac{1}{2}\sqrt{\frac{1}{2}+\frac{1}{2}\sqrt{\frac{1}{2}}}}}\cdots$$

11.5 Exercises

1. In the following, assume p, q, r are all positive. Determine the type of number p, q, r must be (i.e., whether side, plane, and so on).

   (a) $x^2 + px + q = 0$ (c) $x^2 + p = qx$

   (b) $x^2 + px = q$ (d) $x^3 + px^2 + qx = r$

---

[4]Viète, François, *Opera Mathematics*. Franciscus van Schooten, 1646. Page 400.

2. By equating B plane $A - A$ cube $= Z$ solid and B plane $E - E$ cube $= Z$ solid , determine the relationship between the roots A, E of the cubic equation. Use the relationship to construct a cubic with real roots 10 and 2.

3. Consider the synaeresic and diaeresic triangles of Viète. Use the sum and difference identities for sine and cosine to show the synaeresic triangle has a base angle equal to the sum of the two base angles in the original triangle, and the diaeresic triangle has an base angle equal to the difference of the two base angles.

4. Show that *Preliminary Notes*, General Corollary is equivalent to Equation 11.1 and Equation 11.2.

5. Show how the given equations are equivalent to the indicated proportionalities.

    (a) A square $- BA$ equal to Z square; the problem of finding three proportionals with a given mean, and difference between the extremes B, where A is the greater of the extremes.

    (b) $BA - A$ square equal to Z square; the problem of finding three proportionals given the mean Z and the sum of the extremes B.

6. For the inverse equations

$$A \text{ cube } - B \text{ plane } A = Z \text{ cube}$$
$$B \text{ plane } E - E \text{ cube } = Z \text{ cube}$$

determine the relationship between the solutions E and A.

7. Viète knew that there were three solutions to $1N - 3C = \sqrt{2}$. Find the other two solutions.

8. Solve each of the following problems.

    (a) $3N - C = 1$
    (b) $8QQ - 8Q + 1 = \sqrt{2}$

    (c) $1C - 3N = 2$
    (d) $1QQ - 4Q + 2 = 1$

9. Vary Viète's method of solving higher-degree equations to find solutions to other problems.

    (a) Let the first two terms of the sequence in continued proportion be the hypotenuse and the base of the first triangle (i.e., 1 and $1 \cos \theta$); hence the sequence is 1, N, Q, C, ... corresponds to 1, $1 \cos \theta$, $1 \cos^2 \theta$, $1 \cos^3 \theta$, ... Express the bases of the successive triangles in terms of the members of this sequence.

(b) Alternatively, let the first two terms of the sequence be one-third the hypotenuse and the base of the first triangle; hence the sequence 1, N, Q, C, ... corresponds to 1, $3\cos\theta$, $9\cos^2\theta$, $27\cos^3\theta$, ... Express the bases of the successive triangles in terms of the members of this sequence.

10. Use the relationships you found in Problem 9, or relationships you find yourself, to solve the following problems.

(a) $3N - 4C = \frac{1}{\sqrt{2}}$

(b) $16C - 3N = \frac{1}{\sqrt{8}}$

(c) $3N - \frac{4}{9}C = \frac{3}{\sqrt{2}}$

(d) $\frac{1}{4}C - 3N = 4$

11. Pose a ninth-degree "challenge" problem, solvable using Viète's method or a variation of it.

12. Viète did not explain how to find the apotomic sides of the regular polygons. Show that Viète's expressions for the apotomic sides of the octagon, 16-gon, 32-gon, and so on, are correct.

13. Use the first five terms of Viète's product to approximate π. How accurate is the approximation?

14. One of the advantages of a concern with homogeneity is that the existence or nonexistence of real, positive solutions to equations is geometrically obvious. Consider the equations from Problem 1. Prove the following using a geometric argument.

(a) $x^2 + px + q = 0$ has no positive solutions.

(b) $x^2 + px = q$ has a unique positive solution. Hint: $x^2 + px$ is a plane quantity, and x is a line to be determined. What happens if the size of x increases or decreases?

(c) $x^3 + px^2 + qx = r$ has a unique positive solution.

15. Compare the notations of Viète, Stevin, and Chuquet. What are the advantages and disadvantages of each? (Be careful not to criticize the notation from the viewpoint of modern algebraic notation; rather, what are the *internal* strengths and weaknesses of the system?)

11.6 The Development of Logarithms

After his publication of *Regarding the New Star*, Brahe became one of the most famous scientists in Europe, and secured the patronage of the Danish King Frederick II. Frederick gave Brahe the island of Hven, and enough money to establish one of the greatest astronomical observatories in Europe. It became known as Uraniborg ("sky palace"), and attracted many important figures of mathematics and science.

See page 304.

11.6.1   Prosthaphaeresis

One of Brahe's assistants was Paul Wittich (1555-January 9, 1587). Some time before 1574, Wittich and Tycho discovered the formula

$$2\sin A \sin B = \cos(A - B) - \cos(A + B)$$

The method of converting a product of trigonometric functions into a sum of trigonometric functions is referred to as **prosthaphaeresis**, from the Greek words meaning "addition and subtraction."

Since extensive tables of sine and cosine were available, this meant that the product of two sines could be found by subtracting two numbers: in other words, a product could be converted into a sum. For example, to multiply 0.8660 by 0.2588, we could consult a trigonometric table and find $\sin 60° \approx 0.8660$, and $\sin 15° \approx 0.2588$. Applying the prosthaphaeretic formula:

$$
\begin{aligned}
2(0.8660)(0.2588) &= 2\sin 60° \sin 15° \\
&= \cos(60° - 15°) - \cos(60° + 15°) \\
&= \cos 45° - \cos 75°
\end{aligned}
$$

Again consulting the table, we would find the values of $\cos 45°$ and $\cos 75°$, which are 0.7071 and 0.2588, respectively. Hence:

$$
\begin{aligned}
2(0.8660)(0.2588) &= \cos 45° - \cos 75° \\
&= 0.7071 - 0.2588 \\
&= 0.4483
\end{aligned}
$$

so $(0.8660)(0.2588) \approx 0.22415$.

Tycho was an inferior mathematician, so the discovery was probably made by Wittich. Wittich did not publish his discovery until later. In 1584, Nicolai Reymers Ursus visited Uraniborg; four years later he published versions of the prosthaphaeretic formulas in his *Foundations of Astronomy* (1588). Brahe accused Ursus of plagiarism. Ursus claimed in turn that Wittich may have discovered the method, but the first *proof* of the validity of the method was due to Joost (or Jobst) Bürgi. However, Bürgi was a student of Wittich at the University of Kassell, so he may have learned the method there.

To add to the confusion, the prosthaphaeretic rule was implicit in work done by Johannes Werner (1468-1522) between 1505 and 1513, but Werner's work on the subject was not published until 1907. However, a few scientists, notably Rheticus, had learned of it, so in 1611, Jacob Christman attributed the actual discovery to Werner; hence the method of prosthaphaeresis is sometimes called **Werner's method**. Finally, prosthaphaeretic methods were used by the eleventh century Islamic astronomer ibn-Yunis, though his discovery seems to have been unknown in the west until much later.

11.6.2 Logarithms

The value of prosthaphaeresis rests on the fact that addition and subtraction are easier than multiplication and division. However, prosthaphaeresis relied on having a table of chords, and the procedure for creating a table of chords was itself an arduous procedure, involving many multiplications and divisions. What was needed was an easily constructible table of numbers with prosthaphaeretic properties. One possibility was numbers in geometric proportion.

Since the values of $\cos A$ decrease from 1 to 0 while the angle A increases from 0, consider a sequence of numbers decreasing in geometric proportion from 1, and a second sequence increasing in arithmetic proportion from 0.

$$1 \quad \tfrac{1}{2} \quad \tfrac{1}{4} \quad \tfrac{1}{8} \quad \tfrac{1}{16} \quad \tfrac{1}{32} \quad \tfrac{1}{64} \quad \tfrac{1}{128} \quad \tfrac{1}{256} \quad \tfrac{1}{512}$$
$$0 \quad 1 \quad 2 \quad 3 \quad 4 \quad 5 \quad 6 \quad 7 \quad 8 \quad 9$$

Consider two terms in the geometric sequence, say $\tfrac{1}{4}$ and $\tfrac{1}{32}$. The product of the two is $\tfrac{1}{128}$. Now examine the corresponding numbers in the arithmetic sequence:

$$\tfrac{1}{4} \quad \text{times} \quad \tfrac{1}{32} \quad \text{equals} \quad \tfrac{1}{128}$$
$$2 \qquad\qquad 5 \qquad\qquad 7$$

It would seem that the *product* of the numbers in the geometric sequence correspond to the *sum* of the numbers in the arithmetic sequence: thus, the geometric and arithmetic sequences correspond respectively to the cosine and the angle in the prosthaphaeretic formulas.

Napier

A Scottish nobleman, John Napier (1550-April 4, 1617), was among the first to realize that numbers in geometric proportion would satisfy prosthaphaeretic properties and turn a multiplication problem into an addition problem. Today, it is easy to see this, since we note

$$
\begin{array}{cccccccccc}
1 & 2 & 4 & 8 & 16 & 32 & 64 & 128 & 256 & 512 \\
\left(\tfrac{1}{2}\right)^0 & \left(\tfrac{1}{2}\right)^1 & \left(\tfrac{1}{2}\right)^2 & \left(\tfrac{1}{2}\right)^3 & \left(\tfrac{1}{2}\right)^4 & \left(\tfrac{1}{2}\right)^5 & \left(\tfrac{1}{2}\right)^6 & \left(\tfrac{1}{2}\right)^7 & \left(\tfrac{1}{2}\right)^8 & \left(\tfrac{1}{2}\right)^9
\end{array}
$$

and the multiplication $\tfrac{1}{4} \times \tfrac{1}{32} = \tfrac{1}{128}$ is $\left(\tfrac{1}{2}\right)^2 \times \left(\tfrac{1}{2}\right)^5 = \left(\tfrac{1}{2}\right)^7$, so the marvelous property of converting products into sums emerges as part of the rules of exponents. However, the modern notation for exponents did not exist in Napier's time; viewed in this light, Napier's development of logarithms was nothing short of genius.

Napier developed his ideas on how to convert products into sums over a period of about twenty years, and in 1614 published *A Description of the Marvelous Rule of Logarithms*. Napier did not define a logarithm until the middle of his work. The numbers in geometric proportion correspond to the numbers whose products or quotients are desired; the numbers in arithmetic proportion are the logarithms. Thus, Napier defined the logarithm of a sine (i.e., a chord) as:

John Napier. Reprinted with the permission of CORBIS.

First Table	Second Table
10000000.0000000	10000000.0000000
−1.0000000	−100.0000000
9999999.0000000	9999900.0000000
−0.9999999	−99.999000
9999998.0000001	9999800.001000
−0.9999998	−99.998000
9999997.0000003	9999700.003000
⋮	⋮
9999900.0004950	9995001.224804

Table 11.4: Napier's First and Second Tables.

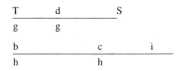

Definition. *Suppose TS is the radius, and dS some sine, and suppose g moves geometrically from T to d in some amount of time. Suppose bi is another line, infinite in length, along which h moves at a constant rate equal to the velocity of g at T. If, in the time during which g moved from T to d, h moved from b to c, then bc is the* **logarithm** *of the sine dS.*

To construct the table, it was thus necessary to find numbers in a geometric sequence. It would be convenient if logarithms existed for as many numbers as possible; hence, the geometric proportion should be a number r very close to 1. However, finding powers of such a number entails a great deal of computation. Napier found an easier way to find the powers of r.

First, emulating his predecessors who found tables of sine values, he used a large radius: $10,000,000$; this allowed him to deal with whole numbers only; indeed, Napier began his *Description* by outlining what we would consider the **rules of significant figures**. Next, rather than multiplying out the powers of r, Napier recognized that if r was equal to $1 - \frac{1}{10^n}$, then the numbers in geometric proportion could be found by subtracting $\frac{1}{10^n}$ of each number. Thus, Napier began with $10,000,000$ and subtracted the ten-millionth part of each number; this formed his **First Table**, consisting of a hundred proportional numbers in sequence, from $10,000,000$ to $9,999,900.0004950$.

Since the last value of the first table is very nearly equal to $9,999,900$, Napier's **Second Table** began with $10,000,000$ and produced successive numbers in geometric proportion by subtracting their hundred thousandth part. Thus, by successively subtracting the hundred thousandth part of the numbers, a second table is produced, consisting of fifty-one numbers from $10,000,000$ to $9,995,001.224804$ (see Table 11.4). Naturally, producing a long table consisting of many arithmetic operations repeated over and over again has its difficulties, and Napier's final result was incorrectly computed to be $9,995,001.222927$.

Again, the last number of this is very nearly equal to $9,995,000$, and so the third table will consist of numbers in proportion, beginning with $10,000,000$, whose $1/2000^{\text{th}}$ are subtracted. The first column of the first table will consist of twenty-one numbers in proportion, from $10,000,000$ to $9,900,473.57808$. Notice

10000000.0000	9900000.0000
9995000.0000	9895050.0000
9990002.5000	9890102.4750
9985007.4987	9885157.4237
⋮	⋮
9900473.5780	9801468.8423

Table 11.5: Napier's Table Three, First and Second Columns

this is very nearly 99/100 the first entry. By making a second column, whose entries are also 99/100 the corresponding entries in the first column, Napier continued the table.

By this construction, the last number in each column will be very nearly equal to $1/100^{\text{th}}$ less than the first number in the column, and thus each number in the table will be 99/100 the number to its immediate left. This produces a table consisting of 69 columns, the last entry of the last column being 4998609.4034, or very nearly half of ten million (see Table 11.5).

In effect, Napier has produced three tables, consisting of numbers in a geometric proportion. For the three tables, the following relationships can be noted:

1. Entries in adjacent rows of Table 1 will have a ratio of 10000000 to 9999999.

2. Entries in adjacent rows of Table 2 will have a ratio of 100000 to 99999.

3. Entries in adjacent rows of Table 3 will have a ratio of 10000 to 9995.

4. Entries in adjacent columns of Table 3 will have a ratio of 100 to 99.

It is only after constructing this table that Napier defined a logarithm. Afterwards, he proved a number of propositions:

Proposition 11.7. *The logarithm of the radius is 0.*

Then

Proposition 11.8. *The logarithm of any given sine is greater than the difference between the sine and the radius, and less than the difference between the radius and the quantity which exceeds the radius in a ratio equal to that of the radius to the given sine.*

Proof. Let the given sine be dS, and let oS be a quantity that exceeds the radius TS in a ratio equal to that of the radius to the given sine; thus $oS : TS = TS : dS$. Then the distances oT, Td, and bc are traversed in equal times. But the velocity of the point from o to T is greater than its velocity at T, and hence oT is greater than bc. Likewise, the velocity of the point from T to d is less than its velocity at T, and hence Td is less than bc. Hence $oT > bc > Td$, which was to be proven, since bc is the logarithm of dS. □

Napier noted that the lower limit is simply the difference between the radius and the sine, while the upper limit can be found by multiplying the radius by the lower limit and dividing by the sine. Thus, the limits of the logarithm of any number on the table can be found very easily.

Example 11.10. *Find the limits on the logarithm of* $9,995,000$. *The difference between the radius and the sine is* 5000, *which makes the lower limit. The upper limit is the product of the radius* $(10,000,000)$ *and the lower limit divided by the number itself:* 5002.5. *Thus, the logarithm is between* 5000 *and* 5002.5.

In practice, Napier required that the upper and lower limits be the same, to the nearest whole number. Thus it was necessary to use:

Proposition 11.9. *The difference of the logarithms of two sines is between two limits. The greater limit is to the radius as the difference of the sines is to the lesser sine; the lower limit is to the radius as the difference of sines is to the greater sine.*

Proof. Let TS be the radius, dS the greater, and eS the lesser sine. Find V so that $TS : TV = eS : de$, and c so that $TS : Tc = dS : de$. Then $eS : dS = TS : VS$, and $eS : de = cS : Tc$, and thus $eS : dS = cS : TS$. Thus, $VS : TS = TS : cS$, and $TS : cS = dS : eS$.

Since $TS : cS = dS : eS$, then the difference between the logarithms of TS, cS will be equal to the difference between the logarithms of dS, eS. But the logarithm of TS is zero; hence the logarithm of cS will be the difference between the logarithms of dS, eS. But the logarithm of cS is between TV and Tc. Hence, the difference of the logarithms of dS, eS is between the greater limit TV and the limit Tc, where the greater limit TV is to the radius TS as the difference of sines de is to the lesser sine eS, and the lower limit Tc is to the radius TS as the difference of sines de is to the greater sine dS. □

Common Logarithms

Though we must credit Napier with being the first to *publish* a work on logarithms, Joost Bürgi had a similar idea at about the same time (being based on the powers of 1.0001, as opposed to Napier's powers of 0.9999999). Bürgi's original work on logarithms may have been done as early as 1588, predating Napier by twelve years. However, Bürgi's work was not published until 1620, by which time logarithms had undergone a major change.

Today, we say that b is the **logarithm to base a of c** if $a^b = c$. Thus, 7 is the logarithm to base 2 of 128. This is written $\log_2 128 = 7$. There are two common bases: e (where the logarithm of c to base e is written $\ln c$) and 10 (where the logarithm of c to base 10 is written $\log c$) 10 is used because it is wellsuited to the decimal system of numeration, but e has many properties that make logarithms to base e particularly elegant. We note that neither form of logarithms given by Napier is the one we use today. Logarithms were put into their present form by Henry Briggs (February 1561-January 26, 1630), who visited Napier in 1615.

See problem 7.

$$\overline{V \qquad T \quad c \qquad d \quad e \qquad S}$$

See problem 8.

Briggs was one of the few scientists of the time who recognized astrology for the superstitious nonsense it is. When Briggs met Napier in 1615, they discussed possible modifications of the method of logarithms. Among the problems were that multiplication of two numbers did not quite correspond to a simple addition, and extra steps were required to recover the final answer. Briggs proposed using 10 as the base for the logarithm, with $\log 10$ being set at ten billion. Napier agreed; he himself had originally considered base 10 logarithms. After some discussion, Briggs came to the conclusion that it would be more convenient if $\log 1 = 0$ and $\log 10 = 1$. To find the values of the logarithms, Briggs found the successive roots of 10: since $\sqrt{10} = 3.162277$, then $\log 3.162277 = 0.5$. By 1617, Briggs had created a table of the logarithms, to base 10 (now called **common logarithms**) of the first 1000 whole numbers. The logarithms were computed to fourteen decimal places (!), and all the usual rules of computation were included.

Briggs's work came to the attention of scholars everywhere. In 1619, Henry Savile (1549-1622), a classical scholar interested in mathematics, donated money to Oxford to fund two professorships, one in geometry and one in astronomy. For his work on logarithms and other contributions to mathematics, Savile himself chose Briggs to become the first Savilian Professor of Geometry.

In 1624, Briggs extended his table to include the logarithms from 1 to 20,000, and from 90,000 to 100,000. In 1627, the gap in the table, from 20,001 to 89,999, was filled by Ezechiel de Decker and Adriaen Vlacq, two Dutch surveyors.

11.6 Exercises

1. Using a calculator to evaluate the appropriate sines, cosines, and inverse sines and cosines, apply the prosthaphaeretic formulas to perform the multiplications:

   (a) 0.4540×0.2756.

   (b) 0.6293×0.6561.

   (c) 0.1908×0.9063.

   (d) 0.7314×0.9962.

   (e) How would you modify the prosthaphaeretic formulas to perform multiplications such as 6293×65.61?

   (f) How would the use of the prosthaphaeretic formulas in the time of Wittich be different from the use in these problems?

2. Using Napier's Proposition 11.8 to find the logarithms of $9,999,999$ and $9,999,900$. Then show that the logarithms found are consistent with the modern definition of a logarithm, where $\log_b a = c$ if $b^c = a$, and $\log_b b = 1$.

3. In Napier's propositions for finding logarithms, he does not use either of the tables constructed. Explain their purpose, and how they could be used to simplify the computation of logarithms.

4. Produce a table of logarithms, according to Napier.

   (a) Produce Table 1 by beginning with 100, and successively subtract the one-thousandth part until you get to a number approximately equal to 99.

   (b) Find the bounds on the logarithms of the remaining numbers in your table.

5. (Calculus) Show that Napier's *definition* of logarithm corresponds to the definition of a logarithm to base $1/e$. What is the *actual* base of the logarithms computed by Napier?

6. Prove, according to Napier's definition, that the logarithm of the radius is 0.

7. Show that the upper limit for the logarithm of a sine may be found by multiplying the radius by the lower limit, then dividing by the sine. This is equivalent to showing that $TS : dS = oT : Td$.

8. Look over the proof of Proposition 11.9.

   (a) Prove that $eS : dS = TS : VS$, and $es : de = cS : Tc$.

   (b) Prove then that $es : ds = cS : TS$.

Chapter 12

The Era of Descartes
and Fermat

By 1600, the Protestant Reformation left Europe divided, but it seemed that Protestants and Catholics might learn to live with each other. In the Holy Roman Empire, the Treaty of Augsburg (1559) made the religion of the Prince the religion of the province (though the choices were limited to Lutheranism or Catholicism). In England, Elizabeth I's Act of Supremacy and Uniformity (1559) was the most liberal, requiring only outward conformity to an Anglican Church that ranged from High Anglican (virtually indistinguishable from the Catholic) to churches virtually indistinguishable from the more radical Protestants. In France, Henry IV's Edict of Nantes (1598) granted Protestants the right to worship as they chose.

Unfortunately, these advanced and liberal policies were far from the norm, and events in Bohemia (modern Czech Republic) would once again tear Europe apart with war. The king of Bohemia, Ferdinand II, was Catholic, but ruled over many Protestant subjects, called *Utraquists*. On May 23, 1618, the Utraquists threw out their Catholic governors—literally, tossing them from a castle window. The governors survived the seventy-foot fall, being born by the wings of angels (Catholic viewpoint) or landing in a pile of dung (Protestant viewpoint), but the "defenestration of Prague" began the Thirty Years' War. The tide of battle would sweep back and forth across central Europe until 1648, and involve all the major powers but England.

12.1 Algebra and Geometry

The factors that gave geometry a two thousand year domination of mathematics began to erode during the course of the seventeenth century. The invention of algebra meant that many powerful problem-solving algorithms were available

Figure 12.1: France in the Seventeenth Century.

to mathematicians. The development of notation meant that no longer was geometry superior in its ability to discuss general cases. Only in the area of logical validity did geometry still reign supreme, for algebraic algorithms oftentimes produced paradoxical results, such as the square roots of negative numbers. A fruitful union of the two methods, the algebraic and the geometric, would result in the invention of **analytic geometry**.

12.1.1   Descartes

René du Perron Descartes (March 31, 1596-February 11, 1650) was educated by Jesuits, but by profession was a mercenary soldier. Descartes joined the army of Maurice of Nassau, student of Simon Stevin (though there is no indication that Stevin and Descartes ever met), fighting on the side of the Protestants to keep Ferdinand of Bohemia deposed. In 1618, Descartes met Isaac Beeckman, who interested him in a variety of scientific and mathematical topics. On March 26, 1619, Descartes wrote to Beeckman, announcing that he had gotten the "first glimpse of a new science": what would eventually become analytic geometry.

Descartes then joined the army of Maximilian, Duke of Bavaria, fighting on the side of the Catholics to restore Ferdinand. On November 10, 1619, Descartes had an important revelation: a proper system of philosophy must begin by doubting everything, and accepting only those things that can be derived logically. The starting point is the acceptance of your own existence, since if you did not exist, you could not even ask the question of whether or not you existed. Descartes summarized this in the immortal phrase, "I think, therefore, I am." This is an early version of the **anthropic principle**, an important (if controversial) element of modern cosmology. Descartes eventually tired of soldiering, and in 1628, went to the Netherlands to develop his philosophical system. *Discourse on the Method* appeared in 1637, and would become one of the most important works in the history of science and mathematics, mainly for its appendices. Two of these dealt with physics: one described the laws of refraction, earlier discovered by Snell, and the other explained the shape and formation (but not colors) of the rainbow.

The third appendix, *The Geometry*, would forever link the name of Descartes with mathematics. The opening sentence reads:

> All problems in geometry can be easily reduced to some terms, for whose construction it is only necessary to know the lengths of certain straight lines.[1]

René Descartes. The text around the figure reads, "René Descartes, Lord of Perron. Born in La Haye, Touraine on the last of March, 1596." Boston Public Library/Rare books Department. Courtesy of the Trustees.

See Section 5.6.

See page 327.

Descartes proceeded to show his readers the geometric analogs of addition, subtraction, multiplication, division, and extracting square roots, all taken from various propositions in the *Elements*. To avoid having to draw lines on paper, Descartes referred to line segments using letters. Other authors had done so before, but it was always a method of "shorthand"; Descartes used letters systematically and consistently. Recall Viète had used vowels to represent unknowns and consonants to represent parameters; it was Descartes who began the practice of using letters near the end of the alphabet, x, y, z, and so on, to represent unknowns and letters near the front of the alphabet, a, b, c, and so on, to represent parameters. Descartes also introduced the modern notation for exponents, though he concerned himself only with positive, integral exponents, and for typographical reasons, a^2 would be written as aa well into the eighteenth century. Descartes used the square root symbol in the modern way, but still retained notation like $\sqrt{Ca^3 - -b^3 + abb}$ to indicate $\sqrt[3]{a^3 - b^3 + ab^2}$: note his use of $--$ to indicate a subtraction.

Descartes clearly outlined his general procedure; we may fruitfully compare it with Pappus's description of analysis and synthesis. First, we suppose the problem already solved, and assign "names" (letters) to all the lines used in the construction. Then relationships are found between the lines until a single quantity can be expressed in two ways, which allows an equation to be constructed and solve algebraically. The arithmetic operations in the solution correspond to geometric procedures, which can be applied to the original problem.

For example, consider the problem

---

[1] Descartes, Rene. *Geometria*. Van Schooten, 1637. Page 1.

Problem 12.1. *Given two lines, to find a third so the square on the second plus the rectangle on the first and third equals the square on the third.*

Let a, b be the two given lines and z the unknown; the problem is equivalent to the algebraic equation

$$z^2 = az + b^2$$

Solving for z, and keeping in mind only positive values of z have any geometric meaning, we have

$$z = \sqrt{b^2 + \frac{1}{4}a^2} + \frac{1}{2}a$$

Descartes pointed out that solving this equation was equivalent to the following construction procedure: begin with a circle centered at N with radius NL equal to $\frac{1}{2}a$; extend LM perpendicular to NL and let $LM = b$; then MO is equal to z. Thus, to find z and solve the geometric construction problem, construct right triangle NLM with $NL = \frac{1}{2}a$ and $LM = b$; draw the circle centered at N with radius NL, and extend MN to O on the circle; MO is the desired line.

Descartes applied his methods to find normals.

Rule 12.1. *Given curve AEC, find the normal to the curve at C. Suppose this is done, and CP is perpendicular to the curve; extend it to P on GA, and let CM be perpendicular to GA. Let $MA = y$, $CM = x$. Let $PC = s$, $PA = v$, and thus $PM = v - y$. Thus*

$$s^2 = x^2 + v^2 - 2vy + y^2$$

and thus $x = \sqrt{s^2 - v^2 + 2vy - y^2}$, or $y = v + \sqrt{s^2 - x^2}$. Either of these expressions can be substituted into the equation relating the points on the curve to eliminate one of the variables.

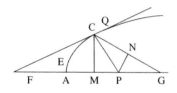

The line PM is called the **subnormal**.

If there is any serious criticism of *The Geometry*, it is that Descartes never chose simple examples; in fact, he stated explicitly that he intended the reader to gain proficiency by working through the problems on their own. Thus, his first example of finding the normal involved an ellipse. We will use a simpler example, using the familiar curve $y = x^2$.

Example 12.1. *Let the curve AEC, with $MA = y$, $CM = x$ be given by $y = x^2$. Then, eliminating x using the preceding equations, we have $y = s^2 - v^2 + 2vy - y^2$. Since the point is given, the goal is to determine s or v, which will determine how to draw the normal.*

Given any point on the parabola, this is an equation with two unknowns, s and v. To find the normal, note that the circle with center P passing through C will, if PC is the normal, touch the curve exactly once, while if P is slightly closer or slightly farther away, the circle will touch the curve twice. Thus, it is necessary that the equation $y = s^2 - v^2 + 2vy - y^2$ to have two equal roots. Hence the equation must be of the form obtained by setting $y = e$ and multiplying $(y - e)$ by itself, or $y^2 - 2ey + e^2 = 0$.

See Problem 3.

See Problem 4.

See Problem 2.

Setting the equation $y = s^2 - v^2 + 2vy - y^2$ equal to zero and equating coefficients, we have $-2e = 1-2v$, and thus $v = \frac{1+2e}{2}$. Or, since $y = e$, then $v = \frac{1+2y}{2}$. Thus, v is found in terms of y, and thus the location of P is determined.

It should be pointed out that Descartes *began* with curves and derived their equations; at no point did Descartes graph a curve from its equation since, to Descartes, it would be pointless to simply write down an equation and expect it to mean anything.

12.1.2 Fermat

Descartes was not alone in his development of analytic geometry; in fact, at least as much credit must be apportioned to Pierre de Fermat (August 20, 1601-January 12, 1665), a lawyer and politician by profession. In his free time, Fermat studied classical literature, including science and mathematics. Since so much of Greek and pre-Greek literature has been lost through the passage of time, a common preoccupation (then, as now) was attempting to reconstruct lost treatises from references in other works. The *Plane Loci* of Apollonius intrigued Fermat, and he began its reconstruction. Around 1636, he realized the fundamental principle of analytic geometry: the relationship between two unknown quantities translated into the solution of a locus problem. This is closer to our own concept of analytic geometry, and especially our idea of the graph of an equation, so perhaps it is more appropriate to call Fermat the real inventor of analytic geometry.

Thus, Fermat began by analyzing various cases of loci. Given a line NZM, with N a fixed point; let segment ZI be constructed from a point on the locus to the line, making a given angle NZI (*not* assumed to be a right angle) with NZM (see Figure 12.2). Let NZ be an unknown quantity A, and the segment ZI be another unknown quantity E (notice that here, Fermat is using Viète's notational system). By assuming various conditions on products of A, E, and other fixed quantities B, D, different loci are generated. The first is that of a line.

Proposition 12.1. *When $D \cdot A = B \cdot E$ (for some constants D, B), the point I describes a straight line, since $B : D = A : E$.*

Fermat did not explain why $B : D$ as $A : E$ guaranteed I was on a straight line, though a proof is relatively easy to construct. Fermat continued with other loci.

See Problem 5.

Proposition 12.2. *If A times E is some constant D, then the point I describes a hyperbola.*

Proof. Draw NR parallel to ZI; through M draw line MO parallel to ZI. Construct rectangle NMO whose area is D. Construct the hyperbola through O with asymptotes NM, NR. The hyperbola satisfies the locus statement, and passes through the point I. $\qquad\square$

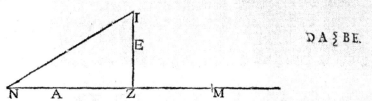

Figure 12.2: Fermat's First Locus. Boston Public Library/Rare books Department. Courtesy of the Trustees.

12.1.3 Theory of Equations

The *Geometry* of Descartes was not just about analytic geometry, but covered a broad range of topics. Because the solution to an algebraic problem was the solution to a geometric problem, Descartes felt it necessary to further investigate the properties of equations, which he did in the Third Book. There he noted

Conjecture 12.1 (Fundamental Theorem of Algebra). *The unknown quantity in any equation can have as many roots or different values as its dimension [degree].*

Descartes provided a demonstration of this conjecture.

Demonstration. If an equation has two distinct roots, say $x = 2$ and $x = 3$, then $(x - 2)(x - 3) = x^2 - 5x + 6 = 0$, and thus x has two values and the equation is of the second degree. If a third root is added, say $x = 4$, we can multiply $x^2 - 5x + 6$ by $(x - 4)$ and obtain the equation $x^3 - 9x^2 + 26x - 24 = 0$, and the equation is of the third degree and x has three values. Thus, it is evident that an equation is always divisible by a binomial consisting of the unknown minus a root. In this way, the degree of an equation can be lowered. Conversely, if an equation is not divisible by a binomial consisting of the unknown minus or plus a quantity, then the quantity is not a root of the equation. $\square$

This proof is not considered valid, and in fact, the fundamental theorem of algebra would remain unproven until the eighteenth century.

Without any justification, Descartes claimed:

Conjecture 12.2 (Descartes's Rule of Signs). *There can be as many positive roots of an equation as there are changes in sign between adjacent terms; there can be as many negative roots of an equation as there are adjacent terms with the same sign.*

Example 12.2. *The equation $x^4 - 4x^3 - 19x^2 + 106x - 120 = 0$ has (taking the first term as $+x^4$) three changes in sign; one pair of adjacent terms ($4x^3$ and $19x^2$) have the same sign; hence the equation has at most three positive roots and at most one negative root.*

Some of these roots, Descartes noted, might be neither negative nor positive, but **imaginary**: it was the first use of this term to refer to the complex numbers.

To get a better idea of what the roots actually were, Descartes suggested a procedure by which one transformed the given equation into another one. The relationship between the positive and negative roots of the new equation gave an indication of the limits of the roots of the equation. As with his analytic geometry, Descartes gave no simple examples, so we will construct one.

Example 12.3. *Applying Descartes's Rule of Signs to $x^2 - 5x - 3 = 0$, we note it has at most one positive root and at most one negative root. If we apply the substitution $x = y - 3$, (and thus $y = x + 3$), we obtain:*

$$
\begin{array}{rrcl}
y^2 & -6y & 9 & = x^2 \\
& -5y & +15 & = -5x \\
& & -3 & = -3 \\
\hline
y^2 & -11y & +21 & = 0
\end{array}
$$

The new equation has no adjacent terms with the same sign, so it has no negative roots. Now, suppose x is a root of the original; then $x + 3$ is a root of the new equation, which must be nonnegative. Thus, if x is real, then $x > -3$.

Other geometers produced similar results; for example, in 1657, Florimond de Beaune (1601-1652), a lawyer, published *On the Nature, Constitution, and Limits of Equations.* A typical analysis was in Chapter I, Proposition I:

Proposition 12.3. *If $x^2 - lx + mm = 0$, then x is less than l but greater than $\frac{mm}{l}$.*

Proof. By transposition, we have $mm = lx - x^2$. Hence lx is greater than x^2, so dividing each term by x, it must be that l is greater than x. Alternately, we have $x^2 = lx - mm$. Hence lx must be greater than mm, and thus, dividing by l, we must have x greater than $\frac{mm}{l}$. $\square$

DeBeaune gave similar propositions for other types of equations. For cubics, he had:

Proposition 12.4. *If $x^3 - mmx + n^3 = 0$, then x is greater than $\frac{n^3}{mm}$ and less than m.*

12.1.4   The Peace of Westphalia

In 1630, Protestant Sweden, led by her charismatic king, Gustavus II Adolphus, entered the Thirty Years' War. Adolphus, considered one of the greatest kings in Swedish history, was one of the first generals to master the use of artillery. Under his direction, Swedish forces swept as far south as the Danube. Unfortunately, Gustavus Adolphus was killed at the battle of Lützen (near modern-day Leipzig) in 1632.

He was succeeded by his daughter, Christina, but she was only 6 at the time of her father's death; hence, Sweden was in fact ruled by Gustavus's Chancellor, Axel Oxenstierna. In 1644, Christina turned 18. She realized the Swedish treasury was being depleted by a war that was, at best, peripheral to Sweden's interests, and helped end the war. The Peace of Westphalia (1648) established the shape of modern Europe. Two key provisions assured the independence of the Low Countries (the Netherlands and Belgium) and the neutrality of Switzerland.

Christina herself was somewhat eccentric. Interested in all aspects of learning, she became known as the "Minerva of the North" (after Minerva, the Roman goddess of wisdom and warfare). In 1649, she invited Descartes to be her tutor in philosophy and to establish an Academy of Sciences in Stockholm. When Descartes arrived, he found one of Christina's habits was waking at five every morning—when she expected her philosophy lessons. Between the rigors of early waking and the harsh Scandinavian winter Descartes, whose health had never been good, died in early 1650.

12.1 Exercises

1. In Problem 12.1, show $MO = z$.

2. Consider Example 12.1

    (a) Explain why the equation relating s, v, and y must have two equal roots if the circle is to touch the parabola exactly once.

    (b) Explain, in terms of synthesis and analysis, why Descartes says the equation will be of the form obtained by setting $y = e$, and then multiplying $y - e$ by itself.

    (c) Explain some of the difficulties with applying the procedure to a general curve AEC.

3. Find the normal to an ellipse, using Descarte's method. If if CE is an ellipse, MA a section of the axis for which CM is an ordinate, r the latus rectum of the ellipse, and q the transverse axis, then Apollonius's *Conics* I-13 tells us $x^2 = ry - \frac{r}{q}y^2$.

4. Explain how Descartes's determination of s or v can be used to draw the normal to the curve AEC. Phrase your answer as a geometric procedure that can be followed.

5. Prove Proposition 12.1, using Euclidean geometry.

6. What are the differences between modern coordinate geometry and the versions of Descartes and Fermat?

7. Determine the maximum number of positive and negative roots the following equations can have.

   (a) $x^4 - 5x^3 + 4x^2 + 3x + 7 = 0$

   (b) $x^5 - 4x^4 - 4x^3 - x^2 + 11x + 15 = 0$

   (c) $x^4 + x^3 + x^2 + x + 1 = 0$

8. Consider an equation whose coefficients are all positive. What happens when you apply the transformation $y = x + a$ to such an equation? What does this say about the negative roots of the equation?

9. What does DeBeaune's method for finding the limits of the equations imply about the assumed nature of solutions to equations? Note that you will have to assume the same things for the following problems.

10. Prove Proposition 2, Chapter I from DeBeaune: if $x^2 - lx - mm = 0$, then x is greater than the greater of $\sqrt{ll + mm}$ and $\sqrt{lm + mm}$ and less than $l + m$.

11. Prove Proposition 3, Chapter I from DeBeaune: if $x^2 + lx - mm = 0$, then x is greater than $\frac{mm}{l+m}$ and less than m.

12.2 Number Theory

In 1621, a Latin version of Diophantus was published, and soon afterwards, Fermat acquired a copy. By 1636, he began developing his own ideas on what would become the modern theory of numbers. What we know of Fermat's achievements in number theory comes primarily through his letters to other mathematicians. In these letters, Fermat claimed a number of important results, often saying that if the recipient was interested, Fermat would supply the proof. Unfortunately, the recipients were almost never interested.

12.2.1 Fermat's Lesser Theorem

One of Fermat's correspondents was Marin Mersenne (September 8, 1588-September 1, 1648), a Minimite friar. Mersenne corresponded with many of the mathematicians and philosophers of the age, serving as a central clearinghouse for letters between them. In mid-June 1640, Fermat sent one of his most useful results to Mersenne. Recall that a Euclidean perfect number is one of the form $2^{p-1}(2^p - 1)$, where $2^p - 1$ is prime. Fermat began with a list of the numbers one less than a power of 2 (which he called the *roots*, as they were in some sense the roots of the perfect numbers) and their corresponding exponents:

See Proposition 5.53.

Exponent	1	2	3	4	5	6	7	8	9	10	...
Root		1	3	7	15	31	63	127	255	511	1023	...

Fermat observed:

1. If the exponent was composite, so was the root.

2. If the exponent was prime, then twice the exponent divided the root minus 1. For example, exponent 7 corresponded to root 127, and $2 \cdot 7 = 14$ divided $127 - 1$.

3. Moreover, if the root was composite, the only factors were primes one more than a multiple of twice the exponent. Thus, Fermat noted, the root 2047, corresponding to the exponent 11, was divisible only by 23 and 89, both of which were 1 more than a multiple of twice 11.

Fermat expanded his result in a letter to Bernard Frenicle de Bessy (1605-January 17, 1675), dated October 18, 1640. This conjecture is now known as Fermat's little (or lesser) theorem.

Conjecture 12.3 (Fermat's Lesser Theorem). *Given any prime number and any geometric progression, the prime number will divide one of the terms of the sequence minus one. Moreover, the exponent of this term will divide the prime number minus one. Finally, any other term whose exponent is a multiple of the first such exponent will also satisfy the proposition.*

[Given prime p, p will divide $r^n - 1$ for some n; moreover, n will divide $p - 1$. Furthermore, p will also divide $r^{kn} - 1$ for all k.]

Fermat included no proof, and omitted a key requirement, probably assuming it was obvious: the ratio of the geometric progression must not be a multiple of the prime number.

Example 12.4. *The successive powers of 3 and their corresponding exponents are:*

Exponent	1	2	3	4	5	6	...
Progression Minus 1	2	8	26	80	242	728	...

Given the prime number 13, we note the third term of the progression minus 1, 26, is divisible by 13; moreover, the corresponding exponent, 3, is a divisor of $13 - 1$.

Fermat noted a few consequences of his result.

Conjecture 12.4. *In order for a prime to divide the power of a progression plus one, the smallest power minus one divisible by the prime must be even. Moreover, the smallest power of the progression plus one so divisible by the prime will be half the even exponent.*

Example 12.5. *Given the prime 23, we note 11 is the smallest power for which $2^m - 1$ is divisible by 23. Since 11 is odd, then 23 will never divide $2^m + 1$, for any m.*

Example 12.6. *Given the prime 5, we note that 4 is the smallest power for which $2^m - 1$ is divisible by 5. Since 4 is even, and half of 4 is 2, then 5 divides $2^2 + 1$.*

Fermat applied the property to factorization. For example, consider the number $2^{37} - 1$. If this number had a prime factor p, it was necessary that 37 be a factor of $p - 1$, and as p had to be odd (since, obviously, 2 does not divide $2^{37} - 1$), then 2 was also a factor. Hence, $2 \times 37 = 74$ had to be a factor of $p - 1$, so the possible factors are the primes in the sequence 75, 149, 223, ... After trying the first candidate, 149 (since 75 is not prime), Fermat found the second one, 223, to be a factor.

Example 12.7. *Find a prime factor p of $2^{11} - 1$. 11 must be a factor of $p - 1$; p must be odd (since 2 obviously cannot divide $2^{11} - 1$); hence 22 must divide $p - 1$, so the candidates are the primes in the sequence 23, 45, 67, ... We find 23 divides $2^{11} - 1$.*

Obviously, if a is odd, then $a^m \pm 1$ has a factor of 2; hence $a^m \pm 1$ can be prime only if a is even. Concentrating on the simplest case, $a = 2$, it can be shown that if m has an odd divisor, then $2^m + 1$ has a factor; hence $2^m + 1$ can be prime only if m is itself a power of 2. Numbers of the form $F_n = 2^{2^n} + 1$ are now called **Fermat numbers**. The first few numbers in the sequence are

$$3, \ 5, \ 17, \ 257, \ 65537, \ 4294967297, \ 18446744073709551617$$

and the first five are prime. In a letter to Pascal, dated August 29, 1654, Fermat conjectured they were all prime, but did not claim then to have a proof. However, in a 1659 letter to Huygens, Fermat claimed to have proven his conjecture. We will see the subsequent history of this conjecture.

Mersenne, perhaps based on Fermat's investigations, announced in 1644 the conjecture

Conjecture 12.5. $N = 2^p - 1$ *is prime for p equal to 2, 3, 5, 7, 13, 17, 19, 31, 67, 127, 257 and for no other values of p less than 257.*

Primes of this form $2^p - 1$ are now called **Mersenne primes**. Mersenne's conjecture It would later be determined that Mersenne was incorrect about 67 and 257, and omitted three other values less than 257 that yielded prime, namely 61, 89, and 107.

Fermat communicated another key result to Mersenne on December 25, 1640:

Conjecture 12.6. *All primes one more than a multiple of four could be expressed as the sum of two squares in a single way.*

It is in the investigation of Conjecture 12.6 that Fermat described the **method of infinite descent**, a specialized form of proof by contradiction. In 1659, Fermat described his proof as follows: if a prime of the form $4n + 1$ was not the sum of two squares, he could show that a smaller prime, also of the form $4n + 1$, was not the sum of two squares, and so on, resulting in a decreasing sequence of primes, none of which can be written as the sum of two squares. But this implied an infinite decreasing sequence of whole numbers, which was impossible. Hence, all primes of the form $4n + 1$ must be expressible as the sum of two squares. However, Fermat omitted a crucial step, that of how to find the smaller prime of the form $4n + 1$ that was inexpressible as the sum of two squares.

12.2.2   The Challenges of 1657

See page 366.

Fermat found so little interest in number theory among his correspondents that he stopped work in the field for most of the 1640s and early 1650s, and instead worked on problems that would eventually form his version of calculus. But Fermat soon returned to the theory of numbers after his belief in its importance was rekindled by correspondence with Pascal over probability (which will be dealt with in Section 12.5). Once again, Fermat tried to interest other mathematicians, with little to no success. Two of Fermat's correspondents, John Wallis and Christian Huygens, felt that number theory was not worth their time. Indeed, Huygens wrote to Wallis about Fermat's interest in number theory, and dismissed the subject, saying "There is no lack of better topics to spend our time on"; Wallis agreed. In 1654, Fermat tried unsuccessfully to interest Pascal and Pierre de Carcavi (1600-April, 1684) into co-authoring a work on the theory of numbers; again, neither was interested.

Still, Fermat tried to interest his correspondents in number theory. One tactic was to dole out enticing results. In a letter to Pascal in 1654, Fermat claimed every number could be written as the sum of three triangular numbers, or four squares, or five pentagonal numbers, and so on; this could be proven, *provided* every prime of the form $4n + 1$ could be written as the sum of two squares (this is the conjecture Fermat sent to Mersenne, above). Moreover, Fermat claimed, every prime one more than a multiple of three was the sum of a square and three times another square; every prime one or three more than a multiple of eight was the sum of a square and twice another square; and that no triangular number was equal to a square number (with the exception of 1). To these Pascal replied, diplomatically, that the results were quite impressive, but that he, Pascal, could not hope to extend them: perhaps Fermat could find someone more qualified to continue research on number theory?

Thus, in 1657 Fermat sent two challenge problems to Paris to be distributed to all mathematicians everywhere; a special copy was sent to England, for John Wallis:

> To be proposed (if you please) to Wallis, and the rest of the English mathematicians, the following questions about numbers.
>
> To find a cube, which, added to all its aliquot parts, makes a square. For example. The number 343 is a cube, with side 7. Its aliquot parts are 1, 7, 49; together with 343 makes the number 400, which is the square of the side of 20.

To find some square number which, added to its aliquot parts makes a cube.

We await the solution; if England, Belgic and Celtic Gaul do not find them, it will be given by Narbonese Gaul. ...[2]

"Gaul" was the Roman name for France; Belgic and Narbonese Gaul were two of its sub-divisions, while the inhabitants of the Gaul in general were known as Celts. Narbonese Gaul included Toulouse—Fermat's home town. Thus, Fermat's final statement was that if the English and the rest of the French could not find the solutions, the Narbonese—Fermat himself—would provide them.

Wallis did not receive the challenge until March; his response was a very brief note: once again, he said he was too busy to concern himself with number theory, but in any case, 1, by itself, answered both questions. Wallis's letter caused a debate between Wallis and Frenicle de Bessy over whether 1 could even be considered a number, which showed that, half a century after Stevin had boldly declared that 1 is a number, mathematicians were still undecided about its status.

See page 322.

Another solution was offered by Brouncker, an English mathematician: 343 divided by any sixth power, for example, $\frac{343}{64}$. Fermat rejected this solution: "Solving this problem by fractions ... does not satisfy me." [3] Here is a key element of modern number theory, and a crucial break with Diophantus: Fermat is insisting on a whole number solution. In any case, Brouncker's solution is invalid. Fermat eventually showed that 7 was the only non-trivial solution to the first problem; he dropped the second problem entirely. Fermat's hope of finding a kindred spirit failed.

12.2.3 The Last Conjecture

Fermat's most famous claim, unfairly designated Fermat's last *theorem*, made its first appearance in 1670 when Fermat's son, Samuel, published his father's copy of Diophantus, complete with his notes. In the notes, the elder Fermat claimed he had a "wonderful proof" that

Conjecture 12.7 (Fermat's Last Conjecture). *The equation $x^n + y^n = z^n$ has no non-trivial integer solutions for $n > 2$.*

However, he did not record his wonderful proof. Nor did this conjecture appear in any of Fermat's letters, which is significant, for in almost every case in which Fermat claimed to a correspondent that a result had a proof, a proof has been found that Fermat could reasonably have obtained. Thus, it seems likely that Fermat thought he had a proof but, upon re-examination, found the proof to be flawed.

However, Fermat apparently proved (or believed he could prove) the case of $n = 4$. In a letter to Huygens, Fermat claimed a proof for

---

[2]Wallis, John, *Opera Mathematicorum* (three volumes). L. Richfield and T. Robison, 1656. Volume II, p. 759.

[3]Ibid., 769

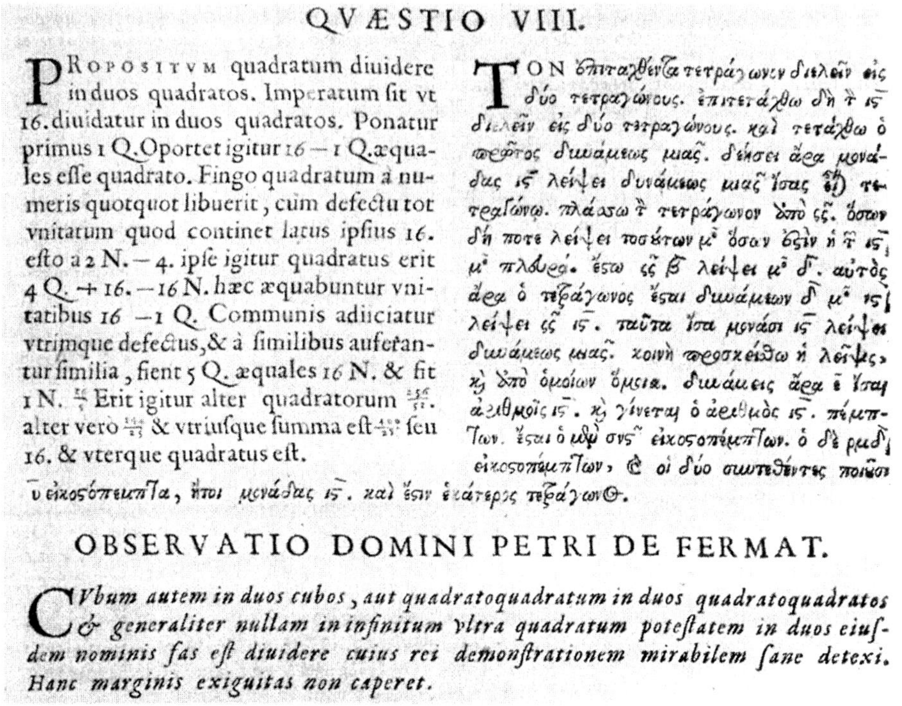

QVÆSTIO VIII.

PROPOSITVM quadratum diuidere in duos quadratos. Imperatum fit vt 16. diuidatur in duos quadratos. Ponatur primus 1 Q. Oportet igitur 16 − 1 Q. æquales esse quadrato. Fingo quadratum à numeris quotquot libuerit, cùm defectu tot vnitatum quod continet latus ipsius 16. esto a 2 N. − 4. ipse igitur quadratus erit 4 Q. + 16. − 16 N. hæc æquabuntur vnitatibus 16 − 1 Q. Communis adiiciatur vtrimque defectus, & à similibus auferantur similia, fient 5 Q. æquales 16 N. & fit 1 N. ⁴⁄₅ Erit igitur alter quadratorum ²⁵⁶⁄₂₅. alter verò ¹⁴⁴⁄₂₅ & vtriusque summa est ⁴⁰⁰⁄₂₅ seu 16. & vterque quadratus est.

ΤΟΝ ἐπιταχθέντα τετράγωνον διελεῖν εἰς δύο τετραγώνους. ἐπιτετάχθω δὴ ὁ ις διελεῖν εἰς δύο τετραγώνους. καὶ τετάχθω ὁ πρῶτος δυνάμεως μιᾶς. δεήσει ἄρα μονάδας ις λείψει δυνάμεως μιᾶς ἴσας ᾖ τετραγώνῳ. πλάσσω τὸν τετράγωνον ἀπὸ ςς. ὅσων δή ποτε λείψει τοσούτων μ᾽ ὅσων ἐστὶν ἡ τ ις μ᾽ πλῆθος. ἔστω ςς β λείψει μ᾽ δ᾽. αὐτὸς ἄρα ὁ τετράγωνος ἔσται δυνάμεων δ᾽ μ᾽ ις λείψει ςς ις. ταῦτα ἴσα μονάσι ις λείψει δυνάμεως μιᾶς. κοινὴ προσκείσθω ἡ λείψις, κ᾽ ἀπὸ ὁμοίων ὅμοια. δυνάμεις ἄρα ε ἴσαι ἀριθμοῖς ις. κ᾽ γίνεται ὁ ἀριθμὸς ις. πέμπτων. ἔσται ὁ μὲν σνς εἰκοστοπέμπτων. ὁ δὲ ρμδ εἰκοστοπέμπτων, ὃ οἱ δύο συντεθέντες ποιοῦσι

υ εἰκοστόπεμπτα, ἤτοι μονάδας ις. καὶ ἔστιν ἑκάτερος τετράγωνος.

OBSERVATIO DOMINI PETRI DE FERMAT.

CVbum autem in duos cubos, aut quadratoquadratum in duos quadratoquadratos & generaliter nullam in infinitum vltra quadratum potestatem in duos eiusdem nominis fas est diuidere cuius rei demonstrationem mirabilem sane detexi. Hanc marginis exiguitas non caperet.

Figure 12.3: The First Appearance of Fermat's Last Conjecture. Boston Public Library/Rare books Department. Courtesy of the Trustees.

Proposition 12.5. *The area of a right triangle whose sides were rational numbers could not be a square number.*

Fermat did not give the proof to Huygens. However, a proof appeared in Fermat's copy of Diophantus. Fermat provided an outline of the proof of this proposition, though omitted many details. Fermat's original outline is difficult to follow, since he expressed it totally without notation; we will add comments in brackets.

Proof. If the area of a right triangle with whole number sides is square, there exists two biquadrates whose difference is a square.

[If the sides of a right triangle are $2xy$, $x^2 - y^2$, and $x^2 + y^2$, and the area of this right triangle is a square, then x and y must be squares, say $x = u^2$ and $y = v^2$, and moreover, $u^4 - v^4$ is a square.]

Hence, there are two squares whose sum and difference are also squares.

[$u^2 + v^2$ and $u^2 - v^2$ are also squares, say

$$u^2 + v^2 = p^2$$
$$u^2 - v^2 = q^2]$$

Thus, there is a square that is the sum of a square and twice a square.
$[p^2 = 2v^2 + q^2.]$
Moreover, the sum of these squares is another square.
$[v^2 + q^2 = u^2.]$
Now, if a square is the sum of a square and twice a square, its side is also made up of a square and twice another square.
[Since $p^2 = 2v^2 + q^2$, there must be numbers m, n where $p = m^2 + 2n^2$.]
From this we can conclude that the side is equal to the sum of the sides of another right triangle, whose base is the square and whose perpendicular is twice the square.
[m^2 and $2n^2$ are the sides of another right triangle with rational area]
This right triangle will thus be formed from two squares, whose sum and difference is also a square. Hence it will be a right triangle whose sides are whole numbers, smaller than the original triangle, whose area is also a square. Thus, if there exists a right triangle whose sides are whole numbers and whose area is square, there exists a smaller right triangle, also whose sides are whole number sides whose area is square, and so on. But it is impossible to find an infinite sequence of decreasing whole numbers; hence, there cannot exist a right triangle whose sides are whole numbers and whose area is a square. □

Fermat claimed to have sent a complete proof to de Bessy, and the proof appeared in Frenicle's *Treatise on Whole Number Right Triangles* (1676). Frenicle gave no credit to Fermat, though this may have been an oversight: Frenicle died before he could complete the *Treatise*.

The proof, as given by de Bessy, is as follows: a right triangle is *primitive* if its sides have no common factor; in this case, one of its legs will be odd, the other even, and the hypotenuse odd. Hence:

Proposition 12.6. *If the odd side of a primitive right triangle is square, the hypotenuse of this triangle will be the sum of two squares, whose roots will be the hypotenuse of a second primitive right triangle and its even side; the root of the odd side of the first right triangle will be the odd side of the second primitive right triangle.*

Example 12.8. *The sides of length 9, 40, and 41 form a primitive right triangle; 9 is a square. We note $41 = 25 + 16$; hence a second right triangle will have hypotenuse 5; even side 4; and odd side 3.*

Proposition 12.7. *If, in a primitive right triangle, the hypotenuse and even side are both square numbers, the root of the hypotenuse will be the hypotenuse of another primitive triangle, whose odd side is a square and whose even side is twice a square.*

From these two propositions, de Bessy concluded

Proposition 12.8. *There is no whole number right triangle whose area is a square or a double square.*

and thus no fourth power could be the sum of two fourth powers.

12.2 Exercises

1. Consider the methods of proof used by Fermat. Why might Fermat have chosen to restrict the theory of numbers to whole numbers only?

2. Explain how Fermat's conjecture in the letter to Frenicle de Bessy is equivalent to Fermat's little theorem as it is given today.

3. Explain why, if $2n$ is the smallest number for which a prime p divides $a^{2n} - 1$, then p will divide $a^n + 1$.

4. Prove that if m has an odd factor, then $a^m + 1$ is composite; hence if $m \neq 2^n$, then $a^n + 1$ cannot be prime.

5. Factor, by finding a, m where $a^m + 1 = N$. Hint: $8^m + 1 = 2^{3m} + 1$, so $a = 8$ need not be considered as a separate possibility.

   (a) $N = 4097$ (c) $N = 1,000,001$ (e) $N = 1,048,577$

   (b) $N = 217$ (d) $N = 46,657$ (f) $N = 6,0466,177$

6. Find the potential factors of $2^{67} - 1$.

7. The claim of Fermat that numbers of the form $2^{2^n} + 1$ were all prime is now known to be false, and Fermat could have found it to be false using his own methods.

   (a) Show that if p divides $2^m + 1$, it also factors $2^{2m} - 1$.

   (b) Show that factors of a number of the form $2^{2^n} + 1$ must be primes of the form $k2^{n+1} + 1$.

   (c) Find the potential factors of F_5.

   (d) Factor F_5.

8. How does Proposition 12.5 follow from Proposition 12.8?

9. Factor, or prove prime:

(a) $6^{2^2} + 1$ (c) $10^{2^2} + 1$ (e) $14^{2^3} + 1$

(b) $6^{2^3} + 1$ (d) $10^{2^3} + 1$ (f) $18^{2^3} + 1$

12.3 The Infinite and Infinitesimals

The study of the infinite in the modern era began with the Italian physicist Galileo Galilei (February 15, 1564-January 8, 1642). In *Dialog Concerning Two New Sciences* (1638), Galileo discussed notions of the infinite and the use of infinitesimals. First, Galileo noted that the number of square numbers is "obviously" less than the number of squares and nonsquares together. But at the same time, there must be as many squares as there are numbers, since each square corresponds to a number, its root. This leads to the conclusion that there are as many squares as there are numbers. Galileo felt this conclusion was absurd, and concluded that it was not possible to apply "greater," "equal," or "lesser" to infinite quantities.

Dialog was primarily concerned with establishing Copernicus's heliocentric universe as a viable alternative to Ptolemy's geocentric one. There are ways of disagreeing with people without making enemies—and then there is Galileo's way. Galileo chose to portray the supporters of the geocentric system as ignorant fools, always citing Aristotle for support, never being willing to think for themselves, and frequently saying they were *unable* to do so. When Pope Urban VIII brought Galileo before the Inquisition on charges of heresy, the only thing that saved Galileo from imprisonment was the fact that Urban VIII was a friend of Galileo's. Still, Galileo was forced to recant, and spent the last eight years of his life under house arrest.

12.3.1 Kepler

Galileo's *Dialog* contrasted the Ptolemaic, Earth-centered universe with the Copernican, sun-centered one. He ignored two other systems proposed at the time. The first, that of Tycho Brahe, was a compromise whereby all planets *except* the Earth revolved around the sun, and the sun, carrying the planets along with it, orbited the Earth. Tycho's system was mathematically equivalent to the Copernican system, but retained the all-important feature of the immovable Earth. The Tychonic system might be seen as the last, dying attempt of an Aristotelian (Brahe) to reconcile the growing weight of evidence that the planets do, in fact, orbit the sun.

More importantly, Galileo ignored the system of Johann Kepler (December 27, 1571-November 15, 1630), who spent most of his life in the Holy Roman Empire. Kepler envisioned the planets moving about the sun in elliptical, not circular, orbits. Galileo knew Kepler and his model of the solar system, though he probably dismissed it: Kepler cast horoscopes, and so Galileo probably considered Kepler's elliptical orbits another pseudoscientific idea.

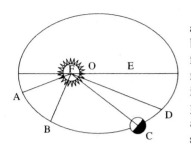

One of the problems with the Copernican system was that, overall, it was *less* accurate than the Ptolemaic system. Kepler, working with the data accumulated by Brahe (who by this time had moved from Uraniborg to Prague), attempted for years to make the path of Mars correspond to a combination of circular motions. Finally, he realized the position could best be modeled if Mars moved in an *elliptical* orbit; the ellipse gave a near perfect match between observed position and predicted position. Interestingly, Kepler had originally theorized an elliptical orbit, but an arithmetical error caused him to discard the correct solution.

One of the important consequences was that the sun was not at the center of the conic section, but rather at what is now called the **focus**. In Latin, the focus is a hearth or fireplace, so the name is particularly apt; moreover, with Kepler's work, the focus became central to the study of the conic sections. **Kepler's three laws** are:

1. Planets move in elliptical orbit, with the sun at one focus.

2. The radius vector from the planet to the sun sweeps out equal areas in equal times. Thus, if the time it takes the planet to move from A to B is the same as the time it takes to move from C to D, then areas AFB, CFD are equla.

3. The cubes of the semimajor axes are proportional to the squares of the periods.

The first two appeared in his *New Astronomy* (1609); the third, sometimes referred to as **Kepler's Harmonic law**, appeared in *Harmony of the World* (1618).

The second law required a consideration of the arc lengths and areas of ellipses. Kepler was able to derive the area formula, $A = \pi ab$ for an ellipse whose semimajor axes were a and b, but could only approximate the perimeter as $L \approx \pi(a + b)$. In fact, determining the arc length of an ellipse would become one of the more fruitful problems of mathematical physics. To find the area of an ellipse, Kepler used the idea of dividing it into a series of infinitely thin rectangles. Likewise, he found volumes by considering a figure to be made up of a series of infinitely thin slices, and in this way he found the volumes of solids of revolution. These results appeared in his *New Stereometry* (1615).

Unfortunately, the Thirty Years' War would begin three years later. Most of the fighting occurred in the empire, and its effect on mathematics is stark: though both the Italian peninsula and the empire were centers of mathematical research before 1600, the war meant the inhabitants of the empire were hardpressed to do more than concentrate on survival. As a result, Kepler's *New Stereometry* had little impact, and it was up to the Italians and the French to develop infinitesimal methods into a powerful tool.

12.3.2   Cavalieri

In 1635 appeared *Geometry, Advanced in a New Way by Indivisibles of the Continua*, by Buenaventura Cavalieri (1598-November 30, 1647), a student of Galileo. Called one of the "most unreadable books" in history, the seven books of *Geometry* included ideas Cavalieri had developed by the 1620s. Cavalieri began with the idea of a tangent, which he did not formally define until Book VII, as:

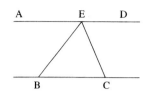

Definition. *A line is **tangent** to a curve at a point if the curve is completely on one side of the line, or if the tangent coincides with the curve over a line and there is one side of the tangent line on which there is no part of the curve.*

An important difference between Cavalieri's definition of a tangent and our is own is that he would consider the lines *AD* and *BC* tangent to triangle *EBC*. Given a plane figure *EFGH* and an arbitrary line *RS* called the **rule**, Cavalieri claimed that parallel to the rule were two tangent lines, which he called the **opposite tangents**, and all other lines either intersected the figure in a line, for example *FH*, or not at all.

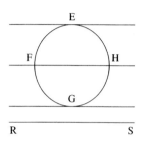

In Book II, Cavalieri introduced an important concept, that of *all lines* of a figure:

Definition. *If, through the opposite tangents of a plane figure are drawn parallel planes, perpendicular or oblique to the plane of the figure, and if one of the planes moves towards the other, always remaining parallel, the totality of the lines of intersection of the plane and the figure form **all lines** of the figure, and one of these lines is taken to be the **rule**.*

Although it would appear that Cavalieri is considering the plane figure to be made of an infinite collection of lines, he specifically denied this, for he knew the difficulties that would arise by comparing infinite collections. For example, consider the parallelograms *ABCD*, *ABFE* in Figure 12.4, with planes α and β. As α moves towards β, it cuts off segments *GH*, *JK* on the two parallelograms, and even though every line *GH* in *ABFE* is equal to every line *JK* in *ABCD*, the two parallelograms are obviously unequal: this seeming equality of unequal areas is known as **Cavalieri's paradox**. It first appeared in a letter from Cavalieri to Evangelista Torricelli (October 15, 1608-October 25, 1647), a student of Galileo's. To resolve this paradox, Cavalieri argued that in *ABFE*, the lines were closer together.

Cavalieri's main principle was

Proposition 12.9 (Cavalieri's Principle). *Given two plane figures ACM, CME. If, at any height RM, the lines BR, RD parallel to the rule always satisfy $BR : RD = AM : ME$, then $ACM : MCE = AM : ME$.*

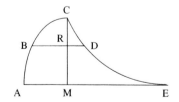

Cavalieri's interest was in finding a *general* method of quadratures (and cubatures). To make the method acceptable, of course, it was necessary to show that it gave the same results as proper, non-infinitesimal methods. In Book II, Theorem 19, Cavalieri used his "all lines" technique of indivisibles to prove:

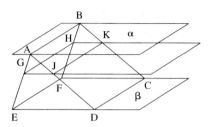

Figure 12.4: Cavalieri's Planes.

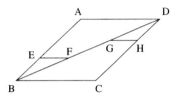

Proposition 12.10. *The diagonal of a parallelogram divides it into two equal triangles.*

Proof. Given parallelogram $ABCD$ and diagonal BD, consider lines EF, GH at equal heights above AD, BC, respectively. They are equal. Hence all lines in ABD are equal to all lines in BDC, and thus the two triangles are equal.   □

By extension, Cavalieri could talk about "all squares," and thus compare the volumes of solid figures (it is in this sense that Cavalieri's principle is most frequently invoked).

An important proposition is:

Proposition 12.11. *Given two parallelograms $ABCD$ and $EFGH$, with altitudes BL, FM, and rules DC, HG, then*

$$\begin{pmatrix}\text{all squares}\\ABCD\end{pmatrix} : \begin{pmatrix}\text{all squares}\\EFGH\end{pmatrix} = \left\{\begin{pmatrix}\text{square}\\DC\end{pmatrix} : \begin{pmatrix}\text{square}\\HG\end{pmatrix}\right\} \cdot (BL : FM)$$

Obviously, this also holds for the figures whose bases are the triangles formed by half the parallelograms. This was essential for proving

Proposition 12.12. *Given a parallelogram and its diagonal, all squares of the parallelogram will be triple all squares of one of the triangles, using one side of the parallelogram as the rule.*

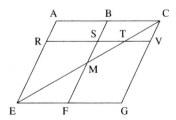

Proof. Given parallelogram $ACGE$, with diagonal CE; let BF bisect the parallelogram and let M be the midpoint; take EG the rule. Consider any line RV parallel to the rule. Then

$$\text{sq. } RT + \text{ sq. } TV = 2 \text{ sq. } RS + 2 \text{ sq. } ST$$

(by Euclid II-9). Thus

$$\begin{pmatrix}\text{all sq.}\\ACE\end{pmatrix} + \begin{pmatrix}\text{all sq.}\\GEC\end{pmatrix} = 2\begin{pmatrix}\text{all sq.}\\ABFE\end{pmatrix} + 2\begin{pmatrix}\text{all sq.}\\BCM\end{pmatrix} + 2\begin{pmatrix}\text{all sq.}\\FEM\end{pmatrix} \quad (12.1)$$

But FEM is equal to BCM, and ACE is equal to GEC, thus all the squares on these figures are equal; hence

$$\begin{pmatrix}\text{all sq.}\\GEC\end{pmatrix} = \begin{pmatrix}\text{all sq.}\\ABFE\end{pmatrix} + 2\begin{pmatrix}\text{all sq.}\\FEM\end{pmatrix}$$

By the previous theorem

$$\binom{\text{all sq.}}{GEC} : \binom{\text{all sq.}}{FEM} = \left\{ \binom{\text{sq.}}{EG} : \binom{\text{sq.}}{EF} \right\} \cdot (CG : MF) = 8 : 1$$

Hence

$$\binom{\text{all sq.}}{FEM} = \frac{1}{8} \binom{\text{all sq.}}{CGE}$$

Moreover

$$\binom{\text{all sq.}}{ACGE} : \binom{\text{all sq.}}{ABFE} = \binom{\text{sq.}}{EG} : \binom{\text{sq.}}{EF} = 4 : 1$$

Hence

$$\binom{\text{all sq.}}{ABFE} = \frac{1}{4} \binom{\text{all sq.}}{ACGE}$$

Hence $3 \binom{\text{all sq.}}{CEG} = \binom{\text{all sq.}}{ACGE}$. $\qquad \square$

Cavalieri's work corresponds to the integration formula $\int_0^a x^2 dx = \frac{1}{3}a^3$. By a straightforward (though tedious) modification, Cavalieri found equivalent formulas for $\int_0^a x^n dx$ for $n = 3, 4, 5, 6,$ and 9. Each formula required the previous one (i.e., the formula for $n = 3$ used the formula for $n = 2$), and for $n = 9$, Cavalieri assumed that a corresponding formula could be found for $n = 7$ and $n = 8$. Perhaps Cavalieri's most remarkable achievement was the determination, around 1643, of the volume of the infinitely large solid formed by revolving a hyperbola about the x-axis. This work of Cavalieri remained unprinted until 1919.

12.3.3 Mengoli

One of Cavalieri's students was Pietro Mengoli (1625-1686). Mengoli, like Oresme before him and Jakob Bernoulli after him, would show the harmonic series diverged by considering that the sum of any *finite* number of its terms could be made as large as possible. From this, Mengoli concluded that the sum of the reciprocals of *any* arithmetic sequence would diverge. A key discovery appeared in his *New Arithmetic Quadratures* (1650), where he noted the infinite sum of the reciprocals of the triangular numbers was 1, since the *finite* sum of the reciprocals of the first N triangular numbers was $\frac{N}{N+2}$.

Mengoli clearly anticipated the modern notion of a limit in his *Examples of Geometric Elements* (1659): if a variable quantity could be made larger than any given finite number, Mengoli called the variable quantity "quasi infinite"; while if the variable quantity could be made smaller than any given positive number, it was "quasi-nil." Unfortunately Mengoli's work, though well known to Wallis and Leibniz, would be forgotten by later mathematicians.

12.3 Exercises

1. Explain how Proposition 12.12 corresponds to $\int_0^a x^2 dx = \frac{1}{3}a^3$.

2. Derive Equation 12.1.

3. State and prove, using Cavalieri's notation and methods, a proposition equivalent to $\int_0^a x^3 dx = \frac{1}{4}a^4$.

12.4 The Tangent and Quadrature Problems

Calculus was originally invented to solve two main problems. First, the tangent problem: given a curve, how can one find the tangent to the curve? Second, the quadrature problem: given a curvilineal region, what is the area of the region? The ancient Greeks solved both of these problems for the curves they knew, but the invention of analytic geometry meant that curves unknown to the Greeks could be created. Moreover, the methods the Greeks used to solve the tangent and quadrature problems had to be specially crafted for each problem. A simple method that could be, in the words of a later mathematician, "learned by a novice" was desirable.

See page 410.

12.4.1 Fermat

From the work of Diophantus, Fermat became acquainted with the notion of **adequating**, whereby one solution to a problem was modified very slightly to give a different solution. Fermat applied a variation of this method to find extreme values. If we suppose the extreme value is found at a and another value—assumed to be very close—at $a+e$, Fermat would then adequate the two values by, essentially, setting them equal to each other. Then, he would collect like terms and divide the remainder by e or some power of e so that one term of the expression did not contain e; finally, any terms still containing e are removed, and the remaining terms are equated.

For example, to divide the line segment AC at some point E so that $AE \times EC$ is a maximum, Fermat supposed $AC = b$, a the length of one of the segments, and $b - a$ the remainder. The product $AE \times EC$ is $ba - a^2$. Now, suppose $a + e$ be the one segment, so the other is $b - a - e$, and their product is $ba - a^2 + be - 2ae - e^2$. Adequating $ba - a^2$ and $ba - a^2 + be - 2ae - e^2$ and canceling the similar terms, we have $2ae - e^2$ adequated with be; dividing all the terms by e we have $2a - e$ adequated to b; dropping the remaining e terms and equating the remainder, $b = 2a$.

Tangents

Fermat used adequation to find the tangents to curves. If BD is a parabola with axis CE, and ordinate BC, let BE be the tangent to the curve at B.

METHODUS

Ad difquirendam maximam & minimam.

OMNIS de inventione maximæ & minimæ doctrina, duabus pofitioni-bus ignotis innititur, & hac unica præceptione; ftatuatur quilibet quæftio-nis terminus effe A, five planum, five folidum, aut longitudo, prout pro-pofito fatisfieri par eft, & inventa maxima aut minima in terminis fub A, gradu ut libet inuolutis; Ponatur rurfus idem qui prius effe terminus A, + E, iterumque inveniatur maxima aut minima in terminis fub A & E, gradibus ut libet coefficientibus. Adæquentur, ut loquitur Diophantus. duo homogenea maximæ aut minimæ æqualia & demptis communibus (quo peracto homogenea omnia ex parte alterutra (ab E, vel ipfius gradibus afficiuntur) applicentur omnia ad E, vel ad elatio-rem ipfius gradum, donec aliquod ex homogeneis, ex parte utravis affectione fub E, omnino liberetur.

Elidantur deinde utrimque homogenea fub E, aut ipfius gradibus quomodolibet in-voluta & reliqua æquentur. Aut fi ex unâ parte nihil fupereft æquentur fane, quod co-dem recidit, negata ad firmatis. Refolutio ultima iftius æqualitatis dabit valorem A, quâ cognita, maxima aut minima ex repetitis prioris refolutionis veftigiis innotefcet.

Exemplum fubijcimus

Sit recta A C, ita dividenda in E, ut rectang. A E C, fit maximum; Recta A C, di-catur B.

A   E   C

ponatur par altera B, effe A, ergo reliqua erit B, — A, & rectang. fub fegmentis erit B, in A, — A² quod debet inueniri maximum. Ponatur rurfus pars altera ipfius B, effe A, + E, ergo reliqua erit B, —, A — E, & rectang. Sub. fegmentis erit B, in A, —, A² + B, in E, ²E in A, — E, quod debet adæquati fuperiori rectang. B, in A, — A², demptis communibus B, in E, adæquabitur A, in E⁴ + E², & omnibus per E, divifis B, adæquabitur ²A + E, elidatur E, B, æquabitur ²A, igitur B, bifariam eit dividenda, ad folutionem propofiti, nec poteft generalior dari methodus.

De Tangentibus linearum curvarum.

AD fuperiorem methodum inventionem Tangentium ad data puncta in lineis quibufcumque curvis reducimus.

H 4

Figure 12.5: Fermat's Method of Optimization. Boston Public Library/Rare books Department. Courtesy of the Trustees.

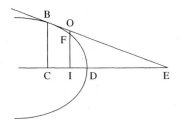

Then O on the tangent line BE must be exterior to the curve; hence by the property of the parabola, $CD : DI = BC^2 : FI^2$; hence $CD : DI > BC^2 : OI^2$. But since BCE, OIE are similar triangles, then $BC : OI = CE : IE$; hence $BC^2 : OI^2 = CE^2 : IE^2$. Hence $CD : DI > CE^2 : IE^2$.

Now, let $CD = d$; since this is the abscissa for the ordinate BC, this is a known quantity. Let $CE = a$, and $CI = e$. Then

$$\frac{d}{d-e} > \frac{a^2}{a^2 + e^2 - 2ae}$$

Hence

$$da^2 + de^2 - 2ade > da^2 - a^2e$$

Adequating the two sides and canceling the like terms, we have $de^2 - 2ade$ adequated to $-a^2e$. Dividing by e, dropping the remaining e terms and setting the remainders equal, we have $2d = a$.

Quadratures

Fermat also tackled the quadrature problem, and by about 1636, he had developed a method of generalizing Cavalieri's technique. Fermat's method of finding the area between a curve and an axis may be summarized as follows: one of the axes is partitioned into intervals in a geometric proportion, and rectangles constructed, bounded by the axis and a point on the curve. The rectangles, if their areas are in geometric proportion, may be summed. As the geometric ratio between the intervals tends to 1, the areas of the rectangles *become* the area under the curve, since more of them are required to partition the axis. Fermat applied the procedure to generalized hyperbolas and parabolas.

In modern terms, Fermat's approach is the following: suppose we wish to find the area under $y = \frac{1}{x^2}$ over the interval $0 \leq x < \infty$. Take $a > 1$, and partition the x-axis into the subintervals $[1, a)$, $[a, a^2)$, $[a^2, a^3)$, ... At the endpoint of each interval, a rectangle is constructed. We choose to use the right endpoint, giving us circumscribed rectangles (though inscribe rectangles will give the same final result). The first rectangle, with sides EG and GH, has area $1(a - 1)$. The second rectangle, with sides HI and HO, has area $\frac{1}{a^2}(a^2 - a)$. The third rectangle has area $\frac{1}{a^4}(a^3 - a^2)$, and so on, so the total area of all the rectangles is

$$\text{Area} = 1(a - 1) + \frac{1}{a^2}(a^2 - a) + \frac{1}{a^4}(a^3 - a^2) + \ldots \qquad (12.2)$$

$$= a \qquad (12.3)$$

The closer a is to 1, the more closely the circumscribed figure approximates the area under the curve. Hence, Fermat let $a = 1$ and obtained 1 for the area under the hyperbola.

Fermat's actual method began by considering the hyperbola $DSEF$ with asymptotes AR, AC; suppose the hyperbola is defined in such a way that the

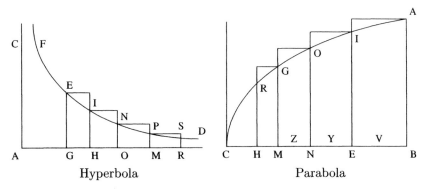

Figure 12.6: Quadrature of the Hyperbola and Parabola.

ratio of some power of AH to the same power of AG was equal to some power of EG to the same power of HI (in our terms, hyperbolas of the form $x^n = ky^m$). For example, suppose $AH^2 : AG^2 = EG : HI$ (i.e., $x^2y = k$). It is worth noting that Fermat here is taking the first steps towards *creating* new curves out of pure algebra, rather than defining them geometrically and deriving an algebraic equation for them.

To perform the quadrature, Fermat began with the summation formula for geometric series given in Euclid (IX-35): in any decreasing geometric sequence, the difference between successive terms is to the smaller as the greater term is to the sum of all the following terms. From this, any hyperbola, save that of Apollonius (i.e., the hyperbola $xy = 1$) could be rectified. What made Fermat's quadrature unusual is that it was the quadrature over an *infinite* region, for he determined the area of the region bound by the hyperbola ESD, the line EG, and the asymptote GR. Although Oresme had also considered infinitely long or infinitely high regions, Fermat was the first to consider nonrectilineal figures.

Fermat began by drawing EG, HI, NO, MP, RS, and so on, parallel to one of the asymptotes, AC in such a way that AG, AH, AO form a geometric sequence whose terms were "close enough" so the parallelograms EH, IO, NM, PR and so on, could be adequated. In particular, AG, GH, HO, and so on, were very nearly equal. Then, because the points in the interval were in a geometric sequence:

$$AG : AH = AH : AO = AO : AM \dots$$

Elements V-17 implied
See page 113.

$$AG : AH = GH : HO = HO : OM \dots$$

We have $\frac{EG}{HI} = \frac{AH^2}{AG^2}$ by the equation of the hyperbola. Moreover, because $AG : AH = AH : AO$, then $\frac{AH^2}{AG^2} = \frac{AO}{AG}$. Hence:

If $\frac{a}{b} = \frac{b}{c}$, then $\frac{a^2}{b^2} = \frac{a}{c}$.

$$\frac{EG \times GH}{HI \times HO} = \frac{EG}{HI} \times \frac{GH}{HO}$$

$$= \frac{AH^2}{AG^2} \times \frac{GH}{HO}$$

$$= \frac{AO}{AG} \times \frac{AG}{AH}$$

$$= \frac{AO}{AH}$$

Hence, the rectangles also form a geometric sequence, since

$$\frac{EG \times GH}{HI \times HO} = \frac{HI \times HO}{ON \times OM}$$

so $\frac{HI \times HO}{NO \times MO} = \frac{AO}{AH}$.

Now, $\frac{AO}{AH} = \frac{AH}{AG}$. Since AG is less than AH, this means the areas of the successive rectangles form a decreasing geometric sequence with ratio $AH : AG$.

Since GH is the difference between the first and second terms, and AG the smaller, then $GH : AG$ is as the first term of the progression, $GE \times GH$, is to the sum of the rest of the parallelograms. Since

$$\frac{GH}{AG} = \frac{GH \times EG}{AG \times EG}$$

then $AG \times EG$ is equal to the sum of the rectangles from IO onward. Adding to this the rectangle EH, which, since the differences were assumed to be nothing, has an area of 0; hence the area of the region $DEGR$ is equal to $AG \times EG$.

Example 12.9. *Suppose $AH^3 : AG^3 = EG : HI$. Find the area $DEGR$. We note*

$$\frac{EG \times GH}{HI \times HO} = \frac{EG}{HI} \times \frac{GH}{HO}$$

$$= \frac{AH^3}{AG^3} \times \frac{GH}{HO}$$

$$= \frac{AM}{AG} \times \frac{AG}{AH}$$

$$= \frac{AM}{AH}$$

$$= \frac{AO}{AG}$$

Likewise, $\frac{HI \times HO}{ON \times OM} = \frac{AO}{AG}$, and thus the areas of the successive rectangles are in a geometric proportion with ratio $AO : AG$. By the proposition, the difference GO is to the smaller AG as the first, $EG \times GH$, is to the sum of all the rectangles, from IO onwards. Since by assumption GH, HO, are very nearly equal, then

$GO = 2GH$. *Thus:*

$$\frac{GO}{AG} = \frac{2GH}{AG}$$
$$= \frac{GH \times EG}{\frac{1}{2}AG \times EG}$$

Hence, the sum of all the rectangles from HN onwards is equal to $\frac{1}{2}AG \times EG$. Adding in the rectangle GI, which, since the width GH, HO, ... were assumed to be 0, has an area of 0, we find area $DEGR$ is equal to $\frac{1}{2}AG \times GE$.

Fermat used a similar method to find the area bound by a generalized parabola of the form $y^m = ax^n$ over a finite interval. Given a parabola COA, where $CE : CB = EI^2 : AB^2$, Fermat made use of circumscribed rectangles; again, dividing the axis so that CB, CE, CN, ... formed a decreasing geometric sequence with BE, EN, NM assumed "nearly equal." The circumscribed rectangles have areas $BE \cdot BA$, $EN \cdot EI$. Here, however, Fermat would have run into difficulties, for

$$\frac{BE \cdot BA}{EN \cdot EI} = \frac{BE}{EN}\frac{BA}{EI}$$
$$= \frac{BC}{CE}\frac{BA}{EI}$$
$$= \frac{BC}{CE}\frac{\sqrt{BC}}{\sqrt{CE}}$$

Thus the quadrature would require determining the sum of a geometric series with ratio $r^{3/2}$ from the geometric series with ratio r, something Fermat and his contemporaries were incapable of calculating.

To avoid this difficulty, Fermat introduced a new set of division points, V, Y, Z, which further divided the axis at the geometric means between CB, CE, ...; in other words

$$BC : CV = CV : CE = CE : VY = \ldots$$

Again, near equality between BV, VE, EY, ... was assumed. Through the use of the auxiliary points, Fermat showed

$$\frac{BE \cdot BA}{EN \cdot EI} = \frac{CB}{CY}$$

Hence the areas of the rectangles formed a geometric sequence with ratio $CB : CY$. Hence

$$\frac{BE \cdot BA}{\text{area } CIE} = \frac{BY}{CY}$$

Since BE, EN, ... are assumed small, then, adequating, $CY = BC$ and thus

$$\frac{BE \cdot BA}{\text{area } CIE} = \frac{BY}{CY}$$
$$= \frac{BY}{BC}$$
$$= \frac{BY \cdot AB}{BC \cdot AB}$$

Alternating

$$\frac{BC \cdot AB}{\text{area } CIE} = \frac{BY \cdot AB}{BE \cdot BA}$$

Hence $AB \cdot BC$ is to the area CIE as $BY : BE$. Adding the zero area of $BA \cdot BE$, then the area under the parabola COA is to the area of the rectangle $AB \cdot BC$ as $BY : BE$. Since BV, VE, EY are nearly equal, Fermat concluded $BY : BE = 3 : 2$.

12.4.2 Gregory of St. Vincent

The one quadrature Fermat's method could not handle was that of the Apollonian hyperbola, $xy = 1$. However, this problem was dealt with by Gregory of St. Vincent (September 8, 1584-1667). St. Vincent was the first to use the term **method of exhaustion** for the Euclidean method of inscribing larger and larger rectilineal figures within a curvilinear figure; the term appeared in St. Vincent's 1647 work, *Geometrical Work on the Squaring of the Circle and Conic Sections.* St. Vincent, like Fermat, divided the axis into segments in geometric proportion, and proved

See Problem 7.

Proposition 12.13. *If $AG : AH = AO : AM$, then $GEIH$ is equal to $ONPM$.*

A friend of his, Alfons A. de Sarasa (1618-1667) noticed then that the area under the Apollonian hyperbola satisfied the same relationships that logarithms did, thus making explicit the integral $\int \frac{1}{x} dx = \ln x$.

Though St. Vincent probably conducted his researches in the years 1622 to 1629, as tutor to Philip IV of Spain, he found himself traveling quite a bit, and was often separated from his papers. Thus, his work was not published until 1647, by which time Fermat's own work on quadrature had appeared. Moreover, the title problem of St. Vincent, the squaring of the circle, was solved incorrectly, tarnishing St. Vincent's reputation: most of those who bought the work did so mainly to analyze the faulty quadrature, rather than to benefit from the flawless ones.

12.4.3 Roberval

Finally, some words should be said about Gilles Persone de Roberval (1602-1675), one of the few professional mathematicians of seventeenth century continental

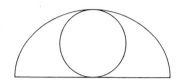

Europe. Roberval taught at the Royal College in Paris. In a curious twist of fate, this meant that he contributed very little to the development of mathematics, for Roberval's position had to be defended every three years by competitive examination, with questions posed by the incumbent. Thus, it was in Roberval's best interest to pose questions that he alone could answer — and the best way to ensure that was to keep his methods of solution secret.

Several of Roberval's main results involve the **cycloid**, which Mersenne brought to the attention of French mathematicians in 1615. The cycloid, already known to Galileo, was defined by Mersenne as the curve described by a point on a circle that rolls along the ground. Roberval determined the area under the arch of the cycloid was three times that of the generating circle (1634) and the method of drawing a tangent to the cycloid at any point (1638), but his need for secrecy was such that he did not communicate the methods of these discoveries to anyone else, and thereby lost credit for them. Consequently, when further developments were made, they followed the methods of Descartes and Fermat, who *did* communicate their discoveries, and not the methods of Roberval, who did not.

12.4 Exercises

1. Show how Equation 12.3 follows from Equation 12.2.

2. Use Fermat's method to find the tangent to the curve where the cube of the ordinate is equal to the solid formed by the abscissa and some fixed area (i.e., where $x^3 = ay$).

3. Use Fermat's method to find the quadrature for the hyperbola $x^4 y = 1$.

4. Fermat calls his method of equating two "very nearly equal" quantities to each other adequating. How does Fermat's method differ from that of Diophantus?

5. For Fermat's quadratures, he used Euclid IX-35. However, Euclid IX-35 applies only to *finite* sums. Prove, using Euclidean methods, that it can also be applied to infinite series. Hint: see Archimedes.

6. What difficulties do you run into using Fermat's method for the hyperbola $x^3 y^2 = 1$?

7. Why does Fermat's method fail for the hyperbola $xy = 1$?

8. Apply Fermat's procedure for the quadrature of the "standard" parabola $y = kx^2$ over the finite interval $[0, 1]$, and generalize it to the curve $y = kx^n$.

9. Examine Fermat's method for finding the quadrature of the parabola.

   (a) Show that $\frac{BE \cdot BA}{EN \cdot EI} = \frac{CB}{CY}$.

   (b) How would the method be changed for the generalized parabola where $CE : CB = EI^3 : BA^3$?

   (c) Find the quadrature of the generalized parabola where $CE : CB = EI^3 : BA^3$.

10. Consider the area under the Apollonian hyperbola $xy = 1$ over the interval $[1, \infty)$.

   (a) Using Fermat's method, show that if the axis is divided into segments in geometric proportion, the corresponding areas under the hyperbolic segments are equal.

   (b) Explain why this suggests a connection between the area under the hyperbola and logarithms, by showing how the areas under a hyperbolic segment obey the same rules that logarithms do.

12.5   Probability

The earliest treatise on probability was one written by Cardano, *On Games of Chance*. As the title suggests, Cardano's main interest was gambling. However, Cardano's work did not appear until 1663, long after his death, so it had little impact on the early history of probability. Around 1654, the Chevalier de Méré, a French nobleman with an interest in mathematics, philosophy, and gambling, posed the following question to an acquaintance of his, also interested in mathematics and philosophy: suppose a game was interrupted before its completion. How should the stakes be divided, if one player is ahead?

12.5.1   Pascal

The acquaintance of Méré was Blaise Pascal (June 19, 1623-August 19, 1662), the third of the three great French amateurs. One may only speculate what Pascal would have done had he lived longer; his mathematical activity was sporadic, confined to two brief periods in his life, but during those periods he produced some first-rate results. Pascal's brilliance in mathematics appeared early; one story, almost certainly untrue, is that he deduced for himself the first thirty-two propositions in the *Elements*, Book I, in *exactly* the same order given by Euclid! On a more reliable note, Pascal was one of the first to construct a calculating machine capable of mechanically performing the operations of addition and subtraction. Between 1642 and 1645, Pascal constructed over 50 different models, and on May 22, 1649, he was granted, by the French King Louis XIV, a monopoly on the manufacture and sale of the adding machines.

Pascal's adding machine (reproduction). ©Bettmann/ CORBIS.

The Problem of Points

In 1654, Pascal wrote to Fermat and described Méré's problem, now called the **problem of points**. In order to win many dice games (such as craps) a player must roll a number, called his point, within a specified number of throws; for example, a player might have eight throws in which to roll a six. If the game is interrupted after three *unsuccessful* throws, how should the stakes be divided?

Pascal's first letter to Fermat is, unfortunately, lost. Fermat's undated reply was probably written around July of 1654, and began by examining what it would be worth to the player to *not* make the next throw. Since there was a one-sixth chance that the player might win on the first throw, then it was worth one-sixth of the stakes *not* to make the first throw. By the same argument, to not make the second throw would be worth one-sixth of the stakes remaining, or $\frac{5}{36}$ of the original stakes, and so on. Thus, if the game was interrupted after the third throw, the player who had not yet made his point should receive $\frac{125}{1296}$ of the stakes.

Fermat's analysis introduced two key principles of early probability theory: first, the examination of **expectations** (the amount of the stakes that one player would earn), rather than the probabilities themselves; and secondly, that all events were equally probable.

Pascal's reply, dated July 29, 1654, agreed that Fermat's analysis was valid, but Pascal himself thought it too complex to apply in general. Instead, Pascal used the idea of reducing a problem to a previous case. Suppose the stakes are 64 *pistoles*, and one player has two of three games needed to win, and the other has one game. Pascal argued as follows: if the first player loses the next game, the score will be two to two, and the players should separate with 32 *pistoles* apiece. But if the first player wins the next game, he will win everything. Thus, he is certain to get 32 *pistoles*, and might or might not get the remaining 32. Thus, he should get the 32 *pistoles* he is certain of, and half of the remaining 32, for a total of 48 *pistoles*.

From here, Pascal worked recursively:

Example 12.10. *Suppose the game count were two games to nothing. How should the stakes (64 pistoles) be divided?*

Solution. If the player who is winning loses the next game, the count will be two to one, and by the previous analysis, he should take 48 *pistoles*. If he wins the next game, he will win the entire amount. Thus, he should receive the 48 *pistoles* he is certain of, plus half the remainder, another 8 pistoles, for a total of 56 *pistoles*. □

Example 12.11. *Suppose the game count was one game to nothing. How should the stakes be divided?*

Solution. Even if the player loses, he will have 32 *pistoles*. If he wins, the game count will be 2 to 0, and he will take 56 *pistoles* (*not* 64, since the match won't be finished). Thus, he is certain to get 32 *pistoles*, and might win an additional 24 *pistoles*, so he should receive the 32 *pistoles* he is certain of, plus half the 24, for a total of 44 *pistoles*. □

Pascal then reduced the analysis to the value of each game. If no games have been played, then the stakes (say 64 *pistoles*) should be divided equally, 32 to 32. If one game has been played, then the player who won it should receive 44 *pistoles*, and the other 20. Hence, winning the first game is worth 12 *pistoles*.

Focusing on the winner of the first game, if he then wins the second game, the stakes will be divided 56 to 8, and thus the second game will be worth 12 more *pistoles* to him. Finally, if he wins the third game, he wins the remaining 8 *pistoles.*

Without explanation Pascal gave a simple procedure for finding the value of the first game. Suppose both players put in equal stakes. If one game has been played, out of five necessary to win, then the outcome of the match will be decided within eight games. Write down, separately, the even and odd numbers from 1 to 8; the product of the odd numbers divided by the product of the even numbers will be the proportion of the other player's stakes that the winning player is owed.

Example 12.12. *If one game out of five necessary to win has been played, then the winner of the first game should be given $\frac{1\cdot3\cdot5\cdot7}{2\cdot4\cdot6\cdot8}$, or $\frac{105}{384}$ of his opponent's stakes.*

Pascal then gave, without proof, two results. In contrast with Stevin, who expressed complex mathematical ideas in French and Dutch, Pascal felt the French language to be "good for nothing" when it comes to logical expression; thus, he switched to Latin to claim:

Conjecture 12.8. *If the combinations of any number of letters, say 8, are taken, then half the combinations of four of the letters, plus all the combinations of five, six, seven, or eight of the letters, will be the fourth term in the quaternary progression beginning with 2.*

In other words, the fourth term in the progression 2, 8, 32, 128, 512, ...; Pascal noted that one used the fourth term because four was half of 8 (the number of letters).

Example 12.13. *Consider the combinations on six letters. Find half the number of combinations on 3, plus the number of combinations on 4, 5, and 6. Half of six is three, so the total number of combinations will be the third term of the sequence 2, 8, 32, 128,..., or 32.*

This restricted form of finding the number of combinations was what Pascal needed to solve the problem of points. He noted that if a player has one (out of five) games necessary to win, the match will be decided in 8 more games; thus the amount of the stakes of the other player that he is owed is equal to half the number of combinations of 4 out of 8 letters, 35, divided by the sum of half the combinations of 4 out of 8, plus the number of combinations of 5, 6, 7, or 8 out of eight. Thus, the winner of the first game is entitled to $\frac{35}{128}$ of his opponent's stakes, which is equal to $\frac{105}{384}$, obtained previously. Pascal then claimed the value of the second throw was always equal to the value of the first throw. Without indicating how he obtained the numbers, he provided a table, showing the value of each throw for a wager of 256 *pistoles* apiece (see Table 12.1).

Throughout this process of calculating how the stakes should be divided, the problem of determining the number of combinations arises. Pascal referred to

Throws Needed to Win						
	6	5	4	3	2	1
First Throw	63	70	80	96	128	256
Second Throw	63	70	80	96	128	
Third Throw	56	60	64	64		
Fourth Throw	42	40	32			
Fifth Throw	24	16				
Sixth Throw	8					

Table 12.1: Value of a Game with Stakes 256 *pistoles*.

"one of my old tables" that he used; it was too long to copy and send to Fermat, but given the context, it was probably a table of combinations — what is now known as **Pascal's triangle**.

The Arithmetic Triangle

Other authors had described the arithmetic triangle before Pascal. It was certainly known to the Chinese, Indians, and Islamic mathematicians of medieval times, though their work was probably unknown to European authors. However, See pages 209 and 243. even among European mathematicians, Pascal was not the first to describe the triangle. Peter Apian (1495-1552) included it in his work on algebra, *Calculation* (1527), and it also appeared in the *Arithmetic of the Integers* (1544) of Michael Stifel (1487-1567). However, most previous authors had simply presented the triangle and its uses; only Yang Hui and ibn Mun'im made any important observations about its general properties. Pascal was the first to provide a compre- See page 244. hensive analysis of the properties of the triangle, and because of this, it is not unreasonable to continue naming the triangle after him. His work, *The Arithmetic Triangle*, was printed around 1654, though not distributed until 1665.

To construct an arithmetic triangle, Pascal began with a large vertical and horizontal array of cells. The numbers in the cells themselves were designated (rather haphazardly) using a combination of Greek and Roman letters: thus, Pascal labeled the entries of the first row G, σ, π, λ, μ, δ, ζ; the entries in the second row were ϕ, ψ, θ, R, S, N; and so on. Pascal gave the following procedure for generating the entries:

1. All entries in the first row are one. Thus G is 1, σ is 1, π is 1, and so on.

2. For the second row: the first entry, ϕ, is 1. The second entry is the sum of the first two entries in the previous row; thus ψ is 2. The third entry is the sum of the first three entries in the previous row, so θ is 3, and so on.

3. For the third row: the first entry, A, is 1. The second entry, B, is the sum of the first two entries in the previous row, namely 1 and 2; thus B is

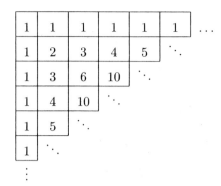

Table 12.2: The Arithmetic Triangle

$1 + 2$ or 3. The third entry, G, is the sum of the first three entries in the previous row, hence G is $1 + 2 + 3$, or 6.

4. The remaining rows are generated in a similar manner.

Pascal then stated and proved 18 propositions about the arithmetic triangle. The first one gives our usual method of generating the triangle:

Proposition 12.14. *Any entry is equal to the sum of the entry immediately to its left and immediately above it.*

Some of the rest are:

Proposition 12.15. *An entry is equal to the sum of the entries of the column preceding it, from the first row to the row of the entry.*

Proposition 12.16. *Each entry, minus one, is the sum of all the entries in the previous rows and columns.*

Pascal called the diagonals that cut across the array, from upper right to lower left, the **bases**; cells on the same base, equidistant from the ends of the base were **reciprocal cells**. The symmetry property of the triangle was thus stated as:

Proposition 12.17. *Reciprocal entries are equal.*

In Consequence 11, Pascal brought **mathematical induction** to the attention of European mathematicians (it had already been used in the work of Jewish and Arab mathematicians). In Pascal's terminology, the **root** of an entry was the number of the column it was in, while the **exponent** was the number of its row.

See page 291.

Proposition 12.18. *Given two adjacent entries on the same base, the lesser is to the greater as the root of the lesser is to the exponent of the greater.*

Before proving this proposition, Pascal noted that the proposition relied on two lemmas:

Lemma 12.1. *The proposition is true for the second base.*

Lemma 12.2. *If the proposition is true for any base, it is true for the next base.*

The first lemma was obviously true, since ϕ is to σ as 1 is to 1. Thus, Pascal noted, all the difficulty of proving the proposition was in proving the second lemma.

To find the entries in the arithmetic triangle, Pascal described a formula equivalent to the binomial coefficient $\binom{n}{r}$, though without the modern factorial notation.

Pascal used the table primarily for finding combinations. For example, Pascal noted

Lemma 12.3. *The sum of the cells of any row is equal to the number of combinations of the number expressing the row, out of the number of objects expressing the base of the triangle.*

What Pascal meant by this is that given any row in the triangle, say the fourth one, the sum of the portion of the row that is within the fourth triangle will be the number of combinations out of four objects.

Example 12.14. *Find the number of combinations of six objects taken four at a time. The fourth row of the triangle consists of the numbers 1, 4, 10, 20, 35, ... The portion of this row contained by the sixth triangle is 1, 4, 10; hence the number of combinations is $1 + 4 + 10$, or 15.*

Since the sum of the row entries is also the method of forming the next entries in the triangle, this means that the *fifth* cell of the *seventh* base will also give the desired value; this can be computed directly.

12.5.2 Christian Huygens

No mathematician, however brilliant, can make a contribution to mathematics unless his or her results are made known. Like Cardano's treatise on probability, the correspondence between Fermat and Pascal would not be published for a long time. However, Christian Huygens (April 14, 1629-June 8, 1695), a Dutch physicist and astronomer, learned of the correspondence and its specifics during a visit to Paris in 1655. He incorporated the ideas outlined by Fermat and Pascal into the first published treatise on probability, *On the Calculations in Games of Chance* (1656). The work mostly duplicated results Cardano had arrived at a hundred years before, but since Huygens was the first to publish, it was as if Cardano's work never existed.

Huygens's work was very elementary, including only fourteen propositions and five exercises. He defined the **chance of a game** as its expected value (which he referred to, confusingly enough, as the chance of an equitable game).

He gave the example of a man holding 3 shillings in one hand, and 7 shillings in the other; the chance to pick one of the hands and receive what was in it was worth 5 shillings, that being the expected value. A few simple propositions followed. For example, if a player has an equal chance of winning sums a or b, their expectation is $\frac{a+b}{2}$ (Proposition I); if a, b, and c are equally likely to be obtained, then the expectation is $\frac{a+b+c}{3}$ (Proposition II). Generalizing this to a player having p opportunities to win a stake of a and q opportunities to win a stake of b, their expectation is thus $\frac{pa+qb}{p+q}$.

12.5.3   Credibility of Witnesses

An anonymous work, "A Calculation of the Credibility of Human Testimony," appeared in the *Philosophical Transactions of the Royal Society of London* of October 1699. The author had a clear understanding of probability; moreover, the problems were stated in a manner that used basic business concepts, which suggests the author was Edmond Halley. The author began by considering an event, such as the arrival of a ship, that would provide a benefit of some amount, say 1200 pounds. If it was *absolutely* assured that the ship had arrived, then not even the smallest amount of insurance would be worthwhile. On the other hand, if there was a chance the report was incorrect, then to pay some amount for insurance would then be worthwhile. Thus, if the event was $\frac{5}{6}$ certain, then there was an "absolute certainty" of 1000 pounds; it would then be worth paying 200 pounds for insurance.

Since an event is invariably reported by a witness, then it is necessary to examine the credibility of the witness. This credibility is governed by two factors: the integrity of the witness (i.e., whether the witness tells the truth or not); and by the "ability" of the witness, which consisted of the witness being able to comprehend the event, and to retain the memory of the event. Had the author dealt with all these factors, he or she would have anticipated the work of Condorcet and Laplace; however, only the first factor was dealt with.

The author considered two cases. The first was n witnesses, all of equal integrity, telling the truth in a cases and not telling the truth in c cases. If these n witnesses were successive (thus, the first witness telling the second witness, who told a third, and so on), then the credibility of the report of the last witness is $\left(\frac{a}{a+c}\right)^n$. On the other hand, if the n witnesses were concurrent, the author gave their combined credibility as $1 - \left(\frac{c}{a+c}\right)^n$.

The author then applied the credibility question to the successful transmission of material through oral or written sources, and considered how long it would be before a particular story was only half reliable: in other words, how many years before the successive retransmission made the probability greater than $\frac{1}{2}$ that the material was no longer reliable. In the case of oral transmission, it was assumed that the story was passed on every 20 years, while in the case of written sources, which might last an indefinite period, the author assumed 100 years between recopying. Moreover, the oral sources could be assumed as less reliable, say with probability $\frac{100}{106}$ that any one witness correctly transmitted the story, compared with $\frac{100}{101}$ that a copy was accurately made. Under these assumptions, the author calculated that an oral tradition might last only 240 years before the probability was greater than $\frac{1}{2}$ that it was inaccurate, compared with 7000 years for written sources.

12.5 Exercises

1. Suppose two games are necessary to win a match. How should a stakes of 64 *pistoles* be divided if the score is 1 to 0? Explain your reasoning.

2. Suppose four games are necessary to win. For each of the following game counts, determine how the stakes (assume 64 *pistoles*) should be divided. Explain your reasoning.

   (a) 3 to 2.

   (b) 3 to 1.

   (c) 2 to nothing.

   (d) 1 to nothing.

3. Pascal's Wager is the following: either God exists or does not, and one can believe or not. If God exists and one is a believer, one "wins" eternal bliss, while if one is an atheist, one "wins" eternal damnation; in all other cases, one "wins" nothing.

   (a) Show that, under these conditions, one should be a believer.

   (b) What is the flaw in Pascal's argument?

4. Prove Pascal's method of determining the value of any entry in the arithmetic triangle: multiply the numbers from 1 up to (but not including) the root (column number); multiply the same number of numbers starting from the exponent (row number); divide the latter by the former, and the result will be the desired entry. For example, to find the entry in the fifth column, third row, multiply $1 \cdot 2 \cdot 3 \cdot 4$, and divide this into $3 \cdot 4 \cdot 5 \cdot 6$, which is 15.

12.6   The Sums of Powers

There are many parallels between the lives and works of Pascal and Leibniz, and an especially intriguing question in the history of mathematics is: had Pascal not died so young, and devoted so much of his too-short life to philosophical, rather than mathematical, pursuits, might he have anticipated Leibniz's development of calculus? Certainly Pascal began to develop a method of quadratures that clearly anticipated some of Leibniz's results.

The method of Pascal came about as a consequence of his investigation into the sums of powers. The results were included in *The Arithmetic Triangle*, though they were written somewhat earlier. To find the sum of the cubes of the arithmetic sequence 5, 8, 11, 14, Pascal began by expanding $(A + 3)^4$ as

$$A^4 + 12A^3 + 54A^2 + 108A + 81$$

The difference between 17^4 and 14^4 is the difference between $(14 + 3)^4$ and 14^4, or just

$$12 \cdot 14^3 + 54 \cdot 14^2 + 108 \cdot 14 + 81$$

Likewise, the difference between 14^4 or $(11 + 3)^4$ and 11^4 is

$$12 \cdot 11^3 + 54 \cdot 11^2 + 108 \cdot 11 + 81$$

and so on. Since

$$17^4 = 17^4 - 14^4 + 14^4 - 11^4 + 11^4 - 8^4 + 8^4 - 5^4 + 5^4$$

and

$$17^4 - 14^4 = 12 \cdot 14^3 + 54 \cdot 14^2 + 108 \cdot 14 + 81$$
$$14^4 - 11^4 = 12 \cdot 11^3 + 54 \cdot 11^2 + 108 \cdot 11 + 81$$
$$11^4 - 8^4 = 12 \cdot 8^3 + 54 \cdot 8^2 + 108 \cdot 8 + 81$$
$$8^4 - 5^4 = 12 \cdot 5^3 + 54 \cdot 5^2 + 108 \cdot 5 + 81$$

then

$$17^4 = (5 + 8 + 11 + 14)108 + (5^2 + 8^2 + 11^2 + 14^2)54$$
$$+ (5^3 + 8^3 + 11^3 + 14^3)12 + 81 + 81 + 81 + 81 + 5^4$$

Since

$$5 + 8 + 11 + 14 = 38$$
$$5^2 + 8^2 + 11^2 + 14^2 = 406$$

(in the general case, it may be assumed that formulas for the sums of the lower powers are known) then

$$5^3 + 8^3 + 11^3 + 14^3 = \frac{17^4 - 108(38) - 54(406) - 5^4 - 81 - 81 - 81 - 81}{12}$$

Pascal described the general procedure for finding the sum of any power of any arithmetic sequence. The description is made considerably more difficult because Pascal lacked our modern method of expressing powers and arithmetic sequences. The procedure was:

Rule 12.2 (Summation of Powers of an Arithmetic Progression). *Take a binomial with first term A and second term the difference of the progression, and raise it to a power one more than the powers on the desired sum. Take the term that would follow the last term in the desired sum, and raise it to the same power as the binomial. From this number subtract:*

1. *The first term of the progression, raised to the same power as the binomial.*

2. *The difference of the progression, raised to the same power as the binomial, and multiplied by the number of terms in the sum.*

3. *The sum of all the given numbers raised to powers less than the desired power, and multiplied by the coefficient of the corresponding power of A in the expansion of the binomial.*

The result is the sum of the powers, multiplied by the corresponding coefficient of A in the expansion of the binomial.

Example 12.15. *To find the sum of the squares of the sequence 3, 5, 7, 9, 11. The difference is 2 and the desired power is 2; hence expand $(A + 2)^3$ to obtain*

$$A^3 + 6A^2 + 12A + 8$$

The next term in the sequence would be 13; raise 13 to the same power as the binomial: $13^3 = 2197$. From this, subtract:

The first term of the progression, raised to third power: $3^3 = 27$.

The difference of the progression, raised to the third power, then multiplied by the number of terms in the progression: $5 \cdot 2^3 = 40$.

The sum of the first powers of the numbers, multiplied by the coefficient of A: $3 + 5 + 7 + 9 + 11 = 35$. Multiplying 35 by 12 gives 420.

The result is the sum of the squares, multiplied by the coefficient of A^2:

$$2197 - 27 - 40 - 420 = 6(3^2 + 5^2 + 7^2 + 9^2 + 11^2)$$

hence

$$285 = 3^2 + 5^2 + 7^2 + 9^2 + 11^2$$

Example 12.16. *Find the sum of the cubes of the sequence 3, 5, 7, 9, 11. Again, the difference is 2. The power is 3, so raising $(A + 2)^4$ we obtain*

$$A^4 + 8A^3 + 24A^2 + 32A + 16$$

The next term would be 13, and $13^4 = 28561$. Subtract:

The first term, raised to the fourth power: $3^4 = 81$.

The difference, raised to the fourth power and multiplied by the number of terms: $5 \cdot 2^4 = 80$.

The sum of the first powers, multiplied by the coefficient of A: $32 \cdot 35 = 1120$.

The sum of the squares, multiplied by the coefficient of A^2: $24 \cdot 285 = 6840$.

The result is the sum of cubes, multiplied by the coefficient of A^3:

$$28561 - 81 - 80 - 1120 - 6840 = 8(3^3 + 5^3 + 7^3 + 9^3 + 11^3)$$

hence

$$2555 = 3^3 + 5^3 + 7^3 + 9^3 + 11^3$$

The Cycloid

On the night of November 23, 1654, between 10:30 PM and 12:30 AM, Pascal underwent a religious experience that caused him to abandon science and mathematics for theology. For the next four years, he produced little of interest to the historian of science, though **Pascal's wager** is an early application of expected value. Then in 1658, a toothache kept him from sleeping; to distract himself, he worked on several problems relating to the cycloid. Miraculously, the pain stopped, and Pascal took it as a sign from God that he was to continue work in mathematics.

See problem 3, page 381.

Pascal found a number of results relating to the cycloid, and posed several questions to the mathematical community, offering a prize for their solutions. Due to publication mishaps and delays, only two entries were received: one by John Wallis, and the other by Antoine de Lalouvère. Neither met with Pascal's standards, and thus no prize was awarded; needless to say, they were not pleased with the outcome. The cycloid caused further problems when Pascal published his results between 1658 and 1659, to which a history of the cycloid was prefaced: the history mentioned nothing about Torricelli, which antagonized Italian mathematicians. It was not for nothing that the cycloid came to be called the "Helen of Geometers" for the arguments it caused.

12.6 Exercises

1. Find the sum of the indicated powers of the indicated sequence.

    (a) The cubes of the sequence 3, 7, 11, 15, 19.

    (b) The squares of the sequence 2, 5, 8, 11, 14.

    (c) The cubes of the sequence 2, 5, 8, 11, 14. Note: You will have found the sum of the squares of the sequence 2, 5, 8, 11, 14 in the previous step.

2. Find a formula for the sum of the squares of the first n odd numbers.

3. Find a formula for the sum of the cubes of the first n odd numbers.

4. Beginning with the sequence of natural numbers 1, 2, 3, ..., n, find:

    (a) The sum of the first n natural numbers.

    (b) The sum of the squares of the first n natural numbers.

    (c) The sum of the cubes of the first n natural numbers.

12.7   Projective Geometry

One of the major differences between the art of the Middle Ages and the art of the Renaissance was that Renaissance artists began to use perspective, so their paintings would appear to be "windows on life." The architect Filippo

Brunelleschi (1377-1446) of Florence began a thorough examination of the laws of perspective, but the first published treatise was that of Leon Battista Alberti (1404-1472). Other work was done by Albrecht Dürer (1471-1528), who also wrote a book on geometric construction techniques. Most sophisticated was the work of Girard Desargues (February 21, 1591-October 1661), who wrote a pioneering work on projective geometry, *Rough Draft of an Attempt to Deal with the Outcome of the Meeting of a Cone and a Plane* (1639), usually referred to as *Rough Draft on Conics*.

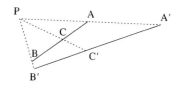

Projective geometry may be defined as the study of the properties of figures that remain unchanged under projections. For example, given a line AB and a point C on the midpoint of AB, consider the projections from P of AB onto $A'B'$. Some of the properties depend on the projection: C', the projection of C onto $A'B'$, will not, in general be at the midpoint of $A'B'$. However, the fact that C is between A and B is preserved under the projection. It was these invariant properties that Desargues proposed to study.

Desargues felt compelled to invent a whole new terminology. Perhaps to make the work more palatable to nonmathematicians, he chose to use many botanical terms: a *trunk* was a line through which other lines, called *branches*, passed; the points of intersection were *knots*, and a set of parallel branches formed a *crown*. Unfortunately, changing the words made the concepts no less difficult for nonmathematicians.

One of the properties of projective figures is known as **Desargues's theorem**. Actually, it appears nowhere in the work of Desargues, but rather in the work of Abraham Bosse (1602-February 14, 1676), whose *The Perspective of Mr. Desargues* (1648) was a detailed commentary on another one of Desargues's works. Three propositions appear at the end, believed to be the work of Desargues, not Bosse. The first is

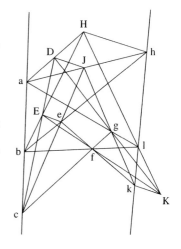

Theorem 12.1 (Desargues's Theorem). *If lines* HDa, HEb, cED, lga, lfb, HlK, DgK, EfK, cba, *which are either in the same plane, or different planes, cut each other in any manner, then the points* c, f, g *lie on the same line.*

The proof of the three-dimensional case is, remarkably, simpler than the proof in the two dimensional case.

Rough Draft on Conics was phenomenally unsuccessful: only 50 copies were made, to be distributed to Desargues's friends. No known copies of *Rough Draft on Conics* were known until one was rediscovered in 1951, over three hundred years after it was published. Until that rediscovery, the only source of information about this work of Desargues was through a handwritten copy made by Philip de la Hire (March 18, 1640-April 21, 1718). De la Hire had already written a work, *New Method of Geometry*, which dealt with projective geometry; like *Rough Draft on Conics*, this, too, was remarkably unsuccessful.

Around 1679, de la Hire came across *Rough Draft on Conics* (de la Hire's father was a friend of Desargues, and probably received one of the copies). In that year, de la Hire published *New Elements of Conic Sections*, dedicated to French finance minister Jean Baptiste Colbert: de la Hire was looking for a rich patron to

support his work. The conic sections were built up using Cartesian methods and
the focus-directrix property. De la Hire used some of Desargues's terminology,
and added a new term: **origin**, used in the modern mathematical sense for the
first time. De la Hire returned to projective methods in *Conic Sections* (1685),
deriving all properties of the conic sections using purely projective methods.

Pascal also wrote a short (1 page) article, "On Conic Sections," which dealt
with projective properties of figures inscribed in conic sections, including a version of the theorem that if a hexagon is inscribed in a circle, the opposite sides
intersect in a line. This theorem highlights the power of projective methods, for
proving the theorem true for a circle is sufficient to prove it true for any hexagon
inscribed in any conic section. But Pascal's work was even less successful than
Desargues's, remaining undiscovered until 1800. As a result the development of
projective geometry would have to wait until the nineteenth century.

12.7 Exercises

1. Prove Desargues's theorem in the three dimensional case by showing c, f,
   g are on two planes that intersect.

2. Desargues's theorem is usually interpreted to mean if one triangle is the
   projection of another, the extensions of corresponding sides intersect along
   a line. Show how the modern interpretation of Desargues's theorem follows
   from the original statement. Hint: let the point of projection be H, and
   the triangles be DKE, *alb*.

3. Show that, if the opposite sides of a hexagon inscribed in a circle meet on
   a line, then the opposite sides of a hexagon inscribed in any conic section
   likewise meet on a line.

Chapter 13

The Era of Newton and Leibniz

Elizabeth I of England died in 1603 without producing an heir (she never married); as a result, the throne of England passed to James VI of Scotland, who became James I of England. James never really understood the English, but the English tolerated his rule. James's son Charles I made an even more serious mistake by not understanding the Scots. A revolt in Scotland blossomed into the English Civil War, fought between the Royalists (also known as Cavaliers), headed by Charles I, and the Parliamentary forces, headed by Oliver Cromwell and known (derisively) as Roundheads. By 1649, the Civil War was over, with Parliament victorious. Cromwell and Parliament took the extraordinary step of sentencing the King to death, and executed him on January 30, 1649. England became a Commonwealth, and in 1653, Cromwell took the title of Lord Protector. Many Royalists fled the country, fearing Parliamentary retribution.

13.1 John Wallis

John Wallis (December 3, 1616-November 8, 1703) was a Parliamentarian during the Civil War. Like Viète, Wallis practiced cryptography, and between 1642 and 1643, decoded Royalist messages intercepted by the Parliamentarians. He was later accused of having decoded the personal letters of King Charles himself, a charge that Wallis adamantly denied. In his old age, Wallis taught what he knew of cryptography to his grandson, William Blencowe, though by then, Wallis admitted, the new French methods of encryption were too complicated to break by the means used by Wallis. In 1649, Wallis became Savilian Professor of Geometry at Oxford.

13.1.1 *Treatise on Conic Sections*

Along with Fermat and Descartes, Wallis should be considered one of the founders of analytic geometry. In *Treatise on Conic Sections: A New Method of Exposition* (1655) Wallis, after introducing the conic sections in the usual way, abandoned their geometric basis and treated them entirely from a coordinate perspective:.

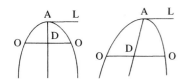

Part I dealt with proving the symptoms of the conic sections, in terms of their origin from the section of a cone. In Part II, Wallis declared:

> I will, therefore, name a *parabola* any thing (whether a curved line in the plane, or a plane figure bound by a curve, such as OAO) the square of whose ordinate is proportional to the intercepted diameter.[1]

Thus, rather than deriving the algebraic property of a parabola from its geometric definition, which Descartes and Fermat did, Wallis defined a curve based on purely algebraic considerations. Using the analytic definition of a parabola, Wallis then proved

Proposition 13.1. *Given a point α on a parabola, a point P on diameter PA with ordinate $P\alpha$, if PA intersects the extension of the diameter at F so that $AF = AP$, then $F\alpha$ touches the parabola.*

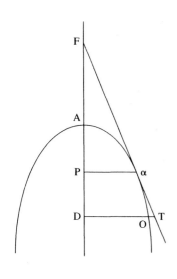

Proof. Take any other point D on PA; draw ordinate DO to T on $F\alpha$. I say that T is beyond the parabola; that is, DT is either equal to DO, in which case D, P coincide, or DT is greater than DO.

Let $P\alpha = p$, $PA = d$, hence $PF = 2d$; let $PD = a$, and so $DA = d \pm a$, $DF = 2d \pm a$. By the property of the parabola, $PA : DA = $ sq. $P\alpha : $ sq. DO. Since $d : d \pm a = p^2 : \frac{d \pm a}{d}p^2$, then $\frac{d \pm a}{d}p^2$ is equal to the square on DO.

By similar triangles, $PF : DF = P\alpha : DT$, hence $2d : 2d \pm a = p : \frac{2d \pm a}{2d}p$, making $DT = \frac{2d \pm a}{2d}p$; thus the square on DT equals $\frac{4d^2 \pm 4da + a^2}{4d^2}p^2$. Thus, DT square minus DO square is equal to $\frac{a^2}{4d^2}p^2$, hence DT must be greater than DO, or if D, P coincide, then $a = 0$ and DT is equal to DO. $\qquad\square$

In the appendix, Wallis included a few new curves, one of which was the cubical paraboloid:

> If a paraboloid is so constituted that the ordinate has a subtriplicate ratio to the intercepted diameter (alternately, that the cube of the ordinate is proportional to the intercepted diameter) it will be called a *cubical paraboloid*. The latus rectum of the cubical paraboloid (designated l), I call the imaginary line whose square, multiplied by the intercepted diameter, equals the cube of the ordinate. In other words, $p^3 = l^2 d$.[2]

Wallis found the tangent relationship for the cubical paraboloid:

---

[1]Wallis, John, *Opera Mathematicorum* (three volumes). L. Richfield and T. Robison, 1656. Vol. I, p. 319.

[2]Ibid., 350

Figure 13.1: England during the Civil War.

Proposition 13.2. *In cubical paraboloid $A\alpha$, if αF is tangent at the point α, and F is on the extension of diameter PA, then AF will be twice the intercepted diameter PA; that is, PF will be three times the diameter PA.*

Wallis used infinitesimals to prove some results, introducing our modern symbol ∞. Citing Cavalieri, Wallis noted plane figures may be considered to be composed of an infinite number of parallelograms of equal height (Wallis considered the parallelograms to run horizontally, rather than vertically). The height of any parallelogram is to the height of the entire figure as $\frac{1}{\infty}$, where Wallis chose the symbol ∞ to represent an "infinite number." Wallis, who was also a classical scholar, may have derived the symbol from the Roman ∞ representing a thousand. Wallis went on to say that the heights of the parallelograms are thus infinitely small.

First, Wallis demonstrated the area of a triangle is half the area of a rectangle on the same base with the same height. Wallis's argument is as follows: given the triangle of base B and height A, the sum of the longitudinal lines in the triangle is equal to the largest, B, taken half an infinite number of times; in

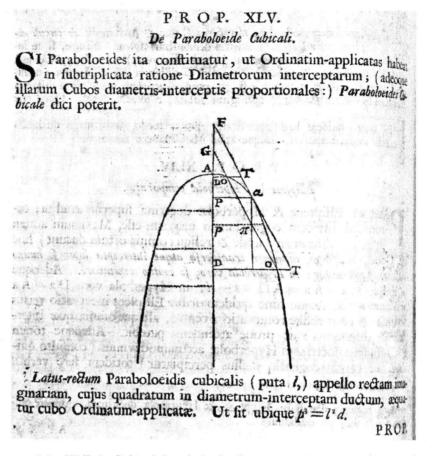

Figure 13.2: Wallis's Cubical Paraboloid. Boston Public Library/Rare books Department. Courtesy of the Trustees.

other words, $\frac{\infty}{2}B$ (this follows from the formula for the sum of an arithmetic progression). Each of the infinitesimal parallelograms has height $\frac{1}{\infty}A$; hence the area will be the product $\frac{1}{\infty}A\frac{\infty}{2}B$, or $\frac{1}{2}AB$.

13.1.2 *Arithmetic of the Infinite*

It was his *Arithmetic of the Infinite, or New Method of Inquiry into the Quadrature of Curves*, published in 1656, which would prove most influential with Newton, ultimately leading to his discovery of the generalized binomial theorem. The key feature of this work is the use of scientific induction and interpolation. Euclid had avoided the question of whether a rectilineal and curvilineal area could be considered equal by comparing the ratio between them; in a similar manner and for a similar reason, Wallis considered the ratio between two infinite series.

Figure 13.3: The First Appearance of the ∞ Symbol. Boston Public Library/Rare books Department. Courtesy of the Trustees.

Series Summation

First, Wallis noted:

$$\frac{0+1}{1+1} = \frac{1}{2}$$

$$\frac{0+1+2}{2+2+2} = \frac{1}{2}$$

$$\frac{0+1+2+3}{3+3+3+3} = \frac{1}{2}$$

$$\frac{0+1+2+3+4}{4+4+4+4+4} = \frac{1}{2}$$

Hence:

Conjecture 13.1. *If we sum a series in arithmetic proportion whose first term is 0, whether the series has a finite or an infinite number of terms, the sum will be to the series consisting of the same number of terms, all equal to the largest term in the first series, as 1 is to 2.*

Wallis then derived a number of quadratures. The first was:

Conjecture 13.2. *The area of a triangle is to the area of a parallelogram that encloses it as 1 to 2.*

Demonstration. The triangle consists of an infinite number of parallel lines in arithmetic proportion, whose largest is the base. The parallelogram consists entirely of equal bases. Since the sum of the lines in arithmetic proportion is to the sum of the equal lines as 1 is to 2, the area of the triangle is to the area of the parallelogram also as 1 is to 2. □

In a similar manner, Wallis demonstrated:

John Wallis. Boston Public Library/Rare books Department. Courtesy of the Trustees.

Conjecture 13.3. *The paraboloid of revolution is to the cylinder that encloses it as 1 is to 2.*

Note that Wallis is not finding the area or volume in our modern sense. Rather, he is finding the ratio between the areas or volumes of two figures.

Wallis used a variety of terms, not always consistently, to refer to the series under consideration. At first, Wallis began with the arithmetic progression 0, 1, 2, ...; the **series of squares** was the sum of the squares of these terms; the **series of cubes** was the sum of the cubes, and so on. For any given series, Wallis called the **series of equals** the series whose terms are the same in number and all equal to the largest term of the series under consideration.

First, Wallis compared the series of squares to the series of equals:

$$\frac{0+1}{1+1} = \frac{1}{2} = \frac{1}{3} + \frac{1}{6 \cdot 1}$$

$$\frac{0+1+4}{4+4+4} = \frac{5}{12} = \frac{1}{3} + \frac{1}{6 \cdot 2}$$

$$\frac{0+1+4+9}{9+9+9+9} = \frac{7}{18} = \frac{1}{3} + \frac{1}{6 \cdot 3}$$

$$\frac{0+1+4+9+16}{16+16+16+16+16} = \frac{3}{8} = \frac{1}{3} + \frac{1}{6 \cdot 4}$$

Hence, Wallis concluded:

Conjecture 13.4. *The ratio of the series of squares to the series of equals will exceed 1 : 3, and the excess will have the ratio of 1 to 6 times the square root of the last term in the series of squares.*

See Problem 7.

A number of quadratures immediately follows.

For the series of cubes, Wallis observed

$$\frac{0+1}{1+1} = \frac{1}{2} = \frac{1}{4} + \frac{1}{4 \cdot 1}$$

$$\frac{0+1+8}{8+8+8} = \frac{9}{24} = \frac{1}{4} + \frac{1}{4 \cdot 2}$$

$$\frac{0+1+8+27}{27+27+27+27} = \frac{36}{108} = \frac{1}{4} + \frac{1}{4 \cdot 3}$$

$$\frac{0+1+8+27+64}{64+64+64+64+64} = \frac{100}{320} = \frac{1}{4} + \frac{1}{4 \cdot 4}$$

hence

Conjecture 13.5. *The ratio of the series of cubes to the series of equals will exceed 1 : 4, and the excess will have the ratio of 1 to 4 times the cube root of the last term in the series of cubes.*

Wallis concluded that in general, the *infinite* series of nth powers are to the series of equals as 1 is to $n + 1$.

Wallis's next few steps can be summarized as follows: the series of squares is 1/3 the series of equals, while the series in arithmetic proportion is 1/2 the series of equals; the series of equals is obviously 1/1 the series of equals (note that the three "series of equals" are, in these three cases, three different series). Comparing the denominators of the ratios, Wallis noted that the arithmetic mean of 1 and 3 is 2, which is the denominator of the ratio for the series whose terms are the *geometric* mean of the corresponding terms in the series of squares and the series of equal. Thus, Wallis concluded the series of square roots must be to the series of equals as 1 is to 3/2, or as 2 to 3, since 3/2 is the arithmetic mean of 1 and 2, while the terms of the series of square roots are the geometric means of the terms in the series of equals and the series in arithmetic proportion. This led Wallis to conclude, again without proof, that:

See Problem 8.

Conjecture 13.6. *If we take an infinite series of nth powers, the ratio of total to the series of equals will be that of 1 to $n + 1$.*

At the end of this proposition, Wallis took a bold leap and concluded the rule holds even if the index is irrational.

It is with Proposition CXXI that Wallis gave the reader the hint of a greater goal: the determination of the area of a quadrant of a circle. Wallis changed his notation and designated the arithmetic sequence as 0, $1a$, $b = 2a$, $c = 3a$, $d = 4a$, ..., R, and let a designate a general term of the sequence; A was the number of terms in the series (hence $R + R + R + \ldots + R = A\,R$). To find the area of a quadrant of a circle of radius R, it is necessary to find the sum of the series whose general term is $\sqrt{R^2 - a^2}$. To do so, Wallis examined a number of other series. He considered series whose terms are powers of $\sqrt{R} - \sqrt{a}$. Adding them termwise, and keeping in mind that the sum of the series of square roots is to the sum of the series of equals as 1 is to $\frac{1}{2} + 1$, then:

Term	Square of Term
$\sqrt{R} - \sqrt{a}$	$R - 2\sqrt{R}\sqrt{a} + a$
$\sqrt{R} - \sqrt{b}$	$R - 2\sqrt{R}\sqrt{b} + b$
$\vdots$	$\vdots$
$\sqrt{R} - \sqrt{R}$	$R - 2\sqrt{R}\sqrt{R} + R$
Sum	$AR - \frac{4}{3}AR + \frac{1}{2}AR$
	$= \frac{1}{6}AR$

An important observation is that the series of equals (i.e., the series consisting of A terms, all equal to $\left(\sqrt{R} - \sqrt{0}\right)^2 = R$) has sum AR; hence, the sum of the series $\left(\sqrt{R} - \sqrt{a}\right)^2$ is to the series of equals as 1 to 6.

Likewise:

Term	Cube of Term
$\sqrt{R} - \sqrt{a}$	$R\sqrt{R} - 3R\sqrt{a} + 3\sqrt{R}a - a\sqrt{a}$
$\sqrt{R} - \sqrt{b}$	$R\sqrt{R} - 3R\sqrt{b} + 3\sqrt{R}b - b\sqrt{b}$
$\vdots$	$\vdots$
$\sqrt{R} - \sqrt{R}$	$R\sqrt{R} - 3R\sqrt{R} + 3\sqrt{R}R - R\sqrt{R}$
Sum	$AR\sqrt{R} - 2AR\sqrt{R} + \frac{3}{2}AR\sqrt{R} - \frac{2}{5}AR\sqrt{R}$
	$= \frac{1}{10}AR\sqrt{R}$

Power of Term

	0	1	2	3	4	...
0	1	1	1	1	1	...
$R - a$	1	2	3	4	5	...
$\sqrt{R} - \sqrt{a}$	1	3	6	10	15	...
$\sqrt[3]{R} - \sqrt[3]{a}$	1	4	10	20	35	...
$\sqrt[4]{R} - \sqrt[4]{a}$	1	5	15	35	70	...
$\vdots$	$\vdots$	$\vdots$	$\vdots$	$\vdots$	$\vdots$	$\ddots$

Table 13.1: Powers of Roots

Again, the series of equals consists of A terms, all equal to $\left(\sqrt{R} - \sqrt{0}\right)^3 = R\sqrt{R}$; hence the sum of the series $\left(\sqrt{R} - \sqrt{a}\right)^3$ is to the sum of the series of equals as 1 is to 10.

Generalizing from these and similar results, Wallis noted that the first series, whose terms are $\sqrt{R} - \sqrt{a}$, has a sum of $\frac{1}{3}A\sqrt{R}$; hence the ratio of the sum of this series to the corresponding series of equals, namely the series of terms all equal to $\sqrt{R}$, is $\frac{1}{3}$, or $\frac{1}{1+2}$. The second series, whose terms consist of the squares of the terms of the first series, or $\left(\sqrt{R} - \sqrt{a}\right)^2$, has a ratio to the series of equals, namely the series of terms all equal to R, of $\frac{1}{1+2+3}$. The third series, with general term $\left(\sqrt{R} - \sqrt{a}\right)^3$, has a ratio to the series of equals $R\sqrt{R}$ of $\frac{1}{1+2+3+4}$.

Next, Wallis considered the series whose terms were powers of $\sqrt[3]{R} - \sqrt[3]{a}$. By the same sort of analysis, Wallis claimed the sums of the powers of the terms was to the series of equals as 1 to 4, 1 to 10, 1 to 20, and 1 to 35. By putting the denominators into a table, Wallis obtained the results in Table 13.1. We might read the table entries as being the ratio between the series of equals to the series of first, second, third, and higher powers of the series whose general term is on the right-hand column.

Interpolation

The series whose sum is desired has general term $\sqrt{R^2 - a^2}$. To find this, Wallis interpolated additional columns and rows between the known columns and rows, generating a new table. To what series should the new rows and columns correspond? Wallis noted that the columns were associated with the powers 0, 1, 2, 3,..., or 0/2, 2/2, 4/2, 6/2..., which suggests the interpolated columns ought to be associated with the powers 1/2, 3/2, 5/2, ... Meanwhile, the indices of the terms are 1, 1/2, 1/3, 1/4,... or 2/2, 2/4, 2/6, 2/8... , which suggests the intermediate rows ought to be associated with the indices 2/1, 2/3, 2/5, ... Thus, the entry indicated $\square$ in Table 13.2 is the desired entry, which corresponds to the series with the power 1/2 of the term with index 2/1: in other words, $\sqrt{R^2 - a^2}$. Moreover, the ratio 1 : $\square$ is the ratio of the series with general term $\sqrt{R^2 - a^2}$ to the series consisting entirely of terms equal to $\sqrt{R^2} = R$. The latter corresponds to the area of a square of side R, while the former corresponds to the quadrant of a circle of radius R. Hence $1 : \square = \square : R^2$.

Now it was necessary to interpolate the values in the rows and columns. The known entries in the second row of the interpolated table are all 1, so Wallis concluded the intermediate entries must also be 1. Because of the symmetry of the table, this further suggested the second column's entries are likewise all 1. The known entries in the fourth row of the interpolated table were 1, 2, 3, ...; thus, Wallis concluded the unknown intermediate entries ought to be $1\frac{1}{2}, 2\frac{1}{2}, 3\frac{1}{2}$, ... The sixth row's known entries, 1, 3, 6, 10,... form a sequence well known to Wallis, the sequence of triangular numbers. The nth triangular number is given by the formula $\frac{n^2+n}{2}$. By assuming this formula also holds for intermediate numbers such as 1/2, 3/2, 5/2, ..., Wallis filled in the missing entries in the sixth row (note that n is one more than the power). For example, the sum of the series $(\sqrt{R} - \sqrt{a})^2$ will be to the series of equals as 1 is to the third triangular number, 6. The eighth row consists of yet another series Wallis knew well, the pyramidal numbers. Wallis noted that the nth pyramidal number is the sum of the first n triangular numbers, and, using his previously obtained results on finite series, the sum is thus $\frac{n^3+3n^2+2n}{6}$.

For the tenth row, Wallis ventures into new territory. The entries in the row are 1, 5, 15, 35, ... Wallis, having noted that the nth triangular number is the sum of the first n whole numbers, and the nth pyramidal number is the sum of the first n triangular numbers, recognized the nth term in the fifth row will be the sum of the first n pyramidal numbers. By the same methods as before, Wallis determined the nth "triangulo-pyramidal" number will be given by $\frac{n^4+6n^3+11n^2+6n}{24}$. These formulas and observations allowed Wallis to interpolate numbers in the columns between known entries.

This left the alternate rows to be determined. To determine these, Wallis

Power of Term

		0	1	2	3	...
0		1	1	1	1	...
	$\square$					
$R - a$		1	2	3	4	...
$\sqrt{R} - \sqrt{a}$		1	3	6	10	...
$\sqrt[3]{R} - \sqrt[3]{a}$		1	4	10	20	...
$\sqrt[4]{R} - \sqrt[4]{a}$		1	5	15	35	...
$\vdots$		$\vdots$	$\vdots$	$\vdots$	$\vdots$	$\ddots$

Table 13.2: Table of Values

		Power of Term						
		0		1		2		$\cdots$
	∞	1	$\frac{1}{2}\square$	$\frac{1}{2}$	$\frac{1}{3}\square$	$\frac{3}{8}$	$\frac{4}{15}\square$	$\cdots$
0	1	1	1	1	1	1	1	$\cdots$
	$\frac{1}{2}\square$	1	$\square$	$1\frac{1}{2}$	$\frac{4}{3}\square$	$1\frac{7}{8}$	$\frac{8}{5}\square$	$\cdots$
$R-a$	$\frac{1}{2}$	1	$1\frac{1}{2}$	2	$2\frac{1}{2}$	3	$3\frac{1}{2}$	$\cdots$
	$\frac{1}{3}\square$	1	$\frac{4}{3}\square$	$2\frac{1}{2}$	$\frac{8}{3}\square$	$\frac{35}{8}$	$\frac{64}{15}\square$	$\cdots$
$\sqrt{R}-\sqrt{a}$	$\frac{3}{8}$	1	$1\frac{7}{8}$	3	$4\frac{3}{8}$	6	$7\frac{7}{8}$	$\cdots$
$\vdots$	$\vdots$	$\vdots$	$\vdots$	$\vdots$	$\vdots$	$\vdots$	$\vdots$	$\ddots$

Table 13.3: Wallis's Final Result.

noted that the entries in each known row were generated by the partial products of $1\times\frac{2n}{2}\times\frac{2n+2}{4}\times\frac{2n+4}{6}\times\ldots$ (i.e., the entries in the first known row were generated by the first factor in the infinite product, or 1; the entries in the second known row were generated by the product of the first two factors, or $1\times\frac{2n}{2}$, and so on). Thus, he proposed, the unknown rows should be generated by the partial products of $A\times\frac{2n-1}{1}\times\frac{2n+1}{3}\times\frac{2n+3}{5}\times\ldots$, where A is some unknown number.

Furthermore, since any entry in the original table is the sum of the entry immediately to its left and immediately above it, Wallis assumed this held for the interpolated table as well; thus, he assumed any entry in the interpolated table was the sum of two entries, the one two spaces to its left, and the one two spaces above it.

Moreover, in the original table, each entry will be the product of the entry above it and the next factor in the infinite product, $1\times\frac{2n}{2}\times\frac{2n+2}{4}\times\frac{2n+4}{6}\times\ldots$ Wallis assumed, again, that this held true for the interpolated table as well, and thus given any entry in one of the interpolated rows of the table, one may find the entry two rows below it by multiplying the entry by the appropriate term in the infinite product $A\times\frac{2n-1}{1}\times\frac{2n+1}{3}\times\frac{2n+3}{5}\times\ldots$ Finally, the table is symmetrical. Thus, the remaining entries may be found. The result is Table 13.3.

See Problem 11.

Now it is necessary to find the value of $\square$. Wallis considered the sequence of values in the third row, namely $\frac{1}{2}\square$, 1, $\square$, $1\frac{1}{2}$, $\frac{4}{3}\square$, $1\frac{7}{8}$, $\frac{8}{5}\square$, $\ldots$ Calling this sequence α, a, β, b, γ, c, $\ldots$, Wallis noted that

$$1:2 \quad \text{as} \quad \alpha:\beta$$
$$2:3 \quad \text{as} \quad a:b$$
$$3:4 \quad \text{as} \quad \beta:\gamma$$
$$\vdots \qquad \vdots$$

Observing that the ratios of the alternate terms decrease, Wallis concluded that the ratio of the successive terms must also decrease. Thus

$$1 : \frac{1}{2}\Box > \Box : 1 \qquad\qquad \Box : 1 > \frac{3}{2} : \Box$$

hence

$$\sqrt{\frac{2}{1}} > \Box > \sqrt{\frac{3}{2}}$$

The next pair of ratios gives

$$\frac{3}{2} : \Box > \frac{4}{3}\Box : \frac{3}{2} \qquad\qquad \frac{4}{3}\Box : \frac{3}{2} > \frac{15}{8} : \frac{4}{3}\Box$$

hence (after some manipulation)

$$\frac{3 \times 3}{2 \times 4}\sqrt{\frac{4}{3}} > \Box > \frac{3 \times 3}{2 \times 4}\sqrt{\frac{5}{4}}$$

Continuing this process, Wallis concluded

$$\frac{3 \times 3 \times 5 \times 5 \times \ldots \times 13 \times 13}{2 \times 4 \times 4 \times 6 \times \ldots \times 12 \times 14}\sqrt{1\frac{1}{13}}$$

$$> \Box > \frac{3 \times 3 \times 5 \times 5 \times \ldots \times 13 \times 13}{2 \times 4 \times 4 \times 6 \times \ldots \times 12 \times 14}\sqrt{1\frac{1}{14}} \quad (13.1)$$

"and so forth to as close an approximation as we wish."

What did Wallis actually find? $\Box$ is the denominator of the ratio of the sum of the series $\sqrt{R^2 - a^2}$ to the sum of the series of equals, or R. Hence, $1 : \Box$ is the ratio of the area of the quarter circle of radius R to the area of the square of side R: in other words, $1 : \Box = \frac{\pi R^2}{4} : R^2$. Thus $\Box = \frac{4}{\pi}$, and Wallis's result implies

$$\frac{4}{\pi} = \frac{3 \times 3 \times 5 \times 5 \times \ldots \times 13 \times 13 \times \ldots}{2 \times 4 \times 4 \times 6 \times \ldots \times 12 \times 14 \times \ldots}$$

13.1.3 Resolution of Impossible Cubics

A minor contribution of Wallis was an early attempt to simplify imaginary quantities. In his 1685 *Algebra*, Wallis noted that Cardano's formulas sometimes resulted in "impossible" cubics. For example, the cubic $r^3 - 63r = 162$ resulted in the solution $r = \sqrt[3]{81 + 30\sqrt{-3}} + \sqrt[3]{81 - 30\sqrt{-3}}$. By assuming $\sqrt[3]{81 + 30\sqrt{-3}} = a - f\sqrt{-e}$, then, cubing both sides and equating the real and imaginary components, Wallis obtained

$$a^3 - 3af^2e = 81$$
$$3a^2f\sqrt{-3} - f^3e\sqrt{-3} = 30\sqrt{-3}$$

By trial and error, Wallis found that $f = \frac{1}{2}$, $e = 3$, and $a = \frac{9}{2}$. He gave no general solution to the problem of reducing the roots of complex numbers to simpler forms. The matter would be taken up again by DeMoivre.

13.1 Exercises

Note: In the following, unless indicated otherwise, assume Wallis's conjectures are valid.

1. Consider the cubical paraboloid, as defined by Wallis in *Conic Sections*

   (a) Examine Figure 13.2, which shows Wallis's sketch of the cubical paraboloid. How does Wallis's definition of a cubical paraboloid and use of coordinates differ from our own?

   (b) Why does Wallis define the latus rectum l as $p^3 = l^2 d$ and not more simply as $p^3 = ld$?

   (c) Show, using Wallis's methods, Proposition 13.2.

2. Find the ratios of the series of fourth powers to the series of equals. What problems would Wallis have run into had he continued to examine the ratios of the series?

3. Can Wallis's results on infinite series be justified? In other words, can the ratio between the sum of a series whose terms have no "largest term" to the sum of a second series consisting entirely of the "largest term" of the first series have any meaning? Explain. Suggestion: what does Wallis actually need in order for him to use the series sums as he does?

4. Wallis may have realized arithmetic operations with infinity did not follow the normal rules. In Proposition CLXXXVIII of *Arithmetic of the Infinite*, he noted that 1 divided by ∞, was 0, as was 2 divided by ∞, and so on.

   (a) Use this to show that multiplication of 0 and ∞ is indeterminate.

   (b) Compare this to Bhaskara's comment about division by 0.

5. Prove Conjecture 13.3 by showing the volume elements of the paraboloid are in arithmetic proportion.

6. Prove Proposition 13.4, and use it to derive a formula for $\sum_{n=0}^{N} n^2$.

7. Demonstrate, using Wallis's methods, the following propositions from Wallis *Arithmetic of the Infinite*:

   (a) The cone is to the enclosing cylinder as 1 is to 3 (Proposition XXII)

   (b) The spiral is to sector of the circle enclosing it as 1 is to 3. (Proposition XXIV) The spiral referred to here is the simple spiral given by the polar equation $r = \theta$.

8. Construct arguments, similar to Wallis's, to demonstrate the following conjectures:

   (a) Proposition LII: The series of square roots is to the series of equals in the ratio of 3 to 2.

   (b) The series of cube roots is to the series of equals as 4 is to 3.

9. Show that the nth pyramidal number is equal to $\frac{n^3+3n^2+2n}{6}$.

10. Show that the nth triangulo-pyramidal number is equal to $\frac{n^4+6n^3+11n^2+6n}{24}$.

11. Complete Wallis's table of interpolated values.

12. Wallis's table of interpolated values, with $\square = \frac{4}{\pi}$, corresponds to a table of definite integrals. To what definite integrals do the table entries correspond?

13. Use Wallis's method to determine the indicated areas.

    (a) The area between the curve $y = x^{1/2}$ and the x-axis.

    (b) The area between the curve $y = x^{1/2}$ and $y = 4$.

14. Use Wallis's method to determine the volumes formed by revolving the indicated regions about the y-axis. Verify your results using integral calculus.

    (a) The region between $y = 4 - x^2$ and the x-axis.

    (b) The region between $y = \sqrt{x}$, the y-axis, $x = 0$, and $x = 4$.

    (c) The region between $y = x^2$, $y = 4$, and the y-axis.

15. Consider Wallis's $\square$.

    (a) Show that $\frac{3}{2} : \square > \frac{4}{3}\square : \frac{3}{2}$ implies $\square > \frac{3\times3}{2\times4}\sqrt{\frac{5}{4}}$.

    (b) Likewise, show how $\frac{4}{3}\square : \frac{3}{2} > \frac{15}{8} : \frac{4}{3}\square$ implies $\frac{3\times3}{2\times4}\sqrt{\frac{4}{3}} > \square$.

    (c) Determine the next pair of inequalities.

    (d) Show how the next pair of inequalities generates the next set of bounds for $\square$.

16. Construct a table, similar to Wallis's, by interpolating *two* rows and columns into Table 13.1.

17. Use the table from Exercise 16 to evaluate the following.

    (a) $\int_0^R \sqrt[3]{R - x}\, dx.$ (b) $\int_0^R \sqrt[3]{\sqrt{R} - \sqrt{x}}\, dx.$ (c) $\int_0^R \sqrt[3]{R^3 - x^3}\, dx.$

18. Suppose $\sqrt[3]{m + \sqrt{-n}} = a + \sqrt{-b}$. Cube both sides and find equations solving for a and b. What problems would Wallis have run into if he assumed he could simplify the cube root of a complex number in this way?

13.2 Isaac Barrow

Isaac Barrow (1630-May 4, 1677) was the son of a linen draper; as a youngster, his inattention and predilection toward fighting caused his father to say (according to one account) that if God had to take one of his children, he would prefer it be Isaac. During the civil war, Barrow was a Royalist. In 1655, after the civil war ended, the position of Regius Professor of Greek at Cambridge opened. Barrow was the best candidate—but ex-Royalists were barred from university positions. He left England shortly thereafter, and traveled through continental Europe.

In 1658, Oliver Cromwell died. Before dying, he had his son, Richard, appointed Lord Protector, and it appeared that the only effect of the civil war was to replace one dynasty with another. Worse yet, Oliver Cromwell and his supporters were Puritans: they closed theaters, prohibited gambling, and banned bear baiting, cockfights, and horse racing; moreover, they prohibited all but religious activities on Sundays. By 1660, the English had had enough of enforced morality, and when Charles, the son of the executed king, agreed to respect the decisions of Parliament, Parliament in turn declared him king, inaugurating the Restoration Era of English history. Charles II kept his promise: he said he had no wish to "resume his travels." In 1663, he granted a charter to the first modern scientific organization: the Royal Society of London (which is still in existence). Barrow was present at the first meeting of the Society, on May 20, 1663.

Isaac Barrow. Boston Public Library/Rare books Department. Courtesy of the Trustees.

In the year of Charles's restoration, 1660, Barrow obtained the position that had been denied him earlier and became Regius Professor of Greek at Cambridge; it would be the first of many honors and positions he would accumulate. In 1662, he became Professor of Geometry at Gresham College in London. In the summer of 1663, Henry Lucas donated money to Cambridge University for a chair in mathematics, and Barrow became the first Lucasian professor that year. He also held a position as a replacement professor of Astronomy, holding all three positions concurrently.

From 1664 to 1666, he delivered a series of lectures that would become his *Geometrical Lectures*, published in book form in 1670. Barrow published the work because he felt it necessary to set a good example for his colleagues: he thought it was the *duty* of an academic to publish his work, so that others may make use of it. This is in sharp contrast from the jealously guarded secret methods of continental mathematicians, like Roberval. Apparently, the duty did not extend to making the work easy to read, and Barrow warned the reader not to expect "anything elaborated, skillfully arranged, or neatly set in order"! It was, indeed, Barrow's intention to publish the lectures "just as they were born." However, he did allow his friends — he names Isaac Newton in particular — to make revisions and corrections so the work might be more comprehensible.

13.2.1 The Fundamental Theorem of Calculus

Barrow deserves credit for proving a version of the **fundamental theorem of calculus**. However, he is a mathematical conservative, and wherever possible he phrased and proved propositions in proper Euclidean fashion. Thus, we have:

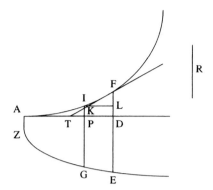

Figure 13.4: The Fundamental Theorem of Calculus, First Part (from Barrow).

Proposition 13.3 (Fundamental Theorem of Calculus, Part 1). *Let ZGE be any curve whose ordinates are increasing and whose axis is AD. Let AIF be a curve, where if any straight line EDF is drawn perpendicular to AD, then the rectangle on DF and a given length R is equal to the area $ADEZ$. Also, let $DE : DF = R : DT$. Then TF is tangent to AIF (see Figure 13.4).*

In modern terms, if $\int_0^x f(t)\, dt = F(x)$, then $F'(x) = f(x)$, though in Barrow's time, $\int$ was not used for integration, $'$ was not used for differentiation, and $f(x)$ did not indicate a function of x.

Proof. Take any point I on AIF, first on the side of F toward A. Draw IG parallel to AZ, and IL parallel to AD. Let IL intersect TF at K. Then $LF : LK = DF : DT = DE : R$. Hence $R \cdot LF$ is equal to $LK \cdot DE$. But from the nature of the lines, $R \cdot LF$ is equal to the area $PDEG$. Hence $LK \cdot DE$ is equal to the area $PDEG$, which is less than the area $DP \cdot DE$. Hence, $LK < DP$ and thus $LK < LI$, so TF does not intersect AIF on the side toward A. Likewise, if I is on the other side of F, $LK > LI$ and does not intersect on the side away from A. Hence, the line TF is tangent to the curve.

Likewise, if ZGE is a curve whose ordinates are decreasing, the tangent TF lies entirely above the curve, which will be concave to the axis. □

Note the analytic character of Barrow's proof: he began by assuming the existence of the curve of area, then deduced the necessary relationship to the original curve.

In Lecture XI, Proposition 19, Barrow then demonstrated the converse of Proposition 13.3. Unlike the earlier proof, Barrow had to rely on infinitely small quantities.

Proposition 13.4 (Fundamental Theorem of Calculus, Part 2). *Let AMB be a curve with axis AD; let BD be drawn perpendicular to AD. Let KZL be a curve such that for any point M on AB, with MT the tangent and*

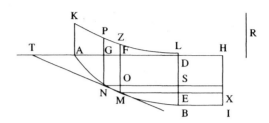

Figure 13.5: The Fundamental Theorem of Calculus, Second Part (from Barrow).

MFZ is parallel to DB, and R a given length, we have $TF : FM = R : FZ$. Then the area ADLK will equal the rectangle on R, DB (see Figure 13.5).

In modern terms, Barrow's second proposition is equivalent to the following: if $F'(x) = f(x)$ (and $F(0) = 0$, something Barrow neither stated nor required), then $\int_0^x f(t)\,dt = F(t)$.

Demonstration. Let $DH = R$, and complete rectangle $BDHI$. Take the indefinitely small arc MN. Draw MEX, NOS parallel to AD. Then $NO : MO = TF : FM = R : FZ$. Hence $NO \cdot FZ = MO \cdot R = ES \cdot EX$, and, since rect. ES, EX is the area $FGPZ$, then $FG \cdot FZ = ES \cdot EX$. But the sum of the rectangles such as $FG \cdot FZ$ differs only in the least degree from the space $ADLK$, and the rectangles $ES \cdot EX$ form the rectangle $DHIB$. Hence the theorem follows. $\qquad\square$

13.2.2   Subtangents to Trigonometric Functions

Barrow was aware of the new methods using infinitesimal quantities, though at first he doubted whether there was any advantage in using them; in addition, his Euclidean predilections made him wary of methods relying on the infinitely small. However, a friend—later determined to be Isaac Newton—convinced Barrow that the method was worth including

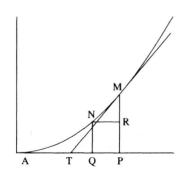

Rule 13.1. *Given a curve ANM with line MT tangent to the curve at M. Set off the indefinitely small arc MN, and draw NQ, NR parallel to MP, AP respectively. Call $MP = m$, $PT = t$, $MR = a$, and $NR = e$. Then to find the subtangent PT:*

1. *Omit all terms containing powers of a or e, or products of these terms.*

2. *Eliminate all terms that do not contain either a or e.*

3. *Substitute m (or MP) for a, and t (or PT) for e, then solve the equation to find the subtangent PT.*

We shall see how Barrow makes use of this rule.

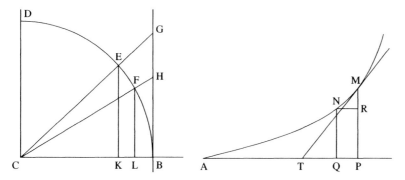

Figure 13.6: Barrow's Example Five.

To complete the problem and find the subtangent PT, Barrow substituted m (or MP) for a, and t (or PT) for e, apparently substituting finite quantities for infinitesimal ones. Barrow's method is virtually identical to the method used nowadays, though he concentrated on finding the value of t (and thus the subtangent PT). He gives five examples; his fifth example is particularly interesting.

See Problem 2.

Example 13.1. *Let DEB be a quadrant of a circle, to which BG is a tangent at B (see Figure 13.6). Let AM be a curve such that if $AP = $ arc BE, and PM is perpendicular to AP, then $PM = BG$. Take $AQ = $ arc BF, very nearly equal to arc BE, and extend CFH. Take EK, FL perpendicular to CB, and let $CB = r$, $CK = f$, $KE = g$, and $EF = e$, $BG = m$, and $BH = a$. Then $CE : EK = $ arc $EF : LK = QP : LK$, and thus $r : g = e : LK$, and thus $LK = \frac{ge}{r}$. Hence $CL = f + \frac{ge}{r}$. Thus:*

$$LF = \sqrt{r^2 - f^2 - \frac{2fge}{r}}$$
$$= \sqrt{g^2 - \frac{2fge}{r}}$$

where the terms containing powers of e have already been eliminated.

Also, $CL : LF = CB : BH$, hence $f + \frac{ge}{r} : \sqrt{g^2 - \frac{2fge}{r}} = r : m - a$. Squaring both sides, and eliminating terms containing higher powers of e or a (by the first part of Rule 13.1), we have $f^2 + \frac{2fge}{r} : g^2 - \frac{2fge}{r} = r^2 : m^2 - 2ma$. Hence

$$rf^2m^2 - 2mrf^2a + 2gfm^2e - 4gfmae = r^3g^2 - 2gfr^2e$$

Now apply the procedure, first eliminating terms containing products of a and e:

$$rf^2m^2 - 2mrf^2a + 2gfm^2e = r^3g^2 - 2gfr^2e$$

Then eliminate terms that do not contain a or e:

$$-2mrf^2a + 2gfm^2e = -2gfr^2e$$

Hence

$$rfma = gr^2 e - gm^2 e$$

Finally, substitute m for a and t for e, we have:

$$rfm^2 = gr^2 t + gm^2 t$$

Solve this equation for t:

$$t = \frac{rfm^2}{gr^2 + gm^2}$$
$$= \frac{BG \cdot CB^2}{CG^2}$$
$$= \frac{BG \cdot CK^2}{CE^2}$$

The last set of equalities needs some explanation: $r^2 + m^2$ is CG^2, while $f : g = CK : KE = CB : BG$. Hence we have

$$\frac{rfm^2}{gr^2 + gm^2} = \frac{\frac{rfm^2}{g}}{r^2 + m^2}$$
$$= \frac{\frac{CB \cdot CB \cdot BG^2}{BG}}{CG^2}$$
$$= \frac{BG \cdot CB^2}{CG^2}$$

What has Barrow found? Since

$$t = \frac{BG \cdot CK^2}{CE^2}$$

then, rearranging

$$\frac{t}{BG} = \frac{CK^2}{CE^2}$$

But, by the definition of the curve AM, $BG = PM$, and by assumption, $t = PT$. Hence:

$$\frac{PT}{PM} = \frac{CK^2}{CE^2}$$

Inverting:

$$\frac{PM}{PT} = \frac{CE^2}{CK^2}$$

where the left hand side is what we would now consider the slope of the tangent line MT, while the right-hand side corresponds to the square of the secant of the angle. Thus, Barrow has shown $y' = \sec^2 x$. Moreover, though Barrow did not do so, the procedure can easily be extended to find the derivatives of *all* the trigonometric functions.

By 1669, Isaac Barrow held two full-time positions: as the Lucasian Professor of Mathematics at Cambridge, and as the Professor of Geometry at Gresham College in London. Today, Cambridge is a short train ride from London, but in the seventeenth century, it would have taken two days for Barrow to travel from one to the other. Not surprisingly, questions arose as to Barrow's ability to hold both positions, and Cambridge University required Barrow to resign one of them. By then, the Cambridge position had become something of a burden to Barrow, for it prevented his advancement to the higher, administrative ranks, so he resigned the Lucasian Professorship and allowed it to be filled by one of his students. Barrow, freed of his teaching duties, went on to become the Royal Chaplain in London, then in 1674 became the Master of Trinity College, and 1675 became its Vice Chancellor. The student who took Barrow's position as Lucasian Professor was Isaac Newton.

13.2 Exercises

1. Consider Proposition 13.3.

   (a) Prove the second half of the first part, namely that if the point *I* is drawn on the "other side" of *F*, then $LK > DP > LI$.

   (b) Prove the second part of the proposition.

2. Consider Barrow's Rule 3. Why does he feel justified in substituting the finite quantities *m* and *t* for the infinitesimal quantities *a* and *e*? Explain.

3. In Example 13.1, show (using infinitesimal geometry) why $CE : EK = $ arc $EF : LK$.

4. The diagram for Example 13.1 can be used to find the derivative of $y = \sin x$. Use Barrow's method to find the derivative of $y = \sin x$.

5. Explain how Propositions 13.3 and 13.4 are equivalent to the fundamental theorem of calculus.

13.3   The Low Countries

The longest reigning monarch in modern European history was Louis XIV of France (1638-1715). When he was very young, the nobility, angry over one of the actions of his regent, Cardinal Mazarin, broke into Louis's bedroom and badly frightened him. When he was declared of age, he vowed that power would be his alone: "I am the state," he said. To history, he would become known as the Sun King, for all of France would revolve around him.

To break the power of the nobility, Louis built a great palace at Versailles, and required the nobles to spend part of the year there. Louis shrewdly realized that, since the nobles could not possibly outdo the king in extravagance, they would attempt to outdo *each other*, and spend themselves into bankruptcy. This they did, leaving Louis to pursue his dream: a united, powerful France. Unity meant one religion, so Louis revoked the Edict of Nantes in 1685, causing many Huguenots to flee France for England or for the Netherlands.

Louis also wanted France to extend to her "natural frontiers," which we might define as a geological structure that clearly separates two regions: an ocean, a mountain range, or a great river. France is surrounded to the north and west by the Atlantic, to the southeast by the Alps and the Mediterranean, and to the southwest by the Pyrenees mountains. To the east and northeast, however, no natural frontier existed—until the Rhine. Thus, it was Louis's dream to conquer all the land from the current border of France to the Rhine River. In particular, the troublesome Netherlands, home to many refugee French Protestants, was a focus of resistance to the French, and Louis invaded the Netherlands many times. Ultimately, Louis's dream was a failure, for no European power wanted France, already the most powerful state in Europe, to become impregnable. Even worse, the costly wars bankrupted the French state, and Louis's descendants would face the consequences.

13.3.1   Hudde

In 1672, Louis XIV invaded the Netherlands. Time and again, the Dutch flooded the polders to slow the advance of invading armies, and in 1672, the breaking of the dikes to let in the sea was directed by Johann Hudde (May 1628-April 15, 1704), one of Amsterdam's most prominent citizens. Before making civic duties his priority, Hudde had been an amateur mathematician, and around 1654, discovered two important rules, known as **Hudde's rules**.

> Half of the Netherlands is **polder**, land recovered from the sea by building dikes to hold back the sea, then pumping out the sea water. The pumps were generally powered by windmills (another invention from Arabia), giving the Netherlands its distinct, windmill-covered landscape.

The rules appeared as two appendices to the 1659 edition of Descartes. The influence of the Renaissance and the interest in classical culture were still apparent, for Hudde gave the date of his "First letter on the Reduction of Equations" as the Ides of July, 1657, and the "Second letter" was dated 6 Calends of February, 1658. The two dates correspond to July 15 and February 6, respectively.

Hudde began with a number of procedures for simplifying equations. Part of Hudde's procedure relied on finding the greatest common divisor of two polynomials, which Hudde solved in the following way: set both polynomials equal to zero (actually, Hudde began with polynomial equations set equal to zero, so this step was unnecessary); "solve" the lower degree equation for the highest power of the variable and substitute this into the other equation, and in this way reduce the power of the other equation. Repeat this procedure until all terms cancel or all the variables are eliminated. The factor corresponding to the last substitution is greatest common divisor; if the terms do not all cancel, then the two polynomials are relatively prime.

Example 13.2. *Find the greatest common divisor of $x^3 + 3x^2 + 3x + 1$ and*

$x^2 - 4x - 5$. *Setting both equal to zero we have*

$$x^3 + 3x^2 + 3x + 1 = 0 \qquad x^2 - 4x - 5 = 0$$

Since the second equation has a lower degree, solve it for x^2, obtaining $x^2 = 4x + 5$. Substitute this expression for x^2 into the first equation, using the fact that $x^3 = x \cdot x^2 = 4x^2 + 5x = 4(4x+5) + 5x = 21x + 20$, and solve for the highest power of x in the result:

$$x^3 + 3x^2 + 3x + 1 = 0$$
$$21x + 20 + 3(4x + 5) + 3x + 1 = 0$$
$$36x + 36 = 0$$
$$x = -1$$

Now substitute this into the expression previously obtained for x^2. Since $x = -1$, then $x^2 = 1$, and the substitution yields:

$$x^2 = 4x + 5$$
$$1 = -4 + 5$$

Since all the terms cancel, the greatest common divisor is $x + 1$, which is the factor corresponding to the last substitution, $x = -1$.

Example 13.3. *Find the greatest common divisor of $x^3 + 4x - 5$ and $x^2 - 3x + 1$. Setting them both equal to zero and solving the second for x^2 gives $x^2 = 3x - 1$; hence $x^3 = x \cdot x^2 = 3x^2 - x = 8x - 3$. Substituting this into the first equation and solving for the highest degree term gives:*

$$x^3 + 4x - 5 = 0$$
$$(8x - 3) + 4x - 5 = 0$$
$$12x - 8 = 0$$
$$x = \frac{2}{3}$$

Substitute this into the equation for x^2 previously obtained:

$$x^2 = 3x - 1$$
$$\left(\frac{2}{3}\right)^2 = 3\left(\frac{2}{3}\right) - 1$$
$$\frac{9}{4} = 1$$

Since the terms do not all cancel, $x^3 + 4x - 5$ and $x^2 - 3x + 1$ have no common divisor.

If the two polynomials have the same degree, it makes no difference which is used first.

Example 13.4. *Find the greatest common divisor of $x^3 - 4x^2 + 5x - 2 = 0$ and $3x^3 - 8x^2 + 5x = 0$. We arbitrarily solve the first equation for the highest degree x-term, obtaining $x^3 = 4x^2 - 5x + 2$; substituting this value into the second equation gives and solving for the highest degree term gives:*

$$3(4x^2 - 5x + 2) - 8x^2 + 5x = 0$$
$$12x^2 - 15x + 6 - 8x^2 + 5x = 0$$
$$4x^2 - 10x + 6 = 0$$
$$x^2 = \frac{5}{2}x - \frac{3}{2}$$

Note that $x^3 = x \cdot x^2$, so:

$$x^3 = \frac{5}{2}x^2 - \frac{3}{2}x$$
$$= \frac{5}{2}\left(\frac{5}{2}x - \frac{3}{2}\right) - \frac{3}{2}x$$
$$= \frac{19}{4}x - \frac{15}{4}$$

Substitute these values into the previously obtained expression for x^3:

$$x^3 = 4x^2 - 5x + 2$$
$$\frac{19}{4}x - \frac{15}{4} = 4\left(\frac{5}{2}x - \frac{3}{2}\right) - 5x + 2$$
$$\frac{19}{4}x - \frac{15}{4} = 10x - 6 - 5x + 2$$

Solving this for the highest power of x (which, in this case, is x) gives $x = 1$. Substitute this into the equation for x^2, previously obtained; since $x = 1$, then $x^2 = 1$ and thus

$$x^2 = \frac{5}{2}x - \frac{3}{2}$$
$$1 = \frac{5}{2} - \frac{3}{2}$$

Since all terms cancel, then $x - 1$, which is the factor corresponding to the equation $x = 1$, is the greatest common divisor.

The ability to find the greatest common divisor of two polynomials was essential for Hudde's main results, expressed in his tenth rule:

Rule 13.2. *To reduce an equation with a double root, multiply each term of the equation by a term in an arithmetic progression. The greatest common divisor of the two equations will correspond to the double root.*

Hudde's example was the equation $x^3 - 4x^2 + 5x - 2 = 0$. Multiplying each term by a term in an arithmetic progression:

x^3	$-4x^2$	$5x$	-2	$=$	0
3	2	1	0		
$3x^3$	$-8x^2$	$5x$		$=$	0

The greatest common divisor of $x^3 - 4x^2 + 5x - 2$ and $3x^3 - 8x^2 + 5x$ is $x - 1$. Hence 1 is a double root of the equation; simple division shows that the third root is 2.

Hudde was quick to point out that multiplying by *any* arithmetic progression would work. For example

x^3	$-4x^2$	$5x$	-2	$=$	0
1	0	-1	-2		
x^3		$-5x$	4	$=$	0

would also have $x - 1$ as its greatest common factor with $x^3 - 4x^2 + 5x - 2$. He gave no explanation of why the rule worked.

Example 13.5. *Find the roots of $x^3 - x^2 - 8x + 12 = 0$. Multiplying the terms by the arithmetic progression 0, 1, 2, 3 gives*

x^3	$-x^2$	$-8x$	12	$=$	0
0	1	2	3		
	$-x^2$	$-16x$	36	$=$	0

We can find the greatest common factor, but it is easier to note that the second equation factors as $-(x+18)(x-2)$, with roots 2 and -18; trial and error shows 2 is a double root of the first equation, and the third root is -3.

In his second letter, Hudde applied his rule to finding maximum or minimum value.

Rule 13.3. *To find the maximum (or minimum) value of an expression, multiply the terms by an arithmetic progression ending with 0, and one of the roots of the new equation will correspond to the maximum (or minimum) of the original.*

Hudde found the maximum of $3ax^3 - bx^3 - \frac{2bba}{3c}x + aab$. Hudde's reasoning is as follows: suppose Z is the maximum value which occurs at x; then $3ax^3 - bx^3 - \frac{2bba}{3c}x + aab - Z = 0$ has a double root at x. Applying the first rule and multiplying the terms by an arithmetic progression ending in 0 gives

$3ax^3 - bx^3$		$-\frac{2bba}{3c}x$	$aab - Z$	$=$	0
3	2	1	0		
$9ax^3 - 3bx^3$		$-\frac{2bba}{3c}x$		$=$	0

Thus, one of the roots of the new equation will be a double root of the original, and thus correspond to the maximum or minimum.

Example 13.6. *Find the least value of $x^3 - 3x^2 - 9x + 27$. Suppose the minimum value is Z; then $x^3 - 3x^2 - 9x + 27 - Z = 0$ has a double root. Multiplying this by the terms in an arithmetic progression ending with 0 gives*

$$
\begin{array}{ccccc}
x^3 & -3x^2 & -9x & +27 - Z & = \ 0 \\
3 & 2 & 1 & 0 & \\
\hline
3x^3 & -6x^2 & -9x & & = \ 0
\end{array}
$$

The second equation has roots 0, -1 and 3. By trial and error, we see that 0 corresponds to neither a maximum or a minimum; -1 corresponds to a maximum, and 3 corresponds to a minimum.

13.3.2   Sluse

The Canon of Liege (in Belgium), Rene François de Sluse (July 2, 1622-March 19, 1685), discovered an algorithmic method for determining subtangents. He communicated his results to the Royal Society in a 1672 letter, "A Method of Drawing Tangents to All Geometrical Curves." His opening statements are quite interesting:

> I send you, Sir, my method of drawing Tangents to any geometrical curve whatever, and submit it to the censure of the learned men of the Royal Society. It appears to me so short and easy, that it may be learned by a novice ... [3]

The method that may be learned by a novice is:

Rule 13.4. *Given a curve DQ, and lines EB, with AD a perpendicular and DC tangent, intersecting EB at C, let $DA = v$, $BA = y$, EB be constant, and $CA = a$.*

1. *Eliminate all terms that contain neither y nor v, then separate y and v to opposite sides.*

2. *On each side, multiply each term by the exponent of v or y, accordingly, and let one y of each term be turned into an a.*

3. *Solve the resulting equation for a.*

Example 13.7. *Suppose the curve is $by - y^2 = v^2$. The y and v variables are already separated, so multiplying each term by the exponent on y (on the left) or v (on the right), and letting one y be turned into an a, we obtain $ba - 2ya = 2v^2$. Hence $a = \frac{2v^2}{b-2y}$.*

If the terms are mixed, Sluse added an extra term corresponding to the mixed term.

---

[3] *Philosophical Transactions of the Royal Society, Abridged with notes,* 1809. C. and R. Baldwin. Vol. II, page 38.

Example 13.8. *If the equation is* $y^3 = bv^2 - yv^2$, *add to the left* yv^2. *Thus:*

$$
\begin{aligned}
yv^2 + y^3 &= bv^2 - yv^2 \\
2av^2 + 3ay^2 &= 2bv^2 - 2yv^2 \\
a &= \frac{2bv^2 - 2yv^2}{v^2 + 3y^2}
\end{aligned}
$$

13.3 Exercises

1. Show how Hudde's method of finding the greatest common factor of two polynomials is equivalent to applying the Euclidean algorithm.

2. Apply Hudde's method to find the greatest common divisors of the two polynomials.

   (a) $x^2 - 3x - 4$ and $x^3 + 1$

   (b) $x^3 - x^2 - 10x - 8$ and $x^3 + 4x^2 + 5x + 2$

   (c) $x^4 + 5x^3 + 2x^2 - 11x - 3$ and $x^3 + 4x^2 - 9$

3. Apply Hudde's rules to find roots to the following equations.

   (a) $x^3 + 8x^2 + 21x + 18 = 0$ (c) $x^3 + 24x^2 - 36x - 864 = 0$

   (b) $x^3 - 3x^2 - 24x + 80 = 0$ (d) $x^4 + 108x + 243 = 0$

4. Verify that Sluse's procedure works to find the subtangent.

5. Apply Sluse's procedure to find the subtangents for the following curves.

   (a) $y = v^2$ (b) $v^2 + y^2 = 4$ (c) $yv = 1$

13.4   Isaac Newton

Isaac Newton (December 25, 1642-March 20, 1727) was born a few months after the death of his father; his mother remarried, but the relationship between Newton and his mother Hannah Ayscough and stepfather Barnabas Smith was strained at best. At one point, he threatened to burn down the house with his parents inside. His stepfather died in 1653, and shortly afterward, his mother decided her eldest son ought to look after the family estate: the Ayscoughs were relatively prosperous farmers. However, Newton showed no ability in estate management, and an uncle, William Ayscough, persuaded Newton's mother that the boy should be sent to a preparatory school for university. He entered Trinity College (part of Cambridge University), his uncle's school, on June 5, 1661.

While in grade school, Newton confessed his primary interests were "money, learning, and pleasure." Thus, it was his intention to become a lawyer. At the time, his interest in mathematics was slight. Euclid was "trifling," though this disdain probably masked a lack of comprehension: Isaac Barrow, who tested Newton on the subject, found Newton's understanding of geometry to be poor. Further evidence for this is that later on Newton became enamored of the nonsense of astrology but found the mathematics too difficult for him, so he began to study trigonometry to make up for his deficiencies. But Newton could not understand trigonometry because he lacked the necessary background in geometry, so he began to work his way through Barrow's edition of the *Elements*; upon reading Proposition I-35 of the *Elements*, Newton realized that Euclid was, in fact, quite profound. Besides the *Elements*, Newton also worked his way through Viète's collected works (published in 1646), Descartes's *Geometry*, and Wallis's *Arithmetic of the Infinite*. Shortly afterward, Newton began to think about the problem of quadrature.

In the 1660s, plague struck England. It devastated London and probably would have killed many more than it actually did but for the Great Fire of London (1666), which burned down much of the city. Amazingly, though 13,000 houses were destroyed and 100,000 people left homeless, only eight people were known to have died in the fire. More importantly, the fire burned the roofs that housed the rats that carried the fleas that harbored the plague bacillus. In 1665, the plague made it to Cambridge, forcing the closure of Trinity College. Newton returned to his family's farm in Lincolnshire. The next few months were, perhaps, the most significant few months in the entire history of science. Newton is known for four major contributions to mathematics and physics made during this time: the binomial theorem; calculus; the theory of gravitation; and the discovery of the nature of light and colors.

See Proposition 5.11.

Isaac Newton. A painting made after Newton's death (no portraits of Newton were made during his lifetime). ©Corbis.

13.4.1   The Binomial Theorem

The use of infinite series would become one of the most powerful tools of mathematicians, but in Newton's time, the study of infinite series was in its infancy. Notwithstanding questions of convergence (the primary concern of modern mathematicians), the actual derivation of a series was a laborious process. Newton's discovery of the **generalized binomial theorem** greatly eased this task, and is perhaps Newton's most important contribution to mathematics.

The binomial expansion of $(a + b)^n$ for integral values of n was well known by Newton's time; what was not known was the corresponding expansion, as an infinite series, for $(a + b)^{m/n}$. Newton's discovery of the generalized binomial theorem was not merely the result of replacing the integral power n with the fractional power m/n but instead followed after extensive analysis of the coefficients, and careful verification of the results; as such, it is a model of mathematical discovery.

Newton claimed to have discovered the generalized binomial theorem before he was acquainted with the procedures of root extraction, suggesting a date of around 1664 or 1665. However, though the use of the generalized binomial expansion would have been quite useful in either *Analysis by Equations of an Infinite Number of Terms*, which he wrote in 1669, or his 1671 *Of the Method of Fluxions and Infinite Series*, it made no appearance in either work. The first written appearance of the binomial theorem was in a letter from Newton to Henry Oldenburg, Secretary of the Royal Society, on June 13, 1676. In a second letter, October 24, 1676, Newton explained to Oldenburg how he happened upon it.

First, Newton considered the areas under the curves

$$(1 - x^2)^{\frac{0}{2}}, \qquad (1 - x^2)^{\frac{2}{2}}, \qquad (1 - x^2)^{\frac{4}{2}}, \qquad (1 - x^2)^{\frac{6}{2}} \quad \ldots$$

The areas under these curves (from 0 to x) form the sequence

$$x, \qquad x - \tfrac{1}{3}x^3, \qquad x - \tfrac{2}{3}x^3 + \tfrac{1}{5}x^5, \qquad x - \tfrac{3}{3}x^3 + \tfrac{3}{5}x^5 - \tfrac{1}{7}x^7 \quad \ldots$$

If one could interpolate between these, one would find the areas of the alternate curves, of which $(1 - x^2)^{\frac{1}{2}}$, the curve giving the half circle, would be the first. Newton, like Wallis, looked for patterns in the terms.

The first term of each is x. The second terms are $-0x^3/3$, $-1x^3/3$, $-2x^3/3$, and so on, which form an obvious progression. Thus, Newton concluded, the first two terms of the interpolated series should be $x - \tfrac{1}{3}(\tfrac{1}{2}x^3)$, $x - \tfrac{1}{3}(\tfrac{3}{2}x^3)$, ...

Next, Newton noted that the denominators of the terms formed a simple arithmetical progression; thus, only the numerators need be found. If the numerator of the second term is m, then the remaining numerators can be found by the successive partial products of

$$\frac{m - 0}{1} \times \frac{m - 1}{2} \times \frac{m - 2}{3} \times \frac{m - 3}{4} \times \ldots$$

where the numerator of the nth term after the first is found by the product of the first n factors. Since a series is desired for the area under $(1 - x^2)^{\frac{1}{2}}$, whose second term has numerator $m = \tfrac{1}{2}$, Newton substituted this in and derived the series

$$x - \frac{\tfrac{1}{2}x^3}{3} - \frac{\tfrac{1}{8}x^5}{5} - \frac{\tfrac{1}{16}x^7}{7} - \frac{\tfrac{5}{128}x^9}{9} - \ldots$$

Interestingly, it was only after Newton found the series for the quadrant of a circle that he realized it was not necessary to consider the areas under the curves, for the expressions themselves could be interpolated. The only necessary change was the elimination of the denominators 3, 5, 7, ..., which were introduced in the determination of the areas. Thus, the coefficients of the binomial expansion of $(1 - x^2)^m$ could be found by the partial products of

$$m \times \frac{m - 1}{2} \times \frac{m - 2}{3} \times \frac{m - 3}{4} \times \ldots$$

Thus, $(1 - x^2)^{\frac{1}{2}}$ could be expanded as $1 - \tfrac{1}{2}x^2 - \tfrac{1}{8}x^4 - \tfrac{1}{16}x^6 - \ldots$ To verify this, Newton multiplied this series by itself, and found the product was $1 - x^2$, with "the remaining terms vanishing by the continuation of the series to infinity."

13.4.2   The Fundamental Theorem of Calculus

Around 1669, Newton first put forth the ideas expressed in his *Analysis by Equations of an Infinite Number of Terms*. Though Newton showed it to his friends, he did not actually publish it until 1711. The reason for this may have been the contentious Robert Hooke (1635-1703). Hooke anticipated (in rough outline) both Newton's work on gravitation and Huygens's theory of light; however, in Hooke's mind, originating an idea was the same as developing it, so when Newton published the *Principia* (1687), Hooke accused Newton of plagiarism. After this and other unfortunate experiences with Hooke, Newton vowed not to publish any more until after Hooke was dead.

In *Analysis by Equations of an Infinite Number of Terms*, Newton began by stating that for a curve whose base $AB = x$ and whose ordinate $BD = y$, then

Proposition 13.5 (Definite Integral of a Rational Power). *If $ax^{\frac{m}{n}} = y$, then area $ABD = \frac{an}{m+n}x^{\frac{m+n}{n}}$.*

The proof is delayed until near the end of his treatise.

Unlike Wallis, Newton did not find the area under a curve by summing an infinite series. Rather, like Barrow, he analyzed the rate of change of an area function for simple curves of the form $y = x^n$, where n is a rational number; his proof is the basis for one version of the modern proof of the fundamental theorem of calculus. First, Newton proved it for a specific case.

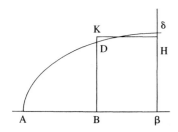

Proof. Let the base AB of any curve AD have BD for its perpendicular ordinate. Let $AB = x$, $BD = y$, and let a, b, c be given quantities, and m, n be whole numbers. Then if $y = ax^{\frac{m}{n}}$, then area $ABD = \frac{an}{m+n}x^{\frac{m+n}{n}}$.

Let the area $ABD = z$. Let $B\beta = o$, $BK = v$, and $B\beta HK = ov$ be the area of $B\beta\delta D$. Therefore, $A\beta = x + o$, and area $A\delta\beta = z + ov$. Assume x, z have any relationship; for example, $\frac{2}{3}x^{\frac{3}{2}} = z$, and so $\frac{4}{9}x^3 = z^2$. Replace x with $x + o$, and z with $z + ov$. Expanding the terms we have $\frac{4}{9}$ of $x^3 + 3x^2o + 3xo^2 + o^3 = z^2 + 2zov + o^2v^2$. Since $\frac{4}{9}x^3$ is equal to z^2, we may eliminate them; dividing the remaining terms by o, we have $\frac{4}{9}$ of $3x^2 + 3xo + o^2 = 2zv + ov^2$. Now suppose $B\beta$ to be diminished indefinitely, so o is nothing. v and y will be equal, and thus $\frac{4}{9}3x^2 = 2zy$, and thus $x^{\frac{1}{2}} = y$. Conversely, if $y = x^{\frac{1}{2}}$, ABD shall be $\frac{2}{3}x^{\frac{3}{2}}$.   □

See Problem 1.

After this example, Newton proved the general case.

Newton then proved the sum rule of integration for positive functions, in much the same way we prove it today. By expanding rational and radical functions as infinite series, and using the sum and power rule of integration, Newton could perform his analysis using equations "of an infinite number of terms." Newton did not use the binomial theorem in any of his series development, strongly suggesting his discovery of the binomial theorem came after he outlined the development in *Analysis by Equations of an Infinite Number of Terms*.

For the curve given by the equation $y = \frac{aa}{b+x}$, Newton first divided $b + x$ into

See Problem 3.

aa, using long division, obtaining

$$y = \frac{aa}{b} - \frac{aax}{b^2} + \frac{aax^2}{b^3} - \frac{aax^3}{b^4} + \ldots$$

Then, using the sum and power rules, the area is determined by integrating this equation term by term; thus, Newton gives the area as

$$\frac{a^2x}{b} - \frac{a^2x^2}{2b^2} + \frac{a^2x^3}{3b^3} - \frac{a^2x^4}{4b^4} + \ldots$$

For functions involving roots, such as $y = \sqrt{aa + xx}$, he follows the procedure for finding square roots to find a series expansion, namely $y = a + \frac{x^2}{2a} - \frac{x^4}{8a^3} + \frac{x^6}{16a^5} + \ldots$ This series is then integrated term by term to yield the desired area. See Problem 7.

13.4.3   Series for Implicit Functions

To construct a series for functions defined implicitly (which Newton called an "affected equation"), such as $y^3 + a^2y - 2a^3 + axy - x^3 = 0$, Newton used a method of successive approximations he invented, now called the **method of reversion**, which is another variant of the much anticipated method of Horner.

First, to solve an equation such as $y^3 - 2y - 5 = 0$, Newton first noted that 2 is an approximate solution; hence, the actual solution can be represented by $y = 2 + p$, where p is a small quantity. Substituting this into the original equation and simplifying, he obtained a new equation $p^3 + 6p^2 + 10p - 1 = 0$.

Since p is small, he felt justified in ignoring the p^2 and p^3 terms, and solved the simpler equation $10p - 1 = 0$, obtaining $p = 0.1$ as the approximate solution to the second equation. By the same argument, the actual solution should be $p = 0.1 + q$, where q is another small number. Substituting $0.1 + q$ for p in the second equation, he generated a third equation, $q^3 + 6.3q^2 + 11.23q + 0.061 = 0$. Again, since q is small, the q^2 and q^3 terms are ignored, resulting in the equation $11.23q + 0.061 = 0$, which has $q = -0.0054$ as a solution. Again, since $q = -0.0054$ is an approximate solution, the actual solution to this equation should be $-0.0054 + r$, where r is very small.

At this point, Newton, chose to ignore the q^3 term in the third equation, so the fourth equation is obtained by substituting $-0.0054 + r$ into the equation $6.3q^2 + 11.23q + 0.061 = 0$. This equation is then solved for r (again ignoring the r^2 terms that result). Since the root is $2 + p + q + r + \ldots$, where the values of p, q, and r were found, the solution 2.09455147 can be found. Newton tabulated his results (see Table 13.4 and notice that in many places Newton wrote only the coefficients of each polynomial).

Newton then applied this method to reducing to a series functions defined implicitly. For $y^3 + a^2y - 2a^3 + axy - x^3 = 0$, Newton began by determining y when $x = 0$. If $x = 0$, then $y^3 + a^2y - 2a^3 = 0$, whose root is $y = a$. Assuming that $y = a + p$, and substituting this into the equation, Newton obtained a second equation, $-x^3 + a^2x + axp + 4a^2p + 3ap^2 + p^3 = 0$.

Assuming x and p small, the largest terms will be those where p or x have the lowest degrees separately. Selecting these terms, he set $4a^2p + a^2x = 0$,

$y = 2 + p$	$+y^3$	$+8$	$+12p$	$+6p^2$	$+p^3$
	$-2y$	-4	$-2p$		
	-5	-5			
	Sum	-1	$+10p$	$+6p^2$	$+p^3$
$p = 0.1 + q$	$+p^3$	$+0.001$	$+0.03q$	$+0.3q^2$	$+q^3$
	$+6p^2$	$+0.06$	$+1.2$	$+6.0$	
	$+10p$	$+1$	$+10$		
	-1	-1			
	Sum	$+0.061$	$+11.23q$	$+6.3q^2$	$+q^3$
$q = -0.0054 + r$	$+6.3q^2$	$+0.000183708$	$-0.06804r$	$+6.3r^2$	
	$+11.23q$	-0.060642	$+11.23$		
	$+0.061$	$+0.061$			
	Sum	$+0.000541708$	$+11.16196r$	$+6.3r^2$	
$r = -0.00004854 + s$					

Table 13.4: Newton's Method of Reversion for $y^3 - 2y - 5 = 0$

$y = a + p$	$+y^3$	$+a^3$	$+3a^2p$	$+3ap^2$	$+p^3$
	$+a^2y$	$+a^3$	$+a^2p$		
	$+axy$	$+a^2x$	$+axp$		
	$-x^3$				
	Sum	$-x^3 + a^2x + axp + 4a^2p + 3ap^2 + p^3$			
$p = -\frac{1}{4}x + q$	$+p^3$	$-\frac{1}{64}x^3$	$+\frac{3}{16}x^2q$	$-\frac{3}{4}xq^2$	$+q^3$
	$+3ap^2$	$+\frac{3}{16}ax^2$	$-\frac{3}{2}axq$	$+3aq^2$	
	$+4a^2p$	$-a^2x$	$+4a^2q$		
	$+axp$	$-\frac{1}{4}ax^2$	$+axq$		
	$+a^2x$	$+a^2x$			
	$-x^3$	$-x^3$			
	Sum	$-\frac{65}{64}x^3 - \frac{1}{16}ax^2 + \frac{3}{16}x^2q - \frac{1}{2}axq + 4a^2q + 3aq^2 + q^3$			

Table 13.5: Newton's Method of Reversion for $y^3 + a^2y - 2a^3 + axy - x^3 = 0$

which gives $p = -\frac{1}{4}x$. Again, assuming that this is close to the actual solution, he assumed $p = -\frac{1}{4}x + q$, and again picking out the terms where x and q are separately of the lowest degree, he sets $-\frac{1}{16}ax^2 + 4a^2q = 0$, and finds $q = \frac{x^2}{64a}$. (Since there is no first-degree x term, Newton used the second degree x term.)

As before, Newton then ignored the q^3 term as being insignificant, and assumed $q = \frac{x^2}{64a} + r$. Substituting this into the expression and again equating the terms of lowest degrees of r and x, separately, and repeating the process one more time, Newton finds the expression for y

$$y = a - \frac{x}{4} + \frac{x^2}{64a} + \frac{131x^3}{512a^2} + \frac{509x^4}{16384a^3} + \cdots$$

One of the difficulties of applying this method in the multivariable case is determining where to begin, since the first step is finding an approximate solution. For example, in the equation $y^6 - 5xy^5 + \frac{x^3}{a}y^4 - 7a^2x^2y^2 + 6a^3x^3 + b^2x^4 = 0$, how can one find an approximate solution for y to begin the process?

Newton provided a solution in a later treatise, the 1671 work *On the Method of Fluxions and Infinite Series*. Though the treatise was not primarily concerned with the method of reversion, it was necessary to use reversion to find many of the functions that Newton would treat with the method of fluxions.

First, indicate on a grid as shown, the grid squares that correspond to a term of the equation. By drawing a line through the grid so that all the squares corresponding to equation terms are on or above the line, one can determine the terms of least order. These are the terms whose sum will initially be set equal to zero. Thus, we note in the equation $y^6 - 5xy^5 + \frac{x^3}{a}y^4 - 7a^2x^2y^2 + 6a^3x^3 + b^2x^4 = 0$, the indicated grid spaces are marked, and the diagonal touches the spaces corresponding to the x^3, x^2y^2, and y^6 terms. Thus, one begins by solving the equation

$$y^6 + 6a^3x^3 - 7a^2x^2y^2 = 0$$

This reduces to the equation

$$v^6 - 7v^2 + 6 = 0$$

using the substitution $y = v\sqrt{ax}$.

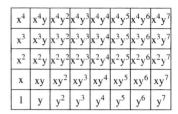

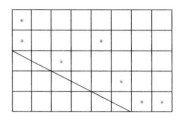

13.4.4 Series for Trigonometric Functions

Newton's process for solving an equation in terms of an infinite series is the basis for his construction of the series for sine, which also appears in *Analysis by Equations of an Infinite Number of Terms*. To begin with, Newton constructed a series for the measure of the length of an arc of a circle. In the following discussion, Newton's term **moments** corresponds to an infinitesimal change in a quantity.

Problem 13.1. *Let $ADLE$ be a circle. Find arc length AD.*

Solution. Draw tangent DHT, and complete the infinitely small rectangle $HGBK$. Let diameter $AE = 1 = 2AC$, and $AB = x$. Then Moment of base BK : Moment of arc HD as $BT : DT$, which is as $BD : DC$, which is as $\sqrt{x - x^2} : \frac{1}{2}$, which is as $BK : DH$. Thus, $BK : DH$ as $1 : \frac{1}{2\sqrt{x-x^2}}$. Thus, the length z of the arc will be $z = x^{\frac{1}{2}} + \frac{1}{6}x^{\frac{3}{2}} + \frac{3}{40}x^{\frac{5}{2}} + \frac{5}{112}x^{\frac{7}{2}} + \ldots$ Alternatively, if CB is supposed to be x, and the radius CA to be 1, then the length z of arc LD will be $z = x + \frac{1}{6}x^3 + \frac{3}{40}x^5 + \ldots$ □

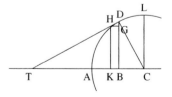

See Problem 10.

The last series corresponds to $z = \arcsin x$. To find a series representation of the length of the arc DL, or in other words, to find a series for the sine of z, Newton relied on his method of reversion. For example, to obtain the series to the fifth degree in z, the arc length, Newton used the arcsine series as far as the fifth degree term in x, then tried to solve the equation

$$0 = x + \frac{1}{6}x^3 + \frac{3}{40}x^5 - z$$

See Problem 11.
See Problem 13.

See Problem 14.

for x in terms of the powers of z, thereby obtaining the familiar power series for $\sin x$. Newton then used this series to find the power series for cosine. In a similar manner, Newton found the area underneath the hyperbola $y = \frac{1}{1+x}$, which would allow him to compute logarithms to any desired degree of accuracy. The inversion of this series produced a series for the exponential function.

13.4.5   James Gregory

James Gregory (1638-1675) found another means of determining the arc length of a circle. In a letter dated February 15, 1671, Gregory noted that if the radius of a circle was r, the arc length a, the tangent t, the secant s (remember the tangent and secant were, at that time, actual lengths), then

$$a = t - \frac{t^3}{3r^2} + \frac{t^5}{5r^4} - \frac{t^7}{7r^6} + \frac{t^9}{9r^8}$$

$$t = a + \frac{a^3}{3r^2} + \frac{2a^5}{15r^4} + \frac{17a^7}{315r^6} + \frac{3233a^9}{181440r^8}$$

$$s = r + \frac{a^2}{2r} + \frac{5a^4}{24r^3} + \frac{61a^6}{720r^5} + \frac{277a^8}{8064r^7}$$

These results can be obtained without the use of Taylor's theorem using either the methods of Newton or those of Barrow, but a mathematical error in the preceding series suggests its relationship to a note written on the back of a letter dated January 29, 1671; the notes in turn suggest that Gregory did use the equivalent of Taylor's theorem.

The back of the document reads:

$$
\begin{array}{llll}
\text{arc} = a & 1^{\text{st}} & 2^{\text{nd}} & 3^{\text{rd}} \\
\text{sine} = s & & & \\
\text{radius} = r & m = q & m = \frac{r^3}{c^2} & m = \frac{2r^4 q}{vc^3} \\
\text{secant} = v & & & \\
\text{cosine} = c & & & \\
\text{tangent} = q & & &
\end{array}
$$

This note suggests that Gregory anticipated the work of Taylor in the discovery of the so-called **Taylor series** of a function. However, Gregory never suggested it might be useful as a general tool, and thus, over the next eighty years, various mathematicians would claim for themselves the discovery of the Taylor expansion.

13.4 Exercises

1. Prove the general case of finding the area under the curve $y = x^{m/n}$, where m and n are whole numbers. Remember that in *Analysis by Equations of an Infinite Number of Terms*, Newton did not use the generalized binomial theorem, so you should not use it in this proof.

2. In the proof of the fundamental theorem for $y = \frac{2}{3}x^{\frac{3}{2}}$, what other theorems does Newton implicitly assume?

3. Consider Newton's division of $b + x$ into aa to obtain Newton's series for the function $y = \frac{aa}{b+x}$.

   (a) Perform the division, and show that if the series converges, the equality of the function and the infinite series is justified.

   (b) For what values of a and b does the series diverge?

   (c) Is there an appropriate transformation of variables that will make the series converge for all values of a and b? Explain.

4. Use Newton's method to integrate the function $y = \frac{1}{1+x^2}$.

5. (Advanced Calculus) Newton dealt primarily with series for functions of the form $y = (1 \pm x^b)^{m/n}$. The series expansion and term-by-term integration work only if the series is uniformly convergent; under what conditions and over what intervals are these series uniformly convergent?

6. (Advanced Calculus) Comment on the validity of the use of the method of reversion to expanding an implicit function in the form of an infinite series.

7. Use Theon's method for finding square roots to develop Newton's series for $y = \sqrt{aa + xx}$. Assume $a > x$.

   (a) For what values of a, x is the series valid?

   (b) Is there an appropriate transformation of variables that make the series convergent for all values of a? Explain.

8. Solve, using Newton's method.

   (a) $y^4 - 4y^3 + 5y^2 - 12y + 17 = 0$. $y = 3$ is an approximate solution.

   (b) $x^3 - 4x^2 + 5x - 8 = 0$

9. Use Newton's method to find the first three terms of the series expansion in x for the function defined implicitly by $y^2 - 5xy + x^2 - 9 = 0$.

10. Analyze Newton's argument for constructing the series for the arc length AD, given $AB = x$.

    (a) Comment on the validity of Newton's assumptions; in particular, note Newton's infinitely small rectangle $HGBK$.

    (b) Show $BD : DC$ is as $\sqrt{x - x^2} : \frac{1}{2}$.

    (c) What quantity does arc AD correspond to in terms of x?

    (d) Use Newton's method of expanding the binomial and integrating term by term to find the first six terms of the series for the length of arc AD.

11. Consider Newton's method of finding the series for arcsine.

    (a) Derive the relationship for the arc length DL to the quantity $AB = x$.

    (b) Expand this relationship as an infinite series.

    (c) Find the first four terms of the series for arc length in terms of x.

    (d) Verify the first three terms of Newton's series by finding the fifth-degree Taylor polynomial for $y = \arcsin x$ around $x = 0$.

12. Use the reversion method to solve $z = x + \frac{1}{6}x^3 + \frac{3}{40}x^5$ for x in terms of a power series in z.

13. Use the series for sine and the identity $\cos x = \sqrt{1 - \sin^2 x}$ to determine the series for $\cos x$.

14. Consider the hyperbola $y = \frac{1}{1+x}$.

    (a) Find, using Newton's method, the area z under the hyperbola. Extend the series to the fifth-degree terms.

    (b) Use the reversion method to solve for z, to the fifth degree in terms of x.

    (c) What function has been found?

15. Verify Newton's binomial series for the expansion of $(1 - x^2)^{1/2}$.

    (a) Use the square root algorithm to find the square root of $1 - x^2$.

    (b) Multiply the binomial series by itself to obtain $1 - x^2$ and show "the remaining terms" will vanish as the series is continued to infinity.

16. Examine the entries in Gregory's note on page 418. How do they support the idea that Gregory was calculating the Taylor series for the tangent function?

17. Calculate the power series for the tangent function, and compare it to Gregory's series. What mathematical error did Gregory make?

13.5 Fluxions

Newton's work helped turn calculus into a generalized, powerful tool with wide applicability. Newton expanded his work on *Analysis by Equations of An Infinite Number of Terms* into *Of the Method of Fluxions and Infinite Series*, written around 1671 but not published until 1736.

Newton began by reiterating his work with infinite series, again showing how to reduce equations to polynomials of infinite degree. He then introduced two general problems:

1. If the total distance a particle travels is given, to find the velocity at any given time.

2. If the velocity of a particle at any given time is known, to find the distance it traveled over a given interval of time.

In Newton's terminology, distance and velocity are simply convenient ways of referring to a quantity that is changing and the rate at which the quantity is changing. The quantity itself Newton referred to as a **fluent**, while the velocity—the rate the quantity is changing—Newton called a **fluxion**. To distinguish between the two, he placed a dot over the variable representing the fluent. Thus, if x is the fluent, then $\dot{x}$ will be its fluxion.

13.5.1 Finding Fluxions and Fluents

To find fluxions, Newton described a simple procedure. First, arrange the terms in order of one of the variables; for example, in order of increasing powers of x. Then multiply the terms by *any* arithmetical progression, and then by $\frac{\dot{x}}{x}$. Repeat the procedure for the other variables, and set the sum of all the terms equal to 0.

For example, in the affected equation $x^3 - ax^2 + axy - y^3 = 0$, set down the x terms in decreasing order, then multiply them by the arithmetic sequence $3, 2, 1$, and then by $\frac{\dot{x}}{x}$.

$$
\begin{array}{ccc}
x^3 & -ax^2 & axy \\
\frac{3\dot{x}}{x} & \frac{2\dot{x}}{x} & \frac{\dot{x}}{x} \\
3\dot{x}x^2 & -2a\dot{x}x & a\dot{x}y
\end{array}
$$

For the y terms, the process results in

$$
\begin{array}{cccc}
-y^3 & & axy & x^3 - ax^2 \\
\frac{3\dot{y}}{y} & \frac{2\dot{y}}{y} & \frac{\dot{y}}{y} & 0 \\
-3y^2\dot{y} & 0 & +ax\dot{y} & 0
\end{array}
$$

Summing the terms and setting the result equal to zero, the fluxional equation is

$$
3\dot{x}x^2 - 2a\dot{x}x + a\dot{x}y - 3y^2\dot{y} + ax\dot{y} = 0
$$

To prove the method is valid, Newton supposed that x increases at a rate of $\dot{x}$ over a short time o, during which time y increases at rate $\dot{y}$. Thus, in time o, y increases by $\dot{y}o$ and x increases by $\dot{x}o$. By substituting these values into the equation $x^3 - ax^2 + axy - y^3 = 0$, one obtains

$$x^3 + 3\dot{x}ox^2 + 3\dot{x}^2o^2x + \dot{x}^3o^3 - ax^2 - 2axox - a\dot{x}^2o^2$$
$$+ axy + a\dot{x}oy + a\dot{y}ox + a\dot{x}\dot{y}o^2 - y^3 - 3\dot{y}oy^2 - 3\dot{y}o^2y - y^3o^3 = 0$$

Since, by supposition, $x^3 - ax^2 + axy - y^3 = 0$, these terms may be eliminated. The equation can then be divided by o and, since o was assumed infinitely small, Newton concluded that the terms multiplied by o would amount to nothing, and thus we obtain

$$3x^2\dot{x} - 2a\dot{x}x + a\dot{x}y + a\dot{x}y - 3\dot{y}y^2 = 0$$

which was to be proven.

Given a fluxional relationship, the relationship between the fluents can be found by reversing the steps, provided the redundant terms are eliminated. Thus, given the fluxional relationship $3\dot{x}x^2 - 2a\dot{x}x + a\dot{x}y - 3\dot{y}y^2 + a\dot{y}x = 0$, the terms are written first in terms of powers of x, which are then divided by $\frac{\dot{x}}{x}$ and then by the terms of any arithmetical sequence.

	$3\dot{x}x^2$	$-2a\dot{x}x$	$+a\dot{x}y$
Divide by $\frac{\dot{x}}{x}$	$3x^3$	$-2ax^2$	axy
Divide by	3	2	1
	x^3	$-ax^2$	$+axy$

The corresponding procedure for y gives

	$-3\dot{y}y^2$		$a\dot{y}x$
Divide by $\frac{\dot{y}}{y}$	$-3y^3$		$+axy$
Divide by	3	2	1
	$-y^3$	0	$+axy$

Since axy appeared in both processes, the second one is redundant; the resulting equation is thus $x^3 - ax^2 + axy - y^3 = 0$.

One of the first uses Newton made of the method of fluxions was finding the maximum or minimum value of a fluent. Newton argued that at the maximum or minimum value of a quantity, it neither increased nor decreased: in other words, the fluxion was equal to zero. From this operation, Newton noted, Hudde's rule could be derived.

See page 408.

Although Newton wrote his *Of the Method of Fluxions and Infinite Series* in 1671, it was not published until 1736, long after the basic procedures of calculus had been established. The first widespread *publication* of Newton's method of fluxions was as one of two appendices to Newton's *Optics* (1704), published the year after Hooke's death. The first appendix, entitled "Enumeration of Lines of the Third Order," classified and graphed 72 types of third-degree equations; it was the first time two axes and negative coordinates were used systematically, and thus the first appearance of our modern version of rectilinear coordinates.

The second appendix, "Treatise on the Quadrature of Curves," introduced and used fluxional quantities in much the same way as Newton's 1671 work, with some refinements. In the introduction, Newton briefly discussed the generation of figures through motion, a concept that would become standard for the next century:

> Lines may be described by the continuous motion of points, surfaces by the motion of lines, solids by the motion of surfaces...[4]

Like Wallis and Euclid before him, Newton avoided the question of a direct comparison of two quantities by examining the *ratio* between them: in Newton's case, he compared two infinitely small quantities. If x increased to $x + o$, then x^n would increase to $(x+o)^n$. Thus, the *ratio* of the increase in x to the increase in x^n was o to $nox^{n-1} + \frac{nn-n}{2}o^2x^{n-2} + \ldots$, or as 1 to $nx^{n-1} + \frac{nn-n}{2}ox^{n-2} + \ldots$. Letting o vanish, the "last ratio" is thus 1 to nx^{n-1}, and hence the fluxion of x is to the fluxion of x^n as 1 to nx^{n-1}.

Newton provided a more formal definition of a fluent or a fluxion:

> Indeterminate quantities which by continuous motion increase or decrease I call *fluents* or *defluents*, and designate them by letters z, y, x, v; their *fluxions* or speeds of increase I note by punctuating the same letters $\dot{z}$, $\dot{y}$, $\dot{x}$, $\dot{v}$.[5]

The rate of change of the fluxion $\dot{x}$ was $\ddot{x}$; the rate of change of $\ddot{x}$ was $\dddot{x}$, and so on. As x might itself be the fluxion of some quantity, Newton designated that x was the fluxion of $\acute{x}$. To find the fluxions, Newton abandoned the complexities of multiplying each term by numbers in arithmetic sequence, and instead used the modern procedure of multiplying by the power of the variable, changing one of the variables to its fluxion. It is important to emphasize that Newton found a *relationship* between the fluxions, and not the derivative in our sense: from the equation $x^3 - xyy + aaz - b^3 = 0$, Newton derived the equation $3\dot{x}x^2 - \dot{x}y^2 - 2xy\dot{y} + a^2\dot{z} = 0$, which gave the relationship of the fluxions.

13.5.2 Tangents

It is, perhaps, significant that Newton showed how fluxions could be used to find maxima and minima before showing how they could be used to find the tangents to curves. One of the problems was there was, as yet, no uniform system of describing curves. Indeed, Newton gave no less than eight methods of finding

[4]Newton, Isaac, *Opticks: or, A treatise of the reflections, refractions, inflexions and colours of light. Also two treatises of the species and magnitude of curvilinear figures.* S. Smith, and B. Walford, 1704. Page 165.

[5]Ibid., 170

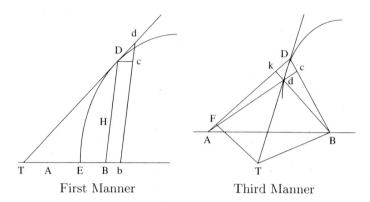

First Manner   Third Manner

Figure 13.7: Two Methods of Finding Tangents

tangents, all depending on how the equation for the curve was described (see Figure 13.7).

The first two methods use Cartesian coordinates. In the **first manner**, where curve EdD is given, suppose TD is tangent to the curve. Let a point on the curve go from D to d. Then $TB : BD = Dc : cd$. If $AB = x$ and $BD = y$, then $Dc = \dot{x}$ and $cd = \dot{y}$, so the length of the subtangent, TB can be found.

The **third manner** uses what are called "bipolar coordinates," where the coordinates of a point are specified by its distance to two fixed points, A and B. Suppose the point D moves in an infinitely small amount of time to d. Make $Ak = Ad$ and $Bc = Bd$, so that kD, cD are the moments of AD, BD. (Notice that here Newton retained a trace of his older terminology.) Make $DF : BD = Dk : Dc$, and drawn BT perpendicular to BD and FT perpendicular to AD. Then $DFTB$ is similar to $Dkdc$, and the diagonal DT will be the tangent.

Newton's **seventh manner**, "for spirals," introduced polar coordinates and the method of finding tangents to curves given the polar equation for them. If BG is the circumference of a circle of radius AG, with D on some curve ADE, let Dd be an infinitesimally small change in the curve. Let $Ac = Ad$, so cD, Gg are the contemporaneous moments of the line AD and the arc BG (i.e., the radius and angle). Draw At parallel to dc, and thus perpendicular to AD, and suppose DT is the tangent to ADE. Then $cD : cd = AD : AT$. Suppose Gt is parallel to DT. Then $cd : Gg = AD : AG$ or, since $Ad = AD$, as $Ad : AG$. Thus $cd : Gg = AT : At$. Therefore $cD : Gg$ is $AD : At$.

Seventh Manner
See Problem 7.

13.5.3 The *Principia*

Even before the appearance of the two treatises in *Optics*, Newton revealed some of his methods of calculus in his masterwork, *Mathematical Principles of Natural Philosophy* (1687), usually known as the *Principia* after its Latin title. Newton sought to establish natural philosophy—what we would now call physics—on a basis as axiomatic and deductive as mathematics itself.

Newton began the *Principia* with some definitions and three axioms, fundamental to physics. Today, these axioms are called **Newton's three laws of motion**:

1. An object at rest will stay at rest, while an object in motion will stay in motion, unless acted upon by an external force.

2. The change in motion of an object is proportional to the force exerted on it.

3. For every action there is an equal and opposite reaction.

Though this is not a text on the history of physics, it is worth noting that today's formulation of Newton's Second Law, that force is proportional to acceleration (i.e., $F = ma$) is not how Newton stated his second law. Rather, Newton's statement is that the force is proportional to the change in motion—what modern physicists call the momentum of an object.

After the introductory definitions, axioms, and a few corollaries, Newton began Book I: On the Motion of Bodies. The first section of the book is devoted to "The Method of First and Last Ratios of Quantities"—Newton's method of fluxions. The introduction to fluxions in the *Principia* is vastly different from the almost modern method in "Treatise on the Quadrature of Curves" or *Of the Method of Fluxions and Infinite Series*. First, though Newton had developed an adequate notation of expressing fluxional quantities in his $\dot{x}$, no trace of this notation appeared in the *Principia*. Second, the method of "first and last ratios" relied heavily on geometrical concepts that make it quite difficult to follow, rather than the simpler algebraic notions used in the method of fluxions and, especially, Leibniz's differential calculus.

Book I, Part 1 opens with an explicit definition of the idea of a limit. The definition appears as part of two lemmas:

Lemma 13.1. *Quantities, and the ratios of quantities, which in any finite time converge continually to equality, and before the end of that time approach nearer to each other than by any given difference, become ultimately equal.*

Though others, notably Stevin, included this idea of convergence of a quantity to a limit, Newton was the first to clearly state it. The next proposition in the *Principia* is:

Lemma 13.2. *If in figure AacE, bounded by straight lines Aa, AE, there is inscribed parallelograms Ab, Bc, Cd, etc., and circumscribed parallelograms aB, bC, cD, etc., all on equal bases AB, BC, CD, etc., and sides parallel to Aa, then as the width of the bases decreases and their number becomes infinite, the ultimate ratios between the inscribed figure, the circumscribed figure, and the figure AacE, will be that of equality.*

Proof. Let the figure be drawn as indicated; the difference between the inscribed and circumscribed figures will be the parallelograms *ab, bc, cd*, etc., which are all parallelograms with equal bases; their total height will be *Aa*. Since the width

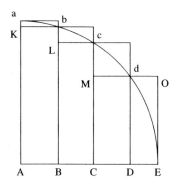

tends to nothing, the difference between the circumscribe and inscribed figures will also tend to nothing, and much more so will the intermediate curvilinear figure be equal to either.   □

Lemma 3 is the corresponding result when the widths are unequal (provided, of course, that the greatest width tends to zero). Several interesting corollaries follow, including:

Corollary. *These ultimate figures (as to their perimeters acE) are not rectilinear, but curvilinear limits of rectilinear figures.*

Recognizable calculus methods make their appearance throughout the *Principia*, but Newton almost invariably resorted to a limit formulation, rather than any algorithmic procedure.

13.5.4   Halley and the Infinite

The publication of the *Principia* was funded by Newton's friend Edmond Halley (November 8, 1656-January 14, 1742), an English astronomer. Halley (pronounced "HOLLY") was also responsible for the publication of the star atlas of John Flamsteed (1646-1719), without Flamsteed's permission: Flamsteed wanted to delay publication until the atlas was perfect, but Halley felt it was too important. It earned Halley an enemy for life.

See page 380.

Halley was also a mathematician, and one of the first to apply the principles of mathematical statistics to the insurance industry: he made up one of the first tables of mortality. Halley might have been the anonymous author of "On the Credibility of Human Testimony." In 1691, Halley published in the *Philosophical Transactions of the Royal Society* an article entitled, "On the Several Species of Infinite Quantity, and the Proportions They Bear to One Another." Galileo suspected infinite quantities might not be comparable in any meaningful sense. Halley suggested otherwise:

> That all magnitudes infinitely great, or such as exceed any assignable quantity, are equal among themselves, though it be vulgarly received for a maxim, is not yet so common as it is erroneous.[6]

As a counterexample, Halley claimed the line stretching from a point in one direction was half as infinite as a line stretching in both directions; the area enclosed by a right angle was a quarter of the infinite area of the plane; and that the area between parallel lines was infinite, but infinitely less than the area between intersecting lines. Halley's views are no longer considered valid, thanks to a rigorous examination of the notion of "infinite" during the nineteenth century.

---

[6]*Philosophical Transactions of the Royal Society, Abridged with notes*, 1809. C. and R. Baldwin. Vol. III, page 465.

13.5.5 Berkeley's Criticism

The loose talk about "ultimate" figures led to an important criticism by George Berkeley (1685-1753), later to become a Bishop in the Anglican Church and a prominent philosopher. Berkeley advanced the notion that unless an object was perceived by some mind, the object did not exist; a variation of this Berkelian philosophy would become an important part of modern physics.

In 1734, Berkeley published "The Analyst: Discourse Addressed to an Infidel Mathematician"— probably Halley—endeavored to prove that mathematics, no less than Christianity, relied on faith in unproven principles; Berkeley focused on the method of fluxions. Berkeley proposed a lemma which was "so obvious" as to not need proof: one cannot suppose a point to prove something and then, in the course of the proof, reject the original point.

In the *Principia*, Book II, Lemma 2, Newton found:

Lemma 13.3. *The moment of any genitum is equal to the moments of each of the generating sides multiplied by the indices of the powers of those sides, and by their coefficients continually.*

By a "genitum," Newton's discussion makes clear, Newton meant a quantity formed by multiplication, division, exponentiation, or extraction of roots. Newton demonstrates the product rule:

Demonstration. If A is increasing at rate a and B increasing at rate b, then the rectangle AB is increasing at rate $aB + bA$. The side A increases from $A - \frac{1}{2}a$ to $A + \frac{1}{2}a$, while at the same time the side B increases from $B - \frac{1}{2}b$ to $B + \frac{1}{2}b$. Hence the product increases from $(A - \frac{1}{2}a)(B - \frac{1}{2}b)$, or $AB - \frac{1}{2}aB - \frac{1}{2}bA + \frac{1}{4}ab$, to $(A + \frac{1}{2}a)(B + \frac{1}{2}b)$, or $AB + \frac{1}{2}aB + \frac{1}{2}bA + \frac{1}{4}ab$. The difference between the two is $aB + bA$. □

By letting $A = B$, Newton demonstrated that the moment of A^2 was $2Aa$, and in general, A^n had moment $nA^{n-1}a$.

Newton's choice to let A increase from $A - \frac{1}{2}a$ to $A + \frac{1}{2}a$ was clearly meant to allow for the cancellation of the ab term, a fact Berkeley realized. The "direct and true method" was to allow A to increase to $A + a$; in that case, then AB would increase to $AB + bA + aB + ab$. To obtain the correct value, mathematicians had to ignore the quantity ab, which, though it might be "infinitesimally small," was nevertheless not zero. Thus, though Newton claimed "not even the slightest" error was allowable in mathematics, here mathematicians were ignoring terms like ab when it proved convenient.

Even worse was Newton's explanation of "last ratios" in the determination of the ratio of the fluxion of x to the fluxion of x^n: first, x was assumed to change by o, then o was allowed to vanish. But in that case x did not change at all, so how could the ratio between the changes in x and x^n be spoken of in the first place? Berkeley was quick to point out that he had no doubts about the *results*: his problem was that the calculus seemed to rest on the "compensation of errors."

Berkeley's criticisms had a detrimental effect on the development of mathematics in England. English mathematicians, aware that the methods of infinitesimal calculus had little logical foundation, sought alternative means of solution, usually involving more complicated geometrical methods. Meanwhile, continental mathematicians, seeing that, by and large, infinitesimal methods *worked*, continued to develop them, confident that some day the necessary logical basis would be developed. The continental mathematicians proved correct, and it would be the work of Leibniz, and not that of Newton, that would drive analysis during the eighteenth century.

13.5 Exercises

1. After arranging the terms of an affected equation in decreasing order of one of its variables, Newton's method required multiplying the terms by "any arithmetic sequence."

    (a) Newton obviously did not mean *any* arithmetic sequence, in our sense. What did Newton actually mean?

    (b) Explain why this works.

2. Consider the equation $\dot{y}\dot{y} = \dot{x}\dot{y} + \dot{x}\dot{x}x^2$.

    (a) Show $\frac{\dot{y}}{\dot{x}} = \frac{1}{2} \pm \sqrt{\frac{1}{4} + x^2}$.

    (b) Expand the right hand side of $\frac{\dot{y}}{\dot{x}} = \frac{1}{2} + \sqrt{\frac{1}{4} + x^2}$ as a power series in x, then use the power series to solve for $\frac{y}{x}$.

3. Newton considered the problem of $\frac{\dot{y}}{\dot{x}} = \frac{a}{x}$.

    (a) Solve this problem using the method outlined by Newton. What difficulty results?

    (b) To resolve this difficulty, Newton noted that the fluents x and $b + x$ had the same fluxion. Thus, the equation could be solved by solving the related equation $\frac{\dot{y}}{\dot{x}} = \frac{a}{b+x}$. Expand the left hand side as a power series in x, then solve the resulting equation for y.

4. Show how Hudde's rules may be derived using Newton's fluxional methods.

5. Refer to the first manner on page 424. Suppose the curve is given by $x^3 - ax^2 + axy - y^3 = 0$. Find the subtangent TB.

6. Refer to the third manner of Newton on page 424.

    (a) Explain why the perpendicularity of BT and FT make $DFTB$ similar to $Dkdc$.

  (b) Explain how to find the tangent to the curve at point D from finding
      the length BT.

  (c) Let $AD = x$ and $BD = y$, and suppose the curve is $a + \frac{ex}{d} - y = 0$.
      Find the tangent to the hyperbola at any given point.

7. Refer to the seventh manner of finding tangents on page 424.

  (a) Explain why if At is parallel to dc then it must also be perpendicular
      to AD. (Remember that Dd is supposed to be an infinitely small
      portion of the curve.)

  (b) Explain why $cD : Gg$ is $AD : At$.

  (c) Suppose the curve ADE is given by $x^3 - ax^2 + axy - y^3$, where $BG = x$
      and $AD = y$. Use the method of fluxions to find the tangent to the
      curve.

8. Prove Lemma 3 of Book I, Section 1 of the *Principia*: even if the widths of
   the parallelograms are unequal, the difference between the circumscribed
   and inscribed figures will still tend to nothing as the widths of the parallelograms tends to nothing and their number becomes infinite.

9. Consider Lemmas 13.1 and 13.2.

  (a) What assumptions does Newton make in the proof of the lemmas?

  (b) To what modern theorem are these two lemmas equivalent?

10. Did Newton "invent" calculus? Explain.

13.6 Leibniz

The important English mathematicians tended to be professionals: Barrow, Newton, and Wallis all had prominent university positions teaching mathematics
or related disciplines. Meanwhile, the prominent continental mathematicians
tended to be amateurs: Descartes was a soldier, Fermat a lawyer, and Pascal
a philosopher. Gottfried Wilhelm Leibniz (July 1, 1646-November 14, 1716)
was no exception to this trend of part-time mathematicians: he was primarily a
diplomat.

  By the age of 25, Leibniz was an advisor to Johann Philip von Schönborn,
the Elector of Mainz, one of seven who chose the Holy Roman Emperor. Louis
XIV's quest for "natural frontiers" meant conquering the German Rhineland
states, including Mainz. At the urgings of Schönborn, and with the help of the
statesman J. C. von Boyneburg, Leibniz prepared a draft proposal that would
keep Louis occupied elsewhere. Rather than invade the Rhineland, which would
be costly in terms of men and money, Louis was encouraged to form a French
Empire in North Africa. To make it easier to travel to the newly established
French trading colonies in India, Louis was further encouraged to build a canal at
Suez, in Egypt. Leibniz went to Paris in early 1672 with the proposal. However,

Gottfried   Wilhelm   Leibniz.
©Bettmann/CORBIS.

See page 406.

Leibniz failed to see the king, who had other plans: with the help of the English, Louis invaded the Netherlands. It was during this invasion that Hudde directed the flooding of the polders.

The English wanted to eliminate the Dutch as commercial rivals: the Dutch East India Company competed with the English in the "spice islands" (modern Indonesia), and the Dutch West India Company threatened the profitability of England's sugar plantations in the Americas. Unfortunately, England's fleet had been in decline since the defeat of the Spanish Armada in 1588, and the English suffered a number of ignominious defeats by the Dutch. Soon, the English were clamoring for a separate peace. To help mediate an end to the war between the English and the Dutch (also known as the Third Dutch War) Leibniz went to London in January 1673. Eventually, the Treaty of Westminster was signed, in February 1674.

Leibniz was by then back in Paris, having arrived in March 1673. In 1676, after Louis XIV showed no interest in meeting Leibniz or discussing his plan, Leibniz left for Hanover. There he met Hudde, and took a position in the service of the duke of Hanover, Johann Friedrich. After the duke's death in 1679, the duke's brother, Ernst August, retained Leibniz, and commissioned him to write a history of the House of Brunswick (the duke's family). The research for this work sent Leibniz on another round of travels, this time through Italy. In Rome, Leibniz met the biologist Malphigi (discoverer of the capillaries that connect arteries to veins), and a Jesuit, Claudio Filippo Grimaldi, who was to leave for China to serve as court mathematician to the Chinese emperor.

Leibniz's work on the genealogy continued; he threw himself into the work so zealously that he developed eyestrain from overwork. The duke was not merely interested in a family history: the genealogy might reveal additional prerogatives due to him, and in 1692, on the basis of Leibniz's research, the duke of Hanover became an eighth Elector.

13.6.1   Series Summation

Leibniz, like Pascal, invented an adding machine, which he presented to the French Academy of the Sciences during his visit to Paris. This brought Leibniz into contact with a number of scientists and mathematicians living in Paris, including Pierre de Carcavi and Christian Huygens. Huygens suggested that, if Leibniz was truly interested in mathematics, he ought to read the works of Pascal (which Leibniz did). Perhaps as a test, Huygens posed to Leibniz the problem of finding the sums of the reciprocals of the triangular numbers:

$$\frac{1}{1} + \frac{1}{3} + \frac{1}{6} + \frac{1}{10} + \frac{1}{15} + \cdots$$

Leibniz soon found the sum to be 2 (a fact probably known to Huygens), which led Leibniz to consider the sums of other series. After reading about Pascal's arithmetical triangle, Leibniz invented a variation, the **harmonic triangle**. Whereas Pascal began with the infinite sequence 1, 1, 1, 1, . . ., and formed

$$\frac{1}{1} \quad \frac{1}{2} \quad \frac{1}{3} \quad \frac{1}{4} \quad \frac{1}{5} \quad \frac{1}{6} \quad \frac{1}{7} \quad \cdots$$

$$\frac{1}{2} \quad \frac{1}{6} \quad \frac{1}{12} \quad \frac{1}{20} \quad \frac{1}{30} \quad \frac{1}{42} \qquad \ddots$$

$$\frac{1}{3} \quad \frac{1}{12} \quad \frac{1}{30} \quad \frac{1}{60} \quad \frac{1}{105} \qquad \ddots$$

$$\frac{1}{4} \quad \frac{1}{20} \quad \frac{1}{60} \quad \frac{1}{140} \qquad \ddots$$

$$\frac{1}{5} \quad \frac{1}{30} \quad \frac{1}{105} \qquad \ddots$$

$$\frac{1}{6} \quad \frac{1}{42} \qquad \ddots$$

$$\frac{1}{7} \qquad \ddots$$

$$\vdots$$

Table 13.6: The Harmonic Triangle.

the terms in successive rows by summing; Leibniz's harmonic triangle began with the harmonic sequence $\frac{1}{1}$, $\frac{1}{2}$, $\frac{1}{3}$, $\frac{1}{4}$, $\frac{1}{5}$, ..., and formed the terms in the successive rows by the *differences* between adjacent terms (Table 13.6).

Because each term in the harmonic triangle is the difference between the terms above it and to the right, this meant that each term was the sum of the entire series of terms in the row below it and to the right. Thus, each row after the first corresponds to what is now referred to as a **telescoping series** (Leibniz had no special term). For example, the terms in the second row are formed from

$$\frac{1}{2} = \frac{1}{1} - \frac{1}{2}$$
$$\frac{1}{6} = \frac{1}{2} - \frac{1}{3}$$
$$\frac{1}{12} = \frac{1}{3} - \frac{1}{4}$$
$$\frac{1}{20} = \frac{1}{4} - \frac{1}{5}$$
$$\vdots$$

Hence

$$\frac{1}{2} + \frac{1}{6} + \frac{1}{12} + \frac{1}{20} + \ldots = \frac{1}{1}$$

as all the terms except the first cancel. Leibniz also claimed the sum of the series

$$\frac{1}{1} + \frac{1}{2} + \frac{1}{3} + \frac{1}{4} + \frac{1}{5} + \ldots$$

was $\frac{1}{0}$, presumably by extending the entries in the first column up one row.

Leibniz then noted another pattern in the harmonic triangle: if the denominators of the fractions in each row were divided by the denominator of the first fraction, the numbers obtained would be the same as those in the arithmetical triangle. Thus, if the denominators in the third row, 3, 12, 30, 60, ..., were divided by the denominator of the first fraction, 3, the results would be 1, 4, 10, 20, ..., which correspond to the numbers in the third row of the arithmetic triangle. Hence, the sums of the reciprocals of the numbers in any row in the arithmetical triangle (i.e., in any of the main diagonals of Pascal's triangle) could be determined.

Example 13.9. *Since*

$$\frac{1}{2} + \frac{1}{6} + \frac{1}{12} + \frac{1}{20} + \frac{1}{30} + \ldots = \frac{1}{1}$$

we can divide every denominator by 2, the denominator of the first fraction, obtaining

$$\frac{1}{1} + \frac{1}{3} + \frac{1}{6} + \frac{1}{10} + \frac{1}{15} + \ldots = \frac{2}{1}$$

which is the sum of the reciprocals of the third diagonal of Pascal's triangle.

These results impressed Leibniz and led him to conclude that *every* series could be summed according to a simple formula, now referred to as the **closed form expression** of the sum.

For example, consider the sum of the squares of the numbers from 1 to x, and suppose this (finite) sum can be expressed by z, where $z = lx^3 + mx^2 + nx$ (the resulting analysis will show that no term higher than x^3 is necessary to express the sum). Leibniz let d indicate the *finite* difference between successive terms; then

$$dz = d\left(lx^3\right) + d\left(mx^2\right) + d\left(nx\right)$$
$$= ld\left(x^3\right) + md\left(x^2\right) + nd\left(x\right)$$

But $d(x^3) = (x+1)^3 - x^3 = 3x^2 + 3x + 1$, $d(x^2) = 2x + 1$, and $dx = 1$; moreover, the difference between the successive values of z, the partial sum of the series of squares, was just the square number $(x+1)^2$ itself. Thus

$$x^2 + 2x + 1 = l(3x^2 + 3x + 1) + m(2x + 1) + n$$

By equating the powers, Leibniz determined $l = \frac{1}{3}$, $m = \frac{1}{2}$, and $n = \frac{1}{6}$; hence the sum of the first x square numbers could be expressed as $z = \frac{1}{3}x^3 + \frac{1}{2}x^2 + \frac{1}{6}x$.

The key to the method was the fact that the finite differences dx, $d(x^2)$, $d(x^3)$, ... could be expressed as a polynomial in x; if this could not be done, finding a closed form expression would be problematic. Leibniz's faith in the summability of any series was further confirmed by his discovery that even if the finite differences could not be expressed as polynomials, they might still be expressible in a simple manner.

Leibniz was probably inspired by his discovery of the relationship between the harmonic and arithmetic triangles. The use of the harmonic triangle relied on the fact that the entries were the difference between two terms; hence the sum of the entries was simply the first term of the previous row. Thus, to find the sum of a series of fractions, it was necessary to find the series for which the fractions themselves were the differences.

He considered the series

$$\frac{1}{3} + \frac{1}{15} + \frac{1}{35} + \frac{1}{63} + \frac{1}{99} + \cdots$$

whose denominators were the products of two successive odd numbers, starting from 1; thus, they have the general form $\frac{1}{4x^2+8x+3}$. If he could find a sequence whose differences made up the terms of the series, then the first term of the sequence would be the sum of the given series.

Suppose the general terms of this sequence were $\frac{e}{bx+c}$. Then the difference between two successive terms should be a term of the given series; hence

$$\frac{e}{bx+c} - \frac{e}{bx+b+c} = \frac{1}{4x^2+8x+3}$$

Solving, we have $b = 2$, $e = \frac{1}{2}$, and $c = 1$, and the desired sequence is $\frac{1}{4x+2}$ (the reciprocals of twice the successive odd numbers). Thus the terms of the given series could be written as the differences of twice the successive odd numbers:

$$\frac{1}{3} = \frac{1}{2} - \frac{1}{6}$$

$$\frac{1}{15} = \frac{1}{6} - \frac{1}{10}$$

$$\frac{1}{35} = \frac{1}{10} - \frac{1}{14}$$

$$\frac{1}{63} = \frac{1}{14} - \frac{1}{18}$$

$$\frac{1}{99} = \frac{1}{18} - \frac{1}{22}$$

$$\vdots$$

Hence

$$\frac{1}{3} + \frac{1}{15} + \frac{1}{35} + \frac{1}{63} + \frac{1}{99} + \cdots = \frac{1}{2}$$

13.6.2 Calculus

During his visit to London in 1673, Leibniz presented his adding machine to the Royal Society. He was elected a member of the Royal Society, and found himself in the company of Pell, Oldenburg, Collins, Hooke, and many others. Leibniz did his best to learn everything he could. He asked Wallis to teach him how to decode messages, but Wallis refused, on the grounds that it was an art that he could not, in good conscience, teach to a foreigner. However, Leibniz did learn of the method of tangents and quadratures of Isaac Barrow, and picked up a copy of Barrow's *Geometrical Lectures*.

The consideration of and easy transition between the sum and the difference led Leibniz to consider the sums and differences of *geometric* quantities. Leibniz's approach to calculus was so fundamentally different from Newton's that the few similarities between them could easily be ascribed to the fact that both Newton and Leibniz were trying to solve the same problems.

On October 25, 1675, after he had returned to Paris, Leibniz recorded the discovery of what would become the **method of integration by parts**. The discovery was tied up with the concept of the moment of a figure about an axis: this is the product of the distance between the axis and the figure's center of mass. The center of mass of a rectangle or a line is at the geometric center of the figure.

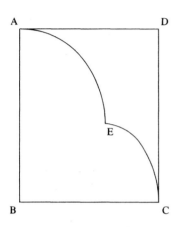

As Leibniz wrote his discovery:

Proposition 13.6. *Given curve AEC, with AD perpendicular to axis AB; let $AB = DC = x$, and let the last $x = b$. Also, let $BC = AD = y$, and let the last $y = c$. Then*

$$\text{all } yx \text{ to } x = \frac{b^2 c}{2} - \text{ all } \frac{x^2}{2} \text{ to } y$$

Leibniz, in 1675, had not yet developed the notation that would make Leibnizian calculus a far more useful tool than Newtonian calculus. What, precisely, was Leibniz saying?

First, by letting $AB = DC = x$, and letting the "last" $x = b$, Leibniz meant that the ordinates of the curve AEC were designated x, and that the maximum value would be b; Leibniz's work implies the minimum value would be $x = 0$; likewise, the abscissas would range from $y = 0$ to $y = c$.

Second, all yx to x meant the sum of all quantities yx (i.e., the product of the ordinate and the abscissa) as far as $x = b$. Any individual yx was the product of the abscissa and its distance from the axis AD; thus yx is the moment of a line about the axis AD. "All" these lines would be the total moment of the figure $AECB$ about axis AD.

In Leibniz's demonstration, AB, BC should be thought of as the ordinate x and the abscissa y, and not the full lengths b and c.

Demonstration. The moment of $ABCE$ about AD is all the rectangles BC, AB [= all yx to x]; and the [moment of the complementary figure] is the sum of the squares on DC halved [= all $\frac{x^2}{2}$ to y], which, if subtracted from the moment of the whole figure $ABCD$ about AD, that is, from c all x, or $\frac{b^2 c}{2}$, will leave the moment of $ABCE$ remains. $\square$

Shortly after this, Leibniz began to use the symbol $\int$, a stylized "S," in place of "all": hence "all x" became $\int x$.

Tangents

On June 26, 1676, Leibniz had a revelation: recognizing that Descartes's method of tangents was unusable in all but the simplest cases, Leibniz declared, in his notebook, that the "true method" of finding tangents relied on the *differences* between successive values of the variables: dx. However, while successive had a clearly defined meaning if x was a discrete sequence of values, such as perfect squares or triangular numbers, what possible meaning might it have if x was a *continuous* variable? Rather than stopping all work and attempting to find a rigorous answer to this question, Leibniz ignored it, and continued to develop his version of calculus. We might compare Leibniz's avoidance of the issue with working on a jigsaw puzzle: if one stops to consider exactly where a particular piece fits into the puzzle, the puzzle will probably never be finished. Instead, it is more productive to piece together separate parts, and ultimately join them together. By 1680, Leibniz had pieced together many separate parts, and had the notation and the methodology for finding tangents, areas, and arc lengths.

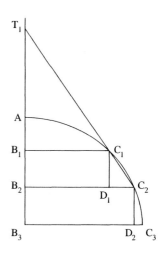

Given a curve AC, with "successive" points C_1, C_2, C_3, ..., with the differences between the abscissas, such as $C_1D_1 = dx$, the differences between the corresponding ordinates $C_2D_1 = dy$; if the differences became infinitely small, then C_1C_2 would be tangent to the curve, considered as a polygon with an infinite number of angles; moreover, the subtangent T_1B_1 would be to the ordinate B_1C_1 as C_1D_1 was to C_2D_1, or $t : y = dx : dy$. Moreover, the last ordinate B_3C_3 is the sum of all the differences, starting from A, or $\int dy = y$. The area of the figure will be the sum of all the rectangles contained by the ordinates, namely $B_1D_1 + B_2D_2 + \ldots$, since the infinitely small triangles such as $C_1D_1C_2$ can be omitted; this is expressed by $\int y \, dx$. Finally, the length of the arc is the sum of the sides C_1C_2, or $\int \sqrt{dx^2 + dy^2}$; in the case of a parabola, where $y = \frac{x^2}{2a}$, then $dy = \frac{x \, dx}{a}$, and thus $C_1C_2 = \frac{dx}{a}\sqrt{a^2 + x^2}$; hence the arc length will be $\frac{1}{a} \int dx \sqrt{a^2 + x^2}$.

To demonstrate the derivative rules, Leibniz returned to the notion that d was the difference between successive values of a quantity. Leibniz's demonstration of the product rule $d(xy) = x \, dy + y \, dx$ is:

Demonstration. $d(xy)$ is the difference between successive values of xy; let one of these be xy, and the next be $(x + dx)(y + dy)$. The difference is thus $y \, dx + x \, dy + dx \, dy$, where $dx \, dy$ is infinitely small in comparison with the rest, and thus may be neglected. □

Leibniz provided demonstrations of the quotient rule and the power rule as well.

New Method of Maxima and Minima

Leibniz summarized his results on the rules of differentiation, though not the demonstrations, in a 1684 publication, "A New Method for Maxima and Minima, and also for Tangents, which is not Obstructed by Irrational Quantities," which we will refer to as "New Method." It was published many years after Leibniz's discovery. One of the reasons for delay in publication may have been that Leibniz was uncomfortable arguing on the basis of "infinitely small" quantities dx; in "New Method," Leibniz made dx a definite quantity, and dy a quantity that had to it the same ratio that the subtangent had to the ordinate.

The publication was forced, in a sense, by Leibniz's fear of being plagiarized by Ehrenfried Walther von Tschirnhaus (April 10, 1651-1708), the Count of Kreslingswalde and Stolzenberg. In 1683, Tschirnhaus published a paper on quadratures that included, without attribution, some work Leibniz had communicated to the Count; Tschirnhaus had not included the method by which the quadratures were obtained, but he was certainly in possession of it, and certainly capable of publishing and claiming it as his own. To forestall this possibility, Leibniz published "New Method," where he gave (without proof) the familiar rules of differentiation; if a is a constant and x, y, z, w, v variable quantities, then:

$$da = 0$$
$$d(ax) = a\ dx$$
$$d(z - y + w + x) = dz - dy + dw + dx$$
$$d(vx) = x\ dv + v\ dx$$
$$d\left(\frac{v}{y}\right) = \frac{\pm y\ dv \mp v\ dy}{y^2}$$

(The $\pm$ and $\mp$ in the quotient rules were necessary because of the treatment of dx as a determinate magnitude) In addition, Leibniz gave the rules for differentiating a power, x^n, as well as the **chain rule**. Leibniz's predecessors had been interested in solving the tangent and quadrature problems, and had set the groundwork for the invention of calculus; in Leibniz's hands, calculus became a powerful algorithmic tool that could then be used to solve a wide variety of problems.

13.6.3　Matrix Algebra

Leibniz was the first European to use matrices to represent and solve a system of equations. On April 28, 1693, in a letter to l'Hôpital, a French nobleman interested in mathematics, Leibniz said he found it convenient to let numbers represent the coefficients in a system of equations. For a system with two variables and three equations, Leibniz wrote:

$$10 + 11x + 12y = 0$$
$$20 + 21x + 22y = 0$$
$$30 + 31x + 32y = 0$$

where a number, such as 32, represents the coefficient of the second variable in the third equation. To eliminate y, he used the first and second equation. Multiplying the first equation by 22 (i.e., the coefficient of the second variable in the second equation) and the second by 12 (i.e., the coefficient of the second variable in the first equation) and subtracting, Leibniz obtained

$$10 \cdot 22 + 11 \cdot 22x - 12 \cdot 20 - 12 \cdot 21x = 0$$

whence the value of x could be determined.

It is tempting to wonder if there is some connection between the matrix algebra of Leibniz and the *fang chang* method of the Chinese; certainly, by Leibniz's time, there was considerable interaction between China and the West. Even more interestingly, one of Leibniz's acquaintances, Claudio Filippo Grimaldi, was a Jesuit mathematician appointed as a representative to the Chinese Imperial court; from Grimaldi (whom he met in Rome in 1668), Leibniz picked up a lifelong interest in things Chinese, evidenced by his 1703 paper on the binary system, where he noted a connection between the trigrams of the *I Ching* (the combination of short and long dashes most easily visible around the edges of the South Korean flag) and the expression in binary numbers. However, there is no evidence that Grimaldi, or anyone else, ever communicated to Leibniz any details of Chinese mathematics.

See page 214.

13.6.4 Imaginary Numbers

In 1702, Leibniz wrote:

> We are led to the important question of whether all quadratures of rational functions can be reduced to the quadratures of hyperbolas and circles ... or [more generally] whether all algebraic equations or rational expressions can be resolved into real factors [i.e., factors with real coefficients] either simple [i.e., linear] or plane [i.e., quadratic].[7]

Leibniz thus posed, for the first time, whether it was always possible to reduce a polynomial into linear and quadratic factors. Leibniz did not believe this was possible; he gave as an example $x^4 + a^4$. Factoring, Leibniz obtained

$$x^4 + a^4 = \left(x^2 + a^2\sqrt{-1}\right)\left(x^2 - a^2\sqrt{-1}\right)$$
$$= \left(x + a\sqrt{-\sqrt{-1}}\right)\left(x - a\sqrt{-\sqrt{-1}}\right)\left(x + a\sqrt{\sqrt{-1}}\right)\left(x - a\sqrt{\sqrt{-1}}\right)$$

no two factors of which yielded an expression with real coefficients. Hence, Leibniz concluded, $x^4 + a^4$ could not be decomposed into linear or even quadratic factors with real coefficients.

To understand the importance of Leibniz's conclusion, consider equations that can be formed using integer coefficients. Some of these, like $5x + 10 = 0$, have solutions among the integers. Others, like $3x - 5 = 0$, do not. In order to solve this equation, one must include the fraction $\frac{5}{3}$ in the number system. However, adding this one fraction does not make it possible to solve all

---

[7]Leibniz, Gottfried Wilhelm, *Mathematische Schriften*, ed. C. I. Gerhardt, 1849-1863. H. W. Schmidt. Vol. V, p. 359.

equations whose coefficients are integers. Clearly, an infinite number of fractions, all different, must be included. Next, consider quadratic equations. Again, some of these, such as $2x^2 + 3x - 2 = 0$, can be solved among the integers and rationals. However, equations like $x^2 - 2 = 0$ require irrational numbers; again, adding just a single irrational number, $\sqrt{2}$, does not allow one to solve all possible equations. Finally, consider an equation like $x^2 + 1 = 0$. By adding the number $\sqrt{-1}$, we are able to solve this equation. Moreover, *any* quadratic is now solvable: although an equation like $x^2 + 4 = 0$ might seem to require a different imaginary number, $\sqrt{-4}$, but this can be thought of as $2\sqrt{-1}$, so we are simply taking an already existing imaginary number and multiplying it by an already existing real number. Leibniz's factorization of $x^4 + a^4$ implied a new type of imaginary, $\sqrt{\sqrt{-1}}$. Indeed, the existence of a single new type of imaginary suggested the possibility of an infinite number of new types. Moreover, it suggested the general quadrature problem might be even more difficult, since it implied that the quadrature of the curves generated by rational functions might require functions other than the trigonometric and logarithmic.

See Problem 9.

13.6.5   The Priority Dispute

One of the more unfortunate events in the history of mathematics was a priority dispute over the inventor of calculus. The dispute was particularly pointless because it focused on whether Leibniz or Newton invented calculus; other contributors, such as Fermat and Barrow, were ignored. Part of the problem was that scholars were still wrestling with the idea of intellectual property. A larger part of the problem was that the notion of "publication" was very different, for there were very few scientific journals. Newton, as Lucasian lecturer, was required to lecture once a week and to put copies of his lectures, in finished form, in the university library, where they might be seen by anyone. Newton also deposited early versions of his works with the Secretary of the Royal Society, where, again, they could be viewed by anyone. However, in neither case would the work be widely distributed.

For this, Leibniz was a victim of unfortunate circumstances. He was a member of the Royal Society, and had been to England on a number of occasion; hence, it was possible for him to have seen many of Newton's manuscripts. As a result, when Leibniz published a version of calculus in 1684 and again in 1686 (where, for the first time, the stylized S is used to indicate a sum: $\int$), without giving any credit to Newton, Newton's supporters in England could reasonably suspect some influence. However, the methods and basis of Leibniz's work were very different from that of Newton's. Moreover, Leibniz's correspondence with Oldenburg seems to indicate that he was unaware of most of Newton's discoveries, and it was through Oldenburg, for example, that Leibniz became aware of the generalized binomial theorem, and Newton's method of infinite series.

The controversy had some unfortunate effects. Leibniz's notation is superior to Newton's. Moreover, Leibniz developed explicit algorithms for finding the differentials dx and dy, while aside from the generalized power rule, Newton tended to rely on the development of a function by means of an infinite series. As a result, Leibniz's calculus was far more usable than Newton's, and it became the method of calculus used outside of England. English mathematicians preferred to use Newton's method, or, after Berkeley's criticisms, to base everything on geometrical methods.

One result was that, after Newton, English contributions to mathematics were in fields other than analysis; meanwhile, great strides in analysis (and its main application, mathematical physics) were made by Leibniz's followers. Another result was that the method of fluxions was associated with Newton's physics, and by rejecting the one, continental mathematicians and scientists rejected the other. Thus, for half a century, European scientists were divided into two groups: the English group, using a valid system of physics (Newton's) and a clumsy system of mathematics (also Newton's), and a continental group, using an invalid system of physics (Descartes's) but a superior system of mathematics (Leibniz's). Something of the importance of mathematics to modern science may be drawn from the fact that when the latter group slowly adopted Newtonian physics, they made great advances in physics, while the English, who continued to use fluxional methods until the 1860s, made almost no significant discoveries in mathematical physics until they abandoned fluxions.

13.6 Exercises

1. Find the sums of the following series.

   (a) $\frac{1}{1} + \frac{1}{4} + \frac{1}{10} + \frac{1}{20} + \frac{1}{35} + \frac{1}{56} + \ldots$ (the sums of the reciprocals of the pyramidal numbers, or the reciprocals of the entries in the fourth row of the arithmetical triangle).

   (b) $\frac{1}{1} + \frac{1}{5} + \frac{1}{15} + \frac{1}{15} + \frac{1}{35} + \frac{1}{70} + \frac{1}{126} + \frac{1}{210} + \ldots$ (the sums of the reciprocals of the entries in the fifth row of the arithmetical triangle)

2. Find a formula for the sum of the cubes of the whole numbers.

3. Find the sum of the series.

   (a) $\frac{1}{8} + \frac{1}{24} + \frac{1}{48} + \ldots$, where the denominators are the products of two consecutive even numbers.

   (b) $\frac{1}{3} + \frac{1}{8} + \frac{1}{15} + \frac{1}{24} + \ldots$, where the denominators are one less than the perfect squares. Hint: use partial fractions.

4. A problem that would later vex Leibniz was finding the sum of the series of reciprocal squares,

$$\frac{1}{1} + \frac{1}{4} + \frac{1}{9} + \frac{1}{16} + \ldots$$

   Apply Leibniz's sum procedure to this series. What difficulties arise?

5. In Leibniz's demonstration of the method of integration by parts, explain why the moment of the complementary figure is all $\frac{x^2}{2}$ to y.

6. Prove, using infinitesimal methods, the quotient rule: $d\left(\frac{x}{y}\right) = \frac{y\,dx - y\,dx}{y^2}$.

7. Prove the power rule $dx^n = nx^{n-1}$ from the product rule.

8. Leibniz claimed $x^4 + a^4$ could not be factored into linear or quadratic factors with real coefficients. This is incorrect; factor $x^4 + a^4$. Hint: $x^4 + 1 = 0$ has for solutions the four fourth roots of 1.

9. What impact would the inability to factor a polynomial into quadratic or linear factors have on calculus? Hint: consider how rational functions are evaluated.

13.7   Johann Bernoulli

Leibniz had a great influence on the development of mathematics. Not only did he publish his work in journals that were read by a wide audience, but he also communicated, via letters, to many of the prominent continental mathematicians. Two of Leibniz's main correspondents were the brothers Bernoulli, Jakob and his young brother Johann.

Johann Bernoulli (August 6, 1667-January 1, 1748) was the tenth child of Nikolaus Bernoulli. It was his father's intent that Johann study to be a merchant, but, like his elder brother Jakob, Johann began to study mathematics and found himself quite adept at it. In 1685, he defended a thesis (with his brother as one of the questioners).

13.7.1   The Brachistochrone Problem

In 1696, Johann proposed the so-called "brachistochrone" problem:

Problem 13.2. *Given two points, A and B on a vertical plane, determine the path AMB taken by a particle M, descending by gravity alone, so it moves from A to B in the shortest amount of time.*

Bernoulli circulated the problem to his friends and professional acquaintances. Leibniz predicted (correctly) there would be only five geometers who would solve the problem: Johann and Jakob Bernoulli, Newton, l'Hôpital, and Leibniz himself; interestingly, Leibniz noted that Hudde would have sent in a solution, if he had still been interested in mathematics. The solution is one of the cycloids that run through points A and B. Since there are an infinite number of cycloids passing through two given points, Bernoulli then gave the method of finding the actual cycloid that is the solution to the brachistochrone problem: draw any cycloid on horizontal base AL that passed through A and intersected the line AB at some point R. Then the diameter of the circle that generates the required cycloid ABL is to the diameter of the circle that generated the cycloid ARS will be as AR to AB.

See page 406.

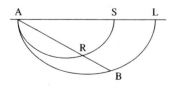

Similar problems were proposed during the course of the next few years, and many important results came out of this early period in what would become the calculus of variations. However, the tools were essentially geometric in nature, and the methods of solution often involved noting a similarity between the problem posed and an equivalent physical situation, so it was hard to generalize the method. Thus, the modern methods of the calculus of variations have little to do with the methods used in the early era.

However, these problems did leave an important legacy. The first definition of a *function* of one variable comes frm a 1718 memoir on an isoperimetric problem, where Johann Bernoulli wrote:

> I call here a function of a variable magnitude a quantity composed in any manner by that variable magnitude and some constants.[8]

This definition would become the standard definition of a function for well over a century.

13.7.2 L'Hôpital's Rule

Guillaume-François Antoine (1661-February 2, 1704), the Marquis de l'Hôpital, was responsible for one of Bernoulli's most significant contributions to mathematics. A nobleman who was also a competent mathematician, l'Hôpital corresponded with all the major mathematicians of the time. On March 17, 1694, l'Hôpital wrote to Bernoulli, offering to pay him three hundred livres each year, on the condition that Bernoulli would work on problems posed by l'Hôpital, and not divulge Bernoulli's solutions to anyone else. Bernoulli accepted the terms.

On June 7, 1694, l'Hôpital asked Bernoulli to evaluate the equation $y = \frac{\sqrt{2a^3x - x^4} - a^3\sqrt{a^2x}}{a - \sqrt[4]{ax^3}}$ when $x = a$. Bernoulli's original response was unsatisfactory to l'Hôpital, who requested him to further elucidate his technique. This Bernoulli did, in a letter dated July 22, 1694.

Rule 13.5 (L'Hôpital's Rule). *Given a curve AEC, with $AD = x$, $DE = y$, AB a constant such that BC is a fraction whose numerator and denominator are both equal to zero; required: to find the magnitude of BC.*

Construct on the same axis adb two curves aeb and αεb so that when the abscissas equal AD, ad, the ordinate de will be in ratio to the numerator of the fraction that expresses the ordinate BC, and dε will be in the same ratio to the denominator of the same fraction; hence de divided by dε will be DE.

Since when de, dε vanish because the numerator and denominator of the fraction vanish, the two curves aeb, αεb intersect at b; thus one has only to find the last differentials βc, βγ and divide one by the other to find the magnitude of BC.

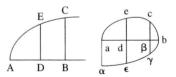

Bernoulli's discovery was incorporated into l'Hôpital's *Analysis of the Infinitely Small*, the first text on differential calculus. L'Hôpital made no claim to originality, and, unusual for his time, gave generous credit to both Johann and Jakob Bernoulli, as well as to Leibniz, for the many discoveries in the book, but Johann felt he deserved even more credit.

---

[8]Bernoulli, Johann, *Opera Omnia*. Reprinted by G. Olms, 1968. Vol. II, p. 241.

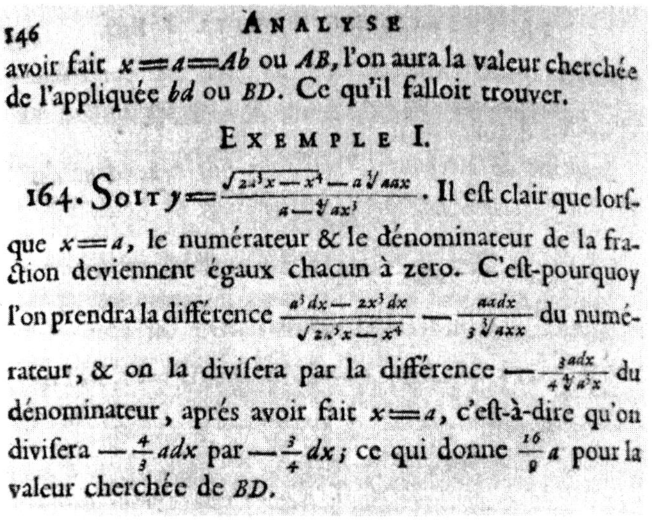

146 ANALYSE

avoir fait $x=a=Ab$ ou AB, l'on aura la valeur cherchée de l'appliquée bd ou BD. Ce qu'il falloit trouver.

EXEMPLE I.

164. SOIT $y=\frac{\sqrt{2a^3x-x^4}-a\sqrt[3]{aax}}{a-\sqrt[4]{ax^3}}$. Il est clair que lorsque $x=a$, le numérateur & le dénominateur de la fraction deviennent égaux chacun à zero. C'est-pourquoy l'on prendra la différence $\frac{a^3dx-2x^3dx}{\sqrt{2a^3x-x^4}}-\frac{aadx}{3\sqrt[3]{axx}}$ du numérateur, & on la divisera par la différence $-\frac{3adx}{4\sqrt[4]{a^3x}}$ du dénominateur, aprés avoir fait $x=a$, c'est-à-dire qu'on divisera $-\frac{4}{3}adx$ par $-\frac{3}{4}dx$; ce qui donne $\frac{16}{9}a$ pour la valeur cherchée de BD.

Figure 13.8: L'Hôpital's Rule, from *Analysis of the Infinitely Small.* Boston Public Library/Rare books Department. Courtesy of the Trustees.

13.7.3 Differential Equations

L'Hôpital was not the only mathematician Johann Bernoulli thought himself slighted by; Johann also felt slighted by older brother Jakob, and during the last few years of Jakob's life, the two communicated mainly through letters published in scientific journals. Jakob was, in general, the better mathematician, but Johann's methods were easier to follow. In December of 1695, Jakob proposed the differential equation

$$a\ dy = y\ p\ dx + b\ y^n q\ dx$$

where a, b were constants, n any power, and p, q both functions of x. The problem was solved by Jakob but, as Johann noted in March, 1697, "the method did not seem to be the best." The alternative method Johann proposed is what we now call the **method of integrating factors**. Johann's solution was:

Solution. Suppose that $y = mz$, and thus $dy = m\ dz + z\ dm$. Substitute this into the differential equation, and we have $a\ z\ dm + a\ m\ dz = m\ z\ p\ dx + b\ m^n z^n q\ dx$. If we suppose $a\ m\ dz = m\ z\ p\ dx$, then $\frac{a\ dz}{z} = p\ dx$. Thus, $z = \zeta$, where ζ is a function of x and constants.

[Specifically, $z = e^{\frac{\int p\ dx}{a}}$.]

Substituting this back into the differential equation, we have $a\ \zeta\ dm = b\ m^n q\ \zeta^n dx$, and hence $a\ m^{-n} dm = b\ \zeta^{n-1} q\ dx$. Hence

$$\frac{am^{-n+1}}{-n+1} = b\int \zeta^{n-1} q\ dx$$

and thus $m = X$ and $y = \zeta X$. □

13.7.4 Infinite Series

Another of Johann Bernoulli's discoveries was a form of the so-called Taylor series. On September 2, 1694, he communicated to Leibniz the discovery that

Rule 13.6. *If an element of space, a curve, or some other thing can be expressed by $n\,dz$, where n is a quantity expressed by variables and quantities, then*

$$\int n\ dz \ = \ nz - \frac{z^2 dn}{1\cdot 2dz} + \frac{z^3 d^2 n}{1\cdot 2\cdot 3dz^2} - \frac{z^4 d^3 n}{1\cdot 2\cdot 3\cdot 4dz^3} + \dots \quad (13.2)$$

Bernoulli obtained the result by expressing $n\,dz$ as the series

$$n\ dz = n\ dz + z\ dn - z\ dn - \frac{z^2 ddn}{1\cdot 2dz} + \frac{z^2 ddn}{1\cdot 2dz}$$
$$- \frac{z^3 d^3 n}{1\cdot 2\cdot 3dz^2} + \frac{z^3 d^3 n}{1\cdot 2\cdot 3dz^2} + \dots \quad (13.3)$$

then integrating the terms pairwise. This discovery, unfortunately, started a new round of controversies over priority that would pit English mathematicians against continental European mathematicians. Leibniz, in a letter to Bernoulli, noted that he had made a similar discovery; De Moivre, too, told Bernoulli of his discovery in a letter of July 6, 1708. However, it was for the English mathematician Brooke Taylor (August 18, 1685-December 29, 1731) that Bernoulli reserved his most acrimonious criticism.

In 1715, Brooke Taylor published *The Method of Increments*, in which appeared **Taylor's theorem** for the expansion of a function. The book was obscurely worded, and used a variation of Newtonian fluxional notation that made it very hard to follow. Bernoulli thought this was deliberate: Taylor, he claimed, had plagiarized his work, and wrote obscurely to hide the fact.

Taylor's main result was:

Proposition 13.7. *Let z, x be two variables, and let z increase uniformly by $\dot{z}$, and let $n\dot{z} = v$, $v - \dot{z} = \acute{v}$, $\acute{v} - \dot{z} = \acute{\acute{v}}$, etc. Then as z varies to $z + v$, x varies to*

$$x + \frac{\dot{x}v}{1\dot{z}} + \frac{\ddot{x}v\acute{v}}{1\cdot 2\dot{z}^2} + \frac{\dddot{x}v\acute{v}\acute{\acute{v}}}{1\cdot 2\cdot 3\dot{z}^3} + \dots \quad (13.4)$$

Taylor derived the series from observing the pattern in the increments. Suppose x changed to $x + \dot{x}$ as z changed to $z + \dot{z}$. Consider the increments:

$$x \quad \dot{x} \quad \ddot{x} \quad \dddot{x} \quad \dots$$

Since the fluxion of x is $\dot{x}$, then as z changed to $z + \dot{z}$, x would change to $x + \dot{x}$. Likewise, $\dot{x}$ will change to $\dot{x} + \ddot{x}$. Hence, we can construct the second row:

$$
\begin{array}{ccccc}
x & \dot{x} & \ddot{x} & \dddot{x} & \dots \\
x + \dot{x} & \dot{x} + \ddot{x} & \ddot{x} + \dddot{x} & \dddot{x} + \ddddot{x} & \dots
\end{array}
$$

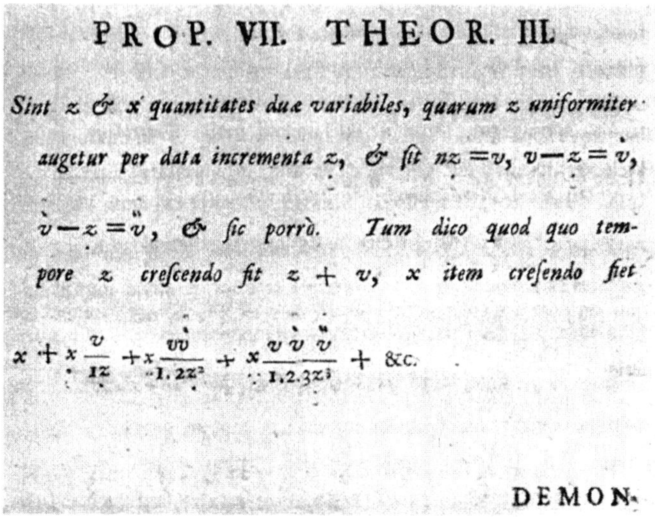

PROP. VII. THEOR. III.

Sint z & x quantitates duæ variabiles, quarum z uniformiter augetur per data incrementa $\dot{z}$, & sit $n\dot{z} = v$, $v - \dot{z} = \dot{v}$, $\dot{v} - \dot{z} = \ddot{v}$, & sic porrò. Tum dico quod quo tempore z crescendo fit $z + v$, x item crescendo fiet

$$x + \dot{x}\frac{v}{1\dot{z}} + \ddot{x}\frac{v\dot{v}}{1.2\dot{z}^2} + \dddot{x}\frac{v\dot{v}\ddot{v}}{1.2.3\dot{z}^3} + \&c.$$

DEMON.

Figure 13.9: The First Appearance of Taylor's Theorem, in *Method of Increments.* Boston Public Library/Rare books Department. Courtesy of the Trustees.

where, as z changes to $z + \dot{z}$, x changes to $x + \dot{x}$. But now, the fluxion of $x + \dot{x}$ is $\dot{x} + \ddot{x}$, so $x + \dot{x}$ changes to $x + 2\dot{x} + \ddot{x}$; likewise, the remaining columns may be filled in:

$$\begin{array}{llll} x & \dot{x} & \ddot{x} & \cdots \\ x + \dot{x} & \dot{x} + \ddot{x} & \ddot{x} + \dddot{x} & \cdots \\ x + 2\dot{x} + \ddot{x} & \dot{x} + 2\ddot{x} + \dddot{x} & \ddot{x} + 2\dddot{x} + \ddddot{x} & \cdots \\ \vdots & \vdots & \vdots & \ddots \end{array}$$

and so on. Hence, as z changes to $z + 2\dot{z}$, x changes to $x + 2\dot{x} + \ddot{x}$; as z changed to $z + 3\dot{z}$, x would change to $x + 3\dot{x} + 3\ddot{x} + \dddot{x}$, and so on. Thus, by letting z change to $z + n\dot{z}$, Taylor obtained Equation 13.4.

13.7 Exercises

1. Integrate the successive pairs in the series in Equation 13.3 to generate the Bernoulli series in Equation 13.2.

2. Use the fact that $dy = \frac{a\ dx}{a+x}$ has the solution $y = \log(a + x)$, to find the series expansion for $\log(a + x)$.

3. What does Bernoulli's formulation of l'Hôpital's rule, on page 441, say about the nature of curves in the time of Bernoulli?

Chapter 14

Probability and Statistics

The region of Switzerland is divided into a number of city-states (called cantons), which were more or less independent of each other, though throughout most of the Middle Ages, they were dominated by the Habsburgs; in fact the family name, originally *Habichtsburg*, was taken from a castle overlooking the Aar River in Switzerland. The Treaty of Westphalia (1648) recognized the independence of Switzerland.

One of the majors exports of Switzerland into the twentieth century was soldiers, both military and intellectual. During the Reformation, the Swiss cantons produced Ulrich Zwingli, a soldier-priest on the side of the Lutherans, and John Calvin, a Protestant philosopher who established Geneva as a "city on a hill," where the precepts of Christian morality formed the entirety of the law. John Knox, who founded the Scottish Presbyterian Church and forced Mary, Queen of Scots to abdicate her throne in favor of her son, James VI (later James I of England) spent some time in Calvin's Geneva. In 1712, in Geneva, was born Jean Jacques Rousseau, whose writings would help bring about both the American and the French revolutions.

14.1 Jakob Bernoulli

The mathematics of the late seventeenth and early eighteenth centuries were dominated by a single family: the Bernoullis of Basel, Switzerland. Modern popular psychology would call the family dysfunctional. Professional rivalry and jealousy caused many harsh words between family members, and two even moved to Russia to escape the family influence.

The father and grandfather of Jakob Bernoulli (December 27, 1654-August 16, 1705) were pharmacists, who intended that Jakob continue the tradition. To study mathematics, Jakob left home and traveled abroad for seven years, going to Geneva, through France, the Netherlands, and England, finally returning to

445

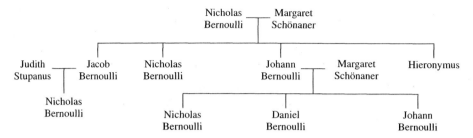

Figure 14.1: The Bernoulli Family.

Jakob Bernoulli.   ©Bettmann/
CORBIS.

Basel in 1683. In 1687, he accepted the chair of mathematics at the University of Basel.

Bernoulli's greatest mathematical work, *The Art of Conjecturing*, began with a commentary on Huygens's *On the Calculations in Games of Chance*; the remainder laid the foundation for the modern theory of probability, and included a number of important original results. Unfortunately, it was unfinished at the time of his death in 1705. Had it been published then, it would have been the first important treatise on probability, but instead, it was given to Jakob's brother Johann to complete. Johann pleaded that he was too busy to undertake such a task, and passed it on to Nicholas, son of another brother. Nicholas, who had been working on probability, intended to add to the work, but found it too difficult a task, and finally had the work published, "as is" in 1713, eight years after Jakob's death. As a result of the delay, Jakob lost credit for some important discoveries in the theory of probability.

14.1.1   The Sums of the Powers

Included in *The Art of Conjecturing* were general formulas for the sums of the *c*th powers of the whole numbers. Wallis had already found some of these formulas by generalizing from specific examples, a method known as **scientific induction**. Though a valid means of *discovery*, it is not a valid means of *proof*, and Jakob (among others) criticized Wallis's work. Bernoulli based the rigorous deduction of these formulas on the theory of combinations and permutations.

The rules for determining the number of permutations had been known to the medieval Muslims and ancient Indians; however, they were still little known to most European mathematicians. The notation of combinatorics was still in its infancy, as Bernoulli's rule for finding the number of permutations shows:

Rule 14.1. *To find the total number of permutations of any given number of things: take all numbers from unity in their natural sequence to the number of objects given. The product will be the number of permutations.*

Example 14.1. *The number of permutations of 9 different objects is* $1 \cdot 2 \cdot 3 \cdots 8 \cdot 9$, *or* $362,880$.

If some of the objects are repeated, then:

Rule 14.2. *To find the number of permutations, where some things are not different: take the number of permutations, where all the objects are different, and divide by the number of permutations, which the number of repeated objects can undergo, or by the product of the number of permutations that the repeated objects can undergo, if there are many repeated objects.*

Example 14.2. *The number of permutations of the letters in the word* studiosus. *The number of permutations of 9 objects is* 362,880. *Since* u *is repeated twice and* s *is repeated three times, the number of distinct permutations is thus* $\frac{362,880}{1 \cdot 2 \cdot 1 \cdot 2 \cdot 3} =$ 30,240.

Bernoulli defined a **combination** in the modern sense.

Definition. *The **combinations** of things are those collections of objects taken from a given multitude, taken without respect to the order.*

Nowadays we speak of "combinations of n objects taken m at a time," but Bernoulli defined:

Definition. *A **binion** is a combination of two objects; a **trinion** is the combination of three; **quaternion** of four, and so on. A **union** is the combination of a single object, and a **nullion** is that of nothing.*

To analyze the number of unions, binions, and so forth, Bernoulli began a well-organized analysis, very similar to that of ibn Mun'im. If there is but a single object, a, then there can be only a single combination, the union a. If a second object, b, is added, then it produces another union, b, and a binion, ab, which is the same as ba. If a third object, c, is added, the combinations are c by itself, c with the previous unions (making ca and cb), and c with the previous binion (making abc). Continuing in this fashion, Bernoulli produced the table

See page 243.

$$
\begin{array}{c}
\hline
a \\
\hline
b, ab \\
\hline
c, ac, bc, abc \\
\hline
d, ad, bd, cd, acd, bcd, abcd \\
\hline
\vdots
\end{array}
$$

A key feature is that Bernoulli organized the table by the *new* combinations created when an additional item is added. Thus, the number of unions follows the sequence 1, 1, 1, ... The number of binions, generated from the unions, form the sequence 0, 1, 2, 3, ... The number of trinions, each generated from the binions, form the sequence 0, 0, 1, 3, 6, ... Organizing the data into a table, Bernoulli produced a variation of Pascal's triangle (see Table 14.1).

Bernoulli then noted and proved a number of "wonderful properties" of the table of combinations. Property 4 is that the sum of the first n entries in any column is the next entry in the adjacent column. He also showed that the column entries could be found using a formula we would recognize today as the

Number	Number in Combination								
of Elements	1	2	3	4	5	6	7	8	9
1	1	0	0	0	0	0	0	0	0
2	1	1	0	0	0	0	0	0	0
3	1	2	1	0	0	0	0	0	0
4	1	3	3	1	0	0	0	0	0
5	1	4	6	4	1	0	0	0	0
6	1	5	10	10	5	1	0	0	0
7	1	6	15	20	15	6	1	0	0
8	1	7	21	35	35	21	7	1	0

Table 14.1: Bernoulli's Combination Table

binomial coefficient, though without notation, Bernoulli is forced to write them out explicitly, such as

$$\frac{n \cdot n - 1 \cdot n - 2}{1 \cdot 2 \cdot 3}$$

Notice here Bernoulli's use of the $\cdot$ to separate groups of terms: Bernoulli's $n \cdot n - 1 \cdot n - 2$ would be written today as $n(n-1)(n-2)$. To avoid confusion, we will include the parentheses.

The use of the combination table allowed Bernoulli to find explicit formulas for the sum of the powers of the first n whole numbers. Since the nth term of the second column is $(n-1)$, the sum of all these terms, which Bernoulli wrote as $\int (n-1)$ is, by Property 4 of the combination table, the next entry in the following column, calculated by $\frac{n(n-1)}{1 \cdot 2} = \frac{n^2-n}{2}$. Hence,

$$\int (n-1) = \int n - \int 1 = \frac{n^2 - n}{2}$$

Thus,

$$\int n = \int 1 + \frac{n^2 - n}{2}$$
$$= n + \frac{n^2 - n}{2}$$
$$= \frac{n^2 + n}{2}$$

which gives the formula for the sum of the series of whole numbers.

The nth term in the third column is $\frac{(n-1)(n-2)}{1 \cdot 2}$ or, expanding, $\frac{n^2-3n+2}{1 \cdot 2}$ so the sum of these terms will be the next term in the fourth column, or $\frac{n(n-1)(n-2)}{1 \cdot 2 \cdot 3}$. Hence,

$$\int \frac{n^2 - 3n + 2}{1 \cdot 2} = \frac{n(n-1)(n-2)}{1 \cdot 2 \cdot 3}$$

and thus,

$$\int n^2 = \frac{1}{3}n^3 + \frac{1}{2}n^2 + \frac{1}{6}n$$

The formulas are not new with Bernoulli, though Bernoulli's derivation differed from that of Pascal and of Leibniz. Bernoulli then gave a general rule for the sum of the cth powers of the whole numbers:

Proposition 14.1. *The sum of the cth powers of n is given by*

$$\int n^c = \frac{1}{c+1}n^{c+1} + \frac{1}{2}n^c + \frac{c}{2}An^{c-1} + \frac{c(c-1)(c-2)}{2\cdot 3\cdot 4}Bn^{c-3}$$
$$+ \frac{c(c-1)(c-2)(c-3)(c-4)}{2\cdot 3\cdot 4\cdot 5\cdot 6}Cn^{c-5}$$
$$+ \frac{c(c-1)(c-2)(c-3)(c-4)(c-5)(c-6)}{2\cdot 3\cdot 4\cdot 5\cdot 6\cdot 7\cdot 8}Dn^{c-7} + \dots$$

where the exponents continue to decrease by 2 until n or n^2 is reached, and the capital letters A, B, C, D, and so on, denote the coefficients of the last terms of $\int n^2$, $\int n^4$, $\int n^6$, ... The successive coefficients are found because they complete the coefficients of the terms of the expression to unity.

The coefficients A, B, C, ... are now called **Bernoulli numbers**.

Example 14.3. *Find the sums of the fourth powers. The coefficient of the last term of the series for $\int n^2$ is $\frac{1}{6}$, so $A = \frac{1}{6}$. Thus:*

$$\int n^4 = \frac{1}{4+1}n^{4+1} + \frac{1}{2}n^4 + \frac{4}{2}\frac{1}{6}n^3 + \frac{4\cdot 3\cdot 2}{2\cdot 3\cdot 4}Bn^1$$

The sum of the coefficients is $\frac{1}{5} + \frac{1}{2} + \frac{1}{3} + B$; setting this equal to 1, we find $B = -\frac{1}{30}$.

At this point, Bernoulli displayed some of the personality traits that caused so much friction between the members of the family. After spending "half of a quarter of an hour" showing that the sum of the tenth powers of the first 1000 numbers was

$$91,409,924,241,424,243,424,241,924,242,500$$

Bernoulli then noted, "It is apparent from this how useless is the work of Ismael Bullialdus," who had written an earlier work, *Arithmetic of the Infinite*, where he had calculated by hand and with a great deal of effort the sums of the powers of the whole numbers.

14.1.2 The Multinomial Theorem

Bernoulli also proved the **multinomial theorem**, of which the binomial theorem is a special case. Again, owing to the delays in publication, another proof, that of Pierre Remond de Montmort (October 27, 1678-October 7, 1719) appeared first, in 1708. Both Bernoulli and Montmort proceeded by identifying the multinomial coefficients with the number of ways that a particular event could occur. This is one of the more powerful tools of combinatorics, and would play a major role in the analysis of error. Bernoulli's proof of the multinomial theorem proceeded as follows: to find the various terms of the powers of the multinomial $a + b + c + d$, Bernoulli noted that the binions were formed by placing each letter before every other letter (including itself); the trinions were formed by placing each letter before every one of the binions; the quaternions were formed by placing every letter before each of the trinions, and so on. However, this was precisely the process by which the power was obtained. Treating all binions, trinions, and so forth, composed of the same number of letters as equal, the coefficient of each term was just the number of permutations of the letters, taken as many times as their powers in the final term.

Example 14.4. *The* $a^5 b^3 c^2$ *term of* $(a + b + c)^{10}$ *is the number of permutations of the set including five* a*'s, three* b*'s, and two* c*'s, or* $\frac{1 \cdot 2 \cdots 9 \cdot 10}{1 \cdot 2 \cdots 5 \cdot 1 \cdot 2 \cdot 3 \cdot 1 \cdot 2} = 2520$.

14.1.3   The Law of Large Numbers

Jakob's major accomplishment in *The Art of Conjecturing* was the first proof of the **law of large numbers**. Bernoulli's proof depends on the binomial expansion of $(r + s)^{nt}$, where $t = r + s$. If an event has r ways to occur and s ways not to occur, then the $r^{nr} s^{ns}$ term of the expansion will be proportional to the probability that the *observed* ratio between occurrence and nonoccurrence of the event is $r : s$, since the actual value of the term (once r and s have been substituted) is the number of ways the event can occur nr ways and not occur ns ways

Lemma 14.1. *Let* M *be the* $r^{nr} s^{ns}$ *term of the binomial expansion of* $(r + s)^{nt}$. *Then the ratio of* M *to the preceding term will be less than the ratio of any two adjacent terms preceding* M, *the ratio being taken of the term nearer to* M *to the term farther from* M. *Likewise, the ratio of* M *to the succeeding term will be less than the ratio of any two adjacent terms following* M, *the ratio being taken of the term nearer to* M *to the term farther from* M.

Proof. M is the term

$$\frac{nt(nt - 1)(nt - 2) \cdots (nt - ns + 1)(nr + 1)}{1 \cdot 2 \cdots ns} r^{nr} s^{ns}$$

or, equivalently,

$$\frac{nt(nt - 1)(nt - 2) \cdots (nt - nr + 1)(ns + 1)}{1 \cdot 2 \cdots nr} r^{nr} s^{ns}$$

Let the term to the left of M, or

$$\frac{nt(nt - 1)(nt - 2) \ldots (nr + 2)}{1 \cdot 2 \cdot 3 \ldots ns - 1} r^{nr+1} s^{ns-1}$$

be designated F; let the term G precede F, H precede G, and so on. Then $M : F = (nr + 1)s : nsr$. Likewise, $F : G = (nr + 2) \cdot s : (ns - 1) \cdot r$. But

$\frac{nr+1}{ns} < \frac{nr+2}{ns-1}$, hence, multiplying both by $\frac{s}{r}$, we have $\frac{(nr+1)s}{nsr} < \frac{(nr+2)s}{(ns-1)r}$, and thus $M : F < F : G$.   □

Next, Bernoulli considered the ratio of M to the nth term to its left, L, and the nth term to its right, Λ.

Lemma 14.2. *The ratio of $M : L$ tends to infinity as n tends to infinity.*

The ratio $M : L$ is a measure of how many times more likely it is to obtain an observed ratio of $r : s$ than $r + 1 : s - 1$. This is not quite the law of large numbers, for it does not yet deal with the possibility of observing a ratio between $r : s$ and $r + 1 : s - 1$. Bernoulli dealt with this in a later lemma. See Problem 6.

Designating the terms H, G, F the terms preceding (in that order) M, and Q, P, L the terms (in that order) following M, then

Lemma 14.3. *The ratio of the sum of the n terms preceding M to the sum of the n terms preceding L is greater than the ratio of M to L.*

Proof. By Lemma 14.1, we have

$$\frac{M}{F} < \frac{L}{P} \qquad\qquad \frac{F}{G} < \frac{P}{Q} \qquad\qquad \frac{G}{H} < \frac{Q}{R}$$

Hence

$$\frac{M}{L} < \frac{F}{P} < \frac{G}{Q} < \frac{H}{R}$$

Summing the antecedents and consequents:

$$\frac{M}{L} < \frac{F + G + H + \dots}{P + Q + R + \dots}$$

□

Since, as the number of trials tends to infinity, the ratio $M : L$ tends to infinity as well, this means that the probability of observing a ratio outside of the interval between $r : s$ and $r + 1 : s - 1$ tends to zero.

Bernoulli did not, however, see the proof as an end by itself, since the *fact* of the law of large numbers was something "known to all." Instead, he was interested in how the ratio approached the probability. In particular, how many trials are necessary before the observed frequency is within a specified distance of the actual probability? It was for this reason that Bernoulli had to examine the terms of the binomial expansion of $(r + s)^{nt}$.

The ratio $M : L$ is

$$\frac{M}{L} = \frac{(nrs + ns)(nrs + ns - s)(nrs + ns - 2s) \cdots (nrs + s)}{(nrs - nr + r)(nrs - nr + 2r)(nrs - nr + 3r) \cdots (nrs)}$$

$$= \left(\frac{nrs + ns}{nrs - nr + r}\right)\left(\frac{nrs + ns - s}{nrs - nr + 2r}\right) \cdots \left(\frac{nrs + s}{nrs}\right)$$

Let c be the level of certainty desired (e.g., if one wished to be 1000 times as certain of the ratio as not, then $c = 1000$), and suppose

$$\left(\frac{r+1}{r}\right)^m \geq c\,(s-1)$$

Solving for m, we have

$$m \geq \frac{\ln(c(s-1))}{\ln(r+1) - \ln r}$$

Since each factor in the ratio $M : L$ is greater than the one preceding it (and all are greater than 1), suppose the mth factor of the ratio $M : L$ is equal to $\frac{1+r}{r}$. So:

$$\frac{1+r}{r} = \frac{nrs + ns - (m-1)s}{nrs - nr + mr}$$

and thus the ratio $M : L$ will be greater than $c(s-1)$. Solving for n we obtain:

$$n = m + \frac{ms - s}{r + 1}$$

Performing the same analysis for the ratio $M : \Lambda$, Bernoulli came up with the other bound, by letting $(\frac{s+1}{s})^m \geq c(r-1)$,

$$m = \frac{\ln(c(r-1))}{\ln(s+1) - \ln s}$$
$$n = m + \frac{mr - r}{r + 1}$$

The number of trials needed to make it c times more likely to observe a frequency between $\frac{r-1}{r+s}$ and $\frac{r+1}{r+s}$ than any other frequency would thus be the greater of the two possible values of nt.

Example 14.5. *Suppose $r = 30$, $s = 20$, $t = r + s = 50$, and $c = 1000$. Our two possible values are*

$$m = \tfrac{\ln(c(s-1))}{\ln(r+1)-\ln r} < 301$$
$$n = m + \tfrac{ms-s}{r+1} < 24728$$

Or

$$m = \tfrac{\ln(c(r-1))}{\ln(s+1)-\ln s} < 211$$
$$n = m + \tfrac{mr-r}{s+1} < 25550$$

Thus, at least $25,550$ trials must be performed.

This result surprised Bernoulli, for it suggested that certainty was much more difficult than it appeared; it may have been one of the reasons why *Art of Conjecturing* did not appear in his lifetime.

14.1 Exercises

1. Find formulas for the indicated sums, *without* using Bernoulli's proposition for the sum of the cth powers.

   (a) $\int n^3$ (b) $\int n^4$

2. Find the Bernoulli numbers C, D, E.

3. Use Proposition 14.1 to find formulas for the following.

   (a) $\int n^5$ (b) $\int n^6$ (c) $\int n^7$ (d) $\int n^8$

4. Find the indicated sums.

   (a) $\sum_{n=1}^{100} n^3$ (b) $\sum_{n=1}^{50} n^5$ (c) $\sum_{n=1}^{20} n^7$

5. Find the indicated term of the given multinomial expansion.

   (a) The $a^5 b^{12}$ term of $(a+b)^{17}$

   (b) The $a^2 b^2 c^2$ term of $(a+b+c)^6$

   (c) The $a^3 b^2 cd$ term of $(a+b+c+d)^7$

6. Prove Lemma 14.2. Why can the ratio $M : L$ be interpreted as the number of times more probable it is to observe the ratio $r : s$ than the ratio $r - 1 : s + 1$?

7. Comment on Bernoulli's proof of the law of large numbers.

   (a) What are its limitations, compared with the modern form of the theorem?

   (b) Bernoulli's proof is complicated by its use of expectations as opposed to probabilities. Modify the proof by letting $r + s = 1$. Hence, r and s are the probabilities of the event occurring or not occurring.

8. Find the number of trials necessary to find the frequency ratios within the given level of certainty.

   (a) $r = 10$, $s = 15$, $c = 20$ (b) $r = 100$, $s = 200$, $c = 100$

9. Forty red balls and sixty white balls are in a vase. A ball is withdrawn, its color is noted, and then it is replaced. Find how many trials are necessary to observe a frequency ratio of red balls that is:

   (a) Fifty times more certain to be between $\frac{39}{100}$ and $\frac{41}{100}$, than any other ratio.

   (b) One hundred times more certain to be between $\frac{3}{10}$ and $\frac{5}{10}$ than any other ratio.

   (c) One thousand times more certain to be between $\frac{1}{5}$ and $\frac{3}{5}$ than any other ratio.

14.2   Abraham De Moivre

In 1685, Louis XIV revoked the Edict of Nantes, causing an estimated 50,000 Huguenots to flee France. One of them was Abraham De Moivre (May 26, 1667-November 27, 1754), who went to England in 1688, after spending two years in a French prison. In England, De Moivre made a name for himself as one of the foremost English mathematicians; in later years, Newton would advise young mathematicians, "Go to Mr. De Moivre; he knows these things better than I."

14.2.1   De Moivre's Formula

The name of De Moivre is most frequently associated with **De Moivre's formula**

$$(\cos a + i \sin a)^n = \cos na + i \sin na$$

though the actual statement of the formula in this form would be left to Euler. There is evidence that De Moivre possessed the formula attributed to him as early as 1707, but its first appearance was in *Miscellaneous Analysis* (1730), where De Moivre stated it as:

Proposition 14.2. *If l, x are cosines of arcs A, B of the unit circle, and $A : B = n : 1$, then*

$$x = \tfrac{1}{2} \sqrt[n]{l + \sqrt{l^2 - 1}} + \frac{\frac{1}{2}}{\sqrt[n]{l + \sqrt{l^2 - 1}}} \tag{14.1}$$

Corollary. *Let $z = \sqrt[n]{l + \sqrt{l^2 - 1}}$. Then*

$$z^{2n} - 2lz^n + 1 = 0 \tag{14.2}$$

$$z^2 - 2zx + 1 = 0 \tag{14.3}$$

De Moivre noted that by eliminating z, a relationship could be found between the cosines of the arcs A and the arcs $\frac{C-A}{n}$, $\frac{C+A}{n}$, $\frac{2C+A}{n}$, $\ldots$

De Moivre also noticed the value of x was related to the nth root of an "impossible" binomial, of the kind Wallis attempted to simplify. On April 29,

See page 397.

1740, a friend, Nicholas Saunderson (who was, at the time, writing a textbook on algebra), wrote to De Moivre, posing the question of how to reduce cube roots of "impossible" binomials, like $\sqrt[3]{-5 + \sqrt{-2}}$. Saunderson was aware of Wallis's procedure, but no general method of reduction existed.

De Moivre had actually solved this problem already as part of a letter he sent to the Royal Society in 1739, entitled *On the Reduction of Radicals to More Simple Terms*, and communicated the result to Saunderson on April 29, 1740—unfortunately after Saunderson's death. De Moivre's solution is as follows.

Problem 14.1. *Reduce the cube root of $a + \sqrt{-b}$.*

Solution. Let the cube root of $a + \sqrt{-b}$ be $x + \sqrt{-y}$. Then $x^3 - 3xy = a$ and $(3x^2 - y)\sqrt{-y} = \sqrt{-b}$ or, expanding

$$a = x^6 - 6x^4y + 9x^2y^2$$
$$-b = -9x^4y + 6x^2y^2y^3$$

Suppose $m = \sqrt[3]{a^2 + b} = x^2 + y$. Then $y = m - x^2$, and so $y = \frac{x^3 - a}{3x}$, or $4x^3 - 3mx = a$.

However, if r is the radius of a circle, l the cosine of an arc, and x the cosine of one-third the arc, then it is true that $4x^3 - 3r^2x = r^2l$. Thus, if we let $r^2 = m$, $r^2l = a$, then $r = \sqrt{m}$ and $l = \frac{a}{m}$. Hence, if we draw of a circle of radius $\sqrt{m}$, take the arc whose cosine is $\frac{a}{m}$, and find the cosine of one-third the arc, this will be x; from which y can be found. □

De Moivre then noted that there are three different possible values of x: if the arc is A and the circumference of the circle is C, then the cosines of $\frac{A}{3}$, $\frac{C-A}{3}$, and $\frac{C+A}{3}$ are all equal to x.

In general, De Moivre concluded, to determine x, y so that $\sqrt[n]{a + \sqrt{-b}} = x + \sqrt{-y}$, first let $m = \sqrt[n]{a^2 + b}$, and find the circle of radius $\sqrt{m}$. Let $p = \frac{n-1}{2}$, and find the arc A whose cosine is $\frac{a}{m^p}$. If C is the circumference, then the required arcs are $\frac{A}{n}$, $\frac{C-A}{n}$, $\frac{C+A}{n}$, ...

14.2.2 Recurrent Series

Also in *Miscellaneous Analysis*, De Moivre included results on **recurrent series**. De Moivre posed the problem

Problem 14.2. *Given a recurrent series $a + bx + cxx + dx^3 + ex^4 + \ldots$, whose scale of relation is $f - g$, to find the sum of the series.*

In a recurrent series, every term is related to the previous terms according to some rule, which De Moivre called the **scale of the relation**. What De Moivre meant by the scale being $f - g$ is made clear by his solution:

Solution. Suppose the terms of the series are

$$P = P$$
$$Q = Q$$
$$R = fQx - gPxx$$
$$S = fRx - gQxx$$
$$T = fSx - gRxx$$

$\square$

In other words, after the first two terms, any term of the series is formed by multiplying the previous term by fx and subtracting the term before it multiplied by gxx.

Solution. [Continued] Let

$$z = P + Q + R + S + T + \ldots \tag{14.4}$$

Then, subtracting P and multiplying through by fx yields

$$fxz - fPx = fQx + fRx + fSx + fTx + \ldots \tag{14.5}$$

If $P + Q$ is added to both sides, this gives

$$P + Q + fxz - fPx = P + Q + fQx + fRx + fSx + fTx + \ldots \tag{14.6}$$

Multiplying Equation 14.4 by $-gxx$ yields

$$-gzxx = -gPxx - gQxx - gRxx - gSxx - \ldots \tag{14.7}$$

Adding Equations 14.6 and 14.7, and remembering that $R = fQx - gPxx$, $S = fRx - gQxx, \ldots$, gives

$$P + Q + fxz - fPx - gzxx = R + S + T + \ldots \tag{14.8}$$
$$= z \tag{14.9}$$

Hence

$$z = \frac{P + Q - fPx}{1 - fx + gxx} \tag{14.10}$$

$\square$

In the *Doctrine of Chances* (see Section 14.2.4), De Moivre gave the modern method for finding the general term of the recurrent series, for the purpose of finding any required term of the series. His procedure is purely algorithmic. If the series is

$$a + br + cr^2 + dr^3 + cr^4 + \dots$$

and the scale of the relationship is $fr - gr^2$ (note that De Moivre's notation has changed), then let m, p be two *distinct* roots of the equation $1 - fx + gx^2 = 0$. If $A = \frac{br - pa}{m - p}$ and $B = \frac{br - ma}{p - m}$, and l is the interval between the first term and the desired term, then the term will be $Am^l + Bp^l$.

14.2.3 *The Measurement of Chance*

De Moivre's primary contribution to mathematics was in the field of probability. In "The Measurement of Chance," which appeared in the January-February-March 1711 issue of *Philosophical Transactions*, De Moivre began by stating that if an event can happen p times and fail to happen q times, then the ratio of the probability of its occurring to the probability that it does not occur is as p to q; this corresponds to the statement that the **odds of an event occurring** are p to q. For the first time, De Moivre added an important stipulation that most writers had assumed but not stated: all the cases must be equally likely to occur.

A key technique used by De Moivre was the use of what we now call **generating functions** to calculate the number of possible cases. For example, if there were two events, and the first one could occur p ways and not occur in q ways, and the second could occur in r ways and not occur in s, then the product of $p + q$ and $r + s$, or $pr + qr + ps + qs$, would give the number of ways the event could occur or not occur.

For example, if a game was being played where A would win only when both events occurred, the ratio of the chance of A winning to that of B winning was pr to $qr + ps + qs$. But if A won whenever either event happened, then the ratio of their chances was $pr + qr + ps$ to qs, and if A won only when the first event happened and the second event did *not* occur, his chances were as ps to $pr + qr + qs$.

The value of this method could be seen in De Moivre's second problem, which considered a game where A needed to win four games, and B needed six games to win the match. This problem broke new ground, for it was the first time in a *published* work that unequal probabilities of winning were considered: De Moivre supposed the probability of A winning to the probability of B winning was $3 : 2$. To find the ratio, De Moivre noted the match would be decided within nine games. First, expanding $(a + b)^9$, he obtained

$$\begin{aligned}(a + b)^9 = {}& a^9 + 9a^8b + 36a^7b^2 + 84a^6b^3 + 126a^5b^4 + 126a^4b^5 \\ & + 84a^3b^6 + 36a^2b^7 + 9ab^8 + b^9\end{aligned}$$

The terms where the power on a is 4 or greater correspond to the number of times A will win four or more games, and thus A will win the match; the remaining terms correspond to the number of times B will win six or more games, and thus B will win the match. By letting $a = 3$ and $b = 2$, De Moivre was able to calculate the odds of A winning. The use of the generating function to calculate the number of cases where an event occurred (A winning) or failed to occur was a portent of things to come, and has become one of the most powerful tools of combinatorics and probability.

De Moivre's third problem was one of the first problems of **inverse probability**: finding the probability of an event, given only the number of times it has occurred or failed to occur. De Moivre considered two players, A and B, and a match where a player must win three games to win the match. From experience, A knows that he can give B a two-game lead, and still be as likely to win as B; in other words, about half the time A could win three games before B could win one more. To solve it, De Moivre supposed the odds of their winning were as z to 1; then the expansion of $(z + 1)^3$ would determine the number of cases where A won the next three games and the number of cases where B won at least one of the next three, namely z^3 and $3z^2 + 3z + 1$. Since the players were equally likely to win the match, then $z^3 = 3z^2 + 3z + 1$. Hence $2z^3 = z^3 + 3z^2 + 3z + 1 = (z + 1)^3$, and $z = \frac{1}{\sqrt[3]{2} - 1}$.

14.2.4   *The Doctrine of Chances*

De Moivre greatly expanded the original "Measurement of Chance" into a complete textbook of probability, *The Doctrine of Chances* (1718, 1738, 1756). In it, he gave, for the first time, a clearly recognizable definition of the probability of an event:

Definition. *The **probability of an event** is greater or less, according to the number of chances by which it may happen, compared with the whole number of chances by which it may either happen or fail.*

Huygens and Jakob Bernoulli had both analyzed games in terms of the expectation; though De Moivre also used the concept of expectation, he just as frequently computed the actual probabilities.

A key problem in inverse probability De Moivre dealt with is

Problem 14.3. *In how many trials will it be equally likely that an event will occur or not occur, if there are a chances of it happening and b chances of it failing in any one trial?*

This is a special case of the Law of Large Numbers.

De Moivre found the number of trials to be $x = \frac{\log 2}{\log(a+b) - \log b}$. Letting $a : b = 1 : q$, and, using the series expansion

$$\log\left(1 + \frac{1}{q}\right) = \frac{1}{q} - \frac{1}{2q^2} + \frac{1}{3q^3} - \dots$$

THE
DOCTRINE
OF
CHANCES.

****

The INTRODUCTION.

1.   H E Probability of an Event is greater or lefs, according to the number of Chances by which it may happen, compared with the whole number of Chances by which it may either happen or fail.

     2. Wherefore, if we conftitute a Fraction whereof the Numerator be the number of Chances whereby an Event may happen, and the Denominator the number of all the Chances whereby it may either happen or fail, that Fraction will be a proper defignation of the Pro-

            B   bability

Figure 14.2: First Page of De Moivre's *The Doctrine of Chances* (Third Edition). Boston Public Library/Rare books Department. Courtesy of the Trustees.

De Moivre found $\frac{x}{q} - \frac{x}{2q^2} + \ldots = \log 2$. For large values of q, we have $x = q \log 2$ or, since $\log 2 \approx 0.7$, then $x = 0.7q$.

De Moivre gave the example of a lottery, where the ratio of blanks to prizes was 39 to 1. If there were only 40 tickets altogether, then 20 tickets would suffice, since the prize-winning ticket would as likely be in one group of 20 as in another group of 20. De Moivre then calculated that with 80 tickets, between 23 and 24 tickets were necessary, and as the number of tickets grew, by multiples of 40, the number of ticket required to have an even chance of winning a prize would never exceed $\frac{7}{10}$ of 39, or between 27 and 28.

Around 1730, through methods that are unknown, De Moivre determined that the ratio of the middle term of the binomial expansion of $(1+1)^n$ for n very large was to the whole binomial, 2^n, as $\frac{2A}{\sqrt{n-1}} \frac{(n-1)^n}{n^n}$, where A is the number with natural logarithm

$$\frac{1}{12} - \frac{1}{360} + \frac{1}{1260} - \frac{1}{1680} + \ldots$$

De Moivre calculated this number to be about $2\frac{21}{125}$.

If n is very large, then the value of $\frac{(n-1)^n}{n^n}$ is "very nearly given" (it is approximately $\frac{1}{e}$) and thus the quantity will be $\frac{2B}{\sqrt{n-1}}$ or $\frac{2B}{\sqrt{n}}$, where B is equal to $\frac{A}{e}$, and hence $\ln B = -1 + \frac{1}{12} - \frac{1}{360} + \frac{1}{1260} - \frac{1}{1680} + \ldots$ Or, changing the signs of the series to $1 - \frac{1}{12} + \frac{1}{360} - \frac{1}{1260} + \ldots$, the expression becomes $\frac{2}{B\sqrt{n}}$. In 1730, James Stirling showed that $B = \sqrt{2\pi}$, from which we may derive **Stirling's formula**:

$$n! \approx \left(\frac{n}{e}\right)^n \sqrt{2\pi n}$$

De Moivre's work was the most thorough treatment of probability to date, and the third edition, that of 1756, is still an excellent introduction to the subject. But the value of the work was missed by his contemporaries. A number of reasons contributed to this fact. One of them was capitalized upon by Thomas Simpson (whose work will be dealt with later): the book was too expensive. Simpson published shorter, cheaper versions of De Moivre's work. Though Simpson praised De Moivre greatly, De Moivre was unhappy, for the existence of the cheaper editions threatened his own sales. A new edition of *Annuities Upon Lives* warned its readers against other editions that "mutilated" his propositions and "confounded" them with a "crowd of useless symbols."

14.2 Exercises

1. Demonstrate that for a recurring series $a + bx + cx^2 + dx^3 + \ldots$ with scale $fx - gx^2$, the sum of the series will be $\frac{a+bx-fax}{1-fx+gx^2}$.

   (a) Let $S = a+bx+cx^2+dx^3+\ldots$ Show that $(1-fx+gx^2)S = a+bx-fax$, and hence $S = \frac{a+bx-fax}{1-fx+gx^2}$.

   (b) What conditions are necessary to make the proof rigorous?

2. Use the procedure in Problem 1 to find the summation for the recurrent series $a + bx + cx^2 + dx^3 + ex^4 + \ldots$ with the indicated law of the relationship.

   (a) $fx - gx^2 + hx^3$.

   (b) $fx - gx^2 + hx^3 - kx^4$.

3. Given a recurrent series $a + bx + cx^2 + dx^3 + \ldots$ with a given law of relationship, explain how you would find the sum of a *finite* number of its terms.

4. Solve.

   (a) If a player must win two games to win a match, and A can give B a one game lead and still be equally likely to win the match, what is the probability A can win any one game?

   (b) If a player must win four games to win a match, and A can give B a three game lead and still be equally likely to win the match, what is the probability A can win any one game?

   (c) If a player must win three games to win a match, and A can give B a one game lead and still be as likely to win as B, what is the probability that A can win any single game?

14.3   Daniel Bernoulli

At the beginning of the eighteenth century, little original research was done at universities. Some scholars, like l'Hôpital, were independently wealthy, but for most, scientific societies, like the Royal Society in England, provided the best source of sponsorship. One of the most important of the scientific societies would be the Russian Academy of the Sciences, founded by Peter the Great in 1724, though it would not open its doors until 1725, after Peter's death.

One of the key figures in the early academy was Christian Goldbach (1690-1764), who, during his travels through Europe, met with Nicholas Bernoulli (1695-1726), the son of Johann Bernoulli (and *not* the Nicholas who edited *The Art of Conjecturing*) in Venice. Goldbach encouraged the Academy to offer positions to Nicholas and his younger brother Daniel (February 8, 1700-March 17, 1782). Both accepted: Johann, whose jealousy poisoned the relationship with his brother Jakob in the last years of his life, had grown jealous of the successes of his own sons, making life in Basel uncomfortable. Nicholas, unfortunately, died shortly after arriving in St. Petersburg. Daniel stayed on for eight years, returning to Basel in 1733, though he would continue to publish important papers through the St. Petersburg Academy journal.

Daniel Bernoulli's greatest contributions were to mathematical physics, for he was one of the first continental physicists to reject Cartesian physics in favor of Newtonian mechanics. By combining a valid system of physics with a useful system of mathematics, physicists in continental Europe would make great strides that would leave their English counterparts, using antiquated Newtonian concepts and notations, far behind.

14.3.1   Observations on Recurrent Series

In an article written in 1728, "Observations on Recurrent Series," Daniel Bernoulli established several key theorems on recurrent series. Some of these had been established by De Moivre, but Bernoulli's formulation was simpler and easier to follow, and aided by useful examples. Moreover, unlike De Moivre's exposition in *The Doctrine of Chances*, Bernoulli included proofs of his results. Finally, De Moivre's method would work only if the roots of the characteristic equation were distinct; Bernoulli removed this limitation.

First, Bernoulli gave (without proof) a method of converting a sequence generated by a polynomial into a recurrent sequence:

Proposition 14.3. *Given a recurrent sequence whose terms are generated according to a $n - 1$st degree polynomial. Take n terms in reverse sequence, A, B, C, D, E, ... (i.e., A is the last of the n terms). Then*

$$A = nB - \frac{n(n-1)}{2}C + \frac{n(n-1)(n-2)}{2 \cdot 3}D - \frac{n(n-1)(n-2)(n-3)}{2 \cdot 3 \cdot 4}E + \dots$$

Example 14.6. *Find a recurrence relationship generating the triangular numbers*

$$1, 3, 6, 10, 15, 21, \dots$$

where the kth triangular number is $\frac{1}{2}k^2 + \frac{1}{2}k$. Since the triangular numbers are generated by a second degree polynomial, $n = 3$, and thus each triangular number is three times the preceding, minus three times the one before, plus the one before that; for example

$$21 = 3 \cdot 15 - 3 \cdot 10 + 6$$

Bernoulli gave the important theorem

Proposition 14.4. *Given a recurrent sequence, where A depends on the $N - 1$ previous terms B, C, D, ..., E according to*

$$A = mB + nC + pD + \dots + qE$$

a series of numbers in continued proportion will satisfy the relationship.

To prove this, Bernoulli assumed that the nth term a_n of a recurrent series had the form $a_n = a_0 r^n$. Since the term also satisfied the recurrence relationship, a polynomial equation in r could be set up and solved.

Bernoulli began with the Fibonacci sequence, though did not give it a specific name:

See page 282.

Example 14.7. *Consider the sequence of terms*

$$1, 1, 2, 3, 5, 8, 13, 21, 34, \dots$$

where each term is the sum of the two preceding it. The primary equation is thus

$$a^2 = a + 1$$

The roots are $a = \frac{1+\sqrt{5}}{2}$ and $a = \frac{1-\sqrt{5}}{2}$. Thus, the xth term of all sequences that satisfy the primary equation will be of the form

$$\beta\left(\frac{1+\sqrt{5}}{2}\right)^x + \gamma\left(\frac{1-\sqrt{5}}{2}\right)^x$$

For this sequence, if $x = 0$, the term of the sequence is 0, and if $x = 1$, the sequence term is 1. Hence

$$0 = \beta + \gamma$$

$$1 = \beta\left(\frac{1+\sqrt{5}}{2}\right) + \gamma\left(\frac{1-\sqrt{5}}{2}\right)$$

Solving, we have $\beta = \frac{1}{\sqrt{5}}$ and $\gamma = -\frac{1}{\sqrt{5}}$, and thus the xth term of the sequence will be given by $\frac{1}{\sqrt{5}}\left\{\left(\frac{1+\sqrt{5}}{2}\right)^x - \left(\frac{1-\sqrt{5}}{2}\right)^x\right\}$.

In the case of a repeated root, a case not dealt with by De Moivre, the power of the root is prefixed by a polynomial of a degree one less than the order of the root. Bernoulli gave the example of the sequence

$$0, 0, 0, 0, 1, 0, 15, -10, 165, -228, \ldots$$

generated by the recurrence relationship

$$A = 0B + 15C - 10D - 60E + 72F$$

The primary equation

$$a^5 - 15a^3 + 10a^2 + 60a - 72 = 0$$

has roots $a = 2$ (order 3) and $a = -3$ (order 2), so the general term is

$$\left(b + cx + dx^2\right)2^x + (e + fx)(-3)^x$$

In this particular case, the general term of the sequence is

$$\frac{\left(1026 - 1035x + 225x^2\right)2^x + (224 - 80x)(-3)^x}{90000}$$

Bernoulli used the recurrence relationship to find approximate solutions to polynomial equations. Given the polynomial equation

$$1 = ax + bx^2 + cx^3 + ex^4 + \ldots$$

form a recurrence relationship with arbitrary initial terms that satisfies

$$E = aD + bC + cB + eA + \ldots$$

where D is the term preceding E, C is the term preceding D, and so on. Then the quotient $\frac{M}{N}$ of two consecutive terms, M, N, approximates a solution to the original equation.

Example 14.8. *The equation*

$$1 = -2x + 5x^2 - 4x^3 + x^4$$

corresponds to the sequence generated by $E = -2D + 5C - 4B + A$. Setting our initial values to be 1, 1, 1, 1, we obtain the sequence

$$1, \ 1, \ 1, \ 1, \ 0, \ 2, \ -7, \ 25, \ -93, \ 341, \ -1254, \ \ldots$$

Thus an approximate solution to the equation is $-\frac{341}{1254}$.

14.3.2 Inoculation

One of the characteristics of the mathematics of the Enlightenment was its concern for "practical" application. Probability was among the most practical. A section of Jakob Bernoulli's *The Art of Conjecturing* showed how to apply probability to ethical and moral dilemmas. One dilemma in the eighteenth century was the question of inoculation against smallpox, perhaps the most feared disease of the day: it killed one-tenth or more of those it infected, and scarred for life most of the survivors. However, two key observations were made: first, those who had the disease (and survived) were immune to it afterwards; and second, it was possible to have a mild case of the disease that left no scars.

Thus began the practice of **inoculation**, which consisted of transferring pus from one person (who had the disease) to a healthy person, who hoped to get a mild case. The practice may have begun in the Ottoman Empire (Turkey), but was first described to the Royal Society in England in 1713. An eccentric English noblewoman, Lady Mary Wortley Montagu, was an early proponent. In 1721, a smallpox epidemic hit the English colonies in the New World, and minister Cotton Mather, of the Massachusetts Bay Colony, urged a local doctor, Zabdiel Boylston, to attempt widespread inoculation. Boylston was the first to inoculate patients in the New World.

It should be stressed that almost all modern inoculation procedures use "killed virus" vaccines, and there is virtually no risk of developing the disease from such an inoculation.

The problem was that inoculation itself carried a high risk, for one could not control the severity of the disease: in modern medical terms, inoculation was a "live virus" treatment. Thus, the person volunteering to be inoculated took the risk of developing a disease he or she might have avoided. To this end, the supporters of inoculation appealed to probability. In 1760 Daniel Bernoulli published "On the Mortality Caused by Smallpox, and the Advantage of Inoculation." Bernoulli's key question was:

Problem 14.4. *Determine the number of years of life, on average, gained by inoculation.*

To solve this problem, Bernoulli let ξ be the number of persons still alive at age x, and s the number of these persons who had not contracted smallpox and are hence susceptible to the disease. Bernoulli assumed that $\frac{1}{n}$ of those susceptible caught the disease in some time dx, and $\frac{1}{m}$ of those who caught the disease actually died from it. Based on empirical studies, Bernoulli determined $n = 8$, $m = 8$.[1] Thus

$$-d\xi - \frac{s\,dx}{mn}$$

is the decrease in the population from causes *other* than smallpox. Since those who contract smallpox are no longer susceptible, then

$$-ds = \frac{s\,dx}{n} - \frac{s\,d\xi}{\xi} - \frac{s^2 dx}{mn\xi}$$

Rearranging

$$\frac{s\,d\xi}{\xi} - ds = \frac{s\,dx}{n} - \frac{s^2 dx}{mn\xi}$$

Multiplying both sides by $\frac{\xi}{ss}$, Bernoulli obtained

$$\frac{s\,d\xi - \xi\,ds}{s^2} = \frac{\xi\,dx}{ns} - \frac{dx}{mn}$$

Making the substitution $q = \frac{\xi}{s}$, then

$$dq = \frac{q\,dx}{n} - \frac{dx}{mn}$$

Solving this equation and substituting back the original variables, Bernoulli ended with

$$s = \frac{m}{e^{\frac{x+c}{n}} + 1}\xi$$

where c is a constant of integration. s is a function of a type now known as a **logistic function**, and functions of this form play a key role in modeling biological populations.

If $x = 0$, $s = \xi$, and (by assumption) $m = n = 8$, so

$$s = \frac{8}{7e^{\frac{x}{8}} + 1}\xi$$

Bernoulli then calculated how many additional years of life, on average, inoculation would give a person, arriving at a value of three years.

14.3 Exercises

1. Find a recurrence relation that generates the sequence of cubes.

2. Suppose $A = a^x$, $B = a^{x-1}$, $C = a^{x-2}$, ... Show that if a satisfies

$$a^{N-1} = ma^{N-2} + na^{N-3} + \ldots + q$$

   then a^x, a^{x-1}, a^{x-2}, ... a^{x-N+1} will satisfy the recurrence relationship.

---

[1]These numbers should be sobering: one eighth of the susceptible population catches a disease from which one eighth will die.

3. Suppose P, Q, R, ... are all the roots of the characteristic equation, assumed all different. Prove

$$\beta P^x + \gamma Q^x + \delta R^x + \ldots + \epsilon S^x$$

also satisfies the recurrence relationship.

4. Verify Bernoulli's result for the formula for the general term of the sequence

$$0, 0, 0, 0, 1, 0, 15, -10, 165, -228, \ldots$$

5. Explain why Bernoulli's method of approximating solutions to equations works.

6. Use Bernoulli's method to find approximate solutions to the following equations. Continue until the approximation is accurate to three decimal places.

   (a) $1 = x + x^2 + x^3$ (b) $1 = x + 2x^2 + 3x^3 + 4x^4 + 5x^5$

14.4 Paradoxes and Fallacies

It would be easy to believe that mathematics, as a deductive science, ought to be free of controversies over the results. However, when the results contradict "common sense," questions are raised over whether or not the *foundations* of the mathematical arguments are sound. Nowhere is this more evident than in probability which, during the eighteenth century, began to be applied to all aspects of society.

14.4.1 The St. Petersburg Paradox

One place where probability contradicted "common sense" occurred in the **St. Petersburg paradox**, worked out between Daniel and his brother Nicholas some time before Nicholas died in 1726 from a cold caught after plunging into the freezing Neva River.

The St. Petersburg paradox first appeared in a problem posed by Nicholas Bernoulli to Montmort; Nicholas and Daniel began discussing the problem somewhat later. The problem revolves around the following game between two players: player A tosses a coin until it lands heads. If it lands heads on the first toss, B pays him A one shilling; if it lands heads on the second toss, B pays A two shillings; if it lands heads on the third toss, B pays A four shillings, and so on; in general, if it does not land heads until the nth throw, B pays A 2^{n-1} shillings. How much should A pay B to play the game? Common sense says the amount should be finite, but the expectation of the game is infinite; hence the paradox. Various explanations were advanced to explain why the expectation was not, in fact, infinite.

14.4.2   D'Alembert

Jean le Rond d'Alembert (November 17, 1717-October 29, 1783) was the illegitimate son of Chevalier (Knight) Destouches, an artillery general, and Madame de Tencin, a wealthy socialite who abandoned d'Alembert when he was an infant, near the church of Jean le Rond (hence d'Alembert's name). He was soon placed in the care of a glazier and his wife, named Rousseau (no relation to the philosopher). Destouches paid for the boy's education. According to one story, when it became clear he would become famous, Madame de Tencin tried to claim him as her son; d'Alembert replied, "You are only my step-mother; the glazier's wife is my mother." D'Alembert never married, but when his longtime love Julie de Lespinasse fell ill in 1765, he moved in to take care of her and stayed until her death in 1776.

Jean le Rond d'Alembert. Used with permission of CORBIS.

D'Alembert was part of a circle of thinkers known as the *philosophes*; other *philosophes* included Denis Diderot and Voltaire. Diderot had embarked on an ambitious project to make a multivolume compendium of human knowledge. It would become the first modern encyclopedia. Unlike modern encyclopedias, Diderot made no attempt to be impartial, and many of the articles included criticisms of the existing order. Most of the time, these critiques escaped the censors because they were buried in longer articles on less interesting subjects; occasionally, they were caught and the encyclopedia was banned, time and again.

D'Alembert worked with Diderot between 1751 and 1772, and wrote many of the mathematical articles in the encyclopedia. Thus, d'Alembert reached a great audience with his views on probability. Unfortunately, his views on probability were invariably incorrect. In the article on *Croix et Pile* (Heads And Tails), which appeared in 1754, d'Alembert considered the probability of obtaining at least one head on two flips of a coin. D'Alembert argued:

> The solution that all authors give, following the ordinary theory, is this: there are four combinations:

First Throw	Second Throw
croix	croix
pile	croix
croix	pile
pile	pile

> Of these four combinations, only one is a loss, and three are a win; there are thus 3 to 1 odds in favor of winning...

> But is this correct? Because if the first throw is *croix*, the game is finished and the second throw is irrelevant. Thus there are properly only three possible combinations:

croix, first throw

pile, croix, first and second throws

pile, pile, first and second throws

whence the odds are only 2 to 1.[2]

Likewise, the standard theory gave the odds of throwing a head on three throws as 7 to 1, while d'Alembert claimed it should in fact only be 3 to 1.

By 1761, d'Alembert had changed his mind, only to make additional mistakes. Noting that the three events might not be equally probable, d'Alembert suggested that the event HT was more likely than the event HH, since this second event required heads to fall twice in a row. However, he continued to argue, the probability was *closer* to $\frac{2}{3}$ than to $\frac{3}{4}$.

See page 464.

D'Alembert also saw an early version Daniel Bernoulli's paper supporting inoculation. Even before the article was published, d'Alembert published a critique of it. Though he concurred with the result, he felt Bernoulli's methods were flawed. D'Alembert's criticisms, though valid, are of little help in improving the analysis: for example, he argued that the latter years of one's life should be considered as having lesser value, due to the infirmities of age, than the earlier ones. He also pointed to a critical problem in dealing with probabilities that affect people: the fact that inoculation, on the whole, lengthens the average life is of little comfort to the parent whose child dies from an inoculation.

14.4 Exercises

1. Show the expected value in the St. Petersburg game is infinite.

2. What is the flaw when d'Alembert claims the probability of obtaining one head on two flips is 2 to 3, and not 3 to 4?

14.5 Simpson

Thomas Simpson (August 20, 1710-May 14, 1761) was a self-taught mathematician who, like Newton, came to mathematics through his interest in the nonsense of astrology. At 19, Simpson married a 50-year-old widow with two children, and when he was 25, the family moved to London where Simpson supported the family by weaving during the day and teaching mathematics at night. His contemporaries accused him of partying with "low company," drinking gin and cheap wine, though Simpson's successor at Woolwich Academy, Charles Hutton, defended Simpson's behavior on the grounds that his family background kept him from consorting with gentlemen in class conscious England.

In Simpson's time there were those who argued that it was better to take a single, accurate observation than to take many observations and average them. Simpson attempted to show the method of averaging multiple observations gave

---

[2]D'Alembert, Jean le Rond, "Croix ou Pile." *Encyclopedia*, ed. D. Diderot. Briasson, David, le Breton, and Durand, 1754. Vol. IV, p. 513.

more accurate results. On April 10, 1755, Simpson presented to the Royal Society, "A Letter to the Right Honorable George Earl of Macclesfield, President of the Royal Society, on the Advantage of Taking the Mean of a Number of Observations in Practical Astronomy."

Simpson made a critical advance. Rather than focusing on the mean of the observations, he focused instead on how the observations differed from the "true" value: in other words, Simpson considered the **deviations from the mean**. Overall, his analysis was limited—it had to be, given the extremely complex nature of the problem—but was a portent of things to come.

To analyze the problem, Simpson began with the sequence

$$r^{-v}, \ldots, r^{-3}, r^{-2}, r^{-1}, r^0, r^1, r^2, r^3, \ldots, r^v$$

Let the exponent represent the size of an error (either positive or negative), and let the coefficients (all 1's in this case) be the number of ways an error could be made. In the first case, Simpson considered that only whole number errors between $-v$ and v were possible, and that all errors were equally likely.

Consider, first, the problem of finding the probability that the mean of n observations differs from the true value by exactly $\frac{m}{n}$ (i.e., the error of the mean is $\frac{m}{n}$). Then the sum of the errors must be m; hence the problem becomes finding the probability that the n observations have a total error of m. To find this, Simpson noted that the coefficient of the r^m term of the expansion of

$$(r^{-v} + \ldots + r^{-1} + r^0 + r^1 + \ldots + r^v)^n$$

was equal to the number of chances of making a total error of size m. To find this coefficient, Simpson noted

$$r^{-v} + \ldots r^{-1} + r^0 + r^1 + \ldots + r^v = r^{-v}\left(\frac{1 - r^{2v+1}}{1 - r}\right)$$

Hence

$$(r^{-v} + \ldots r^{-1} + r^0 + r^1 + \ldots + r^v)^n = r^{-nv}(1 - r^w)^n (1 - r)^{-n}$$

where $w = 2v + 1$; Simpson then expanded $(1 - r^w)^n$ and $(1 - r)^{-n}$ using the binomial theorem. Through tedious but not difficult algebra, Simpson determined the coefficient of the mth term of the multinomial. Then by summing the coefficients of the r^{-m} to the r^m terms, Simpson found the number of chances of making a total error between $-m$ and m, and thus the probability that the mean of n observations was within $\frac{m}{n}$ of the true value.

In the second case, Simpson considered a linear error distribution; hence, he considered the multinomial whose terms were

$$1r^{-v}, 2r^{1-v}, 3r^{2-v}, \ldots, 3r^{v-2}, 2r^{v-1}, 1r^v$$

and thus the coefficient of the r^m term of the expansion of

$$(1r^{-v} + 2r^{1-v} + \ldots + 2r^{v-1} + 1r^v)^n$$

was, again, the number of ways of making a total error of size m in n observations.

Simpson then applied the method to the problem of observational astronomy. Supposing that $\pm 5''$ was the maximum error in any observation, and that the chances of making an error of the indicated size were

Error	-5	-4	-3	$\ldots$	$+3$	$+4$	$+5$
Chances	1	2	3	$\ldots$	3	2	1

Simpson computed that the probability that the mean of 6 observations would result in an error less than 1 second of arc (hence the total error must be 6 or fewer seconds of arc) was $\frac{788814800}{1088391168}$, or odds of about $2\frac{2}{3}$ to 1, while the odds that a single observation made an error less than 1 second of arc were 16 to 20, significantly worse. Thus, the method of taking the mean was more reliable.

14.5 Exercises

1. Suppose the only errors possible are $+2$, $+1$, 0, -1, or -2. Under the assumption that all errors are equally probable:

   (a) What is the probability that a single observation will have an error of no more than ± 1?

   (b) What is the probability that two observations will have an error of no more than ± 1?

   (c) Redo the preceding questions, under the assumption that the error distribution is linear.

2. Consider the following "argument" that it is better to have a single observation than a number of them. Suppose the error distribution is linear, and the possible errors are -1, 0, and 1.

   (a) Find the probability of making no error in one observation.

   (b) Suppose five observations are made. What is the probability that the mean has no error?

   (c) It would seem that making more observations increases the chance of an error. Explain why, despite this fact, it is still better to take the mean of several observations, than to make a single observation.

14.6   Bayes's Theorem

Thomas Bayes (1702?-April 1761) was one of the more enigmatic eighteenth century mathematicians. About all that is known about his life is that he attended Edinburgh University, was a Presbyterian minister, and died in the spring of 1761. On December 23, 1763, Bayes's "An Essay Towards Solving a Problem in the Doctrine of Chances" was read before the Royal Society. Bayes had examined and solved the key problem in **inverse probability**:

Problem 14.5. *Given the number of times an event has occurred and failed to occur. To find the probability that the probability of the event is in a given interval.*

Bayes also enunciated (but did not prove) what is now known as **Bayes's theorem**.

Bayes's work is an interesting blend of the modern methods of probability and the older methods of the seventeenth century. For example, Bayes defined two events as inconsistent if they could not happen simultaneously, though we now use the term **mutually exclusive**. Contrary events were events where one or the other (but never both) had to occur; we now use the term **complementary**. With Bayes, we can see probability working toward a consideration of subsets of the sample space.

However, his definition of the probability of an event is rooted in the past. For Bayes, the probability is the ratio between the value of the expectation, and the value of the event if it should occur. Suppose a game is to be played; if the player wins, he will win 1200. Moreover, suppose the expectation of a game is 1000. Hence, the probability of winning the game will be the ratio 1000 to 1200, or $\frac{5}{6}$. It is helpful to think of the value of the expectation as the amount that should be wagered to play a game with a specified stakes: thus, to play a game in which one has a 5/6 chance of winning 1200, one should have to wager 1000.

Using the value of the expectation, Bayes proved that the probability of a set of inconsistent events was the sum of the probabilities: if the probabilities were $\frac{a}{N}$, $\frac{b}{N}$, $\frac{c}{N}$, then the value of the expectation was $a + b + c$, and thus, by definition, the probability that any of the events occurred was $\frac{a+b+c}{N}$. An immediate corollary is that if the probability of an event is $\frac{P}{N}$, then the probability of its failure is $\frac{N-P}{N}$. Bayes's second proposition is rather complexly stated:

Proposition 14.5. *If a person has an expectation depending on the happening of an event, the probability of the event is to the probability of its failure as his loss if it fails to his gain if it happens.*[3]

Bayes was imagining a situation where a game was to be played, and the player had to put in an amount of money (the value of the expectation) in order to play the game. Then, if he won, he would receive a certain amount, and his gain would be the purse *minus* what he had to put in to play; otherwise, he would simply lose his deposit.

Proof. Suppose one will receive N if an event occurs, and the probability of the event is $\frac{P}{N}$. Hence P is the value of the expectation, and if the event fails, an expectation of P is lost. If the event occurs, you gain N but lose the expectation P; hence the gain is $N - P$. But the probability of the event is $\frac{P}{N}$, so the probability of its failing is $\frac{N-P}{N}$, and $\frac{P}{N} : \frac{N-P}{N}$ as $P : N - P$, which was to be proven. □

---

[3] Bayes, Thomas, "An Essay towards solving a Problem in the Doctrine of Chances." *Philosophical Transactions of the Royal Society*, LIII (1763), page 377.

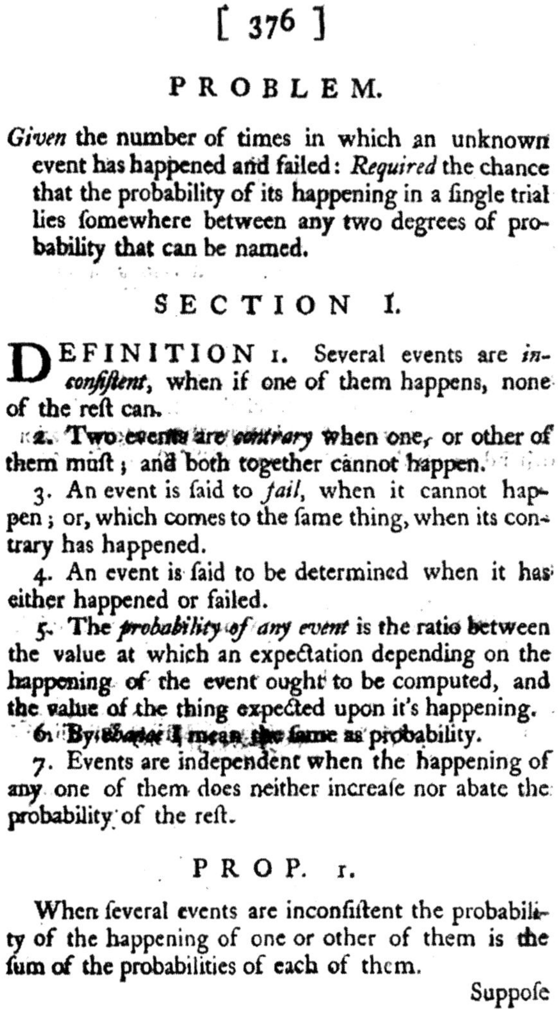

[376]

PROBLEM.

Given the number of times in which an unknown event has happened and failed: *Required* the chance that the probability of its happening in a single trial lies fomewhere between any two degrees of probability that can be named.

SECTION I.

DEFINITION 1. Several events are *inconfiftent*, when if one of them happens, none of the reft can.

2. Two events are *contrary* when one, or other of them muft ; and both together cannot happen.

3. An event is faid to *fail*, when it cannot happen ; or, which comes to the fame thing, when its contrary has happened.

4. An event is faid to be determined when it has either happened or failed.

5. The *probability of any event* is the ratio between the value at which an expectation depending on the happening of the event ought to be computed, and the value of the thing expected upon it's happening.

6. By *chance* I mean the fame as probability.

7. Events are independent when the happening of any one of them does neither increafe nor abate the probability of the reft.

PROP. 1.

When feveral events are inconfiftent the probability of the happening of one or other of them is the fum of the probabilities of each of them.

Suppofe

Figure 14.3: First Page of Bayes's Article in *Philosophical Transactions.* Courtesy of the Internet Library of Early Journals.

The first computations of conditional probabilities appeared in his third proposition:

Proposition 14.6. *The probability that two subsequent events both happen is the product of the probability of the first and the probability of the second on the supposition the first happens.*

This is equivalent to $P(A \cap B) = P(A)P(B|A)$.

Proof. Suppose that if both events occur, one wins N. Let the probability that both occur be $\frac{P}{N}$ and the probability that the first occurs is $\frac{a}{N}$; hence the probability that the first event does not occur is $\frac{N-a}{N}$. Let the probability that the second event occurs if the first one has already occurred be $\frac{b}{N}$.

Since the probability of both events occurring is $\frac{P}{N}$, then by definition P will be the value of the expectation. If the first event happens, then the expectation becomes b and there is a gain of $b - P$; otherwise, P is lost. Thus, by Proposition 14.5, $\frac{a}{N} : \frac{N-a}{N} = P : b - P$. Hence $\frac{a}{N} = \frac{P}{b}$. But $\frac{P}{N} = \frac{P}{b} \cdot \frac{b}{N}$, and thus the probability of both events occurring is the probability that the first event occurs, multiplied by the probability that the second event occurs given that the first event has already occurred. □

Bayes then considered a rather complexly worded case:

Proposition 14.7. *If there be two subsequent events to be determined every day, and each day the probability of the second is $\frac{b}{N}$ and the probability of both $\frac{P}{N}$, and I am to receive N if both events happen the first day on which the second does; I say, according to these conditions, the probability of my obtaining N is $\frac{P}{b}$.* [4]

Consider two events, A and B, of which A, if it occurs, occurs before B. Suppose on a day that B is known to have occurred, you will win N if A has also occurred. What Bayes was trying to find is the probability of winning N. Since the problem is really that of trying to find the probability of A occurring given that the time when B can occur is always *after* the time when A can occur, the problem is one of the so-called **backwards inference** problems. Bayes's proof is:

Proof. Suppose the probability of obtaining N is not $\frac{P}{b}$, but rather $\frac{x}{N}$. By definition, x is the value of the expectation. Take y so that $y : x = N - b : N$. Suppose the second event has occurred. Then, since the probability of both events is $\frac{P}{N}$, the value of the expectation of gaining N is P.

On the other hand, suppose the second event does not occur. Then you will receive x, the value of the expectation (you neither win nor lose, and your wager is returned). The probability of the second event not occurring is $\frac{N-b}{N}$, which is equal to $\frac{y}{x}$. But since we stand to gain x if the second event does not occur, the value of the expectation is y.

Hence, the value of the expectation is $P + y$. But this must be x; hence $P + y = x$. However, since $y : x = N - b : N$, then $x : P = N : b$ and thus $\frac{x}{N} = \frac{P}{b}$. Thus, the probability of obtaining N is $\frac{P}{b}$. □

Bayes's theorem, in its first form, appeared as his fifth proposition; Bayes gave no proof:

Conjecture 14.1 (Bayes's Theorem). *Given two events, the probability of the second being $\frac{b}{N}$ and the probability of both together $\frac{P}{N}$, and given that the second*

---

[4]Ibid., 379

has occurred, the probability that the statement, "The first event has occurred" is correct is $\frac{P}{b}$.

This probability is, of course, the same as the probability that the event has occurred, though it is interesting to see that Bayes had phrased it as the probability of the correctness of an assertion.

In his next proposition, Bayes used the term **independent events** in the modern sense as events whose probabilities were not altered by the occurrence or nonoccurrence of the others.

Bayes's last proposition of the first section is that if the probability that an event fails is a and the probability it occurs b, then the probability the event has occurred p times and failed to occur q times is Ea^pb^q, where E is the binomial coefficient of the a^pb^q term in the expansion of $(a + b)^{p+q}$.

Bayes had not yet answered the question of determining the probability that the probability fell within certain limits. To do so, he began by considering a table, onto which a ball was dropped; he supposed that the ball was equally likely to fall on any portion of the table. First, suppose the ball is thrown onto the table, landing at some point; draw oS parallel to AD through the landing point (see figure in margin). Then, let the ball be thrown $p + q$ more times, and let the ball's resting between oS and AD be the happening of the event M. Bayes first proved

Lemma 14.4. *The probability o lies between any two points on AB is the ratio of the distance between the points to the whole of AB.*

Proof. Let any two points, f, b be taken; draw fF, bL parallel to AD. First, suppose rectangles Cf, Fb, LA are commensurable; if so, they can be divided into a number of parts of equal size, and thus the probability that the ball falls into any one of the equal parts is the sum of the probabilities that it lands in any one of them; hence the probabilities that the ball falls into Cf, Fb, LA are to each other as rectangles; however, the rectangles are to each other as the ratios of the sides Bf, fb, and ba. $\square$

Bayes went on to prove the lemma for the case where the rectangles were incommensurable. Hence, the probability of the event M occurring on any one throw of the ball is the ratio $Ao : AB$.

On the line AB, Bayes then drew a curve $BghikmA$, with the property that if AB was divided at b and bm perpendicular to AB drawn, and

$$bm : AB = y$$
$$Ab : AB = x$$
$$Bb : AB = r$$

and E the a^pb^q coefficient of the expansion of $(a + b)^{p+q}$, then the the ordinate bm is proportional to Ex^pr^q.

Given that M has happened p and failed q times, then

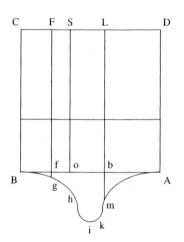

See problem 1.

Proposition 14.8. *Given f, b any two points on AB, and M has occurred p times and failed q times in $p + q$ trials. The probability o is between f and b is the ratio of $fghikmb$ to CA, the square upon AB.*

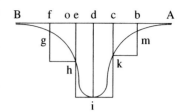

Since the location of o determines the probability of M, what Bayes is actually determining is that the probability of the event M falls between $\frac{Af}{AB}$ and $\frac{Ab}{AB}$: in other words, the probability that the probability falls between two specified values.

Proof. Suppose not. First, let it be the ratio of D, a figure greater than $fghikmb$ to CA; in other words, let the probability be $\frac{D}{CA}$, where D is greater than $fghikmb$. Through e, d, c, draw perpendiculars to AB, and place d so di is the longest perpendicular, and so that the rectangles bk, ci, ei, fh together differ from $fghikmb$ less than D. (In other words, let the rectangles together have an area less than D and greater than $fghikmb$) Since di is the longest of the perpendiculars, then $fg < eh < di$; likewise $di > ck > bm$. If $Ao = Ae$, then the probability of $M = \frac{Ae}{AB} = x$, and the probability that M does nto occur is $\frac{Be}{AB} = r$, and the probability that M occurs p times and fails q times is $Ex^p r^q$, where E is the coefficient of the $x^p r^q$ term in the expansion of $(x + r)^{p+q}$. If $y = eh : AB$, then $y = Ex^p r^q$. Thus, if $Ao = Ae$, the probability M occurs p times and fails q times is $\frac{eh}{AB}$.

Likewise, if $Ao = Af$, then the probability M occurs p times and fails q times is $\frac{fg}{AB}$. But eh is the greatest of the perpendiculars on ef, by the construction of the curve. So if o falls anywhere between f, e, the probability that M occurred p times and failed q times must not exceed $\frac{eh}{AB}$.

There are now two subsequent events: first, that o falls between e and f; and second, that M occurred p times and failed q times. The probability of the first is $\frac{ef}{AB}$. The probability of the second, given that the first occurred, does not exceed $\frac{eh}{AB}$. Hence the probability of both does not exceed the compounded probability, $\frac{ef}{AB} \cdot \frac{eh}{AB} = \frac{fh}{AC}$.

Likewise, we may determine the probability that o falls between e and d, and that M occurred p times and failed q times does not exceed $\frac{ei}{CA}$; the probability that o falls between d and c and that M occurred p times and failed q times does not exceed $\frac{ci}{CA}$, and the probability o fell between c and b does not exceed $\frac{bk}{CA}$.

Hence, the probability that o falls between f and b, and the event M happens p and fails q times, is $\frac{fh+ei+ci+bk}{CA}$. But $fh + ei + ci + bk < D$, so the probability is less than $\frac{D}{CA}$, when it was supposed to be equal. Hence, the probability cannot be as anything greater than $fghikmb$ to CA Likewise, we can show the probability cannot be as anything less than $fghikmb$ to CA, hence it must be equal. $\square$

The ninth proposition, like the fifth, is a statement about the probability of an assertion being correct, namely:

Proposition 14.9. *If M has occurred p times and failed q times, and we guess that o falls between f and b, and thus the probability of M is between $\frac{Ab}{AB}$ and $\frac{Af}{AB}$, the probability we are correct is the ratio of $fghimb$ to AiB.*

Proof. There are two subsequent events: the first is that o is between f and b, and the second is that M has occurred p and failed q times. The probability of the second is the $\frac{AiB}{CA}$, while the probability of both occurring is $\frac{fghimb}{CA}$. Since the second event is known to have occurred, the probability we are correct in saying the first has also occurred is (by Conjecture 14.1) $\frac{fghimb}{AiB}$. $\square$

This solved the inverse probability problem, at least from a theoretical point of view. Bayes noted that the practical application was rather more difficult, since for that, it was necessary to evaluate the area under the section $fghimb$ of curve AiB.

14.6 Exercises

1. Prove Lemma 14.4 by showing that if Cf, Fb, LA are not commensurable, the result still holds. Use the method of exhaustion.

2. Complete the proof of Proposition 14.8 by showing that the probability cannot be as some area less than $fghikmb$ to CA.

Chapter 15

Analysis

Between 1697 and 1698, an enormous man—six and a half feet tall—calling himself "Sergeant Peter Mikhailov" traveled with a large Russian delegation through Europe. For four months he worked in an Amsterdam shipyard, studying carpentry and western methods of naval construction. When the delegation went to England, he worked at the Royal Naval yard in Deptford, continuing his study. In his free time, he visited anyplace that showcased western European culture: schools, museums, and even Parliament.

"Sergeant Mikhailov" was actually Peter I, Tsar of Russia, absolute ruler of the largest—and most backward—European country. Peter intended to drag Russia into the eighteenth century, by force if necessary. For example, since the nobility of "advanced" western European countries were beardless, Peter ordained that Russian nobles must shave. Peter himself walked the streets, shears in hand, looking for nobles who disobeyed his order; those he caught were subject to an impromptu (and none-too-careful) shave.

On territory conquered from Sweden during the Great Northern War (1700-1721), Peter built a new capital: Saint Petersburg. Peter did not long survive his creation of a modern European state. In the autumn of 1724, he dove into the Gulf of Finland to rescue some soldiers whose ship had crashed. The dive into freezing water left him with a chill from from which he never recovered, and Peter the Great died on February 8, 1725.

15.1 Leonhard Euler

Among the giants of the eighteenth century, one stands supreme: the Swiss mathematician Leonhard Euler[1] (April 15, 1707-September 18, 1783), possibly the most prolific mathematician of all time. During his lifetime, Euler published an average of 800 pages a year, and even death was only a minor setback, for

---

[1]Rhymes with "boiler" or, more appropriately, "toiler."

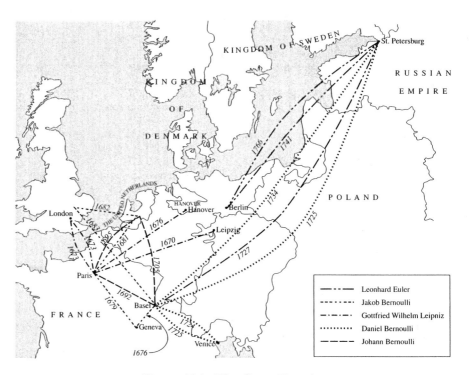

Figure 15.1: The Great Travels.

Leonhard Euler.   ©Bettmann/
CORBIS.

Euler papers were still appearing as late as 1830. A complete bibliography of Euler works runs to 886 items, including some very lengthy books; first drafts of several others remain in manuscript form. Like Euclid, Euler had such an influence on the subsequent development of mathematics that we will examine his work in some detail.

In addition to his mathematical work, Euler translated into German an English work on the principles of gunnery; wrote a popular work, *Letters to a German Princess on Various Scientific and Philosophical Subjects*; wrote *Naval Science: Complete Theory on the Construction and Maneuver of Vessels* for naval cadets; edited a major scientific journal; and communicated extensively with mathematicians and scientists throughout Europe: over 3000 Euler letters exist. Altogether, Euler's output might fill seventy or eighty books the size of this one.

We might envision Euler as a man totally dedicated to his mathematics—and, as we shall see, we would be wrong. In 1720, Euler entered the University of Basel. The university was small, having only about a hundred students and 19 professors—but one of these was Johann Bernoulli. Bernoulli proved too busy to give Euler private lessons, but he did recommend that Euler read what advanced works on mathematics he could find, and they would meet every Saturday afternoon to work through the difficult parts.

At the age of 15, Euler served as opponent for disputants in the history of law and in logic, a position equivalent to serving on a thesis committee for Master's degree students; in 1723, he joined the department of theology at Basel. Euler was also deeply interested in the classics and memorized Vergil's *Aeneid* so well that he could still recall it, word for word, at the age of 70. In the meantime, he wrote a dissertation on the virtues of temperance.

In 1726, a mathematics chair opened up at Basel, and Johann Bernoulli enthusiastically supported Euler. His application for the position included a dissertation on the physics of sound, but Euler was rejected, probably because of his age: he was just 19. Fortunately, a position had opened up in far-off Russia.

Christian Goldbach had already recruited Nicholas and Daniel Bernoulli for the Russian Academy of Science, at the newly founded city of St. Petersburg. They successfully urged the academy to extend an invitation to their friend Euler; he accepted, moving to St. Petersburg in 1726. Unfortunately, Tsar Peter the Great had died the previous year, and from 1725 to 1740, Russia was ruled by a succession of weak or incompetent rulers. Still, Euler made himself at home, marrying Katherine Gsell, the daughter of a Swiss teacher of painting at the high school associated with the academy, in 1733. They would have thirteen children, five of whom (three sons and two daughters) survived into adulthood.

15.1.1 Convergence of Series

During this first St. Petersburg period, Euler made significant contributions to the theory of infinite series. Much of Euler's work on series shows clear traces of the formalism of Johann Bernoulli, his teacher. Euler was not alone in treating series formally, freely rearranging terms and dealing with divergent series: almost every eighteenth century mathematician did so. In fact, Euler was one of the few who actually considered the question of convergence. One of his earliest works, "Observations on the Harmonic Progression" (1734), published in the journal for the St. Petersburg Academy, showed that Euler had a clear conception of convergence in the modern sense.

Euler began by noting that a harmonic sequence was a sequence of fractions with the same numerators, and whose denominators were in an arithmetic progression. Considering the sequence of fractions

$$\frac{c}{a} \quad \frac{c}{a+b} \quad \frac{c}{a+2b} \quad \cdots$$

Euler took i to be an infinitely large number, and considered the $(n-1)i$ terms in the harmonic sequence from $\frac{c}{a+ib}$ to $\frac{c}{a+(ni-1)b}$. The sum of these terms was less than $\frac{(n-1)ic}{a+ib}$, but greater than $\frac{(n-1)ic}{a+(ni-1)b}$. Since i is an infinitely large number, a vanishes in the denominator, and thus the sum of the $(n-1)i$ terms is greater than $\frac{(n-1)c}{b}$. Hence, Euler concluded, the sum could be made as large as desired, and thus the infinite series had a sum that was infinitely great; since the argument did not depend on the form of the arithmetic sequence, this proved

Proposition 15.1. *The series of reciprocals of any arithmetic progression has an infinite sum.*

Next, Euler considered a sequence whose terms were of the form

$$\frac{c}{a+b} \qquad \frac{c}{a+2^{\alpha}b} \qquad \frac{c}{a+3^{\alpha}b} \qquad \cdots$$

By a similar argument, Euler showed that if $\alpha > 1$, the *finite* sum from the ith to the $(n-1)$ith was zero (again, with i being an infinitely large number); hence the series would converge. Euler had formulated what would later become known as the **Cauchy convergence criterion**. Except for Euler's use of actually infinite numbers, which a modern mathematician would replace with limits that tend to infinity, Euler's treatment of series in these propositions very closely approaches our our modern one.

In another early paper, "Universal Method of Finding the Approximate Sum of a Convergent Series" (1736), Euler introduced a variation of the **integral test** for the convergence of series. The sum of the series $a + b + c + \ldots + x$ could be thought of as the sum of a series of rectangles of width 1 and heights a, b, c, $\ldots$; hence if y and Y were two functions that were, respectively, below and above the corresponding rectangles, then $\int y \, dn$ and $\int Y \, dn$ would bound the sum of the series; this allowed Euler to find a method of approximating the sum of the series using the integral.

See Problem 3.

15.1.2   Other Results on Series

More typical of the eighteenth century use of series was Euler's "Various Observations on Infinite Series," written around 1737 but not published until 1744. Christian Goldbach had communicated to Euler the discovery that

$$1 = \frac{1}{3} + \frac{1}{7} + \frac{1}{8} + \frac{1}{15} + \frac{1}{24} + \frac{1}{26} + \frac{1}{31} + \frac{1}{35} + \ldots \tag{15.1}$$

where the denominators were of the form $m^n - 1$, with m, n whole numbers greater than 1, but did not include a proof. Euler demonstrated this result by beginning with the equation

$$x = 1 + \frac{1}{2} + \frac{1}{3} + \frac{1}{4} + \frac{1}{5} + \ldots$$

Since

$$1 = \frac{1}{2} + \frac{1}{4} + \frac{1}{8} + \frac{1}{16} + \ldots$$

then

$$x - 1 = 1 + \frac{1}{3} + \frac{1}{5} + \frac{1}{6} + \frac{1}{7} + \frac{1}{9} + \ldots$$

The largest fraction remaining on the right hand side is $\frac{1}{3}$, and the geometric series beginning with $\frac{1}{3}$ is

$$\frac{1}{2} = \frac{1}{3} + \frac{1}{9} + \frac{1}{27} + \ldots$$

Subtracting from both sides, Euler obtained

$$x - 1 - \frac{1}{2} = 1 + \frac{1}{5} + \frac{1}{6} + \frac{1}{7} + \ldots$$

Continuing this process using

$$\frac{1}{4} = \frac{1}{5} + \frac{1}{25} + \frac{1}{125} + \ldots$$
$$\frac{1}{5} = \frac{1}{6} + \frac{1}{36} + \frac{1}{216} + \ldots$$
$$\vdots$$

Euler "exhausted" the left hand side, and ended with

$$x - 1 - \frac{1}{2} - \frac{1}{4} - \frac{1}{5} - \frac{1}{6} - \ldots = 1$$

Replacing x with $1 + \frac{1}{2} + \frac{1}{3} + \frac{1}{4} + \ldots$ yielded the result. Although Euler's proof is not today considered logically valid, the result is provable.

Theorem 8 of "Various Observations" marks an early appearance of what would become known as the **Riemann ζ-function** in the so-called product form:

See Problem 13.

$$1 + \frac{1}{2^n} + \frac{1}{3^n} + \frac{1}{4^n} + \ldots = \left(\frac{2^n}{2^n - 1} \right) \left(\frac{3^n}{3^n - 1} \right) \left(\frac{5^n}{5^n - 1} \right) \cdots \qquad (15.2)$$

To prove this, Euler once again assumed the sum equal to x:

$$x = 1 + \frac{1}{2^n} + \frac{1}{3^n} + \frac{1}{4^n} + \ldots$$

Multiplying both sides by $\frac{1}{2^n}$ and subtracting gave:

$$\frac{2^n - 1}{2^n} x = 1 + \frac{1}{3^n} + \frac{1}{5^n} + \frac{1}{7^n} + \ldots$$

Repeatedly applying this reduction gave the result.

One of Euler's first great results was in solving a problem posed by Henry Oldenburg, Secretary of the Royal Society, to Leibniz: to find the sum of the series of reciprocal squares. On May 24, 1673, Leibniz wrote back to Oldenburg to report that he had made no progress, but did note that the sum of the terms

See Problem 4.

$$\frac{1}{4} + \frac{1}{9} + \frac{1}{16} + \frac{1}{25} + \ldots$$
$$\frac{1}{8} + \frac{1}{27} + \frac{1}{64} + \frac{1}{125} + \ldots$$
$$\frac{1}{16} + \frac{1}{81} + \frac{1}{256} + \frac{1}{625} + \ldots$$
$$\vdots$$

is equal to 1.

Leibniz communicated the problem to Jakob and Johann Bernoulli. In his November 6, 1696 letter to Johann Bernoulli, Leibniz suggested what he hoped might be a promising track: the sums of the reciprocal squares is the value of

$$y = \frac{x}{1} + \frac{x^2}{4} + \frac{x^3}{9} + \dots$$

when $x = 1$. Differentiating, Leibniz obtained

$$\frac{dy}{dx} = \frac{x}{1} + \frac{x^2}{2} + \frac{x^3}{3} + \dots$$

and thus

$$\frac{d^2 y}{dx^2} = 1 + x + x^2 + x^3 + \dots$$

See Problem 8.

the right hand side of which can be represented as $\frac{1}{1-x}$. Hence by finding $\int (\int \frac{1}{1-x} dx) dx$ when $x = 1$, one could, presumably, find the sum of the series. Unfortunately, Leibniz made several errors in his derivation. Even if he did not, the procedure would not work. A few days later, on November 9, 1696, Leibniz realized his error, and sent off another letter to Johann Bernoulli. Beginning with

$$y = \frac{x}{1} + \frac{x^2}{4} + \frac{x^3}{9} + \dots$$

This time, correct differentiation gives

$$\frac{dy}{dx} = 1 + \frac{x}{2} + \frac{x^2}{3} + \frac{x^3}{4} + \dots$$

Multiplying by x and using the series expansion for $\ln(1 - x)$ gave

$$x \frac{dy}{dx} = \ln(1 - x)$$

where we note that Leibniz omitted a factor of -1. Rearranging, Leibniz obtained

$$y = \int \frac{\ln(1 - x)}{x} \, dx$$

But once again, Leibniz was unable to evaluate the integral.

After having failed with the series expansion communicated to Johann Bernoulli, Leibniz tried a different tactic, which he sent to Jakob Bernoulli on March 15, 1697. The alternating series

$$\frac{1}{1} - \frac{1}{4} + \frac{1}{9} - \frac{1}{16} + \dots$$

was value of the function

$$y = \frac{x}{1} - \frac{x^2}{4} + \frac{x^3}{9} - \frac{x^4}{16} + \ldots$$

when the $x = 1$. Differentiating and multiplying by x, Leibniz obtained

$$x\frac{dy}{dx} = (x - \frac{x^2}{2} + \frac{x^3}{3} - \frac{x^4}{4} + \ldots)$$

the right hand side of which Leibniz recognized as the series expansion for $\log(1 + x)$. This led to the integral equation

$$y = \int \frac{\log(1 + x)}{x} dx$$

which Leibniz was unable to solve. In his reply, dated November 15, 1702, Jakob Bernoulli found the formula for $\int x^c \log x \, dx$, a formula that worked for all values of c—*except* for $c = -1$. After five years, Bernoulli's frustration was evident: "It eludes us yet!" he wrote in his reply.

Neither Leibniz nor the Bernoullis were able to sum the series of reciprocal squares, but sometime before 1736, Euler found a solution, which appeared in "On the Sums of Many Series of Reciprocals" (1740). Johann Bernoulli was quite impressed with the result, noting that his brother Jakob had failed to find the sum of the series, and bemoaning the fact that Jakob was not alive to see Euler's successful summation. (There was probably an element of sibling rivalry: Euler was, after all, one of Johann's students, and not one of Jakob's.)

To find the sum, Euler began with a series for $\frac{1}{\sin x}$.

Conjecture 15.1. *If $y = \sin x$, then*

$$\frac{1}{y} = \frac{1}{A} + \frac{1}{B} + \frac{1}{C} + \ldots$$

where A, B, C, ... are arcs whose sine is y.

Demonstration. From

$$y = \sin x$$

$$= x - \frac{x^3}{3 \cdot 2 \cdot 1} + \frac{x^5}{5 \cdot 4 \cdot 3 \cdot 2 \cdot 1} - \frac{x^7}{7 \cdot 6 \cdot 5 \cdot 4 \cdot 3 \cdot 2 \cdot 1} + \ldots$$

we obtain

$$1 - \frac{x}{y} + \frac{x^3}{3 \cdot 2 \cdot 1 \cdot y} - \frac{x^5}{5 \cdot 4 \cdot 3 \cdot 2 \cdot 1 \cdot y} + \ldots = 0 \qquad (15.3)$$

by dividing by y and moving all terms to the left hand side. We note that if $\sin A = y$, then $x = A$ is a root. Hence

$$\left(1 - \frac{x}{A}\right)\left(1 - \frac{x}{B}\right)\left(1 - \frac{x}{C}\right)\ldots$$

$$= 1 - \frac{x}{y} + \frac{x^3}{3 \cdot 2 \cdot 1 \cdot y} - \frac{x^5}{5 \cdot 4 \cdot 3 \cdot 2 \cdot 1 \cdot y} + \ldots \qquad (15.4)$$

where B, C, and so on, are other values the sine of which is equal to y. Expanding the infinite product and comparing the coefficient of x, we have

$$\frac{1}{y} = \frac{1}{A} + \frac{1}{B} + \frac{1}{C} + \dots$$

$\square$

Next, Euler designated half the perimeter of the circle of radius 1 as p, and A as the least arc whose sine is y. Then the other arcs whose sine is y are A, $p - A$, $2p + A$, $3p - A$, $-p - A$, $-2p + A$, and so on; hence the terms $\frac{1}{A}$, $\frac{1}{B}$, $\frac{1}{C}$, and so on, become $\frac{1}{A}$, $\frac{1}{p-A}$, $\frac{1}{-p-A}$, $\frac{1}{2p+A}$, and so on.

In the specific case where $y = \sin x = 1$, then the least arc whose sine is 1 is $\frac{1}{2}p$. Letting $q = \frac{1}{2}p$, so $p = 2q$, Euler obtained for the terms $\frac{1}{A}$, $\frac{1}{p-A}$, $\frac{1}{-p-A}$, $\frac{1}{2p+A}$ the terms $\frac{1}{q}$, $\frac{1}{q}$, $-\frac{1}{3q}$, $-\frac{1}{3q}$, and so on. To find the sum, Euler needed:

Proposition 15.2. *Consider the series* $a + b + c + d + \dots$ *Let* α *be the sum of this series;* β *be the sum of the products of the terms taken two at a time (i.e.,* $\beta = ab + ac + ad + \dots + bc + bd + \dots$*);* γ *be the sum of the products of the terms taken three at a time, and so on. Let*

$$P = a + b + c + \dots$$
$$Q = a^2 + b^2 + c^2 + \dots$$
$$R = a^3 + b^3 + c^3 + \dots$$
$$\vdots$$

Then

$$P = \alpha$$
$$Q = P\alpha - 2\beta$$
$$R = Q\alpha - P\beta + 3\gamma$$
$$\vdots$$

Euler let $\alpha = \frac{1}{y}$; hence $a = \frac{1}{q}$, $b = \frac{1}{q}$, $c = -\frac{1}{3q}$, $d = -\frac{1}{3q}$, and so on. Since $\beta = 0$ (as there is no x^2 term in the expansion of Equation 15.4), then $Q = \alpha^2 = \frac{1}{y^2}$. But Q is also, by definition, the sum $a^2 + b^2 + c^2 + \dots$ Hence:

Conjecture 15.2.

$$1 + \frac{1}{4} + \frac{1}{9} + \dots = \frac{p^2}{6}$$

Demonstration. The sum of the series whose terms are the squares of $\frac{1}{q}$, $\frac{1}{q}$, $-\frac{1}{3q}$, $-\frac{1}{3q}$, and so on, we have designated Q; hence

$$Q = \frac{2}{q^2}\left(1 + \frac{1}{9} + \frac{1}{25} + \frac{1}{49} + \dots\right)$$

But by the foregoing, $Q = \frac{1}{y^2} = 1$. Hence

$$1 + \frac{1}{9} + \frac{1}{25} + \frac{1}{49} + \ldots = \frac{q^2}{2} = \frac{p^2}{8}$$

Now

$$1 + \frac{1}{9} + \frac{1}{25} + \frac{1}{49} + \ldots = 1 + \frac{1}{4} + \frac{1}{9} + \ldots - \frac{1}{4}\left(1 + \frac{1}{4} + \frac{1}{9} + \ldots\right)$$
$$= \frac{3}{4}\left(1 + \frac{1}{4} + \frac{1}{9} + \ldots\right)$$

Thus the series of the reciprocals of the odd squares is $\frac{3}{4}$ the series of the reciprocals of the squares, so

$$1 + \frac{1}{4} + \frac{1}{9} + \ldots = \frac{p^2}{6}$$

$\square$

Euler then proceeded to find the sums of the series of reciprocals of even powers up to the twelfth.

15.1 Exercises

1. Show that a sequence of fractions with the same numerator and denominators in an arithmetic progression forms a harmonic sequence.

2. Show that the series with general term $\frac{c}{a+n^\alpha b}$ converges for $\alpha > 1$, using Euler's criteria.

3. Euler's method of approximating the sum of a series begins with a series $a + b + c + \ldots$, and a function y where at the integers, y takes on the values of the successive terms of the series; assume y is concave upward; also, suppose a second function, Y takes on the values of the series b, c, $\ldots$ at the successive integers, and suppose that Y is also concave upward.

    (a) Euler first showed that if x is the nth term of the series, then $\int y \, dn < a + b + c + \ldots + x$. Find the appropriate limits for $\int y \, dn$ and show that Euler's result is correct.

    (b) Euler showed next that $\int Y \, dn > a + b + c + \ldots + x$. Find the appropriate limits for $\int y \, dn$, and show Euler's result correct.

    (c) Suppose w and z are the $n+1$ and $n+2$ terms of the series. Show $a + b = c + \ldots + x > \int y \, dn + \frac{a-w}{2}$, and $a + b + c + \ldots + x > \int y \, dn + \frac{a-b}{12} - \frac{w-z}{12}$.

    (d) Consequently, if the series is summed "to infinity," we have $a + b + c + \ldots \approx \int y \, dn + \frac{7a}{12} - \frac{b}{12}$.

4. Demonstrate Leibniz's result; in other words, show $\sum\limits_{n=2}^{\infty} \sum\limits_{m=2}^{\infty} \frac{1}{n^m} = 1$.

5. Consider Conjecture 15.2.

   (a) Demonstrate it, by assuming that algebraic operations on the finite sum $a + b + c + \ldots + n$ will apply to the infinite sum $a + b + c + \ldots$

   (b) What is the corresponding equation for the sum $a^4 + b^4 + \ldots$?

   (c) What conditions must be satisfied to make the demonstration rigorous?

6. For the series whose terms are the cubes of $\frac{1}{q}, \frac{1}{q}, -\frac{1}{3q}, -\frac{1}{3q}$, and so on; find the sum. Why can this not be used to find the sum of the series of the reciprocals of the cubes?

7. Find the sum of the series of reciprocal fourth powers.

8. Show $\int (\int \frac{1}{1-x} dx) dx$ diverges, and thus the integral cannot be used to find the sum of the series of reciprocal squares.

9. Find $\int x^c \log x \, dx$.

10. Find a series for $\sec x = \frac{1}{\cos x}$, using Euler's method in "On the Sums of Many Series of Reciprocals."

11. Use Euler's method from "On the Sums of Many Series of Reciprocals" to find a product form of $\cos x$ by noting that if $x = 0$, $\cos x = 1$, and the solutions to $\cos x = 0$ are $\pm\frac{\pi}{2}, \pm\frac{3\pi}{2}, \ldots$ Use the product and series forms of $\cos x$ to determine the value of $\frac{4}{\pi^2} + \frac{4}{9\pi^2} + \frac{4}{25\pi^2} + \ldots$ Show that your result is consistent with Euler's sum of the reciprocals of the squares.

12. Show that Euler's procedure in "Various Observations on Infinite Series" will result in terms of the form $\frac{1}{m^n - 1}$.

13. Verify, by examining the partial sums, that Euler's result in Equation 15.1 is probably correct. (A rigorous proof is beyond the scope of this text.)

14. Another result from "Various Observations on Infinite Series" are several new series forms for π.

   (a) Use Euler's method to convert

   $$A = 1 - \frac{1}{3^n} + \frac{1}{5^n} - \frac{1}{7^n} + \ldots$$

   into an infinite product.

   (b) If $n = 1$, then $A = \frac{\pi}{4}$. Use this fact to find a infinite product expression for π.

15. Theorem 2 of "Various Observations on Infinite Series" is

$$\ln 2 = \frac{1}{3} + \frac{1}{7} + \frac{1}{15} + \frac{1}{35} + \dots$$

$$1 - \ln 2 = \frac{1}{8} + \frac{1}{24} + \frac{1}{26} + \frac{1}{48} + \dots$$

where the denominators in the first series are the odd numbers that are one less than any power, and the denominators in the second series are the even numbers that are one less than any power. Demonstrate this conjecture of Euler's.

(a) Begin with

$$x = \frac{1}{2} + \frac{1}{4} + \frac{1}{6} + \frac{1}{8} + \dots$$

Demonstrate

$$x = 1 + \frac{1}{5} + \frac{1}{9} + \frac{1}{11} + \frac{1}{13} + \frac{1}{17} + \frac{1}{19} + \dots$$

(b) Next, show that

$$x = 1 + \frac{1}{3} + \frac{1}{5} + \frac{1}{7} + \frac{1}{9} + \dots - \ln 2$$

(c) Finally, compare the two series.

16. Demonstrate the validity of Equation 15.2. Let A be the left hand side of the equation.

(a) Let $B = \left(1 - \frac{1}{2^n}\right) A$. Show that the denominators of the series B do not contain any factors of 2.

(b) Let $C = \left(1 - \frac{1}{3^n}\right) B$. Show that the denominators of the series C do not contain any factors of 3.

(c) Continue the process, "exhausting" the series until only the first term, 1 is left.

15.2   Calculus Textbooks

In 1740, Anna Leopoldovna became regent for her son, the infant Ivan VI. Euler was prescient enough to realize that the next few years would be even more turbulent than the past few, and had had enough of political uncertainty. He accepted an invitation from Frederick I of Prussia to revitalize the Berlin Society of the Sciences. Euler arrived in Berlin just in time to see the start of a war.

Frederick's father, Frederick William I, turned Prussia into a military state, building a permanent standing army of 83,000 men out of a population of only two and a half million. But Frederick William I loved his army too much to send it into battle and risk its destruction, so the mighty Prussian war machine remained at peace for the entirety of his reign.

Frederick William's relationship to his son Frederick was strained, at best. At one point, Frederick ran away from home with his best friend. They were soon caught, and returned home. To teach his son a lesson, Frederick William had the friend executed and considered disinheriting Frederick. However, he never did so, and when he died in 1740, Frederick took the throne. Within five months, he put the army so beloved by his father to use.

Austria's new ruler was the Empress Maria Theresa, and according to one interpretation of the laws of inheritance, a woman could not rule Austria. Several other monarchs with claims based on ancestry moved to secure the throne for themselves, thus beginning the War of the Austrian Succession. Frederick, who had *no* legitimate claim, took the opportunity to annex a portion of Austria known as Silesia. Maria Theresa, however, proved to be a much more capable leader than anyone suspected. By 1742, Austrian armies had expelled all European armies except the Prussian from Austrian territory. The empress formally recognized the annexation of Silesia, and made peace with Frederick in 1742. Frederick gained Silesia—and an enemy for life.

Under Euler's direction, the Berlin Society was revitalized, and in 1744, became the Royal Academy of Sciences and Fine Letters. The seeds of future conflict were already present. Frederick the Great was a francophile and insisted on making French the official language of the academy, over Euler's objections: most scientific publications were in Latin, a language understood by all scholars of the time.

15.2.1 Euler's Textbooks

In Berlin, Euler wrote two textbooks, from which a generation of mathematicians would learn the tools of calculus. *Introduction to the Analysis of the Infinites* (1748) was intended to serve as a bridge between algebra and calculus; as such, it would prove pivotal to the way we teach and learn both. Euler began with notions of variables and functions, defining a function to be an analytic expression composed "in any way" of the variables and constant quantities. Euler specifically disallowed "constant" functions, such as $f(z) = z^0$.

Almost immediately, Euler began using the expression of a function in terms of an infinite series. Since the text is a "precalculus" book, the series expansions must be found without obviously using such calculus tools as Taylor's Theorem, though calculus concepts, particularly limits, lie just below the surface. It was in his derivation of the infinite series for a^x that Euler first used the symbol e for the base of the natural logarithms.

Euler also popularized the use of π in its modern sense, though the first to use the symbol ittself was the English mathematician William Jones, in 1706. However, it was only with Euler that π became the universally recognized symbol it is today.

Notice that Euler had previously used p for this purpose.

Another symbol introduced by Euler was i for $\sqrt{-1}$, though this took several years. First, Euler began to systematically factor out $\sqrt{-1}$; thus, he wrote $\sqrt{-4}$ as $2\sqrt{-1}$. It was not until a paper presented to the St. Petersburg Academy in 1777 that Euler used $i = \sqrt{-1}$; the paper itself did not see publication until 1794.

Like the series for exponentials, Euler developed the series for the trigonometric functions without an obvious use of calculus.

Conjecture 15.3. $\cos v$ *and* $\sin v$ *may be respectively represented as*

$$\cos v = 1 - \frac{v^2}{1 \cdot 2} + \frac{v^4}{1 \cdot 2 \cdot 3 \cdot 4} + \dots$$

$$\sin v = v - \frac{v^3}{1 \cdot 2 \cdot 3} + \frac{v^5}{1 \cdot 2 \cdot 3 \cdot 4 \cdot 5} + \dots$$

Demonstration. Since in general $(\cos z + \sqrt{-1} \sin z)^n = \cos nz + \sqrt{-1} \sin nz$, we have

$$\cos nz = \frac{(\cos z + \sqrt{-1} \sin z)^n + (\cos z - \sqrt{-1} \sin z)^n}{2}$$

$$\sin nz = \frac{(\cos z + \sqrt{-1} \sin z)^n - (\cos z - \sqrt{-1} \sin z)^n}{2}$$

Expanding these using the binomial theorem, we obtain

$$\cos nz = (\cos z)^n - \frac{n(n-1)}{1 \cdot 2}(\cos z)^{n-2}(\sin z)^2$$
$$+ \frac{n(n-1)(n-2)(n-3)}{1 \cdot 2 \cdot 3 \cdot 4}(\cos z)^{n-4}(\sin z)^4$$
$$- \frac{n(n-1)(n-2)(n-3)(n-4)(n-5)}{1 \cdot 2 \cdot 3 \cdot 4 \cdot 5 \cdot 6}(\cos z)^{n-6}(\sin z)^6 + \dots$$

Let z be infinitely small, so $\cos z = 1$ and $\sin z = z$. Then let n be infinitely large, so that nz is a finite quantity v, and $z = \frac{v}{n}$. Since n is infinitely large, then $\frac{n-1}{n}$, $\frac{n-2}{n}$, and so forth, are equal to 1; hence

$$\cos v = 1 - \frac{v^2}{1 \cdot 2} + \frac{v^4}{1 \cdot 2 \cdot 3 \cdot 4} + \dots$$

Likewise

$$\sin v = v - \frac{v^3}{1 \cdot 2 \cdot 3} + \frac{v^5}{1 \cdot 2 \cdot 3 \cdot 4 \cdot 5} + \dots$$

$\square$

Infinite series turn into a powerful tool at the hands of Euler. For example, Euler derived

$$\frac{1}{\left(1 - \frac{1}{2}\right)\left(1 - \frac{1}{3}\right)} = 1 + \frac{1}{2} + \frac{1}{3} + \frac{1}{4} + \frac{1}{6} + \frac{1}{8} + \dots$$

by multiplying termwise the series expansions for $\frac{1}{1-\frac{1}{2}}$ and $\frac{1}{1-\frac{1}{3}}$.

The second of the Berlin textbooks, *Elements of Differential Calculus* (1755), could be used today as a modern calculus text but for two reasons. First, it is written in Latin. Second, Euler's development of differential calculus, including the use of infinite series, is today considered nonrigorous. For example, to find the derivative of $\ln x$, Euler wrote:

$$y = \ln x$$
$$y + dy = \ln(x + dx)$$

Subtracting $y = \ln x$ from both sides, we have:

$$dy = \ln(x + dx) - \ln x$$
$$= \ln\left(1 + \frac{dx}{x}\right)$$

Expanding the right hand side as an infinite series in dx (!), Euler obtained:

$$= \frac{dx}{x} - \frac{(dx)^2}{2x^2} + \frac{(dx)^3}{3x^3} - \dots$$

Since $(dx)^2$, $(dx)^3$, ... are "infinitely small" compared with dx, Euler concluded that $dy = \frac{dx}{x}$ or, dividing both sides by dx, $\frac{dy}{dx} = \frac{1}{x}$.

Despite the nonrigorous reasoning, most of Euler's results are accurate. One might claim that Euler was "lucky," but when considering how much *computational* mathematics appears in Euler's works, a better explanation is that Euler had developed an advanced "number sense" that kept him from making serious errors. On rare occasions, however, Euler's intuition failed him. The most notorious example was where Euler concluded, by substituting $x = 1, 2, 3, \dots$ into

$$\frac{1}{1-x} = 1 + x + x^2 + x^3 + x^4 + \dots$$

that

$$A = 1 + 1 + 1 + \dots = \tfrac{1}{1-1} = \infty$$
$$B = 1 + 2 + 4 + \dots = \tfrac{1}{1-2} = -1$$
$$C = 1 + 3 + 9 + \dots = \tfrac{1}{1-3} = -\tfrac{1}{2}$$

Since every term of the series for B is at least as great as the corresponding term of the series for A, Euler reasoned that the sum of series B ought to be greater than the sum of series A, and thus "in some sense" negative numbers were greater than infinity. It was in part the contradictory nature of results like this that prompted nineteenth century mathematicians to reexamine the basis of calculus.

INSTITUZIONI
ANALITICHE
AD USO
DELLA GIOVENTU' ITALIANA
DI D.ᴺᴬ MARIA GAETANA
AGNESI
MILANESE
Dell' Accademia delle Scienze di Bologna.
TOMO I.

IN MILANO, MDCCXLVIII.
NELLA REGIA-DUCAL CORTE.
CON LICENZA DE' SUPERIORI.

Figure 15.2: Title page of Agnesi's *Foundations of Analysis*. Boston Public Library/Rare books Department. Courtesy of the Trustees.

15.2.2   Maria Agnesi

We might compare Euler's *Introduction to the Analysis of the Infinites* and *Elements of Differential Calculus*, written in Latin, with another text, *Foundations of Analysis for the Use of Italian Youth* (1748), by Maria Gaetana Agnesi (May 16, 1718-January 9, 1799), probably the first great female mathematician in Western history. Agnesi's father, Pietro Agnesi, was professor of mathematics at the University of Bologna, and, recognizing her genius early, established at his own home a salon where she could present and defend theses against vari-

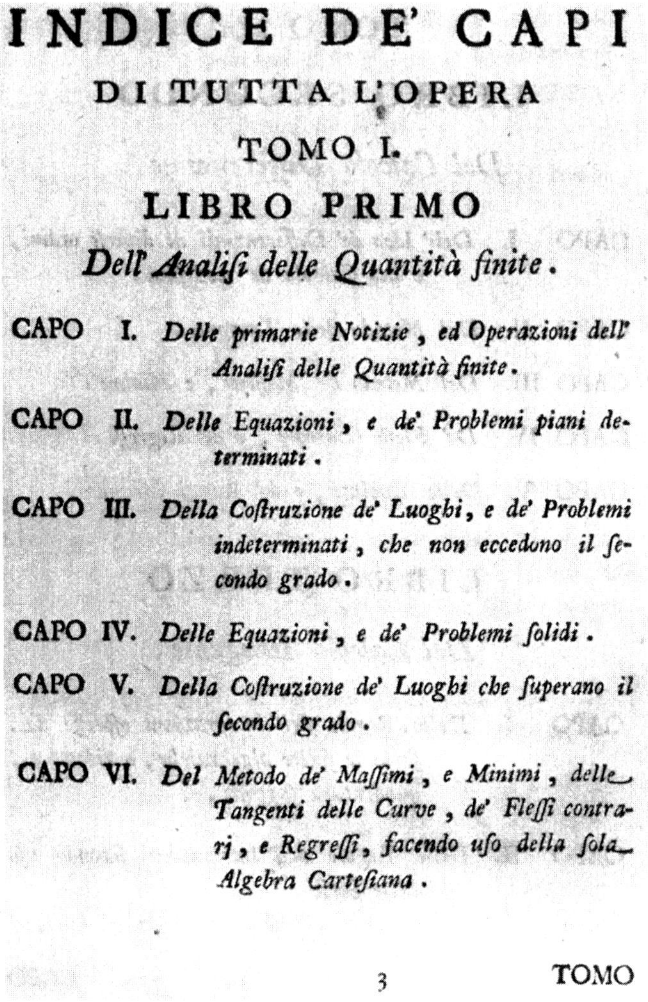

INDICE DE' CAPI

DI TUTTA L'OPERA
TOMO I.
LIBRO PRIMO

Dell' Analisi delle Quantità finite.

CAPO   I.   *Delle primarie Notizie, ed Operazioni dell' Analisi delle Quantità finite.*

CAPO   II.   *Delle Equazioni, e de' Problemi piani determinati.*

CAPO   III.   *Della Costruzione de' Luoghi, e de' Problemi indeterminati, che non eccedono il secondo grado.*

CAPO   IV.   *Delle Equazioni, e de' Problemi solidi.*

CAPO   V.   *Della Costruzione de' Luoghi che superano il secondo grado.*

CAPO   VI.   *Del Metodo de' Massimi, e Minimi, delle Tangenti delle Curve, de' Flessi contrarj, e Regressi, facendo uso della sola Algebra Cartesiana.*

3   TOMO

Figure 15.3: Table of Contents of Agnesi's *Foundations of Analysis*. Note that finding maximum and minimum values, as well as finding tangents, is covered in Chapter VI (CAPO VI) of Volume I (TOMO I), whereas infinitesimal methods are not covered until Volume II. Boston Public Library/Rare books Department. Courtesy of the Trustees.

ous experts. Like many great mathematicians, she was a linguist, and by the age of 11, she could speak French, Latin, Greek, German, Spanish and Hebrew. Though the disputations in her home were usually conducted in Latin, she could (and frequently did) respond to questioners in their native language.

Foundations of Analysis for the Use of Italian Youth was written in Italian, not Latin. Thus, like Stevin, Agnesi found the vernacular quite suitable for abstract thought and did not feel the need, like Euler, to write in Latin so scholars *everywhere* could make use of the work: the title alone indicated its national character. Agnesi dedicated the work to another great woman of her time—the Empress Maria Theresa. In two volumes and 1000 pages, Agnesi covered nearly the whole of mathematics, from algebra through infinitesimal analysis. A French reviewer called it the best and most complete text on the subject, and an English reviewer learned Italian to translate the work, so that its usefulness might not be limited to just Italian youth.

In the first book, she discussed the graphing of curves in a manner that would be familiar to any student of mathematics: from the equation, a few points (notably the x and y-intercepts) are found; then asymptotes are determined; then the curve is sketched. In Chapter V, she discussed curves that arose from equations of higher than the second degree or, alternatively, curves that generated equations of the second degree or higher.

Maria Agnesi.   ©Bettmann/ CORBIS.

Though Agnesi included a number of results original with herself, her name is inextricably linked to something she did not discover (nor claim credit for discovering). In Problem III of Chapter V, she described the following curve: given a semicircle ADC, with diameter AC. The points M where $AB : BD = AC : BM$ describe a curve whose equation can be found as follows: if $AC = a$, $AB = x$, $BM = y$ (note that the x represents a vertical length and y a horizontal length), then, by the property of the circle, $BD = \sqrt{ax - x^2}$ and by the definition of the curve, $y = \frac{a\sqrt{ax - x^2}}{x}$. "This [curve] is called the **turning**."

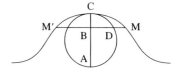

The turning was actually invented by Guido Grandi (October 1, 1671-July 4, 1742), who took up the study of mathematics because he was appointed to teach it at his monastery. "The turning" in Italian is *la Versiera*. The English translator of Angesi's work misread this as *l'avversiera*, "she-devil," or "witch": hence the curve became known as the **witch of Agnesi**.

In 1752, Agnesi's father died, and gradually she distanced herself from academia. Thus, in 1762, when the University of Turin asked if she would serve as a reader for a paper on the calculus of variations, she declined the honor, and missed the opportunity to examine the work of a fellow Italian mathematician: Joseph Louis Lagrange.

15.2 Exercises

1. Derive the series for a^x as Euler did, using the following steps. Let a be some positive number, and ω be infinitely small.

   (a) Show $a^\omega = 1 + k\omega$, for some k.

   (b) Suppose $j = \frac{z}{\omega}$, where z is finite. Show for any j, $a^{j\omega} = 1 + \frac{j}{1}(k\omega) + \frac{j(j-1)}{1\cdot 2}(k\omega)^2 + \ldots$, and hence $a^z = 1 + kz + \frac{k^2 z^2}{2\cdot 1} + \ldots$ Hint: for A any real number, $\lim_{n\to\infty} \frac{n+A}{n} = 1$.

   (c) Find a in terms of k, and then find the "natural" value of a that will make k equal to 1; in other words, the value that will make $a^\omega = 1 + \omega$, for ω infinitely small.

2. Show, using Euler's method in *Introduction to the Analysis of the Infinite*, how the series for $\sin v$ may be developed.

3. Comment on the validity of Euler's methods of infinite series in the derivation of the series for transcendental functions.

4. Demonstrate

$$\frac{1}{\left(1 - \frac{1}{2}\right)\left(1 - \frac{1}{3}\right)} = 1 + \frac{1}{2} + \frac{1}{3} + \frac{1}{4} + \frac{1}{6} + \frac{1}{8} + \ldots$$

by writing the left hand side as a product of two geometric series, then multiplying them term by term.

5. If a, b, c, ... are whole numbers, and

$$\frac{1}{\left(1 - \frac{1}{a}\right)\left(1 - \frac{1}{b}\right)\left(1 - \frac{1}{c}\right)\ldots} = 1 + \frac{1}{A} + \frac{1}{B} + \frac{1}{C} + \frac{1}{D} + \ldots$$

what, in general, is the relationship between a, b, c, ... and A, B, C, ...?

6. Use Problem 5 to determine the sums of the following series.

   (a) $1 + \frac{1}{3} + \frac{1}{5} + \frac{1}{9} + \frac{1}{15} + \frac{1}{25} + \ldots$, where the denominators are all numbers that are a product of powers of 3 and 5.

   (b) $1 - \frac{1}{2} - \frac{1}{3} + \frac{1}{4} + \frac{1}{6} - \frac{1}{8} + \frac{1}{9}$, where the denominators are of the form $2^n 3^m$, and the terms are positive if $n + m$ is even and negative otherwise. Hint: $\frac{1}{1+x} = 1 - x + x^2 - x^3 + \ldots$

7. Demonstrate, using Euler's methods on infinite series, that the number of primes must be infinite: show that $\frac{1}{\left(1 - \frac{1}{2}\right)\left(1 - \frac{1}{3}\right)\left(1 - \frac{1}{5}\right)\left(1 - \frac{1}{7}\right)\ldots}$, where the fractions in the denominator are the reciprocals of the primes, results in the harmonic series; hence, the divergence of the harmonic series implies the number of primes is infinite.

8. What are some of the key differences between Euler's definition of a function and our own?

9. From the definition of the *versiera*, determine its Cartesian equation. Suppose that the circle is situated so that it is tangent to the x-axis at A.

15.3 Mathematical Physics

By applying the tools of mathematics to the principles of physics, one is able to derive powerful results. However, unless both the mathematics and the physics are sound, the results will have nothing whatsoever to do with "reality." Though Newton had described, in his *Mathematical Principles of Natural Philosophy*, an essentially correct system of nature, and Leibniz had established differential calculus as a powerful mathematical tool, the practitioners of both the correct Newtonian physics and the powerful Leibnizian calculus were very few, and for a century, progress in mathematical physics languished.

It should be pointed out that Newtonian notation is, by itself, not inferior to Leibnizian notation; what made the application of Newtonian methods to physics so difficult was the insistence, by English mathematicians, of a return to the geometrical basis. As a crucial example, we will examine the motions of the Moon, a problem Newton described as one that "made my head ache." If the Earth and Moon were alone in the universe, the Moon would trace out a perfect and unchanging ellipse, and its exact position at any future point could easily be determined.

However, the sun's gravitational force causes the ellipse's shape to vary over time. There are several important changes. The **line of the nodes**, where the plane of the Moon's orbit about the Earth intersects the plane of the Earth's orbit about the sun, changes by approximately 19° a year. Newton, using a purely geometrical approach, was able to account for this variation entirely, and included his work in the *Principles*. Unfortunately, the geometrical method was ill-suited for extension to other problems in celestial mechanics; it is not much of an understatement to say that Newton's work in the *Principles* pushed geometrical methods in mathematical physics as far as they could go.

Meanwhile, continental physicists were enamored of a system of the universe put forth by Descartes. Descartes had imagined space to be everywhere filled with particles, which whirled about the sun in a constantly moving vortex. The planets were carried in the vortex like leaves in a stream. While visually attractive, the vortex theory lacked a mathematical basis that would allow useful predictions to be made on any but an *ad hoc* basis. For example, to explain the rotation of the sun, astronomers using Cartesian vortices assumed, with no supporting evidence whatsoever, that the sun's atmosphere extended far beyond its visible surface so that the edge of the atmosphere would move at the proper rate to make the sun's surface (assumed to move at the same angular velocity as the atmosphere) move at the observed rate. As a result, continental physicists labored under an inaccurate description of the universe, while English physicists labored under a difficult and awkward system of mathematics.

15.3.1 The Triumph of Newton

There were some supporters of Newton in France, notably Voltaire, who wrote a popular work on Newton's physics. In 1736, an expedition led by Pierre de Maupertuis verified one of Newton's predictions, also derived geometrically in the *Principles*: the Earth would have an equatorial bulge. Despite the success of Newtonian gravitation in predicting the equatorial bulge, the idea that a force could make itself felt across a vacuum was hard to swallow for most physicist; in fact, as late as the twentieth century, no less a physicist than Einstein concerned himself with gravity's "spooky action at a distance."

To overcome the resistance to action at a distance, a succession of triumphs was necessary. One of these came in 1748, when the Paris Academy of the Sciences offered a prize for a mathematical theory of Jupiter and Saturn, explaining the variations in their motions. Euler won the prize with "Researches on the Question of the Inequalities of the Motion of Jupiter and Saturn."

Euler began with a discussion of the state of celestial mechanics, taking great pains to point out that universal gravitation was *not* an established scientific fact. For example, since Saturn was distinctly nonspherical (it is the most oblate of the major planets), Euler questioned whether its gravitational attraction toward Jupiter could be described by a simple law. However, in the end, Euler consigned himself to using gravitation by the inverse square.

Euler wrote down the differential equations of motion governing the distance between the planets Jupiter and Saturn and the sun, and attempted to solve them. In the expansion of the trigonometric series giving the distance between Saturn and the sun, Euler came across a term of the form $T'' ep \sin(\omega - p)$, where T'', e are constants, and p is the "eccentric anomaly," which increases linearly with time. If T'' is nonzero, then this term would increase without bound, and the distance between Saturn and the sun would undergo wild oscillations, swinging in very close to the sun or very far away, depending on whether $T'' ep \sin(\omega - p)$ was positive or negative. Thus, Euler's analysis suggested the possibility of large variations in the orbits of the planets and, consequently, the end of the world. This did not concern Euler, who was of the opinion that the solar system must, in time, come to an end. In any case, he consoled his readers by noting that if the distance between Saturn and the sun varied too greatly, the approximations he was using would no longer be valid, and the solution itself would thus be compromised.

Euler went on to mathematically derive the observed variations in the motions of Jupiter and Saturn using only Newtonian gravitation. It was a great triumph and might have overcome all resistance to the idea of action at a distance. However, one final piece of observational evidence remained unexplainable by the simple law of gravitation. The point in an elliptical orbit farthest from the body being orbited is called the **apse** (in the case of a planet orbiting the sun, the **aphelion**, or in the case of an object orbiting the Earth, the **apogee**); the Moon's apogee was observed to move at a rate of approximately 3° per orbit. Newton himself was able to account for almost all of the variations in the Moon's orbit—with the exception of the motion of the apogee.

The problem of the motion of the apogee was taken up by Alexis-Claude Clairaut (May 7, 1713-May 17, 1765) even before Euler's 1748 article. Clairaut was one of 20 children (!) most of whom did not survive to adulthood. Clairaut was home-schooled, and at the age of 10, learned infinitesimal calculus from l'Hôpital's work. He was a member of Maupertuis's expedition, and thus had good reason to believe in the validity of Newtonian gravitation. To his chagrin, Clairaut discovered that, at best, only half the motion of the apogee could be accounted for by Newtonian gravitation. He suggested several possible explanations, including that the inverse-square law was only an approximation; this was especially attractive, since the inverse-square law was obtained by assuming the Earth was perfectly spherical, which Clairaut's own experience contradicted.

The Russian Academy of the Sciences offered, as its first prize question, the explanation of the progression of the Moon's apogee (Euler posed the question anonymously). In 1749, Clairaut, by means of more extensive calculations, was finally able to account for the entirety of the motion of the Moon's apogee using only Newtonian gravitation. In the same year, d'Alembert scored another victory for the Newtonian model. By treating the Earth's equatorial bulge as a number of small satellites about the Earth's equator, d'Alembert was able to account for the precession of the equinoxes. It is worth noting that the English, who had an accurate description of the universe from the time of Newton, were so hampered by their system of mathematics that they could make little progress, but the French adopters of Newton's mechanics made great strides in mathematical physics. Though there would still be ardent Cartesians as late as 1765, the work of d'Alembert and Clairaut marked the true beginning of the wholehearted acceptance of Newtonian physics in continental Europe.

15.3.2 Complex Integration

D'Alembert and Clairaut were frequently rivals, working on the same problem and usually critiquing the work of the other (which provided new insight into mathematical structures). D'Alembert was more physicist than mathematician. Like all great mathematical physicists, he forged the tools necessary to solve certain problems in mathematical physics; the tools might or might not be useful to mathematicians, and might or might not be considered rigorous, but so long as they worked, these philosophical considerations mattered little. For example, the first appearance of partial differential equations was in d'Alembert's "Reflections on the General Cause of the Winds" (1747).

In 1740, Clairaut had noted if

$$P \, dx + Q \, dy$$

was the *complete* differential of some function z of x and y, then

$$\frac{dP}{dy} = \frac{dQ}{dx} \tag{15.5}$$

(where d indicates a partial derivative; the notation $\frac{\partial P}{\partial y}$ would not be invented until 1786, by Adrien Marie Legendre). Today, we would say that $P \, dx + Q \, dy$ is an **exact differential**.

In a 1743 paper, Clairaut claimed that if $P \, dx + Q \, dy$ was an exact differential, then $\int P \, dx + Q \, dy = 0$ if the integral is taken around a closed path. In 1768, d'Alembert provided a counterexample to Clairaut's claim, noting that

$$\int \frac{y \, dx - x \, dy}{x^2 + y^2} = -2\pi$$

if the integral is taken around a closed clockwise path about the origin.

Earlier, in "Essay on the Resistance of Fluids" (1752), d'Alembert was led to a problem where $A\ dx + B\ dz$ and $zB\ dx - zA\ dz$ were exact differentials. He noted the solution was related to the similar problem where

$$dq = M\ dx + N\ dz$$
$$dp = N\ dx - M\ dz$$

were exact. If a solution to this equation existed, it would have to satisfy

$$d\left(q + p\sqrt{-1}\right) = \left(M + N\sqrt{-1}\right)\left(dx - \sqrt{-1}\ dz\right) \qquad (15.6)$$
$$d\left(q - p\sqrt{-1}\right) = \left(M - N\sqrt{-1}\right)\left(dx + \sqrt{-1}\ dz\right) \qquad (15.7)$$

Hence, $M + N\sqrt{-1}$ would have to be a function of $F + x - \sqrt{-1}\ z$, and $M - N\sqrt{-1}$ would have to be a function of $G + x + \sqrt{-1}\ z$, where F and G are constants.

Alternatively, p and q must satisfy

$$\frac{dp}{dz} = -\frac{dq}{dx} = -M$$
$$\frac{dp}{dx} = \frac{dq}{dz} = N$$

(since, by assumption, dp and dq were exact differentials). Thus, $q + p\sqrt{-1}$ is a function of $F + x - z\sqrt{-1}$ and $q - p\sqrt{-1}$ is a function of $G + x + z\sqrt{-1}$. This marked the first appearance of the **Cauchy-Riemann equations** of complex function theory.

15.3.3   Limits

See page 467.

In sharp contrast to his views on probability, which were invariably incorrect, d'Alembert's views on limits were very close to the modern view, though he lacked the notation to express them as we do. In the article on *Limit* for the *Encyclopedia*, he wrote:

> One says that a quantity is the *limit* of another quantity, when the second can be nearer to the first than any given quantity, however small, without exceeding the quantity approached.

> For example, given two polygons, one inscribed and the other circumscribed about a circle; it is obvious that one can increase the number of sides as much as one wants; and in this case, each polygon always approaches more and more closely the circle, the perimeter of the inscribed polygon increasing, and that of the circumscribed polygon decreasing, but the perimeter of the first never surpasses the circumference, and that of the second is never smaller than the same circumference.[2]

---

[2]D'Alembert, Jean le Rond, "Limite." *Encyclopedia*, ed. D. Diderot. Briasson, David, le Breton, and Durand, 1765. Vol. IX, p. 542.

Elsewhere in the *Encyclopedia*, d'Alembert defined the differential nearly as we do, as the limit of the ratio of two quantities. D'Alembert had not quite come to the modern conception; like Newton, he still included the proviso that the one quantity not "exceed" the limiting quantity.

15.3 Exercises

1. [Advanced Calculus] Consider d'Alembert's integral $\int \frac{y\,dx - x\,dy}{x^2 + y^2}$. Evaluate it over the contour C, where

   (a) C is the straight line from $(1, 0)$ to $(2, 0)$.

   (b) C is the *closed* path consisting of the straight lines from $(1, 0)$ to $(0, 1)$, then from $(0, 1)$ to $(0, 2)$, then from $(0, 2)$ to $(2, 0)$, and from $(2, 0)$ back to $(1, 0)$. (You will need to parametrize each line segment and evaluate four different integrals).

   (c) C is the circle parametrized by $x = \cos t$, $y = -\sin t$, for $0 \leq t \leq 2\pi$. (D'Alembert's result).

   (d) C is the circle parametrized by $x = \cos t$, $y = \sin t$, for $0 \leq t \leq 4\pi$ (i.e., the circle that wraps around the origin twice).

   (e) What is qualitatively different about the closed paths in 1b and the closed paths in 1c and 1d?

2. Consider Equations 15.6.

   (a) Demonstrate their validity.

   (b) Show that $M + N\sqrt{-1}$ and $M - N\sqrt{-1}$ must have the form claimed by d'Alembert.

15.4   Trigonometric Series

One of the most exciting (and exasperating) events is when the solution to a problem points to flaws in the fundamental basis of a science; mathematics is no exception. The **vibrating string problem** forced mathematicians to give serious consideration to the very foundations of analysis. The problem involved the determination of the shape of a string, attached at its two endpoints, when it was allowed to vibrate in accordance with the laws of physics. The problem involves solving second order partial differential equation, so we will not discuss it in detail. D'Alembert's main contribution to the problem was to show that if y is the displacement of the string above the horizontal, then y satisfies

$$y = \frac{1}{2}f(t + x) + \frac{1}{2}f(t - x) \tag{15.8}$$

where x represents the distance from the endpoint and t the amount of time elapsed from the beginning.

15.4.1   Definition of a Function

Difficulties arise upon more careful consideration of the function $f(x)$. If $t = 0$, then the curve $y = f(x)$ represents the initial shape of the string. To establish Equation 15.8, d'Alembert was forced to impose several conditions on $f(x)$: in particular, it was necessary for it to be twice differentiable. Moreover, the expression of $f(x)$ had to consist of a "single analytic expression"; in other words, it had to be expressible in terms of x according to a single, algebraic rule. In the terms of eighteenth century mathematics (but not in modern terms), $f(x)$ was a **continuous function**.

Euler tackled the vibrating string problem shortly after d'Alembert. In a 1748 paper on the subject, published in the journal for the Berlin Academy, Euler proved:

Proposition 15.3. *If A is a function of t and u, then differentiating A with respect to t first, then u will give the same result differentiating with respect to u first, then t.*

In other words, the mixed partial derivatives of a function of two variables are equal. Euler's proof, reproduced next, shows the state of notation in mid-century.

Proof. Let B be A with t replaced by $t + dt$; C be A with u replaced by $u + du$, and D be A with both t and u replaced by $t + dt$ and $u + du$, respectively. If t is held constant, then the differential will be $C - A$. If then u is held constant, the differential of this quantity will be $D - B - C + A$.

Conversely, if u is held constant the differential will be $B - A$, and if then t is held constant, the differential of this will be $D - C - B + A$, which is the same. Hence, the differential with respect to u first, then t is the same as the differential with respect to t first, then u. □

Euler pointed to a problem with d'Alembert's conclusions: if $f(x)$ was to represent the shape of the string, then, since the string could be put into any initial position whatsoever, in many cases the requirement that $f(x)$ be twice differentiable would fail, and d'Alembert's method of solution would not work. For example, if the string were drawn upward in the middle, forming a "∧," this shape could not be described by a single algebraic expression and even if it could, it was not even once differentiable. Hence, the question arose: in this case, could the initial position of the string be considered to be given by a function? The vibrating string problem forced mathematicians to reconsider what constituted a "function" of a variable.

Daniel Bernoulli added a factor that further complicated the issue. Based primarily on physical considerations, he asserted that the motion of the string could be expressible as a trigonometric series, consisting of the sums of sines and cosines, such as

$$\sin x + \frac{1}{2}\sin 2x + \frac{1}{3}\sin 3x + \dots$$

But this meant that nondifferentiable functions (like that corresponding to the ∧ shape) could be expressed as the sums of differentiable functions. The conceptual difficulties that arose from this problem would eventually force a reconsideration of the basis of analysis.

15.4.2   Series of Sines

With the problem of the vibrating string, trigonometric series moved to the forefront of mathematical investigation. In 1754 Euler wrote a short article, "Simplification of the Calculus of Sines," which appeared in 1760 in the journal of the St. Petersburg Academy.

To simplify the sums of infinite trigonometric series, Euler used a number of clever algorithmic devices. If we let

$$u = \cos\phi + \sqrt{-1}\sin\phi \qquad\qquad v = \cos\phi - \sqrt{-1}\sin\phi$$

then $\cos\phi = \frac{1}{2}(u+v)$, $uv = 1$, and the powers of u and v could be found by De Moivre's formula. Thus, to find the powers of $\cos^n\phi$, Euler expanded $(u+v)^n$:

$$2^n \cos^n\phi = (u+v)^n$$
$$= u^n + nu^{n-1}v + \frac{n(n-1)}{1\cdot 2}u^{n-2}v^2 + \ldots$$

Since $(u+v)^n = (v+u)^n$, then:

$$2^n \cos^n\phi = (v+u)^n$$
$$= v^n + nv^{n-1}u + \frac{n(n-1)}{1\cdot 2}v^{n-2}u^2 + \ldots$$

Adding the two equations together gives

$$2^{n+1}\cos^n\phi = u^n + v^n + \frac{n}{1}\left(u^{n-2} + v^{n-2}\right)uv$$
$$+ \frac{n(n-1)}{1\cdot 2}\left(u^{n-4} + v^{n-4}\right)u^2 v^2 + \ldots$$

Hence

$$2^n \cos^n\phi = \frac{1}{2}\left(u^n + v^n\right) + \frac{n}{2}\left(u^{n-2} + v^{n-2}\right) + \frac{n(n-1)}{1\cdot 2}\frac{1}{2}\left(u^{n-4} + v^{n-4}\right) + \ldots$$
$$= \frac{1}{2}\cos n\phi + \frac{n}{2}\cos(n-2)\phi + \frac{n(n-1)}{1\cdot 2}\frac{1}{2}\cos(n-4)\phi + \ldots$$

Thus the powers of $\cos\phi$ could be expressed easily. Using the fact that $\cos\phi = \sin\psi$, where $\phi = 90° - \psi$, Euler also found series expressions for the powers of $\sin\phi$. Euler then assumed that this expansion held true if n was negative or fractional and thus claimed

$$\frac{1}{\cos\phi} = \cos\phi - \cos 3\phi + \cos 5\phi - \ldots$$

At the end of the paper, Euler considered the expressions

$$U = \frac{u^m}{1 - au^n} \qquad\qquad V = \frac{v^m}{1 - av^n}$$

Thus

$$\frac{U + V}{2} = \frac{\cos m\phi - a\cos(m - n)\phi}{1 + a^2 - 2a\cos n\phi} \tag{15.9}$$

$$\frac{U - V}{2} = \frac{\sin m\phi - a\sin(m - n)\phi}{1 + a^2 - 2a\sin n\phi} \tag{15.10}$$

Since U and V could also be expressed as a geometric series in u or v (which in turn could be expressed in terms of sines and cosines of multiples of ϕ), Euler obtained

$$\frac{\cos m\phi - a\cos(m - n)\phi}{1 + a^2 - 2a\cos n\phi} = \cos m\phi + a\cos(m + n)\phi + a^2\cos(m + 2n)\phi + \ldots$$

Letting $m = 1$, $n = 1$, and $a = 1$, Euler arrived at

$$-\frac{1}{2} = \cos\phi + \cos 2\phi + \cos 3\phi + \ldots \tag{15.11}$$

Daniel Bernoulli considered the derivatives and integrals of this equation (Euler did so as well), and made a remarkable discovery. In a 1772 paper on the subject, Bernoulli began with the equation's indefinite integral:

$$C - \frac{x}{2} = \sin x + \frac{\sin 2x}{2} + \frac{\sin 3x}{3} + \ldots$$

To find C, he first supposed that $x = q$, where q is the quadrant of a circle (in our terms, $q = \frac{\pi}{2}$). This gave $C = q$, and thus

$$q - \frac{x}{2} = \sin x + \frac{\sin 2x}{2} + \frac{\sin 3x}{3} + \ldots \tag{15.12}$$

Since the left hand side decreased as x went from 0 to $4q$, the series must also decrease. But when $x = 4q$ (i.e., $x = 2\pi$), then the sum of the series jumps from $-q$ to q. Bernoulli thus found the remarkable result that a *discontinuous* function could be represented as a infinite sum of continuous trigonometric functions.

15.4.3   The Seven Years' War

Frederick I of Prussia had a way of antagonizing powerful women; he was certainly a misogynist, and may have been homosexual. Besides Maria Theresa, he angered the Tsarina Elizabeth of Russia, and infuriated Madame Pompadour, the mistress of the king of France, Louis XV. In 1756, then, Prussia found itself in a war facing not one, but three of the most powerful countries in Europe: France, Austria, and Russia. Because of its length, the war would become known as the

Seven Years' War. Frederick was the most brilliant tactician of his age—history remembers him as Frederick the Great—but the forces against Prussia seemed too powerful, and it appeared that Prussia might be ground into the dust.

In 1760, the Russian General Tottleben captured Berlin, and during the invasion, Euler's farm in Charlottenburg, which he had bought in 1753, was burnt down. When Tottleben heard that Euler's farm had been destroyed by his own troops, Tottleben (perhaps remembering that Marcellus was known primarily as the general whose troops killed Archimedes) declared he was not waging war against mathematicians, and indemnified Euler for the damage. Tsarina Elizabeth also sent Euler a gift of 4000 florins. In the meantime, Euler continued his mathematics, and despite the war, scientific communication between Berlin and St. Petersburg was unhindered; in fact, several of Euler's publications appeared in the St. Petersburg journal during the war.

Meanwhile, Frederick benefited from a fantastic stroke of luck: in 1762, Tsarina Elizabeth died, leaving the throne to her nephew Peter. Peter was fanatically pro-German (he himself was the Duke of Holstein-Gottorp), and hero-worshipped Frederick; hence, he withdrew Russia from the alliance against Prussia. Without Russia, Austria and France could not defeat Frederick, and in 1763, the war ended. For his pains, Peter was deposed by a palace coup, then assassinated a few days later. Peter's wife, Catherine II, succeeded him, but she saw no point in continuing the war, since Russia had enough internal troubles, which she would concentrate upon. She would be known to history as Catherine the Great.

15.4 Exercises

1. Find the series expression for $\sin^n \phi$.

2. Prove Equations 15.9.

3. Demonstrate that the value of the series in equation 15.12 decreases as x goes from 0 to $4q$, by considering the value of the series when x equals $\frac{2}{3}q$, $\frac{4}{3}q$, $\frac{6}{3}q$, $\frac{8}{3}q$, $\frac{10}{3}q$.

Chapter 16

Algebra

In 1685, Charles II of England and Scotland died, and his brother James became king. James II rapidly proved unpopular. He was an outspoken Catholic, and no one in England or Scotland wanted a return to Catholicism. Fortunately, it seemed that the sickly James would never produce a male heir, and the throne would pass on to James's eldest daughter, the Protestant Mary, who had married William of Orange. But in 1687, a son was born to James, and it looked like England might become a Catholic country again. William of Orange was called on to save England from Catholicism, and in the bloodless "Glorious Revolution" (1688), James fled England for France when it became obvious that no one supported him. Parliament declared that James had abdicated, then offered the throne to William and Mary, who accepted.

William ruled until 1702 (Mary died in 1694), but had no children, so the throne passed to Mary's sister Anne. During Anne's reign, the Act of Union (1707) was passed, an important piece of legislation that formally joined the kingdoms of England and Scotland to create the United Kingdom of Great Britain. An important provision was that, should Anne die without heirs, the throne would pass to the Protestant Sophia, Princess of Hanover (and granddaughter of James I), or her heirs, and not to the children of James II, who were Catholic. Sophia died before Anne, so after Anne's death in 1714, the throne of England passed into the House of Hanover. Thus, the lands of Newton and Leibniz were united. Unfortunately, the mathematical community continued to be divided.

16.1 Newton and Algebra

Part of James II's attempt to return England to Catholicism was filling vacant university positions with Roman Catholics. When he insisted that a Catholic Benedictine monk with no particular qualifications be given a Cambridge degree, Newton encouraged the vice chancellor of Cambridge University to defy the king.

The vice chancellor was dismissed, but Cambridge University got its revenge: the university could send two representatives to Parliament, and on January 15, 1689, Newton was elected as one of them. This Parliament declared James II deposed, and offered the crown of England to William and Mary.

Meanwhile, Newton continued his scientific work and was knighted by Queen Anne in 1705 (the first scientist to be knighted for his scientific work). He was dissatisfied with Descartes's "rule of signs": though it could be used to find the maximum number of positive or negative roots of an equation, Newton wanted a corresponding rule for the number of complex roots of an equation. In his *Universal Arithmetic* (1728), which, despite its name, was actually an algebra text, Newton gave a new rule: See page 350.

Rule 16.1. *Make a series of fractions whose denominators are 1, 2, 3, ... up to the highest power in the equation, and whose numerators are the same numbers in the reverse order. Divide the second by the first, and third by the second, and so on, placing the results over the terms in the equation, beginning with the second. If, under any term, if the square of the term, multiplied by the fraction, exceeds the product of the terms on either side, place a "+" sign underneath; otherwise, place a "−" sign. Under the first and last terms place a "+." The number of alterations of signs, from "+" to "−" or from "−" to "+," will be the minimum number of complex roots.*

Newton provided no proof of this result, and he may have discovered it empirically.

Example 16.1. *For the equation $x^3 + px^2 + 3p^2x - q = 0$, with p, q assumed positive, form the fractions $\frac{3}{1}, \frac{2}{2}, \frac{1}{3}$. Dividing the second by the first, we obtain $\frac{1}{3}$; likewise, dividing the third by the second, we obtain $\frac{1}{3}$. Comparing the squares of each term, multiplied by the fraction above it, with the product of the terms on either side, and writing a "+" if the square is greater and a "−" if the term is less, we obtain:*

$$
\begin{array}{ccccccccc}
 & & \frac{1}{3} & & \frac{1}{3} & & & & \\
x^3 & + & px^2 & + & 3p^2x & - & q & = & 0 \\
+ & & - & & + & & + & &
\end{array}
$$

There are two alterations of signs; hence there are at least two complex roots.

If there is a single missing term, it should be treated as a 0 term.

Example 16.2. *Applying Newton's method to the equation $x^4 - 6x^2 - 3x - 2 = 0$, we determine*

$$
\begin{array}{ccccccccc}
 & & \frac{3}{8} & & \frac{4}{9} & & \frac{3}{8} & & \\
x^4 & + & ** & - & 6x^2 & - & 3x & - & 2 \\
+ & & + & & + & & - & & +
\end{array}
$$

where the missing term is indicated with a ∗∗. There are two alterations in sign, and consequently at least two imaginary roots.

If there are two or more terms in a row that are missing, below the missing terms should be placed a "$-$," then a "$+$," then a "$-$," alternating, but below the last missing term should always be "$+$."

Example 16.3. *Applying Newton's method to $x^5 - 1 = 0$, Newton's procedure produces*

$$
\begin{array}{ccccccccccc}
 & & \frac{2}{5} & & \frac{1}{2} & & \frac{1}{2} & & \frac{1}{2} & & \frac{2}{5} \\
x^5 & + & ** & + & ** & + & ** & + & ** & - & 1 \\
 & + & & - & & + & & - & & + & + \\
\end{array}
$$

There are four changes in sign, so there are at least four complex roots.

Newton also found formulas for the sums of the powers of the roots of an equation, which would become important in the later theory of equations. Designating by p, q, r, s, and so on the "quantity of terms" of an equation with its signs changed (i.e., the negatives of the coefficients), where p is the second term, q is the third term, and so on, and letting $p = a$, $pa + 2q = b$, $pb + qa + 3r = c$, $pc + qb + rc + 4s = d$, $pd + qc + rb + sa + 5t = e$, $pe + qd + rc + sb + 6a = f$, and so on, then a will be the sum of the roots, b will be the sum of the squares of the roots, c will be the sum of the cubes of the roots, and so on.

Example 16.4. *The equation $x^4 - x^3 - 19x^2 + 49x - 30 = 0$. $p = 1$, $q = 19$, $r = -49$, and $s = 30$. Thus $a = 1$, $b = 39$, $c = -89$, $d = 723$. Hence the sum of the roots is 1, the sum of the squares of the roots is 39; the sum of the cubes of the roots is -89, and the sum of the fourth powers of the roots is 723.*

The information about the sums of the powers of the roots could be used to find the limits of the roots; thus, Newton pointed out, the square root of b was greater than the (absolute value) of the largest root; the fourth root of d was greater than the (absolute value) of the largest root and, moreover, closer to the largest root than the square root of b.

Example 16.5. *The sum of the squares of the roots of $x^4 - x^3 - 19x^2 + 49x - 30 = 0$ is 39; thus, the roots all lie between $\pm\sqrt{39} \approx 6\frac{1}{2}$. The sum of the fourth powers of the roots is 723, hence all the roots lie between $\pm\sqrt[4]{723} \approx 5\frac{1}{4}$.*

16.1 Exercises

1. Let M be the largest absolute value of the roots of an equation. Recall that Newton's terminology had b be the sum of the squares of the roots, c be the sum of the cubes, d the sum of the fourth powers, and so on.

   (a) Show that $\sqrt{b}$ and $\sqrt[4]{d}$ is greater than M.

   (b) Show that the $M \leq \sqrt[4]{d} \leq \sqrt{b}$.

2. Suppose none of the roots of an equation are equal to 0 (obviously, any equation with roots of 0 can be reduced to a simpler equation with no 0 roots). Find *lower* bounds for the absolute values of the roots. Hint: replace x with $\frac{1}{v}$, and note that the roots of the resulting equation are the reciprocals of the roots of the original equation.

3. Use Newton's method to determine the minimum number of complex roots.

(a) $x^2 + 3x - 5 = 0$ (c) $x^3 - 3x^2 - 7 = 0$

(b) $x^4 + x^3 + x^2 + x + 1 = 0$ (d) $x^3 - 5x + 4 = 0$

4. For the equations in Problem 3, find bounds on the real roots.

16.2 Maclaurin

Colin Maclaurin (February, 1698-January 14, 1746) popularized Newton's method of fluxions in his *Treatise on Fluxions* (1742); the **Maclaurin series** is so named because it appeared in *Treatise*, though it had been known in a more general form to Taylor and to Gregory before him. However, Maclaurin's primary original contributions to mathematics were in algebra.

16.2.1 Newton's Rule

In the *Philosophical Transactions of the Royal Society* for 1726, Maclaurin attempted to prove Newton's method of determining the number of complex roots. After proving a number of introductory lemmas, Maclaurin showed: See page 505.

Proposition 16.1. *In a quadratic with two real roots, the square of the second term is greater than four times the product of the first and third.*

Like Newton, Maclaurin considered the products of the *terms*, as opposed to the coefficients. A corresponding result for the cubic equation is:

Proposition 16.2. *In a cubic with three real roots, the square of the second term is greater than three times the product of the first and third.*

Hence, if the square of the second was not greater than three times the product of the first and third, the cubic must have two complex roots. Maclaurin pointed out that this was equivalent to Newton's rule. Applying it to the cubic See page 505 $x^3 + px^2 + qx + r = 0$ gave

$$x^3 \quad + \quad \overset{\frac{1}{3}}{px^2} \quad + \quad \overset{\frac{1}{3}}{qx} \quad - \quad q \quad = \quad 0$$
$$+ \qquad\qquad\qquad +$$

If the square of the second, times $\frac{1}{3}$, did not exceed the product of the first and third, a $-$ would be written below the second term; regardless of what sign ended below the third term, there would be two changes of sign, which agreed with the result that there were two complex roots. Maclaurin proved Newton's rule for the fourth degree equation as well.

16.2.2   Cramer's Rule

See page 213.

Cramer's rule made its first appearance in the West in Maclaurin's *Treatise on Algebra*, published in 1748. The rule was named after the Swiss mathematician Gabriel Cramer, who published a work containing it in 1750, though a Chinese form of Cramer's rule existed nearly two thousand years before.

Rule 16.2 (Cramer's Rule for Two Variables). *For the system of equations*

$$ax + by = c$$
$$dx + ey = f$$

$y = \frac{af - dc}{ae - db}$.

In Maclaurin's terminology, the "orders" are the equation coefficients (in modern matrix terminology, the "orders" correspond to the column entries); the "opposite coefficients" are the coefficients of different variables in different equations; hence, "Cramer's" rule is, as described by Maclaurin:

> [The] numerator is the difference of the products of the opposite coefficients in the orders in which y is not found, and the denominator is the difference of the products of the opposite coefficients taken from the orders that involve the two unknown quantities.[1]

For three equations in three unknowns, Maclaurin gave:

Rule 16.3 (Cramer's Rule for 3 Variables). *For the system of equations*

$$ax + by + cz = m$$
$$dx + ey + fz = n$$
$$gx + hy + kz = p$$

$z = \frac{aep - abn - dbm - dbp + gbn - gem}{aek - abf + dbc - dbk + gbf - gec}$, *where the numerator is formed from all the different possible products of "opposite" coefficients from the orders in which z is not found.*

Unfortunately, Maclaurin did not explain the rule for the alternation of the signs, making his method very hard to use and impossible to generalize.

In 1745, the supporters of the descendants of James II, known as Jacobites (from the Latin form of James) tried to place Charles Edward, the grandson of James II, on the throne. Maclaurin helped with the defense of Edinburgh when it was besieged by the Jacobites, but the city was taken on September 11, 1745. However, when the Jacobites tried to move beyond Scotland, they found no support, and were soundly defeated. "Bonnie Prince Charlie," as Charles Edwards was known, was forced to flee over the ocean to France, giving rise to nostalgic poems and songs. Maclaurin, unfortunately, exhausted himself during the defense of the city and died soon after.

---

[1]Maclaurin, Colin, *A treatise of algebra*, fourth edition. J. Nourse, 1779. Page 82

16.2 Exercises

1. Prove Maclaurin's fourth lemma: if A is the sum of the squares of m numbers, and B is the sum of the products of the numbers taken two at a time, then $\frac{(m-1)A}{2} > B$.

2. Explain how Proposition 16.2 is equivalent to Newton's rule for cubics.

3. Prove Newton's rule for the fourth degree equation.

16.3   Euler and Algebra

In the eighteenth century, analysis was the newest, and most attractive, of all the fields of mathematics, so it is not surprising that Euler devoted much time and effort to advancing analysis. Still, the multitalented Euler made important contributions to algebra as well.

16.3.1   Continued Fractions

In algebra, Euler revitalized the study of continued fractions, which had been introduced by Bombelli, Brouncker, Wallis, and others. Euler's systematic study of continued fractions began with "Dissertation on Continued Fractions," presented to the St. Petersburg Academy in 1737 but not published until 1744. Euler considered continued fractions of the form

$$a + \cfrac{\alpha}{b + \cfrac{\beta}{c + \cfrac{\gamma}{d + \cfrac{\delta}{\ddots}}}}$$

where Euler called a, b, c, d the **denominators** and α, β, γ the **numerators**. The successive partial quotients, now called **convergents**, are

$$a = \frac{a}{1}$$

$$a + \frac{\alpha}{b} = \frac{ab + \alpha}{b}$$

$$a + \cfrac{\alpha}{b + \cfrac{\beta}{c}} = \frac{abc + \alpha c + \beta a}{bc + \beta}$$

$$a + \cfrac{\alpha}{b + \cfrac{\beta}{c + \cfrac{\gamma}{d}}} = \frac{abcd + \beta ad + \alpha cd + \gamma ab + \alpha \gamma}{bcd + \beta d + \gamma b}$$

To make the pattern of the successive partial quotients more obvious, Euler arranged them as follows:

$$
\begin{array}{ccccc}
a & b & c & d & e \\
\frac{1}{0} & \frac{a}{1} & \frac{ab+\alpha}{b} & \frac{abc+\alpha c+\beta a}{bc+\beta} & \frac{abcd+\alpha cd+\beta ad+\gamma ab+\alpha \gamma}{bcd+\beta d+\gamma b} \\
\alpha & \beta & \gamma & \delta & \epsilon
\end{array}
\quad
\begin{array}{c}
\cdots \\
\cdots \\
\cdots
\end{array}
$$

where the top row consists of the successive denominators and the bottom row consists of the successive numerators, and the middle row consists of the partial quotients. Then he gave:

Rule 16.4. *The numerator of the partial quotient is the numerator of the previous partial quotient, multiplied by the index above it, plus the numerator of the partial quotient before that, multiplied by the index below it; the denominator is the product of the previous denominator by the index above it, and added to the product of the denominator of the partial quotient before that by the index below it.*

Example 16.6. *The third partial quotient is the numerator of the previous partial quotient, a, times the index above it, b, plus the numerator of the partial quotient before $\frac{a}{1}$, which is 1, multiplied by the index below it, which is α. The denominator is the product of the previous denominator, 1, by the index above it, b, plus the product of the partial quotient before, 0, multiplied by the index below it, b. Thus, the third partial quotient is $\frac{ab+\alpha}{b}$.*

Euler then noted the differences between the successive approximations, which were

$$
\frac{1}{0} \quad \frac{\alpha}{1 \cdot b} \quad -\frac{\alpha\beta}{b(bc+\beta)} \quad +\frac{\alpha\beta\gamma}{(bc+\beta)(bcd+\beta d+\gamma b)} \quad \cdots
$$

Hence, the actual value of the continued fraction could be expressed by the alternating series

$$
a + \frac{\alpha}{b} - \frac{\alpha\beta}{b(bc+\beta)} + \frac{\alpha\beta\gamma}{(bc+\beta)(bcd+\beta d+\gamma b)} - \cdots
$$

By combining pairs of terms, Euler obtained the series of positive terms

$$
a + \frac{\alpha c}{1(bc+\beta)} + \frac{\alpha\beta\gamma e}{(bc+\beta)(bcde+\beta de+\gamma be+\delta bc+\beta\delta)} + \cdots
$$

Together, these could be used to turn an infinite series into a corresponding continued fraction.

Next, Euler tackled the problem of determining the real number to which a continued fraction corresponds. If all the denominators but the first are the same, as in

$$
x = a + \cfrac{1}{b + \cfrac{1}{b + \cfrac{1}{\ddots}}}
$$

then, Euler noted

$$x - a = \cfrac{1}{b + \cfrac{1}{b + \cfrac{1}{b + \cfrac{1}{\ddots}}}}$$

$$= \cfrac{1}{b + x - a}$$

Hence $x^2 - 2ax + bx + a^2 - ab = 1$, so $x = a - \frac{b}{2} + \sqrt{1 + \frac{b^2}{4}}$.

To obtain a continued fraction expansion of a given rational number x, Euler began by letting $b = 2a$, so $x = \sqrt{1 + \frac{b^2}{4}}$. For $x = \sqrt{5}$, $b = 4$, $a = 2$, so the continued fraction expansion is

$$\sqrt{5} = 2 + \cfrac{1}{4 + \cfrac{1}{4 + \cfrac{1}{\ddots}}}$$

Likewise, other roots can be found. Similar formulas can be found if the denominators are periodic, and in general, Euler noted, continued fractions with periodic denominators correspond to the solutions of quadratic equations. See Problem 5.

16.3.2 The Fundamental Theorem of Algebra

The first proof of the fundamental theorem of algebra was due to Euler, though in a rather roundabout way. In the journal for the Berlin Academy, Euler mentioned he possessed a proof, but did not publish one. This prompted d'Alembert to present his own proof. D'Alembert's proof began by chiding Euler for claiming results he had nowhere published:

> Mr. Euler mentioned in Volume VII of *Berlin Miscellany* a work where he demonstrated the proposition [that every polynomial factors into quadratic and linear factors] in general. But it seems to me that Mr. Euler hasn't yet published any of his work on the subject. At least I haven't found any trace of the work...[2]

To prove the fundamental theorem, d'Alembert proved

Proposition 16.3. *Suppose $y = 0$ when $z = 0$. If z is infinitely small, then y is a real number when z is positive, and y is the sum of a real number and a real number multiplied by $\sqrt{-1}$ if z is negative.*

---

[2]D'Alembert, Jean le Rond. "Recherches sur le Calcul Integral." *Histoire de l'Académié de Berlin, 1746*, 1748. Page 183.

Proof. Suppose $y = 0$ when $z = 0$. Then by the theory of infinite series, y can be expressed in terms of z as

$$y = az^{\frac{m}{n}} + bz^{\frac{p}{q}} + ez^{\frac{r}{s}} + \dots$$

If z is positive, y is clearly positive. If z is negative, then the terms are powers of the roots of negative numbers, which can be expressed as the sum of a real number and a real number multiplied by $\sqrt{-1}$. Hence y is a real number and a real number multiplied by $\sqrt{-1}$. □

Using this, d'Alembert proved a polynomial function evaluated for complex numbers had no maximum or minimum value. Hence

Proposition 16.4. *Given a multinomial*

$$x^m + ax^{m-1} + bx^{m-2} + \dots + fx + g$$

which is not equal to 0 for any real number x, there is a number of the form $p + \sqrt{-1}q$ for which it is equal to 0.

Prompted by d'Alembert's work, Euler responded with "Researches on the Imaginary Roots of Equations" (1751). D'Alembert's work relied on infinitely small quantities, which Euler preferred to avoid in this case. The research was motivated in part by a question faced by the Renaissance cossists: when solving the cubic equation using the Cardano-Ferrari-Tartaglia methods, one often ended

See Section 11.3.

with the roots of imaginary quantities. For example, the equation $x^3 = 6x + 4$ results in the solutions $\sqrt[3]{2 + \sqrt{-4}} + \sqrt[3]{2 - \sqrt{-4}}$. These could be reduced, using De Moivre's formulas, to quantities of the form $M + N\sqrt{-1}$, but did every root of *every* equation, particularly those involving trigonometric or logarithmic expressions, result in imaginary numbers that could be reduced to this form?

See page 437.

Leibniz had supposed this was not possible in general.

Euler did not consider the situation where a polynomial had a multiple root, but well-known techniques, such as Hudde's rules, existed in Euler's time for reducing a polynomial with a multiple root to one where all the roots were

See Problem 15.

distinct. Euler first proved

Proposition 16.5. *All polynomial equations of an odd degree have at least one real root, and if there is more than one, the number of real roots is odd.*

Euler's proof follows.

Proof. Consider the graph of the curve expressed by $y = x^{2m+1} + Ax^{2m} + \dots + N$. Obviously, any value of x corresponds to a unique value of y, and if y vanishes for some x, that value of x corresponds to a root. When this happens, the curve crosses the axis; hence the number of real roots is the number of intersections of the curve with the axis.

To determine this number, suppose x is positive and infinitely large, that is, $x = \infty$. Obviously $y = \infty^{2m+1} = \infty$ and thus this branch of the curve is at $+\infty$.

Likewise, if $x = -\infty$, $y = -\infty$, and this branch of the curve is at $-\infty$. Now, this branch is continuous with the branch on the other side of the axis, and thus the curve must cross the axis at some point; and if it crosses at more than one point, the number of crossings must be odd. □

As written, this proof would not today be considered rigorous, though only a few changes are necessary to make it logically valid. For even degree equations, Euler proved

Proposition 16.6. *All equations of an even degree, where the constant term is negative, must have two real roots.*

Euler's first key result is

Proposition 16.7. *All equations of the fourth degree can be decomposed into a product of two quadratic factors with real coefficients.*

Proof. We may restrict ourselves to the consideration of fourth degree polynomials of the form

$$x^4 + Bx^2 + Cx + D$$

since the cubic term may always be eliminated by an appropriate substitution. The factorization must necessarily be of the form

$$x^4 + Bx^2 + Cx + D = \left(x^2 + ux + \alpha\right)\left(x^2 - ux + \beta\right)$$

whence $B = \alpha + \beta - u^2$, $C = (\beta - \alpha)u$, and $D = \alpha\beta$. Hence

$$\alpha + \beta = B + u^2$$

$$\beta - \alpha = \frac{C}{u}$$

Therefore

$$2\beta = B + u^2 + \frac{C}{u} \qquad\qquad 2\alpha = u^2 + B - \frac{C}{u}$$

and $4\alpha\beta = 4D$. Thus

$$u^6 + 2Bu^4 + (B^2 - 4D)u^2 - C^2 = 0 \tag{16.1}$$

But this is an equation of an even degree with a negative constant term; hence, it must have a real root, and thus the equation may be solved for u. Consequently, α and β may be found, and the fourth degree polynomial may be factored. □

In a like manner, it can be shown that an eighth degree equation must factor into two fourth degree factors with real coefficients, and in general, a 2^nth degree equation can be factored into two 2^{n-1}th degree factors with real coefficients. But since, for example, a fifth degree polynomial can be turned into an eighth degree polynomial by multiplying through by a third degree polynomial, this implies See Problem 10.

Theorem 16.1 (Fundamental Theorem of Algebra). *All polynomial expressions with real coefficients can be factored into a product of linear and quadratic terms with real coefficients.*

Since the linear terms give real solutions, and the irreducible quadratic factors give rise to complex conjugate solutions of the form $M \pm N\sqrt{-1}$, Euler concluded

Proposition 16.8. *The complex roots of any polynomial equation are of the form $M + N\sqrt{-1}$.*

Hence, all complex numbers that might be produced as solutions to a polynomial equation are of the form $M + N\sqrt{-1}$. Euler goes on to draw, as a corollary:

Proposition 16.9. *Given an equation of degree $n = \alpha + 2\beta$, it will have α real roots and 2β complex roots, which must all be of the form $M + N\sqrt{-1}$.*

The second half of "Researches on the Imaginary Roots of Equations" showed, first, that the set of complex numbers of the form $M + N\sqrt{-1}$ was closed under addition, subtraction, multiplication or division, as well as exponentiation to integral powers; second, Euler, using the trigonometric form of complex numbers, showed that the rational roots of complex numbers were, likewise, of the form $M + N\sqrt{-1}$. It was in this article that Euler put De Moivre's formula into its modern form. First, Euler showed that if ϕ had tangent $\frac{b}{a}$, and $\sqrt{a^2 + b^2} = c$, then

$$a + b\sqrt{-1} = c(\cos\phi + \sqrt{-1}\sin\phi) \tag{16.2}$$

Today, the right hand side is referred to as the **polar form** of the complex number.

Without proof, Euler then stated

Conjecture 16.1 (De Moivre's Formula).

$$(\cos\phi + \sqrt{-1}\sin\phi)^n = \cos n\phi + \sqrt{-1}\sin n\phi$$

It is not known whether Euler knew of De Moivre's work, but he was certainly the first to put it into the modern form.

To deal with the case of a real number raised to an imaginary power, Euler noted that if

$$a^{m+n\sqrt{-1}} = x + y\sqrt{-1}$$

then

$$(m + n\sqrt{-1})\ln a = \ln(x + y\sqrt{-1})$$

By treating a, x, y as variables and differentiating, Euler obtained

$$\frac{m\, da}{a} + \frac{n\, da\sqrt{-1}}{a} = \frac{dx + dy\sqrt{-1}}{x + y\sqrt{-1}}$$
$$= \frac{x\, dx + y\, dy}{x^2 + y^2} + \frac{x\, dy - y\, dx}{x^2 + y^2}\sqrt{-1}$$

Equating the real and imaginary parts, Euler found

$$\frac{m\,da}{a} = \frac{x\,dx + y\,dy}{x^2 + y^2} \qquad\qquad \frac{n\,da}{a} = \frac{x\,dy - y\,dx}{x^2 + y^2}$$

Solving the first differential equation gives us $a^m = \sqrt{x^2 + y^2}$; solving the second gives us $\arctan \frac{y}{x} = n \ln a$. Hence

$$x = a^m \cos(n \ln a) \qquad\qquad y = a^m \sin(n \ln a)$$

In a like manner, Euler showed that the logarithm of a complex number was another complex number. Like the proof of the fundamental theorem itself, this was prompted by d'Alembert's work. D'Alembert, among others, argued that the logarithm of a negative number had to be a real number, or, at least, could be treated as a real number. He gave several justifications for the idea that the logarithm of a negative number was a real, not complex, number.

The first was that the logarithm of a number, by definition, was the number in *any* arithmetic sequence that corresponded to a number in *any* geometric sequence. In particular, d'Alembert gave the example of the geometric and arithmetic sequences See page 339.

$$\begin{array}{ccccc} 1 & -2 & 4 & -8 & \ldots \\ 0 & n & 2n & 3n & \ldots \end{array}$$

Since -2 is the geometric mean between 1 and 4, d'Alembert concluded that the logarithm of -2 was the arithmetic mean between the logarithm of 1 and the logarithm of 4: in other words, $\log -2 = \frac{\log 4}{2}$.

D'Alembert also argued that it was inconceivable that a function, like the logarithmic, that produced real values for all positive numbers could suddenly begin to generate complex values. D'Alembert wrote:

> At least, this transition [from real to imaginary values] cannot occur in geometric curves. I admit the logarithmic curve is not geometric, and this conclusion isn't completely rigorous, but at least it raises the question of how a quantity represented by the abscissa of a curve can pass from $-\infty$ to an imaginary amount.[3]

A third objection was based on the fact that $(-1)^2 = 1$, so $\log(-1)^2 = \log 1$, and thus $2\log -1 = \log 1 = 0$. Thus $\log -1 = 0$, and consequently, the logarithm of any negative number was the same as the logarithm of the corresponding positive number. A fourth objection of d'Alembert relied on the fact that $\ln a = \int_1^a \frac{dy}{y}$. Since the integral was clearly an area, it had to have a value corresponding to a real number, regardless of the value of a; thus even if a was negative, $\ln a$ had to be a real number.

---

[3] D'Alembert, *Opuscules Mathematiques*, Vol. I, p. 184.

The trigonometric functions of an imaginary number were dealt with using the power series for sine or cosine, and substituting in the imaginary $b\sqrt{-1}$. Thus, Euler obtained the relationships

$$\cos(b\sqrt{-1}) = \frac{e^b + e^{-b}}{2}$$

$$\sin(b\sqrt{-1}) = \frac{e^b - e^{-b}}{2}\sqrt{-1}$$

See Problem 13.

Since all the other trigonometric functions could be expressed in terms of sine or cosine, this implied that the trigonometric functions of a complex number of the form $a + b\sqrt{-1}$ were likewise complex numbers of the same form. This can then be used to show that even the inverse trigonometric functions of complex numbers yielded complex numbers of the same form. Euler's article concluded with

Theorem 16.2 (Closure of Complex Numbers). *All imaginary quantities that are algebraic or transcendental functions of $M + N\sqrt{-1}$ are reducible to the form $A + B\sqrt{-1}$, where A, B, M, N are real numbers.*

16.3 Exercises

1. Determine the law of progression for α, β, γ, etc. for the continued fraction

$$\cfrac{\alpha}{b + \cfrac{\beta}{c + \cfrac{\gamma}{d + {}^{\cdot\cdot\cdot}}}}$$

in terms of the infinite series supposed equal to it, $A - B + C - D + \ldots$ Then determine the values of b, c, d, $\ldots$ that will make α, β, γ whole numbers. Repeat the procedure for the infinite series $\frac{1}{A} - \frac{1}{B} + \frac{1}{C} - \cdots$

2. Show that a continued fraction expansion

$$a + \cfrac{1}{b + \cfrac{1}{c + \cfrac{1}{{}^{\cdot\cdot\cdot}}}}$$

corresponds to a rational number $\frac{p}{q}$ if and only if there are a finite number of (nonzero) terms.

3. Find the continued fraction expansion for the indicated numbers using the given series.

    (a) $\frac{1}{e} = \frac{1}{2 \cdot 1} - \frac{1}{3 \cdot 2 \cdot 1} + \frac{1}{4 \cdot 3 \cdot 2 \cdot 1} + \cdots$

    (b) $\frac{2}{3} = 1 - \frac{1}{2} + \frac{1}{4} - \frac{1}{8} + \cdots$

4. Find the value of the given continued fraction.

    (a)

    $$2 + \cfrac{1}{3 + \cfrac{1}{3 + \cfrac{1}{\ddots}}}$$

    (b)

    $$1 + \cfrac{1}{1 + \cfrac{1}{1 + \cfrac{1}{\ddots}}}$$

5. Suppose

    $$x = a + \cfrac{1}{b + \cfrac{1}{c + \cfrac{1}{b + \cfrac{1}{c + \cfrac{1}{\ddots}}}}}$$

    (a) Determine the irrational number it is equal to.

    (b) Show that any continued fraction with periodic denominators of any period is equal to an irrational number of the form $p + \sqrt{q}$.

6. For rational numbers, the continued fraction expansions may be found using the following principle: if $\frac{p}{q} = m + \frac{r}{q}$, with $0 < r < q$, then $\frac{p}{q} = m + \cfrac{1}{\cfrac{q}{r}}$.

    Then r may be divided into q to obtain the next stage of the process.

    (a) Find the continued fraction expansions for the following.

        i. $\frac{29}{17}$ ii. $\frac{2783}{891}$ iii. $\frac{355}{113}$

    (b) Prove that the continued fraction expansion of a rational number is finite.

7. Euler assumed, but did not prove, that if $x = a$ is a root of a polynomial, then $x - a$ is a factor of the polynomial. Prove this.

8. Consider Euler's proof of Proposition 16.5. What features of this proof are unacceptable today? Modify the proof so that it is acceptable; what must be assumed?

9. Prove Proposition 16.6 by considering the value of the function at $x = 0$ and $x = \infty$, $x = -\infty$.

10. Prove Theorem 5 of "Researches on the Imaginary Roots of Equations": an eighth degree equation must be factorable into two fourth degree factors. Remember the $n-1$ term of a nth degree equation can always be eliminated.

11. Use Euler's method to prove that the logarithm of a complex number of the form $M + N\sqrt{-1}$ produces another complex number of the same form.

12. Examine d'Alembert's arguments that the logarithm of a negative number is a real number. Identify the flaws in the four arguments.

13. Prove

$$\cos(b\sqrt{-1}) = \frac{e^b + e^{-b}}{2} \qquad \text{and} \qquad \sin(b\sqrt{-1}) = \frac{e^b - e^{-b}}{2}\sqrt{-1}$$

14. Prove the inverse sine or cosine of a complex number of the form $p + q\sqrt{-1}$ is a complex number of the form $a + b\sqrt{-1}$.

15. Explain how to reduce a polynomial with repeated roots to one with no repeated roots. Hint: use Hudde's rules. See page 408.

16. Consider the proof of Proposition 16.7.

    (a) Explain why the cubic term in the quartic can always be eliminated.

    (b) Why is it insufficient to note that the roots of a fourth degree equation can be found by the Cardano-Ferrari method?

    (c) Why does the ability to find a real solution to Equation 16.1 imply the proposition?

Chapter 17

Number Theory

Euler lived during the era in European history known as the Enlightenment, which emphasized the rational pursuit of human happiness, freedom, and knowledge. A characteristic development was the so-called **enlightened despot**: an absolute ruler who sought to better his or her subjects. Peter the Great of Russia is considered one of the earliest of the enlightened despots, and his methods were typical: his subjects would be brought into the eighteenth century, by force if necessary, because it was for their own good. Catherine the Great continued Peter's reforms in Russia. She was only one of several like-minded monarchs, including Maria Theresa and her son and successor, Joseph II, and Frederick the Great of Prussia, whose revitalization of the Berlin Society of the Sciences was but a single step in fostering education among his people.

Not all monarchs felt that it was their duty to ensure the betterment of their people, and some adhered to the "Divine Right of Kings": a monarch is God's representative on Earth, and thus responsible to no earthly power. The most vocal adherent of this doctrine was James II of England, and it ultimately cost him his throne and sent him into exile at the court of Louis XIV. An even worse fate was in store for Louis's descendants, who continued to treat the French state as if it was their own personal property, to do with as they pleased.

17.1 Euler and Fermat's Conjectures

The lives of many mathematicians divide neatly into intervals in which they almost exclusively investigate a single subject. Not so with Euler, who pursued all his interests throughout all his career. Thus, at the same time Euler was developing the infinite series into one of the most powerful tools in the mathematician's arsenal, he was also raising number theory to the status of a major field in mathematics. Even today, Euler's work on number theory is cited in professional journals by working mathematicians.

Shortly after Euler arrived in St. Petersburg in 1726, Christian Goldbach informed him of some of Fermat's results. Like so many of Fermat's correspondents, Euler believed number theory not worth his time. But soon he changed his mind, and if Fermat is the father of number theory, Euler raised it (or at least brought it to its adolescence).

In "Observations on Some Theorems by Fermat and Others Regarding Primes," published in 1738, but written around 1732, Euler began with some preliminary observations, stated without proofs. First, $a^n + 1$ always has a factor if n is odd, or if n can be divided by an odd number besides 1. Hence, if $a^n + 1$ is to be prime, n must be a power of 2. Moreover, if a was odd, then $a^n + 1$ was obviously composite (since it would have a factor of 2); and if a was even, sometimes $a^n + 1$ was composite and sometimes it was prime; for example, if $a = 5b \pm 3$, then $a^2 + 1$ was divisible by 5. Finally, Euler disproved Fermat's conjecture regarding the so-called Fermat primes, by showing $2^{2^5} + 1$ has a factor of 641. Thus Euler began the revival of number theory by refuting one of Fermat's conjectures.

17.1.1 Fermat's Lesser Theorem

See page 353.

Euler gave the first proof of Fermat's lesser theorem in "Proof of a Theorem Regarding Prime Numbers" (1736). The proof is a very elegant example of a proof by mathematical induction.

First, Euler proved that $2^{p-1} - 1$ is divisible by p. He did so by writing 2 as $1 + 1$, then expanding $(1 + 1)^{p-1}$ as

$$(1+1)^{p-1} = 1 + \frac{p-1}{1} + \frac{(p-1)(p-2)}{1 \cdot 2} + \frac{(p-1)(p-2)(p-3)}{1 \cdot 2 \cdot 3} + \ldots$$

where all the terms are integers. If p is a prime greater than 2, then $p - 1$ is even and thus the number of terms is odd. Subtracting 1 and collecting the remaining terms pairwise, Euler obtained

$$(1+1)^{p-1} - 1 = \frac{p(p-1)}{1 \cdot 2} + \frac{p(p-1)(p-2)(p-3)}{1 \cdot 2 \cdot 3 \cdot 4} + \ldots$$

This is obviously divisible by p; hence $2^{p-1} - 1$ is divisible by any odd prime p. Obviously, $2^p - 2$ is also divisible by p.

Euler then included what, in modern texts, might be considered an unnecessary proof: if p is a prime not equal to 3, then $3^{p-1} - 1$ is divisible by p, and thus p must also divide $3^p - 3$ and conversely. This follows because 3^p can be expanded as

$$(1+2)^p = 1 + \frac{p}{1}2 + \frac{p(p-1)}{1 \cdot 2}2^2 + \ldots + \frac{p}{1}2^{p-1} + 2^p$$

Clearly $3^p - 2^p - 1$ will be divisible by p, since all the remaining terms will have a factor of p. But $3^p - 2^p - 1 = 3^p - 3 - 2^p + 2$; hence, since $2^p - 2$ is divisible by p, it follows that $3^p - 3$ is divisible by p, and thus $3^{p-1} - 1$ is also divisible by p if $p \neq 3$.

Euler's motivation in including this proof may have been to show the process of developing a mathematical proof. Knowing that $2^p - 2$ was divisible by p allowed him to prove $3^p - 3$ was also divisible by p, which suggested, in general, that if $a^p - a$ was divisible by p, so would $(a+1)^p - (a+1)$, a fact that Euler proved. Thus, if a prime p does not divide a, then $a^{p-1} - 1$ is divisible by p.

17.1.2 Fermat's Last Conjecture

Euler was the first to make significant progress toward proving Fermat's last conjecture, though Fermat himself included a partial proof of the case where $n = 4$. The proof that the sum of two fourth powers could not equal another fourth power appeared "Demonstrations of Some Arithmetic Theorems" (1747), written around 1738. See page 358.

The proof is fairly simple. Euler proved

Lemma 17.1. *If $a^2 + b^2$ is a square, and a and b are relatively prime, then $a = p^2 - q^2$ and $b = 2pq$ for some p and q; moreover, p and q have no common factor, and one will be even, and the other odd.*

Then

Lemma 17.2. *If $a^2 - b^2 = c^2$ and a and b have no common factors, then a is odd, and there exist p, q with no common factors, with $a = p^2 + q^2$, and either $b = p^2 - q^2$ or $b = 2pq$. Moreover, p will be odd and q even, and have no common factors.*

This leads to

Proposition 17.1. *The sum of two fourth powers $a^4 + b^4$ relatively prime to one another cannot be equal to a square.*

Proof. Suppose a is odd and b even. By Lemma 17.1, there exist p, q, relatively prime, such that $a^2 = p^2 - q^2$, $b^2 = 2pq$, where p is odd and q is even, and p and q have no common factors. Since $2pq$ is a square, this means that p and $2q$ are separately squares.

Since $p^2 - q^2$ is a square, there are m, n where $p = m^2 + n^2$ and $q = 2mn$, with m, n having no common factors. But $2q$ is a square; hence $4mn$ is a square, and thus m, n are separately squares, say $m = x^2$ and $n = y^2$. Hence $p = x^4 + y^4$. But x and y are less than a and b, so if $a^4 + b^4$ is a square, we can always find a smaller pair of numbers $x^4 + y^4$ that is also a square. But this is impossible; hence, there can be no numbers whose fourth powers together make a square. $\square$

As the first corollary, Euler noted that $a^4 + b^4$ cannot equal c^4 for any non-trivial values of a, b, or c.

Around the time *Demonstrations* was written, Euler was also engaged in a massive cartographic project, mapping out portions of the Russian Empire.

Through overwork, he lost the sight of one eye and begged Goldbach to reduce his duties; this Goldbach did, reassigning Euler to less exhausting tasks.

17.1 Exercises

1. Show that $a^n + 1$ is composite if n is odd, or if n can be divided by an odd number besides 1.

2. Show that if $a = 5b \pm 3$, then $a^2 + 1$ has a factor of 5.

3. Prove that if $a^p - a$ is divisible by p, then $(a+1)^p - a - 1$ is also divisible by p.

4. Show that if a is not divisible by p and p divides $a^p - a$, then p also divides $a^{p-1} - 1$.

5. Euler did not actually prove Fermat's lesser theorem in the manner it is currently stated, but rather that for *any* positive number a, then $a^p - a$ is divisible by p. Complete the proof of the modern form of Fermat's lesser theorem by showing that so long as p does not divide a, then p divides $a^{p-1} - 1$.

6. Prove Lemma 17.1, by first supposing that $a^2 + b^2 = \left(a + \frac{bq}{p}\right)^2$, where p and q have no common factor.

   (a) Show that $a : b = p^2 - q^2 : 2pq$.

   (b) Show that if $p^2 - q^2$, $2pq$ have no common factor, then $a = p^2 - q^2$ and $b = 2pq$.

   (c) Show that if $p^2 - q^2$, $2pq$ have a common factor, it is 2 and $a = \frac{p^2 - q^2}{2}$, $b = pq$; hence there exist numbers r and s with no common factors where $a = 2rs$ and $b = r^2 - s^2$.

   (d) Prove Corollary 1 to this lemma: if the sum of two squares is a square, one must be even and the other odd; hence (Corollary 2), we may assume that $a = p^2 - q^2$ and $b = 2pq$.

17.2   The Berlin Years

Some of Euler's most important results in number theory were written during his stay in Berlin, though most of his papers on the subject appeared in the St. Petersburg journal. One of the characteristics of Euler's work was that he would return, again and again, to subjects he had already examined and "solved." Thus, though he had already given a proof of Fermat's lesser theorem, he would return to prove it four more times in his mathematical career, each time generalizing the result a little more.

17.2.1 *Theorems on Remainders*

Euler gave a second proof of Fermat's lesser theorem in "Theorems on the Remainders of the Division of Powers" (1758). Some of the ideas and methods used in the paper foreshadowed the development of group theory. Given a prime p and a number a relatively prime to p, Euler considered the remainders when the powers

$$1, a, a^2, a^3, a^4, \ldots$$

were divided by p. Euler first proved none of these remainders were divisible by p (theorem 1); hence, all the remainders were nonzero. In quick succession, Euler proved

Proposition 17.2. *If a^μ on division by p leaves a remainder of r, and a^ν on division by p leaves a remainder of s, then $a^{\mu+\nu}$, on division by p, leaves a remainder of rs.*

Proposition 17.3. *In the infinite sequence*

$$1, a, a^2, a^3, \ldots$$

are infinitely many terms whose remainder, on division by p, is 1; the exponents of these terms form an arithmetic sequence.

Proof. Since there are infinitely many terms in the sequence, but only finitely many possible remainders, then at least two of the terms, say a^μ and a^ν, must have the same remainder, r, on division by p; we may assume $\mu > \nu$. Hence, p must divide $a^\mu - a^\nu$, and thus p must divide $a^\nu (a^{\mu-\nu} - 1)$. But p is prime, and cannot divide a^ν; hence $a^{\mu-\nu}$, on division by p, must leave a remainder of 1.

Let $\mu - \nu = \lambda$. Then the sequence a^λ, $a^{2\lambda}$, $a^{3\lambda}$, ..., whose exponents form an infinite arithmetic sequence, must also have a remainder of 1 on division by p □

An important corollary to this is that if λ is the smallest power for which a^λ has a remainder of 1 on division by p, then λ must necessarily be less than p. In the remainder of the discussion, assume λ is the smallest power of a for which a^λ leaves a remainder of 1 on division by p.

Proposition 17.4. *If the remainder of a^μ, on division by p, is r, and the remainder of $a^{\mu+\nu}$ is rs, then a^ν, on division by p, leaves a remainder of s.*

This gives

Proposition 17.5. *The only powers of a that leave a remainder of 1 on division by p are 1, a^λ, $a^{2\lambda}$, $a^{3\lambda}$, ...*

One of the main results is

Proposition 17.6. *The remainders formed by the division of the terms of the sequence*

$$1, a, a^2, a^3, \ldots, a^{\lambda-1}$$

are all distinct.

Proof. Suppose a^μ, a^ν, with $\nu < \mu < \lambda$, have the same remainder on division by p. Then $a^\mu - a^\nu = a^\nu (a^{\mu-\nu} - 1)$ is divisible by p; hence $a^{\mu-\nu}$ will have a remainder of 1. But $\mu - \nu$ is less than λ, contrary to supposition.   □

Theorem 8 and its corollaries are that the remainders formed by dividing

$$1, a, a^2, a^3, \ldots, a^{\lambda-1}$$

are the only ones possible. Hence:

Proposition 17.7. *If all numbers less than p appear among the remainders of the powers of a divided by p, then the least power λ for which a^λ has a remainder of 1 is $\lambda = p - 1$.*

Proof. λ must be less than p; there are (by Theorem 7) λ distinct remainders in the sequence

$$1, a, a^2, \ldots, a^{\lambda-1}$$

which are all the remainders that are possible when a power of a is divided by p. However, by assumption, every number less than p appears as a residue; hence, there must be $p - 1$ possible remainders. Hence, $\lambda = p - 1$.   □

Proposition 17.8. *If the number of possible remainders is less than $p - 1$, then there must be at least as many nonremainders as there are remainders.*

Euler's proof of this last proposition relied, in essence, on the notion of a **coset** in group theory: if one took the set of remainders and multiplied each element of the set by a nonremainder k, one would end with a distinct set of nonremainders; hence, as there are λ remainders, there must be at least λ nonremainders. A corollary is that if $\lambda < p - 1$, then $\lambda \leq \frac{p-1}{2}$. Theorems 12 and 13 extended the proof: if $\lambda < \frac{p-1}{2}$, then $\lambda \leq \frac{p-1}{3}$; and (Theorem 13), if $\lambda < \frac{p-1}{3}$, then $\lambda \leq \frac{p-1}{4}$.

This set the stage for the proof of Fermat's lesser theorem, the proof of which, since all the key results on remainders were already proven, is remarkably simple:

Theorem 17.1 (Fermat's Lesser Theorem). *If p is prime, and a relatively prime to p and λ the smallest power for which a^λ leaves a remainder of 1 on division by p, then λ divides $p - 1$.*

Proof. There are λ distinct remainders and $p - 1 - \lambda$ nonremainders. λ must divide $p - 1 - \lambda$; hence λ divides $p - 1$.   □

17.2.2   *Observation in Mathematics*

One of the great questions that might be asked is how Euler managed to be so prolific. At least part of the answer is that much of Euler's work consists of fairly elementary results, which were significant only because no one else had noticed them before. However, this should not detract from Euler's genius: part of genius *is* noticing things that no other person has noticed. Around 1756, Euler wrote one of his more interesting papers, "Examples on the Use of Observation in Pure Mathematics." Published in 1761, "Examples" gives us great insight into the workings of Euler's mind. Perhaps it is impossible to learn how to be a great mathematician, but if it is possible, surely the best way is to see how the great mathematicians themselves thought and worked.

It was known that if a number was prime, it could be decomposed as the sum of two squares in only one way; elementary algebra is sufficient to prove this. Euler proved, in "Examples," that primes also had a unique decomposition as a sum of the form $2a^2 + b^2$. The proof is fairly elementary, and the proposition is easy to prove—*once* it is conjectured. But how does one come up with the conjecture in the first place?

Euler began by listing all possible sums of the form $2a^2 + b^2$ less than 500. From this list, Euler made a few observations. The first two of Euler's observations were:

1. If a prime p is on the table, it has a unique decomposition on the table.

2. If a prime p is on the table, $2p$ is a number on the table that has a unique decomposition.

The first observation led Euler to suspect that any prime p could be decomposed as $2a^2 + b^2$ in a unique way. The other observations suggested means by which the proposition could be proven. Thus, Euler proved, in rapid succession:

Proposition 17.9. *If $N = 2m^2 + n^2$, then $2N$ can also be expressed as $2k^2 + l^2$.*

Corollary. *If N has several decompositions of the form $2m^2 + n^2$, so does $2N$.*

Proposition 17.10. *If $2N = 2m^2 + n^2$, then N can also be expressed as $2k^2 + l^2$.*

Proposition 17.11. *If M, N are numbers that can be expressed in the form $2m^2 + n^2$, MN is also a number that can be expressed in this form.*

Proposition 17.12. *If $N = 2a^2 + b^2 = 2c^2 + d^2$ are two different decompositions of N, then N is not prime.*

Proof. Suppose $N = 2a^2 + b^2$ and $N = 2c^2 + d^2$. Multiplying the first by a^2 and the second by c^2 and subtracting, we have $(a^2 - c^2)N = (ad - bc)(ad + bc)$. If N is prime, then it must divide one of the factors, either $(ad - bc)$ or $(ad + bc)$. However, $2N = 2a^2 + b^2 + 2c^2 + d^2$. Subtracting $2ad + 2bc$ from both sides, we have

$$2N - 2ad - 2bc = a^2 + (a - d)^2 + c^2 + (c - b)^2$$

The right-hand side is the sum of four squares, and hence must be positive; hence

$$N > ad + bc$$

Thus, N cannot divide $(ad + bc)$; by assumption, then, N must be able to divide $(ad - bc)$. But this is impossible. Hence, N cannot be prime. $\square$

Corollary. *If N is prime with a decomposition of the form $2m^2 + n^2$, the decomposition is unique.*

17.2.3   Results on Prime Numbers

See page 355.

Another conjecture of Fermat was that all primes that were one more than a multiple of 4 were the sum of two squares. Euler proved this in two papers, "Regarding Numbers Which Are the Sum of Two Squares" (1758), and "Demonstration of Fermat's Theorem That All Prime Numbers of the Form $4n + 1$ Are the Sum of Two Squares" (1760).

In the first article, Euler began by listing all the numbers less than or equal to 200 which could be expressed as the sum of two squares, then all those that could not be expressed in this way. Since even squares are of the form $4a$ and odd squares are of the form $8b + 1$, Euler concluded that the sum of two squares was either divisible by 4, or of the form $4n + 1$ or $8n + 2$.

See Problem 3.

Euler then proved three lemmas:

Lemma 17.3. *If p is the sum of two squares, so is $n^2 p$.*

Lemma 17.4. *If p is the sum of two squares, so are $2p$ and $2n^2 p$.*

Lemma 17.5. *If $2p$ is the sum of two squares, so is p.*

Next comes

Proposition 17.13. *If p and q are the sum of two squares, so is their product.*

Proof. If $p = a^2 + b^2$ and $q = c^2 + d^2$, then

$$pq = (a^2 + b^2)(c^2 + d^2)$$
$$= (ac + bd)^2 + (ad - bc)^2$$

$\square$

This leads to

Proposition 17.14. *If pq is the sum of two squares, and p is prime and the sum of two squares, then q is also expressible as the sum of two squares.*

Proof. Suppose $pq = a^2 + b^2$ and $p = c^2 + d^2$; hence $c^2 + d^2$ divides $a^2 + b^2$. Thus $c^2 + d^2$ divides $c^2(a^2 + b^2) - a^2(c^2 + d^2)$, and thus $c^2 + d^2$ divides $(bc - ad)(bc + ad)$.

Since $c^2 + d^2$ is prime, then c and d have no common factors, and $c^2 + d^2$ must divide one of $bc \pm ad$, so $bc \pm ad = mc^2 + md^2$ for some m. Let $b = mc + x$, $a = \pm md + y$; hence $cx \pm dy = 0$, and thus $\frac{x}{y} = \mp\frac{d}{c}$. Let $x = nd$ and $y = \mp nc$. Thus

$$a = \pm md \mp nc$$
$$b = mc + nd$$

Hence $pq = m^2 d^2 - 2mncd + n^2 c^2 + m^2 c^2 + 2mncd + n^2 d^2$, or $pq = (m^2 + n^2)(c^2 + d^2)$. Hence $q = m^2 + n^2$, which was to be proven. $\square$

Hence

Corollary. *If the sum of two squares is divisible by a prime that is the sum of two squares, the quotient is also the sum of two squares.*

This leads easily to

Proposition 17.15. *If pq is the sum of two squares, and q cannot be expressed as the sum of two squares, then either p is a prime not expressible as the sum of two squares, or p has a prime factor that is not the sum of two squares.*

Euler gave the example of 45, which is the sum of two squares, and 3, which cannot be expressed as the sum of two squares: since $45 = 3 \cdot 15$. Since 15 is not prime, one of its prime factors is inexpressible as the sum of two squares.

After this comes

Proposition 17.16. *If the sum $a^2 + b^2$ of two numbers relatively prime to each other is divisible by p, then another sum $c^2 + d^2$ can be found, also divisible by p, which is less than $\frac{1}{2}p^2$.*

An important, but unstated, result is that c and d can always be found that are relatively prime to each other.

This leads to a key proposition:

Proposition 17.17. *The sum of two squares relatively prime to one another can not be divided by a number that is not the sum of two squares.*

The proof uses Fermat's method of infinite descent:

Proof. Suppose p is not the sum of two squares, and $a^2 + b^2$ is divisible by p. We can find another sum, $c^2 + d^2$, also divisible by p and less than $\frac{1}{2}p^2$. Hence $c^2 + d^2 = pq < \frac{1}{2}p^2$. q cannot be the sum of two squares, since if it was, so would p. Hence one of its factors, say r, is not the sum of two squares. Since $pq < \frac{1}{2}p^2$, Thus $q < \frac{1}{2}p$ and thus $r < \frac{1}{2}p$, and r divides $c^2 + d^2$. Hence, by the preceding proposition, we can find $e^2 + f^2 < \frac{1}{2}r^2 < \frac{1}{8}p^2$, which is divisible by r, assumed not to be the sum of two squares. We can obviously continue this procedure indefinitely, finding smaller and smaller sums of two squares, but this is impossible. □

This gives

Corollary. *If the sum of two squares is not prime, then all its prime factors must be the sum of two squares.*

Corollary. *All numbers that are the sum of two squares prime to one another are, except for 2, primes of the form $4n + 1$, or products of such primes.*

This is the converse of Fermat's conjecture; Euler then provides a partial proof.

Proposition 17.18. *All prime numbers of the form $4n + 1$ can be expressed as the sum of two squares.*

Proof. Given a prime of the form $4n + 1$, and a, b not divisible by $4n + 1$, then $4n + 1$ divides $a^{4n} - b^{4n}$. Hence $4n + 1$ must divide either $a^{2n} - b^{2n}$, or $a^{2n} + b^{2n}$.

Suppose $4n + 1$ divides $a^{2n} + b^{2n}$. Let $p = a^n$, $q = b^n$. Then $4n + 1$ divides $p^2 + q^2$, and p, q are not divisible by $4n + 1$. If m is the greatest common divisor of p, q, then $p^2 + q^2 = m^2(r^s + s^2)$, where r, s are some other numbers, and $4n + 1$ must divide $r^2 + s^2$. But any divisor of the sum of two squares must also be expressible as the sum of two squares; hence the prime $4n + 1$ is expressible as the sum of two squares. $\qquad\square$

See Problem 5.

Euler required that the prime $4n + 1$ divide $a^{2n} + b^{2n}$, and not $a^{2n} - b^{2n}$, for some a, b, a fact he was unable to prove in 1758. The remaining portion of the proof appeared in "Demonstration of Fermat's Theorem That All Prime Numbers of the Form $4n + 1$ Are the Sum of Two Squares" (1760). To prove this, Euler began by considering the sequence

$$1, 2^{2n}, 3^{2n}, 4^{2n}, \ldots, (4n)^{2n}$$

If the first differences

$$2^{2n} - 1, 3^{2n} - 2^{2n}, 4^{2n} - 3^{2n}, \ldots, (4n)^{2n} - (4n - 1)^{2n}$$

were all divisible by $4n + 1$, then the second, third, etc., differences would also be divisible, and the $2n$th differences would be constant, and equal to $1 \cdot 2 \cdot 3 \cdot 4 \cdots 2n$. However, the prime $4n + 1$ cannot divide this quantity; hence at least one of the first differences must not be divisible by $4n + 1$. Thus, there always exists an a, b, where $a^{2n} - b^{2n}$ is not divisible by $4n + 1$, and Fermat's conjecture is proven.

17.2.4 Quadratic Residues

Euler's proof that all primes of the form $4n + 1$ can be expressed as the sum of two squares is an example of what is sometimes referred to as **naive number theory**, because the proofs are primarily algebraic. Though most of Euler's proofs are done algebraically, some began to use more advanced number theoretic concepts. Thus, Euler's 1754 partial proof of Fermat's conjecture that every number could written as the sum of at most four squares, began the study of **quadratic residues**: a is a quadratic residue of p if there exists some number n for which a is the remainder when n^2 is divided by p. For example, 5 is a quadratic residue of 11, since 5 is the remainder when 4^2 is divided by 11.

Euler did not completely prove Fermat's conjecture that all numbers are the sum of four or fewer squares; however, he was able to prove

Proposition 17.19. *All numbers are the sum of four or fewer squares, some of which may be fractional.*

Euler would return periodically to the problem, but a final proof of Fermat's conjecture would be up to Lagrange (1770).

17.2.5 Fermat's Lesser Theorem (New Proof)

One of the characteristics of Euler's work in mathematics is that he never abandoned a subject completely and would periodically return to it, deriving new results or providing simpler proofs: Fermat's lesser theorem was proven five times by Euler, and generalized along the way. Perhaps the most important of his proofs was his third, in the 1760 article "A New Demonstration of a Theorem of Arithmetic," which established what is now called the **Euler-Fermat theorem**.

Euler proved

Proposition 17.20. *Given the n terms of an arithmetic progression with common difference d relatively prime to n. The remainders when the terms are divided by n will include all numbers less than n.*

Proof. Let a be the first term of the progression; it is sufficient to consider the n terms

$$a \quad a+d \quad a+2d \quad a+3d \quad \ldots \quad a+(n-1)d$$

Suppose two of these have the same remainder on division by n, say $a + \mu d$ and $a + \nu d$. Then the difference $\mu d - \nu d$, or $(\mu - \nu)d$ is divisible by n. Since n, d have no common factors, n must divide $\mu - \nu$. However, this is impossible, since $\mu - \nu$ is less than n. Hence, each of the n terms must have a different remainder on division by n, and since there are only n possible remainders, every possible remainder must be formed. □

Next, Euler proved

Proposition 17.21. *Given n terms in an arithmetic progression with a difference d, where d, n are relatively prime, among the remainders when the terms are divided by n, there are as many terms that are relatively prime to n as there are numbers less than n that are relatively prime to it.*

Here is the first investigation of what would become designated the **Euler phi function**: $\phi(n)$ is the number of numbers less than n that are relatively prime to n. The notation $\phi(n)$ is due to Gauss, and Euler himself nowhere named this function.

The next few theorems relate to the properties of the unnamed function:

Proposition 17.22. *If n is any power of a prime p, say $n = p^m$, then among the numbers less than n, there will be*

$$p^m - p^{m-1} = p^{m-1}(p - 1)$$

which are relatively prime to n.

Euler used a beautiful counting argument to prove this theorem. A variation of the counting argument was used to prove the next theorem:

Proposition 17.23. *If $n = pq$, where p and q are unequal primes, then the number of numbers less than n that are relatively prime to it is $(p-1)(q-1)$.*

Proof. The numbers less than pq that are relatively prime to p are

$$
\begin{array}{ccccc}
1 & 2 & 3 & \cdots & p-1 \\
p+1 & p+2 & p+3 & \cdots & 2p-1 \\
2p+1 & 2p+2 & 2p+3 & \cdots & 3p-1 \\
\vdots & \vdots & \vdots & \ddots & \vdots \\
(q-1)p+1 & (q-1)p+2 & (q-1)p+3 & \cdots & pq-1
\end{array}
$$

There are $p-1$ columns, and note that each column consists of q entries of an arithmetic sequence with difference p, relatively prime to q. Hence, if the terms of each column are divided by q, their remainders will include all numbers less than q, so exactly one of the terms will be divisible by q, and the remainder will not be; since q is prime, this means that exactly one term in each column will have a factor in common with q. Thus, there are $p-1$ columns containing $q-1$ entries relatively prime to both p and q. $\qquad\square$

Using a similar technique, Euler proved

Proposition 17.24. *If A and B are relatively prime, and the number of parts to which A is relatively prime is a, and the number of parts to which B is relatively prime is b, then the number of parts less than AB that are relatively prime to AB is ab.*

Euler here is using the Euclidean meaning of the word parts, namely that the parts of a number A are the numbers less than A.

The next theorem is

Proposition 17.25. *If x is relatively prime to N, then the remainder when any power of x is divided by N will be relatively prime to N.*

A critical step in proving the generalized version of Fermat's lesser theorem is

Proposition 17.26. *If the series of powers*

$$
x^0, x^1, x^2, x^3, \ldots
$$

is divided by N, among the remainders will be 1; moreover, this remainder will appear repeatedly.

Proof. At least two of the terms will have the same remainder on division by N (since there are only a finite number of possible remainders), say $x^{\mu+\nu}$ and x^{ν}. Thus, N will divide $x^{\mu+\nu} - x^{\nu} = x^{\nu}(x^{\mu} - 1)$. But N cannot divide x^{ν}, hence it must divide $x^{\nu} - 1$, and thus the remainder from the division of x^{ν} by N is 1. Moreover, it is obvious that the remainders of x^{μ} and $x^{\mu+\nu}$ are divided by N are the same. $\qquad\square$

Proposition 17.27. *Suppose x and N are relatively prime; let the remainders of*

$$1 \quad x \quad x^2 \quad x^3 \quad \dots$$

on division by N be

$$1 \quad a \quad b \quad c \quad \dots$$

Then among the remainders of the division of the powers of x by N will be all powers of all products of a, b, c, ...

Euler assumed it to be understood that if the product or power of the product exceeded N, only the remainder need be considered. Finally, the key theorem is

Proposition 17.28. *The remainders when the powers of x are divided by N, assumed relatively prime to x, will either include all the numbers relatively prime to N, or exclude some of the numbers relatively prime to N, in which case the number of numbers excluded will be a multiple of the number of numbers included.*

Hence

Theorem 17.2 (Euler-Fermat Theorem). *The smallest power ν of x that has a remainder of 1 on division by N, assumed relatively prime to x, will either equal the number of numbers relatively prime to N, or be a factor of this number.*

This is Fermat's lesser theorem in its most generalized form. Euler's proof is beautifully simple.

Proof. Let the number of numbers less than N that are relatively prime to N be n, and let there be ν remainders when the powers of x are divided by N. There are thus $n-\nu$ nonremainders, which must be a multiple of ν, say $n-\nu = (m-1)\nu$. Hence $n = m\nu$, or $\nu = \frac{n}{m}$. $\qquad\square$

From this, Fermat's lesser theorem can be deduced directly.

17.2 Exercises

1. Prove the corollary to Proposition 17.3: if λ is the smallest power for which a^λ, on division by p, leaves a remainder of 1, then $\lambda < p$.

2. Prove Proposition 17.8, by showing if k is a nonremainder, then k, ak, a^2k, ..., $a^{\lambda-1}k$ are distinct nonremainders.

3. Prove that the sum of two squares is of the form $4n$, $4n+1$, or $8n+2$.

4. Prove Proposition 17.16, by letting $a = mp \pm c$, $b = np \pm d$, where c, d are less than $\frac{1}{2}p$. Prove that either c and d are prime to each other, or, if they have a common divisor q, then another pair of numbers $r^2 + s^2$, relatively prime to one another, can be divided by p.

5. In "Theorems on the Divisors of Numbers" (1750), Euler proved that if p does not divide a or b, then p must divide $a^{p-1} - b^{p-1}$. Prove this.

6. Make a list of numbers less than 500 that can be expressed as the sum of two cubes. Based on this list, what conjectures can you make, particularly in regard to prime numbers P that can be expressed as the sum of two cubes? Prove your conjecture.

7. Make a list of numbers less than 500 that can be expressed as the sum of two fourth powers. Based on this list, what conjectures can you make, particularly in regards to prime numbers P that can be expressed as the sum of two fourth powers? Prove these conjectures.

8. Based on Problems 6 and 7, what conjecture or conjectures can you make on the expressibility of a prime as the sum of two nth powers? Prove your conjecture.

9. Prove Proposition 17.22. Begin by noting there are $p^m - 1$ numbers less than p^m. Which of these numbers have a factor in common with p^m?

10. Prove Proposition 17.28. Show that if α is not a remainder when a power of x is divided by N, neither is αa, αb, ...

11. Prove Fermat's lesser theorem, as originally stated by Fermat. (See page 353.)

17.3   Return to St. Petersburg

In 1763, the Seven Years' War ended. Frederick was now tired of war and decided he would enjoy the finer things in life, which (for him) involved writing poetry, which he was especially bad at, and playing the flute, which he was actually quite good at. Unfortunately for Euler, it also meant Frederick took a more direct interest in the goings-on at the Royal Academy, and the two did not get along. Frederick often referred to the one-eyed Euler as his "cyclops," and felt Euler to be too unsophisticated for Frederick's court. In fairness, Euler's philosophical notions were primitive and, as a devout Catholic, he found himself at odds with the Enlightenment. By 1766, Euler had had enough of Frederick's intervention in the running of the academy, and he accepted an invitation from the Tsarina Catherine to return to Saint Petersburg. He left the Berlin Academy in the capable hands of one of his protégés—Joseph Louis Lagrange.

17.3.1   Fermat's Last Conjecture

Although most of Euler's great mathematical discoveries had been made by the time he returned to St. Petersburg, he was still capable of producing mathematical gems. *Elements of Algebra* (1770) included a proof of Fermat's conjecture for cubes. Euler began by establishing a number of minor lemmas. First, if $x^3 + y^3 = z^3$ is unsolvable, so is the difference $x^3 - y^3 = z^3$. Second, x and y may be considered relatively prime to each other. Finally, one of x, y, z must be even, and the others must be odd, and it is sufficient to consider only the case where both x and y are odd.

Proposition 17.29. *The equation $x^3 + y^3 = z^3$ has no non-trivial integer solutions.*

Proof. Since x, y are odd, then $x + y$ and $x - y$ are even, and designating $p = \frac{x+y}{2}$, $q = \frac{x-y}{2}$ and so $x = p + q$, $y = p - q$, then one of p or q is even, and the other odd; since x, y have no common factors, p and q must likewise have no common factors. Since $x^3 + y^3 = 2p(p^2 + 3q^2)$, then it remains to be shown that $2p(p^2 + 3q^2)$ cannot be a cube.

Suppose it is. The expression is even; hence, 1/8th of it must be a whole number; hence $\frac{1}{4}p(p^2 + 3q^2)$ must also be a whole number and a cube. $p^2 + 3q^2$ must be odd, and thus cannot be divisible by 4; hence p is divisible by 4 and q is even.

Either p, $p^2 + 3q^2$ have no common factor, or they have a common factor, and it is 3. Suppose first that p is not divisible by 3, and hence p and $p^2 + 3q^2$ have no common factor. Then the factors $\frac{1}{4}p$ and $(p^2 + 3q^2)$ must separately be cubes, so $p^2 + 3q^2$ must be a cube. Let $p \pm q\sqrt{-3} = (t \pm u\sqrt{-3})^3$, so $p^2 + 3q^2 = (t^2 + 3u^2)^3$. Also $p = t^3 - 9tu^2 = t(t^2 - 9u^2)$, and $q = 3t^2u - 3u^3 = 3u(t^2 - u^2)$. Since q is odd, u must be odd, and t even. See Problem 2.

Since $\frac{p}{4}$ is a cube, then $2p$ is likewise a cube, and hence $2t(t - 3u)(t + 3u)$ is a cube. But t is even and not divisible by 3 (since if t is divisible by 3, so is p, contrary to hypothesis), hence $2t$, $t - 3u$, and $t + 3u$ must be relatively prime to each other, so each must individually be a cube. Suppose $t - 3u = f^3$ and $t + 3u = g^3$. Then $2t$, also a cube, will be equal to the sum of two cubes, each less than x, y. Thus, given a sum of two cubes equal to a cube, we can always find a smaller set of two cubes, equal to another cube. But there are no cubes among small numbers that are equal to the sum of two cubes, so there cannot be cubes among large numbers equal to the sum of two cubes. See Problem 3.

Now, suppose the p is divisible by 3, and thus $p = 3r$ and 3 does not divide q. Then $\frac{p}{4}(p^2 + 3q^2) = \frac{3r}{4}(9r^2 + 3q^2)$, or $\frac{9r}{4}(3r^2 + q^2)$. The two factors of this last are prime to one another, since $3r^2 + q^2$ is not divisible by 2 or 3 (having had the factor of 3 removed in the last step). Therefore, the two factors must separately be cubes.

In the same manner as before, we have $q = t(t^2 - 9u^2)$ and $r = 3u(t^2 - u^2)$. Since q is odd, then t is odd and u is even; since $\frac{9r}{4}$ is a cube, then $\frac{2r}{3}$ is also a cube. Thus $2u(t^2 - u^2)$ is a cube and, as before, $2u$, $t - u$ and $t + u$ must separately be cubes. Letting $f^3 = t + u$ and $g^3 = t - u$, we have $f^3 - g^3 = 2u$, a cube, say h^3, and thus $f^3 = g^3 + h^3$. Thus, as before, given the sum of two cubes equal to a cube, we may find a smaller set of two numbers likewise equal to a cube, but there are no such numbers among the smaller numbers, so there cannot be any among the larger. $\qquad\square$ $\frac{2r}{3}$ is also a whole number, since $r = 3u(t^2 - u^2)$ has a factor of 3.

This proof, perhaps, represents the limit of what could be done using purely algebraic techniques in number theory.

17.3.2   Euler's Last Years

In 1771, after the publication of *Elements of Algebra*, an illness cost Euler the use of his remaining eye. A few years later, a surgeon attempted to restore Euler's sight. The surgery worked briefly, but soon after Euler was essentially blind, though he could still make out large written letters. Thus, Euler had a dining room table fitted with a slate top so he could continue to write and calculate. Remarkably, Euler's production actually increased on his blindness: nearly half his works were written after he had become blind, including some sizable works with very involved computations, most of which Euler did in his head. According to one story, two of his students disagreed in the fiftieth decimal place about the sum of the first 17 terms of a series; Euler summed the series in his head, and announced the correct answer.[1]

In 1773, Euler's wife died; three years later, he married his late wife's half sister. Around 5 P.M. on September 18, 1783, after having spent a normal day teaching a lesson in mathematics to one of his grandsons, dictating some correspondences, and working on the problem of the orbit of the newly discovered planet Uranus, Euler said, "I am dying" and lost consciousness; he had suffered a stroke and died later that day.

Euler marked a great transition in mathematics. Before Euler, it was possible for a single person to master the whole of mathematics, and make vital contributions throughout. After Euler, it took an extraordinary individual to make significant contributions in more than one or two specialized areas. It could be done, but after Euler, generalists were rare, and mathematicians were number theorists, or geometers, or analysts.

One apocryphal story about Euler deserves to be told, if only to be refuted; the story is told by Augustus de Morgan, who lived a century after Euler. It was said that while at Catherine's court in 1773, the French encyclopedist Denis Diderot made himself a nuisance by his openly atheist views. Euler, a good Catholic, was called upon to silence him, so he let it be known to Diderot that he had a mathematical proof of the existence of God. This intrigued Diderot; in full view of the court, he asked Euler to prove the existence of God. "Sir," Euler said, "$\frac{a+b_n}{n} = x$, therefore God exists. Refute that!" Diderot (who knew mathematics well enough to write articles in his *Encyclopedia* about it) realized he had been made a fool of, and eventually returned to his native France. There, his *Encyclopedia* was causing people to question the ancient monarchy. Shortly after Diderot's return, Louis XV died. He was succeeded by his grandson, Louis XVI, who had married one of Maria Theresa's daughters: a flighty, insipid girl named Marie Antoinette.

17.3 Exercises

1. Prove that in the equation $x^3 + y^3 = z^3$, that one of x, y, or z must be even and the other two odd; and that the proving $x^3 + y^3 = z^3$ impossible for x and y odd is sufficient to prove that it is impossible in general.

---

[1]The anecdote does not record which, if either, student was correct.

2. Prove that if p and q are relatively prime, and p and $p^2 + 3q^2$ have a common divisor, the divisor is 3.

3. Show, under the conditions Euler assumes, that $2t$, $t - 3u$, and $t + 3u$ must be relatively prime.

Chapter 18

The Revolutionary Era

The expensive and fruitless wars of Louis XIV crippled French economy. Louis's long life did not help; when he died, he had outlived his son and grandson, and it was his five-year-old great-grandson who succeeded him, as Louis XV in 1715. To raise money for the French state, Louis XV allowed John Law, a Scotsman, to launch an ambitious scheme of selling stock in the Mississippi Company, which controlled French Louisiana. Stock prices were wildly inflated, then crashed when it was realized the company was not profitable; the crash bankrupted thousands, further wrecking the French economy. The Seven Years' War added additional burdens. At the same time, France fought Britain during the French and Indian War (the two wars were fought at the same time but for different reasons and were ended by different treaties). The result was a disaster for everyone involved: France lost all her Canadian colonies, and India, and Britain emerged as a global power with an empire she was unprepared to manage.

Louis XV reigned for 59 years, outliving *his* son, and was succeeded, in 1774, by his grandson, Louis XVI. When Britain's American colonies revolted in 1776, Louis thought it would be a fine opportunity to take back some of France's losses. The result was another costly foreign entanglement, and by 1783, when the War of American Independence ended, the French state was completely bankrupt.

18.1 Analysis

One of the consequences of the priority dispute between Leibniz and Newton was that mathematics in England tended to be isolated from the developments in Europe, particularly in France. An unfortunate result was that English developments in mathematics lagged behind those from continental Europe, particularly in the field of analysis.

18.1.1 Polynomial Interpolation

Still, the English made some important contributions, particularly in those areas whose logical foundations were not questioned. One mathematician of note was Edward Waring (1736-August 15, 1798). Waring's 1770 *Meditations on Algebra* was highly regarded by Lagrange and other continental mathematicians, and Waring himself invented what are now called **Lagrange polynomials**. These appeared in "Problems Concerning Interpolations," in the *Philosophical Transactions of the Royal Society* for the year 1779. If $y = a + bx + cx^2 + \ldots + x^{n-1}$ be some polynomial function of x with a, b, c, ... unknown constants; let $x = \alpha, \beta, \gamma, \ldots$ represent n values of x, and let $y = s^\alpha, s^\beta, s^\gamma, \ldots$ represent the corresponding values of y. Then

$$y = \frac{(x-\beta)(x-\gamma)(x-\delta)\cdots}{(\alpha-\beta)(\alpha-\gamma)(\alpha-\delta)\cdots}s^\alpha + \frac{(x-\alpha)(x-\gamma)(x-\delta)\cdots}{(\beta-\alpha)(\beta-\gamma)\cdots}s^\beta + \ldots$$

Waring proved the result in a straightforward manner.   See Problem 1.

18.1.2 Foundations of Calculus

Continental mathematicians were well aware that calculus had, as yet, no sound logical basis. One of the first attempts to establish a rigorous basis for calculus was by Joseph Louis Lagrange (January 25, 1736-April 10, 1813), born Giuseppe Lodovico, in Turin, Italy. His great-grandfather was a French cavalry Captain in the service of Charles Emmanuel II of Savoy. In 1755, Lagrange communicated to Euler a new version of the calculus of variations. At the time, Euler was working on his own version of the calculus of variations, but was so impressed by Lagrange's treatment that he deliberately withheld the publication of his own work, so that Lagrange could reap the credit of being the first. Moreover, Euler soon adopted Lagrange's treatment in place of his own, and Lagrange's methods soon became the standard techniques.

Joseph Louis Lagrange, wearing the medal of the Legion of Honor. ©CORBIS.

Euler was, at the time, still in Berlin, though Frederick's interference was growing day by day. Lagrange's brilliance provided Euler a way out. Working carefully with d'Alembert, who was on good terms with Frederick, Euler and d'Alembert convinced Frederick that Lagrange, not Euler, was temperamentally a better match for the ruler than Euler. Thus, in 1766 when Euler returned to Russia, Lagrange came to Berlin.

One of the issues that troubled Lagrange was the use of "infinitesimal" and "infinitely small" quantities in calculus. Euler had, to some extent, banished these quantities by declaring that they were no different from 0, but calculus still depended on their existence as quantities somehow different from 0.

In 1772, Lagrange wrote "On a New Type of Calculus," which appeared in the journal for the Berlin Academy. Lagrange suggested a means of avoiding infinitesimal quantities. Function notation was still being developed in the eighteenth century; to give the reader an idea of the importance of notation, we will reproduce Lagrange's work as originally stated (with annotations afterward).

Given a finite function u of a variable x, and letting x vary to $x + \xi$, then "by the known theory of series," u would vary to

$$u = u + p\xi + p'\xi^2 + p''\xi^3 + \ldots \tag{18.1}$$

where p, p', ... are derived in some manner from the function u (and are dependent on x); we would write $u(x + \xi) = u(x) + p(x)\xi + p'(x)\xi^2 + p''(x)\xi^3 + \ldots$ The purpose of differential calculus, according to Lagrange, was to recover the functions p, p', p'', from u, while the purpose of integral calculus consisted of recovering u from p, p', p'', and so on.

To make the relationship between u and p, p', p'', ... more clear, Lagrange further noted that if x varied to $x + \omega$, then all the functions would change accordingly:

$$u = u + p\omega + p'\omega^2 + p''\omega^3 + \ldots$$

$$p = p + \varpi\omega + \rho\omega^2 + \sigma\omega^3 + \ldots$$

$$p' = p' + \varpi'\omega + \rho'\omega^2 + \sigma'\omega^3 + \ldots$$

$$\vdots$$

in other words, $p(x + \omega) = p(x) + \varpi(x)\omega + \rho(x)\omega^2 + \ldots$, and so on.

Lagrange then let $x + \xi$ go to $x + \xi + \omega$. Consider $u(x + \xi + \omega)$. Treating it as $u((x + \omega) + \xi)$, Equation 18.1 becomes

$$u + p\omega + p'\omega^2 + p''\omega^3 + \ldots + (p + \varpi\omega + \rho\omega^2 + \sigma\omega^3 + \ldots)\xi$$
$$+ (p' + \varpi'\omega + \rho'\omega^2 + \sigma'\omega^3 + \ldots)\xi + \ldots \tag{18.2}$$

On the other hand, treating it as $u(x + (\omega + \xi))$ gives

$$u + p(\xi + \omega) + p'(\xi + \omega)^2 + \ldots$$

(i.e., $u(x) + p(x)(\xi + \omega) + p'(x)(\xi + \omega)^2 + \ldots$) or, expanding and collecting like terms,

$$u + p\omega + p'\omega^2 + p''\omega^3 + \ldots + (p + 2p'\omega + 3p''\omega^2 + \ldots)\xi$$
$$+ (p' + 3p''\omega + 6p'''\omega^2 + \ldots)\xi + \ldots \tag{18.3}$$

By comparing these Equations 18.2 and 18.3 termwise, Lagrange noted

$$\begin{array}{cccc}
\varpi & = & 2p' & \quad \rho & = & 3p'' \\
\varpi' & = & 3p'' & \quad \rho' & = & 6p''' \\
\varpi'' & = & 4p''' & \quad \rho'' & = & 10p^{iv}
\end{array}$$

$$\vdots \qquad \vdots \qquad \vdots \qquad \vdots$$

hence $p' = \frac{\varpi}{2}$, $p'' = \frac{\varpi}{3}$, ..., or by making the appropriate choices for u', u'', ..., then if x is replaced with $x + \xi$, then u will become

$$u + u'\xi + \frac{u''\xi^2}{2} + \frac{u'''\xi^3}{2 \cdot 3} + \ldots$$

At this point, Lagrange noted that if ξ is regarded as infinitely small, one may neglect ξ^2, and so on, and thus $u'\xi$ will be du, ξ is dx, and thus $du = u'dx$ or $u' = \frac{du}{dx}$. Hence u' will just be the derivative of u; u'' will be the derivative of u', and so on. Lagrange, of course, recognized his result as Taylor's Theorem, though "it seems to me this is the simplest demonstration." What he meant was that this was the demonstration that required the fewest assumptions, not that it would produce a Taylor expansion the most simply.

One outgrowth of Lagrange's attempt to rigorize calculus was the development of new notation and new terminology. In particular, since the functions u', u'', ... were derived from the function u, Lagrange spoke of them as the derived functions, and later as the **derivatives** of u; the use of u' to indicate the derivative also stems from Lagrange's work.

18.1.3   The Stability of the Solar System

The French, despite a late start in accepting Newtonian physics, made rapid strides. Indeed, the greatest mathematical physicist since Newton was French: See page 495. Pierre Simon de Laplace (March 23, 1749-March 5, 1827). Laplace's family were minor country nobility, none of whom had shown any remarkable scientific ability (though an uncle was a teacher of mathematics); indeed, the young Laplace was intended to become a theologian. But at the provincial school in Caen his interest in mathematics was piqued by two instructors, Christopher Gadbled and Pierre le Canu. In 1768, when Laplace was to leave for Paris, Canu wrote him a letter of recommendation to give to d'Alembert. The self-made d'Alembert had no use for men who had only letters of recommendation in their favor; he sent Laplace away with a problem and told him to come back in a week. Legend has it that Laplace solved it overnight. This impressed d'Alembert, who a few days later managed to get Laplace a position teaching mathematics at the Military School.

Pierre Simon de Laplace.
©Bettmann/CORBIS.

Laplace recognized his own brilliance, and as a result, hardly seems to be a likable character. He was always title conscious: from the beginning, he signed his papers *de la Place*, a reminder to the readers of Laplace's noble descent. When he was passed over for election to the Academy of Sciences not once, but twice, Laplace threatened to leave Paris. On Laplace's behalf, d'Alembert asked Lagrange if there were any positions available in Berlin. However, in 1773, Laplace was finally elected to the academy and decided to stay in Paris.

Laplace was primarily a mathematical physicist, and is best known for analyzing the motions of the planets under Newtonian gravitation. It had been known since the time of Kepler that an ellipse is the best, simple, approximate path of the planets; under the assumption of Newtonian gravitation, it is possible to prove that in a system of just two bodies moving under gravitational forces alone, the paths of the bodies are conic sections. A question that intrigued Newton was whether or not the combined gravitational forces of the planets could, somehow, cause sufficiently large perturbations in their orbits as to need "setting right" by some supernatural force—God, in other words.

Problem 18.1 (Stability Question). *Will the planets of the solar system remain in nearly circular orbits about the sun, or will their combined gravitational influences cause arbitrarily large variations in their orbits?*

See page 496.

Euler had already come across indications that the solar system was *not* stable over long periods of time. However, Euler's result was based on an approximate solution to the differential equations of the planetary motions, so it was possible that Euler's result was an artifact of the method, and not an actual physical reality.

The differential equations are most easily set up in terms of the distance to the sun, r; to first approximation, the sun may be considered the motionless center of the solar system. In 1758, Euler pioneered a new approach to the problem. Rather than try and solve the differential equation for the distance r between a planet and the sun, Euler instead rewrote the differential equations in terms of a, the semimajor axis of an elliptical orbit: this was the birth of the **variation of parameters** method of solving differential equations.

In addition to pioneering a new technique, it allowed the stability question to be asked in a useful manner: given that a was not constant, were the variations **secular** (i.e., could a grow without limit), or were they periodic? There are three parameters of interest: the semimajor axis a; the eccentricity of the ellipse e; and the inclination of the orbit to the plane of the ecliptic θ. In 1760, the Paris Academy of the Sciences offered a prize for a proof of the stability of the solar system. Euler's son Charles attempted to show that the variations in the semimajor axis were periodic, but the analysis was flawed and no prize was awarded for the question. In 1773, Lagrange submitted a paper to the French Academy that showed the inclinations underwent only periodic variations. The paper was submitted to Laplace to referee, and Laplace applied Lagrange's methods to show that the variations in the eccentricities were also periodic. Laplace read his results to the Academy on December 17, 1774 *before* Lagrange's paper was presented; Lagrange's paper appeared somewhat later.

Not surprisingly, Lagrange submitted no more papers to the French Academy journal, and his next important result appeared in the journal for the Berlin Academy of the Sciences. In 1776, Lagrange (apparently unaware of the younger Euler's result) showed that, in general, $\frac{d\left(\frac{1}{2a}\right)}{dt}$ could be written as a trigonometric series; hence $\frac{1}{2a}$ was periodic, and by extension, a underwent only periodic variations. Lagrange refined the proof in 1783 and 1784.

This was not quite the solution to the stability question, for even periodic variations may become large. In particular, the differential equation for $\frac{1}{2a}$ was expressed in a series including terms such as $\cos(nt+k)$, where k was a constant, and n was some real number that might be very small; upon integration, this term would become $\frac{1}{n}\sin(nt+k)$ and could contribute to a significant perturbation.

Thus on November 23, 1785, Laplace announced his important result. If m, m', m'', ... are the relative masses of the planets (where the mass of the sun is set equal to 1); a, a', a'', ... the semimajor axes; e, e', e'', ... the eccentricities;

and θ, θ', θ'', ... are the inclinations of the orbits of the planets to the plane of the ecliptic, then

$$m(e^2 + \theta^2)\sqrt{a} + m'(e'^2 + \theta'^2)\sqrt{a'} + m''(e''^2 + \theta''^2)\sqrt{a''} + \ldots = \text{constant}$$

Today, we would refer to the left hand side as an **integral**. Because the value of the sum was constant, and all the terms were positive, none of them could exceed a certain value. Hence, the variations were not only periodic (Lagrange's work) but bounded (Laplace's work). For the astronomers of the nineteenth century, the stability of the solar system had been proven, though today, the proof is not considered complete.

To their contemporaries, the glory of proving the stability of the solar system was shared by both Lagrange and Laplace, but by the middle of the nineteenth century, Laplace alone would bear the credit for proving the solar system stable. A large part of the reason was probably Laplace's massive five-volume tome, *Celestial Mechanics*, which made the name of Laplace familiar to every astronomer. The first four volumes were translated into English by the American Nathaniel Bowditch, who doubled its length by adding extensive commentaries. Bowditch complained about Laplace's use of "It is easy to see...," for it invariably meant several hours of hard work to see what Laplace claimed was so easy. It should be pointed out the phrase "It is easy to see" was a common one, appearing throughout the works of Laplace, Lagrange, and d'Alembert as well.

Laplace came to be known as the Newton of France. In seamlessly blending mathematics and Newtonian physics, he came to epitomize the Enlightenment ideal that everything in the universe could be described by a simple set of laws. However, this ideal has a dark side to it: as Laplace suggested, it meant that if one knew the position and momentum of every particle in the universe at any instant in time, then one could predict its entire past and, more importantly, its entire future. Thus, free will was an illusion caused by ignorance.

18.1.4 Computation of Multinomials

One of the major computational problems in probability is the integration of functions, such as $\int x^p (1-x)^q dx$, where p and q were very large numbers. Standard integration techniques would result in an exact answer after a great deal of tedious work; meanwhile, series solutions converged too slowly to make any but a high order expansion useful. In 1785 and 1786 appeared Laplace's two-part paper, "Memoir on the Approximation of Formulas Which Are Functions of Very Large Numbers."

In the case of a binomial, the middle term of $(1+1)^{2n}$ corresponds to $\binom{2n}{n}$. To find this, Laplace gave two methods. The second method began by considering the expansion of $\left(e^{\omega\sqrt{-1}} + e^{-\omega\sqrt{-1}}\right)^{2s}$, whose middle term, y_s, would be independent of $e^{\omega\sqrt{-1}}$, while all the other terms could be expressed in terms of the sine or cosine of multiples of 2ω. Hence

$$\int_0^\pi \left(e^{\omega\sqrt{-1}} + e^{-\omega\sqrt{-1}}\right)^{2s} d\omega = \pi y_s$$

But $\left(e^{\omega\sqrt{-1}} + e^{-\omega\sqrt{-1}}\right)^{2s} = 2^{2s}\cos^{2s}\omega$; thus

$$y_s = \frac{2^{2s}}{\pi} \int_0^\pi \cos^{2s}\omega\, d\omega$$

$$= \frac{2^{2s+1}}{\pi} \int_0^{\pi/2} \cos^{2s}\omega\, d\omega$$

$$y_s = \frac{2^{2s+1}}{\pi} \int_0^1 \left(1 - u^2\right)^{s-\frac{1}{2}} du$$

where the substitution $u = \sin\omega$ has been made. To evaluate the integral, Laplace made the additional substitutions $\alpha = \frac{1}{s-\frac{1}{2}}$, $1 - u^2 = e^{-\alpha t^2}$, and thus $u = \sqrt{1 - e^{-\alpha t^2}}$. Hence

$$\int (1-u^2)^{s-\frac{1}{2}}\, du = \int e^{-t^2}\, du$$

To evaluate this integral, Laplace expanded $\sqrt{1 - e^{-\alpha t^2}}$ as a power series in t, namely,

$$u = \sqrt{1 - e^{-\alpha t^2}} \tag{18.4}$$

$$= \alpha^{\frac{1}{2}} t \left(1 + \alpha q^{(1)} t^2 + \alpha^2 q^{(2)} t^4 + \ldots\right) \tag{18.5}$$

where $q^{(1)}$, $q^{(2)}$, $q^{(3)}$, ... are coefficients to be determined later.

Of course, these coefficients could be found by using Taylor's Theorem; however, the differentiation (after the first step) becomes extremely complex. To find the coefficients, Laplace employed a different technique: the **logarithmic derivative**. We have

$$u = \sqrt{1 - e^{-\alpha t^2}}$$

$$\ln u = \frac{1}{2} \ln\left(1 - e^{-\alpha t^2}\right)$$

$$\frac{1}{u} u' = \frac{\alpha t e^{-\alpha t^2}}{1 - e^{-\alpha t^2}}$$

$$= \frac{1 - \alpha t^2 + \frac{1}{1\cdot 2}\alpha^2 t^4 - \frac{1}{1\cdot 2\cdot 3}\alpha^3 t^6 + \ldots}{t - \frac{\alpha t^3}{1\cdot 2} + \frac{\alpha^2 t^5}{1\cdot 2\cdot 3} - \frac{a^3 t^7}{1\cdot 2\cdot 3\cdot 4} + \ldots}$$

On the other hand, from Equation 18.4, we have

$$\frac{1}{u} u' = \frac{1 + 3\alpha q^{(1)} t^2 + 5\alpha^2 q^{(2)} t^4 + \ldots}{t + \alpha q^{(1)} t^3 + \alpha^2 q^{(2)} t^5 + \ldots} \tag{18.6}$$

By setting the two equal to each other and cross multiplying, we obtain

$$\left(1 + 3\alpha q^{(1)}t^2 + 5\alpha^2 q^{(2)}t^4 + \ldots\right) \times \left(t - \frac{\alpha t^3}{1 \cdot 2} + \frac{\alpha^2 t^5}{1 \cdot 2 \cdot 3} - \ldots\right)$$

$$= \left(1 - \alpha t^2 + \frac{1}{1 \cdot 2}\alpha^2 t^4 - \ldots\right) \times \left(t + \alpha q^{(1)}t^3 + \alpha^2 q^{(2)}t^5 + \ldots\right)$$

If the products are expanded and the coefficients compared, we obtain the relationship between $q^{(1)}$, $q^{(2)}$, ... For example, comparing the t^3 coefficients, we have

$$3\alpha q^{(1)} - \frac{\alpha}{1 \cdot 2} = \alpha q^{(1)} - \alpha$$

Hence

$$2q^{(1)} + \frac{1}{2} = 0$$

Comparing the t^5 coefficients, we obtain

$$\frac{\alpha^2}{1 \cdot 2 \cdot 3} - \frac{3\alpha^2 q^{(1)}}{1 \cdot 2} + 5\alpha^2 q^{(2)} = \alpha^2 q^{(2)} - \alpha^2 q^{(1)} + \frac{1}{1 \cdot 2}\alpha^2$$

Hence

$$4q^{(2)} - \frac{1}{1 \cdot 2}q^{(1)} - \frac{2}{1 \cdot 2 \cdot 3} = 0$$

In general, Laplace noted that $q^{(i)}$ could be found using the recurrence relationship

$$0 = 2iq^{(i)} - \frac{2i - 3}{1 \cdot 2}q^{(i-1)} + \frac{2i - 6}{1 \cdot 2 \cdot 3}q^{(i-2)}$$

$$- \frac{2i - 9}{1 \cdot 2 \cdot 3 \cdot 4}q^{(i-3)} + \frac{2i - 12}{1 \cdot 2 \cdot 3 \cdot 4 \cdot 5}q^{(i-4)} - \ldots \quad (18.7)$$

where $q^{(0)} = 1$ (and if the index is negative, the terms are considered to be 0).

Solving equation 18.6 for u' and substituting in u from equation 18.4, we have

$$du = \alpha^{\frac{1}{2}}\left(1 + 3\alpha q^{(1)}t^2 + 5\alpha^2 q^{(2)}t^4 + 7\alpha^3 q^{(3)}t^6 + \ldots\right) dt$$

Hence

$$\int e^{-t^2} du = \alpha^{\frac{1}{2}}\int e^{-t^2}\alpha^{\frac{1}{2}}\left(1 + 3\alpha q^{(1)}t^2 + 5\alpha^2 q^{(2)}t^4 + 7\alpha^3 q^{(3)}t^6 + \ldots\right) dt$$

Since the original integral was from $u = 0$ to $u = 1$, the new integral will be taken from $t = 0$ to $t = \infty$. However,

$$\int_0^\infty t^{2r}e^{-t^2}\,dt = \frac{1 \cdot 3 \cdot 5 \cdots (2r - 1)}{2^r}\int e^{-t^2}\,dt$$

$$= \frac{1 \cdot 3 \cdot 5 \cdots (2r - 1)}{2^r}\frac{\sqrt{\pi}}{2}$$

Thus, making the proper substitutions, we have

$$y_s = \frac{2^{2s}}{\sqrt{\left(s - \frac{1}{2}\right)\pi}}\left(1 + \frac{1\cdot 3}{2}\alpha q^{(1)} + \frac{1\cdot 3\cdot 5}{2^2}\alpha^2 q^{(2)} + \frac{1\cdot 3\cdot 5\cdot 7}{2^3}\alpha^3 q^{(3)} + \ldots\right)$$

where $\alpha = \frac{1}{s - \frac{1}{2}}$.

18.1 Exercises

1. Prove Waring's method works by showing that if $x = \alpha$, $y = s^\alpha$, and so on.

2. Assuming that every function can be expanded as a finite or infinite series, show that Lagrange's identification of the coefficients p, p', p'', ... with the derivatives of u is correct.

3. Find $q^{(1)}$, $q^{(2)}$, $q^{(3)}$; then use these values to estimate $\binom{100}{50}$, $\binom{200}{100}$, and $\binom{400}{200}$. How much error is there in using y_s to approximate $\binom{2s}{s}$?

18.2   Algebra and Number Theory

See page 510.

Euler gave a method by which a series could be turned into a continued fraction equivalent. Since series for e and π were well-known, this meant that continued fraction equivalents for these numbers could be found. In 1761, the astronomer Johann Heinrich Lambert (August 26, 1728-September 25, 1777), born in Alsace, proved π to be irrational, using the continued fraction expansion

$$\tan v = \cfrac{1}{\frac{1}{v} - \cfrac{1}{\frac{3}{v} - \cfrac{1}{\frac{5}{v} - \cfrac{1}{\ddots}}}} \tag{18.8}$$

Based on this, Lambert proved that if the tangent of an arc was rational, the arc itself had to be incommensurable with the radius; since $\tan\frac{\pi}{4} = 1$, this implied that π was an irrational number.

18.2.1   Lagrange and Continued Fractions

The theory of continued fractions was further extended by Lagrange, who read "On the Numerical Solution of Equations" to the Berlin Academy on April 20, 1769. In the third section, Lagrange applied continued fractions to solving equations. Given

$$Ax^m + Bx^{m-1} + \ldots + K = 0$$

let x be a positive root between two consecutive whole numbers, p and $p + 1$; let $x = p + \frac{1}{y}$, where $y > 1$. Substituting into the original equation, we obtain a new equation in y,

$$A'y^m + B'y^{m-1} + \ldots + K' = 0$$

which has at least one positive real solution $y > 1$. Hence, we may repeat the procedure, letting $y = q + \frac{1}{z}$, with $z > 1$. In this way, we obtain a sequence of positive whole numbers p, q, r, $\ldots$, and the solution x to the original equation may be expressed in terms of a continued fraction

$$x = p + \cfrac{1}{q + \cfrac{1}{r + \cfrac{1}{u + \cfrac{1}{\ddots}}}} \tag{18.9}$$

Lagrange showed that the convergents alternately exceed and fall short of the actual root, and that the (reduced) convergent more closely approximates the root than any fraction of a lower denominator.

See page 509.

18.2.2 Algebraic Resolution of Equations

Since the time of Cardano, little progress had been made on solving fifth- or higher-degree equations, though Tschirnhaus (the count who attempted to plagiarize Leibniz) had discovered algebraic transformations that could reduce a general nth degree equation into a simpler nth-degree equation. Tschirnhaus had shown in 1683 it was possible to eliminate the $n - 1$ and $n - 2$ terms from a general nth-degree equation, and he hoped it was possible to reduce a general nth degree equation to one of the form $x^n - K = 0$, which could be solved using elementary operations. If this was possible, then equations of any degree could be solved algebraically.

Vandermonde

The first steps toward the final answer of this question of solvability were taken by Alexandre Theophile Vandermonde (February 28, 1735-January 1, 1796), who, despite his last name, was a Frenchman; Vandermonde and Monge were such close friends that Vandermonde was known as "femme de Monge." Vandermonde only wrote four papers in mathematics. The Vandermonde determinant, which appears in approximation theory but nowhere in Vandermonde's own work. However, Vandermonde was one of the first authors to systematically investigate and prove the properties of determinants, and his other work impressed his contemporaries enough so that he was elected to the Academy of Sciences in 1771, beating out Laplace.

In November 1770, Vandermonde presented to the Paris Academy "Memoir on the Resolution of Equations," where he analyzed the structure of the algebraic solutions of polynomial equations. Vandermonde began by noting that the solvability of the quadratic equation $x^2 - (a+b)x + ab = 0$ relied on the fact that the roots, a and b, could be expressed in terms of the coefficients, as

$$\frac{1}{2}\left((a+b) + \sqrt{(a+b)^2 - 4ab}\right)$$

The expression gave two solutions, depending on whether one used the positive or negative square root of $(a+b)^2 - 4ab$.

Similarly, if r' and r'' were the cube roots of unity not equal to 1, then the expression

$$\frac{1}{3}\left(a + b + c + \sqrt[3]{(a + r'b + r''c)^3} + \sqrt[3]{(a + r''b + r'c)^3}\right)$$

gave the three roots a, b, c to the cubic equation. Actually, as Vandermonde knew, the expression actually gives six possible solutions, but trial and error will eliminate the three extraneous solutions. With some tedious but not too difficult algebra, it is easy to show that this expression can be rewritten in terms of the coefficients of the cubic. From these expressions, Vandermonde derived formulas, equivalent to the quadratic equation, for the solution to the general cubic.

Lagrange Resolvents

Lagrange was apparently unaware of Vandermonde's work when, a few months later, he presented a massive work, "Reflections on the Algebraic Resolution of Equations." Lagrange's approach to the problem was similar to Vandermonde's, though much more general in some respects. Lagrange noted that the resolution of the cubic and quartic equations depended on solving another equation, which he called the **resolvent equation**. In the case of a cubic, the resolvent *appeared* to be of a higher degree but was in fact reducible to an equation of lower degree. Lagrange took this to be the key factor in the solvability of the cubic.

To fully understand Lagrange's work, take a quadratic equation $x^2 - px + q = 0$, with two roots, a and b. Consider a polynomial function of these roots, such as ab or $a^2 + 2b$. Some of these functions, such as ab, will have the same value, regardless of which root is considered a and which is considered b; others, such as $a^2 + 2b$, will have different values. The former are now known as **symmetric functions**.

The coefficients of a polynomial equation are defined in terms of symmetric functions of the roots; for example, the quadratic equation $x^2 - px + q = 0$ with roots a, b has $p = a + b$ and $q = ab$. Since *some* symmetric functions could be expressed in terms of the coefficients, and no one had found a symmetric function that could not be so expressed, eighteenth century mathematicians believed that *all* symmetric functions could be expressed in terms of the coefficients. This is true, though proof was not available until the nineteenth century.

Example 18.1. *For the quadratic $x^2 - px + q = 0$, find an expression for $(a-b)^2$ in terms of the coefficients. We note that $(a-b)^2$ has the same value, regardless of the order the roots are taken, for $(a-b)^2 = (b-a)^2$; hence, it is conceivable that $(a-b)^2$ may be expressed in terms of the coefficients p and q. Since $p = a+b$ and $q = ab$, we have*

$$(a-b)^2 = a^2 + b^2 - 2ab$$
$$= a^2 + b^2 + 2ab - 4ab$$
$$= (a+b)^2 - 4ab$$
$$= p^2 - 4q$$

The importance of this is that it is possible to evaluate certain functions of the roots of an equation *without* actually knowing the roots.

Lagrange began his analysis with the cubic, but to make his work more clear, we will begin by applying it to the quadratic equation. Suppose we wish to solve the quadratic equation $x^2 - px + q = 0$. Consider the (nonsymmetric) function $a - b$ of the roots of $x^2 - px + q = 0$. We note that this expression has two values, depending on the order the roots are taken: $a - b$ or $b - a$. Let these two values be the roots of a resolvent equation. Writing and expanding the resolvent equation, we obtain:

$$(y - (a-b))(y - (b-a)) = 0$$
$$(y - (a-b))(y + (a-b)) = 0$$
$$y^2 - (a-b)^2 = 0$$

Since $(a-b)^2 = p^2 - 4q$, we may write

$$y^2 - (p^2 - 4q) = 0$$

It might seem that we have gone nowhere, for the degree of the resolvent equation is the same as the degree of the original. *However*, and this is the key to Lagrange's analysis (and our ability to solve the quadratic equation), the resolvent is solvable using elementary operations; in particular, $y = \pm\sqrt{p^2 - 4q}$.

It remains to extract the solutions to the original equation from y. We note that y has two values, depending on whether one takes the positive or negative root. Say $y_1 = a - b$ and $y_2 = b - a$. Then $\frac{1}{2}(y_1 + p) = a$, and $\frac{1}{2}(y_2 + p) = b$. Hence, from the two solutions of the resolvent equation may be found the two solutions of the original equation. Note that it does not matter whether y_1 corresponds to $a - b$ or $b - a$, the two possible values of the roots under permutations.

For the cubic equation $x^3 + mx^2 + nx + p = 0$ with roots a, b, c, Lagrange considered the expression $a + \alpha b + \alpha^2 c$, where $\alpha^3 = 1$ and $\alpha \neq 1$. This expression has six possible values when the roots are permuted, namely

$a + \alpha b + \alpha^2 c$	$a + \alpha c + \alpha^2 b$	$b + \alpha a + \alpha^2 c$
$b + \alpha c + \alpha^2 a$	$c + \alpha a + \alpha^2 b$	$c + \alpha b + \alpha^2 a$

Let these values be the roots of a sixth degree resolvent equation. Although it seems our problem has been magnified, in fact it has been simplified, for the choice of our expression, $a + \alpha b + \alpha^2 c$ means that values of the six roots can, in fact, be expressed in terms of two quantities, r and s, where $r = a + \alpha b + \alpha^2 c$ and $s = a + \alpha c + \alpha^2 b$. The six values of the expression under permutation of the roots are, in fact, r, αr, $\alpha^2 r$, s, αs, $\alpha^2 s$. The equation, with these six values as roots, is:

See problem 5.

$$0 = y^6 - (r^3 + s^3)y^3 + r^3 s^3 \tag{18.10}$$

which, though it is a sixth degree equation, is actually quadratic in y^3, and hence solvable, *provided* the coefficients of the equation can be expressed in terms of the coefficients of the original. But $r^3 + s^3$ and $r^3 s^3$ are symmetric functions of the roots; hence, it is conceivable that they can be expressed in terms of the coefficients of the original equation. A little algebra shows:

$$r^3 + s^3 = -2m^3 + 9mn - 27p \tag{18.11}$$

$$r^3 s^3 = (m^2 - 3n)^3 \tag{18.12}$$

Hence, the resolvent equation can be expressed and then solved; from the solutions, the roots of the cubic may be found.

We note that in the case of the quadratic, the resolvent was the equation whose roots were of the form $a - b$, which can be thought of as $1a + (-1)b$, where 1 and -1 are the two square roots of one. For the cubic, the resolvent was the equation whose roots were of the form $1a + \alpha b + \alpha^2 c$, where 1, α, and α^2 were the three cube roots of one. The obvious extension to the fourth-degree equation, which both Vandermonde and Laplace pointed out, was to use the expression $1a + \beta b + \beta^2 c + \beta^3 d$, where β, β^2, β^3 are the three nonunit fourth roots of unity (i.e., i, -1, and $-i$).

However, Lagrange found not one, but two simpler ways to proceed. Let a, b, c and d be the four roots of the fourth-degree equation $x^4 + mx^3 + nx^2 + px + q = 0$, and consider the expression $ab + cd$, which has only three possible values under the permutation of the roots. Thus, the resolvent equation with these values as root is

$$u^3 - Au^2 + Bu - C = 0$$

where

$$A = n$$
$$B = mp - 4q \tag{18.13}$$
$$C = (m^2 - 4n)q + p^2$$

Consider any one of the roots of the resolvent, say the root corresponding to $ab + cd$ (it does not matter which expression the particular root corresponds to). Designate $u = ab + cd$. Since $q = abcd$, then ab and cd were the two roots of the equation

$$t^2 - ut + q = 0$$

Hence, ab and cd could be determined separately. If $t' = ab$ and $t'' = cd$ (again, it makes no difference which product corresponds to which root), then a and b would be the roots to

$$z^2 - \frac{p - mt'}{t' - t''}z + t' = 0$$

and c, d were the roots to

$$z^2 - \frac{p - mt''}{t'' - t'}z + t'' = 0$$

Hence, the fourth-degree equation could be solved by finding a root of a third-degree resolvent equation; this root was then used to form a second-degree equation, the roots of which were used to form two more second-degree equations. The four solutions that resulted were the four solutions to the original fourth-degree equation.

Lagrange's second method is to consider the expression $c + d - a - b$, which has six possible values under permutations of the roots, and thus gives rise to a sixth-degree resolvent which is, however, reducible to cubic.

The obvious extension is to find some function of the five roots of a fifth-degree equation. The problem is clear: with three roots, there are six possible permutation of the roots, so some care must be taken to ensure the resulting equation is solvable. With four roots, there are twenty-four possible permutations, and considerably more care is necessary to ensure that the resulting resolvent equation can be written as an equation of degree less than four. Five roots have 120 permutations, and it is necessary to find an expression with four or fewer possible values under permutation of the roots. Lagrange thought it "very doubtful" that the methods used to solve the equations of degree less than five could profitably be applied to the fifth degree equation, though he admitted that it had not yet been *proven* impossible.

18.2.3 Waring and Wilson

Among the many fields of mathematics in the eighteenth century, it was in algebra that the English were most active, for in algebra, at least, there was little doubt about the logical foundations. In a 1762 work, *Miscellaneous Analysis of Algebraic Equations and the Properties of Curves*, Waring stated that any number was the sum of 4 squares, 9 cubes, 19 fourth powers, and so on. He neither gave a proof, nor explained how the numbers "4," "9," "19," and so on, were obtained. **Waring's conjecture** remained unproven until the beginning of the twentieth century.

Another important result in number theory, known as **Wilson's theorem**, first appeared in Waring's *Meditations on Algebra* (1770). Waring attributed the result to his student, John Wilson (1741-1793). Without proof, Wilson claimed that if n is prime, then $(n - 1)! + 1$ is divisible by n. Though Waring eventually supplied a proof in the third edition of *Meditations on Algebra* (1782), Lagrange was the first to prove Wilson's theorem, as a corollary to a more general theorem, in 1770.

To prove Wilson's theorem, Lagrange began with

Lemma 18.1. *The product*

$$(x+1)(x+2)(x+3)\cdots(x+n-1)$$
$$= x^{n-1} + A'x^{n-2} + A''x^{n-3} + \ldots + A^{(n-1)} \quad (18.14)$$

where

$$A' = \frac{n(n-1)}{2}$$
$$2A'' = \frac{n(n-1)(n-2)}{2\cdot 3} + \frac{(n-1)(n-2)}{2}A'$$
$$3A''' = \frac{n(n-1)(n-2)(n-3)}{2\cdot 3\cdot 4} + \frac{(n-1)(n-2)(n-3)}{2\cdot 3}A' + \frac{(n-2)(n-3)}{2}A''$$
$$\vdots$$

Hence

Proposition 18.1. *If n is prime, A', A'', A''', ..., $A^{(n-2)}$ are all divisible by n, and n also divides $A^{(n-1)} + 1$.*

A number of corollaries follow.

Corollary (Wilson's Theorem). *The product*

$$1\cdot 2\cdot 3\cdot 4\cdots(n-1) + 1$$

is always divisible by n, when n is a prime number.

Corollary. *If x is not divisible by n, then $x^{n-1} - 1$ is divisible by n.*

Corollary. *If n is an odd prime, then*

$$\left[1\cdot 2\cdot 3\cdots\frac{n-1}{2}\right]^2 \pm 1$$

is divisible by n, where one uses the $+$ or $-$ depending on whether $\frac{n-1}{2}$ is even or odd. Hence, if $\frac{n-1}{2} = 2m$ and thus $n = 4m+1$, then

$$\left[1\cdot 2\cdot 3\cdots(2m-1)\right]^2 - 1$$

is divisible by n.

Lagrange noted:

The preceding Corollaries are all the more remarkable since, if n is
not a prime number, the expressions that were divisible by n when
n was prime are no longer so [divisible] ... One can thus derive from
this a direct method for determining if any given odd number n is
prime or not.[1]

Lagrange is referring to the fact that $2 \cdot 3 \cdots (n-1) + 1$ will be divisible by n
if and only if n is prime. Of course, the method is impractical if n is a large
number; Lagrange provided two simplifications, and invited other geometers to
find additional ones.   See problem 17.

Waring also included some conjectures involving primes appearing in an arithmetic sequence, which Lagrange also proved.   See problem 18.

18.2 Exercises

1. Demonstration Equation 18.8. Express $\tan v = \frac{\sin v}{\cos v}$ as a quotient of infinite series, then apply the long division to obtain the continued fraction
   expression for $\tan v$. Hint: use the fact that $\frac{a}{b} = \frac{1}{\frac{b}{a}}$.

2. Use Equation 18.8 to find an infinite series expansion for $\tan v$.

3. Consider Lagrange's method of using continued fractions to solve algebraic
   equations.

   (a) Show that it is possible to produce a sequence of positive, whole num-
       bers that approximate the root x.

   (b) Find the first five numbers in the continued fraction expansion of the
       solution to $x^3 - 2x - 5 = 0$. $x = 2$ is an approximate solution.

4. Consider the continued fraction expansion of the form in Equation 18.9.

   (a) Show that the successive convergents $\frac{\alpha}{\alpha'}$, $\frac{\beta}{\beta'}$, ..., alternately fall short
       and exceed the solution x.

   (b) Moreover, the convergents are relatively prime; hence the error in
       using $x \approx \frac{\delta}{\delta'}$ is less than $\frac{1}{\delta'^2}$.

   (c) Finally, each fraction is closer to the root than any fraction with a
       lower denominator. Hint: consider two successive fractions, say $\frac{\gamma}{\gamma'}$
       and $\frac{\delta}{\delta'}$. Show that $\frac{\mu}{\mu'}$, with $\delta' > \mu'$, leads to a contradiction.

5. Show how to express the six values of $a + \alpha b + \alpha^2 c$ under the permutation
   of the roots in terms of r and s.

6. Prove that the coefficients of a polynomial are determined by symmetric
   functions of the roots.

---

[1]Lagrange, Joseph Louis, *Œuvres Complete de Lagrange*. Gauthier-Villars, 1867-1892. Vol.
III, page 432.

7. Show (without using Equations 18.11 or 18.12) that $r^3 + s^3$, and $r^3 s^3$ are symmetric functions of the roots.

8. Verify Equations 18.11 and 18.12.

9. Write down the "cubic formula," that gives the solution to the general cubic $x^3 + px^2 + qx + r = 0$.

10. Show the resolvent equation for the cubic has the form shown in Equation 18.10.

11. Consider the resolvent equation for the quartic.

   (a) Show that the expression $ab + cd$ has only three distinct values under permutation of the roots.

   (b) Show that the resolvent equation with these three values is $u^3 - Au^2 + Bu - C = 0$, where A, B, and C are given by the equations 18.13.

12. Consider the expression $c + d - a - b$.

   (a) Show that this only has six possible values under permutation of the roots.

   (b) Write down the sixth degree resolvent equation that results.

   (c) Show that this is reducible to cubic.

13. Prove that the coefficients A', A'', A''', ..., $A^{(n-1)}$ in Equation 18.14 are all whole numbers.

14. Prove Lemma 18.1, by replacing x with $x + 1$ in Equation 18.14, and comparing the coefficients A', A'', ..., in the two equations.

15. Prove Theorem 18.1.

   (a) First, show A', A'', ..., $A^{(n-2)}$ are divisible by n if n is prime.

   (b) Next, show

$$(n-1)A^{(n-1)} = 1 + A' + A'' + \ldots + A^{(n-2)}$$

   and thus $A^{(n-1)} + 1$ is divisible by n.

16. Prove that Wilson's theorem holds if and only if n is prime.

17. Lagrange provided two simplifications of the method of using Wilson's theorem to verify primality.

   (a) Show that if n is an odd composite number, $2 \cdot 3 \cdots (n-2)$ is divisible by n.

   (b) If $n = 4m + 1$ or $n = 4m - 1$, what further simplifications may be made?

18. Prove the following about primes in arithmetic sequences.

    (a) If three primes form an arithmetic sequence, the common difference must be divisible by 6, so long as 3 is not a term of the sequence.

    (b) If five primes form an arithmetic sequence, the common difference must be divisible by 30, so long as 5 is not a term of the sequence.

18.3 Probability and Statistics

Besides algebra and number theory, several contributions were made by the English in the fields of probability and statistics, as shown by the work of De Moivre, Simpson, and Bayes. The work of the last two was extended by Lagrange and Laplace. Whether Lagrange and Laplace knew the work of Simpson and Bayes is an open question, but it is (especially in the case of Laplace) a possibility. That neither cited any other mathematicians is more indicative of the climate of the time, for citation of sources was still a rarity—as well it might be, for the number of scientific journals was still small, and a single scientist could easily apprise himself of all developments in the field. Thus, plagiarism would be easy to spot.

18.3.1 Lagrange

Lagrange, like Simpson, approached the theory of errors by examining generating functions in "On the Usefulness of Taking the Arithmetic Mean of Several Observations" (1776). Strictly speaking, the work should be considered as one on *direct* probability, since Lagrange spent most of the work computing the probability of making a given error.

Lagrange, like his contemporaries, defined the probability of an event to be the ratio between the number of cases in which the event occurred to the number of cases in which it was possible for the event to occur. He began by considering the case where the observation was either exact (error of 0), or had an error of $+1$ or -1; assuming these occurred in a, b, or b cases, respectively (hence, the probability of making an error of 0 would be $\frac{a}{a+2b}$). What, then, is the probability that the arithmetic mean of the observations is equal to the actual value? Lagrange proceeded:

> It is easy to see that this question reduces itself to this: let there be n dice, each having a faces marked zero, b faces marked by positive one, and b faces marked by negative one, so that the total number of faces is $a + 2b$; find the probability that the total, if all the dice are thrown, is zero.[2]

To solve this problem, Lagrange used a generating function. Lagrange might or might not have gotten the idea from Thomas Simpson or De Moivre, who used

---

[2]Lagrange, Joseph Louis, *Œuvres Complete de Lagrange*. Gauthier-Villars, 1867-1892. Vol. II, page 175.

similar methods, but he took the development of the generating function idea far beyond its original conception.

Consider the expansion of the trinomial $(a + b(x + x^{-1}))^n$. The constant term A corresponds to those products where the total error is 0. Since the total number of possible cases was $(a + 2b)^n$, then the probability that the total error was 0 would be $\frac{A}{(a+2b)^n}$. Through a straightforward (if lengthy) analysis, Lagrange was able to calculate A. If $a = b$, then the probability of making no total error in n observations is

n	1	2	3	4	5	6
Probability	$\frac{1}{3}$	$\frac{1}{3}$	$\frac{7}{27}$	$\frac{19}{81}$	$\frac{51}{243}$	$\frac{141}{729}$

Lagrange noted a seeming paradox:

> It seems one can conclude, in this case, that it is better to make a single observation than to take the average of many observations...[3]

This is, of course, incorrect; Lagrange went on to explain the reason why.

The second problem was to determine the probability that the error of the sample mean was not greater than $\frac{m}{n}$, where $m < n$. Through a tedious but not difficult analysis, Lagrange found, for the case $a = b$

n	$\pm\frac{0}{n}$	$\pm\frac{1}{n}$	$\pm\frac{2}{n}$	$\pm\frac{3}{n}$	$\pm\frac{4}{n}$	$\pm\frac{5}{n}$
	\multicolumn{6}{c}{Probability the error does not exceed}					
1	$\frac{1}{3}$					
2	$\frac{3}{9}$	$\frac{7}{9}$				
3	$\frac{7}{27}$	$\frac{19}{27}$	$\frac{25}{27}$			
4	$\frac{19}{81}$	$\frac{51}{81}$	$\frac{71}{81}$	$\frac{79}{81}$		
5	$\frac{51}{243}$	$\frac{141}{243}$	$\frac{201}{243}$	$\frac{231}{243}$	$\frac{241}{243}$	
6	$\frac{141}{729}$	$\frac{393}{729}$	$\frac{573}{729}$	$\frac{673}{729}$	$\frac{715}{729}$	$\frac{727}{729}$

It is with the third problem that Lagrange began the true study of error analysis. By allowing the possible errors to be 0, -1, and r (assumed positive) with frequency a, b, and c, respectively, then the probability that the mean was within some specified range could be found by examining the appropriate terms in the expansion of the trinomial $(a + bx^{-1} + cx^r)^n$.

Since the errors are no longer distributed symmetrically about the actual value, the question arises of what is the most probable error, an issue Lagrange dealt with in Problem IV. The most probable error would be the one corresponding to the greatest term in the expansion of $(a + \frac{b}{x} + cx^r)^n$. Supposing this to

---

[3]Ibid., 177.

be the term where a is raised to the α, $\frac{b}{x}$ is raised to the β, and cx^r is raised to the γ, then this term is $\pi a^\alpha b^\beta c^\gamma x^{-\beta+r\gamma}$, where

$$\pi = \frac{1 \cdot 2 \cdot 3 \cdots n}{1 \cdot 2 \cdot 3 \cdots \alpha \cdot 1 \cdot 2 \cdot 3 \cdots \beta \cdot 1 \cdot 2 \cdot 3 \cdots \gamma}$$

Lagrange designated the quantity $\pi a^\alpha b^\beta c^\gamma$ as M; the problem then is that of finding the maximum value of M.

Here, Lagrange must confront one of the difficulties that will plague Laplace: differential calculus is a powerful tool that works with continuous quantities, but there is no corresponding tool that will work with *discrete* quantities. Happily, this particular problem can be solved easily: at the maximum value, varying α, β, or γ must necessarily result in a lower coefficient (by assumption). If α varies to $\alpha + 1$, either β or γ must decrease by 1. In the first case, one obtains

$$\frac{\beta}{\alpha+1} \frac{aM}{b} \leq M$$

Hence

$$\frac{\beta}{\alpha+1} \frac{a}{b} \leq 1$$

Likewise, if α is replaced with $\alpha - 1$ and β with $\beta + 1$, then

$$\frac{\alpha}{\beta+1} \frac{b}{a} \leq 1$$

Comparing the two, Lagrange obtained

$$\frac{\alpha}{\beta+1} \leq \frac{a}{b}$$

and

$$\frac{\alpha+1}{\beta} \geq \frac{a}{b}$$

Hence $\frac{\alpha}{\beta} = \frac{a}{b}$. Likewise, $\frac{\alpha}{\gamma} = \frac{a}{c}$. Thus:

$$\alpha = pa \qquad\qquad \beta = pb \qquad\qquad \gamma = pc$$

Since $\alpha + \beta + \gamma$ is the number of trials n, then $p = \frac{n}{a+b+c}$, and thus

$$\alpha = \frac{na}{a+b+c} \qquad\qquad \beta = \frac{nb}{a+b+c} \qquad\qquad \gamma = \frac{nc}{a+b+c}$$

If α, β, and γ are whole numbers, Lagrange concluded, these would be the most likely numbers of errors of size 0, -1, and r; otherwise, they would be the numbers closest to α, β, γ. Thus, the most probable single error would be $\frac{r\gamma-\beta}{n}$, or $\frac{rc-b}{a+b+c}$. As a corollary, if $r = 1$ and $c = b$, the most likely error of the mean would be 0.

The generalization of this is the fifth problem, where the errors of size p, q, r, s, ... were made with frequency a, b, c, d, ...; hence the most probable error will correspond to the largest term in the expansion of the multinomial $(ax^p + bx^q + cx^r + dx^s + \ldots)^n$. "It is easy to demonstrate," Lagrange claimed, that M, the largest coefficient, will occur when

$$\alpha = \frac{na}{a + b + c + \ldots}$$

$$\beta = \frac{nb}{a + b + c + \ldots}$$

$$\gamma = \frac{nc}{a + b + c + \ldots}$$

$$\vdots$$

hence the most probable error is $\frac{\mu}{n} = \frac{ap+bq+cr+\cdots}{a+b+c+\ldots}$. "This quantity represents the correction that should be made to the mean of many observations."[4]

In order to make use of this correction factor, however, it is necessary to determine the unknown frequencies a, b, ... with which the errors p, q, ... are made. Lagrange attempted to answer this question in Problem VI. Suppose it is determined (through some sort of "calibration" step) that the errors p, q, r, ... are made α, β, γ, ... times; further, suppose that the errors p, q, r, ... are made with (initially unknown) frequencies a, b, c, ... In the multinomial expansion of

$$(ax^p + bx^q + cx^r + \ldots)^n$$

the coefficient of the $x^{p\alpha+q\beta+c\gamma+\cdots}$, divided by $(a + b + c + \ldots)^n$, will give the probability that the errors p, q, r, ... are made α, β, γ, ... times, respectively.

This coefficient Lagrange designated $Na^\alpha b^\beta c^\gamma \ldots$, where N is the multinomial coefficient, namely

$$N = \frac{1 \cdot 2 \cdot 3 \cdots n}{1 \cdot 2 \cdot 3 \cdots \alpha \cdot 1 \cdot 2 \cdot 3 \cdots \beta \cdot 1 \cdot 2 \cdot 3 \cdots \gamma \cdots}$$

Lagrange then assumed that the observed error was the most probable one, namely the one that maximized $Na^\alpha b^\beta c^\gamma \ldots$ By the work in the previous problem, he knew this occurred when

$$\alpha = \frac{na}{a + b + c + \ldots}$$

$$\beta = \frac{nb}{a + b + c + \ldots}$$

$$\gamma = \frac{nc}{a + b + c + \ldots}$$

---

[4]Ibid., 199.

Thus, the values of a, b, c, ... could be determined. It follows that the correction that needs to be made to the mean of n observations is $\frac{\alpha p + \beta q + \gamma r + \ldots}{n}$.

Lagrange proceeded in this manner until Problem Ten (which is actually the ninth problem), which deals with the continuous case. Suppose that the possible errors are between p and $-q$, with p and q both positive, and that y is the probability of making an error of size x on any single observation. What is the probability that the error of the mean is between r and $-s$?

The first part of the problem is finding a function that gives the probability that the error of the mean is z, z being some fixed value. To find this probability, Lagrange considered the integral from $-q$ to p of $ya^x dx$ to be a polynomial in a; the coefficient of the a^z term of the nth power of this polynomial will be the desired probability. (a is a dummy variable, serving the same purpose as x in the expansions of $(ax^p + bx^q + cx^r + dx^s + \ldots)^n$ in Lagrange's fifth problem.)

For example, if $y = K$ and thus all errors between $-q$ and p are equally likely, then $\int_{-q}^{p} K a^x dx = \frac{K(a^p - a^{-q})}{\ln a}$. If this is expanded as a power series in a, then the coefficient of the a^z term is the desired probability; this will itself be a function of z, and if this function is then integrated from $-r$ to s, the desired probability can be found. The analysis is exceptionally complicated even for this, the simplest case. Of greater significance is the examination of integrals of the form $\int ya^x dx$, which is of the form of what would become known as a **Laplace transform**.

18.3.2 Laplace

Somewhat earlier than Lagrange's work on the subject, Laplace published his own pioneering articles. One of his key concerns was the inverse probability problem: given that an event has occurred p times in the past and failed to occur on q other occasions, to find the probability that the event will occur in the future. Laplace's "Memoir on the Probability of the Causes of Events" (1774) examined these and other questions.

Laplace began with a principle he gave without explanation (or justification):

> If an event can be produced by a number n of different causes, the probabilities of these causes are to each other as the probability of the event happening from these reasons, and the probability of any one of them is equal to the probability of the event happening from this cause, divided by the sum of all the probabilities of the event caused by each of the other reasons.[5]

This is equivalent to **Bayes's theorem**; it has been suggested that Laplace was in fact aware of Bayes's work. To clarify this rather complex statement, Laplace gave the following example:

> I suppose there are two urns, A and B, where the first contains p white tickets and q black tickets, and the second contains p' white

---

[5] Laplace, Pierre Simon de, *Œuvres complètes*. Gauthier-Villars, 1878-1912. Vol. VIII, page 29.

tickets and q' black tickets; I draw from one of these urns (I know not which) $f + h$ tickets, of which f are white and h black; one asks, what is the probability that the urn from which I drew these tickets is A or B.[6]

The event, "drew f white and h black tickets," could be produced by either drawing from urn A or drawing from urn B. If the urn was A, let the probability of drawing f white and h black tickets be K; likewise, let the probability of drawing f white and h black from urn B be K'. Then, "the probabilities of these causes are to each other as the probability of the event happening from these reasons"; that is, the probability the urn was A is to the probability the urn was B as K to K'. Moreover, "the probability of any one of them is equal to the probability of the event happening from this cause, divided by the sum of all the probabilities of the event caused by each of the other reasons"; hence the probability that the urn was A is equal to the probability that f white and h black were drawn from urn A, divided by the sum of the probabilities that f white and h black tickets were drawn from A or from B; in other words, $\frac{K}{K+K'}$.

To proceed, Laplace supposed that the urn contained an infinite number of white and black tickets, and that p white and q black tickets were drawn in $p + q$ drawings. What is the probability E that the next ticket drawn would be white? Laplace analyzed the problem thus: if x is the probability the ticket is white, then the probability that p white and q black tickets are drawn is $x^p(1-x)^q$. Laplace then applied his principle; here the cause is "the probability of drawing a white ticket is x," hence the probability that the probability *is x* is $\frac{x^p(1-x)^q \, dx}{\int x^p(1-x)^q dx}$, where the integral is taken from $x = 0$ to $x = 1$. Here, Laplace is summing the probabilities through integration.

If the probability of drawing a white ticket is, indeed, x, then the probability of drawing one more white ticket would be $\frac{x^{p+1}(1-x)^q \, dx}{\int x^p(1-x)^q dx}$. Summing over all possible probabilities between $x = 0$ and $x = 1$, Laplace determined $E = \frac{\int x^{p+1}(1-x)^q dx}{\int x^p(1-x)^q dx}$, where the integrals are taken from $x = 0$ to $x = 1$. Since

$$\int x^{p+1}(1-x)^q dx = \frac{q}{p+2} \int x^{p+2}(1-x)^{q-1} dx \qquad (18.15)$$

then, applying this reduction recursively,

$$\int x^{p+1}(1-x)^q dx = \frac{1 \cdot 2 \cdot 3 \cdots q}{(p+2)(p+3)\cdots(p+q+2)}$$

Hence $E = \frac{p+1}{p+q+2}$.

An important point to recognize is that this is not actually the probability that a ticket is white; it is the probability that a ticket is white *given* that the previous draws have included p white and q black tickets. Thus, Laplace next asks what is the probability that the probability the ticket is white lies between

---

[6]Ibid., 29

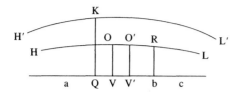

Figure 18.1: Laplace's First Graph.

$\frac{p}{p+q} + \omega$ and $\frac{p}{p+q} - \omega$. If the probability of drawing a white ticket is x, then the probability that the probability is x is

$$\frac{(p+1)(p+2)\cdots(p+q+1)}{1\cdot 2\cdot 3\cdots q}x^p(1-x)^q dx$$

If we let $x = \frac{p}{p+q} + z$, then

$$x^p(1-x)^q = \frac{p^p q^p}{(p+q)^{p+q}}\left(1 + \frac{p+q}{q}z\right)^p\left(1 - \frac{p+q}{q}z\right)^q$$

Integrating this from $z = 0$ to $z = \omega$, and multiplying by $\frac{(p+1)(p+2)\cdots(p+q+1)}{1\cdot 2\cdot 3\cdots q}$, gives the probability that x is between $\frac{p}{p+q}$ and $\frac{p}{p+q} + \omega$. Likewise, we may find the probability that x lies between $\frac{p}{p+q}$ and $\frac{p}{p+q} - \omega$; adding these two values together we obtain the probability that x (which, again, is the probability of drawing a white ticket) is between the bounds $\frac{p}{p+q} - \omega$ and $\frac{p}{p+q} + \omega$. After some exhaustive analysis, Laplace showed that this probability, which he designated E, is approximately 1.

Laplace's real interests appear in Problem III: given three recorded observations of the same phenomenon, to determine the actual time at which it occurred. Laplace is thinking of some astronomical event, which three different observers recorded as happening at three times, a, b, and c; what is the most probable time V for the observed event?

Suppose the event actually occurred at the time corresponding to V. Laplace let the error of the first observation, a, be x (what we would now call the **absolute error**, and always taken to be positive), and supposed the interval between a and b to be p, and the interval between b and c to be q (see Figure 18.1); p and q can, of course, be determined independently of V. He then introduced a function, $\phi(x)$, which was the probability that an (absolute) error of size x was made in recording the true time of V. Thus, according to Laplace, the probability that the three errors were made was $\phi(x)\ \phi(p-x)\ \phi(p+q-x)$; obviously, he is assuming independence of the errors.

If, on the other hand, the event actually occurred at time V', with x' is the distance to a, then the probability of making the three observations (and their corresponding errors) is $\phi(x')\ \phi(p-x')\ \phi(p+q-x')$. Hence (by Laplace's

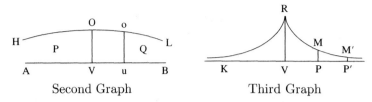

Figure 18.2: Laplace's Second and Third Graphs.

principle), the ratio of the probability that the actual time is at V is to the probability that the actual time is at V' is

$$\frac{\phi(x) \; \phi(p-x) \; \phi(p+q-x)}{\phi(x') \; \phi(p-x') \; \phi(p+q-x')}$$

Laplace then considered the graph HOL, where $y = \phi(x) \; \phi(p-x) \; \phi(p+q-x)$. The problem is to find the point V that represented the *true* time of the phenomenon.

For this, Laplace first considered the nature of $\phi(x)$. Suppose the probabilities that the observations differ from the true time by VP, VP' are the ordinates of some curve RMM', described by some law; let $x = VP$, $y = PM$, and suppose $y = \phi(x)$ (see Figure 18.2). Laplace assumed $\phi(x)$ had the following properties:

1. VR divided the graph into two equal and similar parts, "because it is as likely that the observation strays from the truth to the right as to the left."

2. The line KP is an asymptote, "because the probability the observation is infinitely far from the truth is nothing."

3. The area under the curve RMM' is 1, "because it is certain that an observation falls on some point of KP."

Next, Laplace noted there were two types of "average" being used.

By the average among many observations, one may mean two things of equal importance. The first is the instant at which it is equally probable that the true time of the phenomenon falls before or after: this instant is called the *probable average*.

The second is the instant at which, taking it for the average [i.e., the "true" time], the sum of the errors, multiplied by their probability, is a minimum: one calls this time the *best error* or the *astronomical average*, because it is that which astronomers prefer.[7]

Laplace then gave a beautiful argument that demonstrated the two averages were, in fact, the same.

---

[7]Ibid., 44

Demonstration. Consider the graph HOL of y. For the first type of average, we must find the line OV such that the area HOL is divided equally, so that it is as probable the phenomenon occurs before V as it is that it occurs after V; suppose this is found (see Figure 18.2).

For the second type, we must find the point so that the sum of the ordinates, multiplied by their distance, is a minimum. Let $Vu = dx$, $OV = y$, Q the center of gravity of the region uoL, and M be the area of the region uoL. Let z be the distance to Q from the line OV, P be the center of gravity of the region VOH, N be the area of VOH, and z' be the distance from P to OV.

Take V as the astronomical average, and consider the sum of the ordinates, multiplied by their distance from V: this is $Mz + Nz' + \frac{1}{2}y\,x^2$. Taking u as the astronomical average, then we have the sum of the ordinates, multiplied by their distance from vo is $M(z - dx) + N(z' + dx) + \frac{1}{2}y\,x^2$. The difference between these is $N\,dx - M\,dx$, which must be zero in the case of a minimum value occurring at either V or u; hence $N = M$ and the areas uoL and VOH are equal; VO, the astronomical average, divides the region into two equal parts. $\qquad\square$

Note that Laplace used Fermat's principle of finding an extreme value; not surprisingly, Laplace considered Fermat to be the true inventor of calculus.

The crucial part of the analysis involved the determination of $\phi(x)$, as there are infinitely many functions that satisfy the three criteria given previously. Laplace considered (and rejected) two broad possibilities: $\phi(x)$ could not be a *constant* function, since it is supposed that the errors become less likely the farther they are from the true value; nor could $\phi(x)$ be a function that decreased at a constant rate, since, Laplace felt, the rate of decrease had to itself decrease when the observations were far from the truth. Laplace, falling back to assumed principles, argued:

> As we don't have any reason to suppose a different law for the ordinates as for their differences, it follows that we must, in accordance with the rules of the probabilities, suppose the ratio of two consecutive and infinitely small differences is equal to the ratio of the corresponding ordinates.[8]

Laplace thus considered the ratio between the ordinates at x and $x + dx$, and the corresponding differentials:

$$\frac{d\phi(x + dx)}{d\phi(x)} = \frac{\phi(x + dx)}{\phi(x)}$$

Hence

$$\frac{d\phi(x)}{dx} = -m\phi(x)$$

---

[8]Ibid., 45-6

for some arbitrary m, and thus

$$\phi(x) = \zeta e^{-mx}$$

Since the total area under $\phi(x)$ is 1, then $\zeta = \frac{1}{2}m$, and $\phi(x) = \frac{m}{2}e^{-mx}$, a probability density function now known as the **Laplace Distribution**. It should be noted that for Laplace, x was the distance from a point to the center V; thus, when Laplace writes x, we should be thinking of $|x|$. What remained was to find V so that HOL was divided into two equal parts.

To this end, Laplace began by supposing the true value V to fall in one of the four possible intervals. If V fell in the interval between a and b, then

$$y = \phi(x)\ \phi(p-x)\ \phi(p+q-x)$$
$$= \frac{m^3}{8}e^{-m(2p+q-x)}$$

and the area under y between a and b is $\frac{m^2}{8}e^{-m(2p+q)}\left(e^{mp}-1\right)$.

What is Laplace finding? He did not state it as such, but the area Laplace found is the probability that the errors were made, *given* that the true time fell in the specified interval. Thus, he was finding the "sum of all the probabilities of the event caused by each of the other causes," necessary to apply Bayes's theorem. Hence, the area $\frac{m^2}{8}e^{-m(2p+q)}\left(e^{mp}-1\right)$ is the probability that the three observations are made, *given* that the true time is between a and b.

There are three other possibilities for the actual location of V. If V was between b and c, then

$$y = \phi(x)\ \phi(x-p)\ \phi(q-x+p)$$
$$= \frac{m^3}{8}e^{-m(x+q)}$$

and the area under y between b and c would be $\frac{m^2}{8}e^{-mq}\left(e^{-mp}-e^{-m(p+q)}\right)$.

If V fell in the interval from c to infinity, then the area would be $\frac{m^2}{3\cdot8}e^{-m(p+2q)}$, while if V fell on the interval from a to "infinity on the side of a" (in other words, the absolute error is ∞ but V assumed to the left of a), the area would be $\frac{m^2}{3\cdot8}e^{-m(2p+q)}$. The total of the four areas would thus be:

$$\frac{m^2}{4}e^{-m(p+q)}\left(1-\frac{1}{3}e^{-mp}-\frac{1}{3}e^{-mq}\right)$$

To find the true time, Laplace returned to the idea that it would be the time when it was just as probable to make an error exceeding V as it was to make one falling short of V. Laplace argued this time had to occur between a and b, if $p > q$, or between b and c, if $p < q$; he assumed, for purposes of the calculation, that V was between a and b. This leads to, Laplace noted, the equation

$$m^2 e^{-m(2p+q-x)} = m^2 e^{-m(p+q)}\left(1-\frac{1}{3}e^{-mp}-\frac{1}{3}e^{-mq}\right)$$

Hence

$$x = p + \frac{1}{m} \ln \left(1 - \frac{1}{3} e^{-mp} - \frac{1}{3} e^{-mq} \right)$$

Thus, this is the correction factor needed. Notice that x depends only on the interval p, q, and m, an arbitrary constant.

18.3 Exercises

1. Explain the "paradox" that the more observations one makes, the less likely the mean is to have no error.

2. Find the probability that the mean has 0 error if $a = 2b$, for $n = 1$ to $n = 6$ observations. Find the probabilities if $b = 2a$.

3. Verify the probabilities in the table on page 18.3.1 for $n = 3$.

4. To find the unknown frequencies a, b, c, ..., Lagrange assumed the errors found during the calibration step corresponded to the most probable error that would be made in an actual observation. Is this justified? Explain. Propose a modification of Lagrange's method that might avoid this assumption.

5. Consider Laplace's example of drawing p white and q black tickets from an urn.

   (a) Laplace claimed the probability of drawing p white and q black tickets was $x^p(1-x)^q$. What is the correct probability?

   (b) Explain how Laplace derived the formula $\frac{x^p(1-x)^q dx}{\int x^p(1-x)^q dx}$ for the probability that the probability of drawing a white ticket is x. How would finding the correct probability for drawin gp white and q black tickets affect this derivation?

6. Prove Equation 18.15 and thus derive Laplace's result for E.

7. For each assumption of the location of V (i.e., to the left of a; between a and b; between b and c, and to the right of c) determine:

   (a) The form of y, the probability of making the observed errors.

   (b) The limits of integration and value of $\int y dx$.

8. Use Laplace's principle (i.e., Bayes's theorem) to find

   (a) The probability the true time V is between a and b.

   (b) The probability that the true time is before a.

18.4   Mathematics of Society

The fruitful application of mathematics to physics to determine the motions in the solar system suggested it might be possible to apply mathematics to all aspects of society. If the rulers of France tried to stem the tide of the Enlightenment, the philosophers of France did not, and they began to apply mathematical methods to questions about social injustice.

18.4.1   Borda

Many French thinkers in the years preceding the French Revolution turned their attention to the mathematics of voting, or what is now called **voting theory**. In 1781, Jean Charles de Borda (May 4, 1733-February 19, 1799), a French cavalry officer then serving in the French navy, presented "Memoir on Elections by means of Ballots." In the **plurality method** of voting, each elector votes for one candidate, and the candidate with the most votes wins. But, Borda pointed out, this did not always mean that the person with the most support won; he gave a crucial example. Suppose there are 21 electors and three candidates, A, B, and C. Suppose 13 of the electors preferred B to A and 8 preferred A to B; suppose also that 13 preferred C over A, and again 8 preferred A to C. In the opinion of the electors, A was clearly the worst candidate: 13 voters would rather have B than A, and 13 voters would rather have C than A; in either case, A would have the support of only 8 voters.

Astonishingly, the plurality method of voting might result in A winning an election between A, B, and C. Suppose, of the 13 who preferred B over A, that 7 preferred B over C, while 6 preferred C over B. Then, when restricted to a single choice, 8 electors would choose A; 7 would choose B, and 6 would choose C, and A would win the election: "Which, by hypothesis, is entirely contrary to the opinion of the electors."

Borda proposed a method of voting he called an election by order of merit, but now called a **Borda count**: rather than voting for a *single* candidate, voters would rank *all* candidates. For each last-place vote, a candidate would receive a points; for a next-to-last-place vote, $a + b$; for a third-to-last-place vote $a + 2b$, and so on. The candidate with the greatest total would then win the election.

Example 18.2. *Suppose there are three candidates; thus, each last-place vote would receive a points; each second-place vote would receive a + b points, and each first-place vote would receive a + 2b points. If candidate A is first on the list of 8 voters, second on the lists of 10 voters, and third on the lists of 3 voters, then he would receive* $8 \cdot (a + 2b) + 10 \cdot (a + b) + 3 \cdot (a)$ *points, or* $21a + 26b$.

In the example used by Borda, the twenty-one ballots might be:

$$A\ A\ A\ A\ A\ A\ A\ A\ B\ B\ B\ B\ B\ B\ B\ C\ C\ C\ C\ C\ C$$
$$B\ C\ C\ C\ C\ C\ C\ C\ C\ C\ C\ C\ C\ C\ C\ C\ B\ B\ B\ B\ B\ B$$
$$C\ B\ B\ B\ B\ B\ B\ B\ A\ A\ A\ A\ A\ A\ A\ A\ A\ A\ A\ A\ A$$

where the first ballot shows the elector felt A was the best candidate, B the second-best, and C the worst. This arrangement of the votes is now called a **preference table**, since it shows the preferences of the voters.

Example 18.3. *Using the preference table, determine the winner of the election using the Borda count. Let a last place vote be worth 1 point; a second place 2 points; and a first place 3 points.*

A has 8 first place votes and 13 last place votes, for a total of $8 \cdot 3 + 13 \cdot 1 = 37$ points. B has 7 first place votes, 7 second place votes, and 7 last place votes, for a total of $7 \cdot 3 + 7 \cdot 2 + 7 \cdot 1 = 42$ points. Finally, C, with 6 first place votes, 14 second place votes, and 1 last place vote, has a total of $6 \cdot 3 + 14 \cdot 2 + 1 \cdot 1 = 47$ points. Thus, C is the winner of the election.

Borda noted, in particular, that candidate C, who had the fewest votes according to the plurality method, actually had the greatest support among the electors, while A, who had the most votes according to the plurality method, actually had the least support.

18.4.2   Condorcet

Another contributor to voting theory was Marie Jean Antoinette Nicolas Caritat de Condorcet (September 17, 1743-March 27, 1794), who along with Diderot, d'Alembert, Voltaire, and a number of other *philosophes* helped to bring about the French Revolution. The others did not live to see the fruit of their work; Condorcet did. Condorcet, always interested in social issues, was perhaps the first to apply the powerful tools of mathematical analysis to questions regarding human behavior.

In 1785, Condorcet published a memoir, *Essay on the Applications of Probability to Decisions Rendered by Plurality Voting.* Some of the problems Condorcet considered were the probability that a "correct" decision would be reached by plurality vote; or that the probability that a decision reached by plurality vote was, indeed, the correct one. Condorcet's conclusion was that in order for the correct decision to be reached by plurality, the voters had to be educated: hardly a startling one, but backed by the grim reality of a mathematical analysis.

Marie Condorcet. ©Leonard de Selva/CORBIS.

The Condorcet Winning Condition

More significant were Condorcet's contributions to voting theory. In the introduction to the *Essay*, Condorcet voiced an objection to Borda's method. Suppose one used the Borda method in the case where there were 81 voters and three candidates, A, B, and C. If 30 voters ranked the candidates ABC (that is, A was the highest, B the second, and C the lowest); 1 ranked the candidates ACB; 10 CAB, 29 BAC, 10 BCA, and 1 CBA, then, using Borda's method, candidate A would receive $81a + 101b$ points; candidate B would receive $81a + 109b$ points; and candidate C would receive $81a + 33b$ points; thus, candidate B would be elected. However, Condorcet pointed out that the proposition "A is better than B" was held by $30 + 1 + 10 = 41$ voters, while the converse, "B is better than

A," was held only by $29 + 10 + 1 = 40$ voters. Hence, if an election were held just between A and B, A would win. Likewise, if an election were held between just A and C, A would win again. Since A would win an election against either candidate singly, A should win the election against both candidates; this criterion is now known as the **Condorcet winning condition**. Condorcet's counterexample showed the Borda count does not always satisfy the Condorcet winning condition.

Condorcet proposed an alternative. Suppose there are 60 voters and three candidates, A, B, and C; suppose A received 23 votes, B received 19, and C received 18. Then A would be the winner. However, suppose further that the 23 who voted for A all thought C was a better candidate than B; that the 19 who voted for B all thought C was better than A, and of the 18 who voted for C, then 16 felt B was better than A, and only 2 felt that A was better than B.

Condorcet then examined how the electorate felt about candidate C. Obviously, all 18 who voted for C felt C was better than A or B; in addition, the 23 who voted for A also (by assumption) felt that C was better than B; hence $18 + 23 = 41$ felt C was a better candidate than B. Likewise, $18 + 19 = 37$ voters felt C was better than A. Thus, in both cases, a majority of voters felt C was better than either A or B.

On the other hand, $19 + 16 = 35$ felt B was better than A, and $19 + 0 = 19$ felt that B was better than C. Finally A, the winner of the election by the plurality method, was actually considered to be the worst candidate: a minority of $23 + 2 = 25$ voters felt A was better than B, and a minority of $23 + 0 = 23$ felt A was better than C. Thus, Condorcet concluded, A, who won in the plurality method, actually had the least support, while C, who received the fewest votes, actually had the greatest support. Hence Condorcet suggested that an election be determined by comparing the candidates in pairs, the winner being determined by the candidate who won the most elections this way; this method is now known as the **pairwise comparison method**. Obviously, this method always satisfies the Condorcet winning condition.

Example 18.4. *Determine the winner in the preceding example. There are three possible two-candidate elections: A versus B, A versus C, and B versus C. In A versus B, B would win, 35 to 25. In A versus C, C would win, 37 to 23. In B versus C, C would win again, 41 to 19. Thus, C would be elected.*

Condorcet was not blind to possible difficulties and pointed out a flaw in his system. Suppose again there are three candidates. If the majority prefers A to B, and the majority prefers B to C, it might seem that the majority should prefer A to C. However, Condorcet gave an example where the majority preferred C over A; hence, "preferred" is not a transitive relationship.

The Reliability of Witnesses

Between 1781 and 1784, Condorcet published, in several parts, a long memoir on probability, and in 1783, he applied probability to the reliability of witnesses.

Condorcet's particular bent may be discerned from his title: "On the Probabilities of Extraordinary Events," that is, "miracles." Condorcet's important addition to the theory of testimony was recognizing that if a witness reported the occurrence of an event, there were actually two possibilities: either the event occurred and the witness told the truth; or the event did not occur and the witness lied or was mistaken. Conversely, if a witness reported an event did not occur, then either the event did not occur and the witness told the truth, or the even occurred but the witness, for whatever reason, did not report it. If v and e represent the probability that the event occurred or not, and v' and e' represent the probability that the witness reported the truth or not, then the probability that the event occurred is $\frac{vv'}{vv'+ee'}$, while the probability the event did not occur is $\frac{ee'}{vv'+ee'}$.

18.4.3 The Revolution

The bankruptcy of France forced Louis XVI to summon the Estates General for the first time in over a century; the estates met on May 5, 1789. After some initial discussion, the Third Estate, composed of representatives of the commoners, organized a National Assembly, and invited the First Estate (the nobility) and the Second Estate (the clergy) to join them; they did so, vowing not to disband until they had given France a written constitution.

On July 14, 1789, a mob broke into the Bastille, an old fortress in Paris being used as a prison, because they heard political prisoners were being kept there. The mob murdered the warden but found only a few thieves, drunkards, and prostitutes; nevertheless, the storming of the Bastille is celebrated as the glorious start of the French Revolution.

The Metric System

One of the most important changes made by the French Revolution came from the Committee of Weights and Measures, established on June 22, 1789. The committee underwent a number of changes in membership, and in the end, its main members were Laplace, Lagrange, Borda, Condorcet, and Gaspard Monge.

The fundamental problem facing the committee was simple: too many units of measurement were in use, none of which were easily compatible. For example, consider the ancient riddle, here expressed in two parts: of the following pairs, which weighs more:

1. An ounce of feathers or an ounce of gold?

2. A pound of feathers or a pound of gold?

The surprising answers are: the ounce of gold and the pound of feathers. This is because gold and feathers were weighed according to different scales: gold used the troy scale (named after the medieval trade fair in Troyes, where precious metals were traded), while feathers were weighed in avoirdupois ounces (a shortening of the French phrase meaning "goods having weight"), and the troy ounce was heavier; meanwhile, there are 12 troy ounces to the troy pound, but 16 avoirdupois ounces to an avoirdupois pound, making the avoirdupois pound heavier.

To end this confusion, the committee settled on the use of a *single* fundamental unit of length, and a *single* fundamental unit of mass. For the larger units, much debate followed over whether to use 10 or 12 as the base. The argument in favor of base-12 is that fractions (such as $\frac{1}{3}$, $\frac{1}{6}$, etc.), would have simple forms (just as one-third or one-sixth of a foot correspond to a whole number of inches). Lagrange is said to have argued for base-11 (or any other prime number) on the grounds that all fractions would automatically be in reduced form. One suspects Lagrange was joking to point out the absurdity of using the ease of expressing fractions as a reason to adopt 12 over 10. The committee settled on 10, and the metric system was born.

See page 495.

However, one key question had to be answered: what was to be the fundamental unit of length? Maupertuis's expedition measured the size of the Earth to great accuracy, so the committee adopted the standard unit of length to be one ten-millionth the distance from the north pole to the equator on the meridian running through Paris. This length was named the **meter**, from the Greek word "to measure." The committee made its recommendation on March 25, 1791, and the National Assembly adopted it the next day. Note that the modern definition of a meter is based on a different standard.

The Committe of Public Safety

Some of the more prescient nobility fled France in the early stages of the revolution, and urged the king to do likewise. However, Louis remained until the end of June, 1791: by then, it was too late, and the king, queen, and the royal party got as far as Varennes (near Verdun, close to the Belgian border) before they were intercepted and returned to Paris.

In 1792, the National Convention was formed. Condorcet was a member, representing Paris. The convention was divided into two important factions. Condorcet was a member of the Girondists, a group of moderates. Opposing them was a group called the Mountain (because they sat in the balcony seats in the Convention's meeting place), led by Maximilian Robespierre. France was now under attack, as the nobility who fled during the first few years organized armies to invade. Austria, Prussia, Spain, the Netherlands, and Great Britain united in a coalition (the first of many) that vowed to restore the monarchy to its premier position. On March 18, the Austrians entered Belgium, driving out the French, and it looked like it was only a matter of time before the revolution was undone. On April 6, 1793, the Convention handed nearly dictatorial power to one of the most notorious groups in history: the Committee for Public Safety.

Everyone realized the king was now an important symbol and so long as the king remained the king, he would serve as a rallying point for the reactionaries seeking to restore the monarchy. Condorcet was among the first to vote to depose the king and form a republic; the Mountain wanted to go farther and execute Louis, but Condorcet argued that while Louis might be inconvenient, he had broken no law that would warrant his death. Opposing the Mountain was futile—and dangerous. Louis and Marie Antoinette were executed soon thereafter, and the Mountain would never forget Condorcet's defiance.

The Girondists prepared a constitution, which was rejected; the Mountain prepared another constitution and gave the Convention three days to accept or reject. Once again, Condorcet refused to be intimidated by the radicals, and criticized the new constitution. As a result, the Convention, dominated by Robespierre, issued a warrant for the arrest of Condorcet and other prominent Girondists. Condorcet went into hiding; a revolutionary tribunal declared Condorcet guilty of opposing the unity of the republic and sentenced him to death. Soon afterward, the Mountain swept away all opposition, by intimidation or execution, and the period of the French Revolution known as the Terror began. The radicals, led by Robespierre, executed their opponents by the thousands. The executions were made easier by suspending a suspect's right to a public trial, and by limiting a jury's choice to acquittal or death (with the clear understanding that if the jury chose acquittal, its members would be viewed with suspicion thereafter). On the practical side, the actual process of execution was accelerated by a new invention: the guillotine.

Condorcet, though hiding and under sentence of death, completed one of his greatest philosophical works, *Sketch of the Progress of the Human Mind*, showing that, despite the frightful course the revolution had taken, he still felt optimistic about the future of humanity. He also wrote an elementary work on arithmetic, *A Sure and Easy Method of Learning to Count*. The manuscript was smuggled out, page by page, to his wife from Condorcet's hiding place. Condorcet's wife, Sophie de Grouchy, supported herself through various activities, including painting the portraits of those the revolution had condemned to death; after the chaos of the revolution, she managed to get the book published, and many of the next generation would learn basic arithmetic from Condorcet's work.

Condorcet knew it was only a matter of time before he was found, and he feared the consequences that would be born by those who hid him. Thus, he attempted to leave the country, but was captured at Clamart, by a suspicious innkeeper. According to one story, Condorcet gave himself away by not knowing how many eggs went into an omelet. He was arrested, but before he could be taken to Paris for execution, Condorcet took poison on the night of March 26-7, 1794.

18.4.4 Gaspard Monge

It is around this dark time in French history that Gaspard Monge (May 10, 1746-July 28, 1818) made his appearance. In one sense, Monge was the most important mathematician of the French Revolution, for he helped to preserve the gains of the revolutionaries. Before the revolution, in 1783, Monge evaluated potential naval officers. In the army and navy of France at the time (as well as that of most other countries), noble birth was considered more important than competence.

Monge, whose own opportunities for advancement were limited (he was the son of a peddler) took the unpopular view that a man in charge of an expensive ship and a hundred men ought to have more to recommend him than a noble title. Somehow, Monge managed to avoid dismissal until the coming of the French Revolution. The revolutionaries appointed Monge the Minister of the Navy. Political interference by the new regime made his job unbearable, so Monge quit on February 13, 1793, only to be reappointed five days later. Within a month, France found itself besieged on every front by the armies of the First Coalition. Monge was allowed to resign his commission on April 10, so he could help the French state with war preparations. This he did, and brilliantly.

Monge did more than merely preserve the French Republic: he made provisions for its future. The growing importance of artillery meant it was critical to have combat engineers—which in turn meant it was necessary to have men skilled in applied mathematics. Unfortunately, many schools of higher education closed in the first few years of the revolution, and most of them never reopened. Thus, in 1794 the Commission of Public Works was established and charged with establishing institutions of higher learning, most especially a preparatory school for military engineers. Monge became one of the commission's most prominent members.

The committee wasted no time, and later in 1794, opened the Polytechnical School in Paris. It would train a generation of French soldiers and, more importantly, French mathematicians, and textbooks based on lectures given at the Polytechnical School would teach mathematicians for a century. The workload was heavy: a student of the Polytechnical School was expected to attend over 800 hours of mathematics lectures, given by *Professors*, in addition to time spent with *Repeaters* who explained the lectures and drilled the students. The student would then be tested by *Examiners*.

It was the work of Monge (and the Polytechnical School in general) that revitalized geometry, so long eclipsed by the more glamorous field of calculus. Monge taught a course on descriptive geometry, essentially, the geometry of schematic diagrams, though Monge's course was far more rigorous and included far more material than this simple summary would suggest. Monge also lectured on a new branch of mathematics, one that applied the tools of calculus to geometry: what we would now call **differential geometry**.

On July 27, 1794, four months after Condorcet's suicide, the moderates gathered their courage and deposed Robespierre, who would soon face the guillotine himself. The Terror had cost the lives of over 17,000 "officially" executed, as well as countless others who, like Condorcet, cheated the executioner by committing suicide.

Meanwhile, the coalition armies had been expelled from France and chased back to the Rhine River, ironically achieving the "natural frontiers" sought by Louis XIV. Victorious generals, hoping to spread republican ideals and, more importantly, to establish buffer states, established various republics around France: in the Netherlands, the Batavian Republic, and in Switzerland, the Helvetic Republic.

Marie Antoinette, guillotined in 1793, was an Austrian princess, and not surprisingly, Austria would play a key role in the armies of reaction. Had Austria intervened in the disorganized early days of the revolution, she might have crushed the provisional government and restored the monarchy. However, she had more lucrative matters to attend to: in the latter part of the 18th century, Poland was too weak to defend herself, so Austria, Prussia, and Russia annexed portions of the country, and by 1795, Poland vanished from the face of Europe.

With Poland absorbed, Austria now turned to crushing France. The new government of France, calling itself the Directory, knew the natural invasion route was through northern Italy, so they sent an army there to forestall Austrian action. The army was led by a former artillery officer who had distinguished himself by retaking Toulon from the British in 1794: Napoleon Bonaparte.

Bonaparte thought himself something of a mathematician and would maintain close contact with Laplace throughout his life. In fact, Laplace had been Napoleon's Examiner in mathematics at the Military Academy in September 1785. In 1797, for political reasons, Lazare Carnot lost his position as professor of mechanics at the Military Academy; General Bonaparte was elected to take his place, and Laplace was one of the first to welcome him. Laplace sent Napoleon the first two volumes of his monumental *Celestial Mechanics*. At least one anecdotal story has it that after Napoleon read through them, he remarked, "You make no mention of God in this work." Laplace supposedly replied, "Sire, I have no need for that hypothesis." Lagrange's rejoinder, on hearing Laplace's answer, is worth noting: "Ah, but it is such a beautiful hypothesis, for it explains so many things."

Whether or not this interchange ever took place, we do not know. However, we do know that Napoleon replied to Laplace, in a letter dated October 19, 1799, that he would read them the first six months he had free—an audacious boast, since they would easily occupy a graduate student in mathematics for a year. In the same letter he invited Laplace to dinner the next day "if you have nothing better to do." What happened at dinner we do not know, but on November 12, Laplace became the Minister of the Interior, overseeing all domestic affairs save finance and the secret police. Among his very few official actions was recommending July 14 become a national holiday. Laplace only lasted a few weeks as Minister before being replaced by Napoleon's brother. However, Napoleon continued to shower honors on him: in 1802, Laplace became a Grand Officer of the Legion of Honor, then Chancellor of the Senate in 1803, then a Count of the Empire in 1808. See page 567.

The war with Austria ended in 1797. Austria recognized the northern Italian republics, and ceded Belgium to France. However, the Republic of Venice, which had once dominated the Mediterranean, was given to Austria as a consolation prize. France's possession of the territory to the Rhine River was recognized, and France emerged from the war as the most powerful nation in Europe. One memory still stung: the loss of her overseas possessions in India and Canada. Napoleon sought to restore a French Empire. To this end, he made ready to invade Egypt.

18.4 Exercises

1. Show that in a Borda count, the actual values of a and b are irrelevant: if a candidate wins if the votes are worth a, $a + b$, $a + 2b$, ..., then the candidate will still win if the votes are worth a', $a' + b'$, $a' + 2b'$, ... Thus, the Borda count does not depend on how the points are assigned to a first, second, third, etc., place vote (hence the simplest arrangement, i.e., 1 point for a last place, 2 points for a next to last, etc., will work).

2. Suppose there are 7 voters, and the probabilities a voter chooses A, B, or C are v, e, and i, respectively. What is the probability that:

   (a) Candidates B and C tie, and both beat A.

   (b) Candidate A ties one of B or C, and beats the other.

   (c) Candidate A beats both B and C.

3. Give an example of Condorcet's Voting Paradox: where the majority of voters preferred A to B, and the majority preferred B to C, and the majority preferred C to A.

4. Prove that in a two person race, the plurality method, pairwise comparison method, and Borda count all give the same results.

5. Suppose a jury consists of three persons, each of whom will make the "right" decision in a cases and the "wrong" decision in b cases. What is the probability that the jury, making its decision by plurality voting, will make the "right" decision? (In other words, what is the probability that the majority of the jurors will make the right decision?)

6. Consider Condorcet's analysis of whether or not an event occurred, given that a witness claimed it occurred.

   (a) Show that the probability the event occurred is $\frac{vv'}{vv'+ee'}$, while the probability the event did not occur is $\frac{ee'}{vv'+ee'}$.

   (b) What assumption about the probabilities did Condorcet make in computing the probability that the event actually occurred, or not? Explain, and modify Condorcet's formulas accordingly.

Chapter 19

The Age of Gauss

In 1804, Napoleon crowned himself Napoleon I, Emperor of the French. Two European coalitions against France had already been defeated; a third, between Britain, Austria, Russia, and Sweden, moved to attack. Napoleon defeated his enemies on land and at the Battle of Austerlitz on December 2, 1805, crushed the combined Austrian and Russian armies. Prussia, who a half century before held off the combined forces of France, Austria, *and* Russia, was soundly defeated on October 14, 1806, at the battles of Jena and Auerstadt, and by year's end, all Prussia was under French control. The Holy Roman Empire was disbanded and replaced by a French puppet state, the Confederation of the Rhine.

The thorn in Napoleon's side was Britain, derided by Napoleon as a "nation of shopkeepers." In 1798, Napoleon invaded Egypt and suffered his first serious defeat at the hands of the British, whose fleet was handled masterfully by Admiral Horatio Nelson. Nelson went on to effectively destroy French naval power at the Battle of Trafalgar in 1805 (where he died of his wounds). In 1808, Napoleon invaded Spain, and what was to be a quick campaign bogged down into a long series of battles, thanks to British forces under Arthur Wellesley, later the Duke of Wellington. In every European coalition against Napoleon, the British were prominent. Soon, there would have to be a reckoning.

19.1   The Roots of Equations

Jean Baptiste Joseph Fourier (March 21, 1768-May 16, 1830) stands the middle ground between Condorcet, who lost everything during the revolution, and Laplace, who through the revolutions, managed to maintain and increase his position and wealth. Fourier's misfortunes are so extensive as to be almost comic; his survival attests to the unwavering support of those he had helped and befriended.

573

Fourier's bad luck began when he arrived in Paris for school on the eve of the revolution. His school closed in 1789, when the revolution began. Fourier had prepared a paper for finding the limits of the real roots of an equation, but in the chaos of the revolution, the paper was lost. When Robespierre and the Committee of Public Safety began the Terror, Fourier dared to defend the accused and was himself arrested in 1794. His appeal to Robespierre was rejected, and he would have been executed but for the fall of the Committee of Public Safety. Released, he began to attend the Normal School, which closed the same year it opened (1794).

Fourier made quite an impression at the Normal School, however. Thus, in 1795, he was appointed an assistant lecturer at the Polytechnical School, and things began to look better for him. Unfortunately, later that year he was arrested on the charge of being a *supporter* of Robespierre and the Terror; only the sworn testimony of his friends kept him from a long prison sentence. About this time, Monge (perhaps thinking his friend would be safer outside of France) arranged to have Fourier accompany Napoleon on an overseas expedition, a highly prestigious position.

Unfortunately, the campaign was Napoleon's disastrous expedition to Egypt. The British nearly captured Napoleon as he fled, abandoning his army in the process. Fourier returned to France in 1801, and Napoleon made him a Baron in 1808. The Egyptian fiasco was not a complete disaster: Fourier put together *Description of Egypt*, which helped to found modern Egyptology. In addition, French soldiers found (and British soldiers soon captured) the Rosetta Stone, from which the meanings of the Egyptian hieroglyphics could be deciphered.

See page 623.

Fourier is best known for his development of the Fourier series. However, his lifelong interest was actually the theory of equations, and he made several contributions to the subject. Perhaps the most important was contained in his 1789 paper, which he rewrote in 1820 as "On the Use of the Theorem of Descartes in Finding the Limits of Roots." This contained an extension of Descartes's rule of signs, as well as a new and simple proof of it. Fourier's bad luck continued to plague him, however: the delay in publication meant Budan de Bois-Laurent, another mathematician, could claim priority. Fourier's friends defended him, however, and pointed out that the substance of the paper had been taught by Fourier to his students at the Polytechnical School as early as 1796, making it clear that Fourier had priority of discovery. Still, Fourier could not escape his luck: the result is today known as **Sturm's theorem**.

Let X be an mth degree polynomial with positive first term; consider X and its first m derivatives, X', X'', X''', ..., in reverse order, namely

$$\ldots, X',''\, X,''\, X',\, X$$

Fourier then described his result:

> If one substitutes the number a into the sequence of functions and writes the sign $+$ or $-$ of each result, one makes a sequence of signs which we designate by (α); if one then substitutes a number b, greater

than a, in the same sequence of functions, and notes the resulting signs, a second sequence of signs will be formed which we designate by (β). This being done, determine how many times, in the first sequence of signs (α), there is a change of sign from one term to the next, that is to say how many times in this sequence two adjacent signs are $+$ and $-$. Then find how many times there is a change of sign in the sequence (β). From this method of comparing the two sequences of signs (α) and (β), we will first of all state and then demonstrate the following consequences.[1]

Fourier's main results are:

1. If the number of sign changes in (α) is equal to the number of sign changes in (β), then there is no root in the interval between a and b.

2. There cannot be more sign changes in (β) than there are in (α).

3. If (α) has one more sign change than (β), there is exactly one root between a and b.

4. If (α) has two more sign changes than (β), then there are two real roots between a and b, or two imaginary ones.

5. If (α) has three more sign changes than (β), then there are one or three real roots between a and b.

In general, Fourier concluded, there cannot be more roots between a and b than the excess of the number of sign changes in (α) over (β).

Fourier's proof is a beautiful piece of mathematical analysis, and only slight modifications are needed to make it completely rigorous. Fourier's argument is the following: let X be an mth degree polynomial with positive first term; consider X and its first m derivatives, namely

$$X^{(m)}, X^{(m-1)}, \ldots, X', '' X, '' X', X$$

Note there are $m+1$ terms in this sequence. If an infinitely large negative number is substituted into the sequence, the derivatives will be alternately positive or negative, and the sequence begins with a positive number; hence the sequence will have m changes in sign.

If, on the other hand, one substituted an infinitely large positive number into the sequence of derivatives, then all the terms are positive and there are no changes in sign. Thus, as a increases from $-\infty$ to $+\infty$, the sequence of signs had to lose m sign changes. However, a sign change cannot occur unless at least one of the functions X, X', X'', X''', $\ldots$ equals 0. Suppose a makes one of the functions in the sequence $X^{(m)}$, $X^{(m-1)}$, $\ldots$, X'', X', X equal zero. There are two cases:

---

[1] Fourier, Jean Baptiste, *Œuvres de Fourier*. Gauthier-Villars et fils, 1888-1890. Vol. II, page 292.

1. $X = 0$, and thus a is a root.

2. One of the derivatives X', X'', ... is equal to 0.

Fourier considered these cases separately. In the first case, suppose that none of the derivatives X', X'', ... are equal to 0 at a. X' is either positive or negative; suppose first it is positive. Then as we pass from a number smaller than a by an infinitely small amount to a number larger than a by an infinitely small amount, the signs of X' and X change as indicated:

		X'	X
$< a$	...	$+$	$-$
a	...	$+$	0
$> a$	...	$+$	$+$

Hence, as we pass from less than a, through a, to more than a, the sequence of signs (α) will lose one sign change if X' is positive. The same result follows if X' is negative; hence, every root corresponds to the loss of one of the sign changes.

Next, suppose a makes one of the derivatives equal to 0, say $X^{(n)}$. The two adjacent derivatives, $X^{(n+1)}$ and $X^{(n-1)}$ might be both positive; both negative; or one positive and one negative. In the first case, we have:

		$X^{(n+1)}$	$X^{(n)}$	$X^{(n-1)}$	
$< a$	...	$+$	$-$	$+$	...
a	...	$+$	0	$+$	...
$> a$	...	$+$	$+$	$+$	...

Thus, as we pass from less than a, through a, to greater than a, we lose two sign changes. Alternatively, if $X^{(n+1)}$ is positive and $X^{(n-1)}$ is negative, then

		$X^{(n+1)}$	$X^{(n)}$	$X^{(n-1)}$	
$< a$	...	$+$	$-$	$-$	...
a	...	$+$	0	$-$	...
$> a$	...	$+$	$+$	$-$	...

and the number of sign changes stays the same as we pass from less than a to greater than a. Likewise, for the other two cases, either two sign changes are lost, or the number of sign changes remains the same.

These two together prove the second consequence, which is that (β) always has fewer sign changes than (α); in other words, as a increases from $-\infty$ to $+\infty$, the number of sign changes can only decrease. To prove the remainder, Fourier noted that if all m roots were real, then, since each real root corresponded to a loss of a sign change, and there were only m sign changes to be lost, then each sign change corresponded to a root; thus, even if a made any of the intermediate derivatives equal to 0, no change in the total number of sign changes was possible.

On the other hand, if there were $m - 2$ real roots and 2 imaginary roots, then there would be a loss of a sign change at each of the $m - 2$ real roots. The remaining two sign changes had to be lost at a point when an intermediate function was equal to 0; this corresponded (though not in an obvious manner) to the existence of a pair of imaginary roots.

After this, Fourier proved the corresponding results if a was a multiple root of X, or if a made more than one of the derivatives of X equal to 0.

19.1 Exercises

1. The major flaw in Fourier's proof of "Sturm's" theorem, from a modern standpoint, is his initial step in substituting an "infinitely large" positive or negative number into the sequence of the function and its derivatives. Modify this part of the proof so it is acceptable today.

2. Fourier's proof assumed a number of properties about the derivative of a polynomial. Identify and prove these properties.

3. Prove that if $X = 0$ at a, and none of the derivatives are equal to 0 at a, a sign change must be lost as we move from less than a to more than a.

4. Prove that if $X^{(n)} = 0$ at a and at none of the higher derivatives $X^{(n+1)}$, ..., then there will either be a loss of two sign changes, or a conservation of the number of sign changes, as we move from less than a to more than a.

5. Prove that if X has a real root of multiplicity k at a, then there will be k changes in sign as we move from less than a to more than a.

6. Fourier said that Descartes's rule of signs followed by substituting in $-\infty$, 0, and $+\infty$. Explain.

19.2 Solvability of Equations

Just as the history of the opening years of the nineteenth century was dominated by a single individual, Napoleon, so was the mathematics of the opening era dominated by a single person, Carl Friedrich Gauss (April 30, 1777-February 23, 1855). Gauss was the son of a tradesman and the semiliterate daughter of a peasant stonemason. Stories of Gauss's early mathematical genius abound: when he was 3, he uncovered an error in his father's bookkeeping; when Gauss was 8, Büttner, the local school teacher, gave to the class the problem of summing the first 100 whole numbers, to which Gauss gave the correct answer—5050— almost immediately and without any obvious computation. The teacher was impressed. The mythology of mathematics says the teacher was furious, but Büttner supplied Gauss with more advanced works and encouraged him to work with one of the assistant teachers, Johann Martin Bartels. By the time Gauss was 14, he came to the attention of the duke of Brunswick, Carl Wilhelm Ferdinand. The duke, a disciple of Frederick the Great of Prussia, paid Gauss a stipend so he could keep on learning and someday be a great benefit to the duchy. In 1795, Gauss graduated from the University in Brunswick, and went on to the University in Göttingen, supported by his ducal stipend of 158 thalers a year.

Carl Friedrich Gauss. ©CORBIS.

19.2.1   Construction of the Septendecagon

Göttingen, in Hanover, was a relatively new university, established in 1737 by Duke George August. At Göttingen, Gauss was almost lost to mathematics. There, he was faced with a choice between working with the brilliant classicist G. Heyne, and the mediocre mathematician A. G. Kästner, who wrote well but lectured poorly. In 1796, Gauss had to choose: the classics, or mathematics? It was then he made a monumental discovery.

Since the time of Euclid, compass and straightedge procedures were known for inscribing in a circle regular 3, 4, 5, and 15-sided polygons. Since it is possible to bisect an angle using only compass and straightedge, it is thus possible to produce polygons of $3 \cdot 2^n$, $4 \cdot 2^n$, $5 \cdot 2^n$, and $15 \cdot 2^n$ sides. But these were the only polygons known to be constructible using compass and straightedge alone.

In 1796, at the age of 19, Gauss discovered that it was possible to inscribe a regular 17-sided polygon (a septendecagon) in a circle—the first new compass and straightedge geometric construction to be discovered in nearly two thousand years. According to one story, Gauss approached Kästner with his discovery. Kästner first told Gauss that the discovery was useless, since approximate constructions were "well known"; next, that the construction was impossible, so Gauss's proof had to be flawed; finally, that Gauss's method was something that he, Kästner, already knew about so Gauss's discovery was unimportant. Despite Kästner's discouraging remarks, Gauss could not let a triumph like the construction of a septendecagon remain his only work in mathematics. Gauss later toasted Kästner, who was an amateur poet, as the best poet among mathematicians and the best mathematician among poets.

Gauss included his method in the last section, "Equations Defining Sections of a Circle," of his *Arithmetical Investigations*, written by Gauss between the ages of 19 and 21 and published in 1801. Gauss was not actually attempting to construct the septendecagon; rather, he was investigating the roots of the equation $x^n - 1 = 0$.

One root is real, namely $x = 1$; factoring out $x - 1$ one obtained the equation

$$x^{n-1} + x^{n-2} + \ldots + x + 1 = 0 \qquad\qquad (19.1)$$

now called the **cyclotomic equation**. Gauss called the left hand side X; its "complex" of roots (what we would now call the *set* of all roots) Gauss designated Ω. These roots were expressible as $\cos \frac{kP}{n} + i \sin \frac{kP}{n}$, with k taking on the values $0, 1, \ldots, n-1$, where P, in Gauss's terminology, was the circumference of the circle of four right angles (i.e., $P = 2\pi$).

19.2.2   Properties of the Cyclotomic Equation

First, Gauss proved:

Proposition 19.1. *If X is divisible by an equation*

$$P = x^\lambda + Ax^{\lambda-1} + \ldots Kx + L \qquad (19.2)$$

then the coefficients of P cannot all be rational numbers.

Then, before continuing the discussion, Gauss described his *purpose*: X was to be resolved into equations of lesser degree, and each of these in turn reduced to equations of lower degree, until, finally, one had a set of solvable equations, whose roots would then make up Ω.

Suppose r was a root of Equation 19.1. Gauss needed to consider powers of powers of the roots; since it was difficult, then, to typeset expressions such as r^{λ^2}, Gauss instead wrote this as $[\lambda^2]$; we will follow Gauss's convention. Remember Euler had considered the remainders when the powers of a were divided by some number n (see Section 17.2). Gauss introduced the term **modulus** and invented the modern notation for it: if a divided the difference of the numbers b and c, then Gauss called b and c **congruent** relative to the modulus a, and wrote $b \equiv c(\bmod a)$. Even though he invented the notation, Gauss did not always use it, and frequently wrote "b is congruent to c," with the modulus a being understood.

Consider some number g. If the powers of g modulo n were congruent (in some order) to the numbers $1, 2, 3, \ldots, n-1$, then g is called a **primitive root** under the modulus n.

Example 19.1. *Let $n = 5$. Then $g = 2$ is primitive, since $2^1 \equiv 2 \bmod 5$, $2^2 \equiv 4 \bmod 5$, $2^3 \equiv 3 \bmod 5$, $2^4 \equiv 1 \bmod 5$, and $2^5 \equiv 2 \bmod 5$ again.*

Given a primitive root g; then $[1], [g], [g^2], \ldots, [g^{n-2}]$ were then equal, in some order, to all the roots Ω. Likewise, if n did not divide λ, then $[\lambda], [\lambda g]$, $[\lambda g^2], \ldots, [\lambda g^{n-2}]$ also included all the roots in Ω. Finally, if G was a different primitive root, then $[1], [G], [G^2], \ldots$ were congruent to $[1], [g], [g^2], \ldots$ (again, in some order).

See Problem 1a.

Suppose e divides $n-1$; let $n-1 = ef$ and $g^e \equiv h$, $G^e \equiv H$. Then

$$1, h, h^2, \ldots, h^{f-1} \equiv 1, H, H^2, \ldots, H^{f-1}$$

taken in some order. Hence

$$[\lambda], [\lambda h], [\lambda h^2], \ldots, [\lambda h^{f-1}] \equiv [\lambda], [\lambda H], [\lambda H^2], \ldots, [\lambda H^{f-1}]$$

taken in some order. The set of these f roots was referred to as the **period**, and Gauss designated by (f, λ) the sum of the f roots

$$(f, \lambda) = [\lambda] + [\lambda h] + [\lambda h^2] + \ldots + [\lambda h^{f-1}]$$

Gauss used, as a continuing example, the case of $n = 19$, $f = 6$. If $n = 19$, then $g = 2$ is a primitive root modulo 19, since the powers of 2 modulo 19 are congruent to the numbers (in some order) 1 through $19 - 1 = 18 = e \cdot f$. Since

$f = 6$, then $e = 3$. We note $2^3 \equiv 8$, so $h = 8$. Now consider the powers of 8 modulo 19. The distinct values are:

$$[1], [8], [64], [512], [4096], [32768]$$

or

$$[1], [8], [7], [18], [11], [12]$$

If $\lambda = 1$, we have:

$$(6, 1) = [1] + [7] + [8] + [11] + [12] + [18]$$

If $\lambda = 2$, we have

$$\begin{aligned}
(6, 2) &= [2 \cdot 1] + [2 \cdot 7] + [2 \cdot 8] + [2 \cdot 11] + [2 \cdot 12] + [2 \cdot 18] \\
&= [2] + [14] + [16] + [22] + [24] + [36] \\
&= [2] + [14] + [16] + [3] + [5] + [17] \\
&= [2] + [3] + [5] + [14] + [16] + [17]
\end{aligned}$$

If $\lambda = 3$, we have

$$(6, 3) = [3 \cdot 1] + [3 \cdot 7] + [3 \cdot 8] + [3 \cdot 11] + [3 \cdot 12] + [3 \cdot 18]$$

To simplify this, we note that $[3 \cdot 1] = [3]$, and if two periods have one term in common, they must have all terms in common; hence $(6, 3) = (6, 2)$. The only other distinct period is

$$(6, 4) = [4] + [6] + [9] + [10] + [13] + [15]$$

Example 19.2. *Find the periods for $n = 5$, $f = 2$. We note that $g = 3$ is a primitive root for $n = 5$, and $f = 2$ divides $5 - 1 = 2 \cdot 2$; hence $e = 2$. Thus, $g^e = 3^2 = 9$, and $9 \equiv 4$, so $h = 4$. Now consider 1 and the powers of 4 modulo 5; there are only two distinct values, namely 1 and 4 (since $4^2 \equiv 1$). First, take $\lambda = 1$*

$$(2, 1) = [1] + [4]$$

Next, take $\lambda = 2$:

$$\begin{aligned}
(2, 2) &= [2] + [8] \\
&= [2] + [3]
\end{aligned}$$

Finally, we note that for $\lambda = 3$ or $\lambda = 4$, we have

$$\begin{aligned}
(2, 3) &= [3] + [2] \\
&= (2, 2)
\end{aligned}$$

And similarly

$$(2, 4) = (2, 1)$$

Hence, we note that $(2, 1)$ and $(2, 2)$ are the only distinct periods.

Note that (f, λ) contained only *some* of the roots, and if the values of these expressions could be found, a simpler expression of the roots might be obtained. Gauss's procedure is to find simpler equations whose roots were expressions such as (f, λ). But to find such equations, it is necessary to determine the values of products and sums of these quantities. Thus, Gauss proved:

Proposition 19.2. *If (f, λ) and (f, μ) are similar periods (not necessarily different), with*

$$(f, \lambda) = [\lambda] + [\lambda'] + [\lambda''] + \ldots$$

then $(f, \lambda) \cdot (f, \mu)$ is the sum of f similar periods, namely

$$W = (f, \lambda + \mu) + (f, \lambda' + \mu) + (f, \lambda'' + \mu) + \ldots$$

Moreover, W is independent of the choice of λ.

For $n = 19$, Gauss found $(6, 1) \cdot (6, 1)$, by:

$$(6, 1) = [1] + [7] + [8] + [11] + [12] + [18]$$

Then (with $\lambda = \mu = 1$)

$$(6, 1) \cdot (6, 1) = (6, 2) + (6, 8) + (6, 9) + (6, 12) + (6, 13) + (6, 19)$$
$$= 6 + 2(6, 1) + (6, 2) + 2(6, 4)$$

Example 19.3. *For $n = 5$, find $(2, 1) \cdot (2, 2)$ and verify the product. Recall*

$$(2, 1) = [1] + [4]$$

We have $\mu = 2$, thus

$$(2, 1) \cdot (2, 2) = (2, 3) + (2, 6)$$

Since $(2, 3) = [3] + [2] = (2, 2)$ and $(2, 6) = [1] + [4] = (2, 1)$, we can write

$$(2, 1) \cdot (2, 2) = (2, 2) + (2, 1)$$

We can verify this by noting the product

$$(2, 1) \cdot (2, 2) = ([1] + [4])([2] + [3])$$
$$= [3] + [4] + [6] + [7]$$
$$= [3] + [4] + [1] + [2]$$

(Remember, $[1] = r^1$ and $[2] = r^2$, so $[1][2] = r^1 r^2 = r^3 = [3]$; likewise, $[7] = r^7 = r^5 r^2 = [2]$, since r is one of the fifth roots of unity.) We can see that $[3] + [4] + [1] + [2] = (2, 2) + (2, 1)$. Further, if r is a fifth root of unity, then $r + r^2 + r^3 + r^4 = -1$. Hence, $(2, 2) \cdot (2, 1) = -1$.

Next is:

Proposition 19.3. *If n does not divide λ, and $p = (f, \lambda)$, any similar period (f, μ) where n does not divide μ can be reduced to*

$$(f, \mu) = \alpha + \beta p + \gamma p^2 + \ldots + \theta p^{e-1}$$

where α, β, γ, $\ldots$ are determined quantities.

Proof. Let p', p'', p''', $\ldots$ represent $(f, \lambda g)$, $(f, \lambda g^2)$, $(f, \lambda g^3)$, $\ldots$, up to $(f, \lambda g^{e-1})$. There are thus $e - 1$ periods, one of which must be (f, μ); moreover, p, p', p'', $\ldots$, together with 1, is the sum of all the roots; hence

$$1 + p' + p'' + p''' + \ldots = 0$$

By Proposition 19.2, we can express p^2, p^3, $\ldots$, p^{e-1} in terms of p', p'', $\ldots$, giving us the $e - 1$ equations

$$0 = p^2 + A + ap + a'p' + \ldots$$
$$0 = p^3 + B + bp + b'p' + \ldots$$
$$0 = p^4 + C + cp + c'p' + \ldots$$
$$\vdots$$

where all the coefficients A, B, $\ldots$, a, b, $\ldots$, a', b', $\ldots$ are integers and independent of the choice of λ.

Suppose $(f, \mu) = p'$ (or any other). Solving for p' in terms of p, p^2, $\ldots$, yielding equation (Z):

$$0 = \mathfrak{A} + \mathfrak{B}p + \mathfrak{C}p^2 + \ldots + \mathfrak{M}p^{e-1} + \mathfrak{N}p'$$

If $\mathfrak{N} \neq 0$, the equation may be solved for p' and the theorem is proven.

Suppose $\mathfrak{N} = 0$. Then the resulting equation is of the $e - 1^{\text{st}}$ degree, and there are thus at most $e - 1$ solutions or p. But the equations from which (Z) is derived are independent of λ, hence the form of equation (Z) itself is independent of λ; thus the equation will hold true for any λ not divisible by n. Thus, p has solutions $(f, 1)$, (f, g), (f, g^2), $\ldots$, (f, g^{e-1}); since there are at most $e - 1$ solutions, at least two of theses must be equal. Suppose the equal period consist of the terms $[\zeta]$, $[\zeta']$, $[\zeta'']$, $\ldots$, and $[\eta]$, $[\eta']$, $[\eta'']$, $\ldots$ Obviously, all of the roots in the period will be different and none will be equal to zero. Let

$$Y = x^\zeta + x^{\zeta'} + \ldots - x^\eta - x^{\eta'} - \ldots$$

Clearly, $Y = 0$ when $x = [1]$. But $[1]$ is also a root of $X = 0$; hence, X and Y have a common factor (namely, $x - [1]$). Hence, they have a greatest common factor, which may be found; however, the process of finding this greatest common factor necessarily results in a polynomial with rational coefficients. But this is impossible, for any polynomial that divides X must not have all rational coefficients (Theorem 19.1). Hence our original assumption, $\mathfrak{N} = 0$, cannot hold true. $\qquad \square$

Again, Gauss returned to the $n = 19$, $f = 6$ case. If $p = (6, 1)$, $p' = (6, 2)$, and $p'' = (6, 4)$, then

$$
\begin{aligned}
p^2 &= (6, 1) \cdot (6, 1) \\
&= 6 + 2(6, 1) + (6, 2) + 2(6, 4) \\
&= 6 + 2p + p' + 2p''
\end{aligned}
$$

Moreover, the sum of all the roots is 1; hence

$$
0 = 1 + p + p' + p''
$$

Hence we may express p' and p'' in terms of p:

$$
\begin{aligned}
p' &= 4 - p^2 \\
p'' &= -5 - p + p^2
\end{aligned}
$$

Hence $(6, 2) = 4 - (6, 1)^2$ and $(6, 3) = -5 - (6, 1) - (6, 1)^2$.

Example 19.4. *If $n = 5$, $f = 2$, $p = (2, 1)$ and $p' = (2, 2)$, express p' in terms of p and its powers. We note that*

$$
0 = 1 + p + p'
$$

Hence $p' = -1 - p$.

Finally, Gauss tackles the main questions. Let us apply Gauss's method to $n = 5$; thus, we are seeking to find the four roots of $x^4 + x^3 + x^2 + x + 1 = 0$. We have already noted that the four roots split into two periods, namely $(2, 1)$ and $(2, 2)$. Like Lagrange, Gauss supposed these were the roots of an equation. If these are the roots of a quadratic equation, $(z - (2, 1))(z - (2, 2)) = 0$, then

$$
\begin{aligned}
0 &= (z - (2, 1))(z - (2, 2)) \\
&= z^2 - ((2, 1) + (2, 2))z + (2, 1)(2, 2)
\end{aligned}
$$

From Examples 19.4 and 19.3 we know the sum and product of $(2, 1)$ and $(2, 2)$; hence

$$
\begin{aligned}
0 &= z^2 + z + (2, 1) \cdot (2, 2) \\
&= z^2 + z - 1
\end{aligned}
$$

where $(2, 1) + (2, 2) = [1] + [4] + [2] + [3] = -1$. The solutions to this equation are $z = \frac{-1 \pm \sqrt{5}}{2}$.

To determine which root corresponds to $(2, 1)$ and which root corresponds to $(2, 2)$, Gauss recommended that one find *approximate* values, using a table of sines. Since

$$(2,1) = [1] + [4]$$

$$= \left(\cos \frac{P}{5} + i \sin \frac{P}{5} \right) + \left(\cos \frac{4P}{5} + i \sin \frac{4P}{5} \right)$$

$$\approx 0.61803$$

then $(2,1) = \frac{-1+\sqrt{5}}{2}$ and $(2,2) = \frac{-1-\sqrt{5}}{2}$.

Next, since $(2,1) = [1] + [4]$, it is necessary to write an equation whose roots are $[1]$ and $[4]$, namely

$$0 = (y - [1])(y - [4])$$

$$= y^2 - ([1] + [4]) z + [1][4]$$

$$= y^2 - (2,1)z + 1$$

The roots of this equation are

$$y = \frac{(2,1) \pm \sqrt{(2,1)^2 - 4}}{2}$$

$$= \frac{\frac{-1+\sqrt{5}}{2} \pm \sqrt{\left(\frac{-1+\sqrt{5}}{2} \right)^2 - 4}}{2}$$

which gives $[1]$ and $[4]$, the first and fourth power of the roots.

Likewise, since $(2,2) = [2] + [3]$, the remaining two roots can be found using the equation

$$0 = (y - [2])(y - [3])$$

$$= y^2 - ([2] + [3])y + [2][3]$$

$$= y^2 - (2,2)y + 1$$

For $n = 17$, Gauss found a corresponding decomposition, first into two periods with 8 terms; each of these broke into two periods of 4 terms; each of these four broke into two periods of 2 terms. Since in all cases the periods formed quadratic equations, at each point the solutions were constructible using only compass and straightedge. Thus the straightedge and compass construction of a septendecagon was possible, though Gauss did not give a specific procedure. In general, Gauss claimed (but did not prove) that a regular n-gon, with n prime, was constructible only if n was a prime of the form $2^{2^n} + 1$ (i.e., a Fermat prime). Hence, regular polygons with 3, 5, 17, 257, and 65,537 sides were constructible. Gauss's conjecture was not proven until 1895 by James Pierpont.

See Problem 6.

19.2.3   Gauss's Diary

Gauss prided himself on publishing "few, but polished" works. After he solved the problem of constructing the septendecagon, he began to keep a mathematical diary, in which he recorded his discoveries. A perusal of the diary shows that Gauss was far ahead of his time: he worked out theories of non-Euclidean geometry and elliptic functions more than twenty years before other mathematicians. But his refusal to publish results meant that others would have to independently discover or create these branches of mathematics. To make matters worse, Gauss, though he would admit the priority of other mathematicians in publication, would invariably note that he himself was first in discovery. In the time of Tartaglia and Roberval, refusal to publish was understandable; by the time of Gauss, and into our own time, it is inexcusable.

19.2.4   The Fundamental Theorem of Algebra

Gauss's doctoral dissertation was "A New Demonstration That Every Integral Rational Algebraic Function of One Variable Can Be Decomposed into Real Factors of the First or Second Degree" (1799). As the title suggests, and we have seen, other proofs of the fundamental theorem of algebra existed prior to Gauss's, though today, Gauss's is considered the first *rigorous* one. After critiquing the proofs of d'Alembert, Euler, and Lagrange, Gauss began his proof with

Lemma 19.1. *Let m be any positive integer. Then*

$$(\sin \phi) \, x^m - (\sin m\phi) \, r^{m-1} x + (\sin (m-1) \phi) \, r^m$$

is divisible by $x^2 - 2 (\cos \phi) \, rx + r^2$.

Next Gauss proved

Lemma 19.2. *If r, ϕ are quantities such that*

$$r^m \cos m\phi + A r^{m-1} \cos(m-1)\phi + B r^{m-2} \cos(m-2)\phi + \ldots$$
$$+ K r^2 \cos 2\phi + L r \cos \phi + M = 0 \quad (19.3)$$

$$r^m \sin m\phi + A r^{m-1} \sin(m-1)\phi + B r^{m-2} \sin(m-2)\phi + \ldots$$
$$+ K r^2 \sin 2\phi + L r \sin \phi + M = 0 \quad (19.4)$$

Let

$$x^m + A x^{m-1} + B x^{m-2} + \ldots + K x^2 + L x + M = X \quad (19.5)$$

Then X is divisible either by $x^2 - 2 \cos \phi rx + r^2$, if $r \sin \phi \neq 0$, or $x - r \cos \phi$ if $r \sin \phi = 0$.

Proof. Consider the sequence of expressions

$$\sin \phi \cdot rx^m \quad - \quad \sin m\phi \cdot r^m x \quad + \quad \sin(m-1)\phi \cdot r^{m+1}$$
$$A \sin \phi \cdot rx^{m-1} - A \sin(m-1)\phi \cdot r^{m-1}x + A \sin(m-2)\phi \cdot r^m$$
$$B \sin \phi \cdot rx^{m-2} - B \sin(m-2)\phi \cdot r^{m-2}x + B \sin(m-2)\phi \cdot r^{m-1}$$
$$\vdots \qquad\qquad \vdots \qquad\qquad \vdots$$
$$L \sin \phi \cdot rx \quad - \quad L \sin \phi r \cdot x \quad + \quad *$$
$$M \sin \phi \cdot r \quad - \quad * \qquad\qquad M \sin -\phi \cdot r$$

where $*$ indicates a missing term. By Lemma 19.1, all are divisible by $x^2 - 2\cos\phi rx + r^2$. The sum of the first column is $rX\sin\phi$, while the sum of the other two columns is 0. Hence $rX\sin\phi$ is divisible by $x^2 - 2\cos\phi rx + r^2$.

If $r\sin\phi \neq 0$, then X is divisible by $x^2 - 2\cos\phi rx + r^2$. If $r\sin\phi = 0$, then either $r = 0$, and thus $M = 0$ and X is divisible by $x - r\cos\phi$; or $\sin\phi = 0$, in which case $\cos\phi = \pm 1$, and $\cos n\phi = (\pm 1)^n$. If $x = r\cos\phi$, then $X = 0$ and hence $x - r\cos\phi$ divides X. $\square$

Thus, the theorem follows if it can be shown that for

$$X = x^m + Ax^{m-1} + Bx^{m-2} + \ldots + Lx + M$$

one can find r, ϕ such that equations 19.3 and 19.4 are true.

To show that there is such a r, ϕ, Gauss considered the surfaces generated by the equations

$$T = r^m \sin m\phi + Ar^{m-1}\sin(m-1)\phi + \ldots + Lr\sin\phi$$
$$U = r^m \cos m\phi + Ar^{m-1}\cos(m-1)\phi + \ldots + Lr + M$$

in cylindrical coordinates. The proof thus reduced to proving that the curves $T = 0$ and $U = 0$ intersected for some value of r, ϕ, and hence lemma 19.2 would be applicable and X would have a quadratic or linear factor.

The *first curve* was the level curve $T = 0$; the *second curve* was the level curve $U = 0$. At an infinite distance from the origin, the first curve was asymptotic to $0 = \sin m\phi$, which consisted of a sequence of m lines radiating from the origin, at angles of $0, \frac{1}{m}180°, \frac{2}{m}180°, \ldots$; hence the first curve had $2m$ branches, which divided the circumference of the circle of infinite radius into $2m$ equal parts.

Meanwhile, the second curve was asymptotic to $0 = \cos m\phi$, which consisted of m lines through the origin at angles of $\frac{1}{m}90°, \frac{2}{m}90°, \ldots$ Thus, the second curve consisted of $2m$ branches as well, each branch being between two branches of the first curve. Hence, the two curves must intersect, and thus r, ϕ may be found so that X has a linear or quadratic factor, implying

Theorem 19.1 (Fundamental Theorem of Algebra). *Every polynomial with real coefficients can be factored into linear and quadratic terms with real coefficients.*

Gauss illustrated the case for $m = 4$ (shown in margin). After this, he gave an alternate proof which does not rely on the asymptotic behavior of T and U.

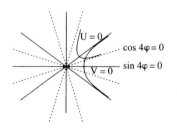

19.2 Exercises

1. Prove that if g is a primitive root of n, and Ω is the set of all roots of $x^{n-1} + x^{n-2} + \ldots + x + 1 = 0$, then:

    (a) The set $[1], [g], [g^2], \ldots$ is equal to Ω.

    (b) If n does not divide λ, then the set $[\lambda], [\lambda g], [\lambda g^2], \ldots, [\lambda g^{n-2}]$ is also equal to Ω.

(c) If G is another primitive root of n, then the set $[1]$, $[G]$, ... is also equal to Ω.

2. From Proposition 19.3, prove that p^2, p^3, ... can be expressed in terms of the equations

$$0 = p^2 + A + ap + a'p' + \ldots$$
$$0 = p^3 + B + bp + b'p' + \ldots$$
$$0 = p^4 + C + cp + c'p' + \ldots$$
$$\vdots$$

where all the coefficients A, B, ..., a, b, ..., a', b', ... are integers.

3. Find all solutions to $x^7 - 1 = 0$.

4. Find all solutions to $x^{13} - 1 = 0$. Hint: first, break down the roots into three groups of four, and solve the corresponding cubic equation. Next, break down each group into two groups of two, and solve the resulting quadratic equations.

5. Find all solutions to $x^{17} - 1 = 0$.

6. Consider Gauss's conjecture; assume n is a prime number; show that it is necessary that n be of the form $2^{2^n} + 1$.

   (a) Suppose m is the product of distinct odd prime factors p_1, p_2, ... and 2^k, where k is a whole number. Explain why the constructibility of a regular m-gon can be reduced to the question of the constructibility of regular p_1, p_2, ...-gons. Why is it necessary to assume the prime factors are distinct?

   (b) Explain why $n - 1$ cannot have prime factors other than 2, if the regular n-sided polygon (with n prime) is to be constructible (using the known compass and straightedge techniques). The essential part of the argument holds for all possible compass and straightedge techniques, though this fact was not known until the middle of the nineteenth century.

   (c) Hence, show that it is necessary that n be a prime of the form $2^{2^n} + 1$.

Though this proves the necessity of the prime being of the form $2^{2^n} + 1$, it does not prove that it is sufficient; for a full proof of Gauss's conjecture, see James Pierpont, "On an Undemonstrated Theorem of the *Disquisitiones Arithmeticae*," *AMS Bulletin* (2), 1895-6, 77-83.

19.3   The Method of Least Squares

Any physical measurement is subject to error. The work of Simpson and Lagrange showed it was better to take the mean of many observations, rather than attempt to make a single, very accurate observation, but there was still no clear way of determining the actual value of a quantity from its measured value.

See page 469.

19.3.1   Legendre

We might pose the problem in the following way: suppose we have a parameter $-A$ that can be calculated using an equation

$$-A = bx + cy + fz + \ldots$$

where b, c, f, ... are known quantities and x, y, z, ... variable quantities. If we observe the parameter to be $-a$, then the error between the "actual" value $-A$ and the "observed" value $-a$ is

$$E = a + bx + cy + fz + \ldots$$

In practice, several observations give several different values of a, so we have a system of equations of the preceding form. The question then becomes how to choose the values of x, y, z, ... so that the value determined for $-A$ is closest to the actual value.

For example, suppose $-A$ is the perimeter of a rectangle with sides of length x and y. Then

$$-A = 2x + 2y$$

Suppose the perimeter is measured three times, to give *measured* perimeters $-a_1$, $-a_2$, and $-a_3$. The problem was to find the *actual* perimeter $-A$, based on these measurements: in other words, to choose the values of the lengths of the sides x and y that were most likely correct. The error in using any one of the values as the actual perimeter will be

$$E_1 = a_1 + 2x + 2y \tag{19.6}$$
$$E_2 = a_2 + 2x + 2y \tag{19.7}$$
$$E_3 = a_3 + 2x + 2y \tag{19.8}$$

Of course, it would be ideal if the errors were all zero, but since this is a system of three equations with two unknowns, it will not, in general, be possible to find such a solution. The question thus arises of finding the values x, y that will determine $-A$ most accurately.

The first published solution was published in 1805 by Adrien-Marie Legendre (September 18, 1752-January 10, 1833). Legendre, like so many of his era, was a universalist, and his best-known work was probably in the theory of numbers, though he wrote an edition of Euclid that became the standard work for over a century. Translations of Legendre's texts into English formed the mainstay of mathematics at advanced schools in the United States, including Harvard University and the Military Academy at West Point.

Legendre's solution appeared as an appendix, dated March 6, 1805, to his *New Method for Determining the Orbit of Comets*. He devoted nine (out of eighty) pages of the appendix to what he called the **method of least squares**. The surprisingly simple solution given by Legendre was that the values of x, y, z, ... that should be chosen to recover $-A$ were the values that minimized the sum of the squares of the errors E. In the preceding case, the values of x and y to be chosen were the ones that made

$$E_1^2 + E_2^2 + E_3^2 \qquad (19.9)$$

a minimum. Legendre showed the best values of x and y could be found using a very simple algorithm. Legendre's publication of the method of least squares in 1805 almost immediately embroiled him in a priority controversy with Gauss, who claimed to have been using the method since 1795. However, he did not publish his version of the method until 1807.

19.3.2 Bowditch and Adrain

Although Legendre's method of least squares provided a very simple and intuitively obvious method of finding values of x and y, he gave no justification for it. The problem is that minimizing the sum of the squares of the errors does not guarantee that the resulting value of the parameter is in fact accurate. The first theoretical justification for the method of least squares would be due to Robert Adrain (September 30, 1775-August 10, 1843), of the fledgling United States; Adrain's justification of the method of least squares was the first significant mathematical contribution to come out of the new country. Adrain himself came from Ireland.

Since the Protestant Reformation, the British feared that Catholic Ireland would be used as a springboard for an invasion by the Catholic powers, first Spain, then as Spain's power declined, France. As a result, the British often treated Ireland as a conquered territory, which caused resentment among both the Catholics and the Protestants. The Irish revolted, time and again, but the revolts were suppressed by the British. Adrain was part of an unsuccessful revolt in 1798. Besides the failure of the revolt, Adrain was injured by being shot in the back by one of his own men. After recovery, Adrain fled to the United States, settling in Princeton, New Jersey.

Adrain was a self-trained mathematician. In 1804, George Baron began publishing the first mathematics journal in the United States, the *Mathematical Correspondent*. Adrain was a frequent contributor, and became its editor in 1807, shortly before it ceased publication. In 1808, he began a new journal, *The Analyst*. The journal was to consist of problems, posed by correspondents, and solutions, sent in by readers. Adrain wrote:

> It would perhaps contribute something to the progress of science if
> the Editor were enabled by the sale of the work to have two Prize
> Questions in each number, a greater and a lesser.[2]

Unfortunately, *The Analyst* only lasted one year before failing and was virtually
unread by the rest of the world.

Shortly before the journal failed, Robert Patterson of Philadelphia posed the
following problem, which appeared in the second issue:

Problem 19.1. *A surveyor walks around a five-sided parcel of land in the directions and for the distances indicated:*

1. *North by 45° east, 40 paces*

2. *South by 30° east, 25 paces*

3. *South by 5° west, 36 paces*

4. *Due west, 29.6 paces*

5. *North by 20° west, 31 paces, back to the starting point*

*However, the path as recorded does not return to the starting point. Find the
most probable area of the field.*

Since the path as recorded does not return to the starting point, there must have
been errors in the recording of the directions and distances. The problem was to
correct the directions and distances. Adrain made this the greater prize question
for the issue, and promised a reward of ten dollars for the best solution; at the
time, the salary of a United States senator was $1500 a year and the president
earned $25,000 a year.

The only solution Adrain judged worthy of the prize was given by Nathaniel
Bowditch (March 26, 1773-March 16, 1838). The self-educated Bowditch was
born in Salem, Massachusetts, and is thus the first mathematician of any significance to be born in the United States, though at the time of his birth, the United
States was still a British colony. Advanced books on science and mathematics
were rare in the infant country, but Bowditch benefited from the Revolutionary
War: in 1791, a privateer brought back to Salem the library of Richard Kirwan,
an Irish chemist and member of the Royal Society of London. From that time,
the eighteen-year-old Bowditch had access to a modern, scientific library.

Bowditch began by supposing $ABCDE$ to represent the figure described by
the survey, where the points A, E should coincide (note that Bowditch solved
the problem for a four-sided plot of land). Since the points A, E do not in fact
coincide, then the difference AE is the total error of the measurements. Let AB'
represent the most probable first side of the field. The first side was actually
measured as AB, so the error in measuring this side was BB'. By substituting
the most probable side, AB', for the measured side, AB, the points C, D, and

A **privateer** is a privately owned
ship and/or its crew, licensed by
a Letter of Marque to attack the
shipping of a hostile power.

---

[2]Adrain, Robert. *The Analyst*, Vol. I, page v.

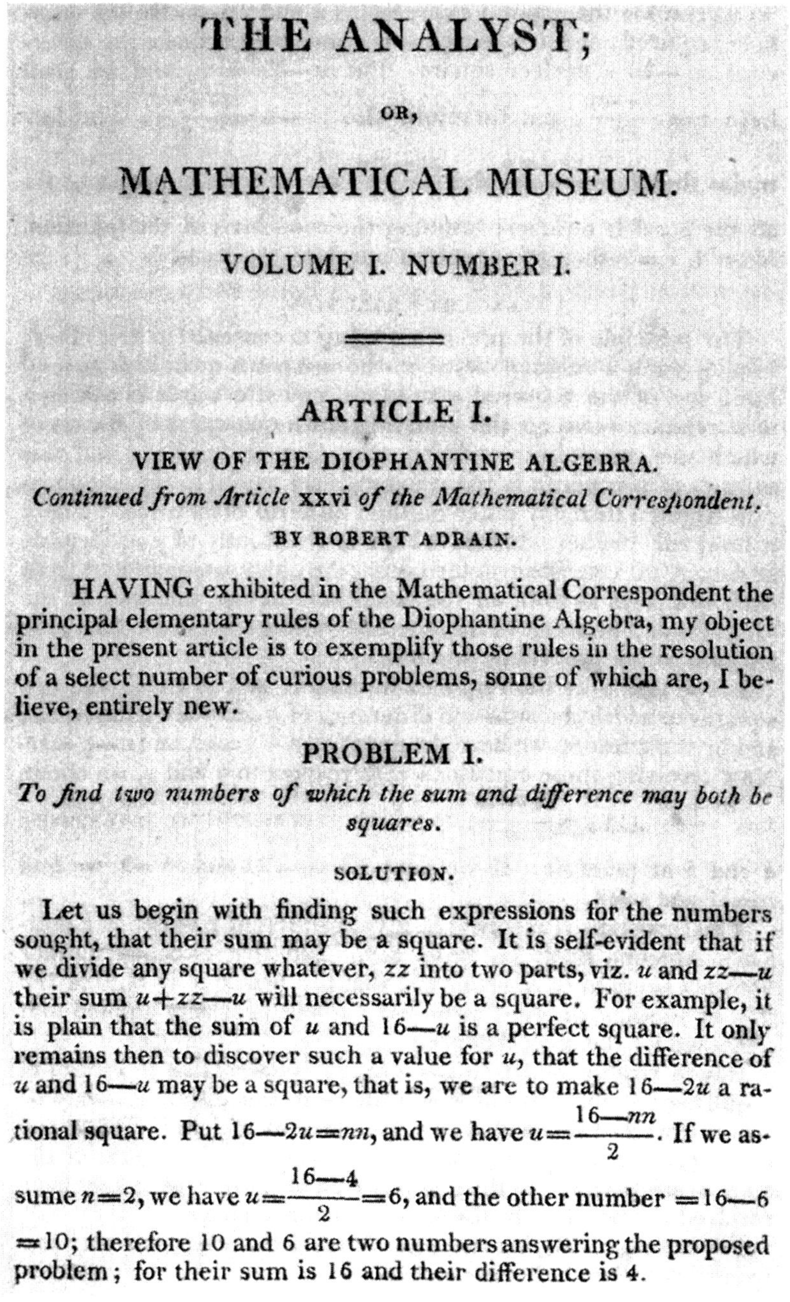

THE ANALYST;

OR,

MATHEMATICAL MUSEUM.

VOLUME I. NUMBER I.

ARTICLE I.

VIEW OF THE DIOPHANTINE ALGEBRA.

Continued from Article xxvi *of the Mathematical Correspondent.*

BY ROBERT ADRAIN.

HAVING exhibited in the Mathematical Correspondent the principal elementary rules of the Diophantine Algebra, my object in the present article is to exemplify those rules in the resolution of a select number of curious problems, some of which are, I believe, entirely new.

PROBLEM I.

To find two numbers of which the sum and difference may both be squares.

SOLUTION.

Let us begin with finding such expressions for the numbers sought, that their sum may be a square. It is self-evident that if we divide any square whatever, zz into two parts, viz. u and $zz-u$ their sum $u+zz-u$ will necessarily be a square. For example, it is plain that the sum of u and $16-u$ is a perfect square. It only remains then to discover such a value for u, that the difference of u and $16-u$ may be a square, that is, we are to make $16-2u$ a rational square. Put $16-2u=nn$, and we have $u=\dfrac{16-nn}{2}$. If we assume $n=2$, we have $u=\dfrac{16-4}{2}=6$, and the other number $=16-6$ $=10$; therefore 10 and 6 are two numbers answering the proposed problem; for their sum is 16 and their difference is 4.

Figure 19.1: The First Page of the First Issue of *The Analyst*. Boston Public Library/Rare books Department. Courtesy of the Trustees.

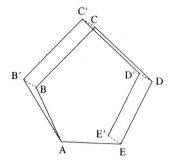

E are moved, respectively, to C', D', and E' by amounts equal to BB' and in the same direction. (This is because the measured side BC no longer begins at B, but instead begins at B' and thus ends at C'; likewise, the measured side CD begins at C' and ends at D', and so on.) Again, let $B'C''$ represent the most probable second side of the field; since it was measured as BC (or $B'C'$, as the two are the same), then the error in the measurement is $C'C''$. Again, substituting the most probable side $B'C''$ for the measured side BC, the points D', and E' are moved to points D'', and E'', respectively. Continuing in this manner, the most probable vertices of the field are at A, B', C'', D''', and E'''', which should coincide with A.

The question then becomes determining the correction factors BB', $C'C''$, $D''D'''$, and $E'''E''''$ needed to make the final point, E'''', coincide with A. Bowditch's principle is that the corrections should all tend to decrease AE. In the simplest case, Bowditch assumed that all the errors should be in the same direction as AE, and the magnitude should be proportional to the measured lengths.

Adrain showed that Bowditch's solution was equivalent to the application of the method of least squares, and considered the problem in more detail. Immediately after printing Bowditch's solution, Adrain posed the problem

Problem 19.2. *Suppose AB is the actual value of a quantity measured to be Ab. What is the probability of obtaining error Bb in measuring AB?*

To solve this problem, Adrain supposed that several measurements were taken in such a way that the total error was known; the surveying problem posed by Patterson resulted in precisely this situation, where the total error, AE, was a known quantity.

Suppose AB, BC are measured as Ab, bc, with total error Cc. Let x, y be the errors in measuring AB, BC, and let the actual measurements Ab, bc be a, b. Adrain assumed that it was most probable that the actual lengths AB, BC were proportional to the measured lengths Ab, bc. This meant that the corresponding errors, x, y, were likewise proportional to the corresponding measurements, a, b; hence

$$\frac{x}{a} = \frac{y}{b} \tag{19.10}$$

if x, y were the errors of greatest probability.

Suppose X, Y were the probabilities of obtaining error x, y in measuring a, b, respectively. The probability of obtaining the errors x and y was thus XY. By assumption, XY is at a maximum, and $x + y = E$. Let $X' = \log X$, and $Y' = \log Y$. If XY is at a maximum, so is $X' + Y'$, hence its fluxion

$$\dot{X}' + \dot{Y}' = 0$$

Under the simplest assumption, X and Y depend only on x and y. At this point, Adrain confusingly uses Lagrange's prime notation to indicate the derivative (whereas $X' = \log X$), writing:

$$X''\dot{x} + Y''\dot{y} = 0$$

Hence

$$X''\dot{x} = -Y''\dot{y}$$

Since $x + y = E$, a known constant, then $\dot{x} + \dot{y} = 0$ and $\dot{x} = -\dot{y}$. Thus $X'' = Y''$.

By assumption, Equation 19.10 must hold. The simplest way to obtain this is to suppose

$$X'' = \frac{mx}{a} \qquad\qquad Y'' = \frac{my}{b}$$

where m is any fixed number. Since

$$\dot{X}' = X''\dot{x}$$
$$= \frac{mx\dot{x}}{a}$$

thus

$$X' = a' + \frac{mx^2}{2a}$$

where a' is some constant, and since $X' = \log X$, then

$$X = e^{a' + \frac{mx^2}{2a}}$$

which is the probability of making error x in measuring a. Since (by assumption) this probability has a maximum value, it is necessary that m be negative; the simplest assumption is that the probability of making an error of size x was e^{-x^2}. Thus the probability of making an error of size x and an error of size y in taking the two measurements is

$$XY = e^{-(x^2 + y^2)}$$

The values of the errors x and y that maximize this probability (and thus correspond to the most probable errors) are the values of x and y that minimize the expression $x^2 + y^2$ in the exponent. In this way, Adrain was the first to provide a rigorous justification for the method of least squares.

Unfortunately for Adrain, *The Analyst* failed to attract the necessary readers; indeed, he and Bowditch were about the only two correspondents of mathematical significance. Thus, it soon ceased publication. Undaunted, Adrain made a second attempt to found a mathematics journal, *The Mathematical Diary*, in 1825. In the first issue, Adrain stated

The principal object of the present work is to excite the genius and
industry of those who have a taste for mathematical studies, by af-
fording them an opportunity of laying their speculations before the
public in an advantageous manner; and thus to spread the knowledge
of mathematics in a way that is both effectual and agreeable.[3]

Like *The Analyst*, *The Mathematical Diary* would consist of questions posed and
answered by correspondents, though this time, the best solution would not be
awarded any prize money; instead, the issue that contained the solution would
be named after the solver. Thus, the second issue of *The Mathematical Diary*
was the "Bowditch" issue, as the prize problem posed in the first issue was solved
by Bowditch. *The Mathematical Diary* fared little better than *The Analyst*, and
a successful mathematical journey in the United States would not appear for
more than half a century.

19.3 Exercises

1. Find the values of x, y that minimize Equation 19.9.

2. What is the total error, in direction and magnitude, of the surveyor's measurements in Problem 19.1? (In other words, if the field is actually walked
   the way it was measured, how far from the starting point do you end?)

3. In the derivation of the method of least squares, Adrain assumed that the
   error was proportional to the size of the quantity being measured. What
   is a more reasonable assumption? Based on your assumption, what is the
   form of the probability function X?

19.4   Number Theory

Gauss, like Euler, was active in many fields of mathematics, but the area where
he had the greatest impact on mathematics was probably in number theory. His
construction of the septendecagon rekindled some interest in the Fermat numbers
$F_n = 2^{2^n} + 1$. A new conjecture, similar to Fermat's, was made by Ferdinand
Gotthold Eisenstein (1823-1852), at the University of Berlin. The **Eisenstein
conjecture**, still unproven, is that the numbers

$$2^2 + 1, 2^{2^2} + 1, 2^{2^{2^2}} + 1$$

and so on are all prime. The first three terms of the sequence are prime; it is
still not known whether any of the remaining numbers are prime. Eisenstein died
quite young, but Gauss ranked him on a level with Archimedes and Newton, so
one wonders what might have become of Eisenstein had he lived.

---

[3] *The Mathematical Diary*, Vol. I, No. 1 (1825), p. iii.

19.4.1 Quadratic Reciprocity

Euler had introduced the study of the residues modulo p of the squares of the whole numbers. In his *Theory of Numbers*, Legendre introduced the modern notation $\left(\frac{m}{q}\right)$ for the quadratic residue, where $\left(\frac{m}{q}\right) = 1$ if m is a quadratic residue mod p (i.e., there is an x such that $x^2 = m \mod p$) and $\left(\frac{m}{q}\right) = -1$ if m is a non-residue. Euler, Legendre, and others had noted a remarkable property: if p and q were distinct odd primes, then

$$\left(\frac{p}{q}\right)\left(\frac{q}{p}\right) = (-1)^{\frac{1}{4}(p-1)(q-1)}$$

Gauss was the first to prove this conjecture and would in fact provide six proofs in the course of his life.

19.4.2 Sophie Germain

Shortly after the publication of *Arithmetical Investigations*, Gauss received a letter from a "Mr. Le Blanc," discussing some aspects of Gauss's work. Thus began a long correspondence between Gauss and "Mr. Le Blanc."

Many years before, another "Mr. Le Blanc" appeared at the Polytechnical School in Paris. There, students were expected to submit end of year reports, describing their accomplishments during the year. The report on mathematics submitted by "Mr. Le Blanc" was particularly good, and Lagrange sought out the identity of the author. To his surprise, he found it was a woman: Sophie Germain (April 1, 1776-June 27, 1831).[4] This impressed Lagrange even more— not because she was a woman, but because the Polytechnical School did not admit women, so her entire knowledge of advanced mathematics was through self-study. Lagrange became one of her greatest supporters.

Germain's interest in mathematics originated when she read a life of Archimedes and decided she, too, would become a mathematician. Like Archimedes, her main interests were in mathematical physics, and she made significant contributions to the mathematical theory of acoustics and of elasticity. In pure mathematics, she worked primarily in number theory. Her correspondence with Legendre was so great that her work was included as a supplement to the second edition of his *Theory of Numbers*.

Her major results in pure mathematics were to prove restricted forms of Fermat's last conjecture. On November 21, 1804, she claimed, in a letter to Gauss, that she could prove that $x^n + y^n = z^n$ had no integer solutions if $n = p - 1$, where p is a prime number of the form $8k + 7$. In the final form (included in Legendre's second edition of *The Theory of Numbers*), Germain's result is the following: if n is an odd prime less than 100, then $x^n + y^n = z^n$ has no integer solutions x, y, z, where x, y, z are prime to n.

---

[4]Thus, Lagrange neatly connects two prominent female mathematicians: Maria Agnesi, who would have refereed one of Lagrange's earliest papers, and Sophie Germain.

Historical forces caused Gauss to learn the identity of his mystery correspondent. One of the casualties at Auerstadt, where Napoleon crushed the Prussians, was Gauss's patron, the Duke of Brunswick Carl Wilhelm Ferdinand. Gauss would never forget the French caused the death of the duke. This did not prevent him from communicating with French mathematicians, though at times, the relationship was forced.

French forces moved on to besiege Hanover in 1807. Germain, worried that Gauss might suffer the same fate as Archimedes, wrote to General Pernety, a close friend of the Germain family, and requested that he protect Gauss. Pernety did as she requested. It was through this act that Gauss ultimately learned the identity of "Mr. Le Blanc," and, like Lagrange and for the same reasons, he was even more impressed with her progress in mathematics.

19.4.3   Gaussian Integers

See page 533.

An important extension of arithmetic was made by Gauss. In his proof of Fermat's last conjecture for $n = 3$, Euler had used properties of "numbers" of the form $p \pm q\sqrt{-3}$, with p and q whole numbers. In his proof, Euler assumed that the properties of the whole numbers also held for numbers of this form, but this had not yet been proven. Other mathematicians had used complex numbers because they were useful, but a "theory of complex numbers" akin to the "theory of arithmetic" had yet to be developed; as Gauss pointed out, complex numbers were not understood, merely tolerated.

As early as October 23, 1813, Gauss (based on his study of biquadratic residues) began to investigate more fully the properties of the complex whole numbers. His first publication on this subject was "Second Memoir on the Theory of Biquadratic Residues" (April 15, 1831). Gauss designated 1, i, -1, and $-i$ to be units; just as Euclid defined number to be a collection of units, Gauss defined the **complex whole numbers**, now called **Gaussian integers**, to be numbers of the form $a + bi$, where a and b are integers. Gauss defined the **conjugate** of $a + bi$ to be $a - bi$, just as we do; his definition of the **norm** of a complex number $a + bi$, however, was the product of the conjugates, $a^2 + b^2$, and not $\sqrt{a^2 + b^2}$, as we define it today. A key property of the norm was that the norm of the product of two complex numbers was the product of the norms.

Among the whole numbers, a number is prime if it has no whole number factors besides itself and 1; hence 5 is prime. By the extension of the numbers to the complex domain, some of these numbers can be factored: $5 = (2 + i)(2 - i)$, and thus 5 is a composite number; on the other hand, $2 \pm i$ cannot be factored into numbers of the form $a + bi$; hence $2 \pm i$ is prime.

Among the whole numbers, factorization into primes is unique. This does not appear to be the case among the complex whole numbers, for $5 = (2+i)(2-i) = (1+2i)(1-2i)$. To retain the useful property of unique factorization, Gauss used the concept of a unit: even though $6 = 2 \cdot 3 = -2 \cdot -3$, we do not consider the second factorization to be different from the first, since the factors differ only by the unit -1. In the same way, since $1 + 2i = i(2 - i)$, the two factorizations of 5 differ only by unit factors. Gauss gave

Proposition 19.4. *All whole number primes of the form $4n + 1$, and 2, have factors of the form $a + bi$.*

as a theorem whose proof was obvious. See Problem 5.

Gauss proved

Proposition 19.5 (Fundamental Theorem of Arithmetic). *A complex whole number M of the form $a + bi$ factors uniquely (up to unit factors) as a product of complex primes of the same form.*

Gauss noted that the theory of cubic residues required an examination of the numbers of the form $a + b\rho$ where $\rho^3 - 1 = 0$, $\rho \neq 1$ and added that a study of the higher powered residues required a similar study of other "complex" numbers. In fact, the Gaussian integers might be properly interpreted as numbers of the form $a + b\rho + c\rho^2 + d\rho^3$, where ρ is a primitive root of $\rho^4 - 1 = 0$. The concept could be and was extended by mathematicians of the nineteenth century.

19.4.4   Gauss's Later Life

In later years, Gauss began to take less notice of the mathematical community. He ignored Abel in person, though he praised him in private. To Dirichlet, who wrote to him in May 1826 asking for advice, Gauss responded—four months later—that he should make sure that he found a job that left him time for research. Not surprisingly, in 1849, when a celebration was organized in honor of the fiftieth anniversary of his thesis (and Gauss presented his fourth and final proof of the Fundamental Theorem of Algebra), it was attended by many scientists but only two mathematicians—Jacobi and Dirichlet.

Gauss and Euler shared many characteristics in their approach to number theory. Both felt that *example* and *scientific induction* were key elements to the discovery of mathematical theorems, which should then be proven as quickly as possible, using whatever tools were necessary. But both felt at the same time that just because a theorem was proven it need not be considered a "finished" problem, and both sought new proofs for results they had found in the earliest years of their careers: Euler and his four proofs of Fermat's Theorem, and Gauss and his four proofs of the Fundamental Theorem of Algebra, as well as six proofs of the law of quadratic reciprocity.

19.4 Exercises

1. Determine $\left(\frac{11}{7}\right)$, $\left(\frac{7}{11}\right)$, and show the law of quadratic reciprocity holds in this case.

2. What are some of the implications of Gauss's proof that all polynomials could be reduced to a product of linear and quadratic factors? In particular, what does this imply about the problem of integrating rational functions?

3. Prove that the product of the norms of two complex numbers is the norm of the product.

4. Prove the fundamental theorem of arithmetic for Gaussian integers.

   (a) Suppose $M = A^\alpha B^\beta C^\gamma \ldots$, where A, B, C, $\ldots$ are distinct complex primes. Let P be a different prime, and designate by p, a, b, c, $\ldots$ the norms of P, A, B, C, $\ldots$ Show that p must equal one of the norms.

   (b) Since p equals one of the norms, we may assume $p = a$, and $A = k - li$. Show that P and A are conjugates, and thus $A \equiv 2k \mod P$.

   (c) Hence p divides $2^{2\alpha} k^{2\alpha} b^\beta c^\gamma \ldots$ Show that p is equal to a second norm, say b, and thus P and some other factor, B are conjugates, which contradicts the assumption that all the factors A, B, $\ldots$ are distinct primes.

5. Prove Proposition 19.4. Hint: the (Gaussian) norm of $a + bi$ is $a^2 + b^2$.

6. Consider numbers of the form $a + b\rho + c\rho^2$, where ρ is a primitive solution to $x^3 - 1 = 0$. What whole number primes p are still primes?

19.5   Geometry

By 1812, Napoleon controlled directly or indirectly France, Spain, Germany, and Italy, while Prussia and Austria were still recovering from their defeats by Napoleon. Britain alone defied Napoleon. To crush the "nation of shopkeepers," Napoleon declared a blockade of Britain as early as 1806. Announced in Berlin after the defeat of Prussia, the Continental System (sometimes called the Berlin Decrees) ordained that no nation in Europe was to trade with the British. While it was effective for the nations under Napoleon's control, the largest European power, Russia, refused to go along with it. Thus, to crush Britain, Napoleon invaded Russia.

19.5.1   Projective Geometry

Napoleon invaded Russia with a "Grand Army" consisting of over 600,000 men. The Russians withdrew, trading distance for time, and by September 1812, the French were in Moscow. The Russians burned the city rather than let it fall into French hands, and the French army was in a precarious position: winter was fast approaching, and they had no place to stay. On October 19, 1812, they began one of the costliest retreats in history. On November 18, the Russians under Marshal Kutuzov attacked the retreating French at the Battle of Krasnoii (near Smolensk, Russia) and dealt them a severe defeat. Over 5000 French soldiers were killed, and 8000 were missing or captured.

Poncelet

Among the captives was a military engineer and product of the Polytechnical School, Jean Victor Poncelet (July 1, 1788-December 22, 1867). He would remain a prisoner of war until June 1814. Poncelet was lucky: of the 600,000 men who went into Russia, fewer than 100,000 would return alive. To pass the time while a prisoner, he taught geometry to fellow prisoners; as he had no access to books, he had to design and teach the course from memory.

Poncelet contributed two important principles to projective geometry. The first is the **principle of duality**. To illustrate the principle of duality, we might note the following sentences:

1. Two *lines* intersect at one *point*.

2. Two *points* lie on one *line*.

Because one can exchange "point" for "line," one can take any theorem in geometry that makes a claim about the properties of "points," and have its dual, a theorem that makes a corresponding claim about the properties of lines. Hence, at a stroke, the number of geometric propositions is doubled.

The second principle used by Poncelet was the **principle of continuity**, which he enunciated in his *Treatise on the Projective Properties of Figures*:

> Is it not obvious that the properties and relationships, found for the first system [of figures], will hold for all successive states of the system, so long as one considers the particular modifications that might need to be required, such as when a magnitude vanishes, or changes directions or signs, etc., modifications that will always be easy to recognize *a priori*, and by certain rules? ... Now this principle, regarded as an axiom by the most sagacious geometers, is called the *principle* or *law of continuity* of mathematical relations on abstract quantities and figures.[5]

For example, consider four points on a circle, A, B, C, and D. If the lines joining the points intersect inside the circle (Figure 19.2, Stage 1), then the *Elements*, III-35 implies $AE \cdot EB = CE \cdot ED$. Now, let A and C move toward each other, until they meet and the intersection point E is on the circle (Stage 2). It is trivially true that $AE \cdot EB = CE \cdot ED$. See page 108.

Suppose A and C now move past each other (Stage 3), so that E is outside the circle. It still remains true that $AE \cdot EB = CE \cdot ED$ (though this is not a Euclidean proposition). Finally, let C continue to move until C, D coincide; then ED is tangent, and we have $AE \cdot EB = CE \cdot ED = ED \cdot ED$, which is *Elements*, III-36.

---

[5] Poncelet, Jean-Victor, *Traité des propriétés projectives des figures*. Gauthier-Villars, 1865-6. Pages xiii-xiv.

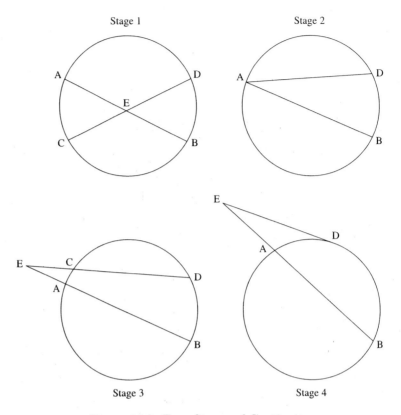

Figure 19.2: Four Stages of Continuity.

Gergonne

Poncelet's rival in projective geometry was Joseph Diaz Gergonne (June 19, 1771-May 4, 1859), a student of Monge. At the beginning of the nineteenth century, there were no mathematical journals for French mathematicians, despite the fact that the French mathematicians led the world in the latter part of the eighteenth century. Thus in 1810, Gergonne founded the *Annals of Pure and Applied Mathematics*, to provide French mathematicians with a suitable outlet.

Gergonne discovered the principle of duality independent of Poncelet, and began publishing, in his *Annals*, theorems and their duals. This led to a priority dispute between Gergonne and Poncelet, though in fact both were anticipated in their enunciation of the principle of duality by Pascal, as well as by a fellow student at the Polytechnical School, Charles Jules Brianchon (1785-1862).

19.5.2   Geometric Representation of Complex Numbers

At the turn of the nineteenth century, two papers appeared on the geometric representation of complex numbers. "On the Analytical Representation of Direction," by Caspar Wessel (June 8, 1745-March 25, 1818) was presented to the Danish Academy of the Sciences on March 10, 1797, and appeared in its memoirs in 1799 (the first paper by a non-member to be printed). Wessel was working his way through the University of Copenhagen as a surveyor for the Danish Academy; it took him fifteen years to complete his degree (in law). He illustrates one of the problems of assigning nationalities to historic figures: Wessel was born in the Kingdom of Denmark, and received his education at the University of Copenhagen, so he was by birth and education Danish. But his birthplace is now part of modern Norway.

Wessel's paper, unfortunately, appeared in Danish and had little impact outside of Denmark. A second, by French geometer Jean Robert Argand (July 18, 1768-August 13, 1822), appeared in 1806 in an anonymous tract. The tract was shown to Legendre, who showed it to François Joseph Français (April 7, 1768-October 30, 1810), another French mathematician. On Français's death, his brother, Jacques Frédéric Français (June 20, 1775-March 9, 1833), found the tract among his brother's papers, and further developed the ideas, publishing them in Gergonne's *Annals* in 1813. At the end, he noted that the ideas were not completely original with him, and that he hoped the unknown author would reveal himself. Argand took the invitation, and soon published a series of articles on geometric representation of complex numbers. This embroiled Argand in controversy with other mathematicians, who wanted to eliminate algebraic notions from geometry.

Argand and Wessel applied arithmetic notions to straight lines. Both gave the same procedure for adding two straight lines: place the beginning of the second line at the end of the first, and then draw a line connecting the beginning of the first line to the end of the second. This corresponds to the modern **tip-to-tail** procedure of adding two vectors.

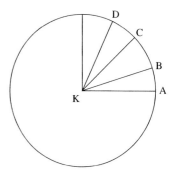

For multiplication, Wessel simply gave a procedure for multiplying two lines, but Argand provided a demonstration of its validity. Begin with two lines, KB and KC, assumed to be of unit length; suppose the unit is KA. Placing them at the common origin K, a circle $ABCD$ could be drawn around them. Let angle CKD equal angle AKB. Since $KA : KB = KC : KD$ (by similar triangles), then $KA \cdot KD = KB \cdot KC$. However, $KA = 1$ by assumption; hence $KD = KB \cdot KC$; note that the angle the product KD makes with the unit KA is equal to the sum of the angles, AKB, AKC that the two factors, KB and KC, make with the unit.

Finally, both Argand and Wessel identified $\sqrt{-1}$ as representing a line drawn perpendicular to the unit. To justify this interpretation, Wessel began by identifying $+1$ with a direction angle of $0°$, -1 with a direction angle of $180°$, and ϵ and $-\epsilon$ with direction angles of $90°$ and $270°$, respectively. The multiplication procedure implies $(+\epsilon)(+\epsilon) = -1$. Hence $\epsilon = \sqrt{-1}$.

Argand's justification is considerably less rigorous. He began by considering the ratios

$$+1 : +1 = -1 : -1$$
$$+1 : -1 = -1 : +1$$

Argand then assumed the proportion held even if $+1$ and -1 represent directions. Then, he considered the direction represent by x where

$$+1 : x = x : -1$$

Thus, the direction of $+1$ to the direction of x would have the same relationship as the direction x to the direction -1. Hence, x is perpendicular (at $+90°$) to $+1$. Solving the proportionality, we find that $x = \sqrt{-1}$.

19.5.3   The Fall of Napoleon

When it became clear Napoleon had suffered disastrous losses in Russia due to "General Winter," Britain, Prussia, and Austria joined Russia in a coalition to bring about Napoleon's final downfall. One defeat may have been particularly galling to Napoleon. At Leipzig on October 16-19, 1813, Napoleon faced a multi-national coalition that included the Swedes, led by Crown Prince Charles John XIV—formerly Jean Baptiste Bernadotte, one of Napoleon's marshals and in power because Napoleon had declined to mediate in a dispute over the Swedish succession.

The victorious allies entered Paris in March, and Napoleon was forced to abdicate on April 11, 1814. Here came another blow that must have incensed Napoleon, for the decree of abdication was signed into law by the Chancellor of the Senate and one who had benefited greatly from Napoleon's patronage: Laplace.

See page 571.

The terms of abdication were generous. Napoleon was given the island of Elba, just off the Italian coast, as his own Principality, and an annual stipend of $2,000,000$ francs. The victorious allies installed Louis XVIII, the brother of Louis XVI, as the king of France. There was no Louis XVII: Louis XVI's son died in 1795.

19.5 Exercises

1. Explain how Wessel's procedure for multiplication gives $(+\epsilon)(+\epsilon) = -1$.

2. Find the figure that is the dual of a triangle.

   (a) Define a triangle.
   (b) Interchange "line" with "point" in your definition.
   (c) What is the resulting figure?
   (d) Which of the theorems regarding triangles hold true about the dual figure? Note that the theorems will have to be restated as their duals.

3. In Figure 19.2, prove that in Stage 3, $AE \cdot EB = CE \cdot ED$.

Chapter 20

Analysis to Midcentury

> Napoleon's abdication was only temporary, and in 1815, he escaped Elba, rallied supporters, and marched on Paris. Louis XVIII fled, but returned with the armies of the allies, which crushed the Bonapartists for the last time at the Battle of Waterloo on June 18, 1815. At the Congress of Vienna, the Great Powers (Austria, Prussia, Russia, and Britain) sought to guarantee that never again would general war wrack the European continent. The Congress redrew the map of Europe: Belgium and the Netherlands were welded together into a single Kingdom; the Confederation of the Rhine was swept away in favor of a Germanic Confederation; Austria received a large portion of the Italian peninsula; Sweden received Norway; and Prussia received the city of Danzig, formerly part of Poland.

20.1   Foundations of Analysis

One of the most important concepts in the philosophy of mathematics is **mathematical rigor**. In a nutshell, the concept of rigor requires that all proofs be logically sound, not only in their arguments, but in the foundations of the arguments. For example, all of Euler's work on infinite series was justified by the fact that the series developments produced the "right answer"; however, they could not be logically justified until a better formulation of the notions of convergence was created.

One might wonder why it was important to establish a logical foundation for the techniques of calculus if they "worked." The answer is that while the techniques of calculus worked *in the cases that had been studied*, this alone was no guarantee they would work for cases that had not yet been analyzed. By establishing a logical foundation, it would be easier to determine when—and if—the techniques of analysis would produce a useful answer.

Figure 20.1: Europe after the Napoleonic Wars.

20.1.1   Bolzano

The monarchs of continental Europe who survived the chaos of the French Revolution saw the great danger in liberal thinking. The universities, in particular,
were hotbeds of revolutionary ideas. However, closing the universities was a
desperate measure that only a few monarchs would take. Franz I of Austria employed another tactic: ensure the liberal thinkers of the university were balanced
by conservative elements, particularly those of the Catholic Church. In 1805 he
decreed that every university would have a chair in religious philosophy. Thus,
in 1805, Bernard Bolzano (October 5, 1781-December 18, 1848) was appointed
to the Chair of Religious Philosophy at the University of Prague.

Bolzano was primarily a philosopher and theologian. However, he had a
long-standing interest in mathematics. Kästner's *Elements of Mathematics* had
impressed upon him the necessity of proving even seemingly self-evident statements. For example, consider the statement "Every closed curve in a plane
divides the plane into two regions, an 'inside' and an 'outside'." Bolzano was
the first to realize that this statement was, in fact, a theorem requiring proof,
though he did not himself prove it. A proof would not be found until 1893, by
Camille Jordan; hence the result is sometimes called the **Jordan curve theorem**.

Good definitions are key elements of an axiomatic system. In 1817, Bolzano
gave the following definition:

Definition. $f(x)$ *varies* **continuously** *if* $f(x + \omega) - f(x)$ *can be made smaller than any given quantity, provided* ω *is as small as necessary.*

In the same paper, Bolzano gave the first proof of the **intermediate value theorem**, which Bolzano stated as:

Theorem 20.1 (Intermediate Value Theorem). *Between any two values that give opposite signs to an expression there lies at least one real root of the corresponding equation.*

See Problem 1.

Previous proofs relied on the assumption that $f(x + n\,\delta x)$ took on every value between $f(x)$ and $f(x + \delta x)$ as n went from 0 to 1; hence if $f(x)$ and $f(x + \delta x)$ had opposite signs, there was necessarily a value of n for which $f(x + n\,\delta x) = 0$. Of course, this was an assumption equivalent to the intermediate value theorem itself, so this "proof" was invalid.

Bolzano began by proving a number of theorems relating to geometric series. Then he proved what is now called the Bolzano-Weierstrass theorem:

Theorem 20.2 (Bolzano-Weierstrass). *If a property M does not hold for all x, but holds for all $x < u$ for some u, there is a U that is the* greatest *value such that M holds for all $x < U$.*

Proof. Given u for which M holds for all $x < u$. Since M does not hold for all x, there must be a $V = u + D$, $D > 0$, such that M is *not* true for all $x < V = u + D$. Consider the numbers $u + \frac{D}{2}$, $u + \frac{D}{2^2}$, $u + \frac{D}{2^3}$, ... We ask the question whether M holds for all $x < u + \frac{D}{2^m}$, whatever the value of m.

By assumption, M does not hold for $x < u + \frac{D}{2^0}$. Suppose that for all m, M does not hold all $x < u + \frac{D}{2^m}$. Then u is the greatest value for which M holds for all $x < u$.

On the other hand, suppose there is some m for which M is true for all $x < u + \frac{D}{2^m}$, and not true for all $x < u + \frac{D}{2^{m-1}}$. Now consider the numbers $u + \frac{D}{2^m} + \frac{D}{2^{m+n}}$ where m is the previously determined value, and n is allowed to vary from $n = 1$ to ∞. Applying the same argument, either $v = u + \frac{D}{2^m}$ is the greatest value for which M is true for all $x < v$, or it is not. Repeating this, we construct a number $u + \frac{D}{2^m} + \frac{D}{2^{m+n}} + \frac{D}{2^{m+n+k}} + \ldots$, which is the greatest number U for which M is true for all $x < U$. $\square$

A corollary is that if the property N holds for all $x > l$, there is an L which is the *least* value such that M holds for all $x > L$. The Bolzano-Weierstrass theorem and its corollary imply the intermediate value theorem, for suppose $f(x)$ changes from negative to positive over some interval. Let M be the property $f(x) < 0$; then there is a U for which if $x < U$, $f(x) < 0$ and if $x \geq U$, $f(x) \geq 0$; likewise, let N be the property $f(x) > 0$; then there is an L for which if $x > L$, $f(x) > 0$ and if $x \leq L$, $f(x) \leq 0$. If $U = L$, then $f(L) = 0$; otherwise $f(x) = 0$ for all $U \leq x \leq L$.

Bolzano can also be considered one of the founders of topology. In 1817, he published a pamphlet, *The Three Problems of Rectification, Arc Measurement, and Cubature*, where he attempted to provide rigorous justifications for the integral formulas for finding areas under curves, arc lengths, and volumes. Bolzano defined a spatial object as a collection of points, which might be infinite or might even be finite in number. Bolzano then defined a line as:

Definition. *If, for every point of a spatial object, there is at least one, but only finitely many, points adjacent to it, each point at a certain distance, the object is a **line**.*

By "finitely many," Bolzano meant that for any given distance, there were only finitely many points at that distance from the point of the object. Likewise:

Definition. *If, for every point of a spatial object, there is at least one, but only finitely many, lines adjacent to it, each at a certain distance, the object is a **surface**.*

20.1.2   Cauchy

Augustin   Louis   Cauchy.
©Bettmann/CORBIS.

Bolzano had little influence, for he published in German, and at Prague, whereas most mathematical work in the nineteenth century was published in French, and at Paris. A mediocre mathematician in Paris could make important contributions; a great one, such as Augustin Louis Cauchy (August 21, 1789-May 22, 1857), could shape mathematics itself.

One of Cauchy's key contributions was in the field of complex function theory. The Bernoullis, d'Alembert, and Euler had freely used complex substitutions to evaluate integrals but without justification. Cauchy's "Memoir on Definite Integrals," presented to the Academy on August 22, 1814. It was meant to establish, in a direct and rigorous manner, various results previously obtained. The **Cauchy-Riemann equations** are derived, as is **Cauchy's theorem** for closed rectangular paths. Legendre and Lacroix reviewed the memoir, and recommended its publication in the Academy's journal.

Before it could be published, Napoleon returned from exile, Louis XVIII fled Paris, and Napoleon had his short-lived "Hundred Days" before Alliance forces crushed the Bonapartists for the last time at the battle of Waterloo on June 18, 1815. After Louis XVIII's return to Paris, he reorganized the Academy because of the political activities of some of its members. Monge and Carnot were expelled, Monge for supporting Napoleon, and Carnot for voting for the death of Louis XVI. Their positions were offered to Cauchy, who accepted immediately, a move that smacked of political opportunism. However, Cauchy quickly proved he deserved the position on merit alone. Incidentally, Laplace, who benefited so much from Napoleon yet signed the decree deposing him, was given more honors by Louis XVIII, and thereafter would be *Marquis* Pierre Simon de Laplace.

The Academy's journal was another casualty of the reorganization; it was not republished until 1827, and thus Cauchy's original article on complex function theory was not published until much later. By then, Cauchy wrote other memoirs on complex analysis, and greatly revised the original article.

Mathematical Rigor

Cauchy, like Bolzano, was concerned about the logical foundations of calculus. Since Euler, infinite series formed the cornerstone of mathematical analysis. In 1822, Cauchy presented "On the Development of Functions by Series," a short article giving an overview of series methods in mathematics. Cauchy noted most mathematicians assumed that the function was completely characterized by its corresponding infinite series. But this was not always true; as a counter-example, Cauchy noted the Maclaurin series for $e^{-\frac{1}{x^2}}$ consists of an infinite series of zeros, despite the fact that the function itself is nonzero. This was not reason enough to abandon series methods, but did imply that careful attention had to be paid to whether or not a series accurately represented a function.

The best exposition of Cauchy's attempt to make calculus rigorous was his *Course in Analysis*, a text written for students at the Royal Polytechnical School in Paris (formerly the Polytechnical School). Because the Royal Polytechnical School was one of the foremost institutions in Europe, it was Cauchy's conceptual development of calculus, not Bolzano's, that would set the standard for mathematical rigor. Indeed, modern undergraduate calculus is largely a product of Cauchy's work. For example, our modern definitions of limit, continuity, and derivative all stem from *Course in Analysis*. The greatest difference between Cauchy's text and modern ones is that infinite series and sequences are used throughout the work, rather than being restricted to an advanced section.

Cauchy defined a **function** as:

> When various quantities are such that, given the value of one quantity, one can find the values of all the rest ... the one that the others can be found by is the *independent variable*; and the other quantities expressed in terms of the independent variable are the *functions* of that variable.[1]

A key difference between Cauchy's notion of a function and ours is that he did not require his functions to return a unique value for a given value of the independent variable; hence, mathematicians still spoke of **multi-valued functions**.

Ideas of limits appear throughout *Course in Analysis*, though a definition does not appear until *Continuation of Lessons Given to the Royal Polytechnic on Infinitesimal Calculus*:

> When the successive values of a given variable indefinitely approach a fixed value, eventually ending by differing as little as one wishes, the last value is called the *limit* of all the others.[2]

Notice that Cauchy's definition, like Newton's, talks of the *last* value. In this sense, Cauchy's definition is rooted in the past. To indicate the limit of a variable or a function, Cauchy introduced the now familiar "lim" notation. Cauchy also gave the definition:

---

[1]Cauchy, Augustin Louis, *Œvrés Complètes d'Augustin Cauchy*, second series, 1897. Gauthier-Villars. Vol. III, page 31.

[2]Ibid., Vol. IV, page 13.

> When the successive values of a given variable decrease indefinitely
> so they fall below any given number, the variable is said to be *in-
> finitesimal* or an *infinitely small* quantity ... [if] the successive values
> of a variable increase beyond any given number, we say the variable
> has *positive infinity* for a limit, indicated by ∞ ...[3]

Cauchy also defined **continuity**:

> If the function $f(x)$ has a unique and finite value for all x between
> two given limits, and the difference $f(x+i) - f(x)$ is always infinitely
> small between those limits [if i is infinitely small], we say $f(x)$ is a
> continuous function of x between those given limits.[4]

Cauchy's definition, combined with his earlier definition of infinitely small, produce our modern definition of a function continuous on an interval.

The **derivative function** is then introduced:

> If the function $y = f(x)$ stays continuous between two given values of
> x, and one takes a value for x between those given limits, an infinitely
> small increase of the variable produces an infinitely small change in
> the function itself. Consequently if one lets $\delta x = i$, the two terms of
> the difference quotient
>
> $$\frac{\delta y}{\delta x} = \frac{f(x+i) - f(x)}{i}$$
>
> will be infinitely small. But since these terms approach indefinitely
> and simultaneously the limit 0, the ratio itself could converge to
> another limit.[5]

As we do, Cauchy designated this limit as y' or $f'(x)$.

Cauchy's definition of the **convergence of a series** is also our modern one:
given a sequence of terms $u_1, u_2, u_3, \ldots$, define $s_n = u_1 + u_2 + \ldots + u_n$. Then

> If for all values of n, the sum s_n approaches indefinitely a certain
> limit s, the series is said to be convergent, and the limit in question
> is called the sum of the series.[6]

Cauchy recognized that defining a series in terms of its limit could cause potential
problems if the limit itself was not known; thus he added:

> It is necessary and sufficient that, for infinitely large values of n, the
> sums
>
> $$s_n, s_{n+1}, s_{n+2}, \ldots$$
>
> differ from the limit s, and consequently from each other, by infinitely
> small quantities.[7]

---

[3]Ibid., Vol. III, page 19.
[4]Ibid., Vol. IV, page 19-20.
[5]Ibid., Vol. IV, page 22.
[6]Ibid., Vol. III, page 114.
[7]Ibid., Vol. III, page 114.

This is known today as the **Cauchy convergence criterion** for infinite series.

Since the determination of the convergence of a series involved proving that the limit of the partial sums tends to some definite number, proving the convergence of any series can be quite a task. Thus, as we do, Cauchy used a number of tests, which would imply the convergence (or divergence) of a series. First, Cauchy introduced the **root test** for series with positive terms:

> Find the limit or limits, as n increases indefinitely, of $(u_n)^{\frac{1}{n}}$, and call k the greatest of these limits, or in other words, the greatest value of the limit of the given expression. The series will be convergent if $k < 1$ and divergent if $k > 1$.[8]

Note that Cauchy's use of the word limit has subtle (but important) differences from the our own use of the word limit: Cauchy's use of the term corresponds to the modern use of $\limsup\limits_{n \to \infty}(u_n)^{\frac{1}{n}}$, rather than to $\lim\limits_{n \to \infty}(u_n)^{\frac{1}{n}}$

Cauchy, apparently unaware of Bolzano's work, gave a second proof that a continuous function that went from negative to positive over some interval had a root within that interval. Cauchy's proof, which he included in one of the appendices of *Course in Analysis*, is the following: if $f(x_0)$ and $f(X)$ had opposite signs, partition the interval at m points, and find x_1, X', which were adjacent values for which $f(x_1)$, $f(X')$ had opposite signs; repeating this procedure, he obtained an increasing sequence

$$x_0, x_1, x_2, \ldots$$

and a decreasing sequence

$$X, X', X,'' \ldots$$

the corresponding terms of which could be made to differ by as little as one desired; hence, they had a common limit, which could be designated $f(a)$; moreover, since $f(x_0)$, $f(x_1)$, $f(x_2)$, $\ldots$ had one sign, while $f(X)$, $f(X')$, $f(X'')$, $\ldots$ had the opposite sign, then $f(a) = 0$.

The **definite integral** is introduced by Cauchy in *Continuation*, the second half of the course in analysis given at the Royal Polytechnical School. Suppose $f(x)$ is a continuous function over the interval $x_0 \leq x \leq X$, and choose x_1, x_2, $x_3, \ldots, x_{n-1}$ so

$$x_0 < x_1 < x_2 < \ldots < x_{n-1} < X$$

Define

$$S = (x_1 - x_0)f(x_0) + (x_2 - x_1)f(x_1) + \ldots + (X - x_{n-1})f(x_{n-1})$$

Then, Cauchy claimed

---

[8]Ibid., Vol. III, page 121.

When the differences that make up $X - x_0$ are infinitely small, the method of partitioning makes an insensible difference in the value of S, and if, as one decreases indefinitely the differences, and increases their number, the value of S ends by being constant or, in other words, it ends by attaining a limit dependent only on the form of the function $f(x)$ and its values at x_0 and X, this limit we call the definite integral.[9]

Cauchy's Character

Cauchy, like Euler, published a great deal of work as rapidly as he could, nearly matching Euler's record with 800 publications of his own, plus many more manuscripts. Cauchy's output was so great that the Academy's own journal imposed a limit on the length of submitted papers, forcing Cauchy to look elsewhere for an outlet. Thus, he founded his own journal, *Mathematical Exercises*, issues of which he was often the sole author.

To this impressive record is a dark side. The Academy received manuscripts from nonmembers, and these were often given to Cauchy to critique. At times, his conduct was negligent, and in two cases, that of Galois and that of Abel, he completely failed to recognize the importance of the work and returned the papers. However, failure to recognize the importance of someone else's work, especially when it is of such a groundbreaking nature, is hardly a blemish on Cauchy's name.

What is more odious were the cases when Cauchy recognized the value of an author's idea. So driven was he by his mathematics that he often took the author's result, reproved it in a different (and often better) manner, generalized it and, for the most part, improved it, announcing his results to the Academy almost immediately while delaying on his report of the original author's work. Finally, he would give his report, always referring back to his previous work on the same subject. While Cauchy never claimed the work of another as his own and, indeed, was one of the first mathematicians to consistently cite the work of others, his delay in reporting the work of others and his haste to publish his own extensions are hardly admirable.

In 1824, Louis XVIII died, to be succeeded by his brother, Charles X. Charles was a reactionary, hoping to restore the monarchy and nobility to the position and power they held before the French Revolution. In 1830, the radicals, including the popular Marquis de Lafayette (of American Revolutionary War fame) marched on Paris. They demanded the abolition of the monarchy and the establishment of a republic (hence they were referred to as *Republicans*) with Lafayette as president. The liberals, who wanted to retain the monarchy but not Charles X, quickly offered the crown to Louis Philippe. In 1830, he accepted it, and the radicals were satisfied, for the moment.

Cauchy refused to take the oath supporting the new king and instead chose exile, accepting a position at Turin, where he proved himself a well-prepared

---

[9]Ibid., Vol. IV, page 125.

council to the king. The king's diary records that on January 16, 1831, Cauchy was asked about his opinion on a subject, and Cauchy had already prepared a written response to the king's question: this happened not once or twice, but five times! In 1833, ex-king Charles called Cauchy to Prague, where he was made a baron. Finally, in 1838, Cauchy returned to France; however, his refusal to take an oath of loyalty to Louis Philippe meant that many positions for which Cauchy was eminently qualified were closed to him.

In 1848, Louis Philippe was driven out of France by a bloodless revolution that installed the nephew of Napoleon Bonaparte, also named Napoleon, as the president of a French Republic (the Second Republic). A few years later, he announced the formation of the Second Empire, and himself as the Emperor Napoleon III (Napoleon II, like Louis XVII, lived but did not reign): the loyalty oaths were reestablished in general, but he made an exception for Cauchy and Arago (a physicist).

20.1.3 Hamilton

Bolzano and Cauchy were hampered in their attempts to rigorize analysis by the lack of a theory of the real and complex numbers. The first tentative attempts at such a theory were made by an Irish mathematician, William Rowan Hamilton (August 4, 1805-September 2, 1865), a child prodigy who, by the age of nine, could speak French, Italian, Latin, Greek, Hebrew, Persian, Arabic, Sanskrit, Chaldean, Syriac, Hindustani, Malay, Maratha, Bengali, "and others," according to his father.

On November 4, 1833, and June 1, 1835, Hamilton read two parts of an essay, "Theory of Conjugate Functions, or Algebraic Couples; With a Preliminary and Elementary Essay on Algebra as the Science of Pure Time." In the first part of the paper, Hamilton contrasted the principles of (Euclidean) geometry, which "no intelligent person can doubt," with the principles of algebra, which had such odd assumptions as "a greater magnitude may be subtracted from a lesser magnitude" or (a difficulty even today), that two negatives may be multiplied to make a positive. Just as Euclidean geometry was developed deductively from a few simple assumptions, Hamilton hoped to derive the properties of the real and complex numbers in the same way.

William Rowan Hamilton. ©Hulton-Deutsch Collection/ CORBIS.

Thus, Hamilton attempted to base the real numbers on a notion of *time*. A quick sketch of Hamilton's ideas follows. Two dates, A and B, were either the same, in which case one wrote $A = B$, or they were different, and one preceded the other. Two other dates, C and D, either represented A and B respectively, or did not. Even if they did not, C might have the same relationship to D that A did to B. If this was the case, then the two pairs, A, B were *analogous* to C, D. Hamilton expressed this relationship as

$$D - C = B - A$$

where the $-$ here should *not* be considered a subtraction; rather, it should be considered a general relationship symbol. If $B - A$ was designated by a, then

$A - B$ could be designated by Θa, which was read as **opposite of a**: in other words, A had to B the *opposite* relationship that B had to A.

In this manner, the symbol "+" could be introduced (again, as a formal relationship): $B = a + A$, where $B - A = a$. Multiples of a (obtained by successive addition) were designated as $1a$, $2a$, $3a$, ... Hamilton continued in this manner, deriving all the usual properties of the real numbers.

The second part of the paper applied the same logical approach to **moment couples**, (A_1, A_2) and (B_1, B_2). In an analogous manner, one could form the **relationship couple** $(B_1 - A_1, B_2 - A_2)$, which was defined as indicating the relationship of the two moment couples to each other; in other words,

$$(B_1, B_2) - (A_1, A_2) = (B_1 - A_1, B_2 - A_2)$$

where, again, the $-$ should not be thought of as a subtraction, but rather as a symbol that indicates the relationship between two objects. Addition of two couples, and the multiplication of a number couple by a whole number could also be derived axiomatically.

The multiplication of two number couples raised some difficulties. To approach the problem, Hamilton considered that the *division* of one number couple by another number couple was clearly defined, for $a(a_1, a_2) = (aa_1, aa_2)$ implied

$$\frac{(aa_1, aa_2)}{(a_1, a_2)} = a$$

However, since the addition or subtraction of two number couples resulted in another number couple, it was desirable to make the division of two number couples likewise result in a number couple; hence, Hamilton designated a as the couple $(a, 0)$. From this, Hamilton defined the multiplication

$$(a, 0) \times (a_1, a_2) = (aa_1, aa_2) \tag{20.1}$$

In a like manner

$$\frac{(aa_1, aa_2)}{(a, 0)} = (a_1, a_2)$$

hence

$$(a_1, a_2) \times (a, 0) = (aa_1, aa_2) \tag{20.2}$$

To try and determine what the product of $(a_1, a_2) \times (b_1, b_2)$ would be, Hamilton noted it was desirable if

$$(b_1 + a_1, b_2 + a_2) \times (c_1, c_2) = (b_1, b_2) \times (c_1, c_2) + (a_1, a_2) \times (c_1, c_2) \tag{20.3}$$

$$(c_1, c_2) \times (b_1 + a_1, b_2 + a_2) = (c_1, c_2) \times (b_1, b_2) + (c_1, c_2) \times (a_1, a_2) \tag{20.4}$$

in other words, the **distributive law** held. Assuming this gave:

$$(a_1, a_2) \times (b_1, b_2) = (a_1, 0) \times (b_1, b_2) + (0, a_2) \times (b_1, b_2)$$
$$= (a_1, 0) \times (b_1, b_2) + (0, a_2) \times (b_1, 0) + (0, a_2) \times (0, b_2)$$

Since the products $(a_1, 0) \times (b_1, b_2)$ and $(0, a_2) \times (b_1, 0)$ were already defined, this could be simplified to:

$$= (a_1 b_1, a_1 b_2) + (0, b_1 a_2) + (0, a_2) \times (0, b_2)$$
$$= (a_1 b_1, a_1 b_2 + b_1 a_2) + (0, a_2) \times (0, b_2)$$

where the product $(0, a_2) \times (0, b_2)$ remained to be determined.

Suppose $(0, x) \times (0, y) = (c_1, c_2)$. Then in order for Equations 20.3 and 20.4 to hold true, c_1 and c_2 must be proportional to the product xy. Thus $\qquad$ See Problem 6.

$$(0, x) \times (0, y) = (\gamma_1 xy, \gamma_2 xy)$$

The values of γ_1, γ_2 may be chosen arbitrarily, but, once chosen, must remain constant; additionally, it would be best if the choice of γ_1, γ_2 were to ensure that each division produced a unique quotient (in the case where the divisor was not zero). The simplest choice is to let $\gamma_1 = -1$, and $\gamma_2 = 0$. In this case, the product rule is

$$(a_1, a_2) \times (b_1, b_2) = (a_1 b_1 - a_2 b_2, a_1 b_2 + a_2 b_1)$$

Since there is now a clearly defined set of operations, a theory of the algebraic operations on the number couples can be derived. Finally, Hamilton made the key identification: $(1, 0)$ could be thought of as the unit 1, while $(0, 1)$ could be thought of as the imaginary unit $\sqrt{-1}$.

Unfortunately, this early attempt to provide a theory of the real and complex numbers was hampered by the fact that mathematicians in the United Kingdom, thanks to their isolation, had little effect on mathematical developments on the continent. For the first half of the century, mathematics would be dominated by the French, and the center would be at Paris and the Polytechnic School.

20.1.4 Poisson

Thus, any Parisian mathematician had an advantage independent of his actual abilities. Cauchy, who was both in Paris *and* brilliant, had an enormous impact on the course of mathematics. Cauchy's nearest rival was Siméon-Denis Poisson (June 21, 1781-April 25, 1840). By all accounts, Poisson was exceptionally good at algebraic manipulation, something that would have made him an eighteenth century mathematician of the same caliber as Euler—unfortunately, Poisson lived in the nineteenth century, by which time algebraic manipulation, though still important, was beginning to take second place to set theoretic considerations. Hence, Poisson's reputation in mathematics has suffered, though he remains a key figure in mathematical physics. According to Arago (the physicist who, like Cauchy, was exempted from the loyalty oath), Poisson once said, "Life is good for only two things: to study mathematics, and to teach it." According to his contemporaries, Poisson was especially good at summarizing, simplifying, and presenting the works of other mathematicians.

Poisson's best-known work in mathematics occurs in the theory of probability, a field that he began to investigate only at the end of his life. Poisson was also the first to use **random variables** in probability applications, and to use the idea of a **cumulative distribution function**, two essential concepts of modern probability. Poisson also named the **law of large numbers**, and invented the **Poisson distribution**.

One particular term, **Poisson stability**, came from Poisson's discovery in 1809 that the solutions to the differential equations for the distance of a planet from the sun included terms of the form $t \sin at$, where t was time: these were terms that could grow infinitely large but also returned to zero infinitely often. Although Euler had made a similar discovery in his 1748 essay on the motions of Jupiter and Saturn, Poisson's discovery came *after* Lagrange and Laplace's "proof" of the stability of the solar system: here, then, was a suggestion that the proof of the stability of the solar system might not be as valid as Laplace had suggested.

20.1 Exercises

1. Compare Bolzano's version of the intermediate value theorem to a version given in a modern calculus text. What is Bolzano assuming about the "expression"?

2. Prove that if the property N holds for all $x > l$, there is an L which is the *least* value such that M holds for all $x > L$.

3. Find the Maclaurin series expansion for $e^{-\frac{1}{x^2}}$ around $x = 0$.

4. Cauchy proved that $\sin x$ was continuous because $\sin(x + \alpha) - \sin x = 2 \sin(\frac{1}{2}\alpha) \cos(x + \frac{1}{2}\alpha)$. Prove this identity.

5. Consider Cauchy's definitions of limit, continuity, derived function, convergence of a series, and definite integral.

   (a) How are they similar to the modern definitions? In what respects are they different?

   (b) Compare Cauchy's definition of continuity to Bolzano's. How do they differ? Does either definition allow for "continuous" functions that are not, by the modern definition, continuous?

6. Show if Equations 20.3 and 20.4 are to hold, then $(0, x)(0, y) = (\gamma_1 xy, \gamma_2 xy)$ for some constant γ_1, γ_2. Hint: suppose $(0, x) \times (0, y) = (c_1, c_2)$. What must $(0, ax) \times (0, y)$ be, where a is a whole number?

20.2   Abel

Norway was originally part of the Kingdom of Denmark, but Napoleon's marshal-turned-king Bernadotte, as crown prince of Sweden, seized Norway from Denmark after the Battle of Leipzig (1814). The seizure was recognized by the

Congress of Vienna, and Norway formally became part of the Kingdom of Sweden. Soon afterward, the Norwegians revolted, but Bernadotte put down the rebels almost bloodlessly.

A leader of the independence movement was Sören Georg Abel, the father of Niels Henrik Abel (August 5, 1802-April 6, 1829). The elder Abel helped to raise the funds to found a Norwegian university, in Christiania (modern-day Oslo). Though the establishment of a university in Oslo in 1813 was a great triumph for the elder Abel and the independence movement, it was bad luck for the young Abel. The cathedral school in Oslo, which dated back to 1250, was a first class institution containing many teachers with excellent credentials—who accepted positions at the university. Thus, when Abel entered school, the remaining teachers were distinctly second rate. One of the teachers, Hans Peter Bader, had a reputation for being an unusually severe disciplinarian: in nineteenth century terms, this meant he beat the students more than the other teachers did.

In 1818, one of Bader's students died eight days after what was perceived as an unusually severe beating. Though the attending physician said the student's death was caused by a "nervous flu," not the beating, and no disciplinary action was taken against Bader, the students staged a massive protest to have the unpopular Bader dismissed. His replacement was Bernt Michael Holmboe.

Holmboe changed the direction of the instruction in mathematics. He encouraged the students to work independently on problems they might find interesting. Almost immediately, he recognized Abel's potential and encouraged him to read the works of Newton, Euler, Lagrange, and other great mathematicians. In later years, Abel would attribute his mathematical success to the fact that he "read the masters, not their students." In his last year at the school, Abel threw himself at solving the general fifth degree equation.

20.2.1   The Solution to the Quintic

Abel thought he had a solution. Unfortunately, no one in Norway was competent to judge the paper. Thus, Abel sent it to Ferdinand Degen at the Danish Academy of the Sciences. Degen made two comments that would profoundly affect Abel's life. The first comment was that though he could find no obvious faults in the paper, Abel could improve its readability by providing an example of its use. (We will deal with Degen's second comment later.)

Abel took the advice, and tried to come up with an example. It was then he discovered that his method was unworkable; soon, he realized that he could, in fact, prove the *impossibility* of solving a fifth (or higher) degree equation.

Unfortunately, Abel had a difficult time trying to get his work published. Somehow, he managed to get enough money to print the paper at his own expense, in 1824. To save printing costs, he reduced the details to a bare minimum, making the original 1824 paper difficult to follow. Later, after his reputation had been made, Abel elaborated some of the key points of the paper.

Let y_1, y_2, y_3, y_4, y_5 be the roots of

$$y^5 - ay^4 + by^3 - cy^2 + dy - e = 0$$

and suppose one of the roots can be expressed in terms of the coefficients a, b, c, d, e in the form

$$y = p + p_1 R^{\frac{1}{m}} + p_2 R^{\frac{2}{m}} + \ldots + p_{m-1} R^{\frac{m-1}{m}} \tag{20.5}$$

where p, p_1, p_2, $\ldots$, R are functions of the same form as y (in other words, they can be expressed in terms of sums of roots of functions of the coefficients); assume $R^{\frac{1}{m}}$ cannot be expressed as a rational function of the coefficients. Abel showed that $R^{\frac{1}{m}}$ must have one of two forms. Either $m = 5$, in which case

$$R^{\frac{1}{5}} = \frac{1}{5}\left(y_1 + \alpha^4 y_2 + \alpha^3 y_3 + \alpha^2 y_1 + \alpha y_5\right)$$

where α is a fifth root of unity. Hence, $R^{\frac{1}{5}}$ can be expressed as a rational function of the roots. However, $R^{\frac{1}{5}}$ can take on only five different values, whereas the right-hand side can take on 120 different values, depending on the order in which the roots are taken; hence, it is impossible for there to be a solution of the form in Equation 20.5 with $m = 5$. The only other possibility is $m = 2$, but in that case, one obtained a similar contradiction.

20.2.2   Crelle's Journal

After graduation, Abel went around the mathematical centers of Europe with several of his friends. His original intention was to go directly to Paris, the center of mathematical activity, but, as the interests of his friends were in mathematical physics, they went to Berlin, and Abel followed them. It was a fateful decision, for there he met August Leopold Crelle (March 11, 1780-October 6, 1855), a construction engineer with an interest in mathematics. In 1826, Crelle would begin publishing what would become one of the most important mathematical journals in history: the *Journal of Pure and Applied Mathematics*, invariably called Crelle's Journal. Crelle was impressed with Abel's work, and strove to give him a wide audience: in the first volume of the journal, he published seven of Abel's papers. Meanwhile, Abel continued his tour and eventually made it to Paris, where he hoped to make his mathematical reputation.

20.2.3   Abelian Rigor

Cauchy may have begun the rigorous examination of the foundations of mathematics, but he was not alone. On January 1, 1826, Abel wrote to his former teacher Holmboe about his current researches. Abel complained:

> Divergent series are in general a terrible thing, and it is a disgrace to base any demonstration whatsoever on them ... can you imagine anything more horrible than to say
>
> $$0 = 1 - 2^n + 3^n - 4^n + \text{ etc.}$$

n being a whole positive number? Moreover I am struck by the fact that except for the simplest cases, for example geometric series, there is in mathematics hardly any infinite series whose sum is determined in a rigorous manner, which is to say that the most essential part of mathematics is without foundation.[10]

A prime example was the expansion of the binomial $(1 + x)^m$. In 1826, Abel's researches on this series were published in Crelle's journal, where he gave the first rigorous proof of the binomial expansion. In the course of the proof, Abel had to show

$$f(x) = v_0 + v_1 \alpha + v_2 \alpha^2 + \dots$$

was a convergent function of x on the interval $x = a$ to $x = b$, where v_0, v_1, ...depended on x and α was chosen so that the series converged. Abel could have "proven" this by appealing to a theorem from Cauchy's *Course in Analysis*, where Cauchy claimed that the limit of a series of continuous functions was itself a continuous function; however, Abel noted this theorem had some exceptions. For example, the function defined by the series

$$\sin x - \frac{1}{2} \sin 2x + \frac{1}{3} \sin 3x - \dots \qquad (20.6)$$

was discontinuous for $x = (2m + 1)\pi$, m being a whole number, despite the fact that the series itself was the limit of a series of continuous functions. Abel came very close to the idea of **uniform continuity**, something that would finally be recognized by Heine.

20.2.4 Elliptic Integrals

Degen, who had suggested Abel provide an example of solving the fifth-degree equation (and thus made Abel discover he had, in fact, shown it was *impossible* to solve the general quintic), also suggested Abel investigate the new field of **elliptic integrals** rather than the tired and overworked field of algebraic solutions to polynomial equations. Soon, Abel was making profound contributions to the theory.

We may divide functions into two broad types. The first, **algebraic functions**, are those expressible using a finite number of elementary operations, such as addition, subtraction, multiplication, division, as well as powers and roots. Other functions, such as the logarithmic and trigonometric functions, cannot be so defined; they are **transcendental functions**.

However, the logarithmic function *can* be expressed in terms of an integral of the reciprocal of a linear factor, while the inverse tangent function could be expressed in terms of the integral of the reciprocal of a quadratic factor; the Fundamental Theorem of Algebra thus implied that the integral of any rational function could be expressed in terms of logarithmic and inverse trigonometric

---

[10] Abel, Niels Henrik, *Œvrés Complètes de Niels Henrik Abel*, 1881. Grøndahl and Søn. Vol. II, page 256-7.

functions. In addition, integrals involving the roots of quadratic polynomials could be evaluated in terms of the other inverse trigonometric functions. De Moivre and Euler pointed out the fundamental relationship between the trigonometric and logarithmic functions, so in some sense, one class of transcendental functions could be generated by integrals involving roots and rational functions of linear and quadratic terms.

Other types of transcendental functions were conceivable. One type were **elliptic functions**, which originated from the problem of finding the arc length of an ellipse in determining planetary orbits. Legendre began a comprehensive study. In addition to his prominence in number theory and geometry, Legendre made many contributions to analysis. His three volume *Treatise of Elliptic Functions and Eulerian Integrals* (1825, 1826, and 1830), became a standard text. Much of the work dealt with reducing elliptic integrals into one of three standard forms:

$$\int \frac{d\theta}{\sqrt{1 - c^2 \sin^2 \theta}} \qquad \int d\theta \sqrt{1 - c^2 \sin^2 \theta} \qquad \int \frac{d\theta}{\left(1 + n \sin^2 \theta\right) \sqrt{1 - c^2 \sin^2 \theta}}$$

Once an elliptic integral was in a standard form, its value could be determined numerically.

Abel thought highly of Legendre, though called him "old as the stones" (he was 74 when the 24-year-old Abel arrived in Paris). Unfortunately, Legendre was also a man of conscience, and when he voted against the government's candidate for the National Institute (the reorganized Academy of Sciences), the government withdrew its financial support of Legendre, and he died in poverty.

Elementary Properties of the Elliptic Functions

Abel further developed the theory of elliptic functions. In "Researches on the Elliptic Functions" (1827), Abel gave a beautiful example of the use of elementary calculus to prove some very fundamental (and surprising) results. Abel considered an elliptic function to be all functions defined by an integral

$$\int \frac{R \, dx}{\sqrt{\alpha + \beta x + \gamma x^2 + \delta x^3 + \epsilon x^4}}$$

where R was a rational function of x. By Legendre's work, Abel knew that these elliptic functions could be reduced to one of the three standard forms, so he could restrict his analysis to these three types.

Abel had the following great insight: consider a function defined by an integral, such as

$$\arcsin x = \int_0^x \frac{1}{\sqrt{1 - \theta^2}} d\theta$$

The arcsine function is inelegant. However, the *inverse* function of arcsine has many nice properties; most usefully, it is periodic, hence it is sufficient to determine its values over a particular interval.

Abel thus considered the functions of the form

$$\alpha(x) = \int_0^x \frac{d\theta}{\sqrt{1 - c^2 \sin^2 \theta}}$$

Under the substitution $x = \sin \theta$, the integral becomes (we abuse notation to follow Abel's development)

$$\alpha(x) = \int_0^x \frac{dx}{\sqrt{(1 - x^2)(1 - c^2 x^2)}}$$

Suppose $c^2 = -e^2$, and, "for reasons of symmetry," write $1 - x^2$ as $1 - c^2 x^2$. This integral becomes

$$\alpha(x) = \int_0^x \frac{dx}{\sqrt{(1 - c^2 x^2)(1 + e^2 x^2)}} \tag{20.7}$$

Abel designated the inverse function of $\alpha(x)$ as $\phi(\alpha)$. Elementary calculus suffices to show:

$$\phi'(\alpha) = \sqrt{\left(1 - c^2 (\phi(\alpha))^2\right)\left(1 + e^2 (\phi(\alpha))^2\right)} \tag{20.8}$$

To simplify the further development, Abel defined two new functions

$$f(\alpha) = \sqrt{1 - c^2 (\phi(\alpha))^2} \quad \text{and} \quad F(\alpha) = \sqrt{1 + e^2 (\phi(\alpha))^2}$$

Some properties of α and $\phi(\alpha)$ can be derived immediately from Equation 20.7 and the properties of definite integrals from elementary calculus. Since the integrand is positive, then $\alpha(x)$ is positive and increasing over the interval $\left(0, \frac{1}{c}\right)$; hence $\phi(\alpha)$ is positive and increasing over the interval $\left(0, \frac{\omega}{2}\right)$, where

$$\frac{\omega}{2} = \int_0^{\frac{1}{c}} \frac{dx}{\sqrt{(1 - c^2 x^2)(1 + e^2 x^2)}}$$

Also, $\phi(0) = 0$ and $\phi\left(\frac{\omega}{2}\right) = \frac{1}{c}$.

Additional properties could be obtained by considering various changes of variable. By replacing x with $-x$ in Equation 20.7, it follows that $\phi(-\alpha) = -\phi(\alpha)$. By (formally) replacing x with xi, Abel obtained the relationship $xi = \phi(\beta i)$, where

$$\beta = \int_0^x \frac{dx}{\sqrt{(1 + c^2 x^2)(1 - e^2 x^2)}}$$

From this, it is clear that β is real and positive from $x = 0$ to $x = \frac{1}{e}$; designating

$$\frac{\tilde{\omega}}{2} = \int_0^{\frac{1}{e}} \frac{dx}{\sqrt{(1 - e^2 x^2)(1 + c^2 x^2)}}$$

we note that x is positive from $\beta = 0$ to $\beta = \frac{\tilde{\omega}}{2}$. Thus, $\phi(\alpha)$ is real for all real values of α between $-\frac{\omega}{2}$ and $\frac{\omega}{2}$, and $\phi(\beta i)$ is pure imaginary when β is between $-\frac{\tilde{\omega}}{2}$ and $\frac{\tilde{\omega}}{2}$. It remained to determine the value of $\phi(\alpha + \beta i)$.

Double Periodicity

To determine these, Abel first showed

$$\phi'(\alpha) = f(\alpha)F(\alpha) \tag{20.9}$$

$$f'(\alpha) = -c^2\phi(\alpha)F(\alpha) \tag{20.10}$$

$$F'(\alpha) = e^2\phi(\alpha)f(\alpha) \tag{20.11}$$

With the derivative formulas, it is possible to determine the addition formulas for the elliptic functions:

$$\phi(\alpha + \beta) = \frac{\phi(\alpha)f(\beta)F(\beta) + \phi(\beta)f(\alpha)F(\alpha)}{1 + e^2c^2\left(\phi(\alpha)\right)^2\left(\phi(\beta)\right)^2} \tag{20.12}$$

$$f(\alpha + \beta) = \frac{f(\alpha)f(\beta) - c^2\phi(\alpha)\phi(\beta)F(\alpha)F(\beta)}{1 + e^2c^2\left(\phi(\alpha)\right)^2\left(\phi(\beta)\right)^2} \tag{20.13}$$

$$F(\alpha + \beta) = \frac{F(\alpha)F(\beta) + e^2\phi(\alpha)\phi(\beta)f(\alpha)f(\beta)}{1 + e^2c^2\left(\phi(\alpha)\right)^2\left(\phi(\beta)\right)^2} \tag{20.14}$$

Although, as Abel noted, these could be determined from the known properties of the elliptic functions, an easier means was available: if r designated the right-hand side of any one of the three formulas (say Equation 20.12), Abel showed that

$$\frac{\partial r}{\partial \alpha} = \frac{\partial r}{\partial \beta}$$

hence r had to be some function of $\alpha + \beta$, say $r = \psi(\alpha + \beta)$. To determine the precise form of ψ, he supposed $\beta = 0$; in which case $r = \psi(\alpha)$. But from the definition of r, we also have $r = \phi(\alpha)$. Hence $\phi(\alpha) = \psi(\alpha)$. Replacing α with $\alpha + \beta$ yielded $r = \psi(\alpha + \beta) = \phi(\alpha + \beta)$, and the result was proven.

From these addition formulas, other properties of the elliptic functions could be determined. The most remarkable property of the elliptic functions, however, was their periodicity. The exponential function has an imaginary period, as the De Moivre's formula demonstrates. The trigonometric functions have a real period. However, the elliptic functions are doubly periodic: they have one real and one imaginary period, since for m, n integers,

$$\phi(m\omega + n\tilde{\omega}i \pm \alpha) = \pm(-1)^{m+n}\phi(\alpha) \tag{20.15}$$

$$f(m\omega + n\tilde{\omega}i \pm \alpha) = (-1)^{m+n}f(\alpha) \tag{20.16}$$

$$F(m\omega + n\tilde{\omega}i \pm \alpha) = (-1)^{m+n}F(\alpha) \tag{20.17}$$

See Problem 2.   which can be derived directly from the addition formulas.

Extensions of Elliptic Integrals

Natural extensions of the transcendental functions might easily be imagined, and on October 30, 1826, Abel presented to the Paris Academy "Memoir on a

General Property of a Very Extensive Class of Transcendental Functions." Unfortunately, though Abel's papers included many fundamental and vital results, the response by the French Academy was near-scandalous. Fourier presented it to the Academy and gave it to Cauchy and Legendre to referee. Cauchy criticized it on several grounds, and the paper was not published until 1841, fifteen years after it was presented.

20.2.5 The Origin of the Modern University

Crelle, unlike Cauchy, recognized Abel's brilliance, and sought to find a position for Abel in Prussia. After enormous efforts, Crelle managed to secure a position at the University of Berlin, and wrote to Abel on April 8, 1829 to tell him the good news. It was too late: Abel had died two days before, from tuberculosis.

The position offered to Abel was part of a growing movement that changed the teaching of mathematics. Thanks to the Polytechnical School and the very active French Academy of the Sciences, Paris was the center of mathematical activity at the beginning of the nineteenth century, and anyone interested in pursuing advanced mathematics had to go to France: Abel's visit was merely one of thousands of similar pilgrimages. Gauss might have been able to start a tradition of mathematics in Confederation of the Rhine, but Gauss hated teaching, and the few students he attracted were taught mathematical astronomy.

After their crushing defeat by Napoleon in 1806, the Prussians spent the next few decades on some much needed internal reform. Some of it was military: the army learned that the past glories of Frederick the Great were of little military value in the present. But much of the reform was social. Education Minister Wilhelm von Humboldt began a sweeping change of the Prussian educational system. At the pre-university (or *gymnasium*) level, mathematics was assigned a primary importance: only the study of Latin had more time devoted to it.

Before von Humboldt, most mathematical and scientific research was done through the various Academies of Science; universities were places to become a lawyer, physician, or theologian. Von Humboldt sought to change this model: Berlin University, founded in 1810, was to be a center of both learning and research.

Another important change was the liberalization of the universities. The University of Halle was a leader in this trend. It had long been teaching classes in German, rather than Latin. Seminars, which encouraged the active participation of all students, were replacing disputations between two advanced students in front of a class of passive listeners. This pioneering liberalism soon spread to other universities. In 1817, the universities at Halle and at nearby Wittenberg, where Martin Luther and Rheticus taught, merged to form a single university.

Besides reform at the university level, Prussia tried to establish an institution to train researchers similar to the Polytechnical School in Paris. Gauss was offered the chance to oversee the entire program: his duties would be primarily administrative and scientific, and he would, most especially, not have to teach. The offer was tempting, but after some consideration, Gauss rejected it. The rejection meant that the idea of founding a school for training researchers had to be shelved, and was replaced with the idea of establishing a special institute for training teachers. It was a position at this institute that was finally offered to Abel. Abel's death meant that this idea, too, was put aside. Out of sheer necessity, the task of training mathematics and science teachers was given to the universities. Through these historical accidents, we see the formation of the modern university in Germany: a multipurpose institution of research, teaching, and teacher training.

20.2.6   Jacobi

An interesting feature of nineteenth century mathematics was that many of the key contributions were made, simultaneously and independently, by two mathematicians. Abel's rival in the development of elliptic functions was Carl Gustav Jacob Jacobi (December 10, 1804-February 18, 1851), the son of a wealthy Jewish banker. Jacobi was one of the first graduates at the newly founded University of Berlin. His Jewish background might have been an impediment to advancement, but Jacobi converted to Christianity, and obtained a position at the University of Berlin, which he left in 1825 for better prospects at the University of Königsburg.

Jacobi's work in the theory of elliptic functions first appeared in the *Astronomical Notices* for December 1827, where he announced many important results without proof. Some time later, he read Abel's "Researches" and with Abel's idea of inverting the functional relationship, was able to prove the results he had earlier announced. Jacobi eventually made inversion a tenet of his mathematical philosophy: "You must always invert," he is reputed to have said. It was Jacobi's claim to success. Soon thereafter, Jacobi was appointed to a professorship at Königsburg. Jacobi was only 24, and some of the faculty objected to promoting him over those with more seniority. However, the Education Department insisted that Jacobi's talent be recognized. In 1829, Jacobi published *Foundations of a New Theory of Elliptic Functions*, and no one thereafter questioned the justice of Jacobi's promotion. What Jacobi might have accomplished in mathematics was tempered by his death, in 1851, from smallpox.

Jacobi's extension to hyperelliptic functions was marred by his discovery of a puzzling problem. By applying the inversion technique to integrals of the form $\int \frac{dx}{\sqrt{X}}$ where X was a sixth degree polynomial, he found quadruply periodic functions that seemed capable of taking on infinitely many values at every point; remember that at this point in time, a "multivalued function" was not a contradiction in terms. The difficulty would be resolved by Riemann.

20.2 Exercises

1. Let $R = 1 + e^2 c^2 \left(\phi(\alpha) \right)^2 \left(\phi(\beta) \right)^2$.

   (a) Find and simplify the formulas for $\phi(\alpha + \beta) \pm \phi(\alpha - \beta)$, $f(\alpha + \beta) \pm f(\alpha - \beta)$, and $F(\alpha + \beta) \pm F(\alpha - \beta)$.

(b) Find and simplify formulas for $\phi(\alpha+\beta)\cdot\phi(\alpha-\beta)$, $f(\alpha+\beta)\cdot f(\alpha-\beta)$, and $F(\alpha+\beta)\cdot F(\alpha-\beta)$.

2. Use the addition formulas 20.12 with $\beta=\pm\frac{\omega}{2}$ and $\beta=\pm\frac{\tilde{\omega}}{2}$ to show

$$\phi\left(\frac{\omega}{2}+\alpha\right)=\phi\left(\frac{\omega}{2}-\alpha\right) \qquad \phi\left(\frac{\tilde{\omega}}{2}+\alpha\right)=\phi\left(\frac{\tilde{\omega}}{2}-\alpha\right)$$

$$f\left(\frac{\omega}{2}+\alpha\right)=-f\left(\frac{\omega}{2}-\alpha\right) \qquad\text{and}\qquad f\left(\frac{\tilde{\omega}}{2}+\alpha\right)=f\left(\frac{\tilde{\omega}}{2}-\alpha\right)$$

$$F\left(\frac{\omega}{2}+\alpha\right)=F\left(\frac{\omega}{2}-\alpha\right) \qquad F\left(\frac{\tilde{\omega}}{2}+\alpha\right)=-F\left(\frac{\tilde{\omega}}{2}-\alpha\right)$$

From these, derive the double periodicity of Equation 20.15.

3. Show that Abel's function in Equation 20.6 is a counterexample for Cauchy's claim that the limit of a series of continuous functions is a continuous function.

20.3 Fourier Series

Fourier, whose bad luck plagued him from the beginning of the revolution to Napoleon's disastrous Egyptian expedition, was made a baron by Napoleon. This honor lasted only until Napoleon fell from power, and was revoked by the new regime of Louis XVIII. During Napoleon's brief return in 1815, he made Fourier a count; however, Fourier, a man of conviction even in dangerous times (witness his defense of those accused during the Terror) resigned his title in protest of the excesses of the Napoleonic regime. The title would have meant little in any case, for soon after, Napoleon fell for the second and final time.

In 1816, Fourier was elected to the Academy of Sciences; however, the newly crowned Louis XVIII vetoed the election because of Fourier's support of Napoleon. In 1820, Fourier rewrote and published the memoir he had written in 1789, and found himself embroiled in a priority dispute with another mathematician, Budan de Bois-Laurent, who had independently discovered the result in 1806. Finally, Fourier was elected to the Academy of Sciences over the king's objections, and became the Academy's Perpetual Secretary in 1822.

20.3.1 Fourier

Fourier, like Newton, developed a general tool for finding an infinite series representation of a function. Whereas Newton's binomial series expanded functions in terms of x^n, the **Fourier series** was an expansion in terms of the sine or cosine of nx. Fourier, like Newton, was anticipated by others, but was the first to express a general, algorithmic means of finding the series coefficients.

In final form, Fourier's work appeared in *The Analytical Theory of Heat* (1822), though ten years earlier, the ideas had won Fourier a prize from the

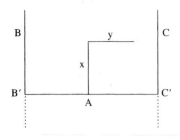

French Academy. As indicated by the title, the primary goal was not to find a trigonometric series expansion. Instead, the series emerged as a by-product of the investigation of a physical problem, that of the equilibrium temperature distribution of a heated solid. Suppose BAC is a solid between two infinite parallel planes B and C, with plane A perpendicular to B and to C. Let $B'AC'$ be a solid, kept at a constant temperature of 1, and suppose that all points on B, C are kept at a constant temperature of 0. Let the length of the base A be $2l$ or, to simplify the computations, π, and let the coordinates of any point m be (x, y), where $y = 0$ corresponds to the middle of the solid. Note that y indicates the *horizontal* axis and x the *vertical* axis. Further, assume that no heat propagates perpendicular to the plane of the drawing (alternatively, that the solid is one of infinitely many slices), and let $v(x, y, t)$ be the temperature at any point (x, y) at time t. At equilibrium, the temperature of a point would not be changing over time.

Fourier argued that the rate of change of temperature could be found by

$$\frac{dv}{dt} = \frac{K}{CD}\left(\frac{d^2v}{dx^2} + \frac{d^2v}{dy^2}\right)$$

where the derivatives on the right hand side are the partial derivatives of v with respect to x or y: Legendre had introduced the use of the symbol ∂ for partial derivatives, but it was not yet in common use; neither was the use of $\bigtriangledown$, to represent the Laplacian operator.

If the temperature is at equilibrium, then $\frac{dv}{dt} = 0$; hence

$$\frac{d^2v}{dx^2} + \frac{d^2v}{dy^2} = 0$$

Suppose the final temperature distribution is given by some function $v = \varphi(x, y)$, which satisfies this differential equation. In addition, Fourier said, $\varphi(x, y)$ must satisfy three more properties: φ must be 0 when $y = \pm\frac{1}{2}\pi$; φ must be 1 when $x = 0$ and y is between $\frac{1}{2}\pi$ and $-\frac{1}{2}\pi$; and finally, φ must have a small value if x is very large.

Fourier began the analysis with a simple case. First, he assumed $v = F(x)f(y)$. Substituting this into the differential equation, he obtained

$$\frac{F''(x)}{F(x)} + \frac{f''(y)}{f(y)} = 0$$

In the simplest case, we may assume $\frac{F''(x)}{F(x)}$ is a positive constant, which can be designated m^2; hence $\frac{f''(y)}{f(y)} = -m^2$. Thus, $F(x) = e^{-mx}$ and $f(y) = \cos my$. Based on physical considerations, Fourier concluded that it is not reasonable for m to be negative, and since $v = 0$ if $y = \pm\frac{\pi}{2}$, then m must be an odd number. Hence, a solution is $\varphi(x, y) = e^{-mx}\cos my$.

See Problem 1.

A more general solution may be found, Fourier claimed, by adding together several terms like the preceding; hence

$$\varphi(x, y) = ae^{-x}\cos y + be^{-3x}\cos 3y + ce^{-5x}\cos 5y + de^{-7x}\cos 7y + \dots$$

where a, b, c, etc., are constants. Since $\varphi(0, y) = 1$ for y between $-\frac{\pi}{2}$ and $\frac{\pi}{2}$, then

$$1 = a \cos y + b \cos 3y + c \cos 5y + d \cos 7y + \ldots$$

Since we have, apparently, the sum of a series of *variable* trigonometric functions equal to a *constant* value, it might seem that such a series could not exist. However, Fourier noted, the solution would make it clear that such a function *was* possible. Here we note an important feature from eighteenth century mathematics: a solution justifies itself.

Before continuing, Fourier noted that the base need not be at a constant temperature of 1 but in fact could be at *any* of a number of different temperatures. For example, it might be at a temperature expressed by $a \cos y$; in this case, $b = c = d = \ldots = 0$. From here, the extension to expressing an *arbitrary* function as a sum of cosines is an easy step. Fourier considered an arbitrary function, $\phi(x)$, whose (Taylor series) expression contained only odd powers of x; he supposed

$$\phi(x) = a \sin x + b \sin 2x + c \sin 3x + d \sin 4x + \ldots$$

Then, taking the series expansion of $\phi(x)$

$$\phi(x) = x\phi'(0) + \frac{x^2}{2}\phi''(0) + \frac{x^3}{6}\phi'''(0) + \ldots$$

which, to make easier to comprehend, Fourier assumed was of the form

$$= Ax - \frac{B}{6}x^3 + \frac{C}{120}x^5 - \ldots$$

Fourier then compared this with the original series, and, expressing A, B, C, ... in terms of a, b, c, ..., obtained

$$A = a + 2b + 3c + 4d + \ldots$$
$$B = a + 2^3 b + 3^3 c + 4^3 d + \ldots$$
$$C = a + 2^5 b + 3^5 c + 4^5 d + \ldots$$
$$\vdots$$

which was a system of infinitely many equations in infinitely many unknowns.

To solve this problem, Fourier considered the sequence of systems of equations

$$a_1 = A_1$$

$$a_2 + 2b_2 = A_2$$
$$a_2 + 2^3 b_2 = B_2$$

$$a_3 + 2b_3 + 3c_3 = A_3$$

$$a_3 + 2^3 b_3 + 3^3 c_3 = B_3$$

$$a_3 + 2^5 b_3 + 3^5 c_3 = C_3$$

He then considered the coefficients a, b, c, . . . to be the limiting values of the solutions a_n, b_n, c_n, . . . After a great deal of effort, the solutions may be found.

However, Fourier noted, there is an easier way. If

$$\phi(x) = a_1 \sin x + a_2 \sin 2x + a_3 \sin 3x + \ldots + a_i \sin ix + \ldots$$

then, multiplying by $\sin ix$ and integrating from 0 to π, one obtained

$$\int_0^\pi \sin ix \phi(x) dx = a_1 \int_0^\pi \sin ix \sin x + a_2 \int_0^\pi \sin ix \sin 2x$$

$$+ a_3 \int_0^\pi \sin ix \sin 3x + \ldots + a_i \int_0^\pi \sin ix \sin ix + \ldots$$

All of the integrals on the right-hand side are 0, except for $a_i \int_0^\pi \sin ix \sin ix dx$; this integral is $\frac{1}{2}\pi a_i$ and thus the coefficients a_i was equal to $\frac{2}{\pi} \int_0^\pi \phi(x) \sin ix dx$.

If $\phi(x)$ was an even function, a similar series could be obtained, namely

$$\phi(x) = a_0 + a_1 \cos x + a_2 \cos 2x + a_3 \cos 3x + \ldots$$

where

$$a_i = \int_0^\pi \cos ix \phi(x) dx$$

Fourier applied the procedure to $\phi(x) = \frac{1}{2}x$ (even though it is not an even function), and obtained the series

$$\frac{1}{2}x = \frac{1}{4}\pi - \frac{2}{\pi}\cos x - \frac{2}{3^2\pi}\cos 3x - \frac{2}{5^2\pi}\cos 5x - \ldots$$

Fourier noted that he now had two values for $\frac{1}{2}x$, one in terms of sines and the other in terms of cosines. These were not equal everywhere, but only for x values between 0 and $\frac{1}{2}\pi$.

20.3.2   Dirichlet

Gustav Peter Lejeune Dirichlet (February 23, 1805-May 5, 1859) began to study mathematics at an early age; by the age of 12, he was using his own allowance to purchase mathematics texts. He studied with the physicist Georg Simon Ohm at the University of Köln, but there was no place in Germany to study higher mathematics, so he went to Paris in May 1822. During the summer of 1822, he suffered an attack of smallpox—the practice of widespread vaccination had not yet been established. Fortunately, he survived. Dirichlet took a position tutoring the children of Maximilian Fay, a former Napoleonic general and a politician. When Fay died in November 1825, Dirichlet, encouraged by Alexander von Humboldt, returned to Germany.

Dirichlet's most famous result was

Proposition 20.1. *In every arithmetic progression* $a + nb$, *where* a, b *have no common factor, there are infinitely many primes.*

He proved this in 1837, generalizing a conjecture Euler made in 1775 that every arithmetic progression with first term 1 contained an infinite number of primes.

Convergence of Fourier Series

With the work of Fourier, it became apparent that many arbitrary functions could be expressed in terms of a trigonometric series. The question naturally arose *which* functions could be expressed in terms of a series (in the sense that the series accurately represented the function). In 1827, Cauchy presented a proof that the Fourier series for a function always converged. However, the proof relied on the assumption that a function $f(x)$ still had a meaningful value when x was a complex number.

Dirichlet took up the problem with "On the Convergence of Trigonometric Series Which Represent an Arbitrary Function Between Given Limits," presented in January 1829. Dirichlet's key result stems from writing the Fourier series for a function $\phi(x)$ as the infinite sum

$$\frac{1}{2\pi} \int \phi(a) \, da + \frac{1}{\pi} \left\{ \begin{array}{l} \cos x \int \phi(a) \cos a \, da + \cos 2x \int \phi(a) \cos 2a \, da + \dots \\ \sin x \int \phi(a) \sin a \, da + \sin 2x \int \phi(a) \sin 2a \, da + \dots \end{array} \right\}$$

which, since all the integrals are definite integrals, is a function of x alone. The sum of the first $2n$ terms of this series could be reduced to

$$\frac{1}{\pi} \int_{-\pi}^{\pi} \phi(a) \left\{ \frac{1}{2} + \cos(a - x) + \cos 2(a - x) + \dots + \cos n(a - x) \right\} \, da$$

or

$$\frac{1}{\pi} \int_{-\pi}^{\pi} \phi(a) \frac{\sin(n + \frac{1}{2})(a - x)}{2 \sin \frac{1}{2}(a - x)} \, da$$

As n went to infinity, this integral converged to the integral of the Fourier series of $\phi(x)$. Dirichlet showed that this integral (and hence the Fourier series itself) converged to the limit

$$\frac{1}{2} \left[\phi(x + \epsilon) + \phi(x - \epsilon) \right] \tag{20.18}$$

for suitably small ϵ. This was the first rigorous determination of the limits over which a Fourier series converged to the function it represented; as an example of a function whose Fourier series nowhere converged to its function, Dirichlet gave his famous **Dirichlet function** $\phi(x)$, which had some value c when x was rational, and a different value d when x was irrational.

Definition of a Function

The consideration of Fourier series led to the modern definition of a function, in Dirichlet's "On the Representation of Completely Arbitrary Functions of Sine and Cosine Series" (1837). It is noteworthy that the article appeared in a physics journal, highlighting the continuing relationship between mathematics and physics that nearly all the significant mathematicians of the nineteenth century cultivated.

> Take a and b two definite quantities and x some other quantity, which lies between a and b. Corresponding to every x let there be a definite, finite y ... it is not necessary that y, in this interval, depend on x by the same law throughout, nor is it even required that the relation be expressible through mathematical operations.[11]

Thus, functions finally emerge in the modern sense as totally arbitrary relationships between two quantities.

Dirichlet's last few years were unfortunate. In 1858, he went to Montreux, Switzerland, to give a speech in honor of Gauss, but had a heart attack and was forced to remain there for some months, recovering. While he was away, his wife died of a stroke, and Dirichlet himself died in 1859, after a year of invalidship.

20.3 Exercises

1. Consider Fourier's derivation of the trigonometric series for the heat equation.

   (a) Explain why it is not reasonable for m to be negative or even.

   (b) Justify Fourier's assumption, in deriving the heat equation, that $\varphi = F(x)f(y)$.

   (c) Show that if $\varphi(x, y) = F(x)f(y)$, then $\frac{F''(x)}{F(x)} + \frac{f''(y)}{f(y)} = 0$.

   (d) Show that $F(x) = e^{-mx}$, $f(y) = \cos my$ is a solution to the differential equation.

2. Fourier found for $\phi(x) = 1$ the series

$$\frac{1}{2}\pi = \sin x + \frac{1}{3}\sin 3x + \frac{1}{5}\sin 5x + \ldots$$

   (a) Show that applying Fourier's methods would produce this series.

   (b) Can $\phi(x)$ be represented as a series of sines?

   (c) What function is actually represented by the preceding equation?

3. Find the Fourier series for $\phi(x) = x$.

---

[11]Dirichlet, G. P. L., *Werke*, 1889. Reimer. Vol. I, page 135.

4. Use Equation 20.18 to show that the Fourier series of a function converges to the function everywhere the function is continuous; moreover, show that, so long as the number of discontinuities of a function are finite, then the Fourier series converges to the function for all values of x except those at which $f(x)$ is discontinuous.

20.4 Transcendental Numbers

Joseph Liouville[12] (March 24, 1809-September 8, 1882) was the editor and founder of one of the more influential mathematical journals of the nineteenth century, the *Journal of Pure and Applied Mathematics*. Liouville began the journal in 1836, in response to a critical need: in 1831, Gergonne's *Annals of Pure and Applied Mathematics* ceased publication.

In the same year, the *Bulletin of the Mathematical, Astronomical, Physical, and Chemical Sciences* also ceased publication, and French mathematicians were left without a French journal in which to present their results. Thus, in January 1836, Liouville began publishing and editing the *Journal of Pure and Applied Mathematics*, and was its editor for almost forty years until he retired in 1874. (The *Journal* itself is still being published).

20.4.1 Transcendental Numbers

One of Liouville's more interesting results involved transcendental numbers. Numbers are said to be **algebraic of the nth degree** if there is an irreducible nth-degree equation with integer coefficients to which they are a root; if there is no such equation to which a number is a solution, the number is said to be **transcendental**. On April 28, 1729, Daniel Bernoulli, writing to Christian Goldbach, claimed that

$$\log \frac{m+n}{n} = \frac{n}{m} - \frac{nn}{2mm} + \frac{n^3}{3m^3} - \frac{n^4}{4m^4} + \ldots$$

in general represented a number that was inexpressible by rationals or even by roots of rational numbers. Goldbach wrote back, saying he suspected Bernoulli was right, but hoped a clear demonstration might be provided. Bernoulli was unable to provide one. Thus, the question of whether there were numbers which could not be the root of any equation remained open.

The beginnings of an answer came in May 1840, when Liouville proved

Proposition 20.2. *Neither e nor e^2 is the root of a quadratic equation.*

---

[12]The name *is* L-i-o-u-ville; this is not a typo.

Proof. Suppose e satisfies $ae + be^{-1} = c$ for some integers a, b, and c. Substitute

$$e = 1 + \frac{1}{1} + \frac{1}{1 \cdot 2} + \frac{1}{1 \cdot 2 \cdot 3} + \ldots$$

$$e^{-1} = 1 - \frac{1}{1} + \frac{1}{1 \cdot 2} - \frac{1}{1 \cdot 2 \cdot 3} + \ldots$$

and multiply by $1 \cdot 2 \cdot 3 \cdots n$; hence

$$\frac{a}{n+1}\left(1 + \frac{1}{n+2} + \ldots\right) \pm \frac{b}{n+1}\left(1 - \frac{1}{n+2} + \ldots\right) = \mu \qquad (20.19)$$

where μ is some integer. One can always ensure that both terms on the left are positive. But the result is absurd, for it claims that μ, an integer, is equal to the sum of two very small numbers. Hence, it cannot be that e satisfies $ae + \frac{b}{e} = c$ for some integers a, b, c; hence e cannot be the root of a quadratic equation. Similarly for e^2. $\qquad\square$

Of course, it would be tedious (and useless) to prove then that e could not be the solution to a cubic, a quartic, and higher-degree equation; the importance of Liouville's result was it suggested that there might be nonalgebraic numbers.

On May 13, 1844, Liouville presented to the French Academy the result of his researches, "On Very Extensive Classes of Quantities Whose Values Are Neither Rational nor Reducible to Algebraic Irrationals."

Let there be an irreducible polynomial function $f(x) = ax^n + bx^{n-1} + \ldots + h$, where a, b, ..., h are integers, and let x satisfy $f(x) = 0$. Let the continued fraction expansion of x be given by

$$x = b_0 + \cfrac{1}{b_1 + \cfrac{1}{b_2 + \cfrac{1}{\ddots \cfrac{}{b_m + \cfrac{1}{x_{m+1}}}}}}$$

In this expansion, x_{m+1} is called the **complete quotient**, b_m is the **incomplete quotient**, and if

$$\frac{p_m}{q_m} = b_0 + \cfrac{1}{b_1 + \cfrac{1}{b_2 + \cfrac{1}{\ddots \cfrac{}{b_{m-1} + \cfrac{1}{b_m}}}}}$$

then the fractions $\frac{p_m}{q_m}$ are called the **convergents**; by the work of Lagrange, it was known that the reduced convergents provided better rational estimates to x than any fraction with a lesser denominator.

See page 544.

Liouville introduced the function

$$F(p, q) = q^n f\left(\frac{p}{q}\right) = ap^n + bp^{n-1}q + \ldots + hq^n$$

If one ignored the sign, then the incomplete quotients $\mu\ [\ = b_m]$, for sufficiently large values of m, must satisfy $\mu < \frac{\partial F(p,q)}{qF(p,q)\partial p}$; Liouville did not include a proof of this statement, though it is true. Clearly, $F(p, q)$ must be a whole number, and unless $x = \frac{p}{q}$, $F(p, q)$ is nonzero. Thus, if x is not a rational number, then $\mu < \frac{\partial F(p,q)}{q\ \partial p} = q^{n-2} f'\left(\frac{p}{q}\right)$. As $\frac{p}{q}$ tends to the root x, then $f'\left(\frac{p}{q}\right)$ tends to $f'(x)$, which may be assumed to be a finite and nonzero; thus for sufficiently large q, we may assume there exists $A > f'(x)$, and thus $\mu < Aq^{n-2}$. This leads to the important proposition:

Proposition 20.3. *If x is algebraic, then the incomplete quotients of the continued fraction representation of x cannot exceed a product of some constant number and the $n - 2$ power of the denominator of the previous convergent.*

As an example of a transcendental, Liouville gave the number expressed by the series

$$\frac{1}{a} + \frac{1}{a^{1\cdot 2}} + \frac{1}{a^{1\cdot 2\cdot 3}} + \frac{1}{a^{1\cdot 2\cdot 3\cdot 4}} + \ldots$$

where a is a whole number.

At the next meeting of the Academy of Sciences, May 20, 1844, Liouville gave a different proof of the existence of transcendentals. Again, letting $f(x)$ be an nth degree polynomial, and $F(p, q) = f\left(\frac{p}{q}\right)$, let x be a real root to $f(x) = 0$, and suppose the other roots $x_1, x_2, \ldots, x_{n-1}$ are distinct. Using the Fundamental Theorem of Algebra and supposing $\frac{p}{q}$ is one of the convergents, Liouville obtained

$$\frac{p}{q} - x > \frac{1}{Aq^n} \tag{20.20}$$

for some A (the absolute value of both sides is assumed), leading to the same result.

These two proofs provide a glimpse into the changing nature of mathematics, from the study of objects to the study of concepts. Liouville's first proof is more difficult and less general, for it required consideration of the continued fraction expansion of x. However, finding a continued fraction expansion of x was a clearly established algorithm that would work for all polynomials $f(x)$ with positive real roots, and moreover, the proof itself showed how to create a transcendental number. The second proof began by *assuming* the existence of a factored form of $f(x)$, which clearly existed but in general could not be found for $n \geq 5$; moreover, though it provided a criterion for determining whether a given number x was algebraic, the proof could in no way be considered constructive: it was a pure existence proof.

It was not until June 8, 1845 that Liouville demonstrated the transcendence of a known number. In "Memoir on the Classification of Transcendentals," he showed that $\ln x$ generated transcendental numbers if x was rational (with the exceptions of the obviously trivial cases such as $\ln 1$). Liouville begins the proof with

Proposition 20.4. $\ln x$ *cannot be expressed as a rational function with integer coefficients.*

From this Liouville derived

Proposition 20.5. *If* $\ln x = y$ *and* x *is rational, then* y *is transcendental.*

Proof. We note $\int \frac{1}{x} dx = \ln x$. Suppose $y = \ln x$ was algebraic and x rational. Hence y is the root to some equation $f(x, y) = 0$, where

$$f(x, y) = Py^n + Qy^{n-1} + \ldots + Ry + S$$

and the coefficients P, Q, $\ldots$ are rational functions of x (and thus rational numbers, since x was assumed rational); we may suppose that $f(x, y)$ is an irreducible polynomial, and that y_1, y_2, $\ldots$, y_n are the n roots of $f(x, y)$.

From $\int \frac{dx}{x} = y$, we obtain $\frac{dx}{x} = dy$. Differentiating $f(x, y) = 0$ with respect to x, and letting $\frac{dy}{dx} = \frac{1}{x}$, we obtain

$$x \, f_x(x, y) + f_y(x, y) = 0$$

This equation must obviously hold for all the roots, y_1, y_2, $\ldots$, y_n. Considering the roots as functions of x, we thus have

$$\frac{dx}{x} = dy_1 = dy_2 = dy_3 = \ldots = dy_n$$

Thus:

$$\int \frac{dx}{x} = \frac{y_1 + y_2 + \ldots + y_n}{n}$$
$$= \frac{-Q}{n}$$

(since $y_1 + y_2 + \ldots + y_n = -Q$, the coefficient of the y^{n-1} term in the polynomial). Hence, $\int \frac{dx}{x}$ is equal to a rational function. But this is impossible, because $\int \frac{dx}{x} = \ln x$, which could not equal a rational function; thus $y = \ln x$ cannot be algebraic if x is rational. $\square$

From this, it is simple to show that e^x cannot be rational if x is algebraic; in particular, e itself is irrational. In a similar manner, Liouville showed $\ln(\ln x)$ could not be expressed as a rational function, and thus (again excepting the trivial cases) $\ln(\ln x)$ produced transcendental numbers.

20.4.2 Fermat's Last Conjecture

Liouville also found himself in the middle of an important, but flawed, attempt to prove Fermat's last conjecture. Progress had been made by Dirichlet (for $n = 5$), and Sophie Germain. In 1839, Gabriel Lamé (July 22, 1795-May 1, 1870) proved the impossibility for $n = 7$; the result appeared in Liouville's journal for 1840. Then in April of 1847, Lamé attempted to prove Fermat's last conjecture in the following way: if $x^n + y^n = z^n$, let the left-hand side be factored among the complex numbers as

$$(x + y)(x + ry)\left(x + r^2 y\right) \cdots \left(x + r^{n-1} y\right) = z^n$$

where -1, $-r$, $-r^2$, ... are roots of $X^n + 1 = 0$. From this result, Lamé derived Fermat's last conjecture. Most French mathematicians hailed this brilliant triumph of their countryman.

Liouville was almost alone among his countrymen in his hesitation to accept Lamé's proof. He pointed out that Lamé assumed the fundamental theorem of arithmetic: in other words, any composite number of the form $a + br$, where r is a primitive root of unity, had a unique factorization into primes.[13] Though Gauss had proven this where r was a primitive third root and a primitive fourth root, a general proof had not been found. Indeed, among the generalized complex numbers, it was easy to find counterexamples to unique factorization; Liouville himself noted that if $r = \sqrt{-17}$, then 13 was prime, since it could not be expressed as a product of numbers of the form $a + \sqrt{-17}b$, but

$$13 \cdot 13 = 169 = \left(4 + 3\sqrt{-17}\right)\left(4 - 3\sqrt{-17}\right)$$

hence 169 did not have a unique prime factorization. This was not necessarily a disproof of Lamé, since $\sqrt{-17}$ was not a root of unity, but it suggested that the unique factorization might fail for general complex whole numbers.

At least part of a proof of unique factorization would involve showing that the Euclidean algorithm for finding the greatest common factor of two (complex) whole numbers terminated; in 1847, Pierre Wantzel, a French mathematician, proved that the Euclidean algorithm worked for the Gaussian integers $a + bi$, and for $a + b\alpha + c\alpha^2$ where α is a primitive third root of unity. Based on these results, Wantzel conjectured that the Euclidean algorithm would always work, but later that year, Cauchy disproved this conjecture. Still, hope was held out for unique factorization.

Liouville was virtually alone among French mathematicians in believing the fundamental theorem of arithmetic did not hold, and his situation is indicative of an important change in mathematics: the growing preeminence of German mathematicians. Liouville, alone among his French contemporaries, maintained close contact with German mathematicians, particularly Dirichlet, whose wife was the cousin of Ernst Eduard Kummer (January 29, 1810-May 14, 1893), born in Sorau, Germany (now Zary, Poland).

---

[13]More generally, the numbers, both composite and primes, were assumed to be of the form $a + br + br^2 + \ldots + br^{n-1}$, where r is a primitive nth root of unity.

As early as 1844, Kummer realized that unique factorization did not hold for the generalized complex whole numbers; indeed, before Lamé's work appeared, Kummer published (1846) a proof to that effect in the proceedings of the Berlin Academy of the Sciences. Kummer's discovery is remarkable, because for numbers of the type $a+b\alpha+c\alpha^2+\ldots+p\alpha^{n-1}$, where α is a primitive nth root of unity, unique factorization *holds* for $n = 2$, 3 and 4; all the counter-examples of unique factorization involved complex numbers of other forms, which might reasonably be said to not represent "proper" complex whole numbers. But Kummer had worked with $n = 23$, for which unique factorization failed even for numbers of the form $a + b\alpha + c\alpha^2 + \ldots + p\alpha^{n-1}$.

Rather than working with $n = 23$, we take an example from the later work of Dedekind: consider numbers of the form $a + b\sqrt{-5}$. For these complex whole numbers, 3 is prime, but $9 = 3^2 = (2 + \sqrt{-5})(2 - \sqrt{-5})$, and thus 9 can be factored in two distinctly different ways. To save unique factorization, Kummer introduced the notion of **ideal factors** (which he never formally defined); in essence, these were new numbers specifically introduced to restore unique factorization. It was not necessary to actually *find* these ideal factors; it was sufficient to assume their existence. Kummer compared the ideals to the way contemporary chemists used the element fluorine. In Kummer's time, chemists knew of the existence of fluorine, by virtue of the compounds it formed, but had not yet succeeded in isolating the highly toxic element (though many chemists had tried, failed—and died).[14] In the preceding case, we may let $a = \frac{\sqrt{10}+\sqrt{-2}}{2}$ and $b = \frac{\sqrt{10}-\sqrt{-2}}{2}$; thus $3 = ab$, $2 + \sqrt{-5} = a^2$, $2 - \sqrt{-5} = b^2$, and thus unique factorization is restored.

See page 449.

In 1847, Kummer proved Fermat's last conjecture for primes λ that did not divide any of the first $\frac{\lambda-3}{2}$ Bernoulli numbers; these primes are now called **regular primes**.[15] Kummer conjectured there were infinitely many such primes, and thus announced his proof of Fermat's last conjecture for "infinitely many" prime numbers. In fact, it is not yet known whether the number of regular primes is finite or infinite.

In 1850, the Paris Academy offered a prize for the proof of Fermat's last conjecture. No proof was forthcoming, but in 1857, the prize committee, including Lamé and Cauchy, awarded the prize to Kummer (who had not submitted an entry) for his work.

20.4 Exercises

1. Complete Liouville's proof that e cannot be the root of a quadratic equation.

   (a) How can we ensure both terms in equation 20.19 are positive?

   (b) Why does the impossibility of $ae + \frac{b}{e} = c$ mean that e cannot be the solution to a quadratic equation with rational coefficients?

---

[14]Elemental fluorine was not isolated until 1886, by Henri Moissan, for which he won the Nobel Prize in 1906.

[15]The smallest irregular prime is 37.

(c) Show that e cannot satisfy $ae^2 + be^{-2} = c$, and thus e^2 is not a quadratic irrational.

2. Show that

$$\frac{1}{a} + \frac{1}{a^{1 \cdot 2}} + \frac{1}{a^{1 \cdot 2 \cdot 3}} + \frac{1}{a^{1 \cdot 2 \cdot 3 \cdot 4}} + \cdots$$

does not satisfy the Liouville criterion for an algebraic number.

3. Show that from a polynomial with rational coefficients $g(x)$ with a root x of multiplicity m, it is possible to form another polynomial with rational coefficients $f(x)$ where the root x has multiplicity 1; in general, given any polynomial, it is possible to reduce it to another polynomial whose roots are distinct. Hint: if x is a root of $f(x)$ with a multiplicity $n > 1$, x is also a root of $f'(x)$.

4. Prove Equation 20.20. Hint: use the Fundamental Theorem of Algebra; remember that $F(p, q)$ is nonzero unless $x = \frac{p}{q}$.

5. Prove Proposition 20.4.

   (a) Suppose it can, and that $\ln x = \frac{X}{Y}$, where X and Y may be supposed relatively prime. Differentiate, and show that $Y = x^n Z$ for some whole number n and some polynomial Z which is not divisible by x.

   (b) Substitute $Y = x^n Z$ into the expression for the derivative of $\ln x$, and show that x must divide X, which contradicts the assumption that X and Y had no common factors.

6. Prove $\sqrt{x}$ cannot be expressed as a rational function of x.

7. Prove e^x cannot be expressed as a rational function of x.

8. Prove that the trigonometric functions cannot be expressed as rational functions.

9. Prove $\ln \ln x$ cannot be expressed as a rational function.

20.5 Chebyshev

Pafnuty Lvovich Chebyshev (May 16, 1821-December 8, 1894) holds, perhaps, the record for the number of variant spellings of his last name. Besides Chebyshev (which is currently the preferred spelling in English) one may find Chebychev, Chebichev, Čebišev, and even Tchebycheff. The variant spellings owe their origin to the attempt to write a Russian name (with an unambiguous spelling in the Cyrillic alphabet) in the letters of the Roman alphabet so readers in English, French, German, or other languages would read it as having a sound similar to the original Russian. Chebyshev, the son of a Russian officer who fought against Napoleon, sucessfully defended his master's dissertation in 1846, "Essay on an Elementary Analysis of the Theory of Probability."

On the strength of his performance, he was invited to Petersburg University in 1847. Euler had been dead for more than half a century, but the St. Petersburg Academy still had a backlog of Euler works. At the instigation of fellow academician Viktor Yakovlevich Bunyakovskii (1804-1889), Chebyshev helped to prepare Euler's works on number theory for publication. Jacobi helped, digging through the archives of the Berlin Academy of Science, and in 1849, two volumes of Euler's works on number theory appeared, with Chebyshev's corrections and reconstructions. In the same year, Chebyshev defended his doctoral dissertation, and in 1850, was appointed professor at the University.

Chebyshev's students came to be known the Chebyshev or St. Petersburg School. Many of them are familiar names from advanced mathematics: Markov, Lyapunov, Voronoi, Krylov. For Chebyshev, the overriding aim of mathematicians was to produce applicable mathematics.

20.5.1 Prime Numbers

For example, it was not sufficient to simply prove a function had a limit or even to find the limit; it was important to find how much the function deviated from the limit, and even to find out how the function itself approached the limit. We see this pragmatic approach in one of Chebyshev's early works, "On the Function Which Determines the Number of Primes Less Than a Given Limit," which appeared in the journal for the St. Petersburg Academy of 1849. In the article, Chebyshev considered the number of primes less than x.

See Problem 2.

Euclid had shown that the number of primes was infinite, but it is easy to show that the distance between prime numbers can be made as large as desired. Hence, the question naturally arises as to the distribution of primes. As early as 1791, Gauss wrote a cryptic note in his diary that can be interpreted as a conjecture that the number of primes less than x was approximately $\frac{x}{\ln x}$. The first to publicly air a conjecture was Legendre in 1798; based on an empirical study of the number of primes, he suggested the formula $\frac{x}{A \ln x + B}$, where $A = 1$ and $B = -1.08366$. In a letter to the astronomer Encke in 1849, Gauss hypothesized that an estimate for the number of primes less than x was $\int_2^x \frac{dx}{\ln x}$. Chebyshev was able to prove

Proposition 20.6. *If $\phi(x)$ is the number of primes less than x, then $\phi(x)$ satisfies the inequalities*

$$\phi(x) > \int_2^x \frac{dx}{\ln x} - \frac{\alpha x}{\ln^n x} \qquad \phi(x) < \int_2^x \frac{dx}{\ln x} + \frac{\alpha x}{\ln^n x}$$

infinitely many times as x tends to infinity, regardless of the smallness of α or the largeness of n.

The proof is very straightforward, using only elementary (albeit somewhat tedious) calculus. From this, Cheybshev derived:

Proposition 20.7. *$\frac{x}{\phi(x)} - \ln x$ cannot have a limit different from -1 as x tends to infinity.*

20.5.2 Probability

An essential part of the mathematical theory of probability is **Chebyshev's inequality**, which first appeared in "Theorem Concerning Mean Values" (1867):

Proposition 20.8 (Chebyshev's Inequality). *If a, b, c, . . . are the expectations of x, y, z, . . ., and a_1, b_1, c_1, . . . are the expectations of x^2, y^2, z^2, . . ., then the probability the sum $x + y + z + \ldots$ is between*

$$a + b + c + \ldots + \alpha\sqrt{a_1 + b_1 + c_1 + \ldots}$$

and

$$a + b + c + \ldots - \alpha\sqrt{a_1 + b_1 + c_1 + \ldots}$$

is greater than $1 - \frac{1}{\alpha^2}$, regardless of the value of α.

To prove this, Chebyshev used the notion that x (y, z, . . .) could take on values x_1, x_2, . . ., with probabilities p_1, p_2, . . .; in other words, x, y, . . . were **random variables** in the modern sense.[16] Though Poisson had used them earlier, Chebyshev made them a key part of probability.

Chebyshev's philosophical inclinations toward the applicable caused him to prove the result using sums. The proof is longer and less general, since x, y, z, . . . can take only discrete values. However, it requires only simple algebra. We present a slightly simplified version of the proof.

Proof. Let a, b be the expectations of x, y, and a_1, b_1 be the expectations of x^2, y^2. Let x_1, x_2 be the only possible outcomes of x with probabilities p_1, p_2 respectively; likewise, let y_1, y_2 have probabilities q_1, q_2; thus

$$p_1 + p_2 = 1 \qquad\qquad q_1 + q_2 = 1 \qquad (20.21)$$
$$p_1 x_1 + p_2 x_2 = a \qquad\qquad q_1 y_1 + q_2 y_2 = b \qquad (20.22)$$
$$p_1 x_1^2 + p_2 x_2^2 = a_1 \qquad\qquad q_1 y_1^2 + q_2 y_2^2 = b_1 \qquad (20.23)$$

The expression

$$\frac{(x_\lambda + y_\mu - a - b)^2}{\alpha^2 (a_1 + b_1 - a^2 - b^2)} p_\lambda q_\mu \qquad (20.24)$$

summed over all λ and μ equals $\frac{1}{\alpha^2}$ (See Problem 1a).

Suppose we reject the terms in the summation where the coefficient of $p_\lambda q_\mu$ is less than 1, and replace the coefficients of $p_\lambda q_\mu$ greater than 1 with 1; obviously, the sum will decrease and be less than $\frac{1}{\alpha^2}$. But the sum will then represent the probability that

$$\frac{(x + y - a - b)^2}{\alpha^2 (a_1 + b_1 - a^2 - b^2)} \geq 1 \qquad (20.25)$$

---

[16] For example, let x be the random variable that represents the number of spots showing upon the rolling of a fair die. Then x could take on the values 1 through 6, each with probability $\frac{1}{6}$.

Let P be the probability that x, y, does not satisfy inequality 20.25; hence P is the probability that $x + y$ is between $a + b + \alpha\sqrt{a_1 + b_1 - a^2 - b^2}$ and $a + b - \alpha\sqrt{a_1 + b_1 - a^2 - b^2}$. But $1 - P < \frac{1}{\alpha^2}$, and thus $P > 1 - \frac{1}{\alpha^2}$, as was to be proven.   □

A few corollaries follow immediately.

Proposition 20.9. *If the quantities x, y, ... are N in number, let $\alpha = \frac{\sqrt{N}}{t}$. Then the probability that the difference of the mean of the quantities $\frac{x+y+z+\cdots}{N}$ and the mean of the expectations $\frac{a+b+c+\cdots}{N}$ will not exceed*

$$\frac{1}{t}\sqrt{\frac{a_1 + b_1 + c_1 + \dots}{N} - \frac{a + b + c + \dots}{N}}$$

is greater than $1 - \frac{t^2}{N}$, whatever the value of t.

The quantity inside the radical is the difference between the mean of the squares of the expectations and the mean of the expectations themselves, and so long as the expectations are below some given finite number, the difference is also finite; hence, by taking a suitably large value of t, the probability that the difference between the arithmetic mean and the mean of the expectations can be made as small as desired.

If the random variables $U_1, U_2, U_3, \dots U_N$ take on only the values of 1 or 0, depending on whether the event E occurs or not on the first, second, third, etc. trials with probabilities $P_1, P_2, P_3, \dots, P_N$, then $U_1 + U_2 + \dots + U_N$ gives the number of times the event occurred in N trials, and thus $\frac{U_1+U_2+\dots+U_N}{N}$ gives the ratio of the number of occurrences to the number of trials. Combined with the previous result, this gives

Proposition 20.10. *The probability that the ratio $\frac{U_1+U_2+\dots+U_N}{N}$ (i.e., the ratio of the number of occurrences to the number of trials) differs from the mean of the probabilities $P_1, P_2, \dots, P_N$ tends to 1 as N tends to infinity.*

If $P_1 = P_2 = P_3 = \dots = p$, then we obtain Bernoulli's Law of Large Numbers.

Chebyshev's result assumed that the events were independent. But in many cases, the events are dependent, and the probability of an event depends on the occurrence or nonoccurrence of another event. The study of these dependent events, and the corresponding law of large numbers, was derived in 1907 by Markov.

20.5 Exercises

1. Consider Chebyshev's inequality (Proposition 20.8.)

   (a) Prove Equation 20.24 by showing the sum of

   $$\left(x_\lambda + y_\mu - a - b\right)^2 p_\lambda q_\mu$$

   over all possible values of λ and μ is equal to $a_1 + b_1 - a^2 - b^2$.

(b) Prove Chebyshev's inequality in general.

2. Show that it is possible to construct arbitrarily long sequences of composite numbers. Hint: $n!$ is divisible by all numbers from 1 to n.

3. Prove Proposition 20.10, and derive Bernoulli's law of large numbers from it.

Chapter 21

Geometry

One of the tenets of political policy in continental Europe after Napoleon was that all the problems of the era were due to liberalism, and autocratic control of all aspects of society was essential. The reaction inevitably caused revolts, but the balance of power was with the autocrats. Austria crushed liberal revolts in Italy, and in 1823, a liberal revolution in Spain was put down—by French troops. Charles X attempted to turn back the clock and restore the nobility and monarchy to the power and prestige they had before the Revolution, but to no avail. He was forced to abdicate in 1830.

Rather surprisingly, the one place on the continent where liberal notions had the greatest foothold was in the Germanic Confederation, particularly Prussia. A consequence of the division of the region of Germany into hundreds of city-states was that each city-state had its own university. During the course of the century, more and more mathematical and scientific discoveries came from the liberalized German universities, and fewer and fewer from more tightly controlled French ones.

21.1   Analytic and Projective Geometry

Analytic and projective geometry had been developed at the beginning of the seventeenth century, but were almost immediately eclipsed by the development of calculus, and they remained in the mathematical background. It was not until the beginning of the nineteenth century that the study of projective geometry returned as an important part of mathematics, mainly through the work of the French and, in particular, through the influence of the Polytechnic School in Paris.

21.1.1 Chasles

Napoleon returned from his Russian campaign with the armies of Europe on his heels. The remaining students of the Polytechnical School (many of whom had already joined the army) joined to help with the defense of Paris. One of these students was Floreál Chasles (November 15, 1793-December 18, 1880). Chasles was named after one of the months in the ill-fated French Republican calendar, which sought to rationalize time by dividing the months into three ten-day weeks. On January 1, 1806, Napoleon I decreed an end to the Republican calendar, to great popular relief, and when he was old enough, Floreál changed his name to Michel, the name by which he is best known.

Chasles was a late bloomer who did not make his first fundamental contributions to mathematics until the age of 44. Chasles helped to further develop projective geometry. In the projection of a figure onto another plane, intersections and straight lines are preserved, but in general angles, areas, and other Euclidean properties are not. However, six quantities, known as **anharmonic** or **cross ratios**, are invariant, and Chasles made them the basis of projective geometry. The first cross ratio Chasles introduced was:

> Four points a, b, c, d, on a straight line, being taken two at a time, give rise to six segments; the expression $\frac{ac}{ad} : \frac{bc}{bd}$ formed by the six segments is called the *anharmonic ratio*
>
> . . .
>
> When four lines A, B, C, D, lying in the same plane, pass through the same point, the *anharmonic ratio* of the four lines is the expression $\frac{\sin(A,C)}{\sin(A,D)} : \frac{\sin(B,C)}{\sin(B,D)}$, formed by the sines of four of the six angles formed by the lines taken two at a time.[1]

By $\sin(A, C)$, Chasles meant the sine of the angle whose sides were lines A and C. These two cross ratios are not essentially different, for:

Proposition 21.1. *If through four points a, b, c, d on a line are passed four lines A, B, C, D, all passing through the same point O, the anharmonic ratio of the four points is equal to the anharmonic ratio of the four lines.*

Hence

Theorem 21.1. *When four points are on a given line, if one takes a projection of these four points onto another plane, the cross ratio is invariant.*

Chasles also began the study of so-called **enumerative geometry**, now a branch of algebraic geometry: in essence, the determination of the number of figures of a given type that satisfy a given relationship, which in turn help to determine the number of solutions to a given equation. Chasles's other contributions included a long-standing belief that that synthetic geometry could be as productive as algebra in its discovery of new results. Moreover, he felt, the

---

[1]Chasles, Michel. *Traité de géométrie supérieure*, 1880. Gauthier, page 7.

proofs of synthetic geometry were more obvious and more clear than those of algebra.

21.1.2   Steiner

This last belief in the supremacy of synthetic geometry over algebraic analysis was echoed by Jakob Steiner (March 18, 1796-April 1, 1863), sometimes called the greatest synthetic geometer of modern times. His will established the Steiner Prize of the Berlin Academy. Steiner, who was so poorly educated he did not learn to read until he was 14, proved that all Euclidean constructions could be done using straightedge alone, provided one is given a single fixed circle.

One of Steiner's unpublished discoveries would find an echo many years later in the work of Cantor: consider a circle of any radius r, however small, with center O. To any other point P inside the circle can be associated a point P', outside the circle and on extension of the line OP, by the relationship $OP \cdot OP' = r^2$; it is easy to find the point P' using simple Euclidean methods; likewise, given any point P' outside the circle, it was always possible to find a point P inside the circle to which it corresponded. Steiner did not publish his result, possibly because it suggested a remarkable paradox: every point inside the circle *except* for the center O was associated with a point outside the circle, and thus, the circle appeared to have *more* points than the entire plane outside of it.

In 1832, Steiner published *Systematic Introduction*, which built up projective geometry based on measurements. Unfortunately, this highlighted a key problem with projective geometry: it still depended on the ideas of distance, perpendiculars, and angles, but these quantities are not invariant. Even though the cross ratio is invariant, it still refers to the distances and angles. Complete freedom from metric notions came with the work of Staudt and, to some extent, Möbius.

21.1.3   Möbius

See page 714.

The history of mathematics is filled with figures like Galois and Gauss, who made profound discoveries by the time they were 20. August Ferdinand Möbius (November 17, 1790-September 26, 1868) shows that mathematical genius need not manifest itself early. Möbius's mother was a descendant of Luther; his father was a dance master at a school for the nobility at Schulpforta, Saxony (now part of Germany). Möbius had the misfortune to study with Gauss, which meant that he learned a great deal of observational astronomy and almost no mathematics. Instead, it was through the study of the French geometers that Möbius acquired an interest in the subject.

Möbius made few important contributions to mathematics until he was 37, when he published *The Barycentric Calculus* (1827). The "barycenter" of the title is simply the center of gravity of an object. Throughout the work, Möbius made use of the idea of a directed line segment: AB was the line segment *from A to B*, and thus $AB = -BA$. Though he did not invent the concept, he was the first to make use of it consistently and fruitfully.

Möbius began by considering two points in the plane, A and B; with a, b any real numbers. Through A, B, let parallel lines be given; it was desired to find points A', B' on the parallel lines so that $a\,AA' + b\,BB' = 0$. To find these lines, Möbius proved:

Proposition 21.2. *Let AB be drawn; find P so that $AP : PB = b : a$. Draw any line through P; its intersection with the parallel lines will be A', B' that satisfy $a\,AA' + b\,BB' = 0$.*

P is known as the **centroid**. We may consider P in another way: if weights a, b are placed at A, B respectively, then the center of gravity of the system will be at P. Möbius extended the notion to as many points A, B, C, ... as desired. If S is the centroid of the points A, B, C with coefficients a, b, $-c$, Möbius wrote the equation

$$aA + bB - cC = (a + b - c)S$$

If

$$aA + bB + cC = 0$$

then the points A, B, C had no finite centroid.

An important idea introduced by Möbius was the **Möbius net**: given any four points A, B, C, D in a plane, connect each pair of points with a line, and let the intersections of the lines formed be A', B', C'. Connect each pair of the points A', B', C', which form six additional intersections, A'', B'', C'', F, G, H. These points of intersection can be connected to each other and to the points already determined, to form new lines, forming new intersections, and so on.

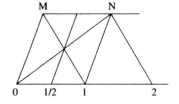

The importance of the Möbius net may be explained as follows. Consider two points on the Cartesian x axis at 0 and 1 (see figure in margin); through these points, draw arbitrary lines that intersect at M; through M draw MN parallel to line 01, and through 1 draw $1N$ parallel to $0M$. From the point N, which can be viewed as being determined by the position of points 0, 1, and M, draw a line parallel to $1M$; this intersects the line 01 at 2. Join $0N$ and through the intersection of $0N$ and $1M$, draw a line parallel to $0M$; this intersects the x axis at 1/2. By repeatedly joining points and drawing parallels, we can locate exactly *any* point corresponding to a rational value of x. If we consider a projection of this "net," the intersections and the ratios of the line segments are preserved.

The best known of Möbius's discoveries was made in September 1858, when he was 68: by taking a strip of paper, and giving it a half-twist before joining the two ends, he created a finite object with *one* edge and *one* side, now called a **Möbius strip**. The idea was certainly "in the air," for in July, Johann Benedict Listing (1806-1882) made the same realization, and published an article on the subject in 1861. It was, however, Möbius's 1865 article, "On the Determination of the Volume of Polyhedra" that familiarized most mathematicians to the existence of polyhedra for which volume and surface area had no clear meaning.

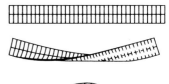

A Möbius strip

21.1.4 Plücker

Julius Plücker (June 16, 1801-May 22, 1868) was a professor at the University of Bonn. In Crelle's Journal, Plücker presented, "On a New Coordinate System" (1830). Cartesian coordinates in two dimensions rely on the distance of a point from two lines. Plücker's new system, which he called **homogeneous variables**, utilized the distances p, q, r of a point from three lines, OO', OO'', and $O'O''$; in a sense, Möbius's barycentric coordinates anticipated Plücker's homogeneous coordinates. The name homogeneous came from the fact that a homogeneous equation of the nth degree in the variables p, q, r (i.e., all terms were of the form $Ap^i q^j r^k$ where i, j, k are positive integers or zero, and $i + j + k$ is a constant) could represent any nth degree equation in the variables ϕ and ψ. In particular, the equation of a line $p + aq + br = 0$ could be reduced to $1 + a\phi + b\psi = 0$.

Plücker also noted another useful feature. The general equation of a line in Cartesian coordinates is $ax + by + c = 0$, which has three parameters but only two variables; hence, the principle of duality is difficult to accept, as it would entail the relationship of *two* numbers (the coordinates of the point) to *three* numbers (the parameters of the line). With homogeneous coordinates, however, we have $ap + bq + cr = 0$, and thus the three homogeneous coordinates (p, q, r) can be matched to the three parameters (a, b, c). Thus, homogeneous coordinates allow for the analytic counterpart to the principle of duality.

21.1.5 Staudt

Karl Georg Christian von Staudt (January 24, 1798-June 1, 1867), like Möbius, studied with Gauss at Göttingen.

Staudt came to Erlangen in October 1835. In 1847, Staudt published *The Geometry of Position* and, in 1856, *Contributions to the Geometry of Position*. According to Felix Klein, whose work will be dealt with elsewhere, these works contained an "extraordinary wealth of ideas ... rigidified almost to the point of lifelessness."

Staudt introduced the "throw": given three fundamental points, the throw that P makes with each of these points is the coordinate of P. The throw, a pure number, is independent of distance; thus, Staudt removed the last connection of projective geometry to the Euclidean metric, and established projective geometry as a pure study of invariant quantities.

21.1 Exercises

1. Explain why $aA + bB + cC = 0$ implies that the points A, B, C have no finite centroid.

2. Prove that the line through N parallel to $1M$ intersects the x-axis at 2.

3. Explain in general how to locate the point at $x = \frac{p}{q}$, for p, q whole numbers.

4. Show how a homogeneous equation of the nth degree in three variables can be reduced to an inhomogeneous equation of the nth degree in two variables.

5. Consider a matrix as representing a linear transformation of 2 variables (p, q). Find a 2×2 matrix representing:

   (a) A reflection across the x axis.

   (b) A reflection across the y axis.

   (c) A rotation of θ degrees.

   (d) Can a 2×2 matrix represent a translation? Explain.

   (e) How can homogeneous coordinates be used to allow a matrix to represent a translation? Remember that the homogeneous coordinates (p, q, r) represents a point in *two*-dimensional space.

6. Explain why "Two lines intersect in exactly one point" implies "Two points are contained on exactly one line," using homogeneous coordinates.

7. Given a circle with center O of radius r and a point P inside the circle and not the center, explain how to construct P' on line OP so that $OP \cdot OP' = r^2$; also, given P' outside the circle, how would you find P inside the circle and on the line OP' to satisfy this same relationship.

21.2 Differential Geometry

Monge had begun what would become differential geometry while teaching at the Polytechnical School, but Gauss's "General Investigations on Curved Surfaces," presented to the Göttingen Royal Society on October 8, 1827, brought differential geometry to a mature form. Gauss, drawing from his work in astronomy, began with a sphere of unit radius situated at an arbitrary point. A point on the sphere represented the same direction as a straight line parallel to the radius drawn to the point. The angle between two lines thus corresponded to the angular measure between the two points on the sphere representing the directions of the two lines; a plane was represented by the great circle parallel to the plane; and the angle between two planes was the spherical angle between the two great circles representing the planes.

An important concept in differential geometry is the **curvature of a surface**. Gauss described a point with **continuous curvature** as one where the tangent plane is "infinitely" close to the surface (which is to say that the tangent plane approximates the surface). The normal to the plane is thus the normal to the surface at the point, and so the point itself can be identified with the point on the unit sphere having the same normal. With the identification of a point on a surface to a point on the sphere, Gauss could then map a curve on the surface to a corresponding curve on the sphere, and then define

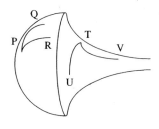

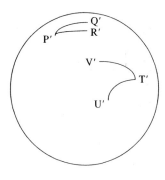

Definition. *The **total curvature** of a region on a curved surface is the area of the corresponding region on the unit sphere.*

Hence, for example, the total curvature of a sphere is 4π, while the total curvature of a cylinder or a plane is 0. To speak of the curvature at a point, Gauss introduced

Definition. *The **measure of curvature** is the curvature of a surface element divided by the area of the element.*

The measure of curvature could be positive (if curves passing through the point P are "similarly placed") or negative (if they were "inversely placed"). For example, the curves PQ and PR on the surface are mapped onto the curves $P'Q'$ and $P'R'$ on the sphere: since the lines are "similarly placed" in the sense that they contain a region that is to the "right" of $P'Q'$ and to the "left" of $P'R'$, the measure of curvature at P is positive. On the other hand, since TU and TV are mapped onto curves $T'U'$ and $T'V'$ of opposite orientation, the measure of curvature at T is negative.

To calculate the measure of curvature at a point, Gauss used the following idea: the section formed by intersecting the surface with a plane through the point was a plane curve whose radius of curvature at the point could be determined. The shape of the plane curve (and hence the radius of curvature) varied, depending on the position of the intersecting plane. Consider the greatest and least of these radii. First, Gauss proved that the planes producing the curves with greatest and least curvature were perpendicular. Then he proved:

Proposition 21.3. *The measure of curvature at a point P is a unit fraction whose denominator is the product of the greatest and least radii of curvature of the sections of the surface through P.*

Cartesian coordinates rely on a grid imposed on the plane; a point is specified by where it lies with respect to the grid lines. In a like manner, Gauss suggested a grid be imposed on the surface, and thus points on the surface received coordinates p and q, corresponding to the particular grid lines on which they happened to lie: along any grid line, p or q would take on a constant value. In Cartesian space, the element of distance was $dx^2 + dy^2 + dz^2$, while on the surface, the distance element was $A\,dp^2 + B\,dp\,dq + C\,dq^2$, where the coefficients A, B, and C depended on the general shape of the surface.

21.2 Exercises

1. Explain from the definition why the total curvature of a sphere of any radius is 4π.

2. Prove the following.

   (a) The total curvature of a plane is 0.

   (b) The total curvature of any cylinder is 0.

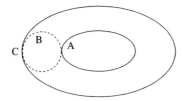

Figure 21.1: Figure for Problems 3 and 4.

    (c) The total curvature of an ellipsoid is 4π.

    (d) The total curvature of the frustrum of a cone is 0.

    (e) The total curvature of a paraboloid $z = x^2 + y^2$ is less than 2π. Hint: map the portion of the paraboloid below $z = M$ to the sphere.

3. What is the sign of the measure of curvature on the indicated surfaces at the indicated points?

    (a) A saddle. For example, the surface given by $z = y^2 - x^2$, at the point $(0, 0, 0)$.

    (b) Point A on the torus in Figure 21.1.

    (c) Point C on the torus in Figure 21.1.

    (d) A point on the surface generated by rotating a hyperbola $xy = 1$ about the x axis.

4. Suppose the torus in Figure 21.1 is formed by revolving the circle of radius r about the origin at a distance of $R > r$. Determine the measure of curvature at points A and C.

21.3 The Fifth Postulate

Since the time of Euclid, the fifth or "parallel" postulate had been a source of contention. Various attempts had been made to prove the fifth postulate, none of which were successful: the proofs that did not have obvious flaws assumed implicitly a postulate equivalent to the fifth. The most sophisticated attempt was that of Nasir Eddin. Wallis had in fact published a translation of Nasir Eddin's See page 266. work, which thus became available to European scholars, and, after criticizing Nasir Eddin's implicit assumption of the fifth postulate, proceeded to give his own proof, based on the axiom

Axiom. *Given any figure, it is possible to find another figure similar to it and having any desired magnitude.*

    Legendre made a similar assumption in the various editions of his *Elements of Geometry*, one of the standard textbooks in Euclidean geometry. However, the variant fifth postulate most commonly used today was introduced in 1795 by John Playfair (March 10, 1748-July 20, 1819), a Scottish mathematician. Playfair's version of the *Elements* included a number of innovations, such as the use of algebraic notation to shorten the proofs. Most importantly, he gave an alternative to Euclid's fifth postulate:

Axiom. *Given a line l and a point P not on the line, there is one and only one line in the plane of P and l that passes through P and is parallel to l.*

Although Playfair himself said he derived the axiom from Proclus, the axiom is now called **Playfair's axiom**.

21.3.1   Saccheri

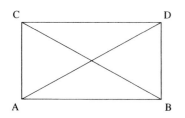

Hope was still held out that the fifth postulate might be provable from the other four. After a perusal of Wallis's translation, Giralamo Saccheri (1667-1733) tried a new tactic. Rather than prove the fifth postulate directly, Saccheri attempted a proof by contradiction; the results appeared in *Euclid Vindicated From All Faults* (1733). Saccheri began with quadrilateral $ABDC$, where AC, BD were equal and made equal angles with the base AB. Saccheri proved a few properties of these quadrilaterals that were independent of the fifth postulate:

Proposition 21.4. *Given quadrilateral $ABDC$, with equal sides AC, BD, at equal angles with AB. Then angles ACD, BDC are equal.*

Proposition 21.5. *In quadrilateral $ABDC$, let AB, CD be bisected at M, H. Then the angles at M and H are all right angles.*

Both of these proofs relied on *Elements* I-4: in two triangles, if two sides and the included angle are equal, then the two triangles are equal.

The first key proposition is

Proposition 21.6. *If $AC = BD$, and AC, BD are perpendicular to AB, then CD is greater, equal, or less than AB according to whether the angles at C, D are acute, right, or obtuse.*

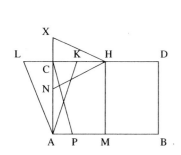

Proof. Suppose ACD and BDC are right angles, and CD not equal to AB. Then one is the greater, say CD. Take DK equal to AB, and join AK. Since $AKDB$ is a quadrilateral with equal sides KD, AB, and equal angles at D, B, then angles DKA, KAB are equal. But this is impossible, since angle KAB is smaller than angle right BAC, while angle DKA, being an external angle of the triangle AKC, must be greater than right angle DCA. Likewise for the acute and obtuse cases.   □

The next proposition is the converse

Proposition 21.7. *Suppose in the quadrilateral $ABDC$ that CD is equal, lesser, or greater than AB; then angles C and D are likewise right, obtuse, or acute.*

Since the angles at C and D are equal if AC, BD are perpendicular to AB, then one of the following must be true about the equal angles ACD, BDC:

1. The angles are right angles. This is Euclid's fifth postulate, which Saccheri called the **hypothesis of the right angle**.

2. The angles are greater than right angles. This is the **hypothesis of the obtuse angle**.

3. The angles are less than right angles. This is the **hypothesis of the acute angle**.

First, Saccheri proved

Proposition 21.8. *If the hypothesis of the right angle holds in a single case, it holds in all cases.*

Proof. Let AC, BD be perpendicular to AB, and suppose angles ACD, BDC are right angles; hence CD is equal to AB. Extend AC to R and BD to X so that CR, DX, AC, and BD are all equal. Join RX. RX is equal to AB, and thus angles ARX, BXR are right angles.

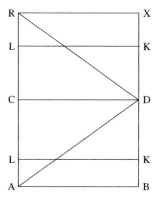

Next, draw any line LK where AL, BK are equal, and thus the angles at L, K are equal. Either both angles on the side toward RX are obtuse and the angles on the side toward AB are acute, or conversely. However, if they are acute on the side toward RX, then LK is greater than RX, and since they are obtuse on the side toward AB, then LK is less than AB. But AB, RX are equal, so LK is greater and less than the same quantity, which is impossible. Hence the hypothesis of the right angle, holding in a single case, must hold in all cases. $\square$

Corresponding results follow:

Proposition 21.9. *The hypothesis of the obtuse angle, if it holds in a single case, must hold in all cases.*

Proof. Suppose AC, BD are equal and perpendicular to AB, and CDB, DCA are equal obtuse angles; hence AB is greater than CD. Extend AC to R, BD to X, so CR, DX are equal; join RX. If the angles CRX, DXR are obtuse, we are done.

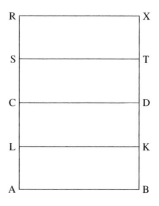

If the angles are right angles, then in every case the angles must be right angles; hence CDB, DCA must also be right angles, which is impossible.

If the angles are acute, then RX must be greater than AB, which is greater than CD. In quadrilateral $CDXR$, draw lines ST where CS, DT are equal; in the transition from CD, which is less than AB, to RX, greater than AB, there must be a line ST, equal to AB. Then in $ABTS$, the angles at S, T are right angles, and in every case, the angles must be right angles, which is impossible. $\square$

The proof of

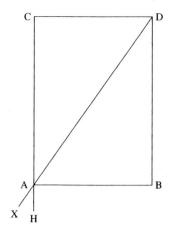

Proposition 21.10. *The hypothesis of the acute angle, if it holds in a single case, must hold in all cases.*

is trivial.

Remember Euclid's fifth postulate concerned the intersection of two lines. In order to prove Euclid's fifth postulate, Saccheri had to consider the conditions of whether two lines intersect or not. Thus he proved

Proposition 21.11. *In right triangle ABD, with HC perpendicular to AB at A, then angle XAH is equal, lesser, or greater than ADB according to whether the hypothesis of the right, obtuse, or acute angle holds true, and inversely.*

Proof. Let AC equal BD, and join CD. In the hypothesis of the obtuse angle, CD is less than AB; hence angle ADB must be greater than angle DAC. But angles DAC, XAH are equal, so angle XAH must be less than angle ADB. Likewise, under the hypotheses of the right or acute angle. □

A key property was

Proposition 21.12. *In a right triangle, the two remaining angles are together equal, greater, or less than a right angle, depending on whether the hypothesis of the right, obtuse, or acute angle is true.*

Next came

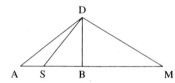

Proposition 21.13. *If DB is perpendicular to AM at B, and DM greater than DA, then BM is greater than BA, and conversely.*

Saccheri then proved that a form of Euclid's fifth postulate followed from the hypothesis of the right angle:

Proposition 21.14. *Let AP be perpendicular to PL at P, and let angle PAD be acute. Then in the hypothesis of the right angle, AD, PL meet.*

Proof. Construct HC perpendicular to AP at A, and extend AD towards X and towards PL (see Figure 21.2). Cut DF equal to AD, and drop DB, FM perpendicular to AP at B, M; join DM. DM must equal DF.

If not, suppose DM is greater than DF. Then angle DFM must be greater than angle DMF. Since angles DFM and XAH are equal (as this is under the hypothesis of the right angle), and angle XAH is equal to angle CAD, then angle CAD is likewise greater than angle DMF.

Since angles CAD, FMH are right angles, and angle CAD is greater than angle DMF, then the remaining angle DAM is less than the remaining angle DMA. Hence DM is less than AD. But this is impossible, DM was supposed equal to DF, and DF, AD were equal. Hence, DM cannot be greater than DF. Likewise, it cannot be less, so DM, DF, DA must be equal, and thus AB, BM are equal, and AM twice AB.

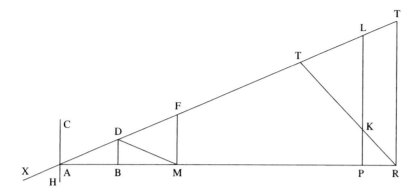

Figure 21.2: Euclid's Fifth Postulate.

We can continue, producing FG equal to AD and extending perpendicular GN, and thus AM, MN are equal, with AN twice AM. Continuing, we eventually arrive at a point T where AT, TV are equal, and perpendicular VR cuts AP at R, where AR is greater than AP, no matter how large AP might be.

This must occur after line AD meets PL. Since if T is before, then TR perpendicular to AP must across PL at some point, say K. Then PKR is a triangle with two right angles, which (under the hypothesis of the right angle) is impossible. Hence AD, PL must meet. $\qquad\square$

In a like manner, Saccheri proved

Proposition 21.15. *Let AP be perpendicular to PL at P, and let angle PAD be acute. Then in the hypothesis of the obtuse angle, AD, PL meet.*

These specialized forms of Euclid's fifth postulate were then used to prove

Proposition 21.16 (Euclid's Fifth Postulate). *If a line XA crosses lines AD, XL and makes the angles XAD, AXL less than two right angles, then AD, XL will meet, under either the hypothesis of the right angle or the hypothesis of the obtuse angle.*

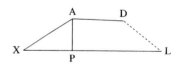

Finally

Proposition 21.17. *The hypothesis of the obtuse angle is false.*

Proof. The hypothesis of the obtuse angle implies Euclid's fifth postulate. But Euclid's fifth postulate implies the hypothesis of the right angle, which must hold in every single case. Hence, the hypothesis of the obtuse angle cannot hold. $\quad\square$

Having disposed of the hypothesis of the obtuse angle, Saccheri then attempted to show that the hypothesis of the acute angle also led to a contradiction. Despite a great deal of effort, Saccheri was unable to find a contradiction. He took refuge in one of the consequences of the hypothesis of the acute angle:

two lines perpendicular to the same line would approach each other asymptotically. This he found "repugnant to the nature of a straight line," and took it to be the contradiction that destroyed the hypothesis of the acute angle.

21.3.2 Lambert

Saccheri's book was almost totally ignored. Some indication of how little known the book was came from the fact that Johann Heinrich Lambert (August 26, 1728-September 25, 1777) also considered non-Euclidean geometries in a manner very similar to Saccheri. Lambert wrote *Theory of Parallel Lines* in 1766, though it was not published until 1786. Like Saccheri, he found the hypothesis of the obtuse angle led to a contradiction. Unlike Saccheri, he realized that the hypothesis of the acute angle did *not* lead to a contradiction, and suspected that it might, in fact, lead to a self-consistent geometry. A consequence of his investigations was the consideration of the trigonometric functions of imaginary angles, which give rise to the **hyperbolic functions**.

21.3.3 Lobachevski and Bolyai

Nikolai Ivanovich Lobachevski. ©Bettmann/CORBIS.

The mathematician who brought non-Euclidean geometry through its birth pains and into the consideration of mathematicians was Nikolai Ivanovich Lobachevsky (November 2, 1793-February 24, 1856). Between 1823 and 1826, he began to consider the possibilities of a geometry in which the fifth postulate of Euclid was replaced with an alternate. By 1826, he became convinced it was feasible to create a self-consistent geometry using a postulate other than Euclid's fifth, and in 1829, took the revolutionary step of publishing, in the Kazan *Messenger*, "On the Principles of Geometry," the first account of a non-Euclidean geometry. Other publications followed: *Imaginary Geometry* was published in Kazan in 1835, and "New Principles of Geometry with a Complete Theory of Parallels" appeared in several parts between 1835 and 1838. Nineteenth century Russia was a backward country, little regarded by western Europe, and Russian scientific discoveries were generally ignored. Lobachevsky risked being another one of history's footnotes, an "also discovered by" in the history of mathematics. However, he had the foresight to prepare a French translation of *Imaginary Geometry* (1836) and a paper for Crelle's journal (1837); a German edition of his work appeared in 1840.

Simultaneously with Lobachevsky's work, another Eastern European mathematician, the Hungarian Janos Bolyai (December 15, 1802-January 27, 1860), began developing his own non-Euclidean geometry. In 1832, Bolyai's father Wolfgang Bolyai published *Essays on the Elements of Mathematics for Studious Youths*, in which he included his son's work as an appendix. When Lobachevky's work appeared in German, it was so similar to Bolyai's work that the younger Bolyai suspected plagiarism. Meanwhile, the lesser side of Gauss's character made an appearance: after the publication of *Essays*, Gauss wrote to the elder Bolyai saying that he could not praise his son's work, for he would be praising himself, as he had earlier investigated non-Euclidean geometry, and Janos

Bolyai's development coincided "almost entirely" with Gauss's own work, conducted for the last "thirty or thirty-five years." Interestingly, Gauss was quite impressed with Lobachevsky's work, recommending him for membership in the Göttingen Academy of the Sciences in 1842, and studying Russian so he could read Lobachevsky's papers in the original.

Imaginary Geometry

Since Lobachevsky's work predated Bolyai's, we will follow Lobachevsky's development. Lobachevsky referred to his geometry as *imaginary* geometry. He began by listing a number of theorems which were independent of the parallel postulate, including

Proposition 21.18. *Two lines perpendicular to a third line never intersect.*

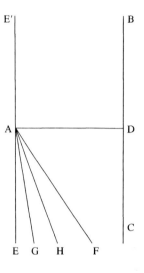

Given a point A, a line BC, and AD perpendicular to BC, and AE perpendicular to AD, then either all lines AG, AH, AF intersect BC if sufficiently extended, or some do and others do not. The **boundary line** AH is the line such that all lines inside angle HAD, such as AF, eventually cross BC; all lines between it and the perpendicular AE, such as AG, do not intersect the line BC. AH is then said to be parallel, and the angle HAD is designated the **angle of parallelism**. The angle of parallelism might depend on the distance AD from the line; if $AD = p$, then the angle of parallelism is designated $\Pi(p)$.

Under the axioms of Euclidean geometry, $\Pi(p) = \frac{\pi}{2}$; thus AE' will also be parallel to DB and all lines through A except EE' will intersect BC, and there is a unique line parallel to BC through A. Otherwise, there will be two distinct parallel lines, namely AH and AK, where angles DAH and DAK have measure $\Pi(p)$; thus, when speaking of parallel lines, it is important to distinguish which *side* of the line is to be parallel to the given line.

Proposition 21.19. *The sum of the angles of a triangle cannot be greater than two right angles.*

Proof. Suppose that in triangle ABC the sum of the angles is equal to $\pi + a$. If the sides are unequal, take the smallest side, say BC, and let AD bisect it. Extend AE so $AD = DE$. Join EC. Triangle ADB is congruent to triangle EDC, and angle ABD is equal to angle ECD, and angle BAD is equal to angle CED; hence the sum of the angles in triangle ACE is also equal to $\pi + a$. Moreover, EAC and AEC are each less than BAC. Continuing in this way, halving the side opposite the smallest angle, we must eventually come to a triangle, the sum of whose angles is $\pi + a$, but with two angles equal and each less than $\frac{1}{2}a$. But the third angle cannot be greater than π; hence a must either be 0 or negative. $\square$

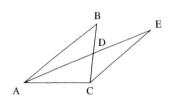

This result corresponds to Saccheri's proof that the hypothesis of the obtuse angle is untenable.

Lobachevsky defined a **boundary line** as a line (or curve) such that the perpendicular bisectors of every chord of the curve were parallel to each other;

one of these lines was the **axis** of the curve, though any of the parallel lines could serve as the axis.

Proposition 21.20. *Let $AA' = BB' = x$ be two parallel lines toward the side from A to A', and let AA', BB' be the axes for two boundary lines $AB = s$, $A'B' = s'$. Then $s' = se^{-x}$, where e is an arbitrary constant.*

Proof. Suppose $s : s' = n : m$, a ratio of whole numbers. Between AA' and BB' draw CC', parallel to AA', BB', with $AC = t$, $A'C' = t'$, and $t : s = p : q$, also a ratio of whole numbers. Thus

$$s = \frac{n}{m}s' \qquad\qquad t = \frac{p}{q}s$$

Divide AB by axes into nq equal parts. There will be mq parts of these size on s' and np parts of this size on t. Moreover, there will be mp such parts on t'; hence

$$\frac{t}{t'} = \frac{s}{s'}$$

But this is true regardless of where CC' is drawn; hence the ratio $s : s'$ is a constant for any fixed distance x. Thus if we let $x = 1$, $s = es'$, and for any x, we have $s' = se^{-x}$. Since e is arbitrary (subject only to $e > 1$, and the unit x is likewise arbitrary, we may let e be the base of the natural logarithms. Hence, parallel lines approach each other asymptotically.   □

The Shape of Space

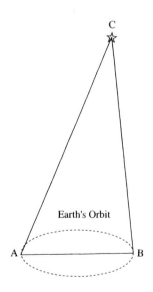

In 1826, Lobachevsky presented to a Kazan scientific society a "rigorous proof" of the Euclidean theory of parallels as they applied to real space. What Lobachevsky apparently meant was that one could not use "inborn" assumptions about the universe; instead, one must rely on factual observations. One such factual observation is that if space were non-Euclidean, the angle sum of the three angles of a triangle would differ from $180°$. By using the latest astronomical observations, Lobachevsky showed that the very large triangle formed by opposite points on the Earth's orbit and the star Sirius differed from $180°$ by less than $0.000372''$; thus, Lobachevski concluded, space was Euclidean. In fact, Lobachevsky made a mistake in his calculations, and the actual difference is a hundred times less than he indicated. It was only in the twentieth century that astronomical instruments became precise enough to measure the very slightly non-Euclidean nature of the universe.

21.3 Exercises

1. Prove the remainder of Proposition 21.6.

2. Prove Proposition 21.10. Use a proof by contradiction.

3. In Proposition 21.11, Saccheri argued angle DAC was less than angle ADB if CD was less than AB. Prove this, then complete the proof of Proposition

21.11 for the hypothesis of the right angle, and the hypothesis of the acute angle.

4. Prove Proposition 21.16. Hint: Drop AP perpendicular to XL.

5. Prove Proposition 21.18 without using the Euclid's fifth postulate.

6. Explain how to construct the boundary line pointwise, given the axis AB.

7. Show that a circle with an infinitely great radius serves as a boundary line.

21.4   Consistency of Geometry

The importance of the discovery of non-Euclidean geometries cannot be underestimated, though it should not be misinterpreted: the value lies not in the discovery, but rather in the *method* of the discovery. Lobachevsky in particular built up a geometry by assuming certain relationships, and deriving the logical consequences of these relationships. In some sense, all mathematics prior to the invention of non-Euclidean geometries dealt with a model of the real universe, where certain rules were implicitly or explicitly assumed. The non-Euclidean geometries suggested mathematics might concern itself with totally abstract relationships between completely arbitrary objects. Thus, whether or not the axioms "made sense" was irrelevant; all that mattered was that they were consistent and that logical conclusions could be derived from them. It is easy to draw logical conclusions, but far more difficult to show that a system of axioms is consistent: the axioms may not contradict each other, nor may any conclusion drawn logically from the axioms contradict any other conclusion.

One way to settle this question was to show that the axioms were not only *assumed* true, but were *in fact* true statements. In the case of Euclidean geometry, the axioms were believed *in fact* to be true statements about actual space, and thus they were "obviously" consistent, as contradictions do not arise in the real universe. But what of the axioms of non-Euclidean geometry? Not only was there no obvious case in which the axioms seemed to be true, but the axioms themselves seemed to contradict common sense.

Consider Saccheri's obtuse geometry, which he proved inconsistent. In 1854, Riemann showed the hypothesis of the obtuse angle could result in a consistent geometry, provided further modifications of Euclid's axioms were made: lines could *not* be extended indefinitely, nor could arbitrarily large circles be created. More importantly, he demonstrated a model in which the axioms *were* true. Rather surprisingly, the model of non-Euclidean geometry had long been studied by geometers: the surface of a sphere, where the "straight lines" correspond to arcs of great circles. For example, the axiom that a straight line could not be extended indefinitely was true about the great circle of a sphere: extended sufficiently far, every great circle meets itself and can no longer be extended. Since the axioms of Riemannian geometry were true statements about the points and "straight lines" on a sphere, then Riemannian geometry was consistent.

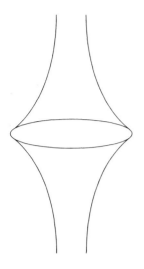

A pseudosphere.

David Hilbert. Used with permission of CORBIS.

What of the geometry of Lobachevsky and Bolyai? In 1866, the Italian Eugenio Beltrami pointed out these could be modeled by using the surface of a pseudosphere, a surface of constant negative curvature; the pseudosphere could be formed by revolving a curve called a tractrix around an axis.

An implication was that the consistency of Euclidean geometry could no longer be assured by appealing to the consistency of the real universe: the geometry on the surface of the Earth, for example, was in fact Riemannian, not Euclidean, and thus the Euclidean axioms were not, in fact, "true." Moreover, the geometry of the surface of a sphere or pseudosphere could be derived from Euclidean geometry, so if Euclidean geometry was inconsistent, so were the geometries of Riemann and Lobachevsky: all three stood or fell on the same judgment.

21.4.1 Pasch

An important step in the direction of proving geometry consistent was its full axiomatization. Although Euclid's *Elements* is usually considered an axiomatic system, in some ways, it falls short of what axiomatics means today. The *Elements* presents geometry as the study of some existing object, whereas a true axiomatic system consists of the study of the relationships between completely abstract objects.

New Introduction to Geometry (1882) by Moritz Pasch (November 8, 1843-September 20, 1930) began the process of axiomatizing of Euclidean geometry. Pasch began with certain undefined notions, such as "point," and certain theorems accepted without proof, such as "On any given straight line segment, a point may be found." Whether or not this conforms with "reality" is immaterial; the proper axiomatic approach would be to concentrate solely on the logical consequences of the assumptions.

21.4.2 Hilbert

Giuseppe Peano continued the axiomatization of geometry in 1894, but it was Hilbert's 1899 work, *Foundations of Geometry*, that established a complete axiomatic basis for Euclidean geometry. David Hilbert (January 23, 1862-February 14, 1943) was born in Königsburg, Prussia whose bridges had, a century before, inspired Euler to begin the study of topology.

To emphasize the point that geometry dealt with relations and not with objects, Hilbert would say that everything said in geometry about points, lines, and surfaces should remain true if the words "points," "lines," and "planes" were replaced with "tables," "chairs," and "beer mugs." Hilbert began with some terms for which he attempted no definition: "point," "line," "plane," "lie," "between," and "congruent." Hilbert then gave 21 axioms. The first eight of these dealt with the undefined terms "point," "line," and "plane":

1. For any two points, there exists a line that contains both.

2. There does not exist more than one such line.

3. There are more than two points on any given line.

4. There exists three points which are not on any one line.

5. For any three points not on one line, there exists a plane containing these three points.

6. If two points A, B are on a line a, and a is in a plane α, then all points of line a are on α.

7. If two planes α, β have a point A in common, they have more than one point in common.

8. There exists four points that are not on any one plane.

To use Hilbert's own analogy, the first axiom might be restated as, "For any two tables, there exists a chair that contains both." Whether or not this is true is irrelevant: what is important is that the truth of the statement is assumed.

Note that Hilbert did not include any axioms relating to circles. This does not mean, however, that Hilbert's geometry excluded circles; rather, it meant that Hilbert felt no need to assume the existence of circles, and felt comfortable deriving the properties of a circle deductively. If circles existed, they would have the properties derived from the axioms; if circles did not exist, the propositions applying to circles applied to nothing, and no harm resulted.

The first two propositions of Hilbert's geometry are:

Proposition 21.21. *Two lines in the same plane either have one point in common, or none at all.*

Proposition 21.22. *Two planes have no point in common, or have a line in common.*

Hilbert, having the advantage of two millennia of mathematical development since the time of Euclid, included what Euclid could not even conceive: a proof of the *consistency* of the mathematical axioms, by showing they were in fact true. By letting the ordered pair (x, y) represent a point and the ordered triple $(u : v : w)$ represent a line (namely the line $ux + vy + w = 0$), Hilbert proved the truth of his axioms. For example, the first axiom can be proven by showing the existence of a line $(u : v : w)$ that contains the two points (a, b) and (c, d). In this way, geometry can be reduced to algebra.

21.4 Exercises

1. The tractrix is formed in the following way: given a line of fixed length AB perpendicular to BC at B. Let B move towards C at a constant rate in such a way that AB, always of fixed length, is tangent to the tractrix at A. Determine the equation of the tractrix.

2. Axioms 7 and 8 are together interpreted to mean that space has exactly three dimensions. Explain.

3. Find and prove the algebraic equivalents of the first four of Hilbert's axioms.

4. What objections might be raised to Hilbert's demonstration of the consistency of Euclidean geometry?

Chapter 22

Analysis After Midcentury

In 1848, liberal revolutions swept across Europe. In February, the king of France, Louis Philippe, was forced to abdicate and the Second Republic was established. In December, Louis Bonaparte, Napoleon's nephew, was elected president. In March of 1848, revolutionary fervor spread to Italy, but the revolt was brutally suppressed by the Austrians. Also in March was an uprising in Frankfurt, which sough to unify many of the north German states into a single German nation. A key action of the Frankfurt Parliament was to elect Friedrich Wilhelm IV of Prussia the emperor, or *kaiser*, of Germany. However, Austria and Russia forced him to refuse the title.

Friedrich Wilhelm's brother, Wilhelm, would consider the refusal coerced. Wilhelm would become an active participant in the drive to form a united Germany, with himself as kaiser. Wilhelm was a figurehead, though (and he was fully aware of it): the real architect of the unification movement was Wilhelm's gifted chancellor, Otto von Bismarck. Slowly, the many states of the German Confederation were combined into a North German Confederation under Prussian leadership.

Meanwhile, Bonaparte's presidency lasted only until 1852—when, with the overwhelming support of the French people, he crowned himself Emperor Napoleon III. Dreaming of the glory days of his uncle, he sought to restore France to the power it had half a century before.

22.1 Analysis in Germany

The Russian Chebyshev and the French Liouville were exceptions to the ever-increasing domination of mathematics by the Germans. Gauss was the first great German mathematician of the nineteenth century, but Gauss hated teaching and, in any case, taught primarily the method of least squares, of great practical importance, but of little theoretical consequence. Still, Göttingen University

Figure 22.1: Germany in the Nineteenth Century.

continued to attract brilliant students, thanks to its reputation—for the study of theology.

22.1.1 Riemann

One such theologically inclined student was Georg Friedrich Bernhard Riemann (September 17, 1826-July 20, 1866). Riemann became interested in mathematics, but besides Gauss, who offered only a single course, on the method of least squares, the rest of the mathematics faculty was otherwise undistinguished. Thus, Riemann went to the University of Berlin, in Prussia.

Riemann Surfaces

Riemann returned to Göttingen in 1851 and presented his dissertation. Here he presented the idea of what is now called a **Riemann surface**, an essential component of modern complex function theory. Some expressions, such as $\sqrt{z}$, can have multiple values: for example, $\sqrt{1} = \pm 1$. Since it is inconvenient to deal with multiple values for a single input, it is preferable to consider just one of them. In the case of $f(z) = \sqrt{z}$, we usually take $\sqrt{z}$ to mean the positive square root of z. Unfortunately, when the function is complex valued, this distinction cannot be made, since positive and negative no longer have clearly definable meanings. However, the additional structure of complex numbers allows for a meaningful way of letting $\sqrt{z}$ take on both values and still be a single-valued function.

This can be accomplished by using the polar form of a number. For example, $1 = e^{0i}$, $1 = e^{2\pi i}$, and so on. When considered in polar form, then $f(e^{0i}) = e^{0i} = 1$, while $f(e^{2\pi i}) = e^{\pi i} = -1$. Hence, the single-valuedness of $f(z) = \sqrt{z}$ is retained, since two "different" values are being input. Moreover, functions like $g(z) = z^2$, which are many-to-one functions (since more than one value of z will give the same value of z^2, for example $g(1) = g(-1) = 1$) can be turned into one-to-one functions, since $g(e^{0i}) = e^{0i}$, while $g(e^{\pi i}) = e^{2\pi i}$. By a similar analysis, Jacobi's problem of dealing with functions that had infinitely many values at a point could be solved. See page 622.

On the Hypotheses That Underlie Geometry

On June 10, 1854, Riemann presented his inaugural lecture (*habilitationschrift*). The inaugural lecture was to display the talents and professional depth of the candidate. It was traditional to present three topics, from which the tenured faculty would choose one for the candidate to talk about. Gauss chose "On the Hypotheses Which Lie at the Foundation of Geometry." This came as no surprise to Riemann, who was fully aware his topic overlapped with Gauss's *General Investigations on Curved Surfaces* of 1827; in fact, "On the Hypotheses. . ." was the only topic for which Riemann had prepared even a rough draft.

Riemann's inaugural lecture displays a rather unfortunate side effect of modern mathematics: specialization. The great mathematicians of the past, like Newton, Euler, and Lagrange, could span the entirety of mathematics, and even lesser mathematicians, such as Poisson, could make important contributions in many fields. The explosive growth of mathematics during the eighteenth and nineteenth centuries meant that specialization was necessary, and more and more frequently, we will see that most of the work of a mathematician is connected with a single work of his or her youth. Riemann, at the age of 27, was about to set the course of his mathematical career in the directions laid out by "On the Hypotheses. . ."

Gauss had envisioned a surface *in* space; Riemann generalized Gauss's idea so that the surface *was* space. An easy way to imagine this is to consider that, in the Cartesian plane, the distance element ds^2 is given by $dx^2 + dy^2$, provided the geometry is Euclidean. But suppose one were to make the distance element $ds^2 = A\ dx^2 + B\ dx\ dy + C\ dy^2$, where A, B, and C are constants or even functions of x, y. For example, the distance between the two points on a flat map of the Earth's surface cannot be accurately calculated using $dx^2 + dy^2$, but instead requires a more complex formula. In this way, all surfaces can be mapped to the Cartesian plane, provided the distance metric is defined appropriately.

Riemann Integrals

See page 627.

Part of Riemann's inaugural lecture discussed the history of trigonometric series representation of functions. Dirichlet had shown the conditions under which the Fourier series of a function would converge to the function itself, and gave a counter example of a function whose Fourier series nowhere converged to the function. Building on Dirichlet's work, Riemann noted that it seemed likely that functions that could *not* be represented by Fourier series did not occur in nature. However, a further investigation into the integrability of functions was warranted because the concept of a Fourier series, though originating with physical problems, was being extended, particularly into the theory of numbers and other aspects of "pure" mathematics.

Since the determination of the Fourier series coefficients relied on the determination of definite integrals, Riemann considered what was meant by the definite integral $\int_a^b f(x)\,dx$. Choosing values x_1, x_2, x_3, ..., such that

$$a < x_1 < x_2 < x_3 < \ldots < x_{n-1} < b$$

and designating $\delta_1 = x_1 - a$, $\delta_2 = x_2 - x_1$, $\delta_3 = x_3 - x_2$, ..., and $0 < \epsilon_i < 1$, Riemann formed the sum

$$S = \delta_1 f\left(a + \epsilon_1 \delta_1\right) + \delta_2 f\left(x_1 + \epsilon_2 \delta_2\right) + \ldots + \delta_{n-1} f\left(x_{n-1} + \epsilon_n \delta_n\right) \qquad (22.1)$$

This sum obviously is dependent on the choices of ϵ_i, δ_i. However, if it has the property that, regardless of how the δ_i and ϵ_i are chosen, the sum approaches a fixed limit A as the δ_i tended to 0, then Riemann called A the value of the definite integral $\int_a^b f(x)\,dx$; conversely, if the sum did not approach a limit, then $\int_a^b f(x)\,dx$ had no value and no meaning.

The question Riemann chose to examine was to find conditions under which a function $f(x)$ is integrable. Since the sum had to converge as the δ_i went to 0, Riemann considered the difference, or "oscillation," between the maximum and the minimum values on each of the intervals (a, x_1), (x_1, x_2), ...; naming these oscillations D_1, D_2, ..., Riemann concluded that the sum

$$\delta_1 D_1 + \delta_2 D_2 + \ldots + \delta_n D_n$$

had to become infinitely small along with δ_i.

Let δ be the greatest of the δ_i, and let Δ be the greatest value of the sum when δ is less than some given d; if $f(x)$ is integrable, Δ becomes infinitely small as d tends to zero. Given some amount, σ, suppose the total length of all the intervals over which the oscillation is greater than σ is s. Hence

$$\sigma s \leq \delta_1 D_1 + \delta_2 D_2 + \ldots + \delta_n D_n \leq \Delta$$

Hence $s \leq \frac{\Delta}{\sigma}$. σ is a fixed number, and if S has a limit (i.e., if $f(x)$ is integrable), then s can be made infinitely small as d tends to nothing. Thus:

Proposition 22.1. *For the sum S to converge as δ becomes infinitely small, it is sufficient that the total length of the intervals over which the oscillation is larger than σ, for any σ, can be made infinitely small by the appropriate choice of d.*

Likewise:

Proposition 22.2. *If $f(x)$ is everywhere finite, and if, by decreasing indefinitely the quantity δ, the greatest value of s of the total lengths of the intervals in which the oscillations of the function are greater than a given σ can be decreased indefinitely, the sum S converges as δ tends to zero.*

Thus, even if a function was discontinuous at an infinite number of places, it might still be integrable, provided the total length of the intervals over which the discontinuities were greater than a given quantity could be made small.

To provide an example, Riemann created the function (x), which was the excess of x over the whole number closest to it, or 0 if x was equidistant from two whole numbers: for example, $(3.3) = 0.3$, $(3.9) = -0.1$, and $(3.5) = 0$. Riemann defined

$$f(x) = \frac{(x)}{1} + \frac{(2x)}{4} + \frac{(3x)}{9} + \dots \tag{22.2}$$

The series representing $f(x)$ is convergent for all x. If $x = \frac{p}{2n}$, where p, n were relatively prime, then (as Riemann wrote it):

$$f(x+0) = f(x) - \frac{\pi^2}{16n^2}$$

$$f(x-0) = f(x) + \frac{\pi^2}{16n^2}$$

In other words, the limit as x approaches $\frac{p}{2n}$ from above is $f(x) - \frac{\pi^2}{16n^2}$, while the limit as x approaches $\frac{p}{2n}$ from below is $f(x) + \frac{\pi^2}{16n^2}$. For all other values of x, the two limits are the same. Hence, the function is discontinuous at an infinite number of points, yet it is integrable, according to Riemann's criterion. Riemann's inaugural lecture was successful, and with Gauss looking on in approval, Riemann became a full professor at Göttingen, earning 300 thalers a year—not quite double Gauss's stipend fifty years earlier.

22.1.2 Darboux

Gaston Darboux (August 14, 1842-February 23, 1917) introduced Riemann's work to French mathematicians. Darboux realized Riemann's reformulation of the integral meant that many notions that seemed self-evident had to be reexamined. For example, it had been assumed that integration and summation could be interchanged. In 1884, Darboux provided a counterexample: the function $f(x) = \sum_{n=1}^{\infty} u_n(x)$, where

$$u_n(x) = -2n^2 x e^{-n^2 x^2} + 2(n+1)^2 x e^{-(n+1)^2 x^2} \tag{22.3}$$

For this function, the integral of the sum of $u_n(x)$ was not equal to the sum of the integrals of $u_n(x)$.

Darboux further modified Riemann's work. In general, Riemann's criteria for integration was too difficult to use in practice, though it was perfectly good for theory. In 1875, Darboux replaced Riemann's $f\left(x_j + \epsilon_{j+1}\delta_{j+1}\right)$ in Equation 22.1 with U_k or L_k, the greatest or least value of the function in the kth interval. This gives our modern definition of the **Riemann integral**.

We might note that Riemann (and Darboux after him) began a serious investigation of what are called **pathological functions**. The Dirichlet function, which took one value at rational points x and a different value for irrational x, is an example of a pathological function; Riemann's integrable but infinitely discontinuous function is a second; Darboux's function is a third. These functions provide counterexamples to long-standing beliefs about the behavior of functions. The production of pathological functions points out some of the changes occurring in mathematics during the nineteenth century: the change from a mathematics of *algorithms* to a mathematics of *concepts*. An algorithm either works or fails, and the only question that arises is whether or not it is applicable. A mathematical concept, on the other hand, is only useful if it is well defined, and part of the rationale of constructing pathological functions is to determine if a concept is well defined and, if not, to see what hidden assumptions are included in its definition.

22.1.3   The Riemann Hypothesis

Not all of Riemann's work stemmed from his inaugural lecture; in fact, one of the most important unsolved problems in mathematics came from "On the Number of Primes Less Than a Given Number," which he presented to the Berlin Academy of Sciences in November 1859. Riemann conjectured

Conjecture 22.1 (Riemann Hypothesis). *The complex zeroes of the function*

$$\zeta(s) = \sum_{n=1}^{\infty} \frac{1}{n^s} \tag{22.4}$$

occur when the real part of s is $\frac{1}{2}$.

Since $\sum_{n=1}^{\infty} \frac{1}{n^s}$ actually diverges if the real part of s is $\frac{1}{2}$, the hypothesis requires some explanation. $\zeta(s)$ should be considered some function, whose series representation is Equation 22.4 for some values of s.

As an analogy, we might consider the following expansions:

$$f(x) = \frac{1}{1-x}$$
$$= 1 + x + x^2 + x^3 + \ldots$$

If we try and evaluate the series at $x = 2$, the series would diverge. However, the function from which it originated would have a value, and $f(2) = -1$. In a like manner, we may consider the value of $\zeta(s)$ at a particular value of s, even if the series in Equation 22.4 diverges.

22.1.4 Weierstrass and Kummer

Riemann turned Göttingen into an important center of mathematical research, but far more important in terms of overall influence was the University of Berlin which, through the influence of Kummer, became one of the premier centers of mathematical research. Kummer could be called the godfather of modern mathematics. Like many other mathematicians, he began his academic career with an interest in theology, but under the influence of his mathematics teacher, Heinrich Ferdinand Scherk, Kummer changed his field. He obtained a position at the Technical High School at Sorau, where he was born. Kummer encouraged the independent investigation of mathematics by his students.

A paper on the hypergeometric series drew the interest of Jacobi and Dirichlet, with whom Kummer began to correspond. Jacobi, with the help of Alexander Humboldt, got Kummer a position at the University of Breslau in 1842. In 1855, Dirichlet left Berlin to become Gauss's successor at Göttingen, and recommended that Kummer become Dirichlet's successor at Berlin; this was done.

This left Kummer's position at Breslau open. There were two prime candidates for the position: Ferdinand Joachimsthal (March 9, 1818-April 5, 1861), one of Kummer's students; and Karl Theodor Wilhelm Weierstrass (October 31, 1815-February 19, 1897). Weierstrass was clearly the better candidate, but Kummer recommended Joachimsthal be given the position. Kummer's reason: he wanted Weierstrass at Berlin, which was arranged the next year.

Karl Weierstrass. Used with the permission of CORBIS.

In 1861, Kummer and Weierstrass organized and ran the first German seminar on pure mathematics. Kummer presented his own research and outlined possibilities for his students to pursue independent investigation of his work. However, Weierstrass did most of the teaching, and modern mathematics is very much a product of Weierstrass and his students, many of whose names are familiar to mathematicians and physicists: Frobenius, Killing, Netto, Hölder, Klein, Lie, Minkowski, Mittag-Leffler, Schwarz, Stolz. It was Weierstrass's work, more than anything else, that turned Berlin into a great center of mathematical studies.

Many of Weierstrass's ideas were never published; like Gauss, he preferred to publish only after a work had reached perfection. Moreover, Weierstrass himself had an aversion to *printed* copies of his lectures (though handwritten copies were permissible). However, Weierstrass enjoyed teaching, and thus had great influence on the development of mathematics. In his courses, he set himself to the goal of methodically building up the whole of mathematics as a sequence of well-ordered thoughts, essentially self-contained: thus, in the course of a series of lectures on a subject, Weierstrass need only cite previous lectures, a pattern that is clearly part of the modern method of teaching mathematics.

Weierstrass also made some profound contributions to the study of continuous functions. On July 18, 1872, Weierstrass presented to the Berlin Academy of the

Sciences the function

$$f(x) = \sum_{n=0}^{\infty} b^n \cos(a^n x \pi) \tag{22.5}$$

where a is an odd whole number and b is a positive constant less than 1 with $ab > 1 + \frac{3}{2}\pi$. $f(x)$ is everywhere continuous and nowhere differentiable. It was the first such function to come to the attention of mathematicians in general.[1] Not long afterward, Darboux produced

$$f(x) = \sum_{n=0}^{\infty} \frac{\sin(n+1)! x}{n!} \tag{22.6}$$

as another example of a continuous, nowhere differentiable function.

22.1.5 Kovalevskaya

One of Weierstrass's more unusual students was Sonya (or Sofia) Vasilyevna (January 15, 1850-February 10, 1891); in Russian naming convention, "yevna" indicates "daughter of," so her name might be read as Sonya, Vasily's daughter. She was a descendant of Mathias Corvinus, the greatest of the medieval kings of Hungary. As a young girl, her room was wallpapered with pages from her father's school text on differential and integral calculus, which provided her with an introduction to the subject. In 1867, she took a course on analysis from Aleksandr N. Stannolyubsky, who recognized her potential as a mathematician and encouraged her to study the subject further.

It was difficult enough for a man to obtain a good education in Tsarist Russia and nearly impossible for a woman to do so. One way out of this difficulty was to marry someone who would be studying out of the country, and in 1868, she did just this, marrying Vladimir Kovalevsky. Kovalevskaya is the female form of Kovalevsky, and it is as Sonya Kovalevskaya that she is better known. The two went to the University of Heidelberg, where she primarily studied physics, with Gustav Robert Kirchoff and Hermann von Helmholtz. In 1871, she came to Berlin, and attracted Weierstrass's attention.

In Berlin, Kovalevskaya began work on partial differential equations and Abelian integrals, especially the reduction of very general Abelian integrals into simpler ones. Like Sophie Germain, she made important contributions to mathematical physics, particularly in the mathematical analysis of the dynamics of the rings of Saturn. Saturn's rings are unique in the solar system (other planets have rings, but none so spectacular as Saturn's), and in the nineteenth century, there was considerable debate over their nature. They were so thin that they vanished when viewed edge on. The English physicist James Clerk Maxwell proved the rings could not possibly be a single, solid body; hence, they had to consist of billions of particles which, at the distance of Saturn, blended into a seemingly continuous structure. But this led to a problem, for every particle

---

[1]Bolzano had discovered a similar function, but his discovery remained in his notebooks until the twentieth century.

should attract every other gravitationally, so what could explain the flatness of the rings? Laplace had attempted to determine a mathematical solution of the ring's nature, but the problem proved very difficult. The complete solution to the dynamics of the ring particles had to await Kovalevskaya, in a work far too complex to discuss here.

In pure mathematics, she completed a proof of a theorem on the uniqueness of the solutions to partial differential equations. A common technique in solving differential equations is to assume the existence of a power series for a function

$$f(x) = a_0 + a_1 x + a_2 x^2 + a_3 x^3 + \ldots$$

The derivatives of this function could be found formally:

$$f'(x) = a_1 + 2a_2 x + 3a_3 x^2 + 4a_4 x^3 + \ldots$$
$$f''(x) = 2a_2 + 6a_3 x + 12a_4 x^2 + 20a_5 x^3 + \ldots$$
$$\vdots$$

By substituting the values of $f(x)$, $f'(x)$, ... into the differential equation and equating the like terms, the coefficients could be found. Since the differential equations dealt with by nineteenth century mathematicians originated in physical problems, it was felt that the corresponding differential equations always had a solution. Weierstrass knew this technique "worked," but the resulting coefficients might not produce a convergent series solution. Cauchy had proven, in 1842, that a wide class of differential equations had solutions, but this work was little known to his contemporaries. Weierstrass eventually proved that existence and uniqueness would hold for first order ordinary differential equations, and attempted to extend his results to partial differential equations. Here, he was unsuccessful, and assigned the problem to Kovalevskaya.

The problem Kovalevskaya faced might be stated as follows: the solution to an ordinary differential equation, such as $y' = y$, includes a constant to be determined by the initial conditions. In this case, $y = Ce^x$, with $C = y(0)$. On the other hand, consider $\frac{\partial y}{\partial x} = y$, where y is a multivariable function. Again, $y = Ce^x$, but in this case C is a function that does not depend on x, and may have an infinite number of forms; thus, even knowing $C = y(0)$ does not help to determine the precise form of C. Hence, it is conceivable uniqueness might not hold for partial differential equations. Fortunately, Kovalevskaya proved that under the right conditions, existence and uniqueness do hold for partial differential equations.

Weierstrass had to teach Kovalevskaya privately, since women were not allowed to take classes at the University of Berlin. Through Weierstrass's influence, she was permitted to be a corresponding student at Göttingen, and in 1874, received her doctorate. Soon thereafter, the Kovalevsky's returned to Russia where, in 1874, Vladimir, having become involved in a financial scandal, committed suicide. Kovalevskaya's financial situation was desperate, but

Mittag-Leffler (another participant in the Kummer-Weierstrass seminar) managed to obtain a position for her at the University of Stockholm. Like Descartes, the Scandinavian winter would prove too difficult for her, and in 1891, she died of pneumonia.

22.1 Exercises

1. What is the difference between Riemann's definition of an integral and Cauchy's definition?

2. Consider Riemann's function in Equation 22.2.

    (a) Show that the series converges for all x.

    (b) If $a = \frac{p}{2n}$ for p, n relatively prime, show that $\lim_{x \to a^+} f(x)$ and $\lim_{x \to a^-} f(x)$ have the form indicated by Riemann.

    (c) Show that $f(x)$ is continuous at all x not of the form $x = \frac{p}{2n}$.

    (d) Show that $f(x)$ is integrable.

3. Prove integration and summation of the function in Equation 22.3 cannot be interchanged.

4. Show that Dirichlet's function (see page 627) is not integrable in Riemann's sense.

5. Consider a flat map of the Earth's surface, and the coordinates of two points, (x_1, y_1) and (x_2, y_2), where the x and y values represent longitude and latitude, respectively. Let R be the radius of the Earth. Derive a formula to calculate the distance between two points on the Earth's surface.

6. Show that the functions in Equations 22.5 and 22.6 are continuous but nowhere differentiable.

22.2   The Real Numbers

One of the issues facing later nineteenth century mathematicians was that the real numbers did not have a proper axiomatic basis. Weierstrass, in his lectures, had suggested one means of handling the real numbers, though he never published his ideas. One of his students, H. Kossak, published *The Elements of Arithmetic* (1872), in which he claimed to present Weierstrass's theory of real numbers, a claim Weierstrass denied. We do know that Weierstrass was concerned over the logical validity of analysis.

The real numbers had invariably been defined as the limit of a sequence of rationals. For example, in Bolzano's proof of the Intermediate Value Theorem, he assumed that the sum

$$u + \frac{D}{2^m} + \frac{D}{2^{m+n}} + \frac{D}{2^{m+n+k}} + \cdots$$

corresponded to some real number, which was the limit of the sequence of partial sums. However, assuming that the sequence had a limit presupposed the existence of the limit. Weierstrass's solution was to consider the sequence *itself* to be the irrational number. Unfortunately, this raised other problems, for any given real number might have many different sequence representations, so the simplest number concept, that of equality, could not be easily defined.

The first to publish an axiomatic basis for the real numbers was Hugues Charles Robert Méray (November 12, 1835-February 11, 1911). In 1869, Méray published, "Remarks on the Nature of Quantities Defined by the Condition of Being the Limit of a Given Variable." The essential idea behind Méray's essay was that if the terms of a sequence v_n were such that $v_{n+p} - v_n$ tended to 0 for all p as n went to infinity, then the sequence could be said to have a limit. In some cases, this limit might be a rational number; in others, the limit was "fictitious" (nonrational).

22.2.1 Cantor and Heine

Méray's theory of the real numbers was well conceived, but Méray published it in an obscure journal, so few knew of it until much later. However, very similar approaches were taken by two colleagues at the University of Halle: Georg Cantor (March, 1845-January 6, 1918) and Eduard Heinrich Heine (May 16, 1821-October 21, 1881). Cantor was born in St. Petersburg, Russia, and attended the University of Berlin, becoming an active participant in the Weierstrass-Kummer seminar and eventually he became one of Kummer's students. In 1869, Cantor accepted a position at the University of Halle, where he stayed for the rest of his life. In 1897, he helped to organize the first international conference of mathematicians, in Zurich, Switzerland.

Georg Cantor. ©CORBIS.

Some preliminary groundwork on a theory of the real numbers was laid on March 20, 1870, when Cantor delivered "Theorems Regarding Trigonometric Series" to Crelle's journal. Cantor found it necessary to give a more formal definition to the idea that a sequence had a limit of 0. In Cantor's words:

> If I speak of an infinite number sequence
>
> $$(G.) \quad \varrho_1 \quad \varrho_2 \quad \varrho_3 \quad \cdots \quad \varrho_n \quad \cdots$$
>
> with $\lim \varrho_n = 0$, I mean that, if δ is a given quantity, one can remove from the sequence $(G.)$ a finite number of values, so the remainder are all less than δ.[2]

In October 1871, Heine, a student of Dirichlet's and Cantor's colleague at Halle, presented "The Elements of Function Theory" for publication in Crelle's journal. The very next month, on November 8, 1871, Cantor presented to another journal, the *Annals of Mathematics*, "Extensions of Theorems Regarding Trigonometric

---

[2]Cantor, George, "Ueber einen die trigonometrischen Reihen betreffenden Lehrsatz." *Journal für die reine und angewandte Mathematik* 72 (1870), page 130.

Series," which included his own theory of the real numbers. As can be judged by their respective titles, it was neither Cantor nor Heine's goal to develop a theory of the real numbers, though both found it necessary. Neither claimed sole (or even partial) authorship of the theory, though both noted that the theory of real numbers had not yet been established in any logical sense.

Heine and Cantor's development of the theory of real numbers is so similar that one suspects it was the joint product of both; the only real difference was that of exact terminology. Since Heine's work was published slightly before Cantor's, and in a somewhat more prestigious journal, we will follow Heine's development. He began with the concept of a **number sequence** which, in Heine's words, was a sequence of numbers, a_1, a_2, a_3, ..., which for every η there is an n so that $a_n - a_{n+\nu}$ is less than η for all positive ν. An **elementary sequence** was one for which a_n falls below any given quantity, for all n greater than some limit. With these two fundamental terms defined, Heine then proved

Proposition 22.3. *If a_1, a_2, ... and b_1, b_2, ... are two number sequences, then so are $a_1 \pm b_1$, $a_2 \pm b_2$, ..., and $a_1 b_1$, $a_2 b_2$, ... Moreover, if a_1, a_2, ... is not an elementary sequence, then $\frac{b_1}{a_1}$, $\frac{b_2}{a_2}$ ... is also a number sequence.*

Two number sequences a_1, a_2, ... and b_1, b_2, ... are said to be **equal** if $a_1 - b_1$, $a_2 - b_2$, ... was an elementary sequence.

To each sequence was associated a symbol; Heine placed the sequence itself in brackets and wrote $[a_1, a_2, a_3, \ldots] = A$. Two symbols A and B were equal if their sequences, $[a_1, a_2, a_3, \ldots]$ and $[b_1, b_2, b_3, \ldots]$ were equal, according to the definition of the equality of two sequences. $A > B$ if $a_n - b_n > 0$ for all n greater than some number.

To the sequence consisting entirely of the *rational number a* would be associated the *symbol a*. For any other sequence, Heine first proved

Proposition 22.4. *The symbol for* any *elementary sequence is* 0.

In other words, all elementary sequences were equal to the elementary sequence $[0, 0, 0, \ldots]$.

The arithmetic operations on the symbols were defined in terms of their sequences: $A \pm B = C$ where C is the symbol for the sequence $[a_1 \pm b_1, a_2 \pm b_2, \ldots]$. Heine gave similar definitions for $AB = C$, and $\frac{B}{A} = C$ for $A \neq 0$. Hence

Proposition 22.5. *If $A \pm B = C$ or $AB = C$, and $A \neq 0$, then $A = C \mp B$, $B = \frac{C}{A}$.*

and thus Heine proved the fundamental laws of arithmetic of the symbols.

Finally, Heine arrived at a key definition: if, for a sequence $[a_1, a_2, \ldots]$, there was some rational number $\mathfrak{U}$ where $\mathfrak{U} - a_n$ was less than any definite value for n larger than some number, then $\mathfrak{U}$ was called the **limit** of the sequence. Moreover,

Proposition 22.6. *If the number sequence a_1, a_2, ... had a limit $\mathfrak{U}$, then $[a_1, a_2, \ldots] = \mathfrak{U}$.*

On the other hand, suppose A was the symbol associated with some sequence $[b_1, b_2, b_3, \ldots]$. Designate B_n to be the sequence $[b_n, b_n, \ldots]$. If $A - B_n$ tended to 0 as n tended to infinity, then A was the limit of the sequence $[b_1, b_2, b_3, \ldots]$. Note that while every sequence had a symbol associated with it, that symbol might or might not be the actual limit of the sequence.

From the set of rational numbers, one can conceive of infinite number sequences and symbols corresponding to the limits of these number sequences. These could be divided into rational numbers, and **irrationals of the first order**. The irrationals of the first order made up a new number class, and one could conceive of number sequences formed from them; the limits would produce the **irrationals of the second order**. One could, in this way, continue, producing irrationals of every order. However, Heine proved:

Theorem 22.1 (Completeness of the Real Numbers). *The irrationals of the second and higher order irrationals are no different from the irrationals of the first order.*

In other words, given a sequence of real numbers with a limit, the limit is another real number. This is a property known as **completeness**.

Uniform Continuity

After introducing a theory of the real numbers, Heine proceeded to the focus of his paper, where he introduced the key property of **uniform continuity**. As defined by Heine, a function was uniformly continuous on the interval $x = a$ to $x = b$

> if for every quantity ϵ, however small, there is a positive quantity η_0, so that for all positive values of η smaller than η_0, $f(x \pm \eta) - f(x)$ is less than ϵ, for whatever value of x, so long as x and $x \pm \eta$ are values between a and b.[3]

Heine then proved a number of propositions regarding uniform continuity, beginning with:

Proposition 22.7. *If $f(x) = x^n$, for n a positive integer, then $f(x)$ is uniformly continuous from $x = a$ to $x = b$, for any values of a and b.*

Heine also gave a proof of

Proposition 22.8. *If $f(x)$ is continuous for every value from $x = a$ to $x = b$, then it is uniformly continuous.*

Heine's proof begins by forming the sequence x_1, x_2, $\ldots$, where x_{m+1} is the first number less than b for which $f(x_{m+1}) - f(x_m) = \frac{\epsilon}{2}$; Heine believed this sequence to be necessarily finite, which is an early version of what would become the **Heine-Borel Theorem**. Hence, one merely takes the width of the smallest of the intervals for η, and the result is proven.

---

[3]Heine, Heinrich, "Die Elemente der Functionenlehre." *Journal für die reine und angewandte Mathematik* 74 (1872), page 184.

The Cantor Set

Both Cantor and Heine considered what we now call the **limit points** of a set: given some set, the limit points are the limits of any convergent sequence of set elements. In essence, a point is a limit point of a set P if there are infinitely many points of the set arbitrarily close to the point. The key discovery was that the limit points of the set of rational numbers were the real numbers, and the limit points of the set of real numbers was simply the set of real numbers.

However, Cantor began to consider other sets, both finite and infinite. Given a set P, Cantor introduced the notion of a **derivative set P′**, which consists of all the limit points of sequences in P.

Example 22.1. *If $P = \left\{ \frac{1}{n} \right\}_{n=1}^{\infty}$, then $P' = \{0\}$.*

P'' is the derivative set of P', and so on; a set P is **perfect** if $P' = P$. A set is **dense in an interval** if, given any two points in the interval, a point of the set may be found between them. It might seem that a perfect set is dense, but in a note in 1883, Cantor gave a counterexample: the set of real numbers on the interval $[0, 1]$ whose ternary expansion

$$\frac{c_1}{3} + \frac{c_2}{3^2} + \frac{c_3}{3^3} + \ldots$$

had $c_i \neq 1$ for all i is perfect, but is dense on no interval. This was the first appearance of what became known as the **Cantor middle-third** set.

22.2.2   Dedekind

Another question that had to be addressed to provide a complete theory of the real numbers was the question of **continuity**. The real numbers form a continuous domain but what, precisely constituted continuity? A solution was presented by Julius Wilhelm Richard Dedekind (October 6, 1831-February 12, 1916) in *Continuity and the Irrational Numbers* (1872), which grew out of lectures he gave in the 1850s.

Dedekind began with the rational numbers, whose existence and properties he assumed. A key property of the rational numbers was that, given a definite rational number a, the rational numbers could be divided into two sets, A_1 and A_2, where all the elements of A_1 were less than a, and all the elements of A_2 were greater than a; a itself could be in either A_1 or A_2. Thus, every rational number a divided the rationals uniquely into two sets, and every element of one set was less than every element of the other set. Dedekind called such a division a **cut** (now called a **Dedekind cut**), and wrote such a cut (A_1, A_2). To indicate the relationship of the number a to the cut it created, Dedekind wrote $a = (A_1, A_2)$.

The property that every rational number produced a cut was analogous to the geometric property of a line, where every point on the line separated the remaining points into two unique sets, one to the left and one to the right. However, Dedekind pointed out, the line had an additional property, that of "continuity." What was the essential feature of continuity? One might suspect that it was the fact that, given any two points of the line, another point on the line could be found between them, but this property, now known as **denseness**, is true for the rational numbers as well.

Rather, Dedekind found the essence of continuity in the following principle, which he discovered on October 24, 1858: not only did every point divide the line into two sets, but every division of the line into two sets was produced by a unique point. The analogous property for the real numbers would be if every cut corresponded to some real number.

To extend the rationals to form the continuous domain of real numbers, Dedekind considered again the division of the rationals into two classes, A_1 and A_2, with *only* the property that every element of A_1 was less than any element of A_2. He referred to this division as the **cut $(\mathbf{A_1, A_2})$**. If the cut was produced by a rational number a, the rational number could be assigned to either set, and would either be the greatest element of A_1, or the smallest element of A_2.

However, some cuts were produced by no rational number. For example, consider the cut where A_1 includes all rationals whose squares are less than 2, while A_2 includes all rationals whose squares are greater than 2. Dedekind said this type of cut *created* a new number, an irrational; the cut was then said to be created by the irrational number. Thus, every real number corresponded to a unique cut in the set of rationals.[4]

Given two cuts, (A_1, A_2) and (B_1, B_2), then either:

1. The cuts were identical, that is, $A_1 = B_1$ and $A_2 = B_2$.

2. The cuts differed by a single number, a, in which case the cuts were "unessentially different," since they were produced by the same number, a.

3. The cuts differed by more than one number, in which case they differed by an infinite number of numbers.

Dedekind then proved some of the properties of the real numbers, including:

Proposition 22.9. *The real numbers form a continuous domain.*

Proof. Consider a cut that divides the real numbers into two sets, $\mathfrak{U}_1$, $\mathfrak{U}_2$, with the property that every number of $\mathfrak{U}_1$ is less than any number of $\mathfrak{U}_2$. This cut in the real numbers corresponds to a cut (A_1, A_2) of the rational numbers; we may assume A_1 consists of the rational numbers in $\mathfrak{U}_1$ and A_2 consists of the rational numbers in $\mathfrak{U}_2$. By assumption, this cut (A_1, A_2) is produced by some number α. Suppose β is any other number. If $\beta < \alpha$, there exists a rational number c so that $\beta < c < \alpha$; hence c belongs to A_1 and consequently to $\mathfrak{U}_1$. Since $\beta < c$, then β also belongs to $\mathfrak{U}_1$. Likewise, if $\beta > \alpha$, then β belongs to $\mathfrak{U}_2$. Hence any real number besides α must belong either to $\mathfrak{U}_1$ or $\mathfrak{U}_2$, so either α is the greatest

---

[4]Two cuts in the case of a rational number, since the rational number itself could be in either A_1 or A_2; however, Dedekind regarded the two "different" cuts here as being essentially the same cut.

number in $\mathfrak{U}_1$ or the least number in $\mathfrak{U}_2$, and thus is the unique real number that produced the cut; hence, every cut is produced by a unique real number, and thus the real numbers form a continuous domain. □

Dedekind then dealt with the problem of arithmetic operations of real numbers. Given two real numbers, a and b, and the two cuts in the set of rationals that they correspond to, (A_1, A_2), (B_1, B_2), then the cut corresponding to the number $a + b$ will be (C_1, C_2), where c is in C_1 if $a_1 + b_1 \leq c$ for some a_1 in A_1 and b_1 in B_1; it is in C_2 otherwise. In a like manner, one could define the multiplication of two real numbers, and, for the first time, provide valid proofs of such statements as $\sqrt{2}\sqrt{3} = \sqrt{6}$.

22.2.3   Transcendental Numbers

After Liouville proved the existence of transcendental numbers, their study was continued by the French mathematician Charles Hermite (December 24, 1822-January 14, 1901) and the German mathematician Carl Louis Ferdinand Lindemann (April 12, 1852-March 6, 1939). Hermite, like so many other mathematicians, began his career by attempting to find a general solution of the fifth degree equation, unaware that Abel had already proven this to be impossible. In 1873, Hermite proved that

$$N + e^a N_1 + e^b N_2 + \ldots + e^h N_n = 0$$

was impossible if N, N_1, N_2, $\ldots$, a, b, $\ldots$, h were integers not all equal to zero; hence e was transcendental. Interestingly, the paper appeared in a British publication, showing that the long isolation of British mathematicians was ending during the nineteenth century.

Hermite believed it would be much more difficult to prove π transcendental. Just nine years later, though, Lindemann would make a slight modification of Hermite's proof, and show that, if z_0, z_1, $\ldots z_r$ were algebraic, and N_0, N_1, $\ldots N_r$ were integers not all equal to zero, then

$$N_0 e^{z_0} + N_1 e^{z_1} + \ldots + N_r e^{z_r} = 0$$

See Problem 6.

was impossible. Hence, it followed that π was not algebraic. Moreover if z was algebraic, e^z was transcendental; hence, the logarithm of an algebraic number was transcendental. An important consequence was that it was impossible to square the circle, not only by compass and straightedge means, but even though the use of solid curves (i.e., conic sections).

Hermite and Lindemann's proofs that e and π were transcendental were simplified by Weierstrass in 1885, who proved further that $\sin \omega$ is likewise transcendental if ω is algebraic (and nonzero); thus, all the trigonometric functions generated transcendental numbers. Rather than being exceptional numbers, it began to appear that transcendental numbers were exceedingly common.

22.2.4 Kronecker

The idea of defining a number in terms of an *infinite* set was hard to swallow; perhaps the greatest objections came from Leopold Kronecker (December 7, 1823-December 29, 1891). Kronecker was one of Kummer's students at the high school in Sorau. He was originally interested in classical philology (like Gauss and Jacobi), but under Kummer's influence, Kronecker began to study mathematics. He followed Kummer to Breslau, and then to Berlin, where, for his dissertation, his thesis examiners consisted not only of the astronomer Encke (application of probability and the method of least squares to astronomy) and mathematician Dirichlet (the theory of infinite series and differential equations), but also August Boeckh (Greek) and Adolf Trendelenburg (the history of legal philosophy).

Kummer's work suggested it was necessary to develop a general theory of numbers of the form $x + y\sqrt{D}$. Kronecker developed such a theory but never published it; later he said that Gauss and Dirichlet had already laid the foundation of the theory, and its development from their work was obvious. Thus, it was left to Dedekind to develop what became the **theory of ideals** and of the **algebraic integers**, an advanced topic we will not deal with further.

Kronecker objected to the free use of infinite sets and irrational numbers. In fact, he argued that *all* mathematics should be based on the notion of natural number alone: to Kronecker is attributed the saying, "God made the whole numbers; all else is the work of man." In "On Number" (1887), he attempted to replace the real and complex numbers entirely with the whole numbers, based on the notion of congruence:

> The concept of negative number can be avoided by replacing the factor -1 with an indeterminate x and using Gaussian congruence modulo $x + 1$. Thus the equation
>
> $$7 - 9 = 3 - 5$$
>
> becomes the congruence[5]
>
> $$7 + 9x \equiv 3 + 5x \mod (x + 1)$$

To avoid fractions, the quotient $\frac{1}{m}$ would be replaced with x_m and an expression would be rewritten as a congruence modulo $mx_m - 1$.

Another key point of Kronecker's mathematical philosophy was that algorithms should include only a finite number of specified steps; "intuitive" algorithms should not be allowed. For example, Kronecker would argue that Liouville's second proof of the existence of transcendental numbers was fundamentally flawed, since the key step, the factorization of an nth degree polynomial, could not be accomplished if $n \geq 5$. Kronecker's reasoning for such a restrictive view was that however the conceptual basis of mathematics shifted, an *algorithm* was

---

[5]Crelle (101), 345

permanent and unchanging. Kronecker contrasted Lagrange's attempt to provide a conceptual basis for calculus, which was no longer accepted, with the Lagrange resolvent, which withstood all changes.

Kronecker's philosophical views placed him at odds with Weierstrass, who freely used infinite sets and processes; the two got into heated debates over the nature of mathematics. The arguments nearly became too much for Weierstrass, who would have left Berlin but for the fact that Kronecker would have chosen his successor.

Kronecker's goal was laudable: the construction of a solid axiomatic foundation for the real numbers, and hence analysis in general. His insistence that infinite processes be eliminated from mathematics was a sensible objection, since by definition, an infinite process never ended, so how could one speak of the end result of such a process? Kronecker's only "sin," in this sense, was to insist that the only possible solution was to abandon the infinite entirely. Hence, he reserved his greatest rancor for the man who would establish the infinite on a sound, axiomatic basis: Georg Cantor.

22.2 Exercises

1. Prove that the division of the rationals into A_1, A_2, where A_1 is the set of all rationals whose square is less than 2, and A_2 is the set of all rationals whose square is greater than 2, is a Dedekind cut that is produced by no rational number.

2. Dedekind claimed that one could find a rigorous basis for the arithmetical operations on the real numbers. Consider two real numbers, $a = (A_1, A_2)$ and $b = (B_1, B_2)$.

   (a) To what cut (C_1, C_2) should the difference $a - b$ correspond?

   (b) To what cut (C_1, C_2) should the product $a \cdot b$ correspond? What difficulties do you encounter?

   (c) To what cut (C_1, C_2) should the quotient $a \div b$ correspond? What difficulties do you encounter?

3. Prove $\sqrt{2}\sqrt{3} = \sqrt{6}$.

4. Dedekind claimed that no one had ever proven $\sqrt{2}\sqrt{3} = \sqrt{6}$. Although this is true, could such a proof been found before the introduction of Dedekind cuts on the rationals? Explain, in terms of the theory of real numbers prior to Dedekind's work.

5. Weierstrass did not think much of Dedekind's definition of a real number, saying that it was no different from Eudoxus's. Explain and evaluate.

6. Use Lindemann's result to prove that π is not algebraic. Explain why this proves the impossibility of squaring the circle, not only by using compass and straightedge alone, but by using any algebraic curves.

7. Show that e^z is transcendental if z is algebraic, hence the logarithm of an algebraic number is transcendental.

8. Show that if $\cos x$ is rational, x is transcendental. Hint: $\cos x = \frac{e^{xi} + e^{-xi}}{2}$.

9. Explain the difference between a function that is continuous and a function that is uniformly continuous.

10. Give an example of a continuous function that is not uniformly continuous.

11. Prove that if $f(x)$ is bounded and continuous over an interval, it must be uniformly continuous over that interval.

12. Prove Proposition 22.8. In particular, explain why the sequence x_1, x_2, ... must necessarily be a finite sequence.

13. Show that the Cantor middle-thirds set is a perfect, nowhere dense set.

14. What equivalencies can be used to eliminate the following numbers: -5, $1 + \sqrt{-2}$? What difficulties would be encountered trying to extend this notion of congruence to all real numbers?

15. Prove, using Kronecker's congruences, that $-5 + -3 = -8$.

16. What is the logical indefensibility in claiming that the real number $\sqrt{2}$ "exists" because it represents the length of the diagonal of the unit square?

22.3   The Natural Numbers

It might be surprising to find out that the theory of the natural numbers was not created until *after* a theory of the real numbers. However, this underscores one of the paradoxes of mathematics: the simpler a concept, the harder it is to define. Since "everyone knows" what a natural number is, they are very difficult to define precisely.

22.3.1   Frege

In the *Elements*, a number was defined to be a "collection of units," a definition that was used for two thousand years. By the mid-nineteenth century, it was realized that this definition was insufficient, and that an axiomatic basis for the natural numbers had to be created. The philosopher and logician Friedrich Ludwig Gottlob Frege (November 8, 1848-July 26, 1925) was the first to publish a theory of the natural numbers in *The Foundations of Arithmetic* (1884).

Frege pointed out the logical difficulties with the previous definitions. Defining number as a *collection* of units implicitly referred to the number concept. For example, if we define 0 as the number associated with the empty set, we are simply saying that 0 is the number associated with the set that has 0 elements, and our definition is circular. After trying several definitions and showing the logical flaws in each, Frege adopted the following definition of 0:

Definition. 0 *is the number associated with the set concept "Not identical with itself."*

Frege then proved an important proposition: all sets "not identical with itself" were identical; in other words, $0 = 0$.

To obtain numbers besides 0, Frege began with

Definition. 1 *is the number of the set F, "Identical to 0."*

0 is, of course, in F, since 0 is identical to 0.

The higher numbers were defined recursively:

Definition. *Given any set K associated with some natural number, and any element x of K. Let a new set L be "In K but not identical to x." The natural number associated with K is the natural number following L, and likewise, L precedes K.*

Hence:

Proposition 22.10. 1 *is the natural number following 0.*

22.3.2 Peano

Frege's reduction of the natural numbers to propositions in pure set theory might, from a logical viewpoint, be viewed as the best of all possible worlds, as it required the fewest axioms. However, from the viewpoint of mathematicians, it would be better to suppose a few simple axioms and derive the properties of the whole numbers from these simple axioms. This work was accomplished by the Italian mathematician Giuseppe Peano (August 27, 1858-April 20, 1932). Like Dirichlet, Peano suffered an attack of smallpox when young. In 1889, Peano presented "The Principles of Arithmetic," doing for arithmetic what Euclid's *Elements* did for geometry. After discussing some of the basic principles of logic, Peano then began to derive the natural numbers axiomatically. A few new symbols were necessary: $a + 1$ meant the **successor to a**, where the "$+1$" should be considered a meaningless symbol with no relationship to the operation of addition. N meant the set of numbers, 1 was unity, and $=$ indicated equality. Peano then introduced nine axioms in symbolic form; translated, these axioms are:

1. 1 is a number.

2. If a is a number, then $a = a$.

3. If a, b are numbers, and $a = b$, then $b = a$.

4. If a, b, c are numbers, and $a = b$ and $b = c$, then $a = c$.

5. If $a = b$ and b is a number, so is a.

6. If a is a number, then $a + 1$ is a number.

7. If a, b are numbers, and $a = b$, then $a + 1 = b + 1$.

8. If a is a number, then $a + 1$ is not 1.

9. If k is in N, and 1 is in k, and if, whenever x is a number in K, then $x + 1$ is also in K, then N is K.

A number of definitions follow: 2 is $1 + 1$ (in other words, 2 is the successor to 1); 3 is $2 + 1$ (3 is the successor to 2), and so on. Peano began with a few simple propositions:

Proposition 22.11. *2 is a number.*

Proof. By assumption, 1 is a number, and the successor of any number is a number; hence $1 + 1$ is a number. But we defined $2 = 1 + 1$, hence 2 is a number. $\square$

A key definition is:

Definition (Addition). *If a, b are numbers, then $a + (b + 1) = (a + b) + 1$.*

Peano noted that this definition means that $a + (b + 1)$ is the number designated by the successor to $(a + b)$, provided $(a + b)$ itself has meaning. In this way, we can define $a + 2$, since $2 = 1 + 1$, and $a + (1 + 1) = (a + 1) + 1$; in other words, $a + 2$ is the successor to the successor of a.

Proposition 22.12. *If a, b are numbers, so is $a + b$.*

To prove this proposition, Peano used the ninth axiom, now called the **axiom of mathematical induction**.

Proof. Since a is a number, then $a + 1$ is also a number, so 1 is an element of the class of numbers T that satisfies the proposition. Suppose b is a number that satisfies the proposition, "$a + b$ is a number." Then $(a + b) + 1$ is a number, since $(a + b) + 1$ is the successor to $(a + b)$, which is a number. But by definition, $(a + b) + 1 = a + (b + 1)$. Hence, $b + 1$ is an element of T. Thus 1 is in T, and if b is in T, so is $b + 1$. Hence N is T, and all natural numbers satisfy the proposition. $\square$

Peano continued in this fashion, proving all the properties of arithmetic. For subtraction, Peano defined

Definition. *If a, b are numbers, then $b - a$ is a number if, for some x, $x + a = b$.*

Definition. *If a, b are numbers, then $a < b$ if $b - a$ is a number.*

The key deficiency of this axiomatization of arithmetic was that, though all the properties of the natural numbers could be derived from these axioms, other sets existed that *also* satisfied all the properties of the natural numbers (for example, the even numbers). In other words, the natural numbers did not form a unique class, according to Peano's axiomatization. Peano was aware of this, and carefully avoided defining what a *number* actually was.

22.3 Exercises

1. Prove Proposition 22.10, using Frege's definition.

2. Using Frege's definitions, how would you define "2"? Based on your definition, show that 2 follows 1 in the number sequence.

3. Prove that if a, b, c are numbers, then if $a = b$, then $a + c = b + c$.

4. If a, b are numbers:

   (a) Prove $a + 1 = 1 + a$.   (b) Prove $a + b = b + a$.

5. Give some examples of other sets that satisfy the axioms of arithmetic according to Peano.

6. Prove that, for a, b, a', and b' numbers, with $a = a'$, $b = b'$, then $b - a = b' - a'$.

22.4   The Infinite

It might seem difficult to believe that the concept of infinity is an important one, since a truly infinite set cannot be created. Calculus avoids dealing with the actual infinite by considering limits: a quantity such as $\frac{1}{x^2}$ is said to have a **limit of infinity** as x tends to 0, since given any finite number N, $\frac{1}{x}$ can be made larger than N for x sufficiently close to 0. In this way, actual infinities are avoided; what one has instead is a potential infinity: a quantity that could *potentially* be larger than any quantity, but is always finite. In the same way, the set of *all* the natural numbers could be conceived to be a potentially infinite set, formed by extending the finite set $\{1, 2, 3, \ldots, n\}$.

But consider a phrase such as, "the set of all rational numbers," essential to the Dedekind definition of a real number. We might suppose this, too, is a potentially infinite set, but how can an existing set be extended so it would include all the rational numbers? For example, if the set is $\left\{1, \frac{1}{2}, \frac{1}{3}, \ldots, \frac{1}{n}\right\}$, then any extension of the set would omit all nonunit fractions. Furthermore, while it is obvious whether or not a set contains all natural numbers, it is not as clear whether a given set will contain all rational numbers. It seems that the only way to create the infinite set of all rational numbers is to begin with the infinite set of all rational numbers. It was questions such as these that Cantor addressed between 1874 and 1895, and in doing so, Cantor founded modern **set theory**.

22.4.1 Countability

One problem was comparing two infinite quantities. Galileo had denied any useful comparisons could be made, but Halley suggested that phrases like "twice as infinite" or "one-fourth as infinite" might have meaning. Cantor used the idea of a one-to-one correspondence between two sets as the basis for determining whether two sets were of the same size, or whether one was larger than the other. On November 29, 1873, Cantor wrote to Dedekind, saying that he had found a proof that the rationals could be put in a one-to-one correspondence with the natural numbers, and speculated on whether or not a correspondence between the real numbers and the natural numbers could also be found. A month later, on December 7, 1873, Cantor wrote that he had found that they could not: the real numbers could *not* be put in a one-to-one correspondence with the natural numbers. His results were published in "On a Property of the Real Algebraic Numbers" (1874).

See page 426.

In general, if an infinite set can be put into a one-to-one correspondence with the natural numbers, it is said to be **countable** or **denumerable**. Thus, Cantor proved first that the rationals were countable, while the real numbers were uncountable. First, Cantor proved

Proposition 22.13 (Countability of the Algebraic Numbers). *The real algebraic numbers can be put in a one-to-one correspondence with the positive whole numbers.*

Proof. Let ω be an algebraic number; then by definition, there is an equation

$$a_0 \omega^n + a_1 \omega^{n-1} + \ldots + a_n = 0$$

where $a_0, a_1, \ldots, a_n$ are integers; we can assume a_0 is a positive whole number; we further assume that n is of the least degree (i.e., that no equation of degree less than n has ω as a solution) and that the a_i's have no common factors. Let the **height N of** ω be defined by

$$N = n - 1 + |a_0| + |a_1| + |a_2| + \ldots + |a_n|$$

Hence, any algebraic number ω has a definite height N associated with it. Clearly, the number of algebraic numbers of a given height N is finite. Let $\phi(n)$ be the number of algebraic numbers of height n. For example, $\phi(1) = 1$, $\phi(2) = 2$, $\phi(3) = 4$, and so on.

Thus, we may order the algebraic numbers. Since there is only one algebraic number of height 1, let ω_1 be that number. There are two algebraic numbers of height 2; let ω_2, ω_3 be these numbers (the order of the numbers does not matter). ω_4, ω_5, ω_6, and ω_7 are the algebraic numbers of height 3, and so on. Hence, all algebraic numbers may be put into the order

$$\omega_1, \omega_2, \omega_3, \ldots$$

which clearly can be put in a one-to-one correspondence with the positive whole numbers. $\qquad\square$

An immediate corollary is that the rational numbers can be put into a one-to-one correspondence with the positive whole numbers. Note this proof is constructive: not only is the countability proven, but a clear method exists for creating the infinite set of rational numbers. Cantor went on to prove

Proposition 22.14 (Uncountability of the Reals). *The real numbers cannot be put in a one-to-one correspondence with the positive whole numbers.*

Proof. Suppose they can, generating the sequence $\omega_1, \omega_2, \ldots$. Consider any interval, (α, β). Let α', β' be the first two numbers of the sequence of real numbers in (α, β). Let α'', β'' be the first two numbers of the sequence in (α', β'), and so on. We construct a nested sequence of intervals, (α, β), (α', β'), $(\alpha,'' \beta'')$, $\ldots$, $(\alpha^{(n)}, \beta^{(n)})$, $\ldots$.

Suppose this sequence of intervals comes to an end, and we have an interval (α^n, β^n) in which there are no numbers of the sequence. Then, obviously, any η in this interval will be a number not included in our list, which is impossible, since the list was assumed to include all real numbers. Alternatively, suppose the sequence of intervals does not come to an end. Consider that α, α', α'', $\ldots$ is an increasing and bounded sequence of real numbers, while β, β', β'', $\ldots$ is a decreasing and bounded sequence of real numbers. Hence, they both have limits, say a^∞ and b^∞. If $a^\infty \neq b^\infty$, then again, any η in (a^∞, b^∞) would serve as a number not in the preceding sequence of real numbers.

But if $a^\infty = b^\infty = \eta$, then η cannot be in the sequence of real numbers, since by its construction, ω_n cannot be in $(a^{(m)}, b^{(m)})$ for $m \geq n$. Thus, in any case, there must be a real number η not in our sequence; hence the sequence is incomplete, and the real numbers *cannot* be put into a one-to-one correspondence with the positive whole numbers.   □

A corollary to this proof was the existence of transcendental numbers.

22.4.2   Space-Filling Curves

Other results regarding the infinite soon followed. On January 5, 1874, in yet another letter to Dedekind, Cantor raised the question of whether or not the unit square could be put in a one-to-one correspondence with the unit interval, and expressed the belief that such a correspondence was not possible. It took over three years, but on June 28, 1877, Cantor conveyed to Dedekind the surprising result that such a correspondence *was* possible: "I see it, but I do not believe it," Cantor wrote. Cantor's proof was remarkably simple: consider the decimal expansion

$$a = a_1 \frac{1}{10} + a_2 \frac{1}{10^2} + a_3 \frac{1}{10^3} + \ldots$$

of any real number a on the unit interval $[0,1]$. a could be mapped to the point in the unit square (often designated $[0,1]^2$) with coordinates

$$x = a_1 \frac{1}{10} + a_3 \frac{1}{10^2} + a_5 \frac{1}{10^3} + \cdots$$
$$y = a_2 \frac{1}{10} + a_4 \frac{1}{10^2} + a_6 \frac{1}{10^3} + \cdots$$

Conversely, any point (x, y) could be mapped to a real number. For example, the real number 0.123456 was mapped to the point $(0.135, 0.246)$, while the point $(0.123, 0.456)$ was in turn mapped to the real number 0.142536. Cantor believed the correspondence was unique: in other words, *every* real number corresponded to a *unique* point, and conversely. Dedekind pointed out an important flaw in Cantor's argument: a terminating decimal, such as 0.5, could be written in two ways, as $0.5000\ldots$ or as $0.4999\ldots$. The flaw, as Cantor remarked, affected "only the proof, not the result," though fixing it sacrificed (in Cantor's viewpoint) the essential simplicity of his earlier argument.

An easy extension was that the unit interval $[0,1]$ could be mapped into any n-dimensional cube $[0,1]^n$. This raised questions about the nature of "dimension," for if a single real number could express the location of a point in two (or three) dimensional space, then the naive connection between dimension and coordinates was impossible to sustain. Was dimension a meaningless concept? Since Cantor's mapping was not continuous, it might seem that continuous mapping was key to giving dimension unambiguous meaning. A flurry of results soon followed. In July 1878, Jakob Lüroth (1844-1910) proved the impossibility of a continuous one-to-one mapping of a line onto a space of dimension greater than 2, or of a plane region into a three or more dimensional space. However, the proof became so complicated that he was unable to generalize it to n-dimensional space. Similar proofs were presented by Enno Jürgens (1849-1907) in August and by Johannes Thomae (1840-1921) somewhat later.

In October 1878, Eugen Netto (1848-1919) addressed a critical point: what *was* dimension? Netto gave a simple example that illustrated his key point. Suppose a continuous, one-to-one mapping exists from the plane M_2 to the line M_1. Take a line segment (a_2, b_2) in M_2, and suppose its image is the line segment (a_1, b_1) in M_1. Choose two points on M_1, p and p', and suppose p is in (a_1, b_1) but p' is outside it. Because of the continuity and the one-to-one nature of the mapping, p and p' correspond to points in the plane, say r and r'. However, in going from p to p' in M_1, it was necessary to cross through one of the endpoints, a_1 or b_1, whereas to go from r to r', it was possible to avoid passing through the endpoints a_2 and b_2. Hence, the one-dimensional line possesses a property that the two dimensional plane does not: a point may serve as a boundary. In general, Netto noted, the key property of dimensionality was that an $n + 1$-dimensional object had an n-dimensional boundary. This ideas echoes Euclid's definition of the "ends" of a line being points, and so on.

Peano, though he is best known for his axiomatization of the natural numbers, considered his contributions to analysis to be more important. In 1886, he

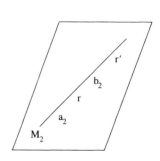

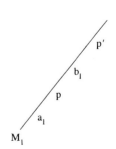

provided the first rigorous proof that the differential equation $y' = f(x, y)$ had a solution, provided $f(x, y)$ was a continuous function. The proof of this involved the so-called **axiom of choice**: given an infinite set of sets, it is possible to create a set consisting of one element from each set. Kronecker would have been aghast at such an axiom, for it clearly and explicitly assumed the existence of a mathematical object that could only be constructed using an infinite number of steps.

Lüroth's work left open the possibility of a continuous mapping of a line onto a plane, and Peano's "On a Curve, Which Fills an Area in the Plane" (1890) showed how to construct a curve that passed through every point in a plane region. Consider a sequence $T = 0.a_1a_2a_3 \ldots$, where the a_n was either 0, 1, or 2. Let $k(a) = 2 - a$, and the notation $k^n(a)$ mean "apply k to a n times in sequence." From this sequence T, let two other sequences be formed:

$$X = 0.b_1b_2b_3 \ldots \qquad\qquad Y = 0.c_1c_2c_3 \ldots$$

where the b's and c's have the relationship

$$b_1 = a_1 \qquad\qquad\qquad c_1 = k^{a_1}(a_1)$$
$$b_2 = k^{a_2}(a_3) \qquad\qquad c_2 = k^{a_1+a_3}(a_4$$
$$b_3 = k^{a_2+a_4}(a_5) \qquad\qquad c_3 = k^{a_1+a_3+a_5}(a_6)$$
$$\vdots \quad \vdots \qquad\qquad\qquad\qquad \vdots \quad \vdots$$
$$b_n = k^{a_2+a_4+\ldots+a_{2n-2}}(a_{2n-1}) \qquad c_n = k^{a_1+a_3+a_5+\ldots+a_{2n-1}}(a_{2n})$$

Given some sequence $T = 0.a_1a_2a_3 \ldots$, define Val T to be

$$\text{Val } T = \frac{a_1}{3} + \frac{a_2}{3^2} + \frac{a_3}{3^3} + \ldots$$

Thus, for every sequence T there is a number t in $0 \leq t \leq 1$. Conversely, the numbers $0 \leq t \leq 1$ divide into two classes: either t, when multiplied by some power of 3, yields a whole number, in which case this value corresponds to two sequences, T and T', where

$$T = 0.a_1a_2a_3 \ldots a_{n-1}a_n 2222 \ldots$$
$$T' = 0.a_1a_2a_3 \ldots a_{n-1}a_n' 0000 \ldots$$

where $a_n' = a_n + 1$; otherwise, t has a unique expression as a sequence T.

Now, if t corresponds to a unique sequence T, then t also corresponds to a unique point (Val X, Val Y). On the other hand, if t is such that there are two sequences, T and T' that correspond to it, and X, Y derived from T and X', Y' derived from T', it can be shown that Val X = Val X' and Val Y = Val Y'. Hence, every real number in $0 \leq t \leq 1$ corresponds to a unique point (x, y) in the unit square; moreover, the relationship is a continuous one.

22.4.3 Orders of Infinity

In 1883, Cantor published the "On Infinite Linear Point Aggregates," the last of five articles of the same name. Over half the article is devoted to justifying the actual existence of the infinite in mathematics. The remainder developed the theory of the infinite ordinals as an extension of the theory of the whole numbers. A key point is the notion of the **power of a set**: two sets had the same power if they could be put in a one-to-one correspondence with each other. Today we say that the two sets have the same **cardinality**. Two sets of the same power can be thought of as having the same number of elements.

Ordinal Numbers

Cantor's development of the orders of infinity is reminiscent of Archimedes's octad notation: in both cases, a set of numbers is constructed, then a new number is generated, based on those already defined. In Cantor's viewpoint, numbers were created by two generating principles. The **first generating principle** was that for every whole number a, there was a number following it, $a + 1$.

See page 155.

The sequence ν of natural numbers could be created by beginning with 1 and applying the first generating principle:

$$1, 2, 3, \ldots$$

Clearly, this sequence could be extended indefinitely; Cantor called this sequence the **numbers of the first class**. To proceed beyond it, Cantor envisioned a **second generating principle**: for any definite succession for which there is no greatest term, a new number could be created, called the **limit of the succession**. This limit was the next higher number in the sequence, and greater than all the numbers preceding it. Thus, the sequence of whole numbers gave rise to a new number, ω, greater than any of them. Cantor described ω as the *first* number that was greater than any of the whole numbers.

The first generating principle could be applied to ω to produce the sequence

$$\omega + 1, \omega + 2, \ldots$$

By the second generating principle, this sequence had a limit that was greater than any term in the sequence. This limit can be written as $\omega + \omega$, or 2ω. Continuing to apply the first and second generating principles, Cantor constructed a set of numbers

$$
\begin{array}{ccccc}
1, & 2, & 3, & \ldots & \omega \\
\omega + 1 & \omega + 2 & \omega + 3 & \ldots & 2\omega \\
2\omega + 1 & 2\omega + 2 & 2\omega + 3 & \ldots & 3\omega \\
3\omega + 1 & 3\omega + 2 & 3\omega + 3 & \ldots & 4\omega \\
\vdots & \vdots & \vdots & \ddots &
\end{array}
$$

In this way, numbers of the form $\mu\omega + \nu$ are constructed. Again applying the second generating principle, numbers of this sequence had a limit, which could be called ω^2. Applying the first generating principle gave rise to the sequence ω^2, $\omega^2 + 1$, $\omega^2 + 2$, ..., which, via the second generating principle, had limit $\omega^2 + \omega$. In this way are built up all numbers of the form

$$\nu_0 \omega^\mu + \nu_1 \omega^{\mu-1} + \ldots + \nu_{\mu-1}\omega + \nu_\mu$$

The second generating principle implied a limit, which could be expressed as ω^ω. Clearly, the application of the first and second generating principles would continue to produce new numbers.

Cantor referred to all the numbers, beginning with ω, that could be created using the first and second generating principle as **numbers of the second class**. Cantor's first remarkable result was that the numbers of the second class of the form

$$\nu_0 \omega^\mu + \nu_1 \omega^{\mu-1} + \ldots + \nu_{\mu-1}\omega + \nu_\mu$$

could be put in a one-to-one correspondence with the numbers of the first class; hence the two sets had the same power.

The numbers of the second class might be written

$$\omega, \omega + 1, \ldots, \omega^\omega, \ldots, \alpha, \ldots$$

Because these numbers have a distinct order, they are called the **transfinite ordinals**. Cantor's second remarkable result was that the power of this set was *not* the same as the power of the whole numbers. The proof is as follows: suppose that the set *can* be put in a one-to-one correspondence with the whole numbers, say in the order $\alpha_1, \alpha_2, \alpha_3, \ldots$ where α_ν represents a definite transfinite number. From this sequence, create a new sequence, defined as follows: let $\alpha_{\varkappa\alpha_2}$ be the first number of the second class greater than α_1; let $\alpha_{\varkappa\alpha_3}$ be the first number of the second class greater than $\alpha_{\varkappa\alpha_2}$, and so on. One then has a sequence of numbers

$$\alpha_1 < \alpha_{\varkappa\alpha_2} < \alpha_{\varkappa\alpha_3} < \ldots$$

where

$$1 < \varkappa\alpha_2 < \varkappa\alpha_3 < \ldots$$

and

$$\alpha_\nu < \alpha_{\varkappa_\lambda}$$

whenever

$$\nu < \varkappa_\lambda$$

Given the sequence $\alpha_1, \alpha_{\varkappa_2}, \alpha_{\varkappa_3}, \ldots$, there are two possibilities. First, it might occur that for some n, $\alpha_{\varkappa_n}$ the greatest of all the ordinals. However, this is impossible, for (by the first generating principle), one can create $\alpha_{\varkappa_n} + 1$, which is a greater ordinal and, by assumption, a number of the second class.

Thus, the sequence $\alpha_1, \alpha_{\varkappa_2}, \alpha_{\varkappa_3}, \ldots$ is infinite. Cantor created a new sequence as follows: consider all the numbers from 1 to α_1; append to this sequence all the numbers from α_1 to $\alpha_{\varkappa_2}$; then all the numbers from $\alpha_{\varkappa_2}$ to $\alpha_{\varkappa_3}$, and so on. In this way, a sequence was formed with no greatest term; hence by the second generating principle, this sequence had a limit β, greater than any term of the sequence. However, β must be a number of the second class, and hence must be a term of the sequence, which is impossible. Thus, the numbers of the second class *cannot* be placed in a one-to-one correspondence with the numbers of the first class; it must have a different power: it was "more infinite" than the set of natural numbers. Cantor went on to prove that the power of the numbers of the second class was, in fact, the next higher transfinite number.

Cardinal Numbers

Cantor refined this early work, and in 1890, he published "On an Elementary Question of Point Aggregates," where he proved anew that the real numbers could not be put in a one-to-one correspondence with the whole numbers. Cantor's proof is very similar to that usually given today and now referred to as the **diagonal proof**: it is simpler than his first proof, but does require consideration of how the real numbers are expressed.

Cantor considered the set M of real numbers, expressed in terms of two characters, m and w (for example, we may think of the binary expansion of a real number); M consists of all sequences that can be formed from these two characters.

For any sequence E_1, E_2, E_3, ... of real numbers expressed in this fashion, there is a number E_0, which corresponds to no E_ν. Suppose, for example

$$E_1 = (a_{1,1}, a_{1,2}, \ldots, a_{1,\nu}, \ldots)$$
$$E_2 = (a_{2,1}, a_{2,2}, \ldots, a_{2,\nu}, \ldots)$$
$$\vdots$$
$$E_\mu = (a_{\mu,1}, a_{\mu,2}, \ldots, a_{\mu,\nu}, \ldots)$$
$$\vdots$$

Construct the sequence b_1, b_2, ... as follows: $b_\nu = w$ if $a_{\nu,\nu} = v$; $b_\nu = v$ if $a_{\nu,\nu} = w$. Then the number

$$E_0 = (b_1, b_2, b_3, \ldots)$$

is not part of the sequence E_1, E_2, ..., since, by construction, $b_n \neq a_{nn}$, thus E_0 is different from each of the E_n for any n. Thus, the real numbers must have a different power from the whole numbers.

In 1895 and 1897, Cantor published "Contributions to the Foundation of the Transfinite Numbers." In these articles, he introduced a few changes that put the theory of the infinite in its modern form. Given a set, $M = \{a, b, c, \ldots\}$, Cantor imagined a "double act of abstraction," $\overline{\overline{M}}$ that removed, from the elements of M, both their order and their identity. The first act of abstraction reduced M to an unordered set, while the second reduced the elements of M to identical units. Hence, $\overline{\overline{M}}$ could be thought of as a collection of units: a number, in the

Euclidean sense. Cantor called $\overline{\overline{M}}$ the **cardinal number** associated with the set M.

Two sets M and N are equivalent, written $M \sim N$, if there was a one-to-one correspondence between the elements of M and N. Obviously, $M \sim M$; moreover, if $M \sim N$ and $N \sim P$, then $M \sim P$. An important theorem is that $\overline{\overline{M}} = \overline{\overline{N}}$ if and only if $M \sim N$.

Having defined equality, Cantor then defined inequality. Suppose $a = \overline{\overline{M}}$ and $b = \overline{\overline{N}}$. Then if no subset of M was equivalent to N, but there was a subset N_1 of N that was equivalent to M, then we say $a < b$; in other words, if M could be put in a one-to-one correspondence with a subset of N, then the cardinality of M was less than the cardinality of N.

From here, the arithmetical operations were defined. Let M and N be two aggregates with no common elements and cardinalities a and b, respectively. Form the union (M, N), and define $a + b = \overline{\overline{(M, N)}}$. The **combining set** $(M \cdot N)$ was the set of all elements (m, n), where m was any element of M and n any element of N; the product $a \cdot b$ was then $\overline{\overline{(M \cdot N)}}$. From these definitions of sum and product, the fundamental laws of arithmetic could be derived.

Exponentiation involved a new definition. Given two sets M and N, with $\overline{\overline{M}} = a$ and $\overline{\overline{N}} = b$, let every element n of N be associated with a specific element of M, designated $f(n)$; it is permissible for two elements of N to be associated with the same element of M. f is called the **covering function of N**; we note that $f(N)$ is a subset of M. For example, if $M = \{1, 2\}$ and $N = \{x, y, z\}$, then one possible covering of N would be $f(n) = 1$ for all n in N. Two covering functions, f_1 and f_2, are equal if for every element n of N, $f_1(n) = f_2(n)$; otherwise they are different coverings. Designate by $(N|M)$ the set of all possible coverings $\{f(N)\}$, and define $a^b = \overline{\overline{(N|M)}}$.

Next, Cantor defined the finite cardinal numbers. For the set E_0 consisting of a single element $\{e_0\}$, Cantor associated the cardinal $1 = \overline{\overline{E_0}}$. The set $E_1 = \{E_0, e_1\}$, and $2 = \overline{\overline{E_1}}$; in this way, the sequence of natural numbers was built up, by forming a new set consisting of the union of the previous set E_n with a new element, e_{n+1}. Based on these definitions, Cantor then proved;

Proposition 22.15. *Every natural number*

$$1, 2, 3, \ldots, \nu, \ldots$$

is different from every other natural number.

Proposition 22.16. *Every number ν is greater than every number preceding, and less than every number following.*

Proposition 22.17. *Given any cardinal number ν, there is no cardinal number between ν and $\nu + 1$ (i.e., between ν and the cardinal number following it in succession).*

Consider $\{\nu\}$, the set of all natural numbers. This was a transfinite aggregate; Cantor associated with $\overline{\overline{\{\nu\}}}$ the cardinal number $\aleph_0$, where $\aleph$ ("aleph") was the first letter of the Hebrew alphabet.[6] Furthermore, $\aleph_0$ was the smallest of the transfinite numbers.

The arithmetic of the transfinite number $\aleph_0$ was different from that of the finite numbers; for example, $\aleph_0 + 1 = \aleph_0$. Indeed, $\aleph_0 + \aleph_0 = \aleph_0$. This follows from the definitions, for let there be two sets, $\{a_\nu\}$ and $\{b_\nu\}$, where $\overline{\overline{\{a_\nu\}}} = \overline{\overline{\{b_\nu\}}} = \aleph_0$. Then by definition, $\aleph_0 + \aleph_0 = \overline{\overline{(\{a_\nu\}, \{b_\nu\})}}$. We have also $\{\nu\} = (\{2\nu - 1\}, \{2\nu\})$; in other words, the set of natural numbers can be considered as the set of odd numbers, together with the set of even numbers. But $(\{2\nu - 1\}, \{2\nu\}) \sim (\{a_\nu\}, \{b_\nu\})$ (since $2\nu - 1$ can be associated with the element a_ν, and 2ν can be associated with the element b_ν. Thus $\overline{\overline{\{\nu\}}} = \overline{\overline{(\{a_\nu\}, \{b_\nu\})}}$, and so $\aleph_0 = \aleph_0 + \aleph_0$.

Alternatively, we may write $\aleph_0 + \aleph_0 = 2 \cdot \aleph_0$. By repeatedly adding $\aleph_0$ to both sides, we have $\aleph_0 = \nu \cdot \aleph_0$ for all finite ν and, in fact, $\aleph_0 \cdot \aleph_0 = \aleph_0$. By definition, $\aleph_0 \cdot \aleph_0$ is the cardinal number of the combining set (μ, ν), where μ and ν are any finite cardinal numbers. It is necessary to show that this is equivalent to another set, λ, which has cardinality $\aleph_0$. Cantor's proof: consider $\rho = \mu + \nu$. Then for any finite ρ, there are $\rho - 1$ elements of (μ, ν) for which $\mu + \nu = \rho$. For example, if $\rho = 4$, then we have the $4 - 1 = 3$ elements $(1, 3), (2, 2), (3, 1)$. We can thus arrange the combining set (μ, ν) by setting down the one element where $\rho = 2$, the two elements where $\rho = 3$, and so on;

$$(1,1), (1,2), (2,1), (1,3), (2,2), (3,1), \ldots$$

and thus any element (μ, ν) of $\{(\mu, \nu)\}$ comes at a determinable place λ; hence the combining set has the same cardinality as λ, namely $\aleph_0$. Incidentally, this shows the rational numbers have cardinality $\aleph_0$. Since $\aleph_0 \cdot \aleph_0 = \aleph_0^2$, we obtain $\aleph_0^\nu = \aleph_0$ for all finite ν.

Cantor had already shown that a higher type of transfinite number existed, which could be derived by consideration of the transfinite ordinals: this was the transfinite number associated with the number of numbers of the second class. This cardinal number he designated $\aleph_1$. Likewise, from $\aleph_1$ arose through some definite law a next higher transfinite, $\aleph_2$, and so on; we may designate the sequence of transfinite cardinals as

$$\aleph_0, \aleph_1, \aleph_2, \ldots$$

Even this did not exhaust the possibilities of transfinite cardinal numbers, for Cantor's second generating principle implied a transfinite number, $\aleph_\omega$, greater than all of them. Thus, Cantor established a hieararchy of the transfinite numbers.

---

[6]Cantor was not the first to introduce letters of the Hebrew alphabet to mathematics; Clairaut, while trying to solve the equation of the Moon's motion, introduced a number of Hebrew letters to represent various constants that appeared in the computations. Incidentally, neither Cantor nor Clairaut were Jewish.

Questions

Two important questions were raised by Cantor's work. The first stemmed from the observation that $\aleph_0$, the cardinality of the whole numbers, was less than the cardinality of the number of real numbers in any interval; this last could be expressed by the transfinite number $2^{\aleph_0}$. It might be tempting to suppose $\aleph_1$, the next higher transfinite number, was in fact equal to $2^{\aleph_0}$: the hypothesis that $\aleph_1 = 2^{\aleph_0}$ is called the **continuum hypothesis** and, despite Cantor's best efforts, he could not prove it.

Another question considered the ordering of the set. A set is ordered if there is an order relationship, $<$, so that for any elements x, y, and z, then either $x < y$ or $y < x$, and if $x < y$ and $y < z$, then $x < z$. The set is **well ordered** if, in addition to these properties, every nonempty subset has a first element. For example, the algebraic numbers are well ordered, since they can be put in a one-to-one correspondence with the natural numbers, and we can define $\omega_n < \omega_m$ if $n < m$. Cantor believed that *any* well-defined set could be well ordered, but was unable to prove this.

22.4.4   The Heine-Borel Theorem

Reaction to Cantor's groundbreaking work was mixed. David Hilbert once compared Cantor's theory of sets to a paradise from which mathematicians would (hopefully) never be expelled. Kronecker, on the other hand, was so opposed to Cantor's ideas that Cantor was forced to seek positions in less prestigious universities (ending in Halle) and eventually he suffered a mental breakdown.

Regardless of how set theory was perceived, it produced a host of new results and new insights into the nature of the real numbers and the theory of functions. Consider a set E of real numbers, and some *finite* set of intervals I_k whose union contains E. In 1882, Hermann H. Hankel (1839-1873) defined the **content of E** to be the lower bound of the sum of the lengths of the I_k, for all possible I_k. At the same time, Paul du Bois-Reymond and Axel Harnack removed the condition that the sets I_k be finite in number.

Even if the sets were infinite in number, a finite subset of I_k might suffice to cover E. In his doctoral thesis (1894), Emil Borel (January 7, 1871-February 3, 1956) included a key theorem in real analysis. The theorem was implicit in Heine's earlier work; hence it is usually called the Heine-Borel theorem:

See page 671.

Theorem 22.2 (Heine-Borel Theorem). *Given an infinite number of intervals whose union contains a closed interval, a finite number of the intervals will contain the same interval.*

Borel proved his result using transfinite induction and Cantorian methods to their fullest. The Paradise alluded to by Hilbert had been entered.

22.4 Exercises

1. Show that if t can be multiplied by a power of 3 to produce a whole number, then $t = \text{Val } T = \text{Val } T'$, where

$$T = 0.a_1 a_2 a_3 \dots a_{n-1} a_n 2222 \dots$$
$$T' = 0.a_1 a_2 a_3 \dots a_{n-1} a_n' 0000 \dots$$

2. Show that Peano's mapping of the interval $[0, 1]$ into the unit square is a continuous mapping.

3. How would you map the unit square onto the unit cube? In other words, given a point in two dimensional space (x, y), to what point in three dimensional space (u, v, w) would it correspond?

4. Consider Cantor's ϕ function.

    (a) Determine the height of the algebraic numbers $\sqrt{2}$, $1 - \sqrt{5}$, $\sqrt[3]{1 - \sqrt{5}}$. Show $\phi(1) = 1$, $\phi(2) = 2$, and $\phi(3) = 4$. What is $\phi(4)$?

    (b) According to Cantor's ordering, what is ω_1? What are the first three algebraic numbers?

5. Modify Cantor's 1874 proof of the countability of the algebraic numbers show that the *rational* numbers can be put in a one-to-one correspondence with the positive whole numbers. What are the first few rational numbers in the sequence?

6. Prove that all numbers of the second class of the form

$$\nu_0 \omega^\mu + \nu_1 \omega^{\mu-1} + \dots + \nu_{\mu-1} \omega + \nu_\mu$$

can be put in a one-to-one correspondence with the numbers of the first class.

7. Explain how to use the first and second generating principles to obtain ω^{ω^ω}.

8. Compare Cantor's first and second proofs of the nondenumerability of the real numbers. What features does the second proof have that alleviate some concern about dealing with the infinite?

9. Prove $\overline{\overline{M}} = \overline{\overline{N}}$ if and only if $M \sim N$.

10. Use Cantor's definitions and prove the commutative and associative laws of addition and multiplication; and the distributive law of multiplication over addition.

11. Let $M = \{1, 2\}$ and $N = \{x, y, z\}$.

    (a) Find $a = \overline{\overline{M}}$ and $b = \overline{\overline{N}}$.

    (b) Prove $a < b$.

    (c) List all elements of $(N|M)$, by indicating the "image" of x, y, z under f. For example, the covering $f(n) = 1$ for all n in N has image $f(\{x, y, z\}) = \{1, 1, 1\}$.

    (d) Determine $\overline{\overline{(N|M)}}$, and hence a^b.

12. What logical objections might be raised about Cantor's definition of 1?

13. Prove that $a^2 = a \cdot a$ (i.e., that Cantor's definition of exponentiation is consistent with the usual definition).

14. Prove the laws of exponentiation: $a^b a^c = a^{b+c}$, $a^c b^c = (a \cdot b)^c$, and $(a^b)^c = a^{b \cdot c}$.

15. Prove that the number of real numbers between 0 and 1 is $2^{\aleph_0}$.

16. Prove $2^{\aleph_0} = 10^{\aleph_0}$ and in general, for any whole number ν greater than 1, $\nu^{\aleph_0} = 2^{\aleph_0}$.

17. Prove that, given any transfinite number T not equal to $\aleph_0$, then $\aleph_0 < T$.

18. Prove $\aleph_0 + 1 = \aleph_0$ and, in general, $\aleph_0 + \nu = \aleph_0$ for any finite ν.

19. Explain how Cantor's proof that the combining set (μ, ν) has cardinality $\aleph_0$ shows that the rational numbers have the same cardinality as the natural numbers.

20. Let E be the set of rational numbers on the interval $[0, 1]$. Determine the content of E. Hint: use an infinite set of intervals I_k, where I_k is an interval around the kth rational number.

22.5   Dynamical Systems

Laplace's application of mathematics to physics implied that it was possible, in theory, to predict the future of the universe based on complete knowledge of its present condition. Of course, it is impossible to do so in practice, but the fact that it was possible in *principle* meant that the entire history of the universe was completely predetermined from its initial condition: free will is an illusion.

Fortunately, there are ways out of this depressing state of affairs. James Clerk Maxwell (the physicist who proved, mathematically, Saturn's rings could not be solid) noted that in some cases, small perturbations in the original state of a system led to small perturbations in the final state but in others, a small perturbation led to a very large difference in the final results. These "watersheds," as

Maxwell called them, accounted for free will, for there are regions where imperceptible differences in the initial state would lead to a vast difference in the final state. The term watershed was not used randomly: in geography, the watershed of a lake is the region where all the water that falls into the region will eventually (if it does not evaporate) drain into the lake. On a grand scale, a **continental divide** separates the watersheds for two oceans: rain falling on one side of the divide will drain into one ocean, while rain falling on the other side will tend to the other ocean. Near the divide, small changes in initial position will frequently lead to very large difference in the ultimate result.

One might argue, and quite reasonably, that while very small changes led to large changes in *some* cases, for the most part, small changes led to small changes. For example, a drop of rain starting in California would eventually drain into the Pacific Ocean. It would take a very large change in the drop's initial position to change the drop's final destination. It would be reasonable to suppose that the watersheds would be easy to find, and the possibilities for small changes leading to great differences in final results would be limited to the boundaries between the watersheds. During the nineteenth century, it became clear that the problem is far more complex than originally believed.

22.5.1   The Fixed Point Theorem

One place where watersheds occur is in the use of iterative techniques, which have been a part of mathematics since the ancient Egyptians. For example, the roots of $f(z) = 0$ can be found using "Newton's" method: begin by choosing a value z_0 so that $f(z_0)$ is an approximate root; then define z_1, z_2, ... iteratively using

$$z_{n+1} = z_n - \frac{f(z_n)}{f'(z_n)}$$

The sequence of z_n's may converge to a root of the equation $f(z) = 0$.

We digress briefly to discuss the history of this method of finding the roots of an equation, which first appeared in *Universal Analysis of Equations* (1691) by Joseph Raphson (1648-1715). Newton had in fact found the same method in his *Method of Fluxions*, which, though written in 1671, was not published until 1736. See page 421. Hence, the method is frequently called **Newton's method** or, less frequently, the **Newton-Raphson method**. However, neither Newton nor Raphson expressed it in the foregoing form. Thomas Simpson was the first to explicitly include the derivative in the formula, and Fourier first expressed it as the preceding algorithm (and, once again, lost credit he rightfully deserved through the vicissitudes of history!). z may take on complex values.

The right hand side can be regarded as a function F of z. In general, if $z = F(z)$, we say z is a **fixed point** of the function F. Points near the fixed point may, under successive iterations, move closer to it; or they may move away from it; or in some regions move towards it and then recede from it. For example, if $F(z) = z^2$, the fixed points are $z = 0$ and $z = 1$. Around $z = 0$, successive iterates move closer to $z = 0$. A point like $z = 0$, near which points move closer to the fixed point, is said to be an **attracting fixed point**.

In 1870, Ernst Schröder (November 25, 1841-June 16, 1902) found a sufficient condition for guaranteeing a fixed point was an attracting fixed point:

Theorem 22.3 (The Fixed Point Theorem). *If* $z' = F(z)$, $z'' = F(z')$, $z''' = F(z'')$, ..., $z^{(r)} = F(z^{(r-1)})$, ..., $F(z_1) = z_1$ *and* $F'(z_1)$ *is less than 1 in absolute value, then for all z sufficiently close to z_1,* $\lim_{r \to \infty} z^{(r)} = z_1$.

Schröder's proof was:

Proof. Suppose $z = z_1 + \epsilon$. Then

$$F(z_1 + \epsilon) = F(z_1) + \epsilon F'(z_1) + \frac{\epsilon^2}{2} F''(z_1) + \dots$$

Hence

$$z' = z_1 + \epsilon F'(z_1) + \frac{\epsilon^2}{2} F''(z_1) + \dots$$

If ϵ is infinitely small, then

$$z' - z_1 = \epsilon F'(z_1)$$

and if $F'(z_1)$ is less than 1 in absolute value, the successive iterates of z_n get closer to z_1.   □

The proof is not considered rigorous by today's standards.

Schröder's theorem is a "local" result, since it applies to points that are infinitely near to the fixed point z_1, and says nothing about points farther away. A "global" result was more desirable, since it would indicate the behavior of any point under the successive iterations. In 1884, Gabriel Koenigs (January 17, 1858-October 29, 1931), a French mathematician, considered points outside the immediate neighborhood D of the fixed point z_1; suppose, for example, there was another point $\tilde{z}$ that converged to the fixed point under iteration. Surrounding this would be a disk $\tilde{D}$ that also converged to the fixed point. The collection of all disks $\tilde{D}$ whose points tend to the fixed point under iteration is now called the **basin of attraction** of the fixed point z. The basins correspond precisely to Maxwell's watersheds.

The basins could become quite complicated for the following reason. Consider the successive iterates of some point. For example, suppose $z = 2$. The successive iterates of $F(z) = z^2$, starting with $z = 2$, form the set

$$\{2, 4, 16, 256, \dots\}$$

which is referred to as the **orbit of z**. The orbit of a fixed point is simply the fixed point itself. However, there are some points that have **closed orbits**; for example, $z = -\frac{1}{2} + \frac{\sqrt{3}}{2}i$, whose orbit under $F(z) = z^2$ is

$$\left\{ -\frac{1}{2} + \frac{\sqrt{3}}{2}i, \quad -\frac{1}{2} - \frac{\sqrt{3}}{2}i, \quad -\frac{1}{2} + \frac{\sqrt{3}}{2}i, \quad -\frac{1}{2} - \frac{\sqrt{3}}{2}i, \dots \right\}$$

where the values repeat themselves periodically. Hence, $z = -\frac{1}{2} + \frac{\sqrt{3}}{2}i$ is said to be a **periodic point**; in this case, it is a **period 2 point**.

If the fixed points are the points where $z = F(z)$, the period n points satisfy $z = F^n(z)$, where $F^n(z)$ is the nth iterate. In the case of $F(z) = z^2$, the period 2 points must satisfy $z = F^2(z) = F(F(z)) = (z^2)^2 = z^4$. The solutions of $z = z^4$ are $z = 0$ and $z = 1$ (which are actually fixed points), and $z = -\frac{1}{2} \pm \frac{\sqrt{3}}{2}i$.

Since a periodic point satisfies $z = F^n(z)$, then the periodic points are fixed points of $F^n(z)$. Suppose z_1 is an attracting fixed point of $F^n(z)$. By Schröder's theorem, there is a disk D around z_1, for which all z tend to z_1 under the iterations of $F^n(z)$. Under the iterations of F, however, these points might not converge to z_1 in the ordinary sense. They *do*, however, periodically return to the vicinity of z_1 (since the nth iterate of z is, by the fixed point theorem, closer to z_1 than z was originally). Points such as these are called **homoclinic points** today, and correspond to points with Poisson stability.

See page 614.

Koenigs recognized an important aspect of iteration: the number of solutions to $F^n(z) = z$ grows as n does. If $F(z)$ is a kth-degree polynomial, there might be k distinct fixed points of $F(z)$, but $F^2(z)$ has all the fixed points of $F(z)$, as well as up to k new ones, corresponding to the period 2 orbits of $F(z)$. In general, $F^n(z)$ might have nk fixed points, some of which are period n orbits of $F(z)$. Hence, there may be arbitrarily many orbits with arbitrarily large periods, and the problem of dividing the plane into basins becomes immensely complicated.

22.5.2 Poincaré

Fixed points and basins of attraction are not limited to iterative functions; they also appear in differential equations. Consider the differential equation $\frac{d^2x}{dt^2} = x$, with $x(0) = 0$. The general solution to this differential equation is $x = ce^t - ce^{-t}$. In 1877, Joseph Valentin Boussinesq (March 15, 1842-February 19, 1929), a physicist, noted if $x'(0) = 0$, then the solution is $x(t) = 0$. However, the smallest perturbation of $x'(0)$ will result in very different behavior for x: if $x'(0) = \epsilon > 0$, then x tends to positive infinity, while if $x'(0) = -\epsilon < 0$, then x tends to negative infinity. Thus, both ∞ and $-\infty$ are in some sense attracting fixed points, with corresponding basins of attraction and the boundary between the two basins at $x'(0) = 0$. Saint-Venant, in 1878, noted this could be used to account for free will: "insensible changes" could lead to very large differences in the final state. However, the intuitive feeling is that at most points, small changes in initial position led to small changes in final conditions. For example, if $x'(0) = \epsilon > 0$, then changing $x'(0)$ to $\epsilon + \delta$, δ assumed very small, still led to the same result: $x(t)$ tended to infinity. The work of Henri Poincaré (April 29, 1854-July 17, 1912) would call this intuition into question.

Poincaré was a graduate of the Polytechnical School in Paris. A student of Hermite's at the University of Paris, Poincaré's thesis on differential equations was praised by his examiners, though they noted that the second half was "a little confused." Poincaré, like Gauss and Euler, was well aware of the role intuition played in mathematics, and in a 1904 article on mathematics education, he wrote, "It is by logic we prove, it is by intuition that we invent."

Poincaré was one of the last mathematicians to be active in all fields of mathematics, not only making important contributions to them, but breaking new ground: it was said he was an explorer, not a colonist. Among other things, Poincaré introduced the idea of an **automorphic function**. The trigonometric functions have one real period, and the exponential function has one imaginary period; the elliptic functions have two periods, one real, and one complex. Poincaré went farther and suggested the existence of functions for which $f(z) = f(z')$, where $z' = \frac{az+b}{cz+d}$. These clearly include the trigonometric and exponential functions, but Poincaré was unsure whether any other automorphic functions existed.

In 1882, Magnus Gustaf (or Gösta) Mittag-Leffler (March 16, 1846-July 7, 1927), a Swedish mathematician and one of Weierstrass's students, established one of today's most important mathematical journals: *Acta Mathematica*. The journal was funded by King Oscar II of Sweden, who also proposed a prize contest, similar to those sponsored by the French Academy of the Sciences. In 1885, a 2500 crown prize was offered in *Acta*, to be awarded to the best answer to one of four questions. By comparison, Mittag-Leffler's annual salary as a university professor was just 7000 crowns. Kronecker objected to the whole contest, calling it a mere attention-getting device for *Acta*: why else, he asked, would the prize announcement be made in *Acta* instead of by the Swedish Academy of the Sciences? Kronecker's distrust of the motives of Mittag-Leffler may have had a lot to do with the fact that Weierstrass was part of the judging committee and Kronecker was not.

Poincaré chose to answer the first question: given the initial positions, masses, and velocities of n bodies, was it possible to predict their future behavior, assuming they moved solely under the influence of gravity? This is called the **n-body problem**. The question related to an interest of Weierstrass, who, in an August 15, 1878 letter to Kovalevskaya, noted that he could find *formal* solutions to the differential equations of the planetary positions but did not know whether these solutions were in fact convergent. Poincaré originally intended to attempt the problem in its full generality, but found it far too complicated. Instead, he attacked the so-called **restricted three-body problem**, where $n = 3$ and one of the bodies is assumed not to affect the motions of the other two. Specific solutions to the restricted three-body problem had been found by Lagrange and Euler, who showed that if a body began in one of five places (now called **Lagrange positions**) with the proper angular velocity about the center of mass, it would maintain its position relative to the other two, and thus its future motions could be completely determined. However, a general solution, even to the restricted three-body problem, had yet to be found.

Poincaré's major contribution was to ignore the overall motions of the system and concentrate on what is now called the **first return map**. Consider the orbit of a planet around the sun and an arbitrary plane not parallel to the plane of the planet's orbit. At some point $t = 0$, the planet is on the plane at some point P_0.

At some later times $t = K$, K', K'', ..., the planet is once again on the plane, at points P', P'', ...; this set of points forms the "return map" of the planet's position. If the orbit of the planet is perfectly stable, the return map consists of a single point; in this case, we say the path is a **closed trajectory**, since the orbit "overlaps" itself. On the other hand, the planet's orbit might vary in such a way that, after some number of orbits, it returns to its starting point after a finite number of orbits; again, this is a closed trajectory. Finally, there is the possibility that the planet's orbit will *never* close up.

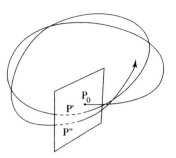

Since the perturbed case is a continuous variation of the stable case, one might suspect that the first return map would vary continuously as well: small perturbations result in small changes in the appearance of the first return map. Poincaré's first solution, to be published in *Acta* in October, 1889, was discovered to have a tremendous flaw in it; as a result, the publication was delayed while Poincaré fixed the flaw. As copies of *Acta* containing the flawed solution had already been printed, Poincaré agreed to pay for the cost of reprinting (which was almost the entire amount of the prize money). While fixing his flawed solution, Poincaré discovered something truly remarkable: suppose there was a closed trajectory through some point P. Since this trajectory was closed, the future position of the particle on the trajectory could be completely determined. Poincaré discovered that arbitrarily close to P were other points, P', and trajectories through P' would stay close to the trajectory through P for a while, but then deviate from it by arbitrarily large distances and then, after some more time, would return arbitrarily close to the trajectory through P. Poincaré originally called these "doubly asymptotic" trajectories, though later, he used the term **homoclinic point** to refer to points such as P. Poincaré expanded the revised paper to *New Methods in Celestial Mechanics*, one of the founding works of the modern theory of celestial mechanics.

22.5 Exercises

1. Let $F(z) = z^2$, and suppose z can take on complex values.

   (a) Show that $z = 0$ is an attracting fixed point.

   (b) Show that the basin is the unit disk $|z| < 1$.

   (c) Show that all period-n points are on the unit disk. Hint: use the polar form of a complex number.

   (d) Suppose z_n is a period-n point. Show that arbitrarily close to z_n is a point which, under iteration, moves away from z_n but also returns arbitrarily close to z_n.

2. Suppose $F(z) = z^2$.

   (a) Find the orbit of $z = i$, $z = -1$, and $z = 3$.

   (b) Find the orbit of $z = -\frac{1}{2} - \frac{\sqrt{3}}{2}i$. Hint: use the polar form of z.

   (c) Find the period 3 points.

   (d) Prove there are period n points, for any value of n.

3. Suppose $F(z) = z^3$. Find:

   (a) The fixed points of $F(z)$.

   (b) The period 2 points.

   (c) The period n points.

4. Show that if two orbits intersect, they are identical from the point of intersection onwards.

5. Suppose a planet is orbiting the sun. Describe the first return map if:

   (a) The planet's orbit is always a perfect circle, but the plane of the orbit changes by $\frac{p}{q}$ radians per orbit, where p and q are whole numbers; assume the plane of the first return map does not include the nodes of the planet's orbit (i.e., the points where the plane of the orbit cross the reference plane).

   (b) The planet's orbit is always a perfect circle, but the plane of the orbit changes by $\frac{p}{q}\pi$ radians per orbit, where p and q are whole numbers. Why is this different from the previous case?

6. Prove that if the first return map has a finite number of points, the orbit is closed.

7. Suppose the initial point P_0 coincides with some future point P' on the first return map. Explain, by way of example, why this does *not* imply a closed orbit.

8. Does the discovery of chaos answer the question of the existence of free will? Explain.

Chapter 23

Algebras

While continental Europe grew more and more reactionary after Napoleon, Britain slowly (and against much internal opposition) liberalized itself. At the same time Louis XVIII of France passed laws restricting the right to vote, Britain was gradually extending the franchise to more and more people. At the same time the French police were busily arresting protesters against the monarchy, British Prime Minister Robert Peel established the first modern police force, designed to protect the population from criminals. The policemen came to be known as "Bobbies," after Peel. In 1833, Britain peacefully abolished slavery at home and in all of her overseas possessions, while the United States would nearly destroy itself in civil war over the issue.

Things were not perfect everywhere, of course. Ireland continued to be a sore point. In 1845, a blight that destroyed the potato crops on which most of the Irish population survived led to the Irish potato famine. The disaster was magnified by economic policies that encouraged the (English) landlords to export food *from* Ireland to England, while corn laws prohibited the importation of corn (grain) unless the domestic price rose above a certain (very high) threshold. Enough British citizens cared about the plight of the Irish to force a repeal of the corn laws in 1846, but by then, one-third of the Irish population had died, or emigrated, mainly to the United States.

23.1 The Algebra of Logic

To first approximation, we can trace the historical development of mathematics by observing the names associated with a field. In analysis, we have Newton's method and Taylor and Maclaurin series; but we also have Euler methods, Lagrange multipliers, Laplace integrals, Fourier series, Cauchy theorems, and Riemann integrals. Clearly, this was a field dominated by the French. In fact, in many fields of mathematics, French (and later, German) names abound. One might wonder what happened to the British during the nineteenth century.

At the time, this "analytical gap" was attributed to the superiority of Leibnizian differential notation over the Newtonian dot notation. Thus, in 1813, three Cambridge students, George Peacock (April 9, 1791-November 8, 1858), John Herschel (1792-1871), and Charles Babbage (December 26, 1792-October 18, 1871) founded the Cambridge Analytical Society. In the words of Babbage, the aim was to "promote the principles of d-ism as opposed to the *dot*-age of the university," a mathematical pun based on the fact that Leibniz's differential notation used d to indicated the differential, while Newton's notation used a dot to indicate the same thing; meanwhile, deism was a religious belief that God created the universe, but was uninvolved in its daily workings. In 1817, when Peacock became the mathematical examiner at Cambridge, he was able to institute this reform by mandating that Leibnizian notation would be used on the Cambridge exams.

While Leibniz's notation has much to commend it, it seems unlikely that it alone was enough to make a difference; after all, the discoveries of Simpson, Maclaurin, and Taylor were all made using dot notation. If we look deeper, we would see that British mathematicians, far more than their continental counterparts, tended to be concerned with making analysis as rigorous as ancient Euclidean geometry. Unfortunately, for the most part, work such as Euler's could be made rigorous, and their desire for rigor merely meant the great discoveries of analysis would be made by others. But at the same time, British mathematical energies were spent productively in the field that rewarded careful concern with logical validity: algebra, and logic itself. The careful work done in the two fields mingled to form modern **abstract algebra**.

23.1.1   De Morgan

After the Seven Years' War, India became part of the growing British Empire, and a contingent of British soldiers were stationed there. The father of Augustus de Morgan (June, 1806-March 18, 1871) was a British army colonel stationed in Madras, where de Morgan was born. He was active in several areas of mathematics: he was probably the first to give a rigorous formulation of Cauchy's definition of a limit though, being an English mathematician, his work was little known by continental mathematicians.

De Morgan's best-known contributions are in logic. One of the problems faced by logicians was that of negation: to say something was "not A" merely meant it could be anything *but* A. De Morgan helped to solve this problem by introducing a **universe U**, from which set elements were to be drawn; thus, not A meant those things in U that were not A, thereby limiting the possibilities. De Morgan also invented a set of symbols for use with logic, which embroiled him in a plagiarism controversy with William Hamilton (March 8, 1788-May 6, 1856), a Scottish philosopher and professor of logic at the University of Edinburgh, who had also invented a set of symbols for logic (William Hamilton had no relation to William *Rowan* Hamilton, who had tried to provide a foundation for the real numbers). However, both systems were soon swept away by the system of Boole.

23.1.2 Boole

George Boole (November 2, 1815-December 8, 1864) was a self-educated mathematician who obtained a position at the newly opened Queen's College in Cork, Ireland, in 1849, even though he had not yet obtained any academic degrees (though in 1844, he won a medal from the Royal Society for his contributions to analysis). By all accounts, Boole was an excellent and conscientious teacher—perhaps too conscientious, for he died soon after walking, through the rain, to deliver a lecture to his students.

In *The Mathematical Analysis of Logic* (1847), Boole showed how logical operations could be treated mathematically, and created what is now called an **operator algebra**. Boole began by letting the universe be represented by "1," and X, Y, Z, ... being individual members of some class; "symbols," such as x, performed the operation of selecting the members X of the class. For example, $x1$ (or simply x, with the "1" understood) meant to select the elements X from the universe 1. Placing two symbols adjacent, such as xy, meant performing the selections in the order y, then x. Boole noted:

1. $x(u + v) = xu + xv$, where $u + v$ is the complete class of which u and v are the component parts: hence the distributive law for logical operators holds.

2. $xy = yx$; hence the commutative law holds.

3. $xx = x$ and, in general, $x^n = x$; this is called the *power* law.

Since "not x" and "x" together formed the universe, Boole assigned the symbol $1 - x$ to "not x"; in the same manner, 0, was to be used to represent "nothing." Boole noted that, because the distributive and commutative laws held, a calculus (i.e., a means of calculating) could be created for logic.

To express "All X are Y," Boole argued as follows: since this states that all Xs could be found in the class that contained all Ys, then first selecting the Ys, then selecting from these the Xs, would give the same result as selecting all the Xs. Hence $xy = x$. Likewise, "No X are Y" meant that selecting the Ys, then the Xs from the set, gave nothing: thus $xy = 0$.

23.1.3 Venn

The most important work of John Venn (August 4, 1834-April 4, 1923) was probably his *Logic of Chance*, which established on a permanent basis the connection between set theory and probability. Venn sought to develop probability on an axiomatic basis. The probability of an event had, before Venn, been defined in terms of equally likely outcomes. Venn chose instead to define it as the limit: if an event occurred s times in n trials, the probability of the event was $\frac{s}{n}$ as n tended to infinity. Venn popularized what are now known as **Venn diagrams**, which were actually invented by Leibniz, though like many of Leibniz's contributions, they were ignored by his contemporaries.

23.1.4   Mechanical Computation

Babbage, who had co-founded the Analytical Society, went on to point out that it was more efficient to charge a flat rate to deliver a letter than to attempt to compute, for every piece of mail, a cost based on the distance the letter was to travel. In part due to Babbage's efforts, the British government began the first modern postal service in 1840, charging a flat rate (one penny) to send a letter anywhere in the United Kingdom. Babbage also invented the railroad "cow catcher," but of all his inventions, he is best known for two that failed to work.

Around 1822, Babbage conceived of a mechanical device that could perform mathematical computations. Leibniz and Pascal had built simple machines, but Babbage envisioned something considerably more complex. The procedures of the calculus of finite differences could easily be translated into mechanical operations, and Babbage conceived of a **difference engine** (in the nineteenth century, an "engine" was any mechanical device), which could compute logarithmic and trigonometric function values.

One problem with the difference engine was that it was a machine that would work (if it worked) for only one problem: if it was necessary to find, say, square roots, a new machine would have to be designed. By 1832, Babbage had envisioned an even more ambitious **analytical engine**, which would work by receiving an external set of instructions, via punched cards (already in use in the textile industry, for creating elaborate tapestries using automatic looms). A new problem could be solved by changing the cards, not the machine.

If it could have been built, the analytical engine would have been the first mechanical computer. Unfortunately, there were severe technical difficulties. To work, gears had to be made with a fine tolerance; in essence, to represent a number, a gear had to be accurately divided into ten equal parts, each tooth corresponding to a digit from 0 to 9. If the tolerances were not precise, the gears would "catch" (causing mechanical jams) or, worse, "slip" (leading to computational errors).

It would be much better if a gear had only to represent two choices. The mathematical study of logic provided such an alternative: in a logical statement, there were only two possibilities, true or false. Thus, one of the most interesting contributions to mechanical methods of computation was made by the logician William Stanley Jevons (1835-1882). Like Boole, Jevons operated with the classes; one of his interests was in being able to express classes in terms of other classes. This could be accomplished easily by writing down all possible combinations, then deleting those that contradicted the premises. A procedure such as this could be performed mechanically, and in 1870, Jevons exhibited a machine that would do so. The device resembled a piano, and Jevons called it his "logical piano."

23.1 Exercises

1. Prove the commutative, distributive, and power laws for logical symbols.

2. Express the following in Boolean notation.

    (a) All X are Y and Z.

    (b) Some X are Y. Hint: X consists of the X that are also Y, and the X that are not Y.

    (c) Some X are not Y.

3. What is $x + x$ equal to? Explain.

4. Consider the rules of algebra and compare them to the rules of logic.

    (a) Solve the following algebraic equations: $x^2 = x$, $x + x = x$, $xy = 0$.

    (b) Solve the corresponding logical equations, and indicate what statements they are equivalent to.

    (c) Which of the rules of algebra hold true for the rules of logic?

23.2   Vector Algebra

The work of Wessel and Argand had made clear that the complex numbers could be viewed as lines in the plane, with the line representing i perpendicular to the line representing 1. The Irish mathematician and linguistic prodigy William See page 600. Rowan Hamilton, who had attempted to base a theory of the real numbers on the notion of time, considered possible extensions to the complex numbers. Hamilton's work is an elegant model of how mathematical structures are ex- See page 611. tended.

Hamilton began by noting the complex numbers $x + iy$ satisfied two key properties: if the complex number $x + iy$ and its square, $(x^2 - y^2) + 2xyi$ were thought of as lines in the plane (namely the lines from the origin to (x, y) and $(x^2 - y^2, 2xy)$, respectively), then the length of the $(x+iy)^2$ was the square of the length of $(x + iy)$ (the **modulus rule**; note that Hamilton is using the modulus in the modern sense, and not Gauss's definition), and the angle between $(x+iy)^2$ and the line represented by 1 was twice the angle between $(x + iy)$ and 1 (the **argument rule**).

Hamilton then considered the possibility of number triples, representing lines in space; for these **hypercomplex numbers**, a third unit, j, was necessary, perpendicular to the plane itself, with $i^2 = j^2 = -1$. If

$$(x + iy + jz)^2 = x^2 - y^2 - z^2 + 2ixy + 2jxz + 2ijyz$$

and the modulus rule held, then ij had to be equal to 0, which seemed paradox-ical. Hamilton wrestled with this problem for a long time. Then, on October 16, 1843, while walking with his wife along the Royal Canal in Dublin, he realized that if $ij = -ji$, the difficulty would vanish, and the law of the modulus could be salvaged. Perhaps the most remarkable feature was that multiplication was not, in general, commutative.

What was ij itself equal to? Hamilton noted the modulus rule itself is a special case of a more generalized rule, namely that the modulus of the product of two complex numbers is the product of the moduli. In general, a hypercomplex number $a + bi + cj + \ldots$ had modulus $\sqrt{a^2 + b^2 + c^2 + \ldots}$ Therefore, Hamilton considered the product

$$(a + ib + jc)(x + iy + jz) = ax - by - cz + i(ay + bx) + j(az + cx) + ij(bz - cy)$$

The product of the moduli of the left hand side is

$$\sqrt{(a^2 + b^2 + c^2)(x^2 + y^2 + z^2)}$$

which, by assumption, should be the modulus of the left hand side. But if $ij = 1$, or $ij = -1$, this would not occur; the modulus rule could be saved only if $ij = k$, a new imaginary. Thus, while algebraic triples of the form $a + bi + cj$ seemed to lead to conceptual difficulties, algebraic quadruples of the form $a + bi + cj + dk$ did not. The multiplication assumptions of this system of *four* units had to be:

$$i^2 = j^2 = k^2 = -1$$
$$ij = -ji = k \qquad jk = -kj = i \qquad ki = -ik = j$$

Hamilton was so impressed that he carved the fundamental equation, $i^2 = j^2 = k^2 = ijk = -1$ on a stone of Brougham Bridge, along the Royal Canal. (Hamil-ton was an inveterate scribbler, writing on anything that was handy: he kept a notebook with him, but he also wrote on his fingernails and on one occasion, according to his son, on a hard-boiled egg.)

Hamilton, classically trained scholar that he was, called the numbers **quater-nions**, after a Roman unit consisting of four soldiers.[1] The concept of quater-nions required the acceptance of some very difficult propositions: not only was multiplication not commutative, but if the units 1, i, and j could be considered mutually perpendicular, what could be made of the *fourth* unit k? "We must admit, in some sense, a *fourth dimension* of space," wrote Hamilton in 1844.

The quaternion $a + bi + cj + dk$ breaks down into a "real part," a, and an "imaginary part," $bi + cj + dk$. Since the real part could be any value on a scale from $-\infty$ to $+\infty$, Hamilton called this the **scalar** part of the quaternion; meanwhile, the imaginary part could be thought of as a straight line having a given length and direction, which Hamilton called the **vector** part; this was the first appearance of these terms in their modern sense. Hamilton used the idea of

---

[1]In Christianity, the Quaternions were the four Roman soldiers who guarded the tomb of Christ after the Crucifixion.

scalar and vector *operators* S and V on the quaternion Q: thus the quaternion could be considered the sum $SQ + VQ$ or $(S + V)Q$.

Hamilton was convinced of the great value of quaternions, and spent the rest of his life developing an algebra of quaternions analogous to the algebra of the real numbers. Although he succeeded in laying the foundations of vector analysis, he was not successful at presenting his results. *Lectures on Quaternions* was a monumental work of 736 pages with a 64 page preface, and the radically different ideas and new notations that filled the work made it difficult reading. Hamilton was at work on his even more monumental *Elements of Quaternions* when he died. A few days before dying, he extracted a promise from Peter Guthrie Tait (April 28, 1831-July 4, 1901), with whom he had corresponded since 1858, to publish a more elementary exposition of the theory of quaternions. Tait did so, and his *Elementary Treatise on Quaternions* (1865) was the first comprehensible account of the theory of quaternions, which, with the associated ideas of vector algebra, soon became of paramount importance in physics.

23.2.1 Grassmann

As Jacobi and Abel developed the theory of elliptic functions independently of one another, Hamilton and Hermann Günther Grassmann (April 15, 1809-September 26, 1877) likewise developed the notion of a vector independently. Like Hamilton, Grassmann was a linguist, writing Latin texts for secondary school students, as well as a German reader for primary school students. Grassmann held the viewpoint that mathematics was the study of *forms*, rather than quantities: this is similar to the modern viewpoint that mathematics is the study of the relations between abstract objects. Grassmann's *Representation Theory*, originating around 1843, provided a treatment of vector algebra similar to Hamilton's. However, like Hamilton's efforts with quaternions, Grassmann's work with vector algebra was likewise ignored. Somewhat later, at the urgings of Möbius (who had ignored the first edition of *Representation Theory*), Grassmann revised this work, though the second edition fared little better than the first. This was due in part to the awkward notation used by Grassmann, but probably more because Grassmann declined to include concrete examples of the value of his vector methods.

23.2.2 Gibbs

In vector analysis we find some of the first well-known American contributions to mathematics. Josiah Willard Gibbs (February 11, 1839-April 28, 1903) is better known as a chemist, working in the field of thermodynamics (Gibbs was responsible for introducing the concept of "entropy" to chemical reactions). However, he maintained a lifelong interest in mathematics as well. He entered Yale University, in his home town of New Haven, Connecticut, in 1854.

In 1861, the American Civil War began over the issue of whether or not a state had the right to allow slavery. The war was a watershed in world and American history, for it was the ultimate test of democracy: what could and should be done if a minority disagrees with the opinions of the majority? In American history, the American Civil War led to the first income tax and the first draft.

The draft was full of exemptions. Some of the more odious were the fact that a rich man could pay for an exemption or have someone else fight in his place. More reasonably, many were exempt for reasons of poor health; Gibbs was in this latter category. Thus, he continued his studies and in 1863, he received from Yale University the first Ph.D. in engineering granted in the United States. In 1865, the American Civil War ended with a total defeat for the South, and the next year Gibbs and two of his older sisters went to Europe. Gibbs toured the academic centers of Europe, spending a year each at the universities of Paris, Berlin, and Heidelberg, studying mathematics. When he returned, he accepted a position at Yale and in 1871, became a Professor of Mathematical Physics.

By then, more and more physicists were aware of the value of quaternions. James Clerk Maxwell's *Treatise on Electricity and Magnetism* used quaternion notation throughout. It was Gibbs who realized that it was the vector part of the quaternion that was of paramount importance, and in 1881 (and again in 1884) printed a pamphlet for his students that gave vector analysis the same form that can be found in every modern textbook.

23.2 Exercises

1. Show that if

$$(x + iy + jz)^2 = x^2 - y^2 - z^2 + 2ixy + 2jxz + 2ijyz$$

   is to satisfy the law of the modulus, $ij = 0$.

2. Show that if $(a + ib + jc)(x + iy + jz)$ is to satisfy the law of the modulus, then $ij = k$, a new imaginary unit. Hint: show $(a^2 + b^2 + c^2)(x^2 + y^2 + z^2)$ is equal to

$$(ax - by - cz)^2 + (ay + bx)^2 + (az + cx)^2 + (bz - cy)^2$$

3. From the assumptions $i^2 = j^2 = -1$ and $ij = -ji = k$, prove the remaining components of the "multiplication table" of the quaternions.

4. Given two quaternions, $\alpha = ix + jy + kz$ and $\alpha' = ix' + jy' + ky'$, consider the product $Q = \alpha\alpha'$. Justify the modern use of the terms scalar product and vector product for the dot and cross product of two vectors, respectively.

5. Show that the system of hypercomplex numbers $a + bi + cj$ has divisors of zero, if $i^2 = j^2 = -1$, regardless of the value of ij.

6. Show that the system of hypercomplex numbers $a + bi + cj + dk$, with multiplication table

$$i^2 = j^2 = k^2 = -1$$
$$ij = ji = k \qquad jk = kj = i \qquad ik = ki = j$$

   is inconsistent.

23.3 Matrix Algebra

In his *Arithmetical Investigations*, Gauss considered numbers of the form $P = ax^2 + 2bxy + cy^2$, where a, b, and c were constants and x, y variables. If x, y underwent a linear transformation of the form

$$x = \alpha x' + \beta y' \qquad\qquad y = \gamma x' + \delta y'$$

then P was transformed into $P' = a'x'^2 + 2b'x'y' + c'y'^2$. Gauss noted that the quantities $D = b^2 - ac$ and $D' = b'^2 - a'c'$ were equal, and called their common value the determinant, though now it is referred to as the **discriminant**. The study of invariants was thus tied up with the study of linear transformations of variables.

Because of their algebraic bent, English mathematicians played a major role in the development of the theory of invariants during the nineteenth century. In particular, Arthur Cayley, James Sylvester, and George Salmon (an Irish theologian and mathematician) were responsible for many major developments in the theory of invariants; the French mathematician Hermite called them "the invariant trinity." Arthur Cayley (August 16, 1821-January 26, 1895), the son of an English merchant living in St. Petersburg, played a major role in developing modern abstract algebra. Like Poisson and Cauchy, Cayley rushed into print whenever he had a new idea (Cayley rivals Cauchy in the number of his publications), not stopping to polish it; as a result, some of Cayley's work has a rough-hewn feel. Occasional mistakes appear, none serious (most are the equivalent of typographical errors).

Arthur Cayley. ©Hulton-Deutsch Collection/CORBIS.

Cayley defined a **quantic** to be a function in several variables; the degree of the function was the order of the quantic. For example, $P = ax^2 + 2bxy + cy^2$, consisting of two variables to the second degree, was a quadratic binary quantic: quadratic because it was of the second degree, and binary because it consisted of two variables. Cayley wrote the general form of binary quadratic quantics as $(a, b, c \, \lozenge \, x, y)^2$. An **invariant** was an expression of the coefficients, such as Gauss's determinant, that did not change under a given linear transformation of the variables. In 1846, Cayley conjectured it was possible, given a binary quantic of any order, and a linear transformation of the variables, to determine *all* invariants.

23.3.1 Sylvester

Cayley's conjecture would be proven by James Joseph Sylvester (September 3, 1814-March 15, 1897). Though English, Sylvester was one of the founders of American mathematics. Like the situation in Europe during the Middle Ages, it was difficult to start a mathematical tradition in the United States, where there were no important centers of mathematical studies. As late as 1818, even Harvard University offered only a modest program consisting of Euclid's *Elements*, trigonometry, and the mathematics required for navigation; what textbooks existed were those of European authors, particularly Legendre's *Geometry*. The military academy at West Point had been established along the lines of the Polytechnical School in Paris, with a rigorous mathematics curriculum, but it produced mainly engineers, not mathematicians.

In 1867, Johns Hopkins, a Baltimore businessman, donated $7 million to found a teaching hospital and a university. At the time, it was the largest single donation for an institute of higher education in the United States. The trustees of the university were advised of the growing importance of science, particularly the practical aspects; moreover, they were advised of the importance of establishing a sound *undergraduate* school. Daniel Coit Gilman became the first president of Johns Hopkins University. On the recommendation of Joseph Henry of the Smithsonian Institution and Benjamin Pierce at Harvard, Gilman offered Sylvester not only a teaching position, but also the chance to build his own mathematics department.

Sylvester was no stranger to the United States. The Jewish Sylvester could not obtain a degree from English universities, which required students to subscribe to the tenets of the Anglican Church. Instead, he graduated from Trinity College in Dublin. His Jewish ancestry prevented him from advancing very far in England, so between November 1841, and February 1842, he taught in the United States at the University of Virginia at Charlottesville. European university students accorded their instructors a healthy dose of respect and reverence. In contrast, the students at the University of Virginia were known for their "drunkenness and lawlessness," and one had in fact murdered a faculty member just before Sylvester's arrival. When a student's threats were met with only a mild censure, Sylvester resigned in protest. After finding no new employment in the United States, he returned to England; among his students there were Florence Nightingale, soon to leave England for the Crimean War. However, Sylvester found the same difficulties with his Jewish ancestry and in 1870 withdrew entirely from mathematics to write poetry and, by most accounts, he was an accomplished amateur poet. By 1876, Sylvester was ready to return to mathematics, and he accepted Gilman's offer.

Sylvester realized that a mathematics journal was essential if any sort of mathematical tradition was to be built. Adrain had twice attempted to found a mathematics journal in the United States, but the country, too recently established, had not yet had time to develop the institutions of higher education or the educated populace that would allow a journal to survive. But by Sylvester's time, the United States had grown considerably.

See page 589.

Thus, in 1878, with the help of William Story (who Sylvester had recruited from Harvard University) and the physicist Simon Newcomb, the *American Journal of Mathematics* began circulation, with Sylvester as editor. The first volume contained not only publications by the English Cayley and the German Lipschitz, but works of American mathematicians as well, including George Hill and Newcomb. The reception of the journal in Europe was excellent, and the United States was on the way to establishing itself as a center of mathematical research.

After withdrawing from mathematics for six years, Sylvester found himself engaged in the problem of invariants. In 1877, Sylvester supplied a proof of

Cayley's conjecture that all invariants of binary quantics could be determined. Then in 1883, he accepted a professorship at Oxford, England, where he spent the rest of his life.

23.3.2 Cayley

The further study of linear transformations led to the development of matrix algebra. Although Leibniz and, centuries before him, the Chinese had used matrix forms as a means of solving equations, they used matrices primarily as a convenient means to record the coefficients of equations, and had not investigated the properties of the matrix itself. Cayley made matrices a subject worthy of study in its own right. Cayley's "On a Theorem in the Geometry of Position" (1841), introduced **determinants** in modern form, though he separated the matrix entries with commas.

Cayley's "Memoir on the Theory of Matrices" (1858) established matrix algebra as an essential component of mathematics. Cayley introduced the **zero matrix**, the unit matrix (our **identity matrix**), as well as the terms **symmetric**, **skew symmetric**, and the operation of transposing a matrix. In addition, he established the rules for the addition of two matrices, as well as the arbitrary seeming rule for the **product of two matrices**. See Problem 3.

For this last, Cayley considered, as do we, the matrix as representing a system of linear equations:

$$X = ax + by + cz$$
$$Y = a'x + b'y + c'z$$
$$Z = a''x + b''y + c''z$$

which can be thought of as the linear transformation of x, y, z into X, Y, Z. If x, y, and z were themselves further defined by linear equations

$$x = \alpha\xi + \beta\eta + \gamma\zeta$$
$$y = \alpha'\xi + \beta'\eta + \gamma'\zeta$$
$$z = \alpha''\xi + \beta''\eta + \gamma''\zeta$$

then X, Y, and Z could be expressed in terms of ξ, η, and ζ as

$$X = A\xi + B\eta + C\zeta$$
$$Y = A'\xi + B'\eta + C'\zeta$$
$$Z = A''\xi + B''\eta + C''\zeta$$

Since the product of two transformations was defined to be the transformation of one, followed by the other, this pair of transformations could be expressed as a single transformation of ξ, η, ζ into X, Y, Z as

$$\begin{pmatrix} A, & B, & C \\ A', & B', & C' \\ A,'' & B,'' & C'' \end{pmatrix} = \begin{pmatrix} a, & b, & c \\ a', & b', & c' \\ a,'' & b,'' & c'' \end{pmatrix} \cdot \begin{pmatrix} \alpha, & \beta, & \gamma \\ \alpha', & \beta', & \gamma' \\ \alpha,'' & \beta,'' & \gamma'' \end{pmatrix}$$

The values of A, B, C, ...could be determined directly.

Example 23.1. *Determine A, B, and C. Since*

$$X = ax + by + cz$$

then, substituting the transformed values of x, y, and z, we have

$$X = a(\alpha\xi + \beta\eta + \gamma\zeta) + b(\alpha'\xi + \beta'\eta + \gamma'\zeta) + c(\alpha''\xi + \beta''\eta + \gamma''\zeta)$$
$$= (a\alpha + b\alpha' + c\alpha'')\xi + (a\beta + b\beta' + c\beta'')\eta + (a\gamma + b\gamma' + c\gamma'')\zeta$$

so $A = a\alpha + b\alpha' + c\alpha''$, $B = a\beta + b\beta' + c\beta''$, and $C = a\gamma + b\gamma' + c\gamma''$.

From this, the **product of two matrices** can be defined. A key fact is that multiplication of two matrices was not a commutative operation, a fact that is clear from the definition.

23.3.3   Dodgson

A story is told that Queen Victoria of England so enjoyed *Alice's Adventures in Wonderland* by Lewis Carroll that she asked for other books by the same author. Among those she received was *An Elementary Treatise on Determinants*, by Charles Lutwidge Dodgson (January 27, 1832-January 14, 1898). "Lewis Carroll" is simply the Latinization of "Charles Lutwidge." Dodgson was an Oxford mathematician keenly interested in logic and logical puzzles (publishing two books on the subject under his pseudonym).

Dodgson might have been a much more important mathematician had he kept abreast of the current state of mathematical knowledge. He wrote on the theory of voting (1873), but his work mainly recapitulated work already done by the French a century before (though it did incorporate some early game theoretic concepts). In *Euclid and his Modern Rivals*, the ghost of Euclid appears before Minos, a college examiner exhausted from grading student papers of lamentable quality. In five acts, Dodgson, through Euclid, criticized the methods of teaching geometry of Legendre, Playfair, and others:

> Euclid: The Modern books on Geometry often attain their much-vaunted brevity by the dangerous process of omitting links in the chain [of reasoning] ...[2]

Not surprisingly, considering its subject, the play was a failure.

One of Dodgson's few original contributions to mathematics was the first proof of one of the fundamental theorems of matrix algebra, which appeared in the *Elementary Treatise* mentioned previously.

Proposition 23.1. *If a system of n inhomogeneous equations in m unknowns is to be consistent, it is necessary and sufficient that the rank of the augmented and unaugmented matrices be equal.*

---

[2]Dodgson, Charles. *Euclid and His Modern Rivals*, 1879. MacMillan. Page 47.

23.3 Exercises

1. Prove that Gauss's determinant of the quadratic form remained unchanged under a linear transformation. Show that it has the form given in the text.

2. Explain why in order to multiply two matrices $A \times B$, the number of columns in A must be equal to the number of rows in B.

3. Derive the rule for the multiplication of two 3×3 matrices.

4. Explain why the multiplication of two matrices is not, in general, commutative.

23.4 Groups

One of the goals of studying the history of mathematics is to understand that mathematics is not a single, imposing edifice, built up by careful accumulation of concepts. Many concepts were introduced haphazardly, and used in different ways by different authors, and it was not until later—sometimes, centuries later—that the axiomatic foundations were finally established and the deductive structure erected. Some central ideas of group theory appeared in the work of Euler and Gauss, but it would take a while before any sort of axiomatic foundation was established.

23.4.1 Cauchy

We can include group theory in the many fields of mathematics pioneered by Cauchy. His "Memoir on the Values That a Function Can Attain Under All Possible Permutations of Its Variables" (1815) began the modern study of **permutation groups**. Suppose K is a function of n variables $a_1, a_2, a_3, \ldots a_n$. A **permutation** of these n variables consists of any rearrangement of them; these permutations Cauchy designated $A_1, A_2, \ldots, A_N$. A **substitution** consisted of replacing the variables whose indices were in one permutation with the variables whose indices were in another permutation. For example, if $A_1 = (1, 2, 3, 4)$ and $A_2 = (2, 4, 3, 1)$, then the substitution $\begin{pmatrix} 1 & 2 & 3 & 4 \\ 2 & 4 & 3 & 1 \end{pmatrix}$ or $\begin{pmatrix} A_1 \\ A_2 \end{pmatrix}$ would replace a_1 with a_2, a_2 with a_4, leave a_3 the same, and replace a_4 with a_1.

In a substitution in the foregoing, Cauchy called the original permutation A_1 the first term, and the final permutation A_2 the second term. The identity substitution was one where the first and second terms were identical. Two substitutions were said to be **contiguous** if the second term of the first substitution was equal to the first term of the second substitution; hence $\begin{pmatrix} A_1 \\ A_2 \end{pmatrix}$ and $\begin{pmatrix} A_2 \\ A_5 \end{pmatrix}$ were contiguous substitutions.

The **product** of two substitutions was the result when the two substitutions were applied in sequence; if the two substitutions were contiguous, the product had a simple form, namely $\begin{pmatrix} A_n \\ A_m \end{pmatrix} \begin{pmatrix} A_m \\ A_k \end{pmatrix} = \begin{pmatrix} A_n \\ A_k \end{pmatrix}$. Finally, one substitution was said to be a power of another if it could be obtained by repeatedly applying the substitution. Cauchy proved

Proposition 23.2. *If p is the largest prime that divides n, then the number of different values a nonsymmetric function of n variables can take on cannot be less than p, unless it is 2.*

This is equivalent to saying that if A is a subgroup of S_n, the group of permutations on n objects, the index of A must be 1, 2, or be greater than p.

23.4.2 Galois

The mathematical career of Evariste Galois (October, 25,1811-May 31, 1832) began in 1828, when he, like Abel, thought he found a solution to the general fifth-degree equation. Like Abel, he found his mistake, and realized that he had in fact proven the impossibility of a general solution. He sent his work to Cauchy, who admired the quality of the work but noted that Abel had already proven the same result. Cauchy recommended Galois read Abel's work, and incorporate some of Abel's theory into Galois's own. Cauchy might have also suggested that Galois submit the paper for consideration for the Academy's Grand Prize in Mathematics. In any case, Galois prepared a new draft, submitting it in February to the secretary, Fourier. Fourier, whose excess of luck (both good and bad) plagued him throughout his life, died shortly thereafter, but his bad luck played a final trick: Galois's paper was never found.

Galois published papers on continued fractions, integration, and other topics, but his main work dealt with the theory of equations. Though Abel had proven that the *general* equation of the fifth or higher degree was unsolvable, there were equations of higher degree that could be solved; for example, $x^{17} - 1 = 0$. Hence the question arose: under what conditions could the roots of an equation be found using elementary operations?

We might summarize the fundamental problem as follows: Lagrange showed the cubic and quartic equations were solvable because there were expressions that, under permutation of the roots, yielded a small number of distinct values. In general, an expression of the n roots of a nth degree equation can have up to $n!$ values under the permutation of the roots. What made the cubic and quartic solvable was that *certain* expressions had fewer than 3! or 4! possible values. Alternatively, there were some permutations of the roots that left the expressions invariant. The problem was that, for the higher degree equations, there was no clear means of producing a resolvent expression that had sufficiently few values.

Galois Theory

In 1831, Galois tackled the problem in his "Memoir on the Conditions for The Solvability of Equations by Radicals," which founded what is now called **Galois**

theory. This time, Poisson was to be the referee. A modern mathematician who knew what the paper was about would find it very difficult to follow Galois's developments; for Poisson, the new ideas were incomprehensible. Still, Poisson felt there was something of value in the paper and asked Galois to write another version, explaining and elaborating the more difficult parts.

Galois's main result was

Proposition 23.3. *There exists a group of permutations [now called the **group of the equation**] for which the following two statements held:*

1. *Every function of the roots invariable under the permutations of the roots was known rationally.*

2. *Every function of the roots that was rationally known was invariable under the permutations of the roots.*

By "invariable," Galois meant that the *actual* value was invariable. For example, if x_1 and x_2 are the roots of the quadratic equation $ax^2 + bx + c = 0$, the expression $x_1^2 - x_2^2$ is not in general invariable. However, for quadratic equations of the form $ax^2 + c = 0$, the expression is invariable, since the form of the equation implies $x_1 = -x_2$. Galois's interpretation of "rationally known" meant that the function could be expressed as a rational (or polynomial) function of the roots and coefficients of the equation. For example, the function $f(a, b, c) = \frac{-b+\sqrt{b^2-4ac}}{2a}$ was a rational function of the roots x_1, x_2, even though it contained a square root, since this could be eliminated using $\sqrt{b^2 - 4ac} = (x_1 - x_2)$.

Galois, unfortunately, gave no concrete example of what he meant (this is part of what makes the paper difficult to read). Let us consider a simple example, the fourth degree equation where $x_1 = -x_2$, and $x_3 = -x_4$. Given this relationship between the roots, it is clear that the equation will be of the form $x^4 - px^2 + q = 0$. Now consider any function $f(x_1, x_2, x_3, x_4)$ of the roots. Some of these might be invariable under some permutations of the roots. For example, consider $f(x_1, x_2, x_3, x_4) = (x_1 + x_2)(x_3 + x_4)$. Since $x_1 + x_2 = x_3 + x_4$, any function of this form will be invariable under a particular set of permutations, namely

$$G = \left\{ \begin{matrix} (1234) & (2134) & (1243) & (2143) \\ (3412) & (4312) & (3421) & (4321) \end{matrix} \right\}$$

Also, by the first part of the proposition, if a function is invariable under this permutation, it can be expressed as a rational function of the roots. For example, $x_1 x_2 + x_3 x_4 - x_1^2 - x_3^2$ is invariant under the permutations in G, and equal to $-p$.

Now consider the expression $x_1^2 - x_3^2$, which is $\sqrt{p^2 - 4q}$. This expression is invariable under a subgroup of G, namely

$$H = \{(1234) \quad (2134) \quad (1243) \quad (2143)\}$$

Let $\theta = \sqrt{p^2 - 4q}$, and consider the expression $\theta_1 = x_1 - x_2 = 2\sqrt{\frac{-p+\theta}{2}}$. This expression is invariable under a subgroup of H, namely

$$K = \{(1234) \quad (1243)\}$$

Finally, let $\theta_2 = x_3 - x_4$. The subgroup of K under which θ_2 is invariable consists only of the identity.

Now, remember that in the method of Lagrange resolvents, the number of distinct values that a function of the roots could have was equal to the degree of the corresponding resolvent equation. θ_2 has two possible values under the permutation of H (the "parent" group of K, under which θ_2 is invariable). Hence, these two values of θ_2 will be the solutions to a second-degree equation, which is, of course, solvable; moreover, the coefficients of this equation can be written in terms of *symmetric* functions of the roots: a little work shows that $x_3 = \sqrt{p + 2q} + \sqrt{p - 2q}$. A key point is that we may thus determine x_3 as a rational expression of the coefficients of the equation, The assumed symmetry of the roots gives $x_4 = -x_3$. Now that x_3 has been found, x_1 may be found, since $\sqrt{p^2 - 4q} = x_1^2 - x_3^2$. In this way, all the roots may be found.

Galois's Death

See page 610.

After Poisson returned the paper with a request for a rewrite, Galois began revising the paper. Near one proposition Galois wrote, "There is something insufficient in this demonstration; I haven't the time to complete it." Galois's lack of time had to do with politics. In July 1830, the Republicans forced Charles X to abdicate in favor of Louis Philippe. Galois supported the Republicans, participating in street riots, and condemning those who would be loyal to the monarchy. At a banquet on May 9, 1831, Galois raised his glass and a knife and toasted, "To Louis Philippe!" It was perceived as a threat, and the next day, Galois was arrested. Galois's lawyer (but not Galois himself) claimed that only the first part of the toast was heard, and that the remainder, "if he defects" (i.e., if he did not live up to the expectations of the Republicans), was drowned by the reaction of the crowd; Galois was acquitted. But on July 14, 1831, he was arrested again and charged with illegally wearing a military uniform. He was found guilty and spent eight months in prison.

For reasons not entirely clear, Galois became involved in a duel over a woman of uncertain background (she may have prostitute, or she may have simply been lower class). The night before the duel, he wrote down yet another version (the fourth) of his discoveries and sent the letter off to a friend, Auguste Chevalier, with two requests: that it be published in the *Encyclopedic Review* (as it was), and that Gauss and Jacobi be *publicly* asked their opinions, "not on the truth, but on the value" of the work (which did not happen). The duel occurred on the morning of May 30, 1832; Galois died of his wounds the next day. He was twenty years old.

23.4.3   Cayley

In the form originally presented by Galois, group theory was nearly indecipherable, so little was done with Galois theory after Galois's death. Then in 1846, Liouville came upon Galois's papers on the theory of equations, and recognized their importance, as well as the fact that they had not yet received a wide distribution. As the editor of his own journal, he could do what Galois could not: he could publish them. Even then, group theory was still hard to comprehend, and there were very few mathematicians who could profitably make use of it; Kronecker was almost the sole exception.

Here, the English made some of their most important contributions during the nineteenth century. Peacock had distinguished between an "arithmetical algebra," where the variables stood for whole positive numbers and the rules were derived from arithmetic, and "symbolic algebra," where the variables could represent anything, and the formulas had to be true for *any* interpretation of the variables. Hence, $a + b = b + a$ is true in arithmetical algebra, since addition of whole numbers is commutative, but such a statement might not be true in symbolic algebra—or what we would now call **abstract algebra**.

A starting point in laying the foundations of group theory was Cayley's "On the Theory of Groups, as Depending on the Symbolic Equation $\theta^n = 1$," which he presented on November 2, 1853. θ was a symbol representing an operation that acted on a system $(x, y, z, \ldots)$, sending it to $(x', y', z', \ldots)$. Cayley stressed:

> it is not even necessary that x', y', $\ldots$ should be the same in number with x, y, $\ldots$ It is not necessary (even if this could be done) to attach any meaning to a symbol such as $\theta \pm \phi$, or to the symbol 0, nor consequently to any equation such as $\theta = 0$ or $\theta \pm \phi = 0$; but the symbol 1 will naturally denote an operation which (either generally or in regard to the particular operand) leaves the operand unaltered, and the equation $\theta = \phi$ will denote that the operation θ is (either generally or in regard to the particular operand) equivalent to $\phi \ldots$[3]

The compounding of two operands $\theta\phi$ was defined as the compound operation that consisted of first performing ϕ, then θ; θ^n was defined, very naturally, as the performance of θ n times in succession.

More generally, Cayley wrote:

> A set of symbols
> $$1, \alpha, \beta, \ldots$$
> all of them different, and such that the product of any two of them (no matter in what order) or the product of any one of them into itself, belongs to the set, is said to be a *group*. It follows that if the entire group is multiplied by any one of the symbols, either as a further [left] or nearer [right] factor, the effect is simply to reproduce the group...[4]

Cayley suggested that the multiplication table for the group be used as a way of representing the group.

---

[3]Cayley, Arthur, "On the Theory of Groups, as Depending on the Symbolic Equation $\theta^n = 1$." *Philosophical Magazine* VII (fourth series), Jan-June 1854. Page 40.

[4]Ibid., page 41.

Since each symbol in a group of n elements satisfied the equation $\theta^n = 1$, then the symbols could be considered to represent the roots of the equation (hence the title of Cayley's work). Cayley considered the following propositions to be elementary

Proposition 23.4. *If for some α the equation $\alpha^r = 1$, with $r < n$, was satisfied, then r divides n.*

Proposition 23.5. *If n is prime, then the group is of the form* 1, α, α^2, α^3, $\ldots$, α^{n-1}.

The second proposition is part of what would eventually become known as the **classification theorem**.

See page 760.

After this, Cayley began a systematic investigation of finite groups. Since Cayley's second proposition implies that all groups with a prime number of elements are the same, there is no need to consider groups with $n = 2$ or $n = 3$ elements; thus he began with the group containing $n = 4$ elements 1, α, β, γ. At least one of the elements, say β, must satisfy $\beta^2 = 1$. Cayley then represented the group as

See Problem 3.

$$1, \alpha, \beta, \alpha\beta, (\beta^2 = 1)$$

Multiplying each term by α, he obtained the elements α, α^2, $\alpha\beta$, $\alpha^2\beta$. (Cayley actually wrote 1, α^2, $\alpha\beta$, $\alpha^2\beta$, an indicator of the haste with which he wrote the article, and the lack of careful editing in nineteenth century journal articles.) Since (by the previous proposition) the product of the elements of a group by any of the elements of the group simply reproduced the group, then α^2 had to be equal to β or else equal to 1. In the former case, the group was simply 1, α, α^2, α^3, while in the latter case, the group's multiplication table could be written:

	1	α	β	γ
1	1	α	β	γ
α	α	1	γ	β
β	β	γ	1	α
γ	γ	β	α	1

Cayley also found all groups of order 6 and 8. One of these groups of order 8 had special properties that Cayley noted: the group whose elements could be listed as

$$1, \alpha, \alpha^2, \alpha^3, \beta, \beta\alpha, \beta\alpha^2, \beta\alpha^3$$

See Problem 6.

where $\alpha^4 = 1$, $\beta^2 = \alpha^2$, $\alpha\beta = \beta\alpha^3$ was equivalent to the quaternions.

23.4 Exercises

1. Explain how Proposition 23.2 is equivalent to the theorem that the index of subgroups of S_n is 1, 2, or not less than p.

2. Show that every element in a group of n elements satisfies the equation $\theta^n = 1$.

3. Prove that in the group of four elements, at least one element β exists such that $\beta^2 = 1$. Hint: if β^2 is not equal to 1, what is it equal to?

4. Find all groups of order 6 by determining their multiplication tables.

5. Find all groups of order 8 by determining their multiplication tables.

6. Show that the group on page 716 is equivalent (isomorphic) to the quaternions. Let $\beta\alpha^3 = \gamma$, and $\alpha^2 = \beta^2 = \theta$.

   (a) Show that $\theta^2 = 1$.

   (b) Show that θ commutes with α, β, and γ.

   (c) Show that $\gamma^2 = \theta$.

   (d) Identify $\theta = -1$, and show that the rules for the multiplication of the group elements correspond to the multiplication rules for the quaternion elements 1, i, j, k.

23.5 Algebra in Paris

Abstract algebra, though originating with ideas put forth by Cauchy and Galois, was developed by British mathematicians. Unfortunately, continental mathematicians were generally unaware of British developments, and as a result, group theory did not become widely used until the French mathematician Camille Jordan (January 5, 1838-January 22, 1921) published a number of papers in the 1860s and 1870s on the subject.

In Liouville's journal of 1867, Jordan published "Memoir on the Algebraic Resolution of Equations," where he clearly laid down the foundations of group theory in an axiomatic form. Jordan defined what is now called a **permutation group**, though he used the term substitution.

> One calls a *substitution* the operation that consists of changing the order of a certain number of things a, b, c, ...

> One designates by AB the substitution which produces the same effect as the two substitutions A and B, executed successively; by 1 the substitution that leaves each thing in its place.

> In the sequence of substitutions 1, A, A^2, ..., all of which are different up to the term A^μ which reduces to 1, from which the others are reproduced periodically, the number μ is called the *order* of the substitution A.

> A system of substitutions A, B, C, ... form a *group* if the product AB of any two is itself part of the system.[5]

---

[5] Jordan, Camille, "Mémoire sur la Résolution Algébrique des Équations." *Journal de Mathématiques Pures et Appliquées, Deuxième Série* 12 (1867), page 109.

From these, Jordan proved many important results. Rather surprisingly to a modern student of abstract algebra, one of the first propositions Jordan proves is:

Proposition 23.6. *The substitution 1 is in every group.*

In modern abstract algebra, this is part of the definition of a group. The existence of **subgroups** is implicit in:

Proposition 23.7. *If a group contains a substitution other than 1, it contains at least one of a prime order.*

23.5.1 Sylow

Jordan's early work on group theory inspired a number of other researchers in the field, such as the Norwegian Peter Ludvig Mejdell Sylow (December 12, 1832-September 7, 1918), who graduated from the same cathedral school in Christiania as Abel. Sylow showed great early promise, going on to the university in Oslo that Abel's father helped found, winning a prize in mathematics in 1853. In 1872, Sylow published a short but important paper, "Theorems on Substitution Groups," containing just eight theorems in ten pages.

To Sylow, a group was a substitution group: the extension of the group concept beyond this very concrete idea had not yet occurred. A key result is now called **Sylow's theorem**, which he stated in the introduction. Cauchy had proven (1845) that if a prime number n divided the order of a group, then the group contained a subgroup of order n. Sylow noted this proposition of Cauchy's was part of a more general theorem:

Theorem 23.1 (Sylow's Theorem). *If the order of a group is divisible by n^α, n being prime, the group contains a subgroup of order n^α.*

Sylow then proved the theorem.

23.5.2 Klein

One of Jordan's students in Paris was Christian Felix Klein (April 25, 1849-June 22, 1925). Klein studied under Plücker at the University of Bonn. By Klein's birth, Bonn was part of the kingdom of Prussia, and in the 1860s, under its dynamic Chancellor Otto von Bismarck, Prussia was fast becoming the nucleus of a German nation. Key to Bismarck's dream was the inclusion of the territories of Alsace and Lorraine. Unfortunately, they were part of France.

In July 1870, the "Iron Chancellor" Bismarck published a letter that goaded the French, under Napoleon III, into declaring war on Prussia; the result was the Franco-Prussian War. Nearly seventy years before, the Prussians had relied on the magic of the name of Frederick the Great, and were crushed at Jena and Auerstadt. In 1870, the French relied on the magic of the name of Napoleon, and were crushed at Sedan. Napoleon III was captured and forced to surrender.

Klein, as an enemy alien during the war, was forced to leave France. He made his way to the University of Erlangen where in 1872 he became a full professor and was expected to give an inaugural lecture. Klein's lecture was on non-Euclidean geometries, which, like group theory, had languished in the back woods of mathematical investigation.

In the longer, published version, Klein mentioned a more ambitious goal, now called the **Erlanger program**, which might be viewed as an attempt to unify the various branches of geometry under the group concept. The key idea of the Erlanger program was that all geometry was a study of properties of figures that were invariant under some spatial transformations. These transformations Klein referred to as the "main group of spatial alterations," though they are now referred to as **transformation groups**. For example, Euclidean geometry consisted of the study of the properties of figures that were unchanged by rigid transformations: rotations, reflections, and translations. In two-dimensional space, these could be described as the transformation of coordinates (x, y) into (x', y') via

$$x' = ax + by + c \qquad\qquad y' = dx + ey + f$$

where $ae - bd = 1$. Notice this makes matrices (and particularly, determinants) of prime importance in the study of higher geometry.

23.5.3   Lie

Klein actually did very little "Erlanger program" research himself, and for the most part, mathematicians ignored Klein's call for unification. The only major exception to this indifference was Marius Sophus Lie (December 17, 1842-February 18, 1899), one of Sylow's students. Lie (pronounced "Lee") was not originally interested in mathematics, and apparently Sylow was an uninspiring teacher.

In 1868, Lie discovered the work of Poncelet and Plücker, and was particularly intrigued by Plücker's use of objects other than points as the basis for geometry. In the winter of 1869-1870, Lie traveled to Paris, where he met Klein. The two began a long correspondence. Since Norway was neutral during the Franco-Prussian War, Lie was allowed to stayed in Paris after Klein had been expelled. Lie studied with Jordan and began the study of what he called "finite continuous" groups, now known as **Lie groups**.

For a transformation group to be finite continuous, Lie required that any transformation in the group differ infinitely little from a family of other transformations (hence continuous), and that the group itself not be decomposable into discrete families (hence finite). In essence, this requires that between any two transformations, there is a continuous chain of transformations that differs from each other by arbitrarily small amounts. For example, any translation can be treated as a sequence of translations each of which move the figure an arbitrarily small amount. Lie gave as an example of a *non*-continuous group the set of transformations in Euclidean geometry: since a reflection cannot be expressed as a continuous chain of transformations within the group, the group is not finite continuous. The study of Lie groups forms a key component of modern mathematical physics, particularly the theory of relativity.

23.5.4   Prelude to the Twentieth Century

Napoleon III's capitulation after the Battle of Sedan meant the end of the Second Empire and the founding of the Third French Republic. German troops marched into Paris and dictated the terms of the peace: all of the French province of Alsace and most of the province of Lorraine were ceded to the German Empire, and a German occupation army was to remain in France until an indemnity of 5 billion francs was paid. No negotiation was allowed: the French were given the choice of signing, or facing continuation of the war. They signed. The Germans announced the formation of the Second German Reich (the Holy Roman Empire being the First Reich) under Kaiser Wilhelm I. The peace treaty was signed and the announcement made at the former palace of the kings of France: Versailles.

23.5 Exercises

1. Read Jordan's definition of a group carefully. How does it differ from the modern definition?

2. Prove Proposition 23.6. Hint: use a counting argument and your results from Problem 1.

3. Prove Proposition 23.7, by explaining how to construct, from the sequence A, A^2, ..., $A^\mu = 1$, an element of prime order p, where p is a factor of μ.

4. In each of the following, a geometry (assumed to be of two dimensional space) is described in terms of the transformations that leave the studied properties invariant. Describe the main group of spatial alterations in terms of matrix properties. For example, Euclidean geometry is the study of properties invariant under rigid transformations; this corresponds to the group M, where M is the set of all adjoined matrices $[A|B]$ where $\det A = 1$.

    (a) Rotations about the origin that preserve length.

    (b) Rotations about the origin; length does not have to be preserved.

    (c) Uniform expansion or contraction of a figure.

    (d) Reflection of a figure about a fixed line.

5. Which of the previous geometries give rise to a Lie group?

Chapter 24

The Twentieth Century

The twentieth century opened with national antagonisms tempered only by national fear. The French were still smarting from the loss of Alsace-Lorraine and sought an alliance with Russia. Russia thought of herself as protector of the Slavs and had gone to war with the Ottoman Empire (modern Turkey) a number of times in the name of Slavic unity, freeing the region of the Balkans into a number of independent states, including Serbia. Serbia considered Bosnia and Herzegovina to be part of her cultural and historical heritage, but these were still part of the Ottoman Empire until 1908, when they were annexed by Austria. Austria, meanwhile, was facing troubles at home, for her empire (the vestige of the Habsburg domination of Europe) was on the verge of disintegration, so she sought help from Germany. In Bismarck's opinion, tying Germany to Austria was dangerous, for Austria's Balkan territories were a powder keg waiting to explode: "The next war," he said, "will start over some damned fool thing in the Balkans."

Unfortunately, Bismarck, Wilhelm I's dynamic chancellor, no longer ran the show. Wilhelm I died in 1888 and was succeeded by his son, Friedrich III, who reigned only 99 days before dying of throat cancer. His son became Kaiser Wilhelm II. One of Wilhelm's first actions was to force Bismarck to resign, and from 1890, *he*, not Bismarck, would run the government, and *he*, not Bismarck, dictated foreign policy.

24.1 The Hilbert Problems

In 1900, in Paris, David Hilbert posed a set of 23 questions for mathematicians of the twentieth century to consider. The problems span the history of mathematics: the oldest concerned Euclid geometry, while the most recent came from the mathematics of Hilbert and his contemporaries. Most of the Hilbert problems deal with advanced mathematics, and even the questions cannot be understood

Figure 24.1: The World in 1900.

without a specialized background, but several relate to "elementary" topics.

The basis of Hilbert's program was something he called the **axiom of solvability**: every problem could either be solved, or proven impossible to be solved. Interestingly enough, Hilbert's intuition was wrong on most of the problems he posed: the solutions were not what he suspected they might be. However, like Fermat's number theory, the value is not so much the goal as it is the path taken toward it.

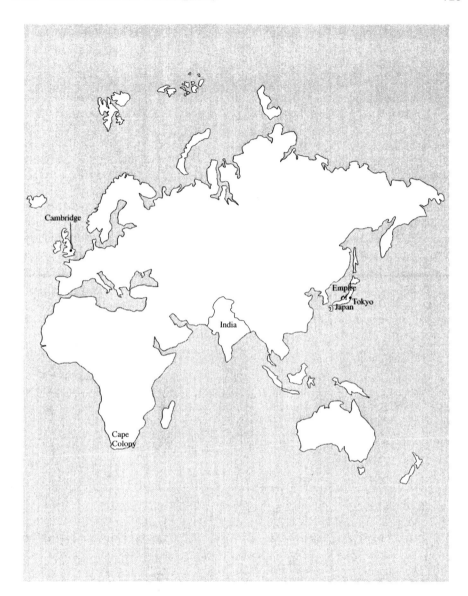

Figure 24.2: The World in 1900.

24.1.1 The Continuum Hypothesis

The very first of Hilbert's questions concerned the continuum hypothesis: was the cardinality of the continuum $\aleph_1$, and could the continuum be well ordered? Hilbert believed the answer to both questions was "Yes." In 1904, Ernst Zermelo (July 27, 1871-May 21, 1953) offered a partial answer to Hilbert's first question by proving that the continuum could be well-ordered. Zermelo relied on the See page 690.

so-called **axiom of choice**: given any set of sets S_n, one can create a set S consisting of one element drawn from each set. There is no difficulty in allowing this for a *finite* sequence of sets, but Hadamard, Peano, and others argued that it was unacceptable to assume that the axiom held for infinite collections of sets.

Kurt Gödel (April 28, 1906-January 14, 1978), an Austrian mathematician, finished answering Hilbert's question about the continuum in 1940, though the answer might appear to be unsatisfactory. We might, like Saccheri, try to disprove the continuum hypothesis by showing that the axioms of set theory, plus the continuum hypothesis, lead to a contradiction. Gödel proved that if such a contradiction arose, then the contradiction could be derived from the axioms of set theory alone. Thus, assuming either the continuum hypothesis or its negation would result in a consistent system, *provided* set theory itself was a consistent system. Gödel also showed that the axioms of set theory were independent of the axiom of choice, and again, it or its negation could be assumed.

24.1.2   The Consistency of Arithmetic

See page 657.

Hilbert's second question arose from the problem of whether or not the axioms of geometry were consistent. His *Foundations of Geometry* had included a proof of the axioms of his geometry, but this proof was based on an appeal to algebra. Hence, geometry is consistent if algebra is consistent. However, the consistency of algebra depended on the consistency of arithmetic, which had not yet been proven. Thus, Hilbert's second question was to prove the laws of arithmetic were consistent: that it was impossible to derive a contradiction from them. The work of Gödel would provide an answer to this question and will be dealt with in Section 24.2.

24.1.3   Volume of a Tetrahedron

Euclid had proven, in *Elements*, XII-7, that a triangular prism could be divided into three equal tetrahedra, which gave the volume formula for pyramids as a consequence. However, Euclid's proof relied on an earlier proposition that pyramids on equal bases with equal heights were equal in volume. The proof of this latter proposition relied on the method of exhaustion, and hence on infinitesimals. Gauss and others felt it should be possible to prove this result using only finite decomposition and reconstitution, but during the nineteenth century, no proof had been found. Hence, Hilbert's third question asked if it was possible to prove the equality of two tetrahedra on equal bases with equal heights *without* the use of infinitesimals. Hilbert himself believed it was *not* possible, and in 1904, Max Dehn (November 13, 1878-June 27, 1952) showed Hilbert to be correct: it was *not* possible, in general, to prove the equality of two tetrahedra without infinitesimals. Dehn's proof is relatively straightforward, essentially relying on the incommensurability of the solid angles of two tetrahedra.

24.1.4 Axiomatization of Physics

Hilbert's sixth question dealt with physics: was it possible to axiomatize physics, and thus derive the laws of physics deductively? Newton's *Mathematical Principles of Natural Philosophy* was an example of how dynamics (the physics of moving objects) could be derived, but since Newton's time, physics had expanded greatly, and the newer parts had not been axiomatized in the way that dynamics had been. Hilbert's original question concerned a branch of physics known as statistical mechanics, but by 1909, Hilbert himself began to work in a branch of physics that did not even exist when he posed the question: the theory of relativity.

Albert Einstein developed the special theory of relativity based on the simple assumption: given a light source, all observers, regardless of their motion or the motion of the source, measure the speed of light to be the same. An important consequence is that if one observer is moving relative to another observer, their measurements of time and distance will be different: one second of time is not the same for all observers, and neither is one meter of length. The **special theory of relativity** dealt with the case where one or another observer was moving with a constant velocity. A more complex problem arises if one or another observer is moving under acceleration; this is the case handled by the **general theory of relativity**. Unfortunately, while the special theory can be handled and derived using simple algebra and geometry, the general theory is significantly more complex.

In 1915, Albert Einstein gave six lectures at the University of Göttingen on his general theory of relativity. One of the main problems Einstein faced centered around the conservation laws, the cornerstones of physics. These conservation laws developed out of very careful observation during the nineteenth century, and are general statements that the total amount of a particular quantity was invariant. For example, the law of conservation of energy states that the energy in a closed system was a constant. These conservation laws are not obvious to the casual observer. For example, a moving hammer has a great deal of kinetic energy (the energy of motion), but when it hits a nail and stops, the kinetic energy appears to have vanished. However, a careful examination shows the nail (and hammer) to be warmer than they were before the impact, so the kinetic energy has not been lost, but has instead been converted into heat energy.

An even more daunting discovery occurred in the last years of the nineteenth century, when radioactive substances were discovered. These substances gave off a tremendous amount of energy with no obvious source. In 1905, Einstein showed that one of the consequences of the special theory of relativity was that mass and energy could be converted: a tiny bit of mass could be converted into a tremendous amount of energy. This could be summarized in the most famous equation of all time, $E = mc^2$.

The problem was that the general theory of relativity appeared to violate the law of conservation of energy. Hilbert had already been working on the theory of relativity, but Einstein's visit seems to have energized him, and he gave the problem of conservation to a remarkable colleague of his: Emmy Amalie Noether

(March 23,1882-April 14, 1935), one of the most brilliant mathematicians of the twentieth century.

Noether's father, Max Noether, was a prominent mathematician in his own right, who taught at the University of Erlangen. Noether herself, like Jacobi and Hamilton, showed an early gift for languages, and in 1900 obtained a certification to teach English and French in Bavarian schools—for girls, as it was considered improper to have women in a male academic environment. However, Noether found mathematics to be her true calling. Thus, in 1900, she obtained the permission to sit in on courses at the University of Erlangen, *provided* the instructor permitted it. In many cases, they did not, though a few were sympathetic, notably Hilbert and Felix Klein. In 1904, she was permitted to study officially, and in 1907, she received her doctorate, working with Paul Gordan: for her dissertation, she found all 331 invariants of ternary biquadratic forms.

See page 707.

In 1915, shortly before Einstein's arrival at Göttingen, Hilbert invited Noether to work with him and his colleagues. At Hilbert's request, she began to investigate general relativity, and eventually, she resolved the difficulty with the conservation of energy. Her results were presented in her inaugural lecture in 1919, "Invariant Variation Problems." The lecture contained a groundbreaking result, which in physics is known as **Noether's theorem**: every symmetry in the laws of nature corresponds to a conservation law, and vice versa. Noether showed, for example, that the symmetry of a system under a translation in time implied the law of conservation of energy. In the general theory of relativity, however, the laws of nature are not symmetric under a translation in time; hence, the failure of the conservation of energy in the general theory. Noether's theorem is central to modern particle physics, where observed conservation laws and their corresponding symmetries play a major role in developing a Grand Unified Theory (GUT).

Noether's inaugural lecture was of a high quality, and had she been a man, she would have become a regular faculty member at Göttingen. However, many of Hilbert's colleagues felt women had no place in an academic environment. Hilbert retorted that the sex of the candidate was irrelevant: "[Göttingen University] is an academic institution, not a bathhouse." The opposition was too great, and the highest position Noether could obtain was that of an Adjunct Extraordinary Professor—a position whose title masked its unimportance ("extraordinary" means "extra" ordinary: untenured) and nonexistent salary. However, on April 22, 1923, she obtained a position as an instructor at Göttingen, which gave her a small stipend, and allowed her to teach officially.

Though the inaugural lecture was, in some sense, supposed to indicate the general future course of the candidate's researches, Noether embarked on a different path that would make her work in physics, important as it was, seem an early tangent to her real career as a mathematician. Since her result implied conservation laws in physics relate to symmetries, the study of these symmetries became important. These symmetries, in general, correspond to invariance under a given spatial transformation (for example, translation, or rotation about

See page 719.

an axis). Hence, they may be described in terms of an associated Lie group. This would lead Noether to become one of the founders of modern abstract algebra.

Noether's main contributions were to what is now called **ring theory**. The real numbers under the two arithmetic operations of addition and multiplication form a system with two many strong properties: both operations are commutative (as $a + b = b + a$, and $a \times b = b \times a$); both have identities (for addition, 0, and for multiplication, 1). In addition, every element has an additive inverse (designated $-a$), and every element except the additive identity has a multiplicative inverse (designated a^{-1}). The real numbers under addition and multiplication form what modern mathematicians called a **field**. Some of these properties do not hold in more general structures. For example, in matrix algebra, addition is commutative but multiplication is not, and not all elements have a multiplicative inverse. Thus, the set of all 2×2 matrices under the usual rules of matrix algebra are an example of a **ring**. If the operation of multiplication is commutative, then the ring is said to be a **commutative ring**: thus, the integers form a commutative ring, but the set of 2×2 matrices does not. If there is an element, designated 1, for which $1 \times a = a \times 1 = a$, then the ring is said to be a **ring with identity**.

A subset I of a ring R is called an **ideal** if it is closed under addition (in other words, the sum of any two elements of I is also in I) and if, for any element r of the ring R, then $rI = Ir = I$ (in other words, multiplying any element of I by r yields an element of I). For example, if R is the set of integers under the operation of addition and multiplication, and I the set of integers divisible by 18, then I is an ideal, since the sum of two integers divisible by 18 is another integer divisible by 18, while multiplying an integer divisible by 18 by any other integer results in an integer divisible by 18. One ideal may be a subset of another ideal; for example, the set of integers divisible by 9 is another ideal, which contains the ideal consisting of the integers divisible by 18. We can imagine a **chain of ideals**, each containing the one before it. For example, if I_n is the ideal consisting of the integers divisible by n, then one such chain might be

$$I_{18} \subseteq I_9 \subseteq I_3 \subseteq R$$

This particular chain has only a finite number of elements, but one can conceive of a chain of ideals with an infinite number of elements. An important component of Noether's main result is

Definition. *A ring R satisfies the **ascending chain condition** if every sequence of ideals C_1, C_2, ... such that*

$$C_1 \subseteq C_2 \subseteq C_3 \subseteq \ldots$$

has a finite number of terms.

We might consider these ideals to be generated by some combination of the two ring operations. I, the ideal consisting of the integers divisible by 18, is generated by $I = \{x | x = 18n, n \in R\}$. In this case, I is said to be **finitely**

generated (the generator is the element 18, which generates the ring by multiplication). What Noether proved was

Theorem 24.1. *In a commutative ring R, each ideal is finitely generated if and only if R satisfies the ascending chain condition.*

Such rings are now called **Noetherian rings**.

24.1.5   Transcendental Numbers

See page 674.

Hilbert's seventh question concerned transcendental numbers. Cantor had proven they were more numerous than the algebraic numbers, but aside from a few cases, there was no general method of determining whether or not a given number was transcendental, nor any general method of creating transcendentals. Thus, Hilbert's seventh question asked whether numbers of the form α^β were transcendental if α was a nonzero algebraic number, and β was an algebraic irrational. The problem is interesting enough by itself, but it probably helped that Hilbert was a student of Lindemann.   In 1934, Aleksandr Osipovich Gelfond (October 24, 1906-November 7, 1968) and T. Schneider independently proved that all such numbers were transcendental.

24.1.6   The Riemann Hypothesis

See page 549.

Hilbert was one of the last of the universalists, and we have already discussed his contributions to geometry and mathematical physics. He also made contributions to number theory, and in 1909, proved proved Waring's conjecture. Thus it is not too surprising that the eighth question dealt with number theory, based on two unsolved problems: the Riemann hypothesis and Goldbach's conjecture. On Hilbert's seventh and eighth questions, his intuition on what problems were "easy" and what problems were "difficult" failed him completely. He believed the seventh question, regarding transcendental numbers, would be very difficult to solve, and the eighth question to be easily solvable. In fact, the seventh was solved in Hilbert's lifetime, but the eighth eluded the best efforts of twentieth century mathematicians.

Hardy and Littlewood

Much work was done on the Riemann hypothesis during the 1910s and 1920s by two English mathematicians, Godfrey Harold Hardy (February 7, 1877-December 1, 1947) and John Edensor Littlewood (June 9, 1885-September 6, 1977), whose lives saw the height and subsequent decline of the British Empire. Hardy's main interests in life were mathematics and cricket, a ball game from which American baseball was probably derived. When Littlewood was young, the British had been gradually acquiring more and more of the region around the Cape of Good Hope, and Littlewood's family moved there. The British expansion brought them into conflict with the Dutch-descended inhabitants of the region, known as

Boers, and the antagonism between the two groups led to the Boer War (1899-1902), which resulted in the British acquisition of what became known as the Union of South Africa.

Littlewood's family, knowing that he would not receive a good education in colonial South Africa, sent him back to England to attend school. At Cambridge, Littlewood's mathematics tutor (a position more akin to a senior graduate advisor) suggested Littlewood investigate the Riemann hypothesis. At the time, Littlewood thought it was merely an interesting problem in the theory of complex functions; he was not aware of its importance to number theory. In 1911, he and Hardy began a fruitful collaboration. It would last thirty-five years and they would write nearly a hundred papers together.

In 1914, Hardy proved that there were infinitely many complex zeroes to the ζ function on the "critical line" $x = \frac{1}{2}$. Later that year, Hardy and Littlewood proved, assuming the truth of the Riemann hypothesis, that if $\Pi(x)$ is the number of primes less than x, and $\mathrm{Li}(x) = \int_2^x \frac{dy}{\ln y}$, then, for an infinite number of values, $\Pi(x) < \mathrm{Li}(x)$, and for an infinite number of values, $\Pi(x) > \mathrm{Li}(x)$, a result See page 636. stronger than a similar one Chebyshev obtained in 1849. However, for all known values of x, $\Pi(x) < \mathrm{Li}(x)$, and the question arises of finding the smallest value of x for which $\Pi(x) > \mathrm{Li}(x)$ is true. For those who enjoy large numbers, the Riemann hypothesis, combined with the prime number theorem, results in some of the largest numbers ever to appear in a mathematical proof. In 1937, S. Skewes, who, like Littlewood, lived in South Africa before coming to school in Britain, proved that x was less than $10^{10^{10^{34}}}$ (now called **Skewes Number**). To get an idea of how large this number is, suppose every atom in the universe could represent a digit from 0 to 9. The largest number that could be so expressed using atoms as digits would be about $10^{10^{68}}$. By comparison, Skewes number is $10^{10^{10,000,000,000,000,000,000,000,000,000,000}}$.

If the Riemann hypothesis was *false*, the Hardy-Littlewood inequality would still hold, but in that case, Skewes showed, x would be less than $10^{10^{10^{10^{10^{1.46}}}}}$. Toward the end of his life, Littlewood began to doubt the truth of the Riemann hypothesis, noting that there was no compelling reason to suspect it was true.

Ramanujan

Another part of the British Empire was India, the so-called "Jewel of the Empire." In January, 1913, Hardy received a letter from a mathematician who had See page 700. already made a reputation there: Srinivasa Ramanujan (December 22, 1887-April 26, 1920). Like Hermite, Abel, and Galois, Ramanujan's early experiences in mathematics included attempting to find a solution to the quintic equation. His exposure to advanced mathematics came from an antiquated text, *Synopsis of Elementary Results in Pure Mathematics*, by G. S. Carr and published in 1856, well before Riemann and Weierstrass had completely reconstructed the foundations of mathematics.

Ramanujan independently discovered the Bernoulli numbers, and in 1911, published a short paper on the subject in the *Journal of the Indian Mathemati-*

cal Society, which accorded him great recognition among other mathematicians in India. One of them, C. L. T. Griffith, sent a collection of Ramanujan's papers to M. J. M. Hill, a professor of mathematics at University College London (where Griffith himself received his education). Hill was encouraging, but suggested Ramanujan study more mathematics before proceeding, since among Ramanujan's results were

$$1 + 2 + 3 + \ldots + \infty = -\frac{1}{12} \tag{24.1}$$

$$1^2 + 2^2 + 3^2 + \ldots + \infty = 0 \tag{24.2}$$

$$1^3 + 2^3 + 3^3 + \ldots + \infty^3 = \frac{1}{240} \tag{24.3}$$

whose sums, Hill pointed out, should all be infinite.

After writing to two other English mathematicians and receiving no answer, Ramanujan finally wrote a letter to Hardy. The letter included about 120 results Ramanujan had derived without proof. These fell into three categories. Some, like

$$\text{If } \alpha\beta = \pi, \text{ then } \sqrt{\alpha} \int_0^\infty \frac{e^{-x^2}}{\cosh \alpha x} dx = \sqrt{\beta} \int_0^\infty \frac{e^{-x^2}}{\cosh \beta x} dx$$

were known results (this one, in fact, had been proven by Hardy himself). Others, like

$$\int_0^\infty \frac{dx}{(1 + x^2)(1 + r^2 x^2)(1 + r^4 x^2) \cdots} = \frac{\pi}{2(1 + r + r^3 + r^6 + r^{10} + \ldots)}$$

could be proven with some difficulty. Still others, like

$$\cfrac{1}{1 + \cfrac{e^{-2\pi}}{1 + \cfrac{e^{-4\pi}}{1 + \cfrac{e^{-6\pi}}{\ddots}}}} = \left\{ \sqrt{\frac{5 + \sqrt{5}}{2}} - \frac{\sqrt{5} + 1}{2} \right\} \sqrt[5]{e^{2\pi}}$$

baffled Hardy completely, though to Hardy, they seemed both new and important.

Perhaps Hardy was impressed by Equations 24.1 through 24.3, which were part of the results of Ramanujan sent by Griffith to Hill. As stated, Ramanujan's claims are incorrect, but the results, $-\frac{1}{12}$, 0, and $\frac{1}{240}$, are in fact the values of $\zeta(-1)$, $\zeta(-2)$, and $\zeta(-3)$. In 1914, Hardy arranged to have Ramanujan brought to Trinity College in Cambridge, where the two began a remarkable collaboration.

Two hundred years before, it might have been sufficient to claim results that could be verified empirically; however, mathematics had changed, and while a good conjecture is still a powerful result, mathematicians are no longer satisfied with the conjecture alone. Hardy noted that without proof, he could not accurately judge the value of Ramanujan's results. Hardy asked Littlewood to teach Ramanujan mathematics; this proved difficult, for as soon as Littlewood introduced an intriguing mathematical concept, Ramanujan would respond with an "avalanche of original ideas," which made it impossible to continue the lesson. The results Ramanujan could be persuaded to prove rigorously were impressive; the results Ramanujan did not prove were of varying quality, some valid and others (like his results on prime numbers) completely wrong.

Unfortunately, Ramanujan's health deteriorated while in England. He was a Brahmin, a member of the highest caste in Indian culture. For religious reasons, Brahmins could only eat certain things, and above all, they could eat no meat. Wartime England was not the place to have a limited diet, and though he returned to India in 1919, he died soon after. Ramanujan left behind a rich legacy of published results that made significant advances in number theory, and unpublished results in his notebooks that inspired other researchers.

24.1.7   Goldbach's Conjecture

The eighth question asked for a proof of **Goldbach's conjecture**, communicated from Goldbach to Euler in 1742:

Conjecture 24.1 (Goldbach's Conjecture). *Any even number can be expressed as the sum of two primes.*

Little progress had been made in proving it. In 1937, Ivan Matveevich Vinogradov (September 14, 1891-March 20, 1983), a Russian mathematician and student of Markov, showed that every "large enough" odd number could be expressed as the sum of three primes, and consequently every "large enough" even number could be written as the sum of four primes. Vinogradov's student, K. V. Borodzin determined (1956) that $3^{3^{15}}$ was an upper bound for "large enough." The best current result, obtained in 1996 by Chen Jing-run and Wang Tian Ze, is that Vinogradov's result is true for all N larger than $e^{e^{9.715}}$.

Meanwhile, progress was also made on reducing the number of primes required to express an even number. The first main result was by the Hungarian mathematician Alfred Rényi (March 20, 1921-February 1, 1970), who is best known for a saying of his: a mathematician is a machine for turning coffee into theorems. In 1948, Rényi proved that every large enough even number was a prime plus a product of at most a primes. Rényi was only able to prove that a was a very large, but nevertheless, finite number. A significant reduction was made in 1965 when A. A. Buhstab and Vinogradov independently proved every even number was a prime plus the product of at most three primes. Finally, in 1966, Chen Jing-run showed that every even number could be written as the sum of a prime and another number, which is either prime or the product of two primes. The final reduction has yet to be made.

24.1.8   Functions of Three Variables

Hilbert's thirteenth question asked for a proof of the impossibility of solving seventh degree equations by means of functions of two variables. This referred to a method, known as **nomography**, of solving equations by considering the intersections of curves. In essence, nomography required the reduction of a function of several variables to a function of two variables. For the method to fail, there would have to be functions of three variables: in other words, Hilbert's thirteenth question asked whether or not there existed any functions of three variables.

This might seem to be an odd question, but consider a function of two variables, such as $f(x, y) = xy$. This can be written as a function of one variable, e^u, where u is the sum of two functions of one variable, namely $u = \ln x + \ln y$. Thus, a function of two variables, $f(x, y) = xy$, can be written as a function of the sum of functions of one variable: in some sense, it is "really" a function of one variable. We might make an analogy with Cantor's discovery of a mapping from $[0, 1]$ into $[0, 1] \times [0, 1]$.

The answer to the thirteenth question was given by the Soviet mathematician Andrei Nikolaevich Kolmogorov (April 25, 1903-October 20, 1987). Kolmogorov published articles on a wide range of topics, including mathematical physics, probability, education, and the literary analysis of poetry. In 1956, he presented, "On the Representation of Continuous Functions of Several Variables as the Superposition of Continuous Functions of a Smaller Number of Variables." If E is the interval $[0, 1]$, and E^n is the n-dimensional cube, Kolmogorov proved that any continuous function $f(x_1, x_2, \ldots x_n)$ on E^n could be written as

$$f(x_1, x_2, \ldots x_n) = \sum_{r=1}^{n} h^r [x_n, g_1^r(x_1, x_2, \ldots, x_{n-1}), g_2^r(x_1, x_2, \ldots, x_{n-1})]$$

where g_1^r and g_2^r are functions of $n - 1$ variables. In effect, Kolmogorov proved that any function on n variables could be written as a sum of functions of three variables. In 1957, the Russian mathematician Arnol'd showed that functions of three variables could, in turn, be written as the sum of at most nine functions of two variables. Finally, on June 20, 1957, Kolmogorov announced the surprising result that *every* continuous function of n variables on E^n could be written as a sum of continuous functions of the sum of continuous functions of one variable on E^1: in the preceding case, we had $f(x, y) = xy$ on E^2 reduced to a function of the sum of continuous functions of one variable. Thus, Hilbert's thirteenth question was answered in the negative: continuous functions of three variables do *not* exist.

24.1 Exercises

1. Show that "Every odd number can be written as the sum of three primes" implies that every even number can be expressed as the sum of four primes.

2. Write the following functions as the sum of continuous functions of the sum of continuous functions of one variable.

   (a) $f(x, y, z) = xyz$ (b) $f(x, y) = \sin(xy)$

24.2 The Consistency of Arithmetic

One of the most common phrases in mathematical proofs is the phrase, "There exists..." However, Kronecker and his followers very reasonably objected to this implied notion of existence. For example, how could the root of a fifth degree polynomial be used as part of a proof, when the root could not be determined in a finite number of steps?

24.2.1 Intuitionism and Formalism

These objections found their twentieth century champion in the Dutch mathematician Luitzen Egbertus Jan Brouwer (February 27, 1881-December 2, 1966). His inaugural address at the University of Amsterdam, entitled, "Intuitionism and Formalism," pointed out a great divide in the philosophy of mathematics.

Despite its name, **Intuitionism** does not imply that mathematics should abandon logic: the name comes from the idea that *some* mathematical concepts must be accepted based on intuition. For example, the natural numbers are accepted. But from these few assumptions, *all* of mathematics must be derived. The strictest interpretation would hold that the derivations must be obtainable within a finite number of steps. Brouwer interpreted "existence" as "constructibility": if a mathematical object cannot be constructed, then it cannot be said to exist, even though it has the *possibility* of existing.

One of the key pillars of mathematics is the **law of the excluded middle**: a statement is either true or false, and there is no other choice (the "excluded middle"). For example, a proof by contradiction relies on the assumption that if a statement cannot be true, it must be false. But an Intuitionist would argue there are three possibilities, illustrated by the following three propositions:

1. *There exists an even prime greater than 2.* This statement is false, and its falsity can be proven.

2. *The number of primes exceeds any given number.* This statement is true, and can be proven in a finite number of steps.

3. *Any even number can be expressed as the sum of two primes.* This statement is *probably* true, but no proof has been found.

The strict intuitionist approach would sweep away most existing mathematics. **Formalism**, an opposing school of thought led by David Hilbert, held that the symbols of mathematics have an existence independent of whether they refer to actual objects. Hence, one can glibly talk of a set S and its subsets S', S'', ..., independent of whether or not the set S actually exists, provided that the deductions based on it are logically sound. An example of this occurred with Hilbert's geometry, where he did not assume the existence of circles. The deductions he made regarding circles were logically valid, and if the space under consideration did not permit the existence of circles, the deductions were *still* valid, even though there was nothing to which they could be applied.

24.2.2   Gödel's Theorem

Support for the intuitionist school, particularly for the notion that one cannot assume existence without constructibility, came from a surprising source: an answer to Hilbert's question about the consistency of arithmetic. In 1931, Kurt Gödel proved that any consistent system of axioms included true, unprovable statements, a result known as **Gödel's theorem**.

A sketch of Gödel's proof follows. Consider a statement such as, "There exists an x such that n divides x." This can be written in the notation of symbolic logic as $(\exists x)(n|x)$. In such a statement, x is a **bound variable** (as it must satisfy certain conditions) while n is a **free variable** (as it can take on any value).

The logical symbols that express mathematical statements must be finite in number. More precisely, new symbols may be invented, but they must be definable in terms of a *finite* sequence using *finitely many* different symbols. Hence, they can be put in a one to one correspondence with the natural numbers themselves, and in this way, a statement, such as $(\exists x)(n|x)$ can be assigned to a natural number. For example, if we let '(' $= 1$, '$\exists$' $= 2$, 'x' $= 3$, ')' $= 4$, 'n' $= 5$, '$|$' $= 6$, then $(\exists x)(n|x)$ can be assigned to the sequence $\{1, 2, 3, 4, 1, 5, 6, 3, 4\}$. This sequence can be turned into a natural number by forming the product of the first n primes, where the nth prime is raised to the nth number in the sequence: thus the formula $(\exists x)(n|x)$ corresponds to the natural number $2^1 3^2 3^3 5^4 7^1 11^5 13^6 17^3 19^4$. Numbers such as these are now called the **Gödel number** of a statement. Thus, *all* statements that can be made in the axiomatic system correspond uniquely to some natural number.

Consider all statements that have a single free variable; Gödel called these **class signs**. Let α be any class sign, and define $[\alpha; n]$ to be the formula obtained by replacing the free variable in the formula α with n: thus, if α is $(\exists x)(n|x)$, then $[\alpha; 5]$ would correspond to the statement $(\exists x)(5|x)$. We note that $[\alpha; n]$ might or might not be true: for example, if β is "There is a number x less than n and more than 1 that divides n," then $[\beta, 5]$ is the statement, "There is a number x less than 5 and more than 1 that divides 5," and this statement is false.

Let these class signs be ordered in some manner, so that $R(n)$ is the nth class sign. Let K be the set of natural numbers n that correspond to the statements of the form $[R(n); n]$ that are unprovable. For example, if $R(1) = \alpha$, then 1 is *not* in K, since the statement $[R(1); 1]$, "There exists an x such that 1 divides x" is provable. Finally, let S be the class sign where $[S; n]$ says that n is in K: i.e., n corresponds to a formula $R(n)$ that is unprovable.

Since S is itself a class sign, then $S = R(q)$ for some q. Now consider $[R(q); q]$, which claims that q is in K. If $[R(q); q]$ is provable, it is true, and q

is in K. However, if q is in K, then $[R(q); q]$ is unprovable, which contradicts the assumption. On the other hand, suppose $[R(q); q]$ is false; hence q is not in K, and thus $[R(q); q]$ is provable, so it is true, which again contradicts the assumption. Thus, $[R(q); q]$ is neither provable, nor false; it must thus be true *and* unprovable.

24.2 Exercises

1. Explain why, "This statement is false" cannot be true or false.

2. Prove that there must exist proofs of arbitrarily long lengths, and thus there are some theorems that cannot be verified practically.

3. Another way of understanding Gödel's proof is the following: imagine a logic machine that can only print true (in the sense of mathematically provable) statements and will, given time, produce all such true statements. Consider the statement, "There are true statements the logic machine cannot print." Show that this implies the existence of unprovable, true statements.

24.3 Real Analysis

Kronecker's counterpart in the physical sciences was the Austrian physicist Ernst Mach (1838-1916). Mach, who studied acoustics (the "Mach number," indicating how many times faster than the speed of sound an object was moving, is named after him), was particularly taken with the idea that quantities must be *measurable* to be meaningful. To the old philosophical question of whether a tree falling in the forest makes a sound if no one hears it, Mach would have answered *no*. In particular, *space* and *time* were concepts freely used by physicists, but no one had yet considered how they were to be measured. Einstein's examination of how space and time were measured led to the special (1905) and general (1915) theories of relativity.

The corresponding examination of how length, area, and volume were measured was begun by the work of Hankel, Harnack, and du Bois-Reymond, who had defined the content of a set. Peano (1887) and Jordan (1892) introduced additional considerations. Consider a set (finite or countably infinite) of intervals I_k that contains a set E. The **inner content** was the content of these sets whose union was entirely within E, while the **outer content** was the content of those sets whose union contained E. Peano and Jordan defined a set as being **measurable** if the inner and outer content were the same as the size of the intervals I_k went to zero. The **measure**, in that case, is the common value of the inner and outer content.

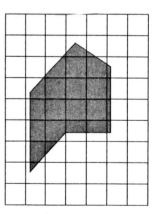

Inner Content

Outer Content

See page 690.

24.3.1 Lebesgue

These ideas were extended by Henri Léon Lebesgue (June 28, 1875-July 26, 1941). Lebesgue put together the measure theoretic ideas developed in the pre-

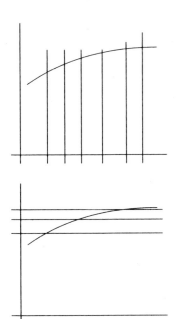

vious half century to form a new (some would say the first) theory of integration. On April 29, 1901, Lebesgue presented, "On a Generalization of the Definite Integral," where he first outlined the idea of what is now known as a **Lebesgue integral**.

The idea of a Lebesgue integral is remarkably simple: in the Riemann integral of $y(x)$ over $[a, b]$, the domain is partitioned into subintervals, and the sum of the products of the widths of the subintervals and a value $y(x)$ takes in the subinterval is found. Lebesgue partitioned the range, rather than the domain. Surprisingly, this has profound effects.

Let m and M be the minimum and maximum values taken by y over the interval $[a, b]$; consider the partition

$$m = m_0 < m_1 < m_2 < \ldots < m_{p-1} < m_p = M$$

Let E be the set of x for which $y = m$; E_i be the set of x for which $m_{i-1} < y \leq m_i$; define the measures λ_0, λ_1, ... of the sets E, E_1, ... Then, considering either of the sums $m_0\lambda_0 + \sum m_i\lambda_i$ or $m_0\lambda_0 + \sum m_{i-1}\lambda_i$, if, as the maximum size of the intervals $[m_i, m_{i+1}]$ tends to zero, the sums tend to the same limit independent of the choice of the m_i, the limit will be called the integral of y, which will be called integrable.

For the concept of "measure," Lebesgue used the following definition: for an interval (a, b) (it is immaterial whether the endpoints are included), the measure of the interval is simply the length $b - a$. For all others:

Definition. *Consider a set of points E in an interval (a, b). If E can be contained in a countably infinite number of intervals, the lower limit of the sum of the lengths of the intervals is the measure of the set.*

Lebesgue defined the set E to be **measurable** if the measure, increased by the measure of the set of points not part of E, is the measure of the entire interval (a, b); however, as this is difficult to apply in practice, one nowadays considers that if E is measurable, $\bar{E}$ is also measurable, and the sum of the two measures is the measure of the interval (a, b).

Example 24.1. *Find the measure of the set $E = \left\{\frac{1}{n}\right\}_{n=1}^{\infty}$. This set is contained within the countably infinite set of intervals*

$$\left\{\left(\frac{1}{n} - \frac{c}{2^n}, \frac{1}{n} + \frac{c}{2^n}\right)\right\}_{n=1}^{\infty}$$

The sum of the lengths of the intervals in this set is $2c$, where c can be any positive number; the lower limit of this sum is 0, which is thus the measure of the set E.

Example 24.2. *Let*

$$f(x) = \begin{cases} 1, & \text{If } x = \frac{1}{n} \text{ for some natural number } n \\ 2, & \text{Otherwise} \end{cases}$$

be defined over the interval $[0,1]$. *Find the Lebesgue integral* $\int_0^1 f(x) \, dx$.
  We note that $m = 1$, $M = 2$. *For any partition*

$$m = m_0 < m_1 < m_2 < \ldots < m_{p-1} < m_p = M$$

the set E *for which* $f(x) = m$ *is* $E = \{\frac{1}{n}\}_{n=1}^{\infty}$, $E_n = \{\emptyset\}$ *if* $n \neq p$; *and* $E_p = \bar{E}$
By the previous, we have $\lambda_0 = 0$, $\lambda_n = 0$ *for* $n \neq p$, *and* $\lambda_p = 1$. *Thus*

$$m_0 \lambda_0 + \sum m_i \lambda_i = 2$$

Likewise, $m_0 \lambda_0 + \sum m_{i-1} \lambda_i$ *tends to 2 as the size of the partition of the range
tends to zero; thus* $\int_0^1 f(x) dx = 2$, *and* $f(x)$ *is Lebesgue integrable.*

24.3.2   Frechet

Lebesgue's theory of integration concerned, essentially, functions of *real numbers*;
in 1904 and 1905, Maurice Frechet (September 2, 1878-June 4, 1973) extended
the notion to functions on some completely abstract set C. Given an infinite
sequence A_1, A_2, $\ldots$, of elements in C, Frechet assumed two axioms:

Axiom. *If the sequence has a limit B, then all infinite subsequences have the
same limit.*

Axiom. *If $A_i = A$ for all i, then the limit of the sequence is A.*

  From these, he defined

Definition. *A subset E is **closed** if every limit sequence of elements of E has
a limit also in E.*

Definition. *C is **compact** if, given any sequence of nonempty subsets E_1, E_2,
$\ldots$, where E_{n+1} is contained in E_n, the intersection of all the E_n is nonempty.*

Definition. *A function F on C is **continuous** over a subset E if, given a sequence in E_i with limit B in E, then the limit of the sequence $F(E_i)$ is $F(B)$.*

  Frechet's thesis (1906) added another element which, combined with his earlier works, may be considered the foundations of modern topology: given any two
elements of a set V, one can define a function (A, B) that satisfies the properties:

1. $(A, B) = (B, A) \geq 0$.

2. $(A, B) = 0$ if and only if $A = B$.

3. For all A, B, C, $(A, C) \leq (A, B) + (B, C)$.

In ordinary Euclidean space, the distance between two points A, B, satisfies
these three properties; the distance function, along with the points for which it
is defined, together form a **metric space**.

24.3.3   Hausdorff and Carathéodory

The idea of a neighborhood, critical to the development of topology, began to be discussed as early as 1912. In a series of lectures given at the University of Bonn, Felix Hausdorff (November 8, 1868-January 26, 1942) defined the **neighborhood** U_x of a point x to be the collection of all points y where the distance from x to y was less than ρ, some positive number. In 1914, he put his ideas together in a crucial work on point-set topology, *Elements of Set Theory*, which gave the first modern definition of a topological space.

Hausdorff's ideas were further extended by Constantin Carathéodory (September 13, 1873-February 2, 1950), a German of Greek descent living in Turkey. Carathéodory succeeded Lindemann at the University of Munich. In 1914, Carathéodory published "On the Linear Measure of Point Sets: A Generalization of the Concept of Length." Given some set A, Carathéodory considered the finite (or countably infinite) collection of sets U_1, U_2, ... whose union contained A. The **diameter d_n** of a set U_n is the greatest distance between any two points of U_n. Carathéodory defined $L_\rho A$ to be the lower limit of the sum of the d_ns for all possible collections U_n that covered A, where if $d_n < \rho$ for all n. Then Carathéodory defined the outer measure $L^* A$ to be given by $L^* A = \lim\limits_{\rho \to 0} L_\rho A$.

24.3.4   World War I

The year Hausdorff and Carathéodory published their groundbreaking work, the "damned fool thing in the Balkans" predicted by Bismarck occurred: Gavrilo Princip, a Serbian fanatic, assassinated Archduke Francis Ferdinand of Austria in Sarajevo. The Austrians demanded the Serbians turn over the murderer for punishment; the Serbians refused. Germany told Austria it would support whatever action Austria decided upon, so Austria invaded Serbia. Serbia's ally was Russia, whose ally was France. Thus, to support Austria in her war against Serbia, Germany invaded France. Though France and Germany share a 200 mile long border, the German war machine went through neutral Belgium, whose ally was Britain. By the end of summer, 1914, the European continent was at war.

The ill-conceived decisions that began the war continued through to the war's end, and beyond. At the Somme (July 1 to November 18, 1916), British and French troops went "over the top," charging at German positions protected by machine guns and barbed wire. Thousands were killed; the Allied forces responded by sending thousands more over the top. By November 18, the British and French had advanced seven miles—and lost over 600,000 men. No nation could stand this loss of men for long. After three years of war, the Russian state disintegrated, helping lead to the Russian Revolution, which established a new nation: the Soviet Union. After four years, the British, French, and Germans were on the verge of collapse, and it was only a question of who would collapse *first*. Britain and France were buoyed by the entry of the United States into the war, so it was Germany that sued for an armistice on November 11, 1918. By then, 8.5 million soldiers and over 13 million noncombatants had been killed.

24.3 Exercises

1. Show that any countable set on an interval has measure 0.

2. Find the Lebesgue integral of the Dirichlet function

$$\phi(x) = \begin{cases} c, & \text{if } x \text{ is rational} \\ d, & \text{if } x \text{ is irrational} \end{cases}$$

   for $c \neq d$, over the interval $[0, 1]$.

3. Prove that if a function, defined over a closed interval $[a, b]$, is Riemann integrable, then it is Lebesgue integrable.

4. Show that the set of rational numbers is not closed.

5. Show, by means of a counterexample, that the set of integers is not compact.

24.4   Topology

The problem of dimensionality had been raised by Cantor's work, showing the existence of one-to-one mappings from the line to the plane, and by Peano's creation of a continuous curve that covered the plane. If a curve could go through every point in the plane, *was* there an essential difference between a line and a plane? Poincaré had suggested the defining characteristic of an $n+1$-dimensional space was that it could be divided into two pieces by an n-dimensional figure: thus, a curve was one dimensional because it could be divided into two pieces by a zero-dimensional point. The notion was used by Euclid and Netto. Poincaré, the "explorer, not colonist," did little further study of the problem of dimensionality. Brouwer, in a lecture to the Dutch Mathematical Society in October, 1910, finally solved the generalized dimensional problem by proving it was impossible to find a continuous, one-to-one map from a space of m dimensions to a space of $m + h$ dimensions.

Brouwer proved that dimension was invariable, but what precisely *was* dimension? Brouwer, the champion of intuitionism, would have demanded that dimension of a space be determinable in some finite fashion. In 1911, a possibility was suggested by Lebesgue, based on the so-called **tiling principle**: if a point in a n-dimensional space D is in at least one of the closed sets E_1, E_2, ..., E_p, assumed finite in number and covering the entire space D, then, if the E_i are sufficiently small, there are points common to $n + 1$ of the sets. For example, we might consider a finite region of the plane to be tiled by the sets E_1, E_2, ... No matter how the plane is tiled, if the E_i are sufficiently small, there must be points that are on the boundary of three of the sets (these are the points where three or more of the tiles meet); thus the plane is two dimensional. Lebesgue thought he could prove the invariance of dimension this way (which infuriated Brouwer, whose paper proving invariance had not yet been published), but Lebesgue's original proof was flawed, and he could not fix the flaw until 1921.

24.4.1   Mathematics in Poland

Most of the fighting in World War I had been in France and Belgium; German territory was virtually untouched. Thus, it came as a shock to most Germans when the peace treaty was announced. The French, remembering history, dictated the terms to the German delegation at Versailles; the Germans were given two choices: to sign, or to face a renewal of the war. They signed, giving rise to a legend that Germany had been "stabbed in the back" by the civilian government. In fact, the delegates were well aware that Germany could not win a war against Britain and France *and* the enormous, untouchable, and barely (in 1918) tapped military potential of the United States. The end of World War I marked the beginning of the United States as a world power.

See page 571.

One of the major provisions of the treaty was the reestablishment of Poland as a nation, which had disappeared at the end of the eighteenth century. The new Polish state was carved mainly out of German territory; in fact, many German mathematicians were born in towns that became part of Poland: Hausdorff and Pasch both saw their birthplaces switch from Germany to Poland, and the birthplaces of Kronecker and Grassmann likewise changed countries. In a move guaranteed to exacerbate ill feelings, a "Polish corridor," including the German city of Danzig (now Gdansk), was established to give Poland access to the Baltic Sea: the Polish corridor separated East Prussia from the rest of Germany. Poland celebrated her newfound existence by going to war against the Soviet Union (formerly Russia) and Lithuania.

In the meantime, however, a Polish school of mathematics began to develop, with interests centered around mathematical logic (which we will not deal with here) and topology. As early as 1910, a Polish school of topology had been founded by Waclaw Sierpinski (March 14, 1882-October 21, 1969), Zygmunt Janiszewski (June 12, 1888-January 3, 1920), and Stefan Mazurkiewicz (September 25, 1888-June 19, 1945). Sierpinski and Mazurkiewicz, his student, began publishing *Fundamenta Mathematica* in 1920: it was the first mathematics journal devoted to one specialized field.

Perhaps one of the most interesting results was found jointly by Stefan Banach (March 30, 1892-August 31, 1945) and Alfred Tarski (January 14, 1902-October 26, 1983). In 1926, in *Fundamenta Mathematica*, they published a joint paper proving the remarkable result that a ball could be decomposed into sets, which could then be reassembled to form two *identical* balls, each equal to the first. The result, now known as the **Banach-Tarski paradox**, implied that volume was a meaningless concept. As the proof relied on the axiom of choice, the paradox caused many mathematicians to reconsider the validity of the axiom of choice.

24.4.2   Fractals

Hausdorff introduced a new means of determining the dimension of an object in 1918. The most remarkable feature of Hausdorff's definition is that it raised the possibility of a figure with fractional dimensionality, from which we get the word

fractal. Hausdorff's definition is actually very difficult to apply in practice, and several other definitions have been used in place. The one most commonly used today was introduced by Georges Bouligand, a French mathematician. On November 17, 1924, Bouligand presented "Dimension, Extension, and Density" to the French Academy of the Sciences. Bouligand extended his ideas in 1928 and 1929.

Bouligand introduced what he called the **Cantor-Minkowski** cover of a set E: let there be taken about every point in E the sphere of radius ρ. The collection of all these spheres, E_ρ, can be thought of as the set E plus all points that are less than distance ρ from an element of E. This collection of sets is sometimes called the **Minkowski sausage**, from its appearance if E is a line or a curve.

Bouligand asserted that the volume of E_ρ has a definite value, $f(\rho)$. If $f(\rho)$ had *infinitesimal order* α, then the dimension of E was $3 - \alpha$ or in general, $m - \alpha$ if E was considered to be in m-dimensional space.

> The **infinitesimal order** of a quantity is the rate at which the quantity tends to zero. For example, x^3 tends to 0 with infinitesimal order 3, while $\sqrt{x}$ tends to zero with infinitesimal order 1/2.

Example 24.3. *Let E be a single point in three-dimensional space. Then $f(\rho) = \frac{4}{3}\pi\rho^3$. As ρ goes to 0, $f(\rho)$ goes to 0 as ρ^3; hence $f(\rho)$ has infinitesimal order $\alpha = 3$, and the point has dimension $3 - 3 = 0$.*

As early as the 1924 article, Bouligand showed certain objects had fractional dimensions. He generalized the Cantor "middle-third" set and considered a line whose middle λth segment was removed, then the middle λth removed from the remaining segments, and so on. Bouligand showed this set has dimensionality $\frac{\ln 2}{\ln\left(\frac{2}{1-\lambda}\right)}$. Bouligand gave no indication of how he arrived at this result, but his 1929 article suggests one possible method. Bouligand considered

Example 24.4. *Let $E = \left\{1, \frac{1}{2}, \frac{1}{3}, \frac{1}{4}, \ldots, 0\right\}$ be a subset of 1-dimensional space. Take about each point the 1-sphere (i.e., an open interval) with radius*

$$\rho = \frac{1}{2}\left(\frac{1}{n} - \frac{1}{n+1}\right) = \frac{1}{2}\frac{1}{n(n+1)}$$

hence ρ tends to 0 as $\frac{1}{n^2}$. Then

$$f(\rho) = \frac{1}{n(n+1)} + \frac{1}{n} + \frac{n-1}{2n(n+1)}$$

hence $f(\rho)$ goes to zero as order $\frac{1}{n}$, or as $\sqrt{\rho}$. Hence $f(\rho)$ has infinitesimal order 0.5, and thus E has dimension $1 - 0.5 = 0.5$.

Bouligand's example suggests a somewhat easier method of approaching the finding of dimension: what is essentially required is to find the volume $f(\rho)$ of the set E when one can cover it with spheres of a particular size. The final comparison, between the infinitesimal order of $f(\rho)$ and the infinitesimal order of ρ can easily be made by noting $f(\rho)$ is of order $\frac{1}{n}$, while ρ itself is order $\frac{1}{n^2}$.

Since $\frac{\ln\frac{1}{n}}{\ln\frac{1}{n^2}} = \frac{1}{2}$, we can determine that $f(\rho)$ is of infinitesimal order $\frac{1}{2}$; in other words, it tends to 0 as $\sqrt{\rho}$.

In comparison, consider the Cantor set. For simplicity, we may assume the Cantor set to be part of a one dimensional line, so the "spheres" are the open intervals $(x - \rho, x + \rho)$; letting $\rho = \frac{1}{3^n}$, we have $f(\rho) = \frac{2^n}{3^n}$; hence the volume tends to zero as $\left(\frac{2}{3}\right)^n$, while ρ itself tends to zero as $\left(\frac{1}{3}\right)^n$. Hence the order of infinitesimal of $f(\rho)$ in terms of ρ is $\alpha = \frac{\ln 3 - \ln 2}{\ln 3}$, and the dimension of the Cantor set itself is $1 - \alpha = \frac{\ln 2}{\ln 3}$.

Other fractal objects made their appearance during the early part of the twentieth century. Since the study of fractals would not become an important part of mathematics until the latter part of the century, these objects were generally introduced for other purposes. In 1906, Niels Fabian Helge von Koch (January 25, 1870-March 11, 1924), a Swedish mathematician, published "An Elementary Geometric Method for the Study of Certain Quantities in the Theory of Plane Curves." Weierstrass had introduced a continuous curve that was nowhere differentiable, but Koch was dissatisfied with Weierstrass's construction, as it involved defining the curve in terms of an analytic expression, and thus hid the geometric nature of the curve. Geometrically, we know that any time a curve has a "corner," it is not differentiable at that point. Thus, Koch created a curve that had corners at every point.

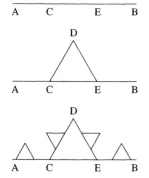

He began with a straight line, AB. Dividing this line into three parts, AC, CE, EB, and construct on CE the equilateral triangle CDE, making curve $ACDEB$. At each of the segments AC, CD, DE, EB, repeat the process: divide each segment into three parts, and on the middle third construct an equilateral triangle; if this process is repeated indefinitely, one ends with a curve P, now called the **Koch snowflake**. Based on this construction, Koch showed:

Proposition 24.1. *The curve P is continuous and has no tangent at any point.*

Proposition 24.2. *The length of the curve P is infinite, but the area enclosed by the curve P is finite.*

24.4.3   Julia and Mandelbrot Sets

See page 693.

Many fractal objects originate from the study of iteration, begun by Schröder and Koenigs. Given some function $F(z)$, which we may assume is a function defined for complex z, we may consider the complex plane to be divided into three regions: first, the periodic orbits or fixed points; second, the basins of any attracting periodic orbit or fixed point; and finally, the boundaries between the regions. For example, if $F(z) = z^2$, the attracting fixed points are $z = 0$ and $z = \infty$; the basins of attractions are the unit disk $|z| < 1$, and $|z| > 1$, and the boundary between the two is the unit circle $|z| = 1$.

In 1918, the French Academy of the Sciences proposed the problem of the study of iteration on a global scale. The winner of the prize was Gaston Julia (February 3, 1893-March 19, 1978), a French mathematician born in Algeria who served as a soldier during World War I; Julia was wounded in an attack and, like Tycho Brahe, lost his nose. At around the same time, another French geometer, Pierre Fatou (February 28, 1878-August 10, 1929) was also working on the problem of iteration (though he did not submit an entry for the Paris Prize). The work of Fatou and Julia overlaps considerably, though Julia's was the more rigorous from a mathematical point of view. In essence, the **Julia set** is the boundary between the basins. The Julia set might be a simple closed curve (like the unit circle in the foregoing example), or it might have a considerably more complex shape.

In 1919, Fatou considered the effects of an arbitrary constant c on the Julia sets for the successive iterations of $\phi(z) = z^m + c$. In the simplest case, $c = 0$, the Julia set is the unit circle; for other values of c, the Julia set is considerably more complicated.

We might consider c to parametrize a function $F_c(z) = z^2 + c$. If c is large, the successive iterates $F_c^n(z)$ tend to infinity for almost all values of z, so ∞ is an attracting fixed point whose basin includes the most of the complex plane. On the other hand, $c = 0$ yields two fixed points, 0 and ∞, with basins $|z| < 1$ and $|z| > 1$, respectively, leading to the Julia set being $|z| = 1$.

In 1945, Szolen Mandelbrojt introduced his newphew, Benoit Mandelbrot (b. November 20, 1924), to Julia's work. Mandelbrot, born in Poland but at that point studying in France, was not originally interested in Julia's work. However, by the 1970s, Mandelbrot was developing some of the earliest software to print computer graphics, and realized that the Julia sets provided a way of producing complex and beautiful graphics. In particular, if one considered the values c for which the successive iterates of $F_c^n(0)$ were bounded, c is now said to be in the **Mandelbrot set**.

Example 24.5. *Determine whether* 1 *is in the Mandelbrot set. We note* $F_1(z) = z^2 + 1$, *and consider the orbit* $F_1^n(0)$, *which is* $\{0, 1, 2, 5, 26, \ldots\}$. *Hence* 1 *is not in the Mandelbrot set.*

Example 24.6. *Determine whether* -1 *is in the Mandelbrot set.* $F_{-1}(z) = z^2 - 1$, *hence the orbit* $F_{-1}^n(0)$ *is* $\{0, -1, 0, -1, 0, \ldots\}$. *Hence* -1 *is in the Mandelbrot set.*

Intuitively, we expect there to be a boundary separating the points in the Mandelbrot set from the points not in the Mandelbrot set. The boundary line is complex, and Mandelbrot and others conjectured that it might have Hausdorff dimensionality two. In 1991, this was proven by the Japanese mathematician Mitsuhiro Shishikura.

24.4 Exercises

1. Show that a point has dimension 0 in 2- or 1-dimensional space.

2. Show that a line is one dimensional.

3. Show that the set formed by removing the middle λth portion from the interval $[0, 1]$, then the middle λth portion from each of the two segments, and so on, has dimensionality $\frac{\ln 2}{\ln\left(\frac{2}{1-\lambda}\right)}$.

4. Show that Bouligand's definition of dimension is equivalent to the following: suppose E is covered by $n(\rho)$ spheres of radius ρ. Then the dimension of E is given by $\lim_{\rho \to 0} -\frac{\ln n(\rho)}{\ln \rho}$.

5. Determine the dimensionality of the Koch snowflake. Hint: consider it to be part of a one-dimensional space.

6. Prove geometrically that the Koch snowflake has no tangent at any point.

7. Show that the dimension of any space filling curve is equal to the dimension of the space the curve fills.

8. Consider $F_c(z) = z^2 + c$. Show that whatever the value of c, there exist points that do *not* tend to ∞ as the number of iterations tend to infinity.

24.5   The Theory of Games

During the eighteenth and nineteenth centuries, mathematics became very complex, and it might seem that there is nothing left to discover; it seems beyond belief that there is anything *simple* to be discovered. However, the twentieth century saw several new fields of mathematical investigation opened up for exploration.

24.5.1   Borel

Emil Borel presented "The Theory of Games," to the French Academy on December 19, 1921. Borel asked the question of whether it was possible to determine a most advantageous method of play, that is, a strategy in which a player stood the best chance of winning. To this end, Borel supposed first that the game was symmetric: if both players made the same moves, each would be equally likely to win. Moreover, both players had the same choices of moves C_1, C_2, ...Suppose that player A made move C_i, and player B made move C_k, and let the probability of A or B winning be, respectively

$$a = \frac{1}{2} + \alpha_{ik} \qquad\qquad b = \frac{1}{2} + \alpha_{ki}$$

where $b = 1 - a$ and α_{ik} was between $-\frac{1}{2}$ and $+\frac{1}{2}$, and $\alpha_{ik} + \alpha_{ki} = 0$. The symmetric nature of the game required that $\alpha_{ii} = 0$.

For example, consider the matrix whose entries $a_{ij} = \alpha_{ij}$ are

$$\begin{pmatrix} 0.0 & 0.2 & 0.0 & 0.0 \\ -0.2 & 0.0 & -0.3 & -0.2 \\ 0.0 & 0.3 & 0.0 & 0.1 \\ 0.0 & 0.2 & -0.1 & 0.0 \end{pmatrix}$$

If A chose move C_2 and B chose move C_3, then $\alpha_{23} = -0.3$ and $\alpha_{32} = 0.3$, and thus A's chance of winning would be 0.2, while B's chance of winning would be 0.8.

Borel noted that certain moves could be eliminated completely as "rational" choices. To analyze the problem, Borel noted that for A, a move C_i would be undesirable (and thus, not a reasonable choice) under three conditions:

1. If $\alpha_{ih} \leq 0$ for all h. In other words, if A made move C_i, then whatever move B made, the probability that A won would be less than $\frac{1}{2}$.

2. If $\alpha_{jk} \leq 0$ for all k such that C_k was not a bad move of the first type for B.

3. If $\alpha_{jk} = 0$ for all moves not already eliminated.

For the preceding game, we note that if A chose move C_2, A would have a 50% or lower chance of winning, regardless of the move B made, so C_2 is not a reasonable choice. Likewise, B would never make move C_2, since this would mean that B would have a 50% or less chance of winning, regardless of the move A made. This means that C_4 is not a reasonable choice for A, since the only case where A has a greater than 50% chance of winning corresponds to the case where B makes a bad move. C_1 is also a poor move, since A does not have greater than a 50% chance of winning (unless B makes a bad move). This means that C_3 is the only reasonable move for A to make. B can apply the same analysis, so for this game, both players will chose C_3, and each will have an equal chance of winning. An interesting, if unappealing, consequence is that free will is, once again, a function of our ignorance: total knowledge implies knowledge of the best possible move; hence, to make any choices other than the best would be pointless.

If there are several possible "good" moves, Borel supposed p_k to be the probability that A chose move C_k, with the corresponding probability q_k that B chose move C_k. (Since the game is symmetric, what is a "good move" for one player is a "good move" for the other player.) The probability that player A wins the game will thus be

$$\sum_{i=1}^{n}\sum_{k=1}^{n}\left(\frac{1}{2}+\alpha_{ik}\right)p_i q_k = \frac{1}{2}+\alpha \tag{24.4}$$

where

$$\alpha = \sum_{i=1}^{n}\sum_{k=1}^{n}\alpha_{ik}p_i q_k = \sum_{i=1}^{n}\sum_{k=1}^{i-1}\alpha_{ik}\left(p_i q_k - p_k q_i\right)$$

In the case where $n = 3$, α can be written as a determinant

$$\alpha = \begin{vmatrix} p_1 & p_2 & p_3 \\ q_1 & q_2 & q_3 \\ \alpha_{23} & \alpha_{31} & \alpha_{12} \end{vmatrix} \tag{24.5}$$

If α_{23}, α_{31}, or α_{12} are nonzero, then it is always possible to choose p_1, p_2, p_3 so that $\alpha = 0$, regardless of the values of q_1, q_2, q_3: in other words, A can always find a method of playing that guarantees he will win half the time.

However, if $n > 3$, Borel believed that it was impossible to predetermine p_1, p_2, p_3, ... so that $\alpha = 0$ regardless of the choice of q_1, q_2, q_3, ...; in fact, α could be given any sign, depending on player B's choices of q_1, ...; in other words, it was impossible to pick a "perfect" strategy. Thus, Borel concluded, mathematics could, at best, eliminate the bad moves, but to decide among the remaining moves was outside the realm of the mathematician.

Borel wrote two more papers on the theory of games. In the last, delivered on January 4, 1927, he posed one of the fundamental questions: a **tactic** is a player's selection of the probabilities x_1, x_2, ..., x_k of choosing the move C_1, C_2, ..., C_k, respectively. Could A choose a tactic in such a manner that even if B finds out what tactic A has chosen, B cannot make his gain positive? In the first paper, Borel had already proven it was impossible for B to make his gain positive if $n = 3$ (since, regardless of player B's choice of q_1, q_2, q_3, the probability that A will win is $\frac{1}{2}$), and in 1924, had shown that it was likewise impossible for B to make his gain positive for $n = 5$; Borel's conclusion was different from his 1921 paper, and he believed now that it *was* possible to pick a perfect strategy.

24.5.2   Von Neumann

Johann (later John) von Neumann (December 28, 1903-February 8, 1957), is generally considered the founder of the modern theory of games, though he began publishing on the theory of games after Borel had completed his work. Von Neumann was born in Hungary of Jewish parents; his father, Max Neumann, purchased a title of nobility (hence the family became "von Neumann," pronounced "NOY-mann") in 1913. Von Neumann taught in the United States from 1929, becoming a Professor at Princeton University in 1931, but periodically returned to Europe until the Nazis came to power in Germany. After that, von Neumann settled in the United States permanently, becoming the youngest permanent member of the Institute of Advanced Study at Princeton. Von Neumann made significant contributions in all areas of mathematics and may have been one of the last to master both pure and applied mathematics. Von Neumann was responsible for answering Hilbert's fifth problem (dealing with Lie groups) and made considerable headway towards solving the sixth problems (axiomatization of physics).

Von Neumann established the modern terminology and methods of the theory of games, but gave no indication he was aware of the work of Borel (echoing the ignorance of French mathematicians of the work of German mathematicians, half a century earlier). Von Neumann's first paper on the subject was delivered at Göttingen on December 7, 1926, where he posed the question: in a game of strategy with n players, how should any individual person play to achieve the optimal result?

Von Neumann began with a very general definition of a game of strategy, where some outcomes depended on chance and others on the actions of the individual players. He then showed how all games of strategy (as defined in his introduction) could be reduced to the simpler case:

Problem 24.1. *Let there be n players, S_1, S_2, ..., S_n. Each player S_m chooses a number x_m from among 1, 2, ..., σ_m (where σ_m is some definite number) without knowing what the other players have chosen. Once all players have made their choices, player S_m receives (or gives out) the amount $g_m(x_1, x_2, ..., x_n)$, subject to the condition that $g_1 + g_2 + ... + g_n = 0$. How should player S_m make his selection of strategies?*

The condition $g_1 + g_2 + ... + g_n = 0$ indicates that of a **zero sum game**, where the players exchange amounts only among themselves. Von Neumann began by examining the case where $n = 2$. Since $g_1 + g_2 = 0$, then $g_1 = g(x, y) = -g_2$. Player S_1 wants to maximize $g(x, y)$, while player S_2 wants to minimize $g(x, y)$. For example, consider the two player game where each player has one of three possible moves, labeled 1, 2, 3. The table for $g(x, y)$ (now called the **payoff matrix**, since it represents the payoffs to one of the players) might look like the following:

$$g(1, 1) = -1 \qquad g(1, 2) = -1 \qquad g(1, 3) = 3$$
$$g(2, 1) = -2 \qquad g(2, 2) = 4 \qquad g(2, 3) = -2$$
$$g(3, 1) = 0 \qquad g(3, 2) = 1 \qquad g(3, 3) = 1$$

First, consider S_1, who wants to make $g(x, y)$ large. By choosing strategy 1, his payoffs will be $g(1, y)$, depending on what S_2 does: the possible payoffs are -1, -1, and 3, and the *minimum* payoff (corresponding to the worst result) would be -1. By choosing strategy 2, his payoffs could be -2, 4, and -2, and the minimum payoff would be -2, while by choosing strategy 3, his payoffs could be 0, 1, 1, and the minimum payoff would be 0. The *maximum* of the *minimum* payoffs, which von Neumann wrote as $\max_x \min_y g(x, y)$, is 0.

Since payoffs represent losses to S_2, this player wants to minimize them. Thus, he wants to find the *minimum* of the *maximums*, or $\min_y \max_x g(x, y)$. If S_2 chose strategy 1, the payoff would be $g(x, 1)$; again depending on what S_1 did, the payoffs might be -1, -2, or 0, with maximum 0; for strategy 2, the maximum result is 4, and for strategy 3, the maximum result is 3. Thus, if S_2 chooses strategy 1, then at worst, he would lose 0, whereas the other strategies might lead to a larger loss.

Since $\min_y \max_x g(x, y) = \max_x \min_y g(x, y) = M$, the outcome of this game will be M. Player S_1 can guarantee that the outcome will be greater than or equal to M, while player S_2 can guarantee the outcome will be less than or equal to M. Notice that for this game, it is irrelevant whether one player knows the other player's move: if S_1 learned that S_2 was going to make move 1, S_1 would not change his strategy, since changing to move 2 or 1 would result in a worse payoff. Thus, neither player has anything to gain by changing strategies.

But consider the following table for $g(x, y)$:

$$g(1,1) = 1 \qquad g(1,2) = -1 \qquad g(1,3) = 3$$
$$g(2,1) = 2 \qquad g(2,2) = 4 \qquad g(2,3) = -2$$
$$g(3,1) = 0 \qquad g(3,2) = 1 \qquad g(3,3) = 1$$

In this case, $\max_x \min_y g(x, y) = 0$, but $\min_y \max_x g(x, y) = 2$. What does this mean? In both games, the best choice is for player S_1 to play move 3 and player S_2 to play move 1, but in the first game, it makes no difference if one player knows the move of the other, since any switch would result in a lower payoff. But in the second game, if S_1 somehow learned that S_2 was making move 1, then S_1 would change his choice to make move 2, increasing his payoff from 0 to 2. Clearly, in such a case, it is essential to keep one's moves secret. But how?

Von Neumann's remarkable answer is the best strategy consists of selecting appropriate *probabilities* ξ_1, ξ_2, ..., of picking moves x_1, x_2, ...: thus, S_1 prevents S_2 from finding out his move by not knowing what move he will make until he actually makes it! In essence, von Neumann shifted the rules of the game, from a player selecting a particular move, to a player choosing a set of probabilities: a tactic, in Borel's terminology.

For example, consider the very simple payoff matrix

$$g(1,1) = 1 \qquad\qquad g(1,2) = -1$$
$$g(2,1) = -1 \qquad\qquad g(2,2) = 1$$

Player A would choose a tactic $(\xi, 1 - \xi)$, where ξ represented the probability of choosing move "1," and $1 - \xi$ the probability of choosing move "2." Likewise, player B would choose a tactic $(\eta, 1 - \eta)$. The expectation $h(\xi, \eta)$ would be

$$h(\xi, \eta) = \xi \eta (1) + \xi (1 - \eta)(-1) + (1 - \xi)\eta(-1) + (1 - \xi)(1 - \eta)(1)$$
$$= 2\xi + 2\eta - 4\xi\eta - 1$$

From this, we can find $\max_\xi \min_\eta h(\xi, \eta)$ and $\min_\eta \max_\xi h(\xi, \eta)$. The **minimax theorem** states that they can be made equal by the appropriate choices of η and ξ. In this case, $\xi = \eta = \frac{1}{2}$, and each player should choose each tactic one-half the time.

In a fair game, $\max_\xi \min_\eta = \min_\eta \max_x = M = 0$ (since neither player should have a "built-in" advantage over the other), and by proving the minimax theorem, von Neumann answered Borel's question: in a fair game, it was *always* possible to choose a tactic that would make it impossible for the other player to make his gain positive. The minimax theorem refuted Borel's supposition that mathematics could do no more than eliminate bad moves.

In 1944, von Neumann and Oscar Morgenstern published *Theory of Games and Economic Behavior*, applying mathematics economics, an area that had traditionally been off limits. Acceptance was slow at first, since before *Theory of Games*, equations in economics papers were virtually unheard of and generally met with incomprehension among economists. Soon, though, mathematics became as much a part of economics as it was of physics.

24.5 Exercises

1. Prove Equation 24.4, then show α can be written as the determinant in Equation 24.5.

2. Use Equation 24.5 to show that, if $\alpha_{23}\alpha_{31}\alpha_{12} \neq 0$, then it is always possible to choose p_1, p_2, p_3 so that $\alpha = 0$, regardless of the value of q_1, q_2, q_3. (In particular, even if player B changes the value of q_1, q_2, q_3, α will still be 0.)

3. For the game "Morra" (better known as "rock, scissors, paper"), the payoff matrix is:

$$g(1,1) = 0 \qquad g(1,2) = 1 \qquad g(1,3) = -1$$
$$g(2,1) = -1 \qquad g(2,2) = 0 \qquad g(2,3) = 1$$
$$g(3,1) = 1 \qquad g(3,2) = -1 \qquad g(3,3) = 0$$

   (a) Find $\max_x \min_y g(x,y)$ and $\min_y \max_x g(x,y)$.

   (b) Determine the tactics (ξ_1, ξ_2, ξ_3) and (η_1, η_2, η_3) that make $\max_\xi \min_\eta$ and $\min_\eta \max_{xi}$ equal. (In this case, $\max_\xi$ should be considered the maximum value over all possible values of (ξ_1, ξ_2, ξ_3), and likewise for $\min_\eta$: remember that $\xi_1 + \xi_2 + \xi_3 = 1$) Does the answer match with your intuition?

24.6 Computer Science

One of the key features of the latter half of the twentieth century was the rapid growth of mechanical and, later, electronic devices that were capable of performing computations with a speed and accuracy unheard of: the computer. One part of **computer science** deals with the engineering task of building computing machines; as discussion of this would be more appropriate in a history of technology or science, we will omit it here. We will instead focus on the mathematical component of computer science: the **theory of algorithms**.

24.6.1 Turing

In mathematical terms, a **computer** is a device that can take a set of instructions, referred to as a **program**, and follow them to produce an **output**. Alan Mathison Turing (June 23, 1912-June 7, 1954) provided the first mathematical analysis of this process, long before computers became common devices. Incidentally, until the 1950s, a "computer" or a "calculator" was a *person* who performed mathematical computations. Turing, though English, was studying at Princeton University, and Gödel's influence is clear in Turing's 1937 article, "On Computable Numbers."

Turing described a machine, now called a **generalized Turing machine**, capable of being in a finite number of conditions, q_1, q_2, $\ldots$, q_R; Turing called

Machine Configuration		Machine Action	
m-configuration	scanned symbol	operation	new configuration
b	blank	$P0, R$	c
c	blank	R	e
e	blank	$P1, R$	f
f	blank	R	b

Table 24.1: Turing's First Machine

these m-configurations (m for machine), though they are now called **states**. Through this machine was fed a "tape," which was divided into "squares," only one of which, say square r, was in the machine at any given time. This square Turing referred to as the "scanned square," containing the "scanned symbol" $\mathfrak{S}(r)$. Turing used quotations to indicate that he was not speaking of an actual physical tape, square, or symbol, but rather anything that would fulfill the same function. In particular, it was assumed that the machine read the symbols one at a time, and in some sort of sequence. The ordered pair $(q_n, \mathfrak{S}(r))$ indicated the current m-configuration of the machine and the scanned symbol.

The machine was assumed capable of four operations: it could write a symbol n to a scanned square (an operation Turing designated Pn); it could erase the symbol on the scanned square (E); it could move the tape one square to the left (L); or it could move the tape one square to the right (R); it could also change its state. A **computing machine** is one that can print two types of symbols: figures (in the simplest case, either a "1" or a "0"), or "symbols of the second type," which include all others.

In the simplest case, where the only figures are either 1 or 0, the sequence of figures can correspond to the binary expression of a number. For example, we might let the sequence 0 1 0 1 0 1 ... correspond to the number

$$0 \cdot 1 + 1 \cdot \frac{1}{2} + 0 \cdot \frac{1}{2^2} + 1 \cdot \frac{1}{2^3} + 0 \cdot \frac{1}{2^4} + \ldots = \frac{2}{3}$$

If the sequence of figures can be output by some machine, the corresponding number is said to be **computable**.

The number $\frac{2}{3}$, corresponding to the sequence 0 1 0 1 0 1 ..., is computable, which Turing proved as follows: consider a machine capable of being in four states, b, c, e, f, with the beginning of an (infinitely long) blank tape fed into it. The machine's actions depended on its current configuration: it would perform some operation and adopt a new m-configuration. Table 24.1 summarizes all possible configurations of the machine and all the actions the machine would perform if it was in a given configuration; Turing referred to this as the **table of the machine**.

Suppose the initial m-configuration is set to b. Since the tape consists entirely of blank squares, then the scanned symbol is blank. According to the table, if the m-configuration is b and the scanned symbol is blank, the action is $P0, R$: the machine prints the symbol 0, then moves the tape one square to the right; moreover, its configuration will change to c. Now the scanned symbol is blank (since the printed symbol, 0, is no longer in the machine), and the m-configuration is c, so the machine moves the tape one space to the right, and adopts m-configuration e. Once again, the scanned symbol is blank, and thus the machine prints the symbol 1 and moves the tape one space to the right, and so on. In this way, the sequence 0 1 0 1 ... is printed.

Turing noted that the machine operations were one of seven forms: E, ER, EL, $P\alpha$, $P\alpha R$, $P\alpha L$, R, or L, where α is one of the symbols the machine can print; if necessary, the machine could be "redesigned" so that the machine action would consist of one of these operations. Designating S_0 a blank space, $S_1 = '0'$, $S_2 = '1'$, and S_3 to S_n the remaining symbols, Turing showed that *any* operation the machine was capable of, from an initial m-configuration q_i and a scanned symbol S_j was one of three forms:

Machine Configuration		Machine Action	
m-configuration	scanned symbol	operation	new configuration
q_i	S_j	$PS_k L$	q_m
q_i	S_j	$PS_k R$	q_m
q_i	S_j	PS_k	q_m

For example, if the machine's action was to erase the symbol in the scanned square, this could be accomplished by the operation PS_0 (print a blank space). Or if the machine's action was to simply move the tape to the right, this could be accomplished by $PS_j R$ (print the scanned symbol in the square, then move the tape to the right). Hence, any row in the table of the machine can be described entirely in terms of four quantities: the machine state q_i, the scanned symbol S_j, the printed symbol S_k, and the new state q_m.

Example 24.7. *The machine for printing out the number* 0 1 0 1 ... *is modified to the machine (remember that S_0 is a blank, S_1 is the symbol '0', and S_2 is the symbol '1'):*

Machine Configuration		Machine Action	
m-configuration	scanned symbol	operation	new configuration
q_1	S_0	S_1, R	q_2
q_2	S_0	S_0, R	q_3
q_3	S_0	S_2, R	q_4
q_4	S_0	S_0, R	q_1

(note that "P" is omitted as being understood).

Now write down all the row entries in the machine table, separated by semicolons and convert the entries into a symbolic sequence in the following way: let q_i be represented by D followed by i A's; S_j be D followed by j C's. This gives the **standard description** (SD) of the machine.

Example 24.8. *The row entries of the machine from Example 24.7 are:*

$$q_1 \ S_0 \ S_1, R \ q_2; q_2 \ S_0 \ S_0, R \ q_3; q_3 \ S_0 \ S_2, R \ q_4; q_4 \ S_0 \ S_0, R \ q_1$$

This gives the SD

$$DADDCRDAA; DAADDRDAAA;$$
$$DAAADDCCRDAAAA; DAAAADDRDA$$

Finally, the SD is converted into the **description number** (DN) by writing 1 for A, 2 for C, 3 for D, 4 for L, 5 for R, 6 for N (never actually used by Turing), and 7 for ';'.

Example 24.9. *The DN for the machine from Example 24.7 is*

$$31332531173113353111731113322531111173111133531$$

The importance of this last step is that any number that is computable can be assigned at least one description number, and every description number corresponds to a unique computable number. Thus

Proposition 24.3. *The number of computable numbers is countable.*

A surprising consequence is that there are numbers that are noncomputable—in effect, there are numbers that cannot be described, except to say that they are indescribable!

Turing went on to discuss a universal machine, $\mathfrak{U}$, which could take the description number of any machine $\mathfrak{M}$ and produce the same number that $\mathfrak{M}$ would produce. Finally, Turing showed that there were some sequences that could not be produced and thus, it was impossible to find a general procedure for determining if a formula was provable. After receiving his Ph.D. from Princeton University in 1938, Turing returned to England. Dark storm clouds were on the horizon.

24.6.2   World War II

In 1933, Adolf Hitler became chancellor of Germany, and between 1933 and 1938, Hitler openly rearmed and remilitarized Germany; publicly repudiated the provisions of the Treaty of Versailles; annexed Austria; and conquered Czechoslovakia without firing a shot, all while France and Britain did nothing. In Italy, Hitler's fellow fascist Mussolini called his political party the *Nazionales*; Hitler appropriated the name Nazi from the Italians, and the swastika from ancient Teutonic symbols. Because the swastika is also a symbol used by the ancient Indians, Hitler claimed a relationship between the ancient Aryans of India and the modern Germans.[1]

Hitler was a remarkably poor student of history. Fourteen hundred years before, Justinian closed the schools of pagan learning and prohibited pagans

---

[1] By a tremendous stretch of the imagination, Hitler rationalized that the swastika, also a Buddhist symbol, implied a connection between the Aryans and the Japanese, hence an alliance with fascist Japan was permissible. The swastika also appears in the symbology of the American Southwest, but this did not prompt Hitler to think of a connection between, say, Navajo and the Germans.

from holding political office or teaching, holding there was a difference between proper "Christian" learning and improper "pagan" learning; the result was the final decline of classical learning and a long "dark age" in the west. Hitler closed many "Jewish" schools and prohibited Jews from holding political office or teaching, and proclaimed a "German science," which was different from the "Jewish science." The idea that the natural universe is amenable to national distinctions is, of course, ludicrous, and the only result of Hitler's persecution was that many scientists fled Europe and came to Britain or America.

We might in passing note some of the scientists who fled: Enrico Fermi, whose wife was an Italian Jew, began experimenting with neutron bombardment of uranium in 1934; Lise Meitner, an Austrian Jew, co-discovered uranium fission in 1938; the next year, based on Meitner's discovery, Leo Szilard, a Hungarian Jew, realized that a nuclear chain reaction might be possible. Szilard and fellow refugee (and Jewish scientist) Eugene Wigner persuaded Albert Einstein, a German Jewish refugee living in the United States, to write a letter to American President Franklin Delano Roosevelt, to begin a large project, subsequently named the Manhattan Project, to build an atomic bomb.[2] One of the chief mathematicians on the project was a third Jewish refugee, Edward Teller. Werner Heisenberg, a prominent and non-Jewish German physicist and Nobel Prize winner, was later put in charge of the project to build a German atomic bomb; after the war, Heisenberg claimed he could have built the bomb but deliberately misdirected the research. While there seems little doubt that Heisenberg was an especially capable scientist, it seems unlikely that he would have been able to overcome the many scientific and technological obstacles necessary to build an atomic bomb when so many scientists who could have helped had already fled Europe.

The Nazis also conceived of a "German mathematics." It was intuitive, not axiomatic; geometrical, not algebraic; and most of all tied to nature and not to "logic chopping." "Aryan students want Aryan mathematics, not Jewish mathematics," said one student with more fanaticism than sense. The Nazis conveniently ignored the highly axiomatic, highly algebraic approaches of German mathematicians like Riemann, Weierstrass, and Dedekind.

Hitler's racist policies forced the dismissal of many prominent mathematicians for no reason other than their ancestry. Noether was dismissed in 1933; wisely, she fled Germany and accepted a position at Bryn Mawr College in Pennsylvania. However, she proved to be an unsuitable undergraduate teacher, interested only in research. She found a new position at the Institute of Advanced Study at Princeton University (where she joined fellow refugees, including von Neumann and Einstein). Unfortunately, Noether contracted a viral infection and died in 1935. Max Dehn, who had solved one of Hilbert's problems, was dismissed from his post in 1935, and was arrested in November 1938. Fortunately, there was insufficient room in the local jail, so Dehn was released and fled Germany the next day. Gödel waited until after the war began to leave, which meant he

---

[2]This was Einstein's *only* contribution to the Manhattan Project.

had to escape through the Soviet Union (then neutral) and Japan before making his way to the United States.

Hausdorff and others stayed. On January 25, 1942, Hausdorff was interned in Endenich, a suburb of Bonn. A friend of his, Hans Wollstein, tried to console Hausdorff with the thought that internment might not be too bad; Hausdorff wrote a letter to Wollstein, noting, grimly, that "Endenich is not the end" (a pun in German: *Endenich ... Ende nicht*): in other words, internment was merely the first step down a terrifying road. After sending off the letter, Hausdorff, his wife, and her sister all committed suicide. A few months later Wollstein, who was also Jewish, would be deported to Auschwitz and die there.

See page 731.

A similar fate would befall the Jews in lands conquered by Hitler. Perhaps the most remarkable story is that of Rényi, who would solve portions of Goldbach's Conjecture. For centuries, European Jews had been restricted to certain portions of cities, called *ghettos*, though by the twentieth century, this practice was more a habit than a law. The Nazis reinstated the practice, herding the Jews to certain portions of a conquered city, and posting armed guards around the ghettos to ensure that the Jews could not leave. Rényi had false papers, so he was not interned, but his parents were. He obtained a soldier's uniform, marched into the ghetto, and demanded—and obtained!—their release.

On September 1, 1939, Hitler (with the consent and support of the Soviet Union) invaded Poland and, finally, France and Britain declared war on Germany: the Second World War had begun. On September 27, the Poles surrendered, and once again, Poland ceased to exist. In Hitler's view, the existence of Poland itself was an affront to Germany, and the Poles suffered grievously. Part of Hitler's campaign was to wipe out Polish academia; fortunately, many Polish mathematicians were able to outwit the Nazis and hide, though a few were murdered and some, like Mazurkiewicz, died from wartime deprivations.

Hitler, the poor student of history, proceeded to repeat Napoleon's mistakes. Napoleon invaded Egypt, to be defeated by the British, and then invaded Russia, and lost his empire and his throne. Hitler invaded Egypt, to be defeated by the British, and then invaded Russia. Even then, a study of recent history—events that occurred within his own lifetime—would have told him that it was vital to keep the United States out of the European war. But when the Japanese attacked the United States at Pearl Harbor on December 7, 1941, Hitler declared war on the United States, even though he was under no obligation to do so. From that point, the outcome was inevitable, though costly: World War II ended in 1945 with the defeat and suicide of Hitler and the partition of Germany. Between 35 and 60 million persons were killed.

During World War II, Turing was part of the British team at Bletchley Park (a country estate fifty miles north of London) that helped break the German "Enigma" codes. Because of their work, the British commanders often knew German orders before the Germans commanders. No doubt the breaking of the codes helped to shorten the war, and some historians go so far as to credit the codebreakers with winning the war for the Allies. However, nations know no gratitude: in the 1950s, homosexuality was still illegal in Britain, and Turing, given the alternative of a medical treatment to "cure" his homosexuality, or prison, chose treatment. The treatment proved too depressing, and Turing committed suicide.

24.6.3   Computability and Complexity

Turing had shown that a computer was capable of performing every operation a human mathematician could, but Turing's machine was designed for generality, not for speed. In a talk given in 1951 (and published in 1953), von Neumann formally considered the *difficulty* of applying a particular algorithm. He considered the so-called "assignment problem," and gave birth to the field of **computational complexity**.

Problem 24.2 (The Assignment Problem). *Given n people and n tasks; let a_{ij} be the value of the ith person doing the jth job. Find the assignment of people to jobs that will maximize the total value.*

A simple algorithm presents itself: examine all $n!$ possible assignments, and determine which one has the greatest total value. But as n grows larger, $n!$ grows very quickly, and this simple algorithm quickly becomes impractical. Von Neumann showed that finding the optimal assignment was equivalent to finding the optimal strategy for the following two-person game:

Problem 24.3. *Suppose there is an $n \times n$ matrix. Player 1 chooses one of the cells; player 2 guesses either a row or a column number (specifying which one he is guessing). If player 2 is correct, he wins α_{ij}; otherwise, he wins nothing. Find the optimal strategy for each player.*

Von Neumann showed that if $\alpha_{ij} = \frac{1}{a_{ij}}$, then the optimal strategy for this game could be used to find optimal assignment. Since player 1 has n^2 possible strategies, and player 2 has $2n$ possible strategies, then finding the optimal strategy for this game consists of examining $2n^3$ possibilities. Thus, rather than examining $n!$ possible assignments, we need only examine $2n^3$ possible strategies.

In 1963, the Canadian mathematician Jack Edmonds further extended the idea of computational complexity by defining an **efficient algorithm** to be one whose difficulty increases polynomially. In the previous case, the algorithm, "Examine all $n!$ possible assignments" is *not* an efficient algorithm, but "Examine all $2n^3$ possible assignments" *is* efficient. Edmonds considered the following problem from graph theory. A **graph** consists of a number of points, called **vertices**, joined by **edges** that run between two points. The problem Edmonds analyzed was:

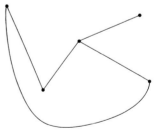

A graph with four vertices and five edges.

Problem 24.4. *Consider a graph G with n edges; a **matching** is a subset of the edges such that no two edges meet at the same vertex. Find the matching with the greatest cardinality (i.e., with the greatest number of edges).*

The matching is a subset of the edges; hence, an obvious algorithm presents itself: examine all subsets of the n edges, and determine which ones are matchings, then choose the one of greatest cardinality. However, this requires an examination of the 2^n subsets of the edges, which means that the difficulty of the task increases exponentially with the number of edges. Edmonds found an algorithm whose difficulty increased as a polynomial in n; we say that the problem can be solved in **polynomial time**, and the problem itself is P-type. Likewise, since a polynomial algorithm exists for the assignment problem, it, too, is P-type.

24.6.4 Cybernetics

Another important branch of mathematics developed in the twentieth century is **cybernetics**, largely begun by the work of Norbert Wiener (November 26, 1896-March 18, 1964), an American mathematician. Wiener entered high school at the age of 9, and graduated two years later; he obtained his Ph.D. at the age of 16 from Harvard University in Cambridge, Massachusetts. A few years later, he joined the faculty of mathematics at a school just down the road from Harvard: the Massachusetts Institute of Technology, universally known as MIT. Wiener's work helped to establish MIT as one of the premier mathematical research centers in the world.

Wiener began work in mathematical physics but after 1940 turned to the theory of communications (see later) and the development of a field he called **cybernetics**, a term used by nineteenth-century engineers, but which Wiener applied to a new field of mathematics. Cybernetics is primarily concerned with **purposeful machines**, which Wiener defined as those having a clearly established final state. For example, Wiener distinguished between a clock, which had a purpose in the ordinary sense, but was not a purposeful machine, since there was no final state to which the clock's motions were directed. On the other hand, a homing torpedo did have a final state, which was to impact the side of a warship at which it was aimed.

Cybernetics is concerned with three important concepts: information, control, and communication. For example, if you reach out to pick up a glass, your eyes pick up information (the relative position of your hand and the glass); your brain processes this information as part of a "control loop," to indicate which way your hand should move to pick up the glass; finally, this information is communicated to the muscles actually necessary to move your hand to the target glass.

24.6.5 Information Theory

We might note that a trend in twentieth century mathematics is an interest in *measurement*: how one measures the area of a region, or the difficulty of a problem, or the appropriateness of a particular strategy. Yet another example is the development of **information theory**. Communication is a key element of cybernetics; we might define communication as the ability to convey information, but what precisely does this mean? In 1948, the Missouri-born Claude Elwood Shannon (April 30, 1916-February 24, 2001), a 1936 graduate of MIT who went to work for Bell Laboratories, suggested one possible answer: **information** consists of the ability to pick out a "real" message from a choice of n possible messages. The amount of information could be found by any function of n, assuming that all messages were equally likely; Shannon pointed out (on the basis of a suggestion by R. V. L. Hartley) that $\log n$ (using any base) was the "natural" method of measuring the amount of information. (The use of the logarithmic function was a standard technique in statistical mechanics, a field pioneered by Gibbs.)

The basic unit of information was the "yes/no" answer, which could be considered a binary digit, a term later abbreviated to **bit** by J. W. Tukey; in this case, it made sense to measure the information content of a message using $\log_2 n$. For example, consider a questionnaire consisting of eight "yes/no" questions; the answers constitute a message. Since there are 256 possible "messages," then any particular message has an information content of $\log_2 256 = 8$ bits. On the other hand, if each question could have one of ten possible answers, then the message would have an information content of $\log_2 10^8 \approx 26.6$ bits. Wiener later defined the amount of information to be proportional to $-\log_2 P$, where P is the probability of any message (all messages assumed to be equally probable).

24.6 Exercises

1. Prove all rational numbers are computable.

2. Devise a simpler machine to print the sequence 0 1 0 1 ...

3. Show that the seven operations E, ER, EL, $P\alpha$, $P\alpha R$, $P\alpha L$, R, or L can be combined to make any sequence of operations of which the machine is capable.

4. Explain why the diagonal proof cannot be used to show the number of computable numbers is *non*denumerable.

5. Redesign the following machine (by adding states, as necessary) so that its standard description and description number may be found. Assume that a blank tape is fed into the machine.

Machine Configuration		Machine Action	
m-configuration	scanned symbol	operation	new configuration
a	blank	$R, P0$	a
a	0	$R, P1$	a
a	1	R	a

6. Show that a decimal digit has an information content of approximately 3.32 bits.

7. How many bits of information does a single letter of a 26-letter alphabet represent?

8. Show Wiener's definition of the amount of information is consistent with Shannon's definition.

9. Why should the *amount* of information have to do with the number of *possible* messages that can be received?

10. Consider the assignment problem. Suppose you had a computer capable of checking 10^{12} different assignments per second. How long would it take to solve the assignment problem for 30 people, if the computer checked every possible assignment to find the assignment of greatest value? How long would it take to solve the assignment problem by reducing it to a game of the type von Neumann envisioned?

24.7   Solved Problems

There have been many great mathematical triumphs during the last half of the 20th century, and three of the great "unsolved" problems have been solved: the four color theorem; Fermat's last conjecture; and the classification theorem.

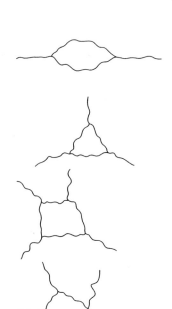

24.7.1   Four Color Theorem

In the course of a lecture given in 1840, Möbius posed the following problem: could a kingdom be divided into 5 regions so that every region touched the other four? Experience shows that this is not possible, which implies that *at most* four colors are required to color any map; this became known as the **four color conjecture**. In 1852, the question was raised again, by Francis Guthrie, a student of Augustus de Morgan. Guthrie had a proof (erroneous) that four colors were sufficient to color any map on the plane, if adjacent regions had to have different colors (thus, the problem is sometimes known as Guthrie's conjecture). De Morgan wrote to William Rowan Hamilton, on October 23, 1852, asking whether any proof of this had been found. Cayley, in an address to a mathematical society on June 13, 1878, noted the problem was unsolved.

In 1879, Alfred Bray Kempe (July 6, 1849-April 21, 1922), one of Cayley's students, showed every map must contain a digon (a map region bordered by two other regions), triangle (a map region bordered by three others), quadrilateral, or pentagon: this set of figures was an **unavoidable set**, since no map could avoid containing at least one of the figures. Consider any one of these figures, say the triangle, and suppose a coloring on the triangle and its neighbors could be extended to a coloring on the whole map; the map is said to have a **reducible configuration**. Kempe showed, to the satisfaction of his contemporaries, that every map was reducible to one of the five elements of the unavoidable set. Since none of these cases required more than four colors, any map could be colored in four or fewer colors. The proof was hailed as a major achievement. But in 1890, Percy Heawood (September 8, 1861-January 24, 1955) gave a counterexample that showed Kempe's proof for a pentagon was flawed, so the problem remained unsolved.

We may describe the subsequent efforts to solve the problem as follows. Kempe's proof, had it worked, would have relied on the fact that every map contained one of a set of unavoidable figures, and each unavoidable figure had reducible configurations associated with it. If every map in the plane could be shown to correspond to one of the reducible configurations, the problem would be solved. Thus, the problem could be approached in two ways: either by increasing the size of the unavoidable set, or by increasing the number of reducible configurations. Progress was made in both directions. In 1904, P. Wernicke increased the size of the unavoidable set to seven, while in 1913, G. D. Birkhoff added a few more reducible configurations. By 1970, H. Heesche conjectured that a finite unavoidable set of reducible configurations must exist; this set might contain as "few" as 9000 elements. Heesch developed a method of creating unavoidable sets, known as the **discharging method**, and all that remained was to show that each was four-colorable. A lengthy task for a human mathematician, but by then, computers were inexpensive enough so that large universities could afford them.

By 1976, Kenneth Appel and Wolfgang Haken of the University of Illinois used a computer to produce an unavoidable set with almost 2000 configurations; 1200 hours of computer time verified that the set was unavoidable, and allowed them to announce a proof of the four color theorem. Despite some qualms about a proof that no single human could verify, there is little argument over result, and the four color conjecture is generally believed proven.

24.7.2 Fermat's Last Theorem

The most hailed achievement of twentieth century mathematics was the proof of Fermat's last conjecture. Kummer's work meant that Fermat's last conjecture was true for a very large (but not necessarily infinite) number of primes. Thus, despite great progress, Fermat's last conjecture remained unproven in the general case.

See page 634.

An important milestone in the proof of Fermat's last conjecture came from two Japanese mathematicians, Yutaka Taniyama (November 12, 1927-November 17, 1958) and Goto Shimura. In September 1955, the two organized in Tokyo a symposium on algebraic number theory. One of the draws of the conference was that many of the results of Japanese number theorists had been published in Japanese journals, and were thus inaccessible to most Westerners. Because the conference itself would have its sessions conducted in English, this meant that anyone who attended the conference would learn about the results of Japanese number theorists before any non-Japanese-speaking mathematician would. In the conference Proceedings, thirty-six problems were posed, and problems 10 through 13, written by Taniyama, included what came to be known as the **Shimura-Taniyama conjecture**, sometimes called the Weil-Taniyama conjecture, though Weil disbelieved the conjecture when it was first proposed. The Shimura-Taniyama conjecture involves the properties of so-called elliptic curves, curves of the form $y^2 = x^3 + ax^2 + bx + c$, where a, b, and c are all integers. The details of the conjecture are too technical to deal with here.

In 1984, Gerhard Frey, a German mathematician, made another important conjecture: if $a^n + b^n = c^n$, for some integers a, b, and c, then the elliptic curve

$$y^2 = x^3 + (a^n - b^n)x^2 - (a^n b^n)x$$

would violate the Shimura-Taniyama conjecture. In other words, Frey conjectured that if a solution to $x^n + y^n = z^n$ existed, it would imply the existence of an elliptic curve that should not exist, by the Shimura-Taniyama conjecture. In 1986, Kenneth Ribet, of the United States, proved Frey's conjecture. In effect, Ribet proved Fermat's last conjecture *provided* the Shimura-Taniyama conjecture was true. The last step, proving the Shimura-Taniyama conjecture, was taken on June 23, 1993, by the Princeton University mathematician Andrew Wiles (b. April 11, 1953). Somewhat later, a flaw was found in the proof, but on September 19, 1994, Wiles, with the help of Richard Taylor from the University of Cambridge, managed to find a way around the difficulty, and the final results were published in May, 1995.

24.7.3   The Classification Theorem

See page 716.

Another achievement of the twentieth century was classification of all finite simple groups (i.e., those groups with a finite number of elements that contained no nontrivial subgroup). For example, Cayley recognized that if p was prime, then all groups with p elements were of the form $1, \alpha, \alpha^2, \ldots, \alpha^p = 1$. These groups are now referred to as Z_p. Another set of simple finite groups is the alternating groups A_n, for $n \geq 5$, where the alternating group A_n consists of the permutations on n letters that can be produced by an even number of transpositions. For example, if $n = 5$, then the permutation (3 1 2 4 5) is in the alternating group, since it is the permutation produced by two transpositions: switching the second and third elements, then the first and second elements. However, (1 2 3 5 4) is not in the alternating group, since it is produced by a single transposition.

Beginning in the 1940s, serious effort was put into classifying all the finite simple groups, and by the mid-1980s, over 500 articles published by mathematicians from around the world classified all finite simple groups into a relatively small number of categories: 18 regular classes, each of which had an infinite number of members (the first two classes were Z_p for p prime, and A_n for $n \geq 5$), plus 26 "sporadic" groups that fit into none of the categories. It has been estimated that the 500 articles of the classification theorem would fill between ten and fifteen thousand pages; hence, the classification theorem is sometimes known as the "enormous theorem."

24.7 Exercises

1. Conjecture how many colors are necessary to color every map that can be drawn on the indicated surface.

    (a) The surface of a cylinder.

    (b) The surface of a torus.

(c) The surface of a sphere.

(d) Prove your result for a sphere.

24.8 Unsolved Problems

There remain a great many questions that mathematicians have yet to answer. Gödel's Theorem suggests that many of these might be true but unprovable. Indeed, we might pose, as one of the problems that have yet to be solved, whether or not a given question is, in fact, unsolvable. On May 24, 2000, the Clay Mathematics Institute presented a set of seven "unsolved problems" of contemporary mathematics. The problems are reminiscent of Hilbert, both in the timing (2000 versus 1900) and the placement (Paris was where the announcement was made in both cases), although there are some crucial differences. The Clay problems are not intended, as Hilbert's were, to shape the direction of mathematics in the twenty-first century. More significantly, solving Hilbert's problems "only" gained one the honor associated with making a brilliant contribution to mathematics, while each of the Clay problems has a $1 million "bounty" for its solution; this has had the unfortunate effect of causing legitimate mathematics journals to be bombarded with submissions of dubious value. We will close with some of the Clay problems, as well as some additional unsolved questions in mathematics.

The first of the Clay problems deals with the so-called **P versus NP problem**. We might describe the difference as follows: a P-type problem is easy to solve, while an NP-type problem is easy to verify. For example, it might be difficult to factor a large number M, but to verify that a given factorization is correct is easy. If P = NP, then the two types of problems are the same, and factorization (in this case) is difficult only because no one has been clever enough to find an "easy" algorithm. This would translate into a disaster of epic proportions, for all computer security and all cryptography assume P $\neq$ NP. In the preceding example, we might suppose that entering one of the factors of the number is the password; the computer can easily verify the password is correct, but to guess the password would be nearly impossible, *provided* P $\neq$ NP. Fortunately, there is no indication whatsoever that P = NP.

The third Clay problem deals with a conjecture posed by Poincaré in 1904. The **Poincaré conjecture** is that any simply connected three-dimensional manifold (i.e., one without any "holes") without a boundary was topologically equivalent to a 3-sphere. (An ordinary sphere is a 2-sphere: it is an object in 3-dimensional space whose surface is a 2-dimensional manifold. A 3-sphere is an object in 4-dimensional space whose surface is a 3-dimensional manifold; in Cartesian coordinates, a 3-sphere is defined by the equation $x^2 + y^2 + z^2 + t^2 = 1$.) The general Poincaré conjecture is that any simply connected n-dimensional manifold is equivalent to an n-sphere; surprisingly, in 1960, Stephen Smale, John Stallings, and Andrew Wallace independently presented proofs for larger values of n: Smale's result, for $n \geq 5$, was the most general (Stallings proved it for $n \geq 7$, and Wallace for $n \geq 6$). The $n = 4$ case was proven in 1982 by M. H.

Freedman, and the $n = 3$ case remains to be proven.

The fourth Clay problem is the familiar Riemann hypothesis, which remains unproven.[3] The fifth and sixth Clay problems deal with questions in mathematical physics; the sixth, in particular, deals with a set of partial differential equations known as the Navier-Stokes equations, involving the flow of fluids. The last of the Clay problems involves the number of rational solutions to Diophantine equations. Hilbert's tenth problem asked whether it was possible to determine, in general, whether a Diophantine equation had any (non-trivial) rational solutions. In 1970, Yuri V. Matiyasevich showed that, in general, it is *not* possible to determine whether a given Diophantine equation has any rational solutions; thus, Hilbert's tenth problem was answered. However, while a general algorithm was thus proven impossible, it might still be possible to determine whether a *particular* Diophantine equation had solutions.

Besides Fermat's Last Conjecture, much progress had been made on Diophantine equations during the twentieth century. In 1922, Louis J. Mordell, an American mathematician, published a conjecture relating to the number of solutions to Diophantine equations. If one considered the surface generated by an equation, the surface might have one or more holes in it. For example, $x^2 + y^2 = 1$ generates a cylindrical surface, with 1 hole, while $x^2 = y$ generated a parabolic cylinder, with no holes. Mordell conjectured that if the surface generated (when the variables were allowed to take on complex values) had more than 2 holes, then the corresponding equation had only finitely many solutions that were relatively prime to one another. In the case of Fermat's Last Conjecture, $x^n + y^n = z^n$ corresponds to a surface containing $\frac{(n-1)(n-2)}{2}$ holes which implied that $x^n + y^n = z^n$ had, at most, finitely many solutions if $n \geq 4$. In 1984, Gerd Faltings (a German mathematician) proved the Mordell Conjecture. The Birch and Swinnerton-Dyer conjecture claims a relationship between the incomplete ζ function and the number of rational solutions to Diophantine equations.

To these seven problems, one more problem may be added, relating to the work of Poincaré, who had shown that the general n body solution was incapable of an exact solution; approximate solutions by means of infinite series were the best that could be hoped for. Unfortunately, these series in general diverged, making them useless for long term predictions. Gerald Jay Sussman and Jack Wisdom established, in 1988, that the outer planets (with the exception of Pluto) had orbits that were stable for the next 875 million years; in 1989, J. Laskar established similar results for the inner planets. However, these results run into a "wall of chaos" that prevents extension into the far future. Important theoretical progress has been made with KAM theory (named after its original authors, Kolmogorov, Arnol'd, and Moser), which showed that, under small perturbations, a large set of initial conditions led to stable solutions. However, in the 1990s, Zhihong Xia and others suggested that arbitrarily near any of these initial conditions were initial conditions that were unstable. Thus, our final question may be asked: can mathematics solve every physical problem?

---

[3]Matti Pikaenen of the University of Helsinki, Finland tentatively announced a proof in early 2000 but later withdrew it.

Appendix A

Answers to Selected Exercises

Chapter 1.1

1b: 213. **1d**: 234. **1e**: $21\frac{1}{3}$. **3a**: ꝯ ∩∩∩ | | | | | | | | | and $100, 30, 7$. **3b**: $2000, 500, 40, 1$. **4a**: $\frac{1}{8}$. **4b**: $\frac{1}{27}$. **5a**: $7, \frac{1}{5}$. **5c**: $9, \frac{1}{4}$. **6e**: $12\frac{1}{10}\ \frac{1}{5}$. **6f**: $12\frac{1}{15}$. **7a**: Possible interpretations are $5\frac{1}{40}$ or $\frac{1}{45}$. $\frac{1}{45}$ of a mile hardly seems worthy of mention: it's about 100 feet. So $5\frac{1}{40}$ seems the better interpretation. **7c**: The number could represent $5\frac{1}{20}$ or $\frac{1}{25}$. "Only" suggests there was a small amount left, so $\frac{1}{25}$ seems the better interpretation.

Chapter 1.2.2

3a: $1, \frac{1}{2}, \frac{1}{4}, \frac{1}{8}, \frac{1}{26}, \frac{1}{104}$. **5a**: $2, \frac{1}{6}$. **5c**: $1, \frac{2}{3}, \frac{1}{6}$. **10a**: $\frac{1}{8}, \frac{1}{22}, \frac{1}{88}$.

Chapter 1.3

3a: $3, \frac{1}{2}, \frac{1}{4}$. **3c**: $5, \frac{1}{3}, \frac{1}{5}, \frac{1}{15}$. **5b**: Breadth 6, length $3, \frac{1}{2}$.

Chapter 1.4

2a: 30 *setat*. **3a**: 224.

Chapter 2.1

1a: $1, 18$. **1b**: $2, 6$. **1c**: $10, 48, 41$. **1d**: $23, 52$. **2a**: 72. **2b**: $11, 711$. **4a**: $10\substack{1\\1}\ \substack{1\\1}\ \substack{110\\10}\ \substack{101\\1}\ \substack{1\\1}\ 1$. **4c**: $1\ 1\ 1\ \substack{10\\10}1$. **5a**: 23 and $36, 003$ are two possible interpretations. **6a**: The simplest interpretation of $10, 1, 1, 1$ would be 13. If each worker consumes $1, 1$ loaves, this is probably 2; for 5 days, 13 workers would then consume 130 loaves, which is $1, 1, 10$. **8a**: Subtract $7, 30, 0$ from $10, 33, 45$ to get $3, 3, 45$. **8c**: Multiply $4; 30$ by 10 to get 45. **14a**: $0; 8, 34, 17, 8, 34, \ldots$ **14b**: $0; 5, 27, 16, 21, 49, \ldots$

Chapter 2.2

1b: A cube 1 GAR on a side has a volume of 90 SAR. **1c**: A volume of 1 SAR is equivalent to 60 GIN.

Chapter 2.3

3a: $A_5 = 1; 43s_5^2$; $A_6 = 2; 35, 53s_6^2$, and $A_7 = 3; 38s_7^2$. The formula for A_5 is within 3% of the correct value, while A_6 and A_7 are within 1% of the correct value. **5**: $\sqrt{3} \approx 1; 43, 30$.

Chapter 3.2

1: 15 is odd-odd and triangular. **2**: Each gnomon is one more than the successive multiples of 3 (i.e., $3n + 1$). **2a**: Each gnomon is one more than the successive multiples of 4. **2b**: Each gnomon is one more than the successive multiples of 5. **2d**: The gnomon of a n-gonal number is one more than the successive multiples of $n - 2$. **6**: If the side of a square is even, it can be divided into two equal parts. Use this to show that the original square is thus even. **5**: $5, 12, 13$; $11, 60, 61$; $13, 84, 85$. **13a**: Harmonic: 6, 3, 2. **13b**: Arithmetic: 6, 4, 2. Geometric: 6, 4, $2 \frac{2}{3}$. Harmonic: 6, 4, 3.

Chapter 3.3

1a: The pairs are $(1,1)$; $(2,3)$; $(5,7)$; $(12,17)$; $(29,41)$. $\frac{41}{29} \approx 1.414$. **1e**: $\frac{3363}{2378} \approx 1.414214$. **2a**: The new side is the diameter plus twice the side; the new diameter is twice the diameters plus five times the side. With initial side and diameter both 1, the square on the diameter will alternately exceed and fall short of five times the square on the side by 4. **2b**: The first four pairs are $(1,1)$, $(3,7)$, $(13,29)$, $(55,123)$, with $\frac{123}{55} \approx 2.236$. **7a**: The new long side is three times the short side plus twice the long side; the new short side is twice the short side plus the long side. **8c**: $(1,1)$, $(4,11)$, $(23,65)$, $(134,379)$, $(781,2209)$. $\frac{2209}{781} \approx 2.828$.

Chapter 5.1

6: Note that there are an infinite number of lines of longitude, so there are an infinite number of straight lines that may be drawn between the two points; moreover, the straight lines cannot be extended indefinitely. Finally, the maximum radius of a circle is half the distance between the north and south poles.

Chapter 5.4

2a: Given two lines, the rectangle on the longer extended by the other, and the longer shortened by the other, is the difference between the square on the longer and the square on the other. **1a**: $6 \frac{9}{60}$. **1b**: $8 \frac{29}{60}$. **1c**: $18 \frac{1}{60}$.

Chapter 5.7

3: It should charge \$6. **4a**: Yes. **4b**: No. **4c**: Yes.

Chapter 5.8

2a: 6, 28, 496, and 8128. **2b**: There seems to be a perfect number in every order of magnitude. **2c**: The next perfect number is $33,550,336$.

Chapter 6.3

4: A is a weight, so it can have a ratio to B. However, A cannot have a ratio to b, since b is a length. **7a**: About $257,000$ stadia. **7b**: Eratosthenes's result is within 3% of the actual value. **7c**: The sun is about $40,000$ stadia from the Earth; this is Anaxagoras's result.

Chapter 6.4

1a: In scientific notation, $M = 10^8$, while $P = 10^{800,000,000}$.

Chapter 7.2

3: crd. $90° = 84^{\mathrm{P}}51'10''$, crd. $120° = 103^{\mathrm{P}}55'23''$. **4**: crd. $15° = 15^{\mathrm{P}}39'47''$. **7**: crd. $1\frac{1}{2}^° = 1^{\mathrm{P}}34'15''$. One method is to begin with crd. $60°$, halve it twice and find crd. $15°$. Then take crd. $72°$, halve it twice and find crd. $18°$. Using the difference formula, crd. $3°$ can be found. Half of this is crd. $1\frac{1}{2}^°$. **10**: Using crd. $1° = 1^{\mathrm{P}}2'50''$, we get the circumference equal to 377 parts, whereas the actual circumference would be 376.99 parts.

Chapter 7.4

2a: $1C$ $4S$ M $6x$ $5U$. **2b**: $3S$ M $4x$ $5U$. **3a**: $2x$ M $3U$; $4S$ $9U$ M $12x$. **3b**: $5x$ $2U$; $25S$ $20x$ $4U$. **3d**: $1S$ $3x$ M $5U$; $1SS$ $6C$ $25U$ M $1S$ $30x$. **4a**: Let the parts be $1x$, $3x$ $4U$. Then $4x$ $4u$ is 80, so $1x$ is 19 and the other part is 61. **4d**: Let number be $1x$; then $1x$ $20U$ is $3x$ M $300U$, so $2x$ is 320 and x is 160.

Chapter 8.1

3a: ‖‖ ⊥‖‖. **3b**: $=$ $\overline{\overline{‖‖ ⊥ ‖}}$.

Chapter 8.2

2: Good harvest is $\frac{35}{26}$ *tou*; mediocre harvest is $\frac{41}{52}$ *tou*. **7**: 15. **9b**: 24, 37.

Chapter 8.3

3a: 120. **3b**: 30.

Chapter 8.4

1a: $\pi \approx 3.1416$. **1b**: $314,160,000$. **1c**: The volume of a sphere with diameter of $20,000$ would be $5,568,000,000,000$, approximately. This is about 30% too large.

Chapter 9.2

1a: $\frac{10}{4} = 2.5$. **1b**: Add four squares of 2.5. **1c**: Take one quarter the number of roots; add four squares of this number.

Chapter 9.4

5: AC corresponds to sine; AD to cosine; BF to tangent; EG to cotangent; OF to secant; OG to cosecant.

Chapter 10.2

1: Answers vary, but one set might be for 36, $\frac{1\ 0\ 0}{4\ 3\ 3}$. For 48, $\frac{1\ 0\ 0}{3\ 4\ 4}$. For 250, $\frac{1\ 0\ 0}{10\ 5\ 5}$.
2a: Using the rule from Problem 1 $\frac{1\ 3\ 2}{3\ 4\ 4}$ 3. **2b**: Answers vary, but using rule $\frac{1\ 0\ 0\ 0}{10\ 3\ 3\ 3}$ we obtain $\frac{4\ 1\ 2\ 2}{10\ 3\ 3\ 3}$ 120.

Chapter 10.3

6: Example 10.4 corresponds to approximating a solution to $f(x) = 24$ when $f(10) = 36$ and $f(8) = 30$. **7a**: $f(x) = 40$; $f(10) = 36$ and $f(16) = 48$. Approximate solution is 12. **7b**: Approximate solution is $9\frac{3}{7}$.

Chapter 10.4

Configurations for Problems 1a-1d:

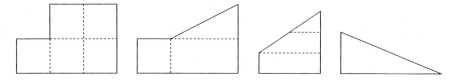

2: For the body in **1a**: $\frac{5}{3}$ of initial velocity. For the body in **1b**: $\frac{5}{4}$ of initial velocity. For the body in **1c**: $\frac{5}{2}$ of initial velocity. For the body in **1d**: half of the initial velocity.

Chapter 11.1

1: The two parts are $\frac{9}{5}$ and $\frac{16}{5}$.

Chapter 11.2.1

2a: $3\frac{2}{11}$. **2b**: $4\frac{4}{7}$.

Chapter 11.5

1b: p is a side, and q is a plane. **2:** B plane $= A$ square $+ AE + E$ square. If $A = 10$ and $E = 2$, then B plane $= 124$ and $Z = 240$. **6:** B plane $= A$ square $- AE + E$square. **13:** $\frac{2}{\pi} \approx 0.6368755$, making $\pi \approx 3.140331$.

Chapter 11.6

1a: $0.4540 \approx \sin 27°$ and $0.2756 \approx \sin 16°$, so $2 \times 0.4540 \times 0.2756 \approx \cos 11° - \cos 43° = 0.2503$. **1b:** $0.6293 \approx \sin 39°$, and $0.6561 \approx \sin 41°$, so $2 \times 0.6293 \times 0.6561 \approx \cos 2° - \cos 80° = 0.8257$. **1e:** Note that $6293 \times 65.61 = 100,000 \times 0.6293 \times 0.6561$. **2:** The logarithm of $9,999,999$ is 1; the logarithm of $9,999,900$ is 100. **4b:** The logarithm of 99.9 is between 0.1000 and 0.1001; the logarithm of 99.501 is between 0.499 and 0.5015.

Chapter 12.1

7a: 2 positive, 2 negative. **7b:** 2 positive, 3 negative. **7c:** 0 positive, 4 negative. **8:** Note that no transformation of the form $y = x + a$, where a is a positive number, will cause any sign changes; hence, if r is a root, there is no value for which $r + a$ will be positive. Hence all the roots must be imaginary.

Chapter 12.2

5a: $N = 2^{12} + 1$, so one factor is 17. **5c:** $N = 10^6 + 1$, so one factor is 101. **9a:** Prime. **9d:** 17 is a factor.

Chapter 12.4

2: Refer to margin figure on page 368. Suppose BD has axis CE with ordinate BC and abscissa CD; suppose the tangent BE intersects CE at E. Then DE is twice CD.

Chapter 12.5

1: The player who is ahead is guaranteed 32 pistoles, for even a loss (which would leave the score tied) would give them to him. As for the rest, he might win them and he might not; thus, he should get the 32 he is guaranteed, plus half the rest: 48. **2a:** 48 to 16. **2b:** 56 to 8. **2c:** 52 to 12. **2d:** 41 to 23.

Chapter 12.6

2: $\frac{8n^3 - 2n}{6}$. **3:** $2n^4 - n^2$. **4a:** $\frac{n^2 + n}{2}$. **4b:** $\frac{n^3}{3} + \frac{n^2}{2} + \frac{n}{6}$. **4c:** $\frac{n^4 + 2n^3 + n^2}{4}$.

Chapter 13.3

5a: $a = 2v^2$. **5b:** $a = \frac{-2v^2}{y}$. **5c:** $a = y$.

Chapter 13.4

8a: $y \approx 3 + 0.06 - 0.00789$. **8b**: $x = 3 + 0.2 + 0.01897$. **9**: $y = 3 + \frac{5x}{2} + \frac{7x^2}{2}$. **12**: $z = x + \frac{1}{6}x^3 + \frac{3}{40}x^5$.

Chapter 13.6

1: $\frac{3}{2}$. **1b**: $\frac{4}{3}$. **3a**: $\frac{1}{4}$. **3b**: $\frac{3}{4}$.

Chapter 14.1

1a: $\frac{n^4 + 2n^3 + n^2}{4}$. **3b**: $\frac{1}{7}n^7 + \frac{1}{2}n^6 + \frac{1}{2}n^5 - \frac{1}{6}n^3 + \frac{1}{42}n$. **2**: $C = \frac{1}{42}$, $D = -\frac{31}{60}$. **4a**: $25,502,500$. **4b**: $2,763,020,625$. **5c**: 420. **8a**: 138. **8b**: 2963.

Chapter 14.2

4a: The odds are $\frac{2+\sqrt{5}}{2}$ to 1. **4b**: The odds are $\frac{1}{\sqrt[4]{2}-1}$ to 1.

Chapter 14.3

6a: $x \approx \frac{193}{355}$. **6b**: $x \approx \frac{71}{183}$.

Chapter 14.5

1a: $\frac{3}{5}$. **1b**: $\frac{11}{25}$. **1c**: $\frac{1}{3}$ and $\frac{46}{81}$. **2a**: $\frac{1}{2}$. **2b**: $\frac{252}{1024}$. **2c**: Compare the probabilities of making a large error in the two cases.

Chapter 15.2

6a: $\frac{15}{8}$. **6b**: $\frac{1}{2}$.

Chapter 15.3

1a: 0. **1b**: 0. **1c**: 2π. **1d**: 0.

Chapter 16.1

3a: 0. **3b**: 2. **3c**: 2. **3d**: 0.

Chapter 16.3

3a: Several possible answers, but using $a = 0$ and $\alpha = \beta = \gamma = \ldots = 1$, then

$$\frac{1}{e} = \cfrac{1}{2 + \cfrac{1}{1 + \cfrac{1}{2 + \ddots}}}$$

4a: $\frac{1}{2} + \frac{\sqrt{13}}{2}$. **4b:** $\frac{1}{2} + \frac{\sqrt{5}}{2}$.

6(a)i: $\frac{29}{17} = 1 + \cfrac{1}{1 + \cfrac{1}{1 + \cfrac{1}{2 + \cfrac{1}{2 + \cfrac{1}{2}}}}}$ **6(a)ii:** $\frac{2783}{891} = 3 + \cfrac{1}{8 + \cfrac{1}{10}}$

Chapter 18.1

3: $q^{(1)} = -\frac{1}{4}$; $q^{(2)} = \frac{5}{96}$; $q^{(3)} = \frac{35}{384}$. Thus $\binom{100}{50} \approx 1.0089 \times 10^{29}$; $\binom{200}{100} \approx 9.04586 \times 10^{58}$; and $\binom{400}{200} \approx 1.029 \times 10^{49}$. The maximum error in using y_s to approximate $\binom{2s}{s}$ is about 0.001%.

Chapter 18.2

3b: $2 + \cfrac{1}{10 + \cfrac{1}{1 + \cfrac{1}{1 + \cfrac{1}{3}}}}$.

Chapter 18.3

1: Although the probability the mean has no error decreases, the probability that the mean is within a certain range of the mean also decreases. **2:** For $n = 1$ to $n = 6$, the probabilities are $\frac{1}{5}$, $\frac{9}{25}$, $\frac{1}{5}$, $\frac{29}{125}$, $\frac{561}{3125}$, and $\frac{2481}{15625}$.

Chapter 18.4

For problem 2a through 2c, consider the multinomial expansion of $(v + e + i)^7$. **2a:** $140ei^3v^3$. **2b:** $140e^3i^3v + 140e^3iv^3$. **5:** $\frac{a^3 + 3a^2b}{(a+b)^3}$.

Chapter 19.2

3: Begin by noting that [2] is a primitive root. Since $n = 7$, e could be 3 or 2, but $e = 3$ does not produce a useful split of the roots. If $e = 2$, then we find $p = [1] + [2] + [4] = \frac{-1+\sqrt{-7}}{2}$ and $p' = [3] + [5] + [6] = \frac{-1-\sqrt{-7}}{2}$. This gives rise to the equation $x^3 + p'x^2 - px - 1 = 0$. Solving this cubic will give us the three values of [1], [2], and [4], and we can use the approximation $[1] = \cos\frac{2\pi}{7} + i\sin\frac{2\pi}{7}$ to determine which of the roots corresponds to [1].

Chapter 19.3

3: It might be more reasonable to suppose that the size of the error is independent of the size of the quantity; thus all errors should be roughly the same size, with $x = y$.

Chapter 22.3

5: Keep in mind that '1' is the symbol associated with the *first* element of N. Thus the even numbers, the prime numbers, and the set of perfect squares all satisfy the axioms of arithmetic.

Chapter 22.2

6: The intersection of two algebraic curves determine a point whose coordinates must necessarily be algebraic numbers.

Chapter 22.4

20: 0. **4a**: $\sqrt{2}$ has height 4; $1 - \sqrt{5}$ has height 8; $\sqrt[3]{1 - \sqrt{5}}$ has height 14. **4a**: $\phi(4) = 20$. **4b**: $\{0, 1, -1\}$.

Chapter 22.5

2a: For $z = i$, the orbit is $\{i, -1, 1, 1, 1, \ldots\}$. **2c**: $F^3(z) = z^8$, so the period 3 points satisfy $z = z^8$, with solutions $z = 0$ and the seven roots of unity, not all of which are period 3 points. **3b**: $\pm \frac{1}{\sqrt{2}} \pm \frac{1}{\sqrt{2}}$ and $\pm i$. **5a**: A circle. **5b**: A finite set of points on the circle.

Chapter 23.1

2a: $xyz = x$. **3**: $x + x = x$. **4b**: $xy = 0$ corresponds to "No X are Y."

Chapter 23.5

4a: $[A|0]$, where $\det A = 1$. These matrices A are also referred to as SU(2) (special unitary 2×2 matrices). In three dimensional space, the matrices would be SU(3). **4b**: $[A|0]$, where $\det A \neq 0$.

Chapter 24.1

2a: $f(x, y, z) = e^{g(x,y,z)}$, where $g(x, y, z) = \ln x + \ln y + \ln z$.

Chapter 24.2

1: Note that assuming the statement to be true leads to a contradiction, but assuming it false *also* leads to a contradiction.

Chapter 24.3

2: *d.*

Chapter 24.6

7: 4.7 bits. **9**: Consider a question for which the only possible answer is 'Yes'. What information is obtained from a message that answers the question?

Appendix B

Select Bibliography

The following is by no means intended as a complete bibliography, but rather a guide to works to consult for further information. Two important journals are the *Archive for the History of the Exact Sciences* (AHES) and *Historia Mathematica* (HM); simply browsing through back issues will inevitably turn up some very interesting bits.

General and Topic Histories

Baron, M. E., *The Origin of Infinitesimal Calculus*. Dover, 1987.

Boyer, Carl B., *History of Analytic Geometry*. Scripta Mathematica, 1956.

Boyer, Carl B., *The History of the Calculus and Its Conceptual Development*. Dover, 1959. One of the classic works in the field.

Boyer, Carl B., *A History of Mathematics*. Princeton University Press, 1985. One of the classics in the field. Earlier editions contain problems not present in the most recent editions.

Cajori, Florian, *A History of Mathematical Notations*. The Open court publishing company 1928-29. A very comprehensive compilation of mathematical notations and their authors, with extensive diagrams.

Calinger, Ronald, *Classics of Mathematics*. Prentice Hall, 1982. A nice selection of translated primary source material, particularly of more recent material.

Dale, Andrew I., *A History of Inverse Probability : from Thomas Bayes to Karl Pearson*. Springer-Verlag, 1991. This is a good general work on the history of the subject.

Dauben, Joseph Warren, *The History of Mathematics From Antiquity to the Present : A Selective Bibliography.* Garland, 1985. A good starting point for any research in the history of mathematics.

Dickson, Leonard E. *History of the Theory of Numbers* (three volumes). Carnegie Institution of Washington, 1919-1923. An comprehensive overview of all the major topics.

Gillespie, Charles, ed., *Dictionary of Scientific Biography* (16 volumes). Scribner 1970-80. Probably the best place to start looking for information about almost any scientist or mathematician before the present era (though they must be dead to be considered for inclusion in the work).

Katz, Victor J., *A History of Mathematics.* HarperCollins, 1993. One of the more recent general histories of mathematics. Katz includes a very large number of problems.

Kline, Morris, *Mathematical Thought From Ancient to Modern Times.* Oxford University Press, 1972. A good, general introduction to the development of the subject, particularly analysis.

Menninger, Karl, *Number Words and Number Symbols*. M.I.T. Press, 1969. An overview of the development of numeration systems around the world.

Newman, James R., *The World of Mathematics.* Simon and Schuster, 1956 (4 volumes). A nice selection of works from a wide variety of authors, both ancient and modern.

Ore, Oystein, *Number Theory and Its History.* Dover, 1988. This is more a text on the theory of numbers, but includes some of the historical background.

Pearson, Karl, *The History of Statistics in the 17th and 18th Centuries Against the Changing Background of Intellectual, Scientific and Religious thought.* Macmillan, 1978. A compilation of Pearson's work in the history of statistics.

Rozenfel'd, B. A., *A History of Non-Euclidean Geometry.* Springer-Verlag, 1988.

Saint Andrews, University of, "The MacTutor History of Mathematics archive." www-groups.dcs.st-and.ac.uk/history/. Contains a wealth of biographical information about mathematicians, as well as lavish visual archives.

Scharlau, Winfried, *From Fermat to Minkowski : Lectures on the Theory of Numbers and Its Historical Development.* Springer-Verlag, 1985.

Smith, David Eugene, *History of Mathematics.* Ginn and Company 1923-25.

Smith attempts to stay true to the original form of the mathematics. His intent in particular is to mine the past for useful resources for the teaching of mathematics.

Smith, David Eugene, *A Source Book in Mathematics*, McGraw-Hill Book Company, 1929. Useful for selections from those areas Struik's lacks (primarily probability).

Stigler, Stephen M., *The History of Statistics : The Measurement of Uncertainty Before 1900*. Harvard University Press, 1986. A good introduction to the history of statistics by one of the leading historians in the field.

Struik, Dirk Jan, *A Source Book in Mathematics, 1200-1800*, Harvard University Press, 1969. Good selections in analysis, algebra, and geometry.

Todhunter, Isaac, *History of the Mathematical Theory of Probability From The Time of Pascal to That of Laplace*. Macmillan, 1865. A good, complete coverage of the subject during the interval stated in the title. For subsequent developments, see Dale, Pearson, and Stigler.

van der Waerden, B. L. *Geometry and Algebra in Ancient Civilizations*. Springer-Verlag, 1983, and *A History of Algebra : from al-Khwarizmi to Emmy Noether*. Springer-Verlag, 1985. These two provide a complete, if somewhat brief, history of algebra

Weil, A., *Number Theory: An Approach Through History*. Birkhäuser, 1984. Covers the period from Hammurabi through Legendre, spending (as it should) a great deal of time on Fermat and Euler.

Ancient mathematics
Chace, A. B., *The Rhind Mathematical Papyrus*, Mathematical Association of America, 1927-29.

Gillings, Richard J., *Mathematics in the Time of the Pharaohs*. MIT Press, 1972. An excellent survey of the available Egyptian mathematical texts.

Hoyrup, Jens, "Investigations of an Early Sumerian Division Problem, c. 2500 BC." *Historia Mathematica* 9 (1) 1982, 19-36. A nice paper that shows how analysis of errors in calculation leads to insights into the methods of calculation.

Neugebauer, Otto, *The Exact Sciences in Antiquity*. Harper and Brothers, 1962.

Neugebauer, Otto, *Mathematical Cuneiform Texts*. American Oriental Society and the American Schools of Oriental Research, 1945. Includes copies of original, translations, and annotations of various Babylonian texts.

van der Waerden, B. L. *Science Awakening.* English translation by Arnold Dresden. P. Noordhoff, 1954.

Greco-Roman era

Aaboe, Asger, *Episodes from the Early History of Mathematics.* L. W. Singer 1964.

Apollonius, *Treatise on Conic Sections.* Translated by Thomas L. Heath. Cambridge University Press, 1896. Hard to find, but the first three books are included in a translation put out by Encyclopedia Britannica in its *Great Books* series.

Archimedes, *Works.* Edited by Marshall Clagett. University of Wisconsin Press, 1964-80. A massive, multi-volume compendium not only of Archimedes, but also of his commentators, imitators, and translators. A very useful sourcebook for anyone interested in the history of higher mathematics from Archimedes to the time of Descartes.

Archimedes, *The Works of Archimedes.* Edited by T. L. Heath. Dover Publications 1953.

Bulmer-Thomas, Ivor, *Selections Illustrating the History of Greek Mathematics.* Translated by Ivor Thomas. Harvard University Press 1939-41. Wide variety of excerpts; English text printed adjacent to Greek text, allowing for useful comparisons of mathematical notations.

Cohen, Morris Raphael, and I. E. Drabkin, *A Source Book in Greek Science.* McGraw-Hill Book Co., 1948. Includes translated selections from various Greek geometers, particularly Heron of Alexandria.

Knorr, Wilbur R, "Arithmêtikê stoicheiôsis: On Diophantus and Hero of Alexandria." *Historia Mathematica* 20 (1), 1993, 180-192. Sheds new evidence on Diophantus's dates.

Heath, Thomas Little, *A History of Greek Mathematics.* The Clarendon Press, 1921. Comprehensive account of all the major geometers from the time of Thales to the final decline of classical civilization.

Heath, Thomas Little, *Diophantos of Alexandria.* Cambridge University press, 1885. A translation of the six books known at the time, plus other available treatises. Heath does translate Diophantus into modern algebraic notation, so a comparison with the Greek text is useful to see how the notation was actually written.

Heath, Thomas Little, *Elements.* E. P. Dutton and Co., 1933. A very thorough account of the *Elements,* including the history of the propositions since Euclid.

Heron of Alexandria, *Metrica. Accedunt partes quaedam selectae Codicis Constantinopolitani Palatii veteris no.1*, edited by E. M. Bruins. E. J. Brill, 1964. Despite the Latin title, volume III is actually an English translation of Heron's *Metrica*. Includes photographs and transcription of the original manuscript, so the form of the original can be seen.

Morrow, Glenn R., *Proclus: A Commentary on The First Book of Euclid's Elements*. Princeton University Press, 1970. Translation and annotation of Proclus's commentary. Useful for seeing what the Greeks actually thought about Euclid.

Nicomachus, of Gerasa, *Introduction to Arithmetic*. Translated into English by Martin Luther D'Ooge; with studies in Greek arithmetic, by Frank Egleston Robbins and Louis Charles Karpinski. Macmillan and Company, 1926.

Russo, Lucio, "The Definitions of Fundamental Geometric Entities Contained in Book I of Euclid's Elements." *Arch. Hist. Exact Sci.* 52 (1998), 195-219. Carefully examines the first seven definitions in the *Elements* and concludes that they are a later interpolation, not original with Euclid. A good example of the historiographical difficulties associated with early Greek mathematics.

Sesiano, Jacques, *Books IV to VII of Diophantus' Arithmetica in the Arabic Translation Attributed to Qusta ibn Luqa*. Springer-Verlag, 1982. A translation of the recovered books of Diophantus. Like Heath's, the notation is translated into a modern form.

Chinese, Indian, and Arab mathematics

Aryabhata, *The Aryabhatiya of Aryabhata*. Translated by Walter Eugene Clark. University of Chicago Press, 1930.

Abu Kamil Shuja'ibn Aslam, *The Algebra of Abu Kamil: Kitab fi al-Jabr wa'l-muqabala, in a commentary by Mordecai Finzi*. Translated by Martin Levey. University of Wisconsin Press, 1966.

Berggren, J. L., *Episodes in the Mathematics of Medieval Islam*. Springer-Verlag, 1986. I highly recommend this work, as it faithfully reproduces the original forms, rather than translating the results into modern technical mathematics.

Bhaskara, *Bijaganita*, translated by Edward Strachey. W. Glendinning, 1813 (?). Difficult to find, but a valuable translated primary source.

Daffa', 'Ali 'Abd Allah., *The Muslim Contribution to Mathematics*. Humanities Press, 1977.

Datta, Bibhutibhusan and Avadhesh Singh, *History of Hindu Mathematics, a*

Source Book Motilal Banarsi Das, 1935, and *History of Hindu mathematics, Part II*, 1938. Motilal Banarsi Das, Lahore.

Hayashi, Takao, *The Bhakshālī Manuscript : An Ancient Indian Mathematical Treatise*. Egbert Forsten, 1995. A translation of the Bhakshālī manuscript, with commentary.

Kangshen, shen, John Crossby, and Anthony W.-C. Lun, *The Nine Chapters on the Mathematical Art*. Oxford University Press, 1999. An English translation of a copy of the *Nine Chapters* with the commentary of Liu Hui.

Khuwarizmi, Muhammad ibn Musa, *Robert of Chester's Latin translation of the Algebra of al-Khowarizmi, with an introduction, critical notes and an English version by Louis Charles Karpinski*. Macmillan, 1915. A comparison with the Rosen translation is instructive.

Khuwarizmi, Muhammad ibn Musa, *The algebra of Mohammed ben Musa*, translated by Frederic Rosen. J. Murray 1831. Very difficult to find, but a translation of the complete work.

Kūshyār ibn Labbān, *Principles of Hindu Reckoning*, translated by Martin Levey and Marvin Petruck. The University of Wisconsin Press, 1965. Parallel Arabic text and English, allowing comparison between how the numbers and operations were performed in the original and how they are done today.

Martzloff, Jean-Claude. *A History of Chinese Mathematics*, 1987. Translated by Stephen S. Wilson, 1997. Springer-Verlag.

Omar Khayyam, *The algebra of Omar Khayyam*. Translated by Daoud S. Kasir. Columbia University, 1931.

Rabinovitch, Nachum, "Rabbi Levi Ben Gershon and the Origins of Mathematical Induction" AHES 6 (3), 1970, 237-248.

Rashdi, Rashed, "L'induction mathématique: al-Karajī, as-Samaw'al" AHES 9 (1), 1972, 1-21.

Swetz, Frank J, *The Sea Island Mathematical Manual*. Pennsylvania State University Press, 1992. Includes the text and annotations.

Yong, Lam Lay, "On the Existing Fragments of Yang Hui's *Hsiang Chieh Suan Fa*," AHES 6 (1), 1969, 82-88.

Yong, Lam Lay, and Ang Tian Se. *Fleeting Footsteps: Tracing the Conception of Arithmetic and Algebra in Ancient China*, 1992. World Scientific Publishing Co. A beautiful introduction to Chinese elementary mathematics.

Medieval and Renaissance era

Flegg, Graham, Cynthia Hay, Barbara Moss, *Nicolas Chuquet, Renaissance Mathematician*. Kluwer Academic Publishers, 1985. Includes a translation of most of Chuquet's *Triparty*.

Grant, Edward, *De Proportionibus Proportionum, and Ad Pauca Respicientes*. University of Wisconsin Press, 1966. Includes a translation of Oresme.

Hughes, Barnabas, *De Numeris Datis*. University of California Press, 1981. A translation of Nemorarius's work, with the original Latin text for comparison.

Hughes, Barnabas, *Regiomontanus: On Triangles. De triangulis omnimodis.*. University of Wisconsin Press, 1967. English translation of Regiomontanus, with a facsimile of the original Latin text. A beautiful example of how translations should be presented.

Kaunzer, Wolfgang. "Über eine frühe lateinische Bearbeitung der Algebra al-Khwārizmīs in MS Lyell 52 der Bodleian Library Oxford," AHES 32 (1), 1985, 1-16. Includes the Latin verses from Pacioli's *Summary* that describe how to solve quadratic equations.

Leonardo of Pisa. *Scritti di Leonardo Pisano* (two volumes). B. Boncompagni, 1857. Contains Leonardo's complete works (in Latin), and the only published version of *Liber Abaci*. For translations of *The Book of Squares*, see Sigler. For translations of short excerpts, see Struik. No translation of the whole has yet been done.

Oystein Ore, *Cardano: The Gambling Scholar*, Princeton University Press, 1953. Good background information on Cardano and his time; also includes a translation of Cardano's *On Games of Chance*.

Sesiano, Jacques, "The Appearance of Negative Solutions in Medieval Mathematics," AHES 32 (2), 1985, 105-150.

Sigler, L. E., *Book of Squares*. Academic Press, 1987. Translation of Leonardo's *Liber Quadratorum*.

Steele, Robert, *The Earliest Arithmetics in English*. Oxford University Press, 1922. Includes translations of Recorde and Villedieu.

Stevin, Simon, *Principal Works*. C. V. Swets and Zeitlinger, 1955-1966. The principal works of Stevin, in an English translation.

Swetz, Frank J. *Capitalism and Arithmetic*, 1987. Open Court Publishing. This includes a translated copy of the Treviso arithmetic (1487), the first printed mathematical work in European history.

Thoren, Victor E., "Prosthaphaeresis Revisisted." *Historia Mathematica* 15 (1) 1988, 32-9.

van Looy, Herman, "A Chronological Historical Analysis of the Mathematical Manuscripts of Gregorius a Sancto Vincentio (1584-1667)," *Historia Mathematica* 11 (1) 1984, 57-75.

Viète, François, *The Analytic Art*, translated by T. Richard Witmer. Kent State University Press, 1983. Here, Witmer retains Viète's original notation as much as possible.

Viète, François, et al, *The Early Theory of Equations*. Golden Hind Press, 1986. Contains translations of three early treatises on equations.

Viète, François, *Opera Mathematics*. Franciscus van Schooten, 1646. Contains all of Viète's works. Recently reprinted.

Witmer, Richard T., *The Great Art; or, The Rules of Algebra*. M.I.T. Press, 1968. A translation of Cardano's *Ars Magna*, though Witmer translates Cardano's mathematics into modern mathematical notation, so it is useful to compare Witmer's translation with the selection in Smith's *Source Book*.

Seventeenth century

Anderson, Kirsti, "Cavalieri's Method of Indivisibles," AHES 31 (4), 1985, 291-367. Anderson points out that Cavalieri is frequently misrepresented; this work includes a number of quotations taken directly from Cavalieri, and presents Cavalieri's developments in a form very close to the original (with only minor changes in notation and style).

Barrow, Isaac, *The Geometrical Lectures of Isaac Barrow*. The Open court publishing company, 1916.

Bernoulli, Johann, *Opera Omnia*. Reprinted by G. Olms, 1968.

Child, J. M., *The Early Mathematical Manuscripts of Leibniz; translated from the Latin texts published by Carl Immanuel Gerhardt with critical and historical notes*, 1920. An English translation of Leibniz's more important works.

Descartes, René, *The Geometry of René Descartes*, The Open court publishing company, 1925. Smith's translation; side-by-side French and English.

Descartes, Rene. *Geometria*, van Schooten, 1637. The Latin edition includes, besides the *Geometry*, appendices by Florimond de Beaune, Jan Hudde, and others. Available via gallica.bnf.fr.

Edwards, A. W. F., *Pascal's arithmetical triangle*. Oxford University Press, 1987

Field, J. V., and J. J. Gray. *The Geometrical Work of Girard Desargues*, 1987. Springer-Verlag. Includes Desargues' *Rough Draft on Conics* as well as Desargues's Theorem from Bosse's work.

Freudenthal, Hans, "Huygens' Foundations of Probability," *Historia Mathematica* 7 (2) 1980, 113-7. Includes a review of Huygen's work.

Gregory, James, *James Gregory. Tercentenary Memorial Volume.* G. Bell and Sons, ltd., 1939.

Hofmann, Joseph Ehrenfried, *Leibniz in Paris, 1672-1676.* Cambridge University Press, 1974.

Leibniz, Gottfried Wilhelm, *Mathematische Schriften* (7 volumes), ed. C. I. Gerhardt, 1849-1863. A. Asher and H. W. Schmidt. The mathematical works and correspondence of Leibniz. In the original language (usually Latin or French).

Napier, John, *Rabdologie.* Translated by William Frank Richardson. MIT Press, 1990.

Newton, Isaac, *The Mathematical Papers of Isaac Newton.* Edited by D. T. Whiteside. Cambridge University Press, 1967-1981. Transcriptions of Newton's original work (in Latin), and English translations of the same.

Newton, Isaac, *Mathematical Principles of Natural Philosophy. [Translated by Andrew Motte. Rev. by Florian Cajori] Optics. By Sir Isaac Newton. Treatise on light by Christiaan Huygens. [Translated by Silvanus P. Thompson].* Encyclopaedia Britannica 1955.

Newton, Isaac, *Mathematical works.* Assembled by Derek T. Whiteside. Johnson Reprint Corp., 1964-67. In two volumes; includes some of Newton's lesser known works, like *Universal Arithmetic* and *Analysis by Equations with an Infinite Number of Terms.*

Newton, Isaac, *Opticks: or, A Treatise of The Reflections, Refractions, Inflexions and Colours of Light. Also Two Treatises of the Species and Magnitude of*

Curvilinear Figures. S. Smith, and B. Walford, 1704. The interest of this work is in the two appendices which are in the original Latin.

Wallis, John, *Opera Mathematicorum* (three volumes). L. Richfield and T. Robison, 1656. Wallis's complete works. Latin.

Eighteenth century
Dunham, William, *Euler: The Master of Us All.* Mathematical Association of America, 1999. I haven't been able to get a copy of this, but the reviews of it are good.

Euler, Leonard, *Introduction to Analysis of The Infinite.* Springer-Verlag, 1988, 1990.

Euler, Leonard, *Elements of Algebra*, Springer-Verlag, 1984. This is the work where Euler included his proof of Fermat's Last Conjecture for $n = 3$.

Katz, Victor J. "The Calculus of the Trigonometric Functions." *Historia Mathematica* (14) 1987, No. 4 (November 1987), p. 311-324.

Kopelevich, Yu. Kh. "The Petersburg Astronomy Contest in 1751." Soviet Astronomy-AJ, Vol. 9, No. 4 (January-February 1966), p. 653-660.

Moivre, Abraham de, *The doctrine of chances; or, A method of calculating the probabilities of events in play.* Chelsea, 1967. A reprint of the third edition.

Suzuki, Jeff, *A History of the Stability Problem in Celestial Mechanics, From Newton to Laplace*, 1996. Doctoral thesis. An overview of one of the issues in mathematical physics during the eighteenth century. A lot of translations of original source material (particularly Clairaut, d'Alembert, Euler, Lagrange, and Laplace). Some of the translations are even correct.

Wilson, Curtis. "The Great Inequality of Jupiter and Saturn: From Kepler to Laplace," AHES 33 (1985) p. 15-290. Discusses in detail one of the key problems in mathematical physics during its formative years.

Nineteenth century
Alexander, Daniel S., *A History of Complex Dynamics: From Schrder to Fatou and Julia.* F. Vieweg, 1994. A very nice account of the history of the mathematics leading up to the Mandelbrot set; the title indicates it is primarily about complex dynamics though, by necessity, it includes a good deal of history of dynamical systems in general.

Birkhoff, Garrett, *A Source Book in Classical Analysis.* Harvard University Press, 1973. Selections of the works of some of the main contributors to real analysis during the latter part of the nineteenth and twentieth century. Foreign language works appear in the original (mainly French). Useful for those important mathematicians who do not have a collected works (or one that is easily available).

Bottazzini, Umberto, *The Higher Calculus: A History of Real and Complex Analysis from Euler to Weierstrass*. Translated by Warren Van Egmond. Springer-Verlag, 1986. A good view of how mathematics grew from the age of Euler to the age of rigor.

Cantor, Georg, *Contributions to the Founding of the Theory of Transfinite Numbers*. Dover Publications 1955. Two of Cantor's later works in English translation. A minor disadvantage is that, because they are later works, they are considerably more polished than his earlier ones, so the papers are of lesser interest historically.

Dauben, Joseph W. *Georg Cantor: His Mathematics and Philosophy of the Infinite*, Harvard University Press, 1979. A good biography of Cantor and his mathematics.

Dedekind, Richard, *Essays on the Theory of Numbers*. Dover Publications 1963. Two of Dedekind's seminal works in English translation.

Dutka, Jacques. "Robert Adrain and the Method of Least Squares," AHES 41 (2) 1990, 171-184.

Edwards, Harold M. "The Background of Kummer's Proof of Fermat's Last Theorem for Regular Primes," AHES 14 (3), 1975, 219-236.

Edwards, Harold, *Galois Theory*. Springer-Verlag, 1984. A good introduction to Galois theory; includes a translation of Galois's key paper on the subject.

Edwards, Harold M. "The Genesis of Ideal Theory," AHES 23 (4), 1980, 321-378.

Fourier, Jean Baptiste Joseph, *The Analytical Theory of Heat*. Dover publications, 1955. Worth examining, particularly for Fourier's development of what we now call a Fourier series.

Frege, Gottlob, *The Foundations of Arithmetic*, English translation by J. L. Austin. Northwestern University Press, 1968.

Gauss, Carl Friedrich, *Disquisitiones Arithmeticae*, English; translated by Arthur A. Clarke ; revised by William C. Waterhouse with the help of Cornelius Greither and A.W. Grootendorst. Springer-Verlag, 1986.

Gauss, Carl Friedrich, *General Investigations of Curved Surfaces of 1827 and 1825*. Princeton University Library, 1902. English translations of two of Gauss's Latin treatises.

Gauss, Carl Friedrich, *Theory of the Motion of the Heavenly Bodies Moving About the Sun in Conic Sections: A translation of Gauss's "Theoria motus"*. With an appendix by Charles Henry Davis. Little, Brown and company, 1857.

Hawkins, Thomas. "The *Erlanger Programm* of Felix Klein: Reflections on Its Place in the History of Mathematics," *Historia Mathematica* 11 (4), 1984, 442-470.

Hawkins, Thomas, *Lebesgue's Theory of Integration*. Chelsea, 1975. An excellent study of the history of analysis during the nineteenth century.

Johnson, D. M. "Prelude to Dimension Theory: The Geometrical Investigations of Bernard Bolzano." *Archive for the History of the Exact Sciences*, 17 (3) 1977, p. 261-295.

Keen, Linda. *The Legacy of Sonya Kovalevskaya*, 1987. American Mathematical Society.

Kiernan, B. M., "The Development of Galois Theory from Lagrange to Artin," AHES 8 (2), 1971, 40-154.

Klein, Felix, *Development of Mathematics in the Nineteenth Century*, 1928. Translated, 1979, M. Ackerman. Math Sci Press. One of the classics in the history of mathematics. Klein focuses on analysis, geometry, and algebra; for the development of other subjects, see Kolmogorov and Yushkevich.

Kolmogorov, A. N., Yushkevich, A. P. (Adol'f Pavlovich), *Mathematics of the 19th century: Mathematical Logic, Algebra, Number Theory, Probability Theory*. Birkhäuser Verlag, 1992. Together with Klein, provide an almost complete overview of the mathematics of the nineteenth century.

Laugwitz, Detlef, *Bernhard Riemann, 1826-1866: Turning Points in the Conception of Mathematics*. Translated by Abe Shenitzer. Birkhäuser, 1999. A great deal of insight into mathematics, particularly in Berlin after the arrival of Riemann.

Lobachevskii, N. I., *Geometrical Researches on the Theory of Parallels*, translated by George Bruce Halsted. Open Court Publishing Company, 1914. Lobachevsky's work, in English.

Lützen, Jesper, *Joseph Liouville, 1809-1882, Master of Pure and Applied Mathematics*. Springer-Verlag, 1990. An excellent biography of Liouville and his mathematics.

Ore, Oystein, *Niels Henrik Abel, Mathematician Extraordinary*. Chelsea Pub. Co., 1974.

Peano, Giuseppe, *Selected works of Giuseppe Peano*. Translated into English and edited, with a biographical sketch and bibliography, by Hubert C. Kennedy. University of Toronto Press 1973.

Russ, S. B., "A Translation of Bolzano's Paper on the Intermediate Value Theorem," *Historia Mathematica* 7 (2) 1980, 156-185. As the title indicates, it includes a translation of Bolzano's paper.

Sheynin, O. B., "Carl Friedrich Gauss and the Theory of Errors," AHES 20 (1) 1979, 21-72.

Sheynin, O. B. "Simeon Denis Poisson's Work in Probability," AHES 18 (3) 1978, 245-300.

Smithies, Frank. *Cauchy and the Creation of Complex Function Theory*, 1997. Cambridge University Press.

Temple, George. *100 Years of Mathematics*, 1981. Springer-Verlag. An excellent source on the history of *recent* mathematics, as its title suggests: covers the latter part of the nineteenth century, and makes references to works as recent as the late 1960s. If read in conjunction with Klein and Kolmogorov/Yushkevich, these provide a good introduction to the recent history of mathematics.

Twentieth century

Anderson, K. G. "Poincaré's Discovery of Homoclinic Points," AHES 48 (2), 1994, 133-147.

Aczel, Amir D. *Fermat's Last Theorem*, 1996. Four Walls Eight Windows. Useful for understanding the basis of Wiles's proof of Fermat's Last Theorem.

Barrow-Green, June, "Oscar II's Prize Competition and the Error in Poincaré's Memoir on the Three Body Problem," AHES 48 (2) 1994, 187-131.

Bell, Eric Temple, *The Last Problem*. Mathematical Association of America, 1990. An examination of attempts to prove Fermat's last conjecture; reviewed and updated by Underwood Dudley.

Clay Mathematics Institute, "Millennium Prize Problems." www.claymath.org/prize_problems/index.html. Includes not only the "informal" description of the seven problems, but also a more formal mathematical description of the problem.

Dimand, Mary Ann, *A History of Game Theory*. Routledge, 1996.

James, I. M., *History of Topology*, Elsevier Science B.V., 1999. A collection of articles by prominent historians of mathematics. Most do not require specializeed background.

Johnson, D. M. "The Problem of the Invariance of Dimension in the Growth of Modern Topology," Parts I and II, AHES 20 (2), 1979, and 25 (2/3), 1981, 85-267.

Monastyrskii, Mikhail Il'ich, *Modern Mathematics in the Light of the Fields medals*. A.K. Peters, 1997. A nice general overview of twentieth century mathematics, though it does require some rather specialized knowledge to follow.

Sussman, Gerald Jay, and Jack Wisdom. "Numerical Evidence that the Motion of Pluto is Chaotic," *Science* 241 (July 22, 1988).

von Neumann, John, "On the Theory of Games of Strategy." Included in *Contributions to the Theory of Games, Volume IV*, A. W. Tucker and R. D. Luce, ed., 1959. Princeton University Press. A translation of von Neumann's original paper (which appeared in German).

Index